Lecture Notes in Computer Science **16299**

The series Lecture Notes in Computer Science (LNCS), including its subseries Lecture Notes in Artificial Intelligence (LNAI) and Lecture Notes in Bioinformatics (LNBI), has established itself as a medium for the publication of new developments in computer science and information technology research, teaching, and education.

LNCS enjoys close cooperation with the computer science R & D community, the series counts many renowned academics among its volume editors and paper authors, and collaborates with prestigious societies. Its mission is to serve this international community by providing an invaluable service, mainly focused on the publication of conference and workshop proceedings and postproceedings. LNCS commenced publication in 1973.

Shiyu Yang · Jiali Mao · Minghe Yu ·
Cheqing Jin · Lemen Chao
Editors

Web Information Systems and Applications

22nd International Conference, WISA 2025
Xi'an, China, October 1, 2025
Proceedings

Editors
Shiyu Yang
Guangzhou University
Guangzhou, China

Jiali Mao
East China Normal University
Shanghai, China

Minghe Yu
Northeastern University
Shenyang, China

Cheqing Jin
East China Normal University
Shanghai, China

Lemen Chao
Renmin University of China
Beijing, China

ISSN 0302-9743 ISSN 1611-3349 (electronic)
Lecture Notes in Computer Science
ISBN 978-981-95-6425-5 ISBN 978-981-95-6426-2 (eBook)
https://doi.org/10.1007/978-981-95-6426-2

This Springer imprint is published by the registered company Springer Nature Singapore Pte Ltd.
The registered company address is: 152 Beach Road, #21-01/04 Gateway East, Singapore 189721, Singapore

Preface

It is our great honor to present the proceedings of the 22nd Conference on Network Information Systems and Applications (WISA 2025). WISA 2025 was jointly hosted by the Information System Technology Committee of the China Computer Federation, Chang'an University, and Northwest University. WISA 2025 provided a premium forum for researchers, professionals, practitioners, and officers closely related to information systems and applications to discuss the theme of intelligent information systems, digital transformation and information system security, focusing on difficult and critical issues in metaverse, knowledge graphs, blockchain, and recommendation systems, and the promotion of innovative technology for new application areas of information systems.

WISA 2025 was held in Xi'an, Shaanxi Province, China from October 31st to November 2nd, 2025.This year, we received 210 papers, and each paper was assigned to at least three Program Committee (PC) members for review. The peer review process was double-blind. The PC carefully discussed each paper and ultimately selected 57 research papers (including 10 short papers). These papers focus on research in areas such as knowledge construction, security, intelligent services, intelligent computing, large language models, and information system applications. WISA 2025 invited famous experts from various fields of artificial intelligence and information systems to give keynote speeches and special lectures, sharing their cutting-edge technologies as well as the latest viewpoints from both academia and industry.

We are grateful to the general chairs, Xing Chunxiao (Tsinghua University), Pei Jianzhong (Chang'an University), and Jiang Lin (Northwest University), as well as all PC members and external reviewers, who contributed their time and professional knowledge to the review process. We would like to express our gratitude to all the members of the organizing committee and the many volunteers for their strong support in the organization of the conference. Especially, we would also like to thank the publication chairs, Shiyu Yang (Guangzhou University), Mao Jiali (East China Normal University), and Yu Minghe (Northeastern University), for their efforts in the publication of the conference proceedings. Last but not least, we would like to express our sincere gratitude to all the authors who submitted papers to the conference.

October 2025

Shiyu Yang
Jiali Mao
Minghe Yu
Cheqing Jin
Lemen Chao

Organization

General Co-chairs

Chunxiao Xing	Tsinghua University, China
Jianzhong Pei	Chang'an University, China
Lin Jiang	Northwest University, China
Xingshe Zhou	Northwestern Polytechnical University, China

Program Committee Co-chairs

Cheqing Jin	East China Normal University, China
Lemen Chao	Renmin University of China, China

Workshop Co-chairs

Xin Wang	Tianjin University, China
Xiang Zhao	National University of Defense Technology, China
Haofen Wang	Tongji University, China

Publication Co-chairs

Shiyu Yang	Guangzhou University, China
Jiali Mao	East China Normal University, China
Minghe Yu	Northeastern University, China

Publicity Co-chairs

Yong Tang	South China Normal University, China
Bin Cao	Beijing SME Online Co., Ltd., China
Chenchen Sun	Chang'an University, China
Xianjia Meng	Northwest University, China

Challenge Co-chairs

Sponsor Co-chairs

Website Co-chairs

Poster Co-chairs

Demo Co-chairs

CCF Liaison

Organizing Committee Co-chairs

Xuequn Shang	Northwestern Polytechnical University, China
Yisheng An	Chang'an University, China
Jinye Peng	Northwest University, China

Program Committee

Baoning Niu	Taiyuan University of Technology, China
Baoyan Song	Liaoning University, China
Bin Li	Yangzhou University, China
Bo Fu	Liaoning Normal University, China
Bohan Li	Nanjing University of Aeronautics and Astronautics, China
Bolin Chen	Northwestern Polytechnical University, China
Buyu Wang	Inner Mongolia Agricultural University, China
Chao Kong	Anhui Polytechnic University, China
Chao Lemen	Renmin University of China, China
Chen Liu	North China University of Technology, China
Chenchen Sun	Tianjin University of Technology, China
Chengcheng Yu	Shanghai Polytechnic University, China
Cheqing Jin	East China Normal University, China
Chuanqi Tao	Nanjing University of Aeronautics and Astronautics, China
Chunyao Song	Nankai University, China
Dan Yin	Beijing University of Civil Engineering and Architecture, China
Derong Shen	Northeastern University, China
Dian Ouyang	Guangzhou University, China
Dong Li	Liaoning University, China
Erping Zhao	Xizang Minzu University, China
Fan Zhang	Guangzhou University, China
Fang Zhou	East China Normal University, China
Feng Qi	East China Normal University, China
Feng Zhao	Huazhong University of Science and Technology, China
Fengda Zhao	Yanshan University, China
Gansen Zhao	South China Normal University, China
Genggeng Liu	Fuzhou University, China
Guan Yuan	China University of Mining and Technology, China

Guanglai Gao	Inner Mongolia University, China
Guigang Zhang	Institute of Automation, Chinese Academy of Sciences, China
Guojiang Shen	Zhejiang University of Technology, China
Guojun Wang	Guangzhou University, China
Haitao Wang	Zhejiang Lab, China
Haiwei Zhang	Nankai University, China
Haofen Wang	Tongji University, China
Hua Yin	Guangdong University of Finance and Economics, China
Huang Mengxing	Hainan University, China
Hui Li	Guizhou University, China
Jiadong Ren	Yanshan University, China
Jiali Mao	East China Normal University, China
Jianbin Qin	Shenzhen Institute of Computing Sciences, Shenzhen, China
Weijin Jiang	Hunan University of Commerce, China
Jianye Yang	Guangzhou University, China
Jiazhen Xi	Huobi Group, China
Jinbao Wang	Harbin Institute of Technology, China
Jinguo You	Kunming University of Science and Technology, China
Jiping Zheng	Nanjing University of Aeronautics and Astronautics, China
Jun Pang	Wuhan University of Science and Technology, China
Jun Wang	iWudao, China
Junying Chen	South China University of Technology, China
Kai Wang	Shanghai Jiao Tong University, China
Kaiqi Zhang	Harbin Institute of Technology, China
Lan You	Hubei University, China
Lei Xu	Nanjing University, China
Li Jiajia	Shenyang Aerospace University, China
Lin Li	Wuhan University of Technology, China
Lin Wang	Wenge Group, China
Lina Chen	Zhejiang Normal University, China
Ling Chen	Yangzhou University, China
Lingyun Song	Northwestern Polytechnical University, China
Linlin Ding	Liaoning University, China
Liping Chen	Tarim University, China
Long Yuan	Nanjing University of Science and Technology, China
Luyi Bai	Northeastern University, China

Lyu Ni	East China Normal University, China
Mei Yu	Tianjin University, China
Meihui Zhang	Beijing Institute of Technology, China
Ming Gao	East China Normal University, China
Minghe Yu	Northeastern University, China
Ningyu Zhang	Zhejiang University, China
Peng Cheng	East China Normal University, China
Qian Zhou	Nanjing University of Posts and Communications, China
Qiaoming Zhu	Soochow University, China
Qingsheng Zhu	Chongqing University, China
Qingzhong Li	Shandong University, China
Qinming He	Zhejiang University, China
Ronghua Li	Beijing Institute of Technology, China
Ruixuan Li	Huazhong University of Science and Technology, China
Shanshan Yao	Shanxi University, China
Shaojie Qiao	Chengdu University of Information Technology, China
Sheng Wang	Wuhan University, China
Shengli Wu	Jiangsu University, China
Shi Lin Huang	Mingyang Digital, China
Shiyu Yang	Guangzhou University, China
Shujuan Jiang	China University of Mining and Technology, China
Shumin Han	Northeastern University, China
Shuo Yu	Dalian University of Technology, China
Shurui Fan	Hebei University of Technology, China
Tianxing Wu	Southeast University, China
Tieke He	Nanjing University, China
Tiezheng Nie	Northeastern University, China
Wei Li	Harbin Engineering University, China
Wei Song	Wuhan University, China
Wei Wang	East China Normal University, China
Wei Yu	Wuhan University, China
Weiguang Qu	Nanjing Normal University, China
Weiyu Guo	Central University of Finance and Economics, China
Weimin Li	Shanghai University, China
Weiwei Ni	Southeast University, China
Xiang Zhao	National University of Defense Technology, China

Xiangfu Meng	Liaoning Technical University, China
Xiangjie Kong	Zhejiang University of Technology, China
Xiangrui Cai	Nankai University, China
Xiaohua Shi	Shanghai Jiao Tong University, China
Xiaojie Yuan	Nankai University, China
Xiaoran Yan	Zhejiang Lab, China
Ximing Li	Jilin University, China
Xinbiao Gan	National University of Defense Technology, China
Xingce Wang	Beijing Normal University, China
Xu Liu	SAP Labs China, China
Xu Lizhen	Southeast University, China
Xueqing Zhao	Xi'an Polytechnic University, China
Xuequn Shang	Northwestern Polytechnical University, China
Xuesong Lu	East China Normal University, China
Yajun Yang	Tianjin University, China
Yanfeng Zhang	Northeastern University, China
Yanhui Ding	Shandong Normal University, China
Yanhui Gu	Nanjing Normal University, China
Yanlong Wen	Nankai University, China
Yanping Chen	Guizhou University, China
Ye Liang	Beijing Foreign Studies University, China
Yi Cai	South China University of Technology, China
Ying Zhang	Nankai University, China
Yinghua Zhou	University of Science and Technology of China, China
Yingxia Shao	Beijing University of Posts and Telecommunications, China
Yong Qi	Xi'an Jiaotong University, China
Yong Tang	South China Normal University, China
Yong Zhang	Tsinghua University, China
Yonggong Ren	Liaoning Normal University, China
Yongquan Dong	Jiangsu Normal University, China
Yu Gu	Northeastern University, China
Yuan Li	North China University of Technology, China
Yuanyuan Zhu	Wuhan University, China
Yue Kou	Northeastern University, China
Yuhua Li	Huazhong University of Science and Technology, China
Yupei Zhang	Northwestern Polytechnical University, China
Yuren Mao	Zhejiang University, China
Zhang Sijia	Dalian Ocean University, China

Zhigang Wang	Ocean University of China, China
Zhiyong Peng	Wuhan University, China
Zhongbin Sun	China University of Mining and Technology, China
Zhongle Xie	Zhejiang University, China
Zhuoming Xu	Hohai University, China
Ziqiang Yu	Yantai University, China
Mingdong Zhu	Henan Institute of Technology, China
Shan Lu	Southeast University, China
Bin Xu	Northeastern University, China
Cao Yang	Tokyo Tech, Japan
Hepu Deng	Royal Melbourne Institute of Technology, Australia
Xiaodong Qi	Nanyang Technological University, Singapore
Jiarui Sun	University of California, Los Angeles, USA
Tianyi Li	Aalborg University, Denmark
Zhuoran Wang	University of New South Wales, Australia
Jianliang Xu	Hong Kong Baptist University, China
Jiujing Zhang	University of New South Wales, Australia
Zihao Zhang	Stony Brook University, USA
Rui Zhou	Swinburne University of Technology, Australia

Contents

Information System Applications

Intelligent Service

Knowledge Construction

Large Language Model

Information System Applications

Multi-relational Context Learning with Cross-Attention Augmentation for Human Object Interaction Detection

Wuyou Wang[1,2], Jiaying Wang[1,2](✉), Jing Shan[1,2], and Xiaoxu Song[1,2]

[1] Shenyang University of Technology, Shenyang 110870, China
jiaying@sut.edu.cn

[2] Shenyang Key Laboratory of Intelligent Technology of Advanced Industrial Equipment Manufacturing, Shenyang 110870, China

Abstract. In recent years, transformer-based human-object interaction (HOI) detection methods have gained widespread attention for their excellent relational modeling capabilities. However, existing methods generally adopt static fusion strategies in the feature enhancement stage, which makes it difficult to fully explore the fine-grained interaction information in multi-relational contexts, limiting the inference ability and generalization performance of the model in complex scenarios. To this end, we propose a cross-attention augmentation module (CAAM) to enhance the selectivity and expressiveness of semantic fusion by introducing a task-aware cross-attention mechanism to establish dynamic interaction paths between task features and multi-relational contexts. The module can be flexibly embedded into existing multi-branch HOI detection frameworks without introducing additional supervision. Experiments on the V-COCO dataset show that the proposed method can stably improve interaction detection accuracy, demonstrating its potential for application in complex interaction understanding tasks.

Keywords: Human Object Interaction · Cross-Attention · Transformer · Multi-Relational Context Learning

1 Introduction

With the continuous evolution of computer vision technology, image understanding tasks are gradually expanding from target recognition and classification to higher-level semantic reasoning [24]. Among them, Human-Object Interaction (HOI) detection, as one of the key vision tasks, aims to recognize the semantic associations between people and objects in an image, and predicts the $\langle human, verb, object \rangle$ triples, to realize the structured representation of complex interactions in an image, HOI detection focuses on recognizing the presence of human and object in an image and inferring the specific interaction between them, such as "riding a bike", "holding a cup", "pushing a door", etc. The enhanced understanding of images enabled by these triples of high-quality

S. Yang et al. (Eds.): WISA 2025, LNCS 16299, pp. 3–14, 2026.
https://doi.org/10.1007/978-981-95-6426-2_1

phrases holds significant practical value for applications including intelligent surveillance, autonomous driving, robotic interaction, and humancomputer interaction [25,32].

HOI detection was first developed under the deep learning framework, which constructs interaction relationship classifiers based on a two-stage mechanism. Methods such as HO-RCNN [2] and iCAN [6] realized human and object detection with the help of Faster R-CNN [17]. Subsequently, researchers have gradually introduced gesture estimation [12], graph modeling [16,19], and attention mechanisms [6] to enhance fine-grained interaction understanding. CNN-based methods perform well in small and medium-sized interaction scenarios. Still, it is difficult to adapt to long-distance reasoning in complex scenarios due to its structural dependence on the combination of explicit candidate frames.

To address these limitations, single-stage HOI detection frameworks were proposed to integrate detection and interaction classification into an end-to-end pipeline. Methods like UnionDet [7] and PPDM [15] directly predict human, object, and interaction triplets from image features, thus avoiding region pairing operations. These approaches offer improved efficiency and inspire subsequent Transformer-based solutions by introducing unified query representations that jointly encode spatial, semantic, and contextual information.

Motivated by the success of Transformers [20] in natural language, vision, forecasting tasks [23,26], Transformer-based HOI models have rapidly emerged. DETR [1] laid the foundation for such models by unifying object detection under a set-based prediction framework. Transformer-based HOI methods can be broadly divided into two categories: (1) Unified models such as QPIC [18], which jointly model detection and interaction prediction in a compact architecture, and (2) Decoupled models such as HOTR [8] and AS-Net [3], which separate tasks into multiple branches and utilize fusion mechanisms to mitigate semantic fragmentation.

Despite their promising performance, current Transformer-based models still face key challenges. Most adopt multi-branch decoders that isolate instance detection from interaction classification, limiting contextual information exchange across tasks. This often results in suboptimal performance in complex scenes involving fine-grained actions or cross-domain interactions. Additionally, common feature fusion strategies, such as concatenation or linear projections, lack adaptive modeling capabilities, impeding the model's ability to dynamically reason over diverse interaction types or long-range dependencies. Consequently, achieving a balance between feature expressiveness, contextual inference, and computational efficiency remains an open problem.

To address these limitations, we propose a multi-relational context learning with cross-attention augmentation for HOI detection. Built upon a compact Transformer architecture, the module introduces a cross-attention mechanism to facilitate dynamic interaction between task features and contextual cues. Unlike traditional MLP-based enhancement, our approach selectively emphasizes task-relevant relational information, thereby enhancing reasoning accuracy and robustness while maintaining low computational overhead and strong model generalizability.

The contribution of this paper can be summarized as follows:

- We introduce a lightweight cross-attention fusion module, named the Cross-Attention Augmentation Module, to model dynamic interactions between task-specific features and multi-relational contexts.
- We enhance reasoning capabilities by leveraging fine-grained cross-attention interactions to accurately capture long-range human-object interaction relationships and subtle action distinctions within complex scenes.
- We conducted extensive experiments on the V-COCO dataset and compared our approach with current mainstream methods. Our approach achieved stable improvements in the Role mAP metric, demonstrating the effectiveness of the proposed method.

2 Methods

2.1 Overall Architecture

This section outlines the overall architecture of the proposed HOI detection model. As illustrated in Fig. 1, the input image is first processed by a backbone network followed by a Transformer encoder to extract global semantic features at the image level. Next, three types of task-specific queries, corresponding to humans, human-object interactions, and objects, are independently decoded through three separate branches to generate their respective candidate representations. Each branch employs a distinct Transformer decoder for semantic modeling and instance perception. The resulting feature representations are denoted as $F^H, F^I, F^O \in \mathbb{R}^{L \times d}$, where L represents the number of queries and d the feature dimension.

To further strengthen inter-branch collaboration, the model incorporates a unified context integration module, referred to as Multi-branch Fusion. This module leverages the MURE [10] mechanism to perform interaction fusion across multiple branches, producing fused interaction representations (contextual feature sets) for downstream tasks. On this basis, we propose a lightweight Cross-Attention Augmentation module (CAAM) to replace the original Attentive Fusion module (AFM). The module enhances the adaptability of context selectivity and fusion by introducing an attention-aware matching mechanism with a feed-forward expression layer.

In the original architecture, the query vectors output from the three-branch task and their corresponding contextual features are spliced and fed into an AFM module, as shown in Fig. 2. The module uses a static gating mechanism to model the interaction feature fusion, and its structure consists of parallel branches, one for generating the augmentation content, i.e., non-linear transformations of the spliced features to mine the potential semantic expressions, and the other for generating the augmentation weights, which are used to measure the importance

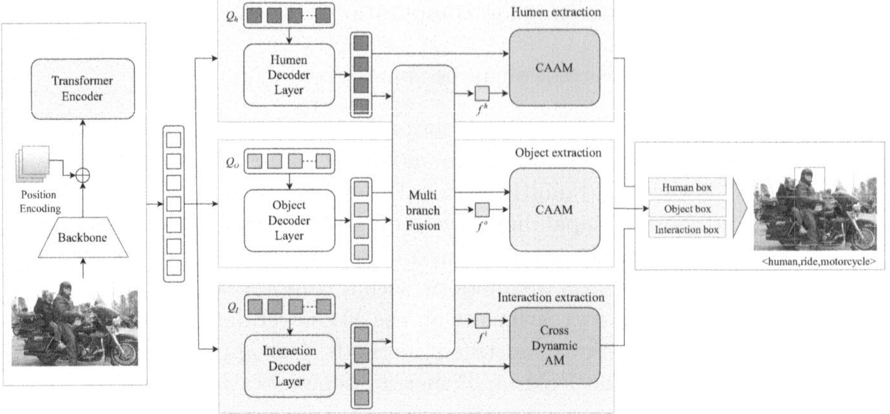

Fig. 1. The overall architecture of the HOI detection model.

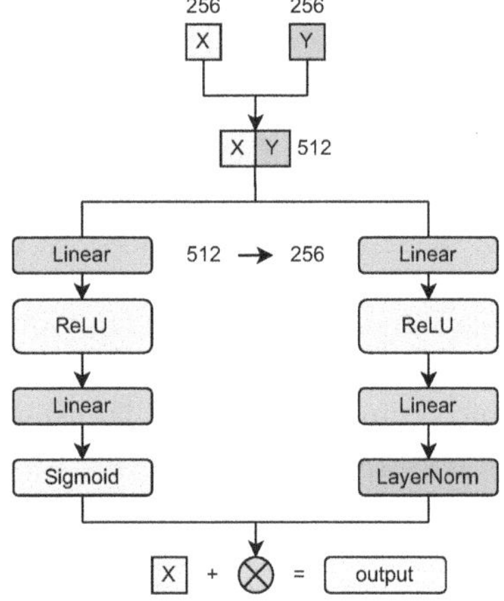

Fig. 2. Attentive Fusion Module.

of the contextual information in the current task. Its inputs are spliced vectors as shown in Eq. 1.

$$\mathbf{z} = \begin{bmatrix} \mathbf{x} \\ \mathbf{y} \end{bmatrix} \in \mathbb{R}^{2d} \tag{1}$$

where x represents the current task feature and y denotes the contextual fusion information from the integration module, the enhancement procedure unfolds in two stages: initially, potential enhanced features emerge through a two-layer MLP transformation. Subsequently, weights $\mathbf{w} \in [0,1]^d$ are derived via a two-layer MLP followed by a Sigmoid function. The ultimate fusion outcome materializes by blending residual connections with the weighted coefficients as shown in Eq. 2.

$$f_{\text{out}} = x + \mathbf{w} \odot f(\mathbf{z}) \tag{2}$$

2.2 Cross-Attention Augmentation Module

Although the current multi-branch structure offers some feature representation capabilities, the inherent task heterogeneity (target detection, action recognition, ternary reasoning) results in semantic disparities among branches. This "task mismatch" challenges the unified augmentation module due to uniform feature dimensions but diverse semantic sources. Different branches focus distinctly within the same context. The human branch emphasizes spatial posture, appearance, and localized character parts, often limited to the character area and action starting point. In contrast, the interaction branch prioritizes action intentions, position changes between action and object, and scene context linkage, with a broader region of interest that may span beyond the detected target boundary. The object branch concentrates on category attributes and interaction regions with less emphasis on character and action semantics.

Previous approaches have employed gating mechanisms to address this issue. Their fusion weights are generated by shallow multi-layer perceptrons (MLPs), which limits their ability to effectively capture contextual discriminative information. In this work, we propose a cross-attention augmentation module (CAAM) to establish context-aware dynamic enhancement pathways via a learnable attention weight matrix. Compared to Self-Attention, the principal advantage of the cross-attention structure lies in its inherent asymmetry: the query originates from the task backbone, while the key and value are derived from the contextual features. This asymmetric design imparts the module with semantic-level directionality and focus, thereby enhancing its capability to selectively integrate relevant contextual information.

Figure 3 shows the structure of the CAAM. The input of the module consists of the feature X and the corresponding context feature Y from the task branch, firstly, X as a query vector and Y as a key-value vector are inputted to the Cross-Attention module to complete the context information extraction based on the task semantics. The output of Cross-Attention is residually connected with the original feature X and then normalized to get the preliminary fusion of the intermediate representation. Subsequently, this intermediate feature is fed into a feedforward sublayer consisting of two layers of linear mapping and nonlinear activation functions for feature enhancement to further improve the representation capability. The enhanced features are again residually concatenated with the backbone features and normalized to output the final result of the module.

Let the input feature vector of the current branch be a query vector, denoted as $f_q \in \mathbb{R}^d$, which represents the local representation of the task (e.g., a human body, an object, or an interaction); meanwhile, let the set of context features obtained from the fusion module be $\mathbf{C} = \{\mathbf{c}_1, \ldots, \mathbf{c}_k\} \in \mathbb{R}^{k \times d}$. Where k denotes the number of context vectors, and d is the dimension of each feature. To achieve

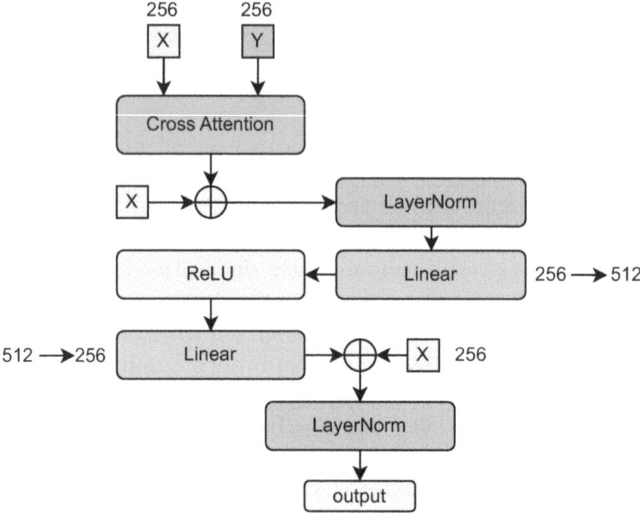

Fig. 3. Cross-Attention Augmentation Modul.

semantic alignment and dynamic fusion between task features and context, this paper introduces the Cross-Attention mechanism in the CAAM module. This mechanism first linearly maps the inputs to get the attention triples: query, key, and value, which is computed as shown in Eq. 3.

$$Q = W_q f_q, \quad K = W_k C, \quad V = W_v C \tag{3}$$

where $W_q, W_k, W_v \in \mathbb{R}^{d \times d}$ is the learnable linear transformation matrix. Next, Scaled Dot-Product Attention is used to compute the matching relationship between the query vectors and the context set to obtain the aggregated context representation as shown in Eq. 4.

$$A = \text{softmax}\left(\frac{QK^{\mathrm{T}}}{\sqrt{d}}\right), \quad f_{\text{att}} = AV \tag{4}$$

where $A \in \mathbb{R}^{1 \times k}$ denotes the attention weight of the current task for each context vector and $f_{\text{att}} \in \mathbb{R}^d$ is the final obtained context aggregation feature. This aggregated feature is summed with the original task vector by residual concatenation and normalized as shown in Eq. 5.

$$f_1 = \text{LayerNorm}(f_q + f_{\text{att}}) \tag{5}$$

Subsequently, two Feed-Forward Neural Networks are introduced for further transformation to enhance feature representation as shown in Eq. 6.

$$f_2 = \text{ReLU}(W_1 f_1), \quad f_3 = \text{LayerNorm}(W_2 f_2 + f_1) \tag{6}$$

where $W_1, W_2 \in \mathbb{R}^{d \times d}$, ReLU is the nonlinear activation function. The final output $f_3 \in \mathbb{R}^d$ is the augmented representation vector of the branch task after fusing the context.

2.3 Loss Function and Optimization Objective

This work adheres to the conventional loss formulation employed in human-object interaction (HOI) detection tasks. The overall optimization objective comprises two principal components: the matching loss and the interaction classification loss. In alignment with prior methods, we utilize the Hungarian matching algorithm to establish a one-to-one correspondence between predicted candidate queries and ground truth interaction instances. To enable joint optimization, the model is trained in a multi-task learning framework. For each predicted human-object-action triplet, denoted as (human, object, verb), the model outputs the following components: the bounding box coordinates for the human and object instances, represented as $b_h, b_o \in \mathbb{R}^4$ and a probability distribution over interaction categories, denoted as $p^a \in \mathbb{R}^C$, where C indicates the total number of predefined action classes. The overall loss function is defined in Eq. 7.

$$L = \lambda_{\text{cls}} \cdot \mathcal{L}_{\text{cls}} + \lambda_{\text{box}} \cdot \mathcal{L}_{\text{box}} + \lambda_{\text{giou}} \cdot \mathcal{L}_{\text{giou}} \tag{7}$$

where $\mathcal{L}_{\text{cls}}$ is the multi-label classification loss for interaction categories, using binary cross entropy; $\mathcal{L}_{\text{box}}$ is the bounding box L1 loss for human and objects; $\mathcal{L}_{\text{giou}}$ is the generalized IOU loss for enhancing the positional fitting of the box; and the weights of each loss term $\lambda_{\text{cls}}, \lambda_{\text{box}}, \lambda_{\text{giou}}$ are set empirically.

The module is not only suitable for human-object interaction detection, but can also be generalized to other multi-branch tasks that require semantic fusion, such as visual question and answer (VQA), multimodal scene graph generation, and relationship retrieval. In the next section, we will further quantify the effectiveness and stability of this paper's method in multiple scenarios through systematic experimental comparisons, ablation validation and generalization evaluation.

3 Experiments

In order to verify the effectiveness of the proposed CAAM module in the human-object interaction detection task, this paper conducts a systematic comparison experiment and ablation analysis on the V-COCO dataset.

3.1 Data Sets and Evaluation Metrics

In this paper, the proposed method is validated on the V-COCO (Verbs in COCO [14]) dataset, which focuses on the Human-Object Interaction (HOI) detection task. The dataset consists of 5,400 images containing 2,533 training images and 2,867 validation images. A total of 16,199 interaction instances are labeled in the dataset, covering 29 types of actions and 80 types of object labels. Each interaction instance consists of a triad (human, action, object).

In this study, Role Mean Average Precision (Role mAP) is adopted as the primary evaluation metric. Following the official protocol, we evaluate model performance under two standard scenarios: Scenario 1 $\text{AP}^{\#1}_{\text{role}}$: The model must generate complete predictions for all interactions in the image. Scenario 2 $AP^{\#2}_{role}$: The evaluation ignores bounding box predictions for occluded targets and only considers visible interaction pairs.

3.2 Experimental Settings

The model is optimized using AdamW with differential learning rates of 1×10^{-5} for the backbone and 1×10^{-4} for the remaining modules. The model is trained for 150 epochs, using a linear warm-up in the first epoch and fixed-percentage decay thereafter. The batch size is set to 4, and all experiments are conducted on two NVIDIA RTX 4090 GPUs. The weighting factors of loss function are set as $\lambda_{\text{cls}} = 1.0$, $\lambda_{\text{box}} = 2.5$, $\lambda_{\text{giou}} = 1.0$.

To ensure fair evaluation of the CAAM module, all ablation experiments are conducted under a unified backbone and training protocol. Each model is trained five times, with the best result used for comparison. ResNet-50 serves as the backbone for global feature extraction, while both the Transformer encoder and decoder comprise 6 layers. The number of query vectors is fixed at 100.

3.3 Model Performance Comparison and Analysis

To comprehensively evaluate the effectiveness of the proposed method, we conduct comparative experiments against state-of-the-art HOI detection approaches. The comparison encompasses two representative categories of methods: those based on convolutional neural networks (CNNs) and those employing Transformer architectures. The corresponding results are summarized in Table 1.

As can be seen from the table, this paper's method outperforms existing methods in $AP^{\#1}_{role}$ and reaches the next highest level in $AP\#2_{\text{role}}$, especially in the premise of using only visual backbone feature inputs, it still outperforms complex methods using additional semantic labels (e.g., action lexicon, gesture map, semantic map, etc.), and demonstrates a very high level of contextual understanding and action inference capability.

3.4 Ablation Experiment

To assess the impact of the proposed CAAM module, we conduct ablation studies on the V-COCO dataset, with results shown in Table 2. The analysis is carried out along two dimensions: (1) Module structure, comparing the baseline AFM, a no-fusion variant, a two-layer MLP without attention, and CAAM; (2) Insertion position, evaluating CAAM individually in the human, object, and interaction branches, as well as its combined effect across all branches.

Table 1. Comparison of HOI detection methods under $AP^{\#1}_{role}$ and $AP^{\#2}_{role}$. "A", "S", "P", "L" and "M" denote appearance, spatial, pose, language and motion features respectively. "×" indicates results not reported.

Year	Method	Backbone	Feature	$AP^{\#1}_{role}$	$AP^{\#2}_{role}$
CNN-based Methods					
2018	GPNN [16]	ResNet-101	A	44.0	×
2018	ICAN [6]	ResNet-50	A+S	45.3	52.4
2019	TIK [12]	ResNet-50-FPN	A+S+P	49.1	×
2020	UnionDet [7]	ResNet-50-FPN	A	47.5	56.2
2020	DRG [5]	ResNet-50-FPN	A+S+L+M	51.0	×
2020	VSGNet [19]	ResNet-152	A+S	51.76	57.03
2020	CHG [22]	ResNet-50-FPN	A+S	52.7	×
2020	ConsNet [15]	ResNet-50-FPN	A+S+L	53.2	×
2021	SCG [30]	ResNet-50-FPN	A+S+M	54.2	60.9
2021	DIRV [4]	EfficientNet-B3	A	56.1	×
Transformer-based Methods					
2021	E2E-HOI [34]	ResNet-101	A	52.9	×
2021	AS-Net [3]	ResNet-50	A	53.9	×
2021	HOTR [8]	ResNet-50	A	55.2	61.0
2021	QPIC [18]	ResNet-101	A	58.8	61.0
2021	CDN [29]	ResNet-101	A	63.91	65.89
2022	UPT [31]	ResNet-101-DC5	A	61.3	67.1
2022	MSTR [9]	ResNet-50	A+M	62.0	65.2
2022	Gen-Vlkt [13]	ResNet-101	A+L	63.6	65.9
2023	NeurIPS [11]	ResNet-50	A+S	64.4	65.6
2023	RLIPv2 [28]	ResNet-50	A	65.9	×
2024	FleVRS [33]	Focal-L	A+S+L	65.2	66.5
2024	MP-HOI-S [27]	ResNet-50	A+S+L	<u>66.2</u>	67.6
2024	BCOM [21]	ResNet-50	A+S+L	65.8	**69.9**
–	Base	ResNet-50	A	65.54	67.89
–	**Ours**	ResNet-50	A	**66.28**	<u>68.42</u>

The ablation results demonstrate that removing the enhancement module leads to a marked performance drop, highlighting the necessity of context modeling in HOI detection. Replacing CAAM with an MLP yields only marginal gains, underscoring the critical role of the attention mechanism in aligning task and context semantics.

Regarding insertion positions, applying CAAM to either the human or interaction branch improves accuracy, with the latter showing the greatest impact–indicating that context-aware interaction features are most crucial for action recognition. Integrating CAAM into all three branches yields the best performance, confirming its effectiveness across semantic paths.

Table 2. Ablation study of different enhancement configurations.

Method	$AP^{\#1}_{role}$	$AP^{\#2}_{role}$
Baseline AFM (original fusion)	65.54	67.89
No Fusion Module	63.40	65.72
Replaced with Two-layer MLP	64.31	66.70
CAAM in Human Branch	65.64	68.07
CAAM in Object Branch	65.71	68.05
CAAM in Interaction Branch	65.87	68.12
CAAM in All Branches (Ours)	**66.28**	**68.42**

4 Conclusion

This paper addresses the challenge of insufficient semantic fusion in multi-branch human-object interaction (HOI) detection by introducing a lightweight Cross-Attention Augmentation Module (CAAM). The proposed CAAM leverages a task-aware Cross-Attention mechanism to facilitate fine-grained contextual interaction and semantic alignment between different task branches. By dynamically adjusting feature fusion based on inter-branch relevance, CAAM significantly improves representational capacity. Extensive experiments on the V-COCO dataset, including both comparative evaluations and ablation studies, demonstrate the superiority of CAAM over existing enhancement modules. The results validate the effectiveness of the proposed approach and highlight its potential as a plug-and-play component for improving multi-branch architectures in HOI detection and other interaction reasoning tasks.

Acknowledgements. This work is partly supported by the Scientific Research Project of Liaoning Provincial Department of Education (No. JYTMS20231226, LJ232410142088), Ministry of Education industry-university cooperative education project (No. 231002108104009) and Open Project of the State Key Laboratory of Integrated Automation in Process Industry (No. SAPI-2025-KFKT-06).

References

1. Carion, N., Massa, F., Synnaeve, G., Usunier, N., Kirillov, A., Zagoruyko, S.: End-to-End object detection with transformers. In: Vedaldi, A., Bischof, H., Brox, T., Frahm, J.-M. (eds.) ECCV 2020. LNCS, vol. 12346, pp. 213–229. Springer, Cham (2020). https://doi.org/10.1007/978-3-030-58452-8_13
2. Chao, Y.W., Liu, Y., Liu, X., Zeng, H., Deng, J.: Learning to detect human-object interactions. In: 2018 IEEE Winter Conference on Applications of Computer Vision (WACV), pp. 381–389. IEEE (2018)
3. Chen, M., Liao, Y., Liu, S., Chen, Z., Wang, F., Qian, C.: Reformulating hoi detection as adaptive set prediction. In: Proceedings of the IEEE/CVF Conference on Computer Vision and Pattern Recognition, pp. 9004–9013 (2021)

4. Fang, H.S., Xie, Y., Shao, D., Lu, C.: Dirv: dense interaction region voting for end-to-end human-object interaction detection. In: Proceedings of the AAAI Conference on Artificial Intelligence, vol. 35, pp. 1291–1299 (2021)
5. Gao, C., Xu, J., Zou, Y., Huang, J.-B.: DRG: dual relation graph for human-object interaction detection. In: Vedaldi, A., Bischof, H., Brox, T., Frahm, J.-M. (eds.) ECCV 2020. LNCS, vol. 12357, pp. 696–712. Springer, Cham (2020). https://doi.org/10.1007/978-3-030-58610-2_41
6. Gao, C., Zou, Y., Huang, J.B.: ican: Instance-centric attention network for human-object interaction detection. arXiv preprint arXiv:1808.10437 (2018)
7. Kim, B., Choi, T., Kang, J., Kim, H.J.: UnionDet: union-level detector towards real-time human-object interaction detection. In: Vedaldi, A., Bischof, H., Brox, T., Frahm, J.-M. (eds.) ECCV 2020. LNCS, vol. 12360, pp. 498–514. Springer, Cham (2020). https://doi.org/10.1007/978-3-030-58555-6_30
8. Kim, B., Lee, J., Kang, J., Kim, E.S., Kim, H.J.: Hotr: end-to-end human-object interaction detection with transformers. In: Proceedings of the IEEE/CVF Conference on Computer Vision and Pattern Recognition, pp. 74–83 (2021)
9. Kim, B., Mun, J., On, K.W., Shin, M., Lee, J., Kim, E.S.: Mstr: multi-scale transformer for end-to-end human-object interaction detection. In: Proceedings of the IEEE/CVF Conference on Computer Vision and Pattern Recognition, pp. 19578–19587 (2022)
10. Kim, S., Jung, D., Cho, M.: Relational context learning for human-object interaction detection. In: Proceedings of the IEEE/CVF Conference on Computer Vision and Pattern Recognition, pp. 2925–2934 (2023)
11. Li, L., Wei, J., Wang, W., Yang, Y.: Neural-logic human-object interaction detection. Adv. Neural. Inf. Process. Syst. **36**, 21158–21171 (2023)
12. Li, Y.L., et al.: Transferable interactiveness knowledge for human-object interaction detection. In: Proceedings of the IEEE/CVF Conference on Computer Vision and Pattern Recognition, pp. 3585–3594 (2019)
13. Liao, Y., Zhang, A., Lu, M., Wang, Y., Li, X., Liu, S.: Gen-vlkt: simplify association and enhance interaction understanding for hoi detection. In: Proceedings of the IEEE/CVF Conference on Computer Vision and Pattern Recognition, pp. 20123–20132 (2022)
14. Lin, T.-Y., et al.: Microsoft COCO: common objects in context. In: Fleet, D., Pajdla, T., Schiele, B., Tuytelaars, T. (eds.) ECCV 2014. LNCS, vol. 8693, pp. 740–755. Springer, Cham (2014). https://doi.org/10.1007/978-3-319-10602-1_48
15. Liu, Y., Yuan, J., Chen, C.W.: Consnet: learning consistency graph for zero-shot human-object interaction detection. In: Proceedings of the 28th ACM International Conference on Multimedia, pp. 4235–4243 (2020)
16. Qi, S., Wang, W., Jia, B., Shen, J., Zhu, S.C.: Learning human-object interactions by graph parsing neural networks. In: Proceedings of the European Conference on Computer Vision (ECCV), pp. 401–417 (2018)
17. Ren, S., He, K., Girshick, R., Sun, J.: Faster r-cnn: towards real-time object detection with region proposal networks. Adv. Neural Inf. Process. Syst. **28** (2015)
18. Tamura, M., Ohashi, H., Yoshinaga, T.: Qpic: query-based pairwise human-object interaction detection with image-wide contextual information. In: Proceedings of the IEEE/CVF Conference on Computer Vision and Pattern Recognition, pp. 10410–10419 (2021)
19. Ulutan, O., Iftekhar, A., Manjunath, B.S.: Vsgnet: spatial attention network for detecting human object interactions using graph convolutions. In: Proceedings of the IEEE/CVF Conference on Computer Vision and Pattern Recognition, pp. 13617–13626 (2020)

20. Vaswani, A., et al.: Attention is all you need. Adv. Neural Inf. Process. Syst. **30** (2017)
21. Wang, G., Guo, Y., Xu, Z., Kankanhalli, M.: Bilateral adaptation for human-object interaction detection with occlusion-robustness. In: Proceedings of the IEEE/CVF Conference on Computer Vision and Pattern Recognition, pp. 27970–27980 (2024)
22. Wang, H., Zheng, W., Yingbiao, L.: Contextual heterogeneous graph network for human-object interaction detection. In: Vedaldi, A., Bischof, H., Brox, T., Frahm, J.-M. (eds.) ECCV 2020. LNCS, vol. 12362, pp. 248–264. Springer, Cham (2020). https://doi.org/10.1007/978-3-030-58520-4_15
23. Wang, J., Cao, L., Shan, J., Song, X., Jiang, J.: Dual learning model of code summary and generation based on transformer. In: International Conference on Web Information Systems and Applications, pp. 41–52. Springer, Heidelberg (2024). https://doi.org/10.1007/978-981-97-7707-5_4
24. Wang, J., Hao, S., Shan, J., Song, X.: Visual language–let the product say what you want. In: Proceedings of the AAAI Conference on Artificial Intelligence, pp. 23841–23843 (2024)
25. Wang, J., Shan, J., Santos, O.E., Bao, J.: High quality error-tolerant phrase mining on text corpus. Expert Syst. Appl. **171**, 114557 (2021)
26. Wang, J., Yang, H., Shan, J., Song, X., Jiang, J.: Attention-based spatial-temporal fusion networks for traffic flow prediction. In: International Conference on Web Information Systems and Applications, pp. 500–511. Springer, Heidelberg (2024). https://doi.org/10.1007/978-981-97-7707-5_41
27. Yang, J., Li, B., Zeng, A., Zhang, L., Zhang, R.: Open-world human-object interaction detection via multi-modal prompts. In: Proceedings of the IEEE/CVF Conference on Computer Vision and Pattern Recognition, pp. 16954–16964 (2024)
28. Yuan, H., et al.: Rlipv2: fast scaling of relational language-image pre-training. In: Proceedings of the IEEE/CVF International Conference on Computer Vision, pp. 21649–21661 (2023)
29. Zhang, A., et al.: Mining the benefits of two-stage and one-stage hoi detection. Adv. Neural. Inf. Process. Syst. **34**, 17209–17220 (2021)
30. Zhang, F.Z., Campbell, D., Gould, S.: Spatially conditioned graphs for detecting human-object interactions. In: Proceedings of the IEEE/CVF International Conference on Computer Vision, pp. 13319–13327 (2021)
31. Zhang, F.Z., Campbell, D., Gould, S.: Efficient two-stage detection of human-object interactions with a novel unary-pairwise transformer. In: Proceedings of the IEEE/CVF Conference on Computer Vision and Pattern Recognition, pp. 20104–20112 (2022)
32. Zhen, R., Song, W., He, Q., Cao, J., Shi, L., Luo, J.: Human-computer interaction system: a survey of talking-head generation. Electronics **12**(1), 218 (2023)
33. Zhu, F., Yang, J., Jiang, H.: Towards flexible visual relationship segmentation. Adv. Neural. Inf. Process. Syst. **37**, 107633–107661 (2024)
34. Zou, C., et al.: End-to-end human object interaction detection with hoi transformer. In: Proceedings of the IEEE/CVF Conference on Computer Vision and Pattern Recognition, pp. 11825–11834 (2021)

PTE-DHNet: Priority-Aware Temporal Embedding and Dynamic Hypergraph Network for Steel Transportation Volume Prediction

Xiaopeng Huang[1], Yitao Dong[1], Jiali Mao[1(✉)], Jiajun Liao[1], Yiming Zhao[1], and Shuaihua Chen[2]

[1] East China Normal University, Shanghai, China
{51265903102,52285903012,52215903003,51275903059}@stu.ecnu.edu.cn,
jlmao@dase.ecnu.edu.cn
[2] Ouyeel Co., LTD., Shanghai, China
chenshuaihua@ouyeel.com

Abstract. Accurate prediction of steel transportation volume is crucial for logistics planning in steel enterprises. However, existing methods often overlook two key challenges: the uncertainty in processing orders due to dynamic prioritization and resource competition, and complex temporal dependencies across multi-stage production processes that drive output fluctuations. To address these issues, we propose a *Priority-Aware Temporal Embedding and Dynamic Hypergraph Network* for steel transportation volume prediction (**PTE-DHNet**), which introduces two key innovations. First, a *Priority-aware Temporal Embedding Network* jointly models historical processing order patterns across products and dynamic order adjustments caused by resource competition, learning the impact of production priorities on output variability through feature fusion. Second, a *Transfer Perception Dynamic Hypergraph* models the time-varying influence of upstream outputs on downstream processing steps, capturing both intra-step product category interactions and inter-step variations in production output dynamics. Extensive experiments on three real-world steel production datasets show that our method outperforms representative baseline models, achieving over 5.7% improvement in WMAPE.

Keywords: Steel Transportation Prediction · Priority-aware · Transfer Perception · Dynamic Hypergraph

1 Introduction

Steel transportation volume prediction is a fundamental task in steel logistics, involving the estimation of finished product weights to be transported in future time windows, categorized by product type and delivery destination. As steel production is order-driven, accurate forecasting enables logistics departments

S. Yang et al. (Eds.): WISA 2025, LNCS 16299, pp. 15–28, 2026.
https://doi.org/10.1007/978-981-95-6426-2_2

to optimize shipping capacity allocation and improve cargo-ship matching efficiency. Crucially, only orders completing all manufacturing processes within the forecast window contribute to the transport volume. For instance, as shown in Fig. 1(a), on May 8, 2024, several previously scheduled orders are in production, but only a subset will be completed due to constrained production capacity - exemplified by the 15 tons of hot-rolled coil destined for Wuhan and 12 tons of hot-rolled sheet for Shanghai ready for shipment on May 9, 2024. The core challenge, therefore, is to estimate completion volumes per product category within the target time frame, which directly determine transportation volume.

Accurate prediction of steel transportation volume requires precise estimation of completed outputs per product category within the forecast window. The manufacturing process, comprising multiple interdependent steps, poses two key challenges to prediction accuracy. First, capacity constraints create resource competition across processing steps, generating queuing effects that drive significant output variability across product categories. Second, each step's output depends on dynamic inflows from multiple upstream steps, introducing complex interdependencies that increase modeling difficulty and output uncertainty. While recent advances in production prediction have employed deep learning to capture long-range dependencies within individual production sequences [2,6,10], these inherent complexities of multi-step manufacturing systems continue to pose significant obstacles for reliable transportation volume forecasting.

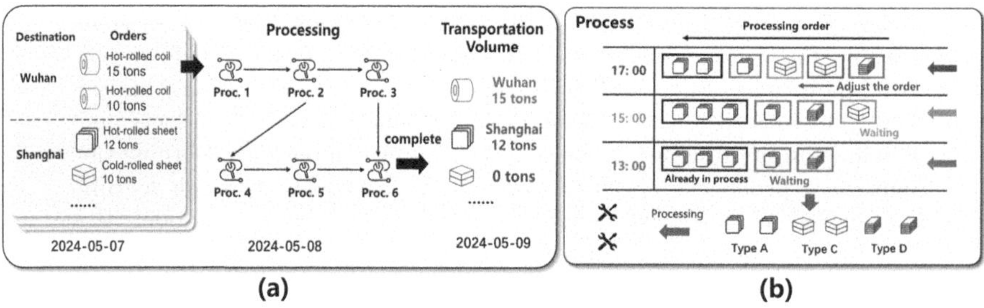

Fig. 1. (a): Illustration of Steel Production Process and Steel Transportation Volume. **(b)**: Illustration of the processing queue evolution over time in the cutting process.

Challenge 1. ***Uncertain processing order of steel products within a processing step affects the output of different categories.*** When multiple production tasks from different orders arrive at a processing step, they must queue due to limited resources and high processing demand. The total processing time of a product at this step includes both waiting and operation times, with the former largely determined by the sequence in which tasks are processed. Whether a product can be completed within a forecast window and contribute to the predicted output depends critically on its total processing time. For instance, products of category A scheduled earlier are more likely to finish on time, thereby contributing more significantly to the predicted output. However, earlier arrival does not always imply earlier processing, as the processing order is influenced by the attributes of preceding tasks. As illustrated in Fig. 1(b), during the queuing phase of a cutting process, both Type A and Type D products enter at 13:00. Due to their greater weight and similarity in type with previously queued items, Type A products receive higher priority to maintain processing stability. Additionally, Type C products arrive at 17:00 and share similar machining requirements with earlier Type C products already in the queue. To minimize tool changes and workpiece handling, the scheduler groups Type C products for batch processing, reducing their waiting time and increasing their contribution to the output within the upcoming time window. Yet, existing methods overlook the impact of processing order on actual output of different steel categories, thereby reducing the accuracy of transportation volume prediction.

Challenge 2. ***Output of a processing step exhibits complex temporal dependencies on the outputs of preceding steps.*** Different steel products often share common processing steps, resulting in each step handling multiple product types and continuously receiving semi-finished products from various upstream steps for further processing. Consequently, the output of any given processing step is influenced by the outputs of preceding steps in multiple ways. First, the quantity and category composition of preceding step outputs have varying degrees of impact on the current step's production. Outputs of the same category are more likely to be processed together, facilitating batch production and increasing yield within the upcoming time window, whereas cross-category outputs may compete for shared resources, indirectly affecting throughput. Second, preceding outputs are inherently time-varying; products with higher processing priority are more likely to enter the current step within the target time window, thereby exerting a stronger influence on its output. Moreover, as each processing step typically receives inputs from multiple upstream steps, these combined inflows contribute collectively to the uncertainty in its final output. Graph-based methods offer a promising approach by capturing entity interactions and spatiotemporal dependencies, making them well-suited for modeling complex production systems. However, recent studies [5,8,9] predominantly focus on pairwise relationships and fall short in modeling the cascading effects that arise among multiple interconnected entities, which leads to inaccuracies in transportation volume prediction.

To address the influence of steel product processing order on transportation volume prediction, we propose a *Priority-aware Temporal Embedding network* that jointly captures the historical processing order features of the target product and its interaction dynamics with co-processed products within the same step. A *cross-attention mechanism* is incorporated to adaptively weigh and integrate these two feature types, enabling the model to better distinguish priority-based output contributions. Additionally, an auxiliary task is introduced to predict the processing order probability distribution using a *ConvLSTM* network [3], where the resulting order-aware embeddings are further integrated into the main prediction task to enhance the model's sensitivity to category-specific outputs within the target time window. To model the temporal dependencies between upstream and downstream processing steps, we further design a *Transfer Perception Dynamic Hypergraph* module. This module first employs a *temporal hypergraph convolution mechanism* to capture intra-step variations in processing volumes across different product categories, thereby enhancing the representation of completed output. Moreover, a *transfer-aware dynamic hypergraph generation component* is introduced to model inter-step fluctuations in completion volumes from multiple preceding steps, allowing the model to learn the evolving impact of upstream outputs and category differences on downstream processing capacity, ultimately improving the accuracy of steel transportation volume prediction.

The main contributions of this work can be summarized as follows:

- To accurately predict steel transportation volume, we propose **PTE-DHNet**, a novel framework comprising a *priority-aware temporal embedding module* and a *transfer perception dynamic hypergraph module.*
- To capture the impact of uncertain processing order on the output of each processing step, we design a *priority-aware temporal embedding module* that enhances the model's ability to perceive category-specific completion volumes within each step by jointly modeling historical processing order patterns and product interaction dynamics.
- To address the complex temporal dependencies between a processing step and its preceding steps, we integrate *temporal hypergraph convolution* with *transfer perception dynamic hypergraph* to model the time-varying influence of upstream outputs on downstream processing step outputs, capturing both intra-step category interactions and inter-step production dynamics.
- Extensive comparative experiments on three real-world steel transportation datasets show that the proposed prediction method achieves a *WMAPE* improvement of over 5.7% compared to existing state-of-the-art approaches, demonstrating its superior accuracy and effectiveness in capturing complex production and logistics dynamics.

2 Related Work

Time series forecasting in manufacturing has evolved significantly from early *CNN* and *RNN* approaches [3,6,10], which faced limitations in modeling long-range dependencies. While recent *Transformer-based* architectures [11,14,21] have improved long-term forecasting through innovations like *ProbSparse* attention [21] and multi-scale patch modeling [14], they fail to address production-specific challenges like processing order effects. Recent advances in external memory integration [7,17], particularly the *Sparse Tandem Hopfield Network*'s memory retrieval mechanism [17], demonstrate the value of incorporating long-term memory of sequential patterns. Building on these insights, our work uniquely combines external memory module with cross-perception to explicitly model processing orders in steel transportation prediction.

Graph-based approaches have shown promise for modeling spatiotemporal dependencies in industrial forecasting [16,18,19], with steel products naturally represented as graph nodes. While early methods like *STGCN* [19] and

STFGNN [9] effectively captured spatial and temporal patterns using *graph convolution networks*, they struggle to capture long-term dependencies across a large number of products. Recent advances incorporating attention mechanisms (e.g., *ASTGNN* [5]'s trend-aware attention and *GBTTE* [15]'s spatiotemporal graph attention) improved dynamic relationship modeling, but remain limited to pairwise interactions. This proves inadequate for manufacturing scenarios where resource competition creates complex, high-order dependencies among multiple products. Though hypergraph networks (*DYHSL* [20]) can model such relationships through similarity-based hyperedges, existing approaches [4,20] fail to model the varying degrees of influence that outputs of different steel product types in upstream processing steps exert on downstream outputs. Our work bridges this gap by introducing a transfer perception dynamic hypergraph module that dynamically generates hypergraphs to capture the complex temporal dependency influences across different processing steps.

3 Problem Definition

Definition 1 (Steel Order). *A steel order S represents a production request from downstream customers in city c, specifying the required type and quantity of steel products from a steel mill. It is formally defined as a tuple $S = (type, w, q, c, t)$, where type denotes the steel product type, characterized by physical properties, weight grade, and shape, w is the unit weight (e.g., in tons) of each steel unit, q represents the total quantity of steel units requested, c indicates the customer's location (city), and t is the order creation timestamp.*

Definition 2 (Processing Record). *For a steel order undergoing sequential processing steps $p_i \in P$, its processing record at time t is defined as $R_i^t = (S_i, p_{pre}, p_{post}, t)$, where p_{pre} refers to last completed processing step, p_{post} is the current step, t represents the timestamp of the record, Steel products may follow distinct combinations of processing steps due to variations in weight, type, and physical specifications.*

Definition 3 (Steel Production Unit). *For logistics and dispatching purposes, in-process steel products are grouped into Steel Production Units ($SPUs$) based on shared destination city, product type, and current processing step. A unit v_i is formally defined as a collection of such products, represented by an attribute vector x_i encoding aggregated features like total product quantity, average unit weight, and mean processing duration.*

Problem Definition (Steel Production Volume Prediction). Given the historical τ Steel Production Unit (SPU) data $X_{t-\tau:t-1} = \{X_{t-\tau}, X_{t-\tau+1}, \cdots, X_{t-1}\}$ and associated processing records R, the objective is to forecast the transportation volume for each product type-city combination over a 24-hour horizon. Formally, we learn a prediction function: $\hat{Y}_{t:t+24} = \mathcal{F}(X_{t-\tau:t-1}, \mathcal{R}), \hat{Y}_{t:t+24} \in R^C$, where C denotes the total number of unique product type-city pairs.

4 Framework

To address the challenges in transportation volume prediction, we propose **PTE-DHNet**, a novel framework comprising two key components: a *Priority-aware Temporal Embedding Network* and a *Transfer Perception Dynamic Hypergraph* module, as shown in Fig. 2. We begin by introducing the Steel Processing Hypergraph Construct as the foundation of modeling the steel production. The Priority-aware Temporal Embedding Network retrieves typical processing order patterns via an external memory unit and models sequential dependencies among product nodes. An auxiliary task predicts processing order distributions, enhancing the perception of output variations across different steel categories. The Transfer Perception Dynamic Hypergraph module employs temporal fusion hypergraph convolution to capture the interactions impact among steel nodes within each processing step. It further extracts multi-scale transfer-aware features to explicitly model the varying influence of upstream product nodes on downstream outputs. By adaptively constructing dynamic hypergraph structures, the model captures complex temporal dependencies among processing outputs, improving the accuracy of transportation volume prediction.

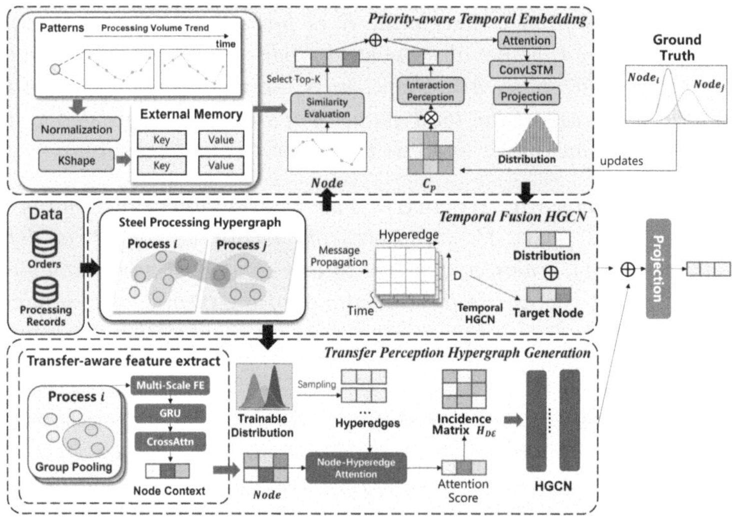

Fig. 2. Overall Framework of PTE-DHNet.

4.1 Steel Processing Hypergraph Construction

To capture the interactions within each processing step and complex temporal dependencies among processing steps, we construct a **Steel Processing Hypergraph** as the foundation, denoted as $G = (V, \mathcal{E}, W)$, where V denotes the set of steel processing units(treated as nodes), $\mathcal{E}$ is the set of hyperedges e, and W is a learnable hyperedge weight matrix. Intra-step hyperedges are constructed among nodes of the same product type to model resource competition. Moreover, steel products transfer across different processing steps during production, we establish additional hyperedges connecting nodes shared the same categories $(type, c)$ across different processing steps. The steel processing hypergraph is encoded via an incidence matrix $\mathbf{H} \in \mathbb{R}^{|V| \times |\mathcal{E}|}$, where $H(v, e) = 1$ if node v belongs to hyperedge e, and 0 otherwise. We further define the node degree matrix $\mathbf{D}_v$, hyperedge degree matrix $\mathbf{D}_e$, and node feature matrix $\mathbf{X}$.

4.2 Priority-Aware Temporal Embedding Network

External Memory Module. To model the impact of processing order within each processing step, we employ the K-Shape clustering algorithm to construct an external memory of typical processing order patterns [12]. For each node, a sliding window of length z is applied to its historical processing volume series to extract subsequences. Given node heterogeneity, clustering is performed in each *type* within a processing step to obtain representative processing patterns $P_{type,p} = \{pat_j | j \in 1, 2, \cdots, J\}$. Furthermore, to standardize the representation, we define a node's processing order distribution in each time window as a discrete probability vector indicating how many processing windows remain before its associated steel products are completed. Each pattern pat_j is paired with its corresponding mean processing order distribution d_j, forming the external memory $EM = \{(pat_j, d_j) | i \in 1, 2, \cdots, J\}$.

To fuse similar historical processing order patterns, we project the processing volume trend $u_{i,t-T:t}$ of node v_i within the window $[t-z, t]$ into a shared embedding space using learnable mapping matrices W_u and W_p, enabling similarity computation with the external memory. The retrieval is formally defined as:

$$Q_u = u_{i,t-z:t} W_u, \quad K_{EM} = P_{type,p} W_p, \quad A_{EM} = Softmax(\frac{Q_u K^\top}{\sqrt{z}}) \tag{1}$$

$$x_{i,EM} = (topk(A) \odot A) D_{type,p} \tag{2}$$

where $topK(\cdot)$ retains the k largest values per row and sets others to zero, $\odot$ represents the element-wise multiplication, $D_{type,p}$ denotes the corresponding processing order distribution matrix. To reduce redundancy, we retain the top-k most relevant patterns during fusion.

Processing Interaction Perception. Due to resource competition, the processing order of steel items is influenced by other nodes within a processing step. We initialize a cross-perception matrix C_p for each step p with all entries set to 1, indicating equal initial correlations. During inference, cross-sequence information is fused via C_p, and the perception process is formally defined as:

$$x_{i,f} = x_i \oplus x_{i,EM} \oplus C_{i,p} X_{p,EM} \tag{3}$$

where $\oplus$ denotes feature concatenation, $X_{p,EM}$ is the feature matrix composed of $x_{i,EM}$ within a processing step. The vector $x_{i,f}$ is then fed into an attention-based ConvLSTM network to predict order distributions. During training, the matrix C_p is iteratively refined by minimizing the Jensen-Shannon (JS) divergence between ground-truth sequence distributions. Smaller divergences indicate stronger joint production tendencies and weaker resource competition. To enable end-to-end optimization, the order distribution loss is integrated into the overall objective $\mathcal{L}$.

$$c_{i,j} \leftarrow (1-\alpha)c_{i,j} + \alpha\,(1 - \mathrm{JS}(d_i \| d_j)) \tag{4}$$

$$\mathcal{L} = \sum_i |y_i - \hat{y}_i|^2 + \beta \cdot \sum_i \mathrm{JS}(\hat{d}_i \| d_i) \tag{5}$$

where $c_{i,j}$ denotes the correlation score between nodes v_i and v_j in C_p, α and β are weighting hyperparameters, d_i and $\hat{d}_i$ represent the ground-truth and predicted processing order distributions of node v_i. The predicted distribution $\hat{d}_i$ is further concatenated into the node feature to enhance the model's awareness of completion volumes across product categories within each time window.

4.3 Transfer Perception Dynamic Hypergraph

Temporal Fusion Hypergraph Convolution. To capture interaction effects among multiple nodes and the complex temporal dependencies between upstream and downstream outputs, we introduce a Temporal Fusion Hypergraph Convolution. Specifically, we adopt the spectral hypergraph convolution framework from Gao et al. [4], where the node representation at layer l ($X^0 = X$) is computed via a learnable filter parameters Θ, formally defined:

$$X^l = \sigma(D_v^{-1/2} H W D_e^{-1/2} H^T D_v^{-1/2} X^{l-1} \Theta) \tag{6}$$

Then, we integrate a hyperedge temporal self-attention mechanism to capture the temporal dynamics of each hyperedge. The incidence matrix H is used to aggregate node features and generate hyperedge representation $X_{\mathcal{E}}$, along with its temporal augmented matrix $X_{\mathcal{E},t-\tau:t}$:

$$X_{\mathcal{E}} = W D_e^{-1} H^{\top} X^l, \quad X_{\mathcal{E},t-\tau:t} = \{X_{\mathcal{E},t-\tau}, X_{\mathcal{E},t-\tau+1}, \dots, X_{\mathcal{E},t}\} \tag{7}$$

Next, we employ learnable weight matrices $W_Q, W_K, W_V \in \mathbb{R}^{d \times d'}$ to project temporal hyperedge features for temporal attention,yielding the temporally attended feature $x_{\mathcal{E}}^{att}$, which is integrated into the final hypergraph convolution layer to enhance spatiotemporal modeling capacity. The spatiotemporal feature X^{ST} is further concatenated with the order distribution for joint learning.

$$Q_{\mathcal{E}} = x_{\mathcal{E},t} W_Q, \quad K_{\mathcal{E}} = x_{\mathcal{E},t-\tau:t} W_K, \quad V_{\mathcal{E}} = x_{\mathcal{E},t-\tau:t} W_V \tag{8}$$

$$A_{\mathcal{E}} = \mathrm{Softmax}\left(\frac{Q_{\mathcal{E}} K_{\mathcal{E}}^{\top}}{\sqrt{d'}}\right), x_{\mathcal{E}}^{att} = A_{\mathcal{E}} V_{\mathcal{E}} \tag{9}$$

$$X^{ST} = X^l + \sigma(D_v^{-1/2} H X_{\mathcal{E}}^{att} \Theta') \tag{10}$$

Transfer Perception Hypergraph Generation. To capture the complex temporal influence of upstream production outputs on the downstream processing, we develop a Transfer Perception Hypergraph Generation module. Specifically, to model the aggregated processing volume trends of different steel types across various processes, nodes are grouped by their product type and processing step. For each group, a temporal representation $g_{type,p}$ is then computed via mean pooling within each time window. To model inter-group temporal interactions, a multi-scale transfer-aware feature extraction module is employed. It applies temporal convolutional blocks to $g_{type,p}$ across multiple time scales. The resulting features are then fused using a GRU layer to capture temporal dependencies across groups, producing the transfer perception feature $GF_{type,p}$ for each group. To model the influence of transfer patterns on each node, we introduce a transfer perception cross-attention mechanism, enabling each node to adaptively extract information from its group representation based on its own temporal dynamics. The mechanism is formally defined as:

$$Q_G = X_{t-\tau:t}^{type,p} W_{QG}, \quad K_G = GF_{type,p} W_{KG}, \quad V_G = GF_{type,p} W_{VG} \tag{11}$$

$$X_{type,p}^{TP} = \text{Softmax}\left(\frac{Q_G K_G^\top}{\sqrt{d}}\right) V_G \tag{12}$$

where $X_{t-\tau:t}^{type,p}$ denotes the input features of nodes within each group over a temporal window $[t-\tau, t]$, and W_{QG}, W_{KG}, W_{VG} are learnable projection matrices for the query, key and value, respectively.

To further capture the varying temporal influence of upstream nodes, we dynamically construct a hypergraph based on transfer-aware node features $X_{type,p}^{TP}$. Following the dynamic hypergraph construction method proposed by Zhou et al. [13], we assume each hyperedge feature follows a trainable Gaussian distribution, i.e., $\mathbf{X}_{\mathcal{DE}} \sim N(\mu, diag(\sigma))$. Using the reparameterization trick, we sample m noise vectors $\varepsilon \in R^{m \times d}$ from $N(0,1)$ and initial the corresponding hyperedge features as: $\mathbf{X}_{\mathcal{DE}} = \mu + \sigma \odot \varepsilon$.

We then apply the attention mechanism between $X_{type,p}^{TP}$ and $\mathbf{X}_{\mathcal{DE}}$ to compute the attention score matrix $A_{\mathcal{N}} \in \mathbb{R}^{|V| \times m}$, which quantifies each node's contribution to its associated hyperedges. To retain the most informative connections, we construct the dynamic hypergraph structure by linking each node to its *top-k* most relevant hyperedges, resulting in the incidence matrix $H_{\mathcal{DE}} = \text{topK}(A_{\mathcal{N}})$. Based on that, the features $X_{type,p}^{TP}$ of each node are encoded via the dynamic hypergraph structure $H_{\mathcal{DE}}$ using Eq. (6) and further integrated with X^{ST} through a projection layer to predict transportation volume.

5 Experiments

5.1 Datasets and Model Settings

We evaluate our approach on three real-world steel transportation datasets collected from different steel plants, spanning the period from January 1, 2024 to July 1, 2024. Dataset statistics are summarized in Table 1. Each dataset is randomly split into 70% for training, 10% for validation, and 20% for testing.

Table 1. Statistics of Steel Transportation Datasets

Dataset	#Orders	#Cities	#Types	#steps
Fac_01	523,720	35	8	15
Fac_02	524,369	40	10	12
Fac_03	102,763	34	6	7

Model Settings. The input time window is set to 6 h, with a historical sequence length $\tau = 16$. For trend pattern extraction, we apply a sliding window of size $z = 8$. In the dynamic hypergraph generation module, $m = 50$ hyperedges are sampled per iteration, and each node connects to up to 10 hyperedges. The cross-perception update rate is set to $\alpha = 0.01$, and the weight of the distribution loss is $\beta = 20$. All models are trained for 150 epochs with a batch size of 100 using Python 3.11 and PyTorch on a single NVIDIA Tesla V100 GPU.

Metrics. We adopt Mean Absolute Error (**MAE**), Weighted Mean Absolute Percentage Error (**WMAPE**) and Root Mean Square Error (**RMSE**) as evaluation metrics to assess the performance. Since the daily transportation volumes vary significantly across different steel product types and destination cities, WMAPE is used to normalize prediction errors. Metrics formally defined as: $\text{MAE} = \frac{1}{N}\sum_{i=1}^{N}|\hat{y}_i - y_i|$, $\text{WMAPE} = \frac{\sum_{i=1}^{N}|\hat{y}_i - y_i|}{\sum_{i=1}^{N} y_i}$, $\text{RMSE} = \sqrt{\frac{1}{N}\sum_{i=1}^{N}(\hat{y}_i - y_i)^2}$.

5.2 Baselines

We benchmark our method against representative time series models and advanced traffic forecasting approaches, which provide transferable techniques for capturing spatiotemporal patterns in manufacturing.

- **Informer** [21] is a Transformer-based time series forecasting model that introduces a ProbSparse attention mechanism and self-attention distillation to efficiently capture long-range dependencies.
- **iTransformer** [11] is an encoder-only architecture for time series forecasting that embeds the full time series of each variable as individual tokens and uses self-attention to model multivariate correlations.
- **xLSTM** [1] is an enhanced LSTM model equipped with exponential gating, matrix memory, and memory mixing mechanisms, designed to improve long-term memory and capture complex temporal patterns.
- **STFGNN** [9] constructs a temporal graph using dynamic time warping to capture latent spatiotemporal dependencies, and employs a gated convolution module for joint temporal-spatial feature fusion.
- **PDFormer** [7] introduces two graph masking modules to learn traffic flow correlations from adjacent space and semantic similarity, with a delay-aware transformation to capture temporal propagation effects.

- **STPGNN** [8] identifies key nodes based on degree centrality and flow aggregation capacity, and separately models key and non-key node subgraphs to extract richer spatial-temporal dependencies.

5.3 Ablation Configuration

- **w/o HC:** Replaces the temporal fusion hypergraph convolution with a standard GCN and removes the dynamic hypergraph module.
- **w/o DHG:** Removes the transfer perception dynamic hypergraph module and uses a static hypergraph throughout training and inference.
- **w/o PA:** Removes the priority-aware temporal embedding network, thus disabling the modeling of processing order distributions.

5.4 Performance Comparison and Analysis

Comparison with Baselines. As shown in Table 2, our model consistently outperforms all baselines across datasets and metrics, demonstrating superior accuracy and robustness. Graph-based methods (STFGNN, PDFormer, STPGNN) generally outperform advanced time series prediction models, highlighting the importance of modeling complex spatial-temporal relationships among products, which critically influence completion volumes. Specifically, STPGNN performs second-best on most datasets, benefiting from its ability to distinguish key nodes and construct targeted subgraphs. However, its performance deteriorates on the sparse dataset Fac_03, indicating its limitations under data-scarce scenarios. In contrast, PDFormer achieves better accuracy on Fac_03 due to its delay-aware transformation module, which leverages external memory to enhance temporal modeling. Among time series forecasting models, Informer–based on an encoder-decoder architecture–underperforms the encoder-only models, likely due to training instability and error accumulation introduced by its decoder.

Ablation Study. To evaluate the contribution of each component in PTE-DHNet, we remove three key modules individually. Replacing the hypergraph with a standard graph in the **w/o HC** variant leads to a 4.60% increase in average WMAPE, indicating that hypergraphs play a crucial role in modeling complex temporal dependencies. Moreover, since the **w/o DHG** variant fails to dynamically generate hypergraph structure, the results show a 2.58% increase in average WMAPE, highlighting the importance of dynamically capturing the impact of transfer patterns. Furthermore, the **w/o PA** variant removes the explicit prediction of processing order, also shows a significant WMAPE degradation, confirming the importance of order modeling for accurate estimation of completion volume. Notably, consistent with baseline comparisons, the priority aware module proves especially beneficial in sparse data scenarios.

Table 2. Performance Benchmarking and Component Analysis

Model	Fac_01			Fac_02			Fac_03		
	MAE	WMAPE(%)	RMSE	MAE	WMAPE(%)	RMSE	MAE	WMAPE(%)	RMSE
Informer	48.22	28.51	56.70	28.97	26.17	40.83	10.65	28.80	34.40
iTransformer	44.29	26.19	51.78	27.22	24.60	39.00	9.59	25.93	32.06
xLSTM	43.91	25.96	48.28	27.53	24.87	39.34	9.52	25.74	35.11
STFGNN	40.90	24.18	50.41	24.27	21.92	38.70	8.55	23.11	30.43
PDFormer	40.35	23.86	51.00	23.65	21.37	38.20	7.97	21.54	32.28
STPGNN	39.58	23.40	51.53	22.09	19.95	37.30	8.17	22.08	31.33
PTE-DHNet	**28.95**	**17.12**	**39.16**	**15.77**	**14.25**	**32.17**	**5.66**	**15.31**	**22.08**
w/o HC	35.70	21.11	46.05	21.71	19.61	36.19	7.30	19.75	30.18
w/o DHG	32.96	19.49	42.31	19.55	17.69	35.32	6.29	17.02	28.01
w/o PA	34.34	20.30	42.78	20.46	18.48	34.49	7.36	19.89	31.36

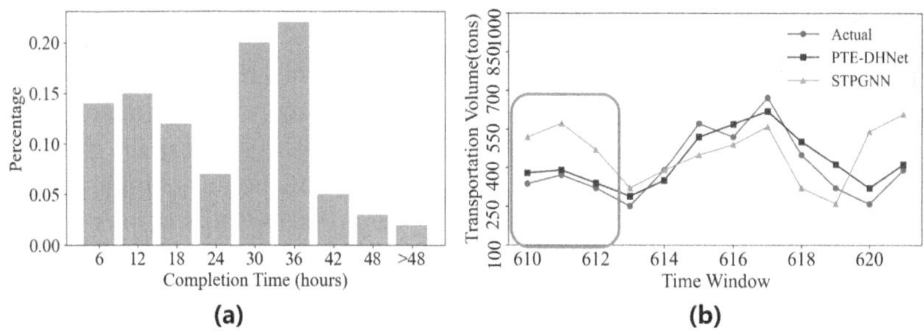

Fig. 3. (a):Completion time distribution of HP-type steel in the final processing step.(b):Transportation volume prediction for HP-type steel destined for Wuxi.

Case Study We conduct a case study to investigate the impact of processing order on transportation volume. Figure 3 shows a representative example from the Fac_01 dataset. In Fig. 3(a), over 40% of HP-type steel waiting for processing within time window 610 due to resource contention, resulting in reduced output in the adjacent window. As shown in Fig. 3(b), PTE-DHNet explicitly models processing order impact, closely matches the actual volume, while STPGNN fails to capture this effect and overestimates the transportation volume.

6 Conclusion

This paper proposes PTE-DHNet, a novel framework that effectively addresses two key challenges in steel transportation volume prediction: the uncertainty in processing order and complex temporal dependencies between processing steps. The framework achieves this through two synergistic innovations: a *Priority-aware Temporal Embedding module* that adaptively models intra-step order dynamics by fusing historical patterns with real-time priority adjustments, and a *Temporal Dynamic Hypergraph Generation mechanism* that explicitly quantifies inter-step process influences through dynamic hypergraph. Extensive experiments on three real-world datasets demonstrate that PTE-DHNet outperforms state-of-the-art methods, with its hypergraph architecture proving particularly effective in capturing nonlinear production cascades. Future work will explore incorporating detailed equipment operation data to further enhance fine-grained modeling of the production process and improve long-term forecasting stability.

References

1. Alharthi, M., Mahmood, A.: xLSTMTime: long-term time series forecasting with xLSTM. AI **5**(3), 1482–1495 (2024)
2. Barzola-Monteses, J., Gómez-Romero, J., Espinoza-Andaluz, M., et al.: Hydropower production prediction using artificial neural networks: an ecuadorian application case. Neural Comput. Appl. **34**(16), 13253–13266 (2022)
3. Essien, A., Giannetti, C.: A deep learning model for smart manufacturing using convolutional LSTM neural network autoencoders. IEEE Trans. Ind. Inf. **16**(9), 6069–6078 (2020)
4. Gao, Y., Feng, Y., Ji, S., et al.: HGNN+: general hypergraph neural networks. IEEE Trans. Pattern Anal. Mach. Intell. **45**(3), 3181–3199 (2023)
5. Guo, S., Lin, Y., Wan, H., et al.: Learning dynamics and heterogeneity of spatial-temporal graph data for traffic forecasting. IEEE Trans. Knowl. Data Eng. **34**(11), 5415–5428 (2021)
6. Han, Y., Du, Z., Geng, Z., et al.: Production prediction model of complex industrial processes based on GRU neural network, pp. 1102–1106 (2020)
7. Jiang, J., Han, C., Zhao, W. X., et al.: PDFormer: propagation delay-aware dynamic long-range transformer for traffic flow prediction. In: AAAI, vol. 37, no. 4, pp. 4365–4373 (2023)
8. Kong, W., Guo, Z., Liu, Y.: Spatio-temporal pivotal graph neural networks for traffic flow forecasting. In: AAAI, vol. 38, no. 8, pp. 8627–8635 (2024)
9. Li, M., Zhu, Z.: Spatial-temporal fusion graph neural networks for traffic flow forecasting. In: AAAI, vol. 35, no. 5, pp. 4189–4196 (2021)
10. Liu, J., Chen, L., Xu, W., et al.: Novel production prediction model of gasoline production processes for energy saving and economic increasing based on AM-GRU integrating the UMAP algorithm. Energy (Oxford) **261**, 125536 (2023)
11. Liu, Y., Hu, T., Zhang, H., et al.: iTransformer: inverted transformers are effective for time series forecasting. In: ICLR, vol. 12 (2024)
12. Paparrizos, J., Gravano, L.: k-shape: efficient and accurate clustering of time series. SIGMOD **45**(1), 69–76 (2016)
13. Peng, Z., Zongqian, W., Xiangxiang, Z., et al.: Totally dynamic hypergraph neural networks. IJCAI **32**, 2476–2483 (2023)
14. Ren, Y., Zhao, Y., Zhang, T., et al.: Pathformer: multi-scale transformers with adaptive pathways for time series forecasting. In: ICLR, vol. 12 (2024)
15. Rong, Y., Yao, J., Liu, J., et al.: GBTTE: graph attention network based bus travel time estimation. In: CIKM, vol. 32, pp. 4794–4800 (2023)
16. Wang, J., Yang, H., Shan, J., Song, X., Jiang, J. Attention-based spatial-temporal fusion networks for traffic flow prediction. In: WISA, pp. 500–511 (2024)
17. Wu, D., Hu, J. Y.-C., Li, W., et al.: STanHop: sparse tandem hopfield model for memory-enhanced time series prediction. In: ICLR, vol. 12 (2024)

18. Wu, Z., Pan, S., Long, G., et al.: Graph WaveNet for deep spatial-temporal graph modeling. IJCAI **28**, 1907–1913 (2019)
19. Yu, T., Yin, H., Zhu, Z.: Spatio-temporal graph convolutional neural network: a deep learning framework for traffic forecasting. IJCAI **27**, 3634–3640 (2018)
20. Zhao, Y., Luo, X., Ju, W.: Dynamic hypergraph structure learning for traffic flow forecasting. In: ICDE, vol. 39, pp. 2303–2316 (2023)
21. Zhou, H., Zhang, S., Peng, J., et al.: Informer: beyond efficient transformer for long sequence time-series forecasting. In: AAAI, vol. 35, no. 12, pp. 11106–11115 (2021)

Optimizing Few-Shot NL2SQL for Sepsis Data Analysis in Medical Databases

Zeming Li[1(✉)], Xinhao Liu[1], Xianbo Liu[1], Yuhang Hu[1], Peng Ren[2], and Chunxiao Xing[2]

[1] School of Computer Science and Technology, Beijing Institute of Technology, Beijing 100081, China
{lizeming,liuxinhao,liuxianbo2024,huyuhang_21}@bit.edu.cn
[2] BNRist, DCST, RIIT, Tsinghua University, Beijing 100084, China
{renpeng,xingcx}@tsinghua.edu.cn

Abstract. Sepsis exhibits high mortality rates, and efficient data exploration is vital for advancing clinical research. However, the complexity of sepsis research poses significant challenges for healthcare professionals, as navigating large and intricate medical databases demands specialized skills and substantial effort. Natural Language to SQL (NL2SQL) frameworks, leveraging few-shot learning, offer a promising solution by allowing clinicians to query databases like MIMIC-IV using natural language, thereby bypassing the need for deep SQL expertise. Yet, current NL2SQL systems, built on language models pretrained on general-domain data, encounter critical issues in the healthcare domain due to data scarcity and insufficient domain-specific training. To address these challenges, we present a customized NL2SQL framework for medical data exploration that integrates data augmentation techniques to mitigate data scarcity and a query-retrieval mechanism to select relevant SQL examples. This approach significantly enhances few-shot learning capabilities, boosting accuracy from 0.76 to 0.94 on unseen queries, and thereby strengthens the utility of NL2SQL in clinical research.

Keywords: NL2SQL · data augmentation · medical databases

1 Introduction

Sepsis is a severe clinical syndrome and a major global public health challenge, distinguished by its substantial incidence and mortality rates. According to estimates, the global incidence of hospital-treated sepsis is significant, with high mortality rates reported across various studies [1]. In hospital settings, mortality rates for sepsis range from 30% to 40%, escalating to as high as 50% in cases of septic shock [2]. These statistics underscore not only the immediate threat to patient survival but also the profound burden on healthcare systems worldwide. The complexity of sepsis arises from its rapid progression and diverse clinical presentations, rendering early detection and timely intervention critical to improving outcomes. However, traditional diagnostic approaches often depend heavily

S. Yang et al. (Eds.): WISA 2025, LNCS 16299, pp. 29–40, 2026.
https://doi.org/10.1007/978-981-95-6426-2_3

on physician expertise and the availability of actionable data. Consequently, the efficient exploration of sepsis-related electronic medical record (EMR) data emerges as a vital tool, enabling real-time patient monitoring while delivering data-driven insights into epidemiology, treatment optimization, and outcome prediction. This data-centric paradigm, a cornerstone of modern medicine, holds particular significance for sepsis—a quintessential high-risk condition.

EMR data serves as the cornerstone for achieving these goals, typically stored in structured databases encompassing longitudinal patient information across multiple tables, such as demographics, diagnostic records, treatment histories, prescriptions, and laboratory results. This wealth of data offers unparalleled opportunities to elucidate sepsis pathophysiology, identify critical biomarkers, and assess therapeutic efficacy. However, its value is tempered by inherent complexity. The multi-table structure, temporal dynamics, and domain-specific terminology render the extraction of meaningful insights technically demanding. For instance, retrieving trends in a sepsis patient's oxygenation index (PaO2/FiO2 ratio) during intensive care unit (ICU) stays requires multi-table joins, time-series filtering, and tailored computations. Such intricacy often precludes clinicians from fully harnessing EMR data, thereby constraining its utility in both clinical practice and research.

Existing medical data query systems aim to mitigate these challenges but remain inadequate. Predominantly rule-based, these systems require clinicians to manually translate clinical questions into SQL queries—a process that demands database knowledge and programming proficiency while suffering from limited flexibility and scalability. For example, investigating the relationship between white blood cell counts and antibiotic administration in sepsis patients over a specific period might necessitate manually configuring multiple filters and crafting intricate query logic. This approach is both time-consuming and inefficient, particularly ill-suited to the fast-paced demands of clinical decision-making. More critically, such systems fail to accommodate clinicians' diverse exploratory needs—such as validating novel hypotheses or uncovering latent disease patterns—revealing a profound disconnect between clinical requirements and current technological solutions.

Natural Language to SQL (NL2SQL) frameworks, powered by advances in natural language processing, provide an effective solution by enabling clinicians to query databases using natural language, eliminating the need for programming expertise. In general domains, NL2SQL systems excel on benchmarks like Spider and WikiSQL [3,4], adeptly parsing complex queries. Recent work, such as MQRLD [5], further demonstrates the potential of query-aware feature representation to enhance retrieval efficiency in complex data environments, informing NL2SQL advancements. However, applying NL2SQL to healthcare, particularly for sepsis, presents distinct challenges, including multi-table joins, time-series analysis, specialized terminology like the Sequential Organ Failure Assessment (SOFA) score—a sepsis severity metric—and constraints like data scarcity and privacy regulations. Existing medical datasets, such as MIMICSQL and

EHRSQL [6], primarily support simple retrieval tasks, failing to address the multidimensional, domain-specific queries required for sepsis research.

To overcome these limitations, we propose a novel NL2SQL framework tailored for sepsis data exploration on the MIMIC-IV dataset. Our approach employs data augmentation to generate diverse query-SQL pairs, addressing data scarcity, and integrates few-shot learning with a query-retrieval mechanism to enhance adaptability to complex, multi-table clinical queries. Experiments demonstrate a significant improvement in query execution accuracy, from 0.76 (baseline medical query systems) to 0.94, underscoring the framework's efficacy in healthcare settings. By aligning technical advancements with clinical needs, our work not only advances sepsis research but also provides a scalable model for data-driven clinical decision-making in critical care conditions, enhancing patient outcomes and healthcare efficiency.

2 Related Work

Natural Language to SQL (NL2SQL) stands as a central pillar in natural language processing, focused on converting human queries into structured database commands. Our long-term exploration of this field highlights its progression from rule-based systems to sophisticated neural architectures. Early approaches predominantly relied on syntactic parsing and slot-filling techniques [7], which offered limited flexibility. The emergence of sequence-to-sequence (Seq2Seq) models, enhanced by attention mechanisms [8], marked a substantial leap forward, achieving execution accuracies of around 60–70% on benchmarks like WikiSQL [4]. The advent of large language models (LLMs) has since reshaped the domain, capitalizing on extensive pre-training and contextual comprehension to adeptly handle complex queries—such as multi-table joins and nested conditions—on datasets like Spider [3], with accuracies exceeding 85%. Innovations like constraint-aware decoding [9] exemplify the cutting-edge progress in general-domain NL2SQL.

Our literature review underscores inherent challenges when extending these techniques to the medical domain. Databases like MIMIC-IV present intricate schemas with multiple interconnected tables, temporal dynamics, and specialized terminology (e.g., "lactate clearance"), imposing demands that surpass standard benchmarks. A frequently cited obstacle in the research community is the scarcity of annotated query-SQL pairs, a problem intensified by privacy regulations such as HIPAA [10], which severely restricts model training. Additionally, the resource-intensive nature of LLMs often necessitates cloud-based deployment, introducing security risks—illustrated by the 2023 HCA Healthcare data breach affecting 11 million patients [11]. Together, these issues drive the need for customized, efficient solutions tailored to clinical settings.

Prior research on adapting natural language to SQL (NL2SQL) for medical applications has established a critical foundation, yet it struggles to meet the specialized demands of sepsis research, where intricate schemas and temporal dynamics challenge general domain advancements. Early efforts, such as TREQS

[12], pioneered Seq2Seq-based architectures with attentive copying mechanisms and domain-specific lookup tables to query electronic medical records, achieving accuracies of 60–70% on the MIMICSQL dataset. While suitable for simple tasks like patient count retrieval, TREQS falters on the multi-table, time-sensitive queries essential for sepsis analysis, such as tracking lactate clearance trends. The advent of large language models (LLMs) has marked a paradigm shift, offering enhanced contextual understanding, though medical domain complexities temper the high accuracies (e.g., 85% on Spider [3]) observed in general benchmarks. The EHRSQL dataset [13] has enabled robust evaluations, with privacy-preserving on-premises LLMs, deployed locally to ensure data security, achieving competitive accuracies of 66–68% [14–16]. Among these, Jo et al. [15] leverage SQLCoder, a pre-trained LLM tailored for NL2SQL tasks, demonstrating the efficacy of such models for medical queries. However, their approach relies solely on fine-tuned capabilities, neglecting advanced techniques like retrieval-augmented generation or domain-specific example selection, which limits its performance on sepsis-specific queries requiring temporal reasoning, such as sequential organ failure assessments.

Drawing from these studies, we identify several persistent shortcomings: limited support for complex, multi-dimensional queries; reliance on scarce or proprietary data; and architectural designs that compromise privacy. Building on these insights, we propose a specialized NL2SQL framework for sepsis data exploration in MIMIC-IV. In contrast to TREQS's Seq2Seq constraints or EHRSQL's API-reliant approach, we leverage localized 7B-scale models—proven effective in recent evaluations—to ensure both privacy and performance. To tackle data scarcity, we expand 236 real sepsis queries into a diverse set of 760 pairs, encompassing intricate clinical scenarios. By integrating few-shot learning with retrieval-based example selection, our method advances beyond prior work, delivering a robust, domain-tailored tool for sepsis research.

3 Methodology

In this section, we delineate the methodology devised to construct and evaluate a Natural Language to SQL (NL2SQL) system tailored for sepsis-related data extraction from the MIMIC-IV database. Our approach integrates two core components: the systematic construction and augmentation of a specialized dataset, and the enhancement of NL2SQL model performance through parameter-efficient fine-tuning and retrieval-based example selection. This framework, illustrated in Fig. 1, addresses the challenges of query complexity, domain-specific terminology, and data sparsity inherent in clinical NL2SQL applications.

3.1 SQL Query Augmentation Pipeline

The cornerstone of our NL2SQL system is a meticulously curated dataset that encapsulates the intricacies of sepsis-related clinical queries. We propose a systematic pipeline encompassing authentic query collection, template extraction,

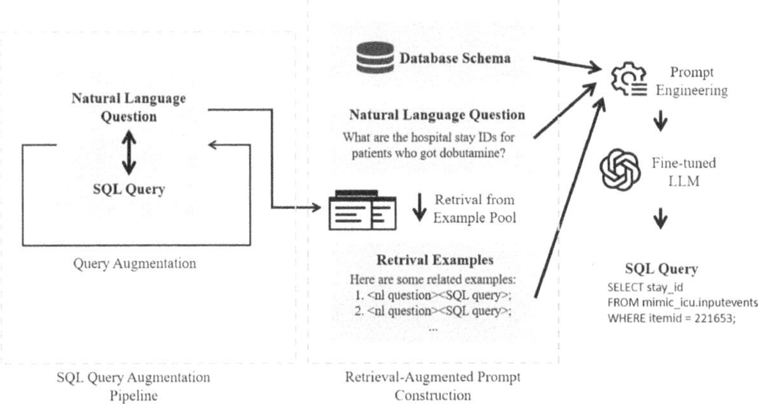

Fig. 1. Overall architecture of the proposed NL2SQL system for sepsis-related data extraction.

algorithmic augmentation, and natural language generation to produce a robust dataset of 760 NL2SQL pairs.

The process begins with the acquisition of 236 real-world SQL queries sourced from hospital interactions with the MIMIC-IV database, each focused on sepsis-related tasks such as patient cohort identification, laboratory result retrieval, and treatment outcome analysis. These queries exhibit significant complexity, incorporating multi-table joins, nested subqueries, and temporal constraints. From this corpus, we extract 236 parameterizable templates that abstract the structural patterns of the original queries while enabling parameterization of numerical and categorical values.

To amplify this foundational dataset, we employ an iterative power method to generate diverse SQL query variants, expanding the initial 236 queries to 760 unique, executable SQL queries. This method starts by identifying predicates in each template—conditions like `WHERE lab_value > 50`—and determining the valid range of values for each parameter based on the MIMIC-IV database. It then iteratively adjusts these parameters by computing incremental changes proportional to the value range, testing both positive and negative directions. For each adjustment, a new query is constructed and validated against the database to ensure it returns non-empty results, preserving relevance to sepsis data extraction. This process enhances structural and contextual diversity, as visualized in Fig. 2.

To finalize the NL2SQL dataset, each of the 760 SQL queries is paired with a corresponding natural language description generated by the Qwen-Instruct-7B model, accessed via its API. We employ carefully crafted prompts to ensure the descriptions are varied, contextually appropriate, and aligned with human linguistic conventions. For example, a query retrieving patients with specific lab results might be paired with "Find all patients with a positive blood culture

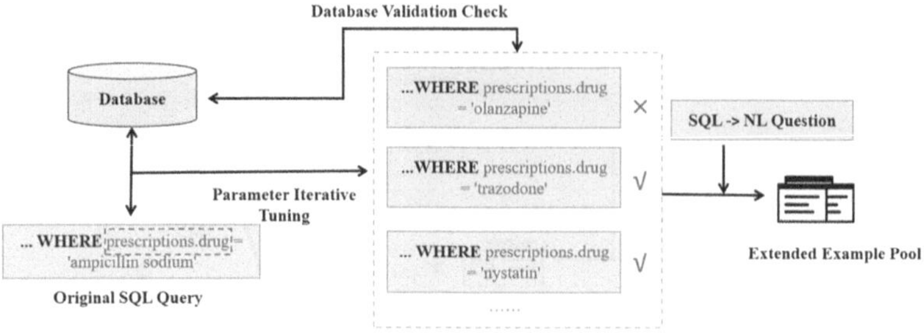

Fig. 2. Flowchart of the power method for SQL query augmentation.

within 24 h of sepsis onset." This results in a comprehensive dataset of 760 NL2SQL pairs, primed to support diverse sepsis-related data extraction tasks.

3.2 Retrieval-Augmented Prompt Construction

Clinical NL2SQL tasks demand models capable of navigating complex queries and sparse data. To this end, we propose a framework that enhances model performance through parameter-efficient fine-tuning and a retrieval-based example selection mechanism, tailored to sepsis queries on the MIMIC-IV dataset. This process is depicted in Fig. 3.

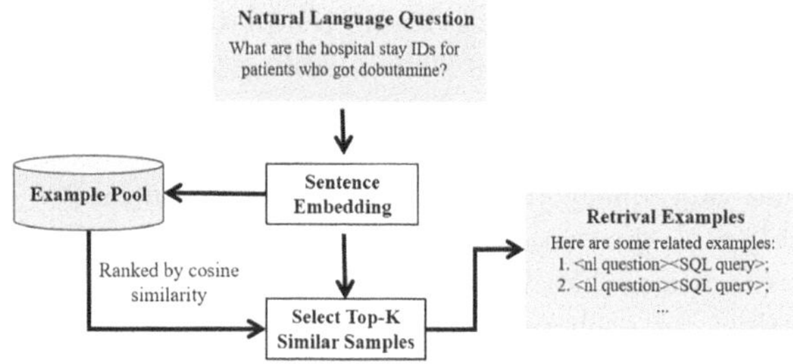

Fig. 3. Schematic of the retrieval-based example selection mechanism for enhancing NL2SQL performance.

We adapt pre-trained NL2SQL models to the sepsis domain by fine-tuning them on our augmented dataset using Low-Rank Adaptation (LoRA) [17] with a rank of 16. This approach updates only 0.1% of the model's parameters, striking a balance between expressiveness and computational efficiency, as determined through empirical tuning. To bolster inference, input prompts are enriched with the MIMIC-IV schema—e.g., `patients: [subject_id, gender, dob]` and `diagnoses_icd: [subject_id, icd_code, icd_version]`—providing critical structural context. Further, we enhance few-shot learning by incorporating k

semantically relevant examples into the prompt. These examples are selected via a top-k retrieval approach inspired by Retrieval-Augmented Generation (RAG) [18]. We encode input and training queries into 384-dimensional normalized embeddings using the Sentence Transformer model (`all-MiniLM-L6-v2`), chosen for its efficiency and robust pre-training. These embeddings are indexed in a vector search engine with an Inverted File Index (IVF) of 100 clusters and Hierarchical Navigable Small World (HNSW) quantization, optimizing retrieval speed and accuracy. For each input query, we retrieve the top-k most similar training queries based on cosine similarity, integrating their SQL-NL pairs into the prompt. This dynamic selection enables the model to leverage contextually pertinent examples, improving its ability to generate accurate SQL queries for sepsis-related tasks.

4 Experiments and Results

To assess the efficacy of our methodology, we conduct experiments on the constructed dataset, executed on a single NVIDIA H5000 GPU with the fine-tuned model deployed on a PostgreSQL server hosting the MIMIC-IV database. Our evaluation comprises three phases: a benchmark of pre-trained NL2SQL models, a comparison against state-of-the-art methods, and an ablation study analyzing the contributions of data augmentation and retrieval-based example selection.

4.1 Model Benchmarking

We evaluate four pre-trained NL2SQL models—DeepSeek-R1-SQL-EXP, natural-sql-7b, SQLCoder-llama3-8B, and sqlcoder-codellama-7b—each with 7âĂŞ8 billion parameters, selected for their scale and architectural diversity. In a few-shot setting, prompts include the MIMIC-IV schema and five fixed examples. Performance is measured on the original 47 test queries using two metrics: Exact Match (EM), the proportion of generated SQL queries that exactly match the ground-truth SQL queries, expressed as a percentage, and Execution Accuracy (EX), the percentage of queries that execute correctly against MIMIC-IV, yielding expected results.

Results, presented in Table 1, reveal that SQLCoder-llama3-8B outperforms its peers, achieving an EM score of 66.73% and an EX score of 76.44%. The higher EX score compared to EM indicates that, while SQLCoder-llama3-8B may not always generate SQL queries identical to the ground-truth, it produces functionally equivalent queries that yield correct results when executed against the MIMIC-IV database. This superior performance underscores its aptitude for sepsis-specific queries, prompting its selection for further analysis.

4.2 Comparison with State-of-the-Art Methods

To validate the effectiveness of our approach, we compare our best-performing configuration (SQLCoder-llama3-8B with data augmentation and top-k = 7

Table 1. Performance of pre-trained NL2SQL models on 47 MIMIC-IV sepsis test queries. SQLCoder-llama3-8B is selected for further analysis.

Model	EM (%)	EX (%)
DeepSeek-R1-SQL-EXP	57.48	51.86
natural-sql-7b	69.33	56.34
SQLCoder-llama3-8B	66.73	76.44
sqlcoder-codellama-7b	64.88	55.44

retrieval) against three recent state-of-the-art methods, implemented and evaluated on our MIMIC-IV dataset tailored for sepsis queries: SCAS [14], LTRC-IIITH [16], and LG AI & KAIST [15]. These methods, originally developed for the EHRSQL 2024 shared task, were adapted to our dataset to ensure a fair comparison under identical conditions. We assess performance on our 152 test queries from the augmented dataset, using two metrics: **Exact Match (EM)**, the percentage of generated SQL queries that exactly match the ground-truth, and **Execution Accuracy (EX)**, the percentage of queries that execute correctly, yielding expected results.

Table 2 presents the results. Our method achieves an EM of 88.00% and an EX of 94.00%, significantly outperforming SCAS (EM: 50.00%, EX: 53.21%), LTRC-IIITH (EM: 52.00%, EX: 55.00%), and LG AI & KAIST (EM: 55.00%, EX: 60.00%). The substantial gains highlight the efficacy of our data augmentation and retrieval-based strategies in handling complex, sepsis-specific clinical queries. Note that EM values for competing methods are approximate, as their primary focus was on execution accuracy; our evaluation provides a comprehensive benchmark for both metrics on MIMIC-IV.

Table 2. Comparison of our best result with state-of-the-art methods, evaluated on 152 MIMIC-IV sepsis test queries. EM values for SCAS, LTRC-IIITH, and LG AI & KAIST are approximate, based on their performance trends.

Method	EM (%)	EX (%)
SCAS [14]	42.52	55.70
LTRC-IIITH [16]	38.16	51.23
LG AI & KAIST [15]	32.62	39.24
Ours (SQLCoder-llama3-8B, $k = 7$)	88.00	94.00

4.3 Ablation Study and Analysis

We conduct an ablation study to dissect the impacts of data augmentation and retrieval-based example selection on SQLCoder-llama3-8B's performance. The model is fine-tuned on two datasets: the original 189 training queries and the augmented 608 training queries, with results averaged over three runs to mitigate variance. We compare fixed prompts (five examples) against dynamic top-k retrieval-based selection ($k = 1, 3, 5, 7$), evaluated on the respective test sets of 47 and 152 queries.

Table 3 summarizes the findings. With fixed prompts, data augmentation alone elevates performance: EM rises from 64.00% to 68.00%, and EX from 58.00% to 79.00%, highlighting the value of a richer training corpus. Integrating top-k retrieval further amplifies these gains. On the augmented dataset, performance peaks at $k = 7$, with an EM of 88.00% and an EX of 94.00%—substantial improvements over the baseline. On the original dataset, EX peaks at 87.00% for $k = 3$, dipping slightly to 86.00% at $k = 7$, suggesting that retrieval benefits scale with dataset diversity.

Table 3. Ablation study results for SQLCoder-llama3-8B, comparing data augmentation and top-k retrieval-based example selection on 47 (original) and 152 (augmented) test queries.

Methods	Example Selection	EM (%)	EX (%)
Original (236 queries)	Fixed (5 examples)	64.00	58.00
Augmented (760 queries)	Fixed (5 examples)	68.00	79.00
Original (236 queries)	Top-k Retrieval ($k = 1$)	72.00	77.00
Original (236 queries)	Top-k Retrieval ($k = 3$)	79.00	87.00
Original (236 queries)	Top-k Retrieval ($k = 5$)	80.00	80.00
Original (236 queries)	Top-k Retrieval ($k = 7$)	82.00	86.00
Augmented (760 queries)	Top-k Retrieval ($k = 1$)	74.00	75.00
Augmented (760 queries)	Top-k Retrieval ($k = 3$)	82.00	88.00
Augmented (760 queries)	Top-k Retrieval ($k = 5$)	84.00	91.00
Augmented (760 queries)	Top-k Retrieval ($k = 7$)	**88.00**	**94.00**

4.4 Analysis

The results affirm the synergistic efficacy of our methodology. Data augmentation enriches the training corpus, enabling the model to generalize across diverse sepsis queries, while retrieval-based example selection dynamically tailors prompts to each input, enhancing precision. The peak performance on the augmented dataset with $k = 7$ (88.00% EM and 94.00% EX) demonstrates the framework's capacity to handle complex clinical NL2SQL tasks. The slight EX dip at higher k on the original dataset suggests an optimal k may depend on dataset size, warranting further exploration. Compared to state-of-the-art methods, our approach excels in execution accuracy, achieving 94.00% EX against 53.21% to approximately 60.00% for EHRSQL methods, particularly for sepsis-specific queries, underscoring its potential for clinical deployment. Collectively, these advancements establish a robust foundation for deploying NL2SQL systems in healthcare settings like sepsis management.

5 Discussion and Future Work

Our NL2SQL framework significantly enhances sepsis data exploration in MIMIC-IV by enabling natural language queries. Integrating data augmentation with retrieval-based example selection, it improves SQL similarity (0.64 to 0.88) and execution accuracy (0.58 to 0.94) on unseen queries. This streamlines access to complex medical data, lowering technical barriers for clinicians and supporting data-driven insights for sepsis, a high-mortality condition with diagnostic challenges.

However, the framework struggles with highly complex queries, such as those requiring extensive multi-table joins or nested temporal conditions (e.g., tracking multi-organ failure trends). These queries may produce syntactic errors or lack semantic accuracy due to MIMIC-IV's intricate schema and limited training data diversity. Despite expanding from 236 to 760 query-SQL pairs, the dataset does not fully capture real-world clinical scenarios. Additionally, fine-tuned models may miss nuanced medical terminology, introducing biases or gaps in query interpretation.

Future research should prioritize expanding the dataset with complex, representative queries, possibly through healthcare collaborations or advanced synthetic data generation. Graph-based schema representations, modeling tables as nodes and relationships as edges, could improve handling of joins and query accuracy. A clinician feedback loop, assessing and refining SQL outputs, would align the model with practical needs and reveal new query patterns. Extending the framework to conditions like heart failure or diabetes requires tailored datasets and retrieval mechanisms, broadening its clinical impact.

Acknowledgments. This work was supported by the National Natural Science Foundation of China (No. 42371480).

References

1. Fleischmann, C., Scherag, A., Adhikari, N.K., et al.: Assessment of global incidence and mortality of hospital-treated sepsis. current estimates and limitations. Am. J. Respiratory Critical Care Medicine, **193**(3), 259–272 (2016)
2. Bauer, M., Gerlach, H., Vogelmann, T., et al.: Mortality in sepsis and septic shock in europe, north america and australia between 2009 and 2019–results from a systematic review and meta-analysis. Crit. Care **24**(1), 239 (2020)
3. Yu, T., et al.: Spider: a large-scale human-labeled dataset for complex and cross-domain semantic parsing and text-to-SQL task. In: Riloff, E., Chiang, D., Hockenmaier, J., Tsujii, J. I., editors, Proceedings of the 2018 Conference on Empirical Methods in Natural Language Processing. Association for Computational Linguistics, pp. 3911–3921, Brussels (2018)

4. Zhong, V., Xiong, C., Socher, R.: Seq2SQL: generating structured queries from natural language using reinforcement learning. https://arxiv.org/abs/1709.00103 (2017)
5. Sheng, M., et al.: MQRLD: a multimodal data retrieval platform with query-aware feature representation and learned index based on data lake. Inf. Process. Manag. **62**(4), 104101 (2025)
6. Lee, G., et al.: EHRSQL: a practical text-to-SQL benchmark for electronic health records. In: Koyejo, S., Mohamed, S., Agarwal, A., Belgrave, D., Cho, K., Oh, A., eds., Advances in Neural Information Processing Systems, vol. 35, pp. 15589–15601. Curran Associates, Inc. (2022)
7. Yaghmazadeh, N., Wang, Y., Dillig, I., Dillig, T.: Query synthesis from natural language. Proc. ACM Program. Lang., **1**(OOPSLA), 1–26 (2017)
8. Dong, L., Lapata, M.: Language to logical form with neural attention. In Proceedings of the 54th Annual Meeting of the Association for Computational Linguistics (Vol. 1: Long Papers), pp. 33–43. Association for Computational Linguistics, Berlin (2016)
9. Scholak, T., Schucher, N., Bahdanau, D.: PICARD: parsing incrementally for constrained auto-regressive decoding from language models. In: Moens, M.F., Huang, X.J., Specia, L., Yih, W.T., eds., Proceedings of the 2021 Conference on Empirical Methods in Natural Language Processing, pp. 9895–9901. Association for Computational Linguistics, Online and Punta Cana, Dominican Republic (2021)
10. U.S. Department of Health and Human Services. HIPAA enforcement: penalties for non-compliance. https://www.hhs.gov/hipaa/for-professionals/compliance-enforcement/index.html (2020)
11. HIPAA Journal. HCA healthcare data breach impacts 11 million patients. https://www.hipaajournal.com/hca-healthcare-data-breach-impacts-11-million-patients/ (2023)
12. Wang, P., Shi, T., Reddy, C.K.: Text-to-SQL generation for question answering on electronic medical records. In: Proceedings of The Web Conference 2020, pp. 350–361. Association for Computing Machinery, Taipei (2020)
13. Lee, G., Kweon, S., Bae, S., Choi, E.: Overview of the EHRSQL 2024 shared task on reliable text-to-SQL modeling on electronic health records. In: Abacha, A.B., Bethard, S., Bitterman, D., Naumann, T., Roberts, K., eds., Proceedings of the 6th Clinical Natural Language Processing Workshop, pp. 644–654. Association for Computational Linguistics, Mexico City (2024)
14. Jabir, M., Kanakarajan, K., Sankarasubbu, M.: Saama technologies at EHRSQL 2024: SQL generation through classification answer selector by LLM. In: Naumann, T., Abacha, A.B., Bethard, S., Roberts, K., Bitterman, D., eds., Proceedings of the 6th Clinical Natural Language Processing Workshop, pp. 655–671. Association for Computational Linguistics, Mexico City (2024)
15. Jo, Y., Lee, S., Seo, M., Hwang, S.J., Lee, M.: LG AI research & KAIST at EHRSQL 2024: Self-training large language models with pseudo-labeled unanswerable questions for a reliable text-to-SQL system on EHRs. In: Naumann, T., Abacha, A.B., Bethard, S., Roberts, K., Bitterman, D., eds., Proceedings of the 6th Clinical Natural Language Processing Workshop, pp. 635–643. Association for Computational Linguistics, Mexico City (2024)

16. Thomas, J., Mishra, P., Sharma, D.M., Krishnamurthy, P.: LTRC-IIITH at EHRSQL 2024: Enhancing reliability of text-to-SQL systems through abstention and confidence thresholding. In Naumann, T., Abacha, A.B., Bethard, S., Roberts, K., Bitterman, D., eds., Proceedings of the 6th Clinical Natural Language Processing Workshop, pp. 697–702. Association for Computational Linguistics, Mexico City (2024)
17. Hu, E.J., et al.: LoRA: low-rank adaptation of large language models. In: The Tenth International Conference on Learning Representations, ICLR 2022, Virtual Event (2022)
18. Meng, X., Zhang, Z., Cui, S., Huo, H.: Top-k collective spatial keyword approximate query. In: Jin, C., Yang, S., Shang, X., Wang, H., Zhang, Y., eds., Web Information Systems and Applications, vol. 14883 of Lecture Notes in Computer Science. Springer, Singapore (2024)

KSDF: A Knowledge-Based Sensitivity Detection Framework for Film Reviews Using Graph Convolutional Network

Ruichen Liu[1,2], Jiaying Wang[1,2](✉), Jing Shan[1,2], Xiaoxu Song[1,2], and Haiwen Feng[1,2]

[1] Shenyang University of Technology, Shenyang 110870, China
Jiaying@sut.edu.cn
[2] Shenyang Key Laboratory of Intelligent Technology of Advanced Industrial Equipment Manufacturing, Shenyang 110870, China

Abstract. The proliferation of user-generated content on film review platforms has heightened reliance on user comments for movie selection. However, the misuse of comment sections to spread illegal, misleading, or offensive content poses serious moderation challenges. Traditional detection methods, primarily based on textual and statistical features, often fail to address evasion tactics such as homophonic substitutions and metaphorical expressions. To overcome these limitations, we propose KSDF, a Knowledge-based Sensitivity Detection Framework that integrates BERT with a Knowledge Graph (KG). KSDF employs a Graph Convolutional Network (GCN) to extract deep knowledge features from a sensitive word KG and utilizes BERT to capture semantic representations of review texts. A semantic association mechanism aligns knowledge-based and text-based features, constructing a multi-dimensional joint discrimination model. Additionally, we introduce an entity embedding enhancement strategy, incorporating named entities from review texts as auxiliary features to refine detection performance. Extensive experiments on real-world datasets demonstrate that KSDF achieves superior sensitivity detection accuracy, outperforming state-of-the-art baselines.

Keywords: Sensitive information detection · BERT · Knowledge graph · Graph convolutional network

1 Introduction

In recent years, with the rapid expansion of the Internet industry, film review aggregation platforms such as Maoyan and Douban have become essential reference sources for audiences making movie-watching decisions [25]. However, the current review landscape faces two major challenges. On one hand, numerous reviews contain illegal remarks involving sensitive topics across various domains [21]. On the other hand, fan communities and commercial interest

S. Yang et al. (Eds.): WISA 2025, LNCS 16299, pp. 41–53, 2026.
https://doi.org/10.1007/978-981-95-6426-2_4

groups frequently manipulate public opinion by employing online trolls to attack competing films. These groups flood review platforms with misleading, low-quality, or offensive content, engaging in malicious competition that not only saturates platforms with invalid information but also distorts user perception.

Current review detection methods primarily rely on the textual and statistical characteristics of reviews [6,22]. However, malicious users frequently employ homophonic substitutions and cryptic expressions to evade detection, posing significant challenges for traditional detection systems. To address the limitations of existing methods, this study proposes an innovative approach: KSDF, a knowledge-based sensitivity detection framework for film reviews using graph convolutional network. Compared with the detection methods that rely on traditional text feature analysis, the KSDF framework has two core advantages: firstly, by integrating knowledge graph, the dual detection barrier of semantic understanding and knowledge reasoning is constructed, which can effectively break through adversarial interference such as homophonic substitution; Secondly, the entity enhancement strategy can help to dig deeper into the hidden metaphorical expressions in the comments. The main contributions of this study are summarized below:

- We propose KSDF, a system that fuses semantic understanding with knowledge association, leveraging a Graph Convolutional Network to embed sensitive word knowledge and effectively counter adversarial strategies such as homophonic substitution.
- We design a knowledge graph-driven entity enhancement mechanism, constructing enriched representation vectors by extracting entity features from comment texts, thereby enhancing the model's sensitivity to implicit and obscure expressions.
- Extensive experiments, including comparative and ablation studies, validate the effectiveness and superiority of the proposed approach.

2 Related Work

2.1 Spam Movie Review Detection

Early methods primarily relied on manual feature extraction and machine learning classifiers [12,15,24], though these approaches are labor-intensive and limited in capturing deep semantic information. To address these limitations, deep learning-based models have emerged [8,18,25,27,29,30], offering improved performance by learning representations directly from data. However, supervised models face challenges related to data labeling costs and dependence on annotation quality, prompting exploration of unsupervised and semi-supervised alternatives [6,22]. Notably, Gao et al. [6] introduced a GAN-based unsupervised method to evaluate review authenticity using textual and statistical features, and Cao et al. [2] proposed a GCN model integrating movie knowledge and user data. However, these works primarily focus on authenticity assessment, with limited attention to detecting sensitive or harmful content.

2.2 Sensitive Information Detection

In 2015, the first significant attempt to identify confidential text was made by Berardi et al. [1] who used keyword-matching techniques. In 2017, a recurrent neural network (RNN) was introduced [21] to detect sensitive information more accurately by analyzing the syntactic and grammatical structure of text. In 2018, a specialized RNN for detecting sensitive phrases was developed [20]; later, Xu et al. [13] integrated the TF-IDF model to classify sensitive information. In 2019, Xu et al. [33] presented a TextCNN-based method and Wang et al. [31] presented a classification model based on BERT and CNN. In 2020, Lin et al. [16] proposed a framework combining BiLSTM and CNN. Another novel framework [10] using network traffic recovery techniques was introduced. Meanwhile, a BERT-based sequence labeling algorithm was successfully employed to identify and remove sensitive data in Spanish clinical literature [7]. In 2021, Ding et al. [4] developed a corpus to train detection models and applied the BERT model. That same year, Gan et al. [5] designed a scalable CNN multi-channel architecture combined with a BiLSTM model (CNN-BiLSTM) to detect sensitive information in Chinese text. In 2022, Cong et al. [3] introduced KGDetector, a Chinese-sensitive information detection framework augmented by knowledge graph-based BERT. In 2023, Mao et al. [19] proposed a lightweight model for detecting sensitive information in images, based on yolov5s. Also, Huo and Jiang introduced the KG-ERNIE model and the THIIP algorithm [11]. In 2024, Zhang et al. [35] proposed an enhanced YOLOv8 model for detecting sensitive content in images. Wang et al. [28] describe a model that is able to generate sensitive content in images.

3 Preliminaries

Knowledge Graph of Sensitive Information. We construct an undirected graph $G = (V, E)$, which contains sensitive information and its corresponding homophone combinations. Sensitive information is sourced from an open-source list of sensitive keywords. On this basis, we used DeepSeek to generate a list of homophones for these sensitive words. Here are the details of how the graph is built: Our undirected graph contains two types of nodes: sensitive vocabulary nodes $S = \{V_1^s, V_2^s, \ldots, V_i^s, \ldots, V_M^s\}$ and homophone nodes $L = \{V_1^l, V_2^l, \ldots, V_j^l, \ldots, V_N^l\}$, where M is the number of sensitive lexical nodes and N is the number of homophone nodes. Specifically, V_i^s Indicates the i-th sensitive word node and V_j^l represents the j-th homophone node, i.e., $V = S \cup L$. In addition, we use $F = \{f_1, f_2, f_3, \ldots, f_M\}$, which indicates a set containing M sensitive words. Since a sensitive word can correspond to multiple homophones, and these homophones have the same meaning as sensitive words and different sensitive words will also have semantic associations with each other.

4 Methods

As shown in Fig. 1, we propose a framework for detecting sensitive film reviews, which leverages a Graph Convolutional Network (GCN) integrated with sensitive information knowledge. Initially, it utilizes a knowledge graph of sensitive words to learn the distributed representations of sensitive concepts. This model employs relationship paths between entities to capture the internal structure and semantic associations of sensitive information. Next, we incorporate a pre-trained BERT model to extract features from the film review. It converts the text into vector representations, capturing the core semantic content of the review. Building on this, we introduce a named entity embedding enhancement strategy. This strategy identifies named entities within the review text and integrates them into the feature space using a knowledge graph embedding approach. It helps to extract more meaningful intermediate information and generate corresponding embedding vectors. Subsequently, the embedding vectors of the named entities and the text's semantic vectors are combined to merge the semantic information from both sources. To further refine the process, we propose a semantic relevance calculation module. This module establishes a mapping between the text representation and the sensitive knowledge graph via bimodal feature comparison. By comparing the two modalities, we calculate a correlation vector that quantifies the semantic association between the film review and sensitive information. Finally, the two feature vectors are concatenated and passed through a fully connected layer, followed by a Softmax layer. After performing nonlinear transformations and normalization, the classification results are output.

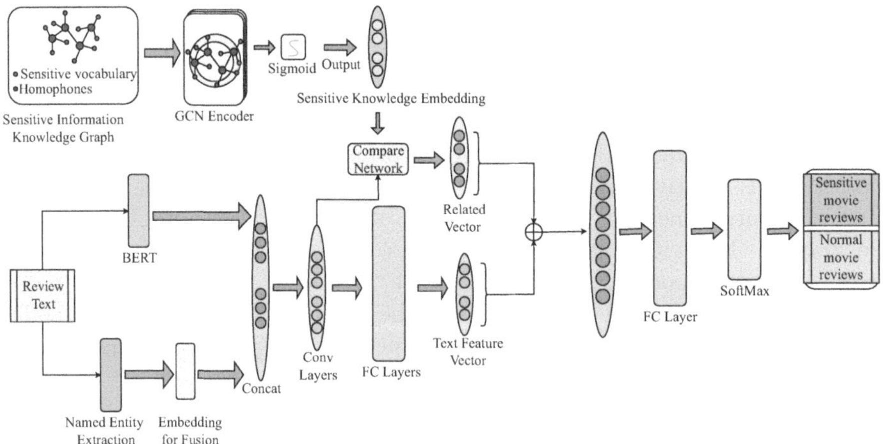

Fig. 1. The overall framework of KSDF model.

4.1 Sensitive Knowledge Embedding Using GCN

We built a GCN model based on an undirected graph of sensitive information knowledge $G = (V, E)$. Firstly, the pre-trained BERT model is used to embed the node features, in which the embedding features of the node and V_s are represented by vectors $x_s \in R^D$. This constructs a feature matrix $X = \{x_1, x_2, x_3, \ldots, x_{|V|}\} \in R^{|V| \times D}$, which contains the eigenvectors of all nodes, where each row of the matrix X represents the eigenvector of one node. We define A as an adjacency matrix and D as a degree matrix, and the graph convolutional layer updates the aggregate feature representation of the $(i+1)$ layer by aggregating the features of its neighbors. The updated formula is:

$$H^{i+1} = \sigma(D^{-\frac{1}{2}}(A+I)D^{-\frac{1}{2}}H^iW^i) \tag{1}$$

where I is the identity matrix of the dimension $|V|$, W^i is the weight matrix of the i-th layer, H^i is the feature matrix of the i-th layer, σ is a nonlinear activation function (i.e. ReLU function), and H^{i+1} is the next layer of the feature matrix after the aggregate feature matrix. The details can be found in Fig. 2.

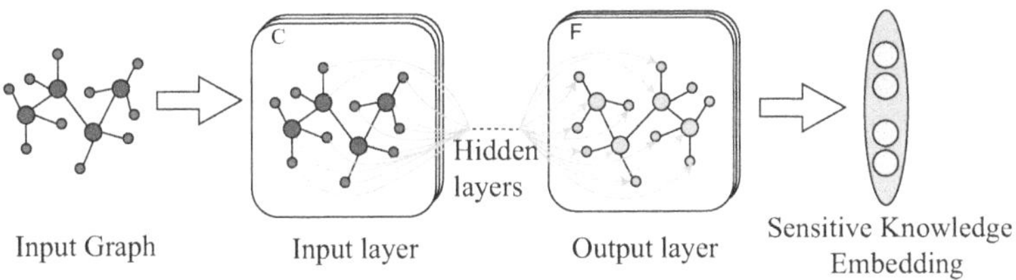

Fig. 2. Sensitive knowledge embedding using GCN.

4.2 Entity Embedding Based on Knowledge Graphs

We take the film review as input data and feed it into our pre-trained BERT model to obtain its semantic vector representation, and then connect with the entity embedding vector to obtain the text feature vector T_n. Although the semantic vectors extracted by the pre-trained BERT are already good at representing the context of the input text, we find that sometimes a subset of entities are ignored, and these entities have high reference values in some cases. Therefore, we propose a named entity embedding strategy based on a knowledge graph and use the relevant knowledge information of named entities in the comment text to enhance the sensitive information detection. For example, in Fig. 3, if the entity "pig" appears in the text, one should be vigilant and analyze based on the contextual information whether it is related to corresponding sensitive information, such as fat, stupid, etc.

For the training of the entity embedding model, we utilized DBpedia, a multilingual knowledge graph. In line with prior studies [3,14], we adopted knowledge

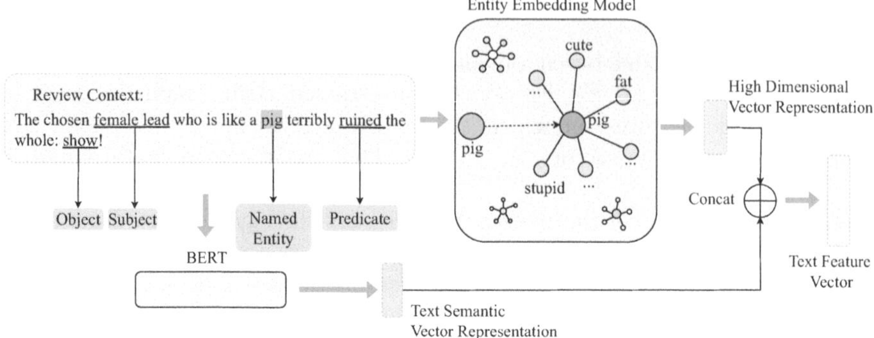

Fig. 3. Entity embedding process.

graph-based entity embeddings of dimension 200. The knowledge graph is composed of triples denoted as $D = (a, r, b)$, where a and b are entities from the entity set E, and r represents a relation from the relation set R. By applying an embedding function $\varepsilon(\cdot)$ to entities and relations separately, we obtained their respective vector representations as $\boldsymbol{a} = \varepsilon(a)$, $\boldsymbol{r} = \varepsilon(r)$, and $\boldsymbol{b} = \varepsilon(b)$. After initializing the embedding vectors randomly, the model was optimized using the following equation:

$$\min_{\varepsilon \in R^d} \sum_{(a,r,b)\in D} \sum_{(\underline{a},r,\underline{b})\in \underline{D}} \mathrm{ReLU}(\eta + \|\boldsymbol{a} + \boldsymbol{r} - \boldsymbol{b}\| - \underline{\boldsymbol{a}} + \boldsymbol{r} - \underline{\boldsymbol{b}}) \tag{2}$$

where $||\cdot||$ denotes ℓ_2 norm. It is important to point out that η is a positive boundary, and $\underline{D}$ is a corrupted dataset that contains mismatched triples, defined as:

$$\underline{D} = \{(\underline{a}, r, b)|\underline{a} \in E\} \cup \{(a, r, \underline{b})|\underline{b} \in E\} \tag{3}$$

We take the optimal solution ε^* of the Eq. 2 as an embedding model for the entity. Specifically, for a given entity $\{e_1, e_2, \ldots, e_M\}$, we embed them into $\{\boldsymbol{e}_1, \boldsymbol{e}_2, \ldots, \boldsymbol{e}_M\}$ based on ε^*. Considering that the entities produce a sparse distribution in the embedding space, we calculate an embedding $\mathbf{e}$ to represent $\{e_1, e_2, \cdots, e_M\}$ as:

$$\mathbf{e} = \sum_{i=1}^{M} \left(\frac{\boldsymbol{e}_i}{M}\right) \tag{4}$$

The model extracts named entities from the film review, calls the trained knowledge graph embedding model, and maps these entities to the corresponding nodes in the knowledge graph to generate embedding vectors. These vectors are fed to the Concat layer to combine with the text semantic vector, forming the text feature vector T_n.

4.3 Feature Extraction for Sensitivity Film Review Detection

After obtaining the embedding representation of sensitive information knowledge, the text feature vector T_n is compared with the embedding vector of sensitive information knowledge x_n, and their correlation vectors $S_n = f_{cmp}(T_n, x_n)$ are obtained, where f_{cmp} is the comparison function. We design it as $f_{cmp}(x, y) = W[x-y, x \odot y]$ where W is the transformation matrix, x and y are the embedding vectors of sensitive information knowledge, and the feature vectors of comments, respectively, and $\odot$ represent element-by-element multiplication operations.

Then we connect the text feature representation vector, T_n and the comment correlation vector S_n to get the vector F_n, and the connection formula is as follows:

$$F_n = concat(T_n, S_n) \tag{5}$$

The vector F_n is then fed to the Softmax classification layer to obtain Z:

$$Z = Softmax(wF_n + b) \tag{6}$$

where w is the parameter of the fully connected layer, and b is the bias.

5 Experiments

5.1 Dataset

In this study, we used four datasets: two Chinese and two English film review databases. To balance the data due to the low occurrence of sensitive reviews, we applied downsampling to non-sensitive reviews. The first dataset, named MaoYan, was collected from the Maoyan platform and consists of 60,778 comments selected from a total of 501,047 reviews across various movies, contributed by 100,038 unique users. Among these, 20,112 were identified as sensitive. The second dataset, Douban, was derived from Douban's online comment section, where 59,420 reviews were chosen from a pool of 487,124 reviews by 110,857 unique users. Out of these, 19,908 reviews were classified as sensitive. The third dataset, RottenTomatoes, was sampled from 52,226 reviews of 538,446 comments posted by 29,800 unique users over the past three years. Among these, 18,438 comments were considered sensitive. Lastly, the fourth dataset, IMDb, was selected from the IMDb Movie Reviews Dataset, with 62,398 comments chosen from 554,237 reviews posted by 31,392 unique users. Of these, 22,604 reviews were labeled as sensitive. The detailed statistics of these datasets are presented in Table 1. Superscripts 1–4 denote data sources: 1 Maoyan; 2 Douban; 3 RottenTomatoes; 4 IMDb.

5.2 Experimental Settings

In this study, all experiments were conducted on a system equipped with an Intel(R) Core(TM) i7-9750F CPU (2.60 GHz), an NVIDIA GeForce RTX 3070 GPU, and 64 GB of RAM. To assess the performance of the proposed KSDF

Table 1. Dataset statistics

Name	Original quantity	Movie	Review	Sensitive	Normal
MaoYan[a]	501,047	420	60,778	20,112	40,666
Douban[b]	487,124	417	59,420	19,908	39,512
RottenTomatoes[c]	538,446	402	52,226	18,438	33,788
IMDb[d]	554,237	466	62,398	22,604	39,794

[a] https://www.maoyan.com/.
[b] https://www.douban.com/.
[c] https://www.rottentomatoes.com/.
[d] https://www.imdb.com/.

model, we employed four evaluation metrics: accuracy, precision, recall, and F1 score. These indicators were used to quantitatively measure the model's effectiveness in detecting sensitive film reviews.

In this study, we evaluated the KSDF model and compared it to three advanced sensitive film review detection models, BERT-CNN [34], BERT-BiLSTM [23] and Prompt Tuning [9]. In addition, to further validate the performance of the model, we also selected several advanced binary classification text classification models as the baseline model, including SSTuning [17], ARTC-LLM [32] and KInIT [26].

5.3 Comparison Analysis

We conducted an extensive experimental evaluation of the proposed model, comparing it with the six advanced methods described above on four datasets, and the results are shown in the Tables 2 and 3, where the bold results indicate the best performance and the underlined results indicate the second best performance. Through comparative analysis, we found that our KSDF method showed excellent performance on four datasets due to the introduction of knowledge embedding. In particular, our method is a significant improvement over other baseline models in terms of accuracy and recall, with an improvement of nearly 10% points over the average of several different models. This is mainly because the model is more effective at identifying sensitive comments that use homophones to evade detection. In terms of precision, while BERT-BiLSTM performed best on the MaoYan dataset, our method came in second with a negligible gap between the two, and we were the best on the other three datasets. However, it is worth mentioning that although our model has a leading position in the rankings, the values are not much different from the second and third models, which is where we have the potential to improve further. In terms of F1 score, our model is significantly better than other models. The experimental results confirm our method's excellent performance and show that our model excels in detecting sensitive movie reviews, particularly those using homophones and metaphors to evade detection.

Table 2. Results of comparative experiments based on the Chinese datasets

Model	MaoYan				Douban			
	Accuracy	Precision	Recall	F1-score	Accuracy	Precision	Recall	F1-score
BERT-CNN	76.94	77.31	<u>77.06</u>	77.18	77.23	77.18	77.21	77.19
BERT-BiLSTM	<u>77.08</u>	**80.97**	75.42	<u>78.10</u>	77.91	<u>80.13</u>	76.55	<u>78.30</u>
Prompt Tuning	74.79	77.61	71.27	74.31	75.53	76.99	73.42	75.16
SSTuning	76.88	79.63	73.98	76.70	76.86	78.74	74.22	76.41
ARTC-LLM	76.77	78.09	76.51	77.29	<u>78.01</u>	79.04	<u>77.23</u>	78.12
KInIT	76.36	76.87	76.07	76.47	76.47	77.16	76.65	76.22
KSDF(our)	**86.11**	<u>80.31</u>	**86.93**	**83.48**	**86.35**	**80.97**	**87.12**	**83.93**

Table 3. Results of comparative experiments based on the English datasets

Model	RottenTomatoes				IMDb			
	Accuracy	Precision	Recall	F1-score	Accuracy	Precision	Recall	F1-score
BERT-CNN	78.26	79.14	<u>78.89</u>	79.01	78.76	80.01	<u>79.63</u>	79.82
BERT-BiLSTM	<u>78.52</u>	<u>82.69</u>	77.26	<u>79.88</u>	79.61	<u>82.16</u>	78.71	<u>80.40</u>
Prompt Tuning	76.33	79.18	72.59	75.74	76.97	79.92	74.17	76.94
SSTuning	77.18	79.52	73.14	76.20	<u>80.10</u>	74.08	77.23	75.62
ARTC-LLM	77.61	80.03	76.93	78.45	78.12	79.78	77.84	78.80
KInIT	76.97	78.76	77.65	78.20	76.62	78.83	79.11	78.97
KSDF(our)	**87.38**	**82.72**	**88.26**	**85.40**	**87.79**	**82.52**	**87.64**	**85.00**

5.4 Ablation Study

The KSDF model we proposed integrates text features, entity association features, and sensitive information knowledge to enhance its ability to detect sensitive comments. To gain a deeper understanding of the specific contribution of each component to the overall performance of the model, we designed and performed ablation experiments on the MaoYan dataset. To evaluate the effectiveness of the model, we utilized four key metrics: accuracy, precision, recall, and F1-score. The results of these experiments are presented in Table 4, where bold results indicate the best performance and underlined results represent the second-best performance. Here, "S" denotes sensitive knowledge embedding, and "E" represents entity embedding, clearly illustrating the impact of each feature module on the model's performance. The experimental results indicate that the sensitive word knowledge embedding method and the entity embedding method based on the knowledge graph significantly outperform traditional methods that rely solely on text features. Furthermore, the experiments reveal that each component of the model plays a crucial role in its overall performance. Specifically, the removal of sensitive knowledge features, entity features, or both individually led to a noticeable decline in the model's effectiveness. This highlights the indispensable role these two modules play in enhancing the performance of the KSDF model, and together, they contribute to a comprehensive improvement in the model's ability to detect sensitive comments.

Table 4. Results of the ablation experiments using the Maoyan database

Model	Accuracy	Precision	Recall	F1-score
KSDF-S	82.70	75.51	79.50	77.47
KSDF-E	81.47	75.05	81.43	78.91
KSDF-SE	80.04	72.31	77.07	71.41
KSDF(our)	**86.11**	**80.31**	**86.93**	**83.48**

6 Conclusion

In this study, we propose a novel detection framework called KSDF, which integrates knowledge graphs and deep learning technology to improve the accuracy of sensitive film review detection, especially those sensitive film reviews that use homophones to avoid detection. The framework uses the graph convolutional neural network to extract the knowledge features of sensitive words from the undirected graph, uses the pre-trained BERT model to extract the features of the network comment text, and combines the named entity information in the comment text as an aid, and finally compares and connects them to evaluate the correlation between the film review and the sensitive information. The framework combines graph convolutional neural networks with entity embeddings based on knowledge graphs to improve the detection ability of sensitive movie reviews. Experimental results on the real-world datasets show that the proposed algorithm has better classification performance than the traditional model. In future work, we plan to explore its application in other domains.

Acknowledgements. This work is partly supported by the Scientific Research Project of Liaoning Provincial Department of Education(No. JYTMS20231226, LJ232410142088), Ministry of Education industry-university cooperative education project(No. 231002108104009) and Open Project of the State Key Laboratory of Integrated Automation in Process Industry(No. SAPI-2025-KFKT-06).

References

1. Berardi, G., Esuli, A., Macdonald, C., Ounis, I., Sebastiani, F.: Semi-automated text classification for sensitivity identification. In: Proceedings of the 24th ACM International on Conference on Information and Knowledge Management, pp. 1711–1714 (2015)
2. Cao, H., Li, H., He, Y., Yan, X., Yang, F., Wang, H.: GCMK: detecting spam movie review based on graph convolutional network embedding movie background knowledge. In: International Conference on Artificial Neural Networks, pp. 494–505. Springer (2022)
3. Cong, K., Li, T., Li, B., Gao, Z., Xu, Y., Gao, F.: Kgdetector: detecting Chinese sensitive information via knowledge graph-enhanced Bert. Secur. Commun. Netw. **2022**(1), 4656837 (2022)

4. Ding, M., Wang, X., Wu, C., Wang, K., Yang, X.: Research on automated detection of sensitive information based on bert. In: Journal of Physics: Conference Series. vol. 1757, p. 012088. IOP Publishing (2021)
5. Gan, C., Feng, Q., Zhang, Z.: Scalable multi-channel dilated cnn-bilstm model with attention mechanism for chinese textual sentiment analysis. Futur. Gener. Comput. Syst. **118**, 297–309 (2021)
6. Gao, Y., Gong, M., Xie, Y., Qin, A.K.: An attention-based unsupervised adversarial model for movie review spam detection. IEEE Trans. Multimedia **23**, 784–796 (2020)
7. García Pablos, A., Perez, N., Cuadros, M.: Sensitive data detection and classification in Spanish clinical text: experiments with BERT. In: Calzolari, N., et al., (eds.) Proceedings of the Twelfth Language Resources and Evaluation Conference, pp. 4486–4494. European Language Resources Association, Marseille (2020). https://aclanthology.org/2020.lrec-1.552/
8. Hajek, P., Barushka, A., Munk, M.: Fake consumer review detection using deep neural networks integrating word embeddings and emotion mining. Neural Comput. Appl. **32**(23), 17259–17274 (2020). https://doi.org/10.1007/s00521-020-04757-2
9. He, X., Zannettou, S., Shen, Y., Zhang, Y.: You only prompt once: on the capabilities of prompt learning on large language models to tackle toxic content. In: 2024 IEEE Symposium on Security and Privacy (SP), pp. 770–787. IEEE (2024)
10. Hong, Q., Zheng, T., Wenli, L., Jianwei, T., Hongyu, Z.: A sensitive information detection method based on network traffic restore. In: 2020 12th International Conference on Measuring Technology and Mechatronics Automation (ICMTMA), pp. 832–836. IEEE (2020)
11. Huo, L., Jiang, J.: Research on intelligent perception algorithm for sensitive information. Appl. Sci. **13**(6), 3383 (2023)
12. Jindal, N., Liu, B.: Opinion spam and analysis. In: Proceedings of the 2008 International Conference on Web Search and Data Mining, pp. 219–230 (2008)
13. Katić, T., Milićević, N.: Comparing sentiment analysis and document representation methods of amazon reviews. In: 2018 IEEE 16th International Symposium on Intelligent Systems and Informatics (SISY), pp. 000283–000286. IEEE (2018)
14. Lerer, A., Wu, L., Shen, J., Lacroix, T., Wehrstedt, L., Bose, A., Peysakhovich, A.: Pytorch-biggraph: a large scale graph embedding system. Proc. Mach. Learn. Syst. **1**, 120–131 (2019)
15. Li, J., Ott, M., Cardie, C., Hovy, E.: Towards a general rule for identifying deceptive opinion spam. In: Proceedings of the 52nd Annual Meeting of the Association for Computational Linguistics (Volume 1: Long Papers), pp. 1566–1576 (2014)
16. Lin, Y., Xu, G., Xu, G., Chen, Y., Sun, D.: Sensitive information detection based on convolution neural network and bi-directional LSTM. In: 2020 IEEE 19th International Conference on Trust, Security and Privacy in Computing and Communications (TrustCom), pp. 1614–1621. IEEE (2020)
17. Liu, C., Zhang, W., Chen, G., Wu, X., Luu, A.T., Chang, C., Bing, L.: Zero-shot text classification via self-supervised tuning. In: Rogers, A., Boyd-Graber, J.L., Okazaki, N. (eds.) Findings of the Association for Computational Linguistics: ACL 2023, Toronto, Canada, July 9-14, 2023, pp. 1743–1761. Association for Computational Linguistics (2023). https://doi.org/10.18653/V1/2023.FINDINGS-ACL.110
18. Liu, Y., Wang, L., Shi, T., Li, J.: Detection of spam reviews through a hierarchical attention architecture with n-gram cnn and bi lstm. Inf. Syst. **103**, 101865 (2022)

19. Mao, Y., Song, B., Zhang, Z., Yang, W., Lan, Y., et al.: A lightweight image sensitive information detection model based on yolov5s. Academic J. Comput. Inf. Sci. **6**(3), 20–27 (2023)
20. Neerbek, J., Assent, I., Dolog, P.: Detecting complex sensitive information via phrase structure in recursive neural networks. In: Phung, D., Tseng, V.S., Webb, G.I., Ho, B., Ganji, M., Rashidi, L. (eds.) PAKDD 2018. LNCS (LNAI), vol. 10939, pp. 373–385. Springer, Cham (2018). https://doi.org/10.1007/978-3-319-93040-4_30
21. Neerbeky, J., Assentz, I., Dolog, P.: Taboo: detecting unstructured sensitive information using recursive neural networks. In: 2017 IEEE 33rd International Conference on Data Engineering (ICDE), pp. 1399–1400. IEEE (2017)
22. Neisari, A., Rueda, L., Saad, S.: Spam review detection using self-organizing maps and convolutional neural networks. Comput. Secur. **106**, 102274 (2021)
23. Nkhata, G., Gauch, S., Anjum, U., Zhan, J.: Fine-tuning Bert with bidirectional LSTM for fine-grained movie reviews sentiment analysis. arXiv preprint arXiv:2502.20682 (2025)
24. Ott, M., Choi, Y., Cardie, C., Hancock, J.T.: Finding deceptive opinion spam by any stretch of the imagination. In: Lin, D., Matsumoto, Y., Mihalcea, R. (eds.) Proceedings of the 49th Annual Meeting of the Association for Computational Linguistics: Human Language Technologies, pp. 309–319. Association for Computational Linguistics, Portland (2011). https://aclanthology.org/P11-1032/
25. Soubraylu, S., Rajalakshmi, R.: Hybrid convolutional bidirectional recurrent neural network based sentiment analysis on movie reviews. Comput. Intell. **37**(2), 735–757 (2021)
26. Spiegel, M., Macko, D.: Kinit at semeval-2024 task 8: fine-tuned LLMs for multilingual machine-generated text detection. In: Ojha, A.K., Doğruöz, A.S., Madabushi, H.T., Martino, G.D.S., Rosenthal, S., Rosá, A. (eds.) Proceedings of the 18th International Workshop on Semantic Evaluation, SemEval@NAACL 2024, Mexico City, Mexico, June 20-21, 2024, pp. 558–564. Association for Computational Linguistics (2024). https://doi.org/10.18653/V1/2024.SEMEVAL-1.84
27. Wang, J., Cao, L., Shan, J., Song, X., Jiang, J.: Dual learning model of code summary and generation based on transformer. In: International Conference on Web Information Systems and Applications, pp. 41–52. Springer (2024)
28. Wang, J., Hao, S., Shan, J., Song, X.: Visual language–let the product say what you want. In: Proceedings of the AAAI Conference on Artificial Intelligence, pp. 23841–23843 (2024)
29. Wang, J., Shan, J., Santos, O.E., Bao, J.: High quality error-tolerant phrase mining on text corpus. Expert Syst. Appl. **171**, 114557 (2021)
30. Wang, J., Yang, H., Shan, J., Song, X., Jiang, J.: Attention-based spatial-temporal fusion networks for traffic flow prediction. In: International Conference on Web Information Systems and Applications, pp. 500–511. Springer (2024)
31. Wang, Y., Shen, X., Yang, Y.: The classification of Chinese sensitive information based on Bert-CNN. In: International Symposium on Intelligence Computation and Applications, pp. 269–280. Springer (2019)
32. Wang, Z., Pang, Y., Lin, Y., Zhu, X.: Adaptable and reliable text classification using large language models. In: IEEE International Conference on Data Mining, ICDM 2024 - Workshops, Abu Dhabi, United Arab Emirates, December 9, 2024, pp. 67–74. IEEE (2024). https://doi.org/10.1109/ICDMW65004.2024.00015

33. Xu, G., Qi, C., Yu, H., Xu, S., Zhao, C., Yuan, J.: Detecting sensitive information of unstructured text using convolutional neural network. In: 2019 International Conference on Cyber-Enabled Distributed Computing and Knowledge Discovery (CyberC), pp. 474–479. IEEE (2019)
34. Zhang, B.: A Bert-CNN based approach on movie review sentiment analysis. In: SHS Web of Conferences. vol. 163, p. 04007. EDP Sciences (2023)
35. Zhang, R., Zhu, D., Wu, C., Xu, J., Wu, C.H.: Sensitive information detection based on deep learning models. Appl. Sci. **14**(17), 7541 (2024)

Cross-Language Summarization Method for Enhancing Factual Consistency Based on Auxiliary Information

Xingyue Li[1], Mengzhu Liu[2(✉)], Yurui Yang[1], and Shenggen Ju[1]

[1] College of Computer Science, Sichuan University, Chengdu 610005, China
[2] College of Computer and Information Engineering, Guizhou University of Commerce, Guiyang, Guizhou, China
mengzhuliu12@163.com

Abstract. Cross-lingual summarization refers to generating a summary in a target language different from the source document. Currently, most work in this field simplistically constructs training samples for cross-lingual summarization by translating mono-lingual summarization corpora, overlooking the potential hallucinations and errors this may introduce. Furthermore, existing cross-lingual summarization methods typically fail to address the enhancement of factual consistency in generated summaries, resulting in outputs with low factual accuracy. To address these limitations, we propose a Factuality Optimization based on Auxiliary Information method (FOAI) for cross-lingual summarization. FOAI first constructs a cross-lingual summarization corpus with reduced hallucination and error rates using a context-guided approach. Subsequently, during the fine-tuning stage, a factuality optimization mechanism is introduced. This mechanism incorporates an auxiliary module that generates factual signals, augmenting the input to the large language model (LLM) to effectively enhance the factual consistency of the generated cross-lingual summaries. Experimental results on cross-lingual datasets derived from CNN/DailyMai and XSum demonstrate that FOAI successfully introduces a factual optimization mechanism into cross-lingual summarization tasks. The study further analyzes the impact of different data scales on experimental outcomes.

Keywords: Text summarization · Cross-Lingual summarization · Factual consistency · Large language model

1 Introduction

Despite significant advances in abstractive summarization driven by developments in deep learning and pre-trained language models [1–3], the majority of research has focused on mono-lingual English scenarios, neglecting the integration of other languages. This limitation has become particularly pronounced with the rapid globalization of the modern internet, for which cross-lingual summarization serves as a key technology to overcome language barriers. Cross-lingual

S. Yang et al. (Eds.): WISA 2025, LNCS 16299, pp. 54–65, 2026.
https://doi.org/10.1007/978-981-95-6426-2_5

summarization(CLS)–defined as generating summaries in a target language distinct from the source document–enables humans to efficiently grasp essential information from lengthy foreign-language texts.

Traditional cross-lingual summarization techniques predominantly employ a pipeline approach [4]to generate target-language summaries, typically involving machine translation models or other translation services. Despite its simplicity, mounting research [5,6] identifies critical limitations including error propagation across processing stages, high dependence on external data/modules, and inherent system brittleness. Current prevalent cross-lingual summarization techniques primarily employ end-to-end approaches [7,8] to directly generate target-language summaries. To train models on sufficient corpora, researchers typically construct cross-lingual summarization samples. A common construction method involves translating mono-lingual summary samples, yet direct translation frequently introduces new hallucinatory errors.

While advancements in enhancing factual consistency for mono-lingual summarization(MLS) have garnered significant attention and rich research progress [9,10], studies addressing factual consistency in cross-lingual summarization remain limited [11]. To imbue cross-lingual summarization with a factuality-aware mechanism, this study proposes the FOAI (Factuality Optimization based on Auxiliary Information) method. Specifically, we introduce an auxiliary factual cue generation module to extract core factual elements from source documents. These elements are then integrated with the source text into a structured prompt as input for the large language model (LLM), guiding it to strengthen factually grounded outputs.

The primary contributions of this study are summarized as follows:

- We construct a factuality-enhanced cross-lingual summarization dataset by translating mono-lingual summarization corpora while jointly leveraging source documents and source-language summaries, subsequently employing the FOAI method to optimize factual consistency in cross-lingual summaries.
- Experiments conducted on our in-house datasets verify the effectiveness of FOAI in enhancing factual consistency, demonstrating its superiority over traditional pipeline approaches and contemporary end-to-end/multi-task joint training methods.
- Extensive experiments use another corpus construction method that does not consider factuality optimization as a comparative baseline and investigate the impact of variable data scales on fine-tuning effectiveness under few-shot settings.

2 Related Work

2.1 Cross-Lingual Summarization Methods

Current cross-lingual summarization techniques primarily encompass pipeline methods, end-to-end approaches, and multitask training frameworks. Early research concentrated on pipeline methods, which rely on mono-lingual summarization models and translation modules to generate cross-lingual summaries. Based on task execution order, pipeline approaches bifurcate into Summarize-Then-Translate [4] and Translate-Then-Summarize [12]. While pipeline methods offer design simplicity, they inherently suffer from error propagation and dependency on external resources, driving recent research towards end-to-end methods. Zhu et al. [5] achieved breakthrough results with the NCLS model by extensively training Transformer-based architectures on large-scale in-house cross-lingual corpora; Zhu et al. [13], inspired by See et al. [14], integrated a translation layer into the decoder to identify pivotal source-language words via attention mechanisms, generate candidate translations through probabilistic bilingual lexicons, and contextually gate output decisions–establishing the ATS model;Cai et al. [15] proposed the CAR model to align cross-attention matrices between cross-lingual and mono-lingual summarizers without added training costs.

Building upon their original NCLS model, Zhu et al. [5] introduced multitask training mechanisms to propose the CLS+MS and CLS+MT models. CLS+MS integrates training objectives from both cross-lingual summarization and mono-lingual summarization tasks, while CLS+MT alternates between cross-lingual summarization and machine translation objectives. This work catalyzed subsequent multitask approaches: Cao et al. [16] combined mono-lingual summarization with context vector alignment, enhancing cross-lingual encoder representations by maximizing similarity with source-language encoders; Liang et al. [17] proposed the VHM model, leveraging conditional variational autoencoders to fuse local latent representations from mono-lingual summarization and machine translation tasks into hierarchical global variables.

2.2 Construction of Cross-Lingual Summarization Datasets

Existing methods to construct CLS datasets fall into two categories: translation-based approaches and automatic alignment using parallel data.

Translation-based methodologies typically render samples from mono-lingual summarization datasets into target languages to construct sample pairs (D_{src}, S_{tgt}), where D_{src} denotes the source-language document and S_{tgt} represents the target-language reference summary. For instance, Zhu et al. [5] employed machine translation systems to automatically translate mono-lingual summaries, augmenting the process with back-translation verification and manual post-editing to build large-scale cross-lingual datasets En2Zh and Zh2En. Automatic alignment using parallel data involves identifying multilingual counterparts of identical content within inherently multilingual resources, subsequently extracting cross-lingual sample pairs. For instance, Ladhak et al. [18] leveraged WikiHow articles aligned via illustrations to create WikiLingua; Pérez-Beltrachini et al. [19] exploited Wikipedia's multilingual infrastructure to align article versions through interlanguage links, yielding XWikis.

3 Methodology

3.1 Constructing Low-Hallucination Cross-Lingual Summarization Corpora

We integrate the English source document and its summary into predefined prompts to guide large language models in translating summaries into Chinese, thereby yielding cross-lingual summarization samples (D^{En}, S^{Zh}) where D^{En} and S^{Zh} denote the English source document and Chinese summary respectively. The following elaborates on corpus construction specifics.

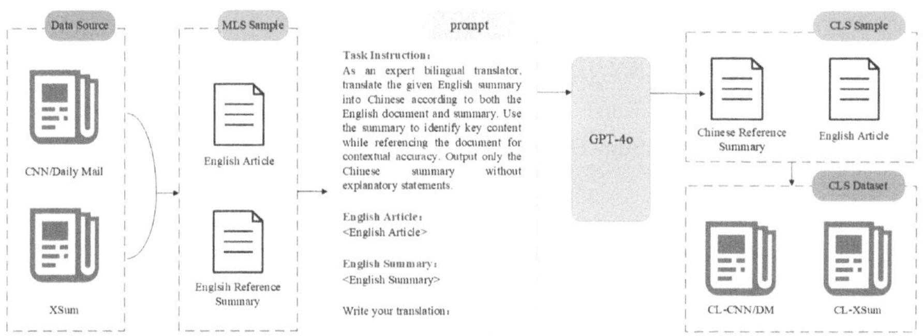

Fig. 1. Translate to Construct CLS Dataset through Large Language Models.

As illustrated in Fig. 1, we utilized CNN/DailyMail and XSum–prominent English mono-lingual summarization benchmark datasets–as data sources, integrating each document-source-summary pair into structured prompts for the large language model GPT-4o to construct cross-lingual corpora. GPT-4o (OpenAI's large language model with massive parameters and multilingual text generation capabilities) received prompts comprising: (1) Task Instruction: "As an expert bilingual translator, translate the given English summary into Chinese according to both the English document and summary. Use the summary to identify key content while referencing the document for contextual accuracy. Output only the Chinese summary without explanatory statements", explicitly requiring dual-source referencing; (2) Source Document: providing full context; (3) English Summary: as both translation target and salient content indicator. This framework enables the LLM to leverage document context for reduced translation errors and hallucinations while exploiting summary salience to enhance content relevance and minimize extraneous details in Chinese outputs.

Using the aforementioned methodology, we randomly sampled CNN/DailyMail and XSum to construct two cross-lingual corpora CL-CNN/DM and CL-XSum, each with identically sized training/validation/test sets (20k/6k/2k). Additionally, we created contrastive datasets CL-CNN/DM-Direct and CL-XSum-Direct by translating the same source summaries without document context: this approach modified prompts to "As an expert bilingual translator, render the given English text into Chinese" while removing original documents. To comparatively evaluate the factual consistency performance of these two construction methodologies, we employed the subsequently described automated evaluation metrics to assess all self-constructed datasets, with per-dataset subset consolidation applied during evaluation and the results are presented in Table 1. Here, "FactScore" quantifies the factual consistency metric while "ErrorScore↓" denotes factual error severity–lower values correspond to fewer factual errors.

Table 1. Factuality Evaluation of Different Datasets

Dataset	FactScore	ErrorScore↓
CL-CNN/DM	8.35	2.59
CL-CNN/DM-Direct	8.13	2.78
CL-XSum	6.88	4.98
CL-XSum-Direct	6.80	5.17

As evident from the table, both the CL-CNN/DM and CL-XSum datasets outperformed the CL-CNN/DM-Direct and CL-XSum-Direct datasets on FactScore and ErrorScore evaluations, respectively, indicating that translating source-language summaries while incorporating the context provided by the source document enables the construction of more accurate cross-lingual summarization datasets. Moreover, it is noteworthy that even the factual accuracy of CL-CNN/DM-Direct, built via direct translation, still exceeded that of CL-XSum, which was built by combining translation with the source document. This phenomenon arises because the XSum dataset itself exhibits a high degree of abstractiveness, inherently containing more errors and hallucinations compared to the CNN/DailyMail dataset, leading to this outcome.

Algorithm 1. Factual Statement Filtering Algorithm

Require: Candidate factual statement set F, FactCC calculation formula $Fact$
Ensure: Filtered candidate factual statement setF_s
1: Initialize an empty set E
2: **for** f in F **do**
3: Compute the FactCC value for the fact: $FactCC \leftarrow Fact(f)$
4: Add this value to set $E:\ \leftarrow E\cup$ FactCC
5: **end for**
6: Determine the maximum value in set $E: MaxFactCC \leftarrow max(E)$
7: **for** f in F **do**
8: **if** $Fact(f) \geq 0.4 \times MaxFactCC$ **then**
9: $F_s \leftarrow F_s \cup f$
10: **end if**
11: **end for**
12: **return** the filtered candidate factual statement set F_s

3.2 Cross-Lingual Summarization Based on Factual Information and LLM

We propose the FOAI method to generate summaries with high factual consistency through large language models. As illustrated in Fig. 2, for an input English source document, it is first incorporated into a specific prompt and processed by the GLM-4-Long model to generate a series of factual statement encapsulations, where each statement captures essential content from the source document and expresses it in concise natural language. Subsequently, each factual statement is treated as a pseudo-summary and evaluated using the FactCC [20] metric to compute its factual consistency score. This score then informs the filtering of factual statements to select the most representative encapsulations, with this screening process detailed in Algorithm 1.

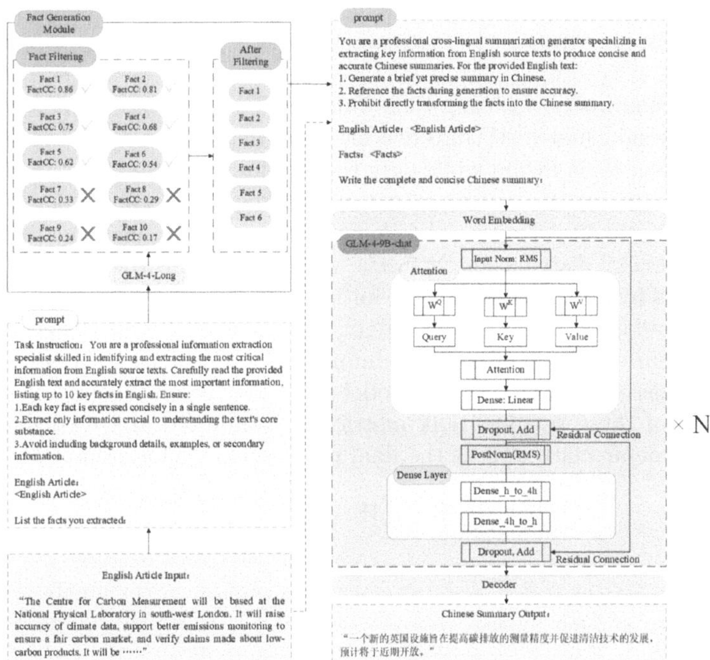

Fig. 2. Cross-Lingual Summarization Process of FOAI.

Subsequently, we incorporate the source document and the selected factual statements into another prompt and utilizes the GLM-4-9B-chat model to generate the Chinese summary. The two large language models involved above are produced by ChatGLM, with the former, GLM-4-Long, supporting ultra-long context input and possessing outstanding long-text processing and reasoning capabilities, while the latter, GLM-4-9B-chat, has been trained with human preference alignment.

As shown in the figure, the GLM-4-9B-chat model generates the Chinese summary based on the English source document and the filtered factual statements, thereby ensuring that the summary accurately and concisely reflects the core information of the article while minimizing errors or hallucinations. Concurrently, an explicit directive – "Prohibit directly transforming the facts into the Chinese summary" – is incorporated into the summary generation prompt to ensure that the factual statement encapsulations serve solely as a reference frame during the model's summarization process.

When fine-tuning the GLM-4-9B-chat model, the traditional full fine-tuning approach updates all model parameters during training. While this enables the model to fully adapt to the specific task and achieve optimal downstream performance, the computational and storage costs become prohibitively substantial for large language models with extremely large parameters. Consequently, we adopt LoRA (Low-Rank Adaptation) [21] to fine-tune GLM-4-9B-chat for the cross-lingual summarization task without significantly increasing computational overhead, while simultaneously maintaining the model's strong performance on this task.

LoRA is a Parameter-Efficient Fine-Tuning (PEFT) method that decomposes the model's weight matrix W into two low-rank matrices A and B with rank r. During training, W is frozen while only the two low-rank matrices are updated, significantly reducing the number of parameters requiring adjustment to achieve parameter-efficient fine-tuning. The elements of matrix A are randomly initialized from a normal distribution $N(0, \sigma^2)$, where σ controls the variance. Both σ and r serve as hyperparameters. This initialization strategy provides the model with flexible adaptation capacity. Matrix B, conversely, is initialized as a zero matrix to ensure stability in early training stages, preventing excessive adjustments from disrupting the original model structure. Upon training completion, the product of these two low-rank matrices is computed and merged into the original parameter matrix W as the final parameters, as formalized in Eq. 1.

$$W' = W + AB \tag{1}$$

4 Experiments

4.1 Metrics

We employ both text quality and factual consistency metrics to evaluate the cross-lingual summaries generated by different methods. For text quality assessment, we select Rouge-1, Rouge-2, and Rouge-L metrics from the Rouge [22] suite. Regarding factual consistency evaluation, popular automatic evaluation methods in monolingual summarization such as those based on text entailment or question-answering paradigms cannot be directly transferred to cross-lingual contexts [11]. Consequently, we leverage the GPT-4o model combined with a specific prompt to outputs corresponding factual consistency scores (FactScore) and error severity scores (ErrorScore) based on the provided English article and Chinese summary.

4.2 Comparative Methods

We compare FOAI against the following methods:

Pipeline. Two distinct pipeline approaches are employed for comparison, namely Sum-Trans and Trans-Sum: the former denotes first generating a source-language summary and then translating it into the target language, while the latter involves first translating the source document into the target language and subsequently using a mono-lingual summarization model to produce the summary. For these methods, we utilize the GPT-4o model for translation.

E2E. We configure the GLM-4-9B-chat model to take the source-language document as input, skipping the generation of facts to directly output the target-language summary, thereby removing guidance for factual consistency. Consequently, relevant components are omitted from the original prompt to enable the model to operate under a vanilla end-to-end training paradigm.

MS+CLS. Drawing inspiration from MCLAS [6], we introduce a multi-task training approach where the large language model first generates a source-language summary for the original document and subsequently produces a target-language summary guided by this summary, with both summaries concatenated.

For all aforementioned methods except the pipeline approaches, we fine-tuned the GLM-4-9B-chat model using LoRA, whereas in pipeline implementations, the model was directly utilized for summary generation without fine-tuning.

4.3 Experiment Results Analysis

The experimental results are presented in Table 2, where R1, R2, RL, FS, and ES denote the Rouge-1, Rouge-2, Rouge-L, FactScore, and ErrorScore metrics, respectively, with the optimal value for each metric highlighted in bold. It can be seen that the FOAI method outperforms other baselines across both factual evaluation metrics.

Table 2. Overall Experimental Results

Method	CL-CNN/DM					CL-XSum				
	R1	R2	RL	ES	FS↓	R1	R2	RL	ES	FS↓
Sum-Trans	32.62	10.89	22.61	8.36	2.21	25.39	6.77	17.99	7.41	3.96
Trans-Sum	27.71	7.64	19.35	8.09	2.74	23.85	5.34	17.79	7.28	3.77
E2E	37.21	13.05	25.73	8.28	2.38	40.51	15.19	32.89	7.34	3.72
MS+CLS	38.77	13.67	28.20	8.16	2.67	**40.89**	**15.90**	**33.87**	6.82	4.66
FOAI(Ours)	**39.82**	**15.85**	**28.36**	**8.49**	**1.76**	40.74	15.41	33.04	**7.50**	**3.63**

On CL-CNN/DM, compared to pipeline approaches, E2E, MS+CLS, and FOAI achieve significantly higher Rouge scores, showing that fine-tuning pre-trained models improves cross-lingual text generation. However, Sum-Trans outperforms both E2E and MS+CLS in factual consistency. This is likely because Sum-Trans translates a distilled source-language summary, helping it preserve core information. The translation model GPT-4o, trained on massive corpora, also helps reduce errors and hallucinations. In contrast, E2E handles both comprehension and generation, which may compromise factual accuracy. Although MS+CLS uses multi-task learning, its simplified source summary may lack factual detail and suffer from task interference. Additionally, FOAI's improvement in factual consistency is more notable in ErrorScore reduction–decreasing by 0.62 compared to a 0.21 FactScore increase–suggesting a stronger focus on reducing factual errors.

On CL-XSum, MS+CLS leads in Rouge scores, while FOAI slightly underperforms. However, FOAI shows better factual consistency with higher FactScore and lower ErrorScore. MS+CLS yields the poorest factual results, likely because XSum's highly abstractive summaries contain inherent errors; these may propagate during target summary generation, reducing factual accuracy. On CL-CNN/DM, compared to pipeline approaches, E2E, MS+CLS, and FOAI achieve significantly higher Rouge scores, showing that fine-tuning pre-trained models improves cross-lingual text generation. However, Sum-Trans outperforms both E2E and MS+CLS in factual consistency. This is likely because Sum-Trans translates a distilled source-language summary, helping it preserve core information. The translation model GPT-4o, trained on massive corpora, also helps reduce errors and hallucinations. In contrast, E2E handles both comprehension and generation, which may compromise factual accuracy. Although MS+CLS uses multi-task learning, its simplified source summary may lack factual detail and suffer from task interference. Additionally, FOAI's improvement in factual consistency is more notable in ErrorScore reduction–decreasing by 0.62 compared to a 0.21 FactScore increase–suggesting a stronger focus on reducing factual errors.

Table 3. Experimental Results on the Contrastive Datasets

Method	CL-CNN/DM-Direct					CL-XSum-Direct				
	R1	R2	RL	ES	FS$\downarrow$	R1	R2	RL	ES	FS$\downarrow$
E2E	37.48	13.17	25.57	8.13	2.41	40.67	14.98	32.81	7.24	4.05
MS+CLS	37.28	13.14	**26.81**	8.11	2.73	**41.20**	**16.10**	**34.23**	6.68	4.89
FOAI(Ours)	**38.71**	**14.89**	26.75	**8.34**	**2.24**	40.93	15.37	33.13	**7.35**	**3.86**

Table 3 presents the automatic evaluation results of three fine-tuning methods on the CL-CNN/DM-Direct and CL-XSum-Direct datasets. It can be observed that, on these cross-lingual summarization datasets constructed by direct translation, all methods exhibit degraded performance on both FactScore and ErrorScore metrics compared to their fine-tuned counterparts on CL-CNN/DM and CL-XSum. This underscores the advantage of our previously described cross-lingual summarization dataset construction method, which–by translating summaries while incorporating source document context–effectively mitigates certain errors and hallucinations, thereby enhancing the overall accuracy of cross-lingual summarization datasets. However, due to the inherently higher prevalence of hallucinations and errors in the XSum source data compared to CNN/DM, FOAI's factual consistency performance remains superior even when fine-tuned on CL-CNN/DM-Direct over fine-tuning on CL-XSum, a phenomenon consistent with the earlier factual evaluation disparities between the datasets. This Experiment demonstrates the inherent factual quality of the dataset is critically important for training outcomes, thus warranting further research into constructing cross-lingual summarization datasets with minimal hallucinations and errors.

4.4 Experiments at Different Data Scales

We investigate the performance of FOAI under zero-shot and few-shot scenarios. Specifically, the training data proportion was adjusted to 0%, 5%, 10%, 25%, and 50% of the original dataset - where 0% corresponds to the zero-shot setting, and the subsequent four proportions represent varying degrees of few-shot learning. Experimental results are presented in Table 4.

Table 4. Experimental Results of Different Data Scales

Data Scale	CL-CNN/DM					CL-XSum				
	R1	R2	RL	ES	FS↓	R1	R2	RL	ES	FS↓
0(0%)	32.79	10.30	22.76	8.40	2.17	24.36	5.97	17.78	7.46	3.98
1k(5%)	36.07	12.66	25.74	8.42	1.96	36.63	10.82	28.66	**7.58**	**3.38**
2k(10%)	37.28	14.07	26.15	8.36	1.93	38.59	11.57	29.97	7.33	3.70
5k(25%)	37.69	14.34	26.93	8.39	2.76	39.27	13.25	31.94	7.38	3.71
10k(50%)	38.28	14.88	28.12	8.43	1.91	40.05	14.77	32.47	7.42	3.67
20k(100%)	**39.82**	**15.85**	**28.36**	**8.49**	**1.76**	**40.74**	**15.41**	**33.04**	7.50	3.63

Even in zero-shot settings, FOAI enables GLM-4-9B-chat to produce factually strong cross-lingual summaries, while a mere 5% training data proportion yields substantial Rouge improvements, which is particularly notable on CL-XSum, with metrics progressively approaching full-data performance as data scales increase. Crucially, at 5% scale, both FactScore and ErrorScore show excellent performance and even outperform full-data results on CL-XSum. This signifies the method's feasibility for low-resource languages in cross-lingual summarization, highlighting FOAI's substantial applicability in data-scarce scenarios.

5 Conclusion

This paper proposes FOAI, a novel method for enhancing factual consistency in cross-lingual summarization. FOAI operates by first extracting factual information from source-language documents and integrating it with original text through specialized prompt. This combined input guides the large language model GLM-4-9B-chat to generate target-language summaries with improved factual consistency. For model fine-tuning, we address hallucination reduction through context-aware translation techniques to construct a low-hallucination cross-lingual summarization dataset. Experimental results demonstrate FOAI's effectiveness in boosting factual consistency and few-shot experiments further validate FOAI's robust feasibility in low-resource scenarios.

Acknowledgments. This work was supported by the Key projects of the National Natural Science Foundation of China (No.62137001) and the Key Research and Development Program of Sichuan Province.(2023YFG0265).

Disclosure of Interests. The authors have no competing interests to declare that are relevant to the content of this article.

References

1. Lewis, M., et al.: Bart: denoising sequence-to-sequence pre-training for natural language generation, translation, and comprehension. In: Proceedings of the 58th Annual Meeting of the Association for Computational Linguistics, pp. 7871–7880 (2020)
2. Zhang, J., Zhao, Y., Saleh, M., Liu, P.: Pegasus: pre-training with extracted gap-sentences for abstractive summarization. In: International Conference on Machine Learning, pp. 11328–11339. PMLR (2020)
3. Liu, Y., et al.: The journey of language models in understanding natural language. In: Web Information Systems and Applications, pp. 331–363 (2024)
4. Wan, X., Li, H., Xiao, J.: Cross-language document summarization based on machine translation quality prediction. In: Proceedings of the 48th Annual Meeting of the Association for Computational Linguistics, pp. 917–926 (2010)
5. Zhu, J., et al.: Ncls: neural cross-lingual summarization. In: Proceedings of the 2019 Conference on Empirical Methods in Natural Language Processing and the 9th International Joint Conference on Natural Language Processing (EMNLP-IJCNLP), pp. 3054–3064 (2019)
6. Teo, T.W., Choy, B.H.: in. In: Tan, O.S., Low, E.L., Tay, E.G., Yan, Y.K. (eds.) Singapore Math and Science Education Innovation. ETLPPSIP, vol. 1, pp. 43–59. Springer, Singapore (2021). https://doi.org/10.1007/978-981-16-1357-9_3
7. Le, T.: Cross-lingual summarization with pseudo-label regularization. In: Findings of the Association for Computational Linguistics: NAACL 2024, pp. 4644–4677 (2024)
8. Huot, F., et al.: Plan: summarizing using a content plan as cross-lingual bridge. In: Proceedings of the 18th Conference of the European Chapter of the Association for Computational Linguistics, vol. 1: Long Papers, pp. 2146–2163 (2024)

9. Balachandran, V., Hajishirzi, H., Cohen, W., Tsvetkov, Y.: Correcting diverse factual errors in abstractive summarization via post-editing and language model infilling. In: Proceedings of the 2022 Conference on Empirical Methods in Natural Language Processing, pp. 9818–9830 (2022)
10. Fabbri, A., Choubey, P.K., Vig, J., Wu, C.S., Xiong, C.: Improving factual consistency in summarization with compression-based post-editing. In: Proceedings of the 2022 Conference on Empirical Methods in Natural Language Processing, pp. 9149–9156 (2022)
11. Zhang, H., Perez-Beltrachini, L.: Leveraging entailment judgements in cross-lingual summarisation. In: The 62nd Annual Meeting of the Association for Computational Linguistics, pp. 14481–14497. Association for Computational Linguistics (ACL) (2024)
12. Leuski, A., Lin, C.Y., Zhou, L., Germann, U., Och, F.J., Hovy, E.: Cross-lingual c* st* rd: English access to hindi information. ACM Trans. Asian Lang. Inf. Process. (TALIP) **2**(3), 245–269 (2003)
13. Zhu, J., Zhou, Y., Zhang, J., Zong, C.; Attend, translate and summarize: An efficient method for neural cross-lingual summarization. In: Proceedings of the 58th annual meeting of the Association for Computational Linguistics, pp. 1309–1321 (2020)
14. See, A., Liu, P.J., Manning, C.D.: Get to the point: summarization with pointer-generator networks. In: Proceedings of the 55th Annual Meeting of the Association for Computational Linguistics, vol. 1: Long Papers, pp. 1073–1083 (2017)
15. Cai, Y., Yuan, Y.: Car-transformer: cross-attention reinforcement transformer for cross-lingual summarization. In: Proceedings of the AAAI Conference on Artificial Intelligence, vol. 38, pp. 17718–17726 (2024)
16. Cao, Y., Liu, H., Wan, X.: Jointly learning to align and summarize for neural cross-lingual summarization. In: Proceedings of the 58th Annual Meeting of the Association for Computational Linguistics, pp. 6220–6231 (2020)
17. Liang, Y., et al.: A variational hierarchical model for neural cross-lingual summarization. In: Proceedings of the 60th Annual Meeting of the Association for Computational Linguistics, vol. 1: Long Papers, pp. 2088–2099 (2022)
18. Ladhak, F., Durmus, E., Cardie, C., Mckeown, K.: Wikilingua: a new benchmark dataset for cross-lingual abstractive summarization. In: Findings of the Association for Computational Linguistics: EMNLP 2020, pp. 4034–4048 (2020)
19. Perez-Beltrachini, L., Lapata, M.: Models and datasets for cross-lingual summarisation. In: Proceedings of the 2021 Conference on Empirical Methods in Natural Language Processing, pp. 9408–9423 (2021)
20. Kryściński, W., McCann, B., Xiong, C., Socher, R.: Evaluating the factual consistency of abstractive text summarization. In: Proceedings of the 2020 Conference on Empirical Methods in Natural Language Processing (EMNLP), pp. 9332–9346 (2020)
21. Hu, E.J., et al.: Lora: low-rank adaptation of large language models. In: ICLR, vol. 1, no. 2, p. 3 (2022)
22. Lin, C.Y.: Rouge: a package for automatic evaluation of summaries. In: Text Summarization Branches Out, pp. 74–81 (2004)

A Multi-level Pruning Method for Medical Redundant Data Based on Data Lake

Tianyi Liu[1], Jiaxu Guo[1(✉)], Chao Xia[2(✉)], Peng Ren[3], and Chunxiao Xing[3]

[1] School of Software, Henan University, Kaifeng 475004, China
{liutianyi,gjx2023}@henu.edu.cn

[2] Kaifeng 155 Hospital, Kaifeng 475003, China
35090741@qq.com

[3] BNRist, DCST, RIIT, Tsinghua University, Beijing 100084, China
{renpeng,xingcx}@tsinghua.edu.cn

Abstract. As the volume of data in the medical field grows exponentially, the issue of data redundancy has become increasingly prominent. For instance, in the case of the large public medical dataset MIMIC-III, the redundancy rate in hospitalization records reaches as high as 97%–98%. This redundancy not only significantly increases storage costs but also reduces data processing efficiency. To date, there has been no proposed method for table-level redundant data pruning specifically tailored to the unique data types found in medical datasets. To address this issue, this paper introduces a multi-level pruning method based on a data lake, which efficiently eliminates redundancy through a hierarchical processing strategy. The method progressively narrows the search space by constructing an inclusion graph (Containment Graph Construction) combined with the multi-level pruning technique, enabling the efficient identification and removal of redundant data. Additionally, to ensure the efficiency, accuracy, and clinical practicality of the pruning results, this paper proposes a "Clinical Decision Making" mechanism. Experimental results demonstrate that this approach can efficiently process large scale medical data and accurately eliminate approximately 23% of redundant data.

Keywords: Data Lake · Medical Redundant Data · Data Pruning

1 Introduction

With the advent of the Industry 4.0 era, the medical field is undergoing an unprecedented digital transformation. The popularization of advanced medical devices and technologies (such as electronic medical record systems and medical imaging equipment) has given rise to massive multi-modal medical data. These data include not only traditional structured data (such as electronic medical records and laboratory test results) but also a large amount of unstructured data (such as X-rays, MRI scans, CT scans, and PET scans) [1].

Wrenn et al. studied the redundancy of inpatient record data and found that the duplicate information gradually increases as the length of hospital stay prolongs. On average, 78% of the information in inpatient records and 54% of

S. Yang et al. (Eds.): WISA 2025, LNCS 16299, pp. 66–77, 2026.
https://doi.org/10.1007/978-981-95-6426-2_6

the information in case records are duplicates of previous texts [2]. However, these studies focused on text processing and did not explore how to solve the problem of medical data redundancy. Other studies have shown that through deduplication technology, the electronic medical records and backup loads in the same department can save 65% and 90% of the storage costs respectively [3,4]. In addition, 99% of human DNA sequences are similar, and deduplication and incremental compression technologies have saved a large amount of storage space for genetic data [5].

These redundant data not only occupy massive storage space and increase the economic costs of storage systems but also consume valuable bandwidth resources during data transmission. Additionally, frequent writes of redundant data can shorten the service life of flash storage devices. Therefore, how to efficiently eliminate redundant data to reduce storage costs and economic expenses has become a critical issue to be urgently addressed in the field of medical data management.

To solve the above problems, this paper proposes a multi-level pruning method for medical redundant data based on a data lake. This method streamlines the processed dataset through "clinical decision-making" by hospital engineers, extracts meaningful columns to construct a Containment Graph (Containment Graph Construction), and combines multi-level pruning strategies for numerical, enumeration, and text data to gradually narrow the search space, achieving efficient identification of redundant data. Finally, "clinical decision making" by physicians is used to safely delete redundant data. Experimental results show that this framework can process terabyte-scale multimodal medical data within hours and significantly reduce storage costs.

The main contributions of this paper include:

- Proposing a scalable redundant data identification and pruning method that can efficiently prune medical redundant data based on different data types;
- Designing a "Clinical Decision Making" mechanism to integrate the prior knowledge of hospital engineers and physicians into the redundant detection and deletion process, improving pruning efficiency and ensuring the accuracy and clinical practicality of pruning results;
- Verifying the efficiency and reliability of this method in large scale medical data scenarios through experiments, providing an innovative solution for medical data management.

2 Related Work

In the early days of medical data management, database technology was adopted [6]. During the period when the amount of medical data was relatively small, databases could efficiently handle online transactions in the medical field thanks to their advantages of fast insertion, deletion, and query operations. However, as the scale and complexity of medical data continued to increase, the limitations of traditional database platform became more and more apparent when dealing with analytical tasks characterized by a large amount of data reading.

To break through this dilemma, Ren et al. proposed an efficient Multi-Source Heterogeneous Data Lake Platform (MHDP), aiming to achieve efficient medical data management [7]. Sheng et al. proposed a multi-modal data retrieval platform MQRLD based on the data lake. This platform makes full use of the transparent storage capabilities of the data lake, breaking through the limitations of low retrieval efficiency and poor compatibility of multi-modal data in the traditional data storage mode [8].

The identification and deletion of duplicate data have been widely applied in various industries due to their advantages in achieving efficient storage [9]. With the vigorous development of big data medical platforms, the problem of medical redundant data has become increasingly severe, and redundant data pruning has also become a research focus. The R2D2 framework proposed by Raunak Shah et al. has the characteristics of scalability and hierarchy, providing a new idea for identifying data containment relationships [10]. Zhang et al. proposed the DCDedupe method, which intelligently selects deduplication or delta compression techniques for different types of data to prune redundant data and improve storage efficiency [11]. Fu et al. designed an n- dimensional parameter space, mapping existing deduplication schemes as points in the space, and weighing storage overhead, backup performance, and recovery performance through various parameters [12]. Zhang et al. analyzed the capabilities and needs of clinicians, designed tasks involving clinicians, and implemented a clinician in the loop toolkit, integrating clinicians' prior knowledge into the construction of medical information platforms. These studies explore effective ways to handle medical redundant data from different perspectives and continuously promote the in depth development of research in this field [13]. Liu et al. proposed a novel model architecture named DiagNet. This architecture leverages both sequential and global analysis of patient information to enhance prediction accuracy [15].

3 Platform Architecture

At the medical data generation layer, the complexity of medical data redundancy is rooted in its multi-source heterogeneity. When medical data analysts query and analyze the same dataset, repeated storage of results in the data lake leads to substantial table level redundancy. In multi-center collaborative systems of medical institutions, data redundancy arises through two mechanisms: horizontally, heterogeneous data standards in independently built hospital systems cause core medical entities (e.g., patient profiles, drug inventory) to be redundantly registered across storage nodes; vertically, temporary data synchronization during new-old system parallel operation and historical data migration—without deduplication—generates temporal redundant copies during system upgrades or architecture transitions.

At the data lake layer, medical data is aggregated to the data lake through pipelines. The unified storage of multi-source heterogeneous data is achieved based on the distributed storage capability of HDFS. The Spark framework is used to perform redundant pruning on the data in HDFS through datasets and in-memory computing engines. The multi-level pruning logic is executed using the distributed computing capability of Spark. After undergoing the data pruning layer, the refined datasets are stored back in HDFS, completing a closed loop processing chain of "lake ingestion redundancy pruning archival optimization". This architecture ensures both storage efficiency and clinical data integrity.

The data pruning layer, as the key research object of this paper, will be described in detail in Sect. 4. The overall architecture is shown in Fig. 1.

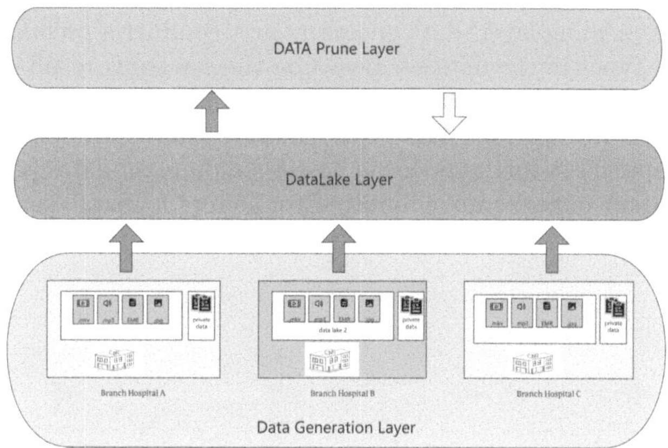

Fig. 1. Architecture Diagram of Hospital Data Platform.

4 Redundant Data Pruning Method

To construct a redundant data identification system in the medical data lake environment, we first establish a formal definition of the containment relationship for table - level data:

Definition 1 (Complete Containment Relationship): Suppose that both dataset P and dataset Q are medical data tables. If every row record $r \in Q$ exists in P, that is, $\forall r \in Q, \exists r' \in P$, such that the values of r and r' are exactly the same in all corresponding columns, then P is said to completely contain Q, denoted as $Q \subseteq P$. This relationship indicates that Q is a redundant copy of P, and theoretically, Q can be directly deleted to eliminate redundancy.

Definition 2 (Column Containment Relationship): For datasets P and Q, if the column set of Q is a subset of the column set of P, that is, $\text{cols}(Q) \subseteq \text{cols}(P)$ (where cols() represents the set of column names of the dataset), then P is said to contain Q in terms of columns, denoted as $Q \to P$. This relationship indicates that P has all the attribute dimensions of Q. The column containment relationship is the basis for constructing the containment graph.

In an ideal scenario, if we can identify all complete containment relationships, we can completely eliminate redundant data. However, this requires a huge amount of computing resources. To strike a balance between computing resource optimization and redundancy identification efficiency, this study aims to construct a redundant and containment relationship mapping among datasets in the data lake in a scalable and efficient manner. Furthermore, the study hopes to select candidate datasets from this relationship set that simultaneously meet the clinical data integrity standards and the risk free conditions of business logic. On the premise of ensuring the medical decision making support ability is not impaired, the study aims to minimize data redundancy and optimize storage costs.

At the data pruning layer, data engineers first conduct a preliminary screening of the data types in the dataset. Based on the screening results, an abstract relationship containment graph is constructed. Then, a multi - level pruning strategy designed for different data types (such as numerical, enumeration, and text types) is applied to further identify possible redundant datasets. Finally, the selected redundant datasets are submitted to doctors for professional judgment to decide which datasets should be deleted. This process is shown in Fig. 2.

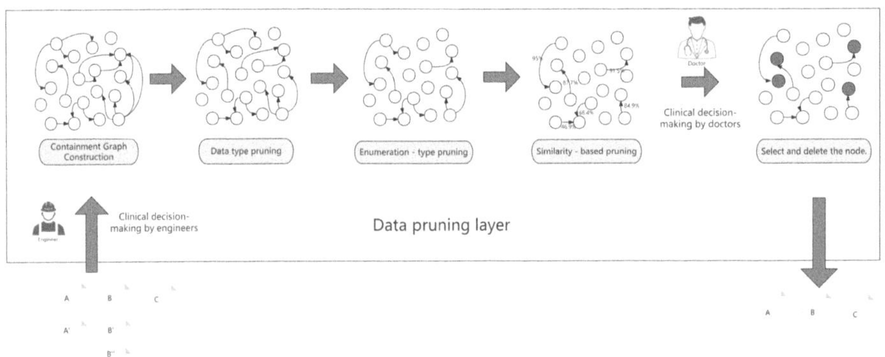

Fig. 2. Flowchart of Data Pruning Layer.

4.1 Containment Graph Construction

The edges of the containment graph represent the column containment relationships between data. In this step, we calculate all the column containment relationships within the dataset (these edges will be pruned in subsequent stages). The brute force approach is to conduct pairwise comparisons for all pairs of schemas. For N datasets, this will result in $O(N^2)$ comparisons and will not generate additional edges. However, we propose to reduce the computational load by sacrificing a certain amount of space to decrease the number of comparisons at this stage.

We abstract the tables and their containment relationships into nodes and directed edges in a graph structure. Specifically, if there is a containment relationship between two tables, a directed edge is constructed from the node representing the smaller dataset to the node representing the larger dataset, so as to intuitively present the hierarchical relationship between the data. This process is realized through the Containment Graph Construction (CGC) algorithm. This algorithm avoids high computational overhead This process is shown in Fig. 3. Next, we describe the CGC algorithm in detail:

1. Sort the dataset S in ascending order according to the number of columns in each dataset to obtain the sorted dataset sequence $S' = \{S_1, S_2, \cdots, S_n\}$.
2. Create an empty directed graph $G = (V, E)$, where V is the set of nodes, initially containing all the datasets in the dataset sequence S'; E is the set of edges, initially empty, and each table S_i corresponds to a node.
3. Start from the first node S_1 in the sorted sequence and compare it with the subsequent nodes one by one. For the current node S_j and the subsequent node S_k ($j < k$), check whether each column name in S_j exists in S_k. If all the columns in S_j exist in S_k, then add a directed edge in the graph G from S_j to S_k, that is, $S_j \rightarrow S_k$.
4. Traverse and compare backward starting from the node S_k. If a directed edge $S_k \rightarrow S_l$ is determined to exist, according to the transitivity of the containment relationship, it can be directly deduced that $S_j \rightarrow S_k \rightarrow S_l$, and the comparison between S_j and S_l can be skipped. After that, use S_l as a new starting point to continue the comparison backward until the end of the sequence, and then return to S_j to compare it with S_{k+1}. Repeat this process until all the comparisons of S_j are completed, and then update the starting node to S_{j+1}. Repeat this cycle until the comparison of S_n is completed, and finally construct the complete containment graph.

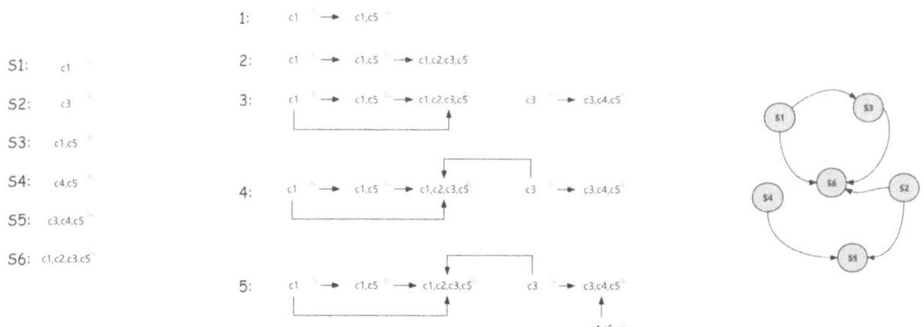

Fig. 3. The Process of Constructing Inclusion Graph by CGC Algorithm.

4.2 Data Type Pruning

We take the constructed containment graph as input and utilize the relationship between the minimum and maximum values of columns to further verify the containment relationships and prune the edges that do not meet the criteria.

Suppose $r(A) \leq r(B)$, that is, the number of rows in table A does not exceed that in table B. Let C be the set of common columns between table A and table B. If $A \subseteq B$, the following two conditions must be satisfied:

1. Minimum value condition: For all common columns $c \in C$, the minimum value of column c in table A must be greater than or equal to the minimum value of column c in table B.
2. Maximum value condition: For all common columns $c \in C$, the maximum value of column c in table A must be less than or equal to the maximum value of column c in table B.

If any column does not satisfy the above conditions, the edge $A \rightarrow B$ can be safely pruned, and we can conclude that A is not contained in B.

For datasets stored in a partitioned form, the minimum and maximum values of columns are usually stored as metadata. This makes it highly efficient to find these values. By directly querying the metadata instead of performing a full table scan, the time complexity of constraint condition verification can be reduced to the O(1) level. This method can significantly improve the efficiency of redundancy verification in scenarios involving large scale data.

4.3 Enumeration Type Pruning

Enumeration type data is central to medical redundant data cleaning systems. As a key enabler for clinical data standardization, it serves as the structured backbone for decision support systems. The high frequency usage of enumeration type data makes it a primary source of redundancy and simultaneously the data type with the greatest potential for storage optimization. Aiming at the special data structure of hospitals, we designed an enumeration type pruning method. For enumeration type data (such as doctor names, blood types, diagnostic codes, etc.), the values of their columns typically come from a finite discrete set. To verify whether table A is contained in table B, we need to ensure that all possible values of the enumeration type columns in table A are included in the set of possible values of the corresponding columns in table B. Let C be the set of common columns between table A and table B. For each common column $c \in C$, define:

1. $V_A(c)$ as the set of all possible values of column c in table A;
2. $V_B(c)$ as the set of all possible values of column c in table B.

If $A \subseteq B$, the following condition must be satisfied:

$$V_A(c) \subseteq V_B(c), \quad \forall c \in C$$

That is, all possible values of column c in table A must be completely contained within the value set of column c in table B. If any column $c \in C$ does not satisfy the above condition, the edge $A \rightarrow B$ can be safely pruned, and it is determined that A is not contained in B.

4.4 Similarity Based Pruning

Hospital datasets exhibit significant particularity. In various types of hospital data such as clinical case records, diagnostic reports, and medical research documents, there are numerous words that are semantically similar but differently expressed. For example, "fever" and "pyrexia" have nearly identical meanings in a medical context, both used to describe the symptom of elevated body temperature in humans. If only the method of directly comparing whether the text is identical is used to determine data redundancy, data with similar semantics but different literal expressions such as "fever" and "pyrexia" will be incorrectly classified as non-redundant, affecting data quality and subsequent analysis results. Therefore, we adopt a semantic similarity based approach to determine the redundancy of hospital datasets, with data alignment required during the calculation process. For instance, when calculating the semantic similarity of cases for the same individual in two tables, it is first necessary to identify the records corresponding to the same person in both tables. For the aligned data, the BERT-Base-Chinese model is used to extract features from the textual data in the hospital dataset, mapping each word to a high dimensional vector space such that semantically similar words are positioned close to each other. Finally, cosine similarity is calculated for the extracted semantic features. A threshold is set for the cosine similarity values: records with values above this threshold are marked as suspected redundant data, which will be reviewed and pruned by physicians during the final Clinical Decision-Making phase.

4.5 Clinical Decision Making

So far, we have introduced the process of constructing dataset relationship graphs and elaborated on the multi-level pruning method for redundant data. To enhance the efficiency of the pruning method and ensure the accuracy of pruning results, we introduced a clinical decision making mechanism. In the clinical decision-making system, engineer clinical decision making and physician clinical decision making form a complete closed loop of complementary collaboration.

The engineer's work is divided into two core steps: Firstly, hospital data engineers undertake the tasks of data structure annotation and pruning method selection in the clinical decision-making process to construct a multi-level pruning workflow. Engineers conduct a comprehensive sorting out of hospital data and accurately annotate the data structure. Through accurate annotation, engineers can clearly grasp the characteristics and uses of the data, laying a foundation for subsequent pruning operations. Subsequently, engineers can comprehensively consider factors such as data scale, complexity, and the importance to medical decision-making, and select appropriate pruning strategies from a variety of methods to ensure that the selected method can effectively reduce data redundancy while completely retaining the key medical information.

Physician clinical decision making focuses on the final review of redundant data. In medical data processing, preliminary pruning operations generate quantitative results of similarity between different datasets. Physicians, relying on professional medical knowledge and clinical experience, deeply analyze these calculation results and determine the redundant attributes of data by evaluating the relevance of data to clinical diagnosis, treatment, and research. Their core goal is to minimize data redundancy without compromising critical information, thereby optimizing the storage and utilization efficiency of medical data while ensuring the accuracy and reliability of medical decisions.

5 Experiment

The experimental environment is built on a three-node distributed cluster with uniform configurations. Each node is equipped with an 11th Generation Intel Core i7-11700F processor and 16 GB of DDR4 memory, and runs on the Ubuntu operating system. Additionally, the Hadoop distributed file system and computing framework are deployed across the cluster, enabling efficient data storage and parallel processing, which provides a stable and high-performance foundation for data processing and algorithm validation in the experiments.

We evaluated the effectiveness of our pruning method on hospital data and synthetic data. For hospital data, we used real-world data from two different hospital campuses. For synthetic data, we constructed it using the publicly available MIMIC-III [14] dataset, a common dataset in the field. The data synthesis process involved sampling for size reduction, adding rows, adding columns, introducing noise, and combinations thereof.

In terms of runtime comparison (as shown in Table 1), both the method proposed in this study and the R2D2 algorithm significantly outperform the brute-force comparison algorithm in terms of time efficiency. Compared with the R2D2 algorithm, our method demonstrates better performance in processing medical datasets. On the one hand, this method fully considers the special data types of medical datasets and designs a targeted pruning strategy, which better meets the needs of the medical field for efficient pruning of redundant data. On the other hand, by comparing the runtime of this method under two scenarios—without clinical decision Making and with clinical decision Making it is found that the runtime is further shortened after incorporating clinical decisions. This indicates that the participation of hospital engineers in clinical decision-making can inject professional knowledge into the redundant pruning process, optimize the pruning logic, and thus significantly improve pruning speed. This strongly demonstrates the critical value of clinical decision-making in medical data redundancy pruning.

In terms of redundant edge pruning effect (as shown in Table 2), the CGC algorithm first constructs a schema containment graph to generate the containment relationships of all columns. Subsequently, a series of targeted pruning

Table 1. Comparison of Different Methods Across Datasets

Datasets	Method	Time
Hospital 1	Brute - Force Algorithm	~days
	R2D2	4.8 h
	Without Clinical Decision	4.28 h
	With Clinical Decision	3.54 h
Hospital 2	Brute - Force Algorithm	~days
	R2D2	4.2 h
	Without Clinical Decision	3.7 h
	With Clinical Decision	2.96 h
MIMIC-III	Brute - Force Algorithm	~days
	R2D2	3.45 h
	Without Clinical Decision	2.45 h
	With Clinical Decision	2.12 h

algorithms process the edges of the schema containment graph, gradually compressing the space of redundant data and significantly reducing the selection scope for doctors when screening data. Each step from data type-sensitive pruning to enumeration pruning and text pruning performs its specific function while collaborating synergistically. Ultimately, this effectively achieves precise control over medical data redundancy, verifying the scientificity and rationality of the multi-level pruning strategy proposed in this study.

Table 2. Comparison of the number of remaining edges after each pruning step

Data	size	Edge after CGC	Edge after DTP	Edge after ETP	Edge after SBP&CDM
Hospital 1	312G	1265	132	83	50
Hospital 2	103G	1187	178	69	42
MIMIC-III	36G	853	664	265	256

In the verification of redundancy pruning effectiveness (as shown in Table 3), after applying this method to prune three medical datasets with different scales and characteristics , the results show that the proportion of deleted datasets in all three exceeds 20%. This result strongly confirms the effectiveness of this method in medical data redundancy pruning, as it can effectively identify and eliminate a large amount of redundant data, contributing to improving medical data storage efficiency and optimizing data quality.

In the final stage of this experiment, the computational workload comparison for containment graph construction (as shown in Fig. 4 and the corresponding figure) analyzed the number of column-wise comparisons during the graph

Table 3. The proportion of redundant data pruned by our method

Data	Size	Deleted datasets	Proportion
Hospital 1	321G	76.8G	24.6%
Hospital 2	103G	22.7G	22.7%
MIMIC-III	50G	11.4G	27.6%

construction process among the current mainstream brute-force algorithm, the R2D2 algorithm, and the algorithm proposed in this study. The results show that on Hospital 1, Hospital 2, and MIMIC-III, the number of column comparisons of our algorithm is significantly fewer than that of the R2D2 algorithm. This demonstrates that our algorithm effectively reduces the computational workload for containment graph construction through optimization strategies, significantly saving the time required for graph generation and improving the execution efficiency of this phase.

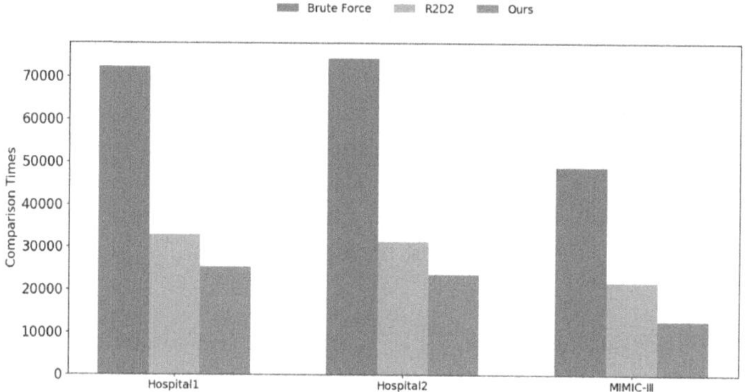

Fig. 4. Comparison Times of Edges in Graph Construction by Different Algorithms.

6 Conclusion

This paper proposes a multi - level pruning method for multi-modal medical redundant data based on data lakes. Through constructing a schema containment graph, implementing a multi - level pruning strategy, and adopting a "doctor in the loop" mechanism, it achieves efficient redundant detection and optimization of storage costs. This method significantly reduces the storage and management burden of medical data, performs well in different types and scales of data, and can be practically applied in clinical environments. However, the proposed method still has limitations in precise semantic recognition and multimodal medical data pruning. Future work will focus on the application of natural language processing and fusion pruning of multi-modal medical data.

References

1. Zhang, Y., Sheng, M., Liu, X., et al.: A heterogeneous multi-modal medical data fusion framework supporting hybrid data exploration. Health Inf. Sci. Syst. **10**(1), 22 (2022)
2. Wrenn, J.O., Stein, D.M., Bakken, S., et al.: Quantifying clinical narrative redundancy in an electronic health record. J. Am. Med. Inf. Assoc. **17**(1), 49–53 (2010)
3. Zhang, Y., Xu, C., Li, H., et al.: HealthDep: an efficient and secure deduplication scheme for cloud-assisted eHealth systems. IEEE Trans. Ind. Inf. **14**(9), 4101–4112 (2018)
4. Xu, L.J., Hao, R., Yu, J., et al.: Secure deduplication for big data with efficient dynamic ownership updates. Comput. Electr. Eng. **96**, 107531 (2021)
5. Cogo, V., Paulo, J., Bessani, A.: Genodedup: similarity-based deduplication and delta-encoding for genome sequencing data. IEEE Trans. Comput. **70**(5), 669–681 (2020)
6. Mohamad, B., Orazio, L., Gruenwald, L.: Towards a hybrid row-column database for a cloud-based medical data management system. In: 1st International Workshop on Cloud Intelligence, pp. 1–4. ACM, New York (2012)
7. Ren, P., et al.: MHDP: an efficient data lake platform for medical multi-source heterogeneous data. In: Xing, C., Fu, X., Zhang, Y., Zhang, G., Borjigin, C. (eds.) WISA 2021. LNCS, vol. 12999, pp. 727–738. Springer, Cham (2021). https://doi.org/10.1007/978-3-030-87571-8_63
8. Sheng, M., Wang, S., Zhang, Y., et al.: MQRLD: a multimodal data retrieval platform with query-aware feature representation and learned index based on data lake. Inf. Process. Manag. **62**(4), 104101 (2025)
9. Tian, G., Hu, Y., Wei, J., et al.: Blockchain-based secure deduplication and shared auditing in decentralized storage. IEEE Trans. Dependable Secure Comput. **19**(6), 3941–3954 (2021)
10. Shah, R., Mukherjee, K., Tyagi, A., et al.: R2D2: reducing redundancy and duplication in data lakes. In: Proceedings of the 2023 Conference on Data Engineering, pp. 1–12. ACM (2023)
11. Zhang, B., Wang, C., Zhou, B.B., et al.: DCDedupe: selective deduplication and delta compression with effective routing for distributed storage. J. Grid Comput. **16**, 195–209 (2018)
12. Fu, M., Feng, D., Hua, Y., et al.: Design tradeoffs for data deduplication performance in backup workloads. In: Proceedings of USENIX Conference on File and Storage Technologies, pp. 331–344. USENIX (2015)
13. Zhang, Y., Sheng, M., Zhou, R., et al.: HKGB: an inclusive, extensible, intelligent, semi-auto-constructed knowledge graph framework for healthcare with clinicians' expertise incorporated. Inf. Process. Manag. **57**(6), 102324 (2020)
14. Johnson, A.E.W., Pollard, T.J., Shen, L., et al.: MIMIC-III, a freely accessible critical care database. In: Proceedings of the Scientific Data, vol. 3, pp. 160035. Nature Publishing Group (2016)
15. Liu, X., Wang, K., Wang, W., et al.: Sepsis mortality prediction with electronic health records based on sequential and attention-based models. In: Jin, C., Yang, S., Shang, X., Wang, H., Zhang, Y. (eds.) Web Information Systems and Applications. WISA 2024. Lecture Notes in Computer Science, vol. 14883, pp. 476–486. Springer, Singapore (2024). https://doi.org/10.1007/978-981-97-7707-5_39

Porcelain-SegNet: A Boundary-Aware Transformer for Semantic Segmentation of Figures on Chinese Export Porcelain

Yanfang Xu[1], Hua Yin[2(✉)], and Zhensheng Hu[3]

[1] School of Data Science and Computer Science, Guangdong Peizheng College, Hua Du District, Guangzhou 510830, People's Republic of China
katexu@suool.net

[2] Guangdong University of Finance and Economics, Guangzhou 510000, China
yinhua@gdufe.edu.cn

[3] Sun Yat-Sen University, Guangzhou 510080, People's Republic of China
huzhsh6@mail2.sysu.edu.cn

Abstract. Human figures on Chinese export porcelain are a vital resource for art-historical study, yet their analysis is hindered by the limitations of manual, subjective methods. The precise segmentation of these figures from intricate backgrounds is a significant technical challenge, complicated by stylistic diversity and ambiguous object boundaries. To automate and scale this process, we propose Porcelain-SegNet, a novel, boundary-aware deep learning framework designed specifically for this task. We first introduce EPF-Seg-1k, the first large-scale benchmark dataset for segmenting figures on porcelain, comprising over 1,000 high-resolution images with meticulous, pixel-level annotations. Our Porcelain-SegNet architecture features a Swin Transformer backbone to effectively model global context, attention-guided skip connections (CBAM) to enhance feature fusion by suppressing background noise, and a composite, boundary-aware loss function to ensure the generation of crisp and accurate outlines. Extensive experiments and ablation studies demonstrate that our method achieves state-of-the-art performance on the EPF-Seg-1k test set, obtaining a mean Intersection over Union (mIoU) of 90.4% and significantly outperforming established segmentation models. The high-fidelity masks produced by our model provide a robust computational foundation for enabling new forms of large-scale, quantitative analysis of iconographic and stylistic features in the field of digital art history.

Keywords: Semantic Segmentation · Chinese Export Porcelain · Vision Transformer · Digital Art History · Cultural Heritage Computing

1 Introduction

Chinese export porcelain stands as a tangible testament to centuries of global trade and cross-cultural aesthetic exchange, particularly from the 16th to the

S. Yang et al. (Eds.): WISA 2025, LNCS 16299, pp. 78–89, 2026.
https://doi.org/10.1007/978-981-95-6426-2_7

19th century [1]. The human figures adorning these wares are of paramount art-historical value, serving as a rich repository of iconographic and stylistic information. A precise, pixel-level analysis of these figures—delineating the exact contours of robes, gestures, and accessories—can unlock profound insights into artistic conventions, workshop practices, and cross-cultural adaptations [2]. However, achieving such fine-grained analysis through traditional scholarly methods is exceptionally challenging. Manual tracing or visual inspection is not only labor-intensive and subjective but also infeasible for large-scale comparative studies. This creates a critical need for an automated method that can precisely segment human figures from complex backgrounds, enabling quantitative analysis of their morphological and decorative features.

The advent of deep learning has catalyzed a paradigm shift in the analysis of cultural heritage. In recent years, a wave of research has demonstrated the profound potential of these methods, moving beyond simple classification towards a more granular, pixel-level understanding of artistic objects. Successful applications now include the semantic segmentation of complex Dunhuang murals [3], the stylistic classification of Tang dynasty tomb paintings [4], and the recognition of ancient scripts such as Naxi Dongba characters and bronze inscriptions [5,6]. This progress highlights a clear trajectory towards leveraging AI for nuanced artistic and historical inquiry, opening up the compelling possibility of applying similar high-precision techniques to the unique challenges presented by ceramic artifacts.

Nevertheless, the semantic segmentation of figures on Chinese export porcelain introduces a unique set of formidable challenges not fully addressed by existing work. First, these figures are often integrated into intricate backgrounds with visually similar patterns and colors, leading to highly ambiguous boundaries. Second, the artistic styles exhibit vast diversity, from the fluid brushstrokes of blue-and-white porcelain to the delicate enamels of *famille rose*, demanding a model with exceptional generalization capabilities across varied material properties. Third, crucial iconographic details are often minute and rendered on curved, glazed surfaces, requiring a model that can preserve fine-grained features throughout its processing pipeline. Standard segmentation models may struggle to address these specific issues simultaneously.

To address this research gap, this paper proposes an innovative and specialized deep learning framework. Our primary contributions are:

1. **The construction of a new benchmark dataset:** We introduce "EPF-Seg-1k," the first large-scale dataset for this task, with high-resolution images and meticulously annotated, pixel-level segmentation masks.
2. **A novel, tailored segmentation model, *Porcelain-SegNet*:** We propose a new architecture featuring a **Swin Transformer** backbone [7] to capture global context, **attention-guided skip connections** to enhance feature fusion, and a **boundary-aware dual loss function** to ensure crisp, accurate outlines.
3. **Enabling new avenues for quantitative art analysis:** We demonstrate that the precise masks generated by our model can serve as a foundation

for novel quantitative analysis in art history, such as the statistical study of posture, costume morphology, and decorative patterns on clothing.

The remainder of this paper is organized as follows. Section 2 reviews related work in semantic segmentation and AI in cultural heritage. Section 3 details our dataset and the proposed *Porcelain-SegNet* architecture. Section 4 presents the experimental results. Finally, Sect. 5 concludes the paper.

2 Related Work

The automated analysis of visual information via semantic segmentation has been fundamentally reshaped by deep learning. A pivotal moment was the development of Fully Convolutional Networks (FCNs), which adapted classification architectures for dense, pixel-level prediction [8]. This gave rise to the influential encoder-decoder paradigm, exemplified by the U-Net architecture, whose skip connections masterfully combine deep semantic context with shallow spatial details [9]. More recently, the field has seen another paradigm shift with the advent of Vision Transformers. The Swin Transformer, in particular, has demonstrated exceptional performance on dense prediction tasks due to its hierarchical structure and efficient self-attention mechanism, making it a powerful backbone for modeling complex imagery [7]. To further refine segmentation, especially for intricate subjects, advanced techniques such as boundary-aware loss functions [10] and attention mechanisms [11] have been introduced to respectively sharpen object contours and focus on the most salient visual features.

These technological advancements are being increasingly leveraged within the field of cultural heritage, enabling analyses of unprecedented scale and detail. Recent studies showcase this trend, applying deep learning to a wide array of artifacts. For instance, researchers have developed specialized networks for the semantic segmentation of Dunhuang murals [3] and for the stylistic classification of Tang dynasty tomb paintings [4], demonstrating the ability to analyze complex, painted surfaces. In parallel, other works have focused on deciphering symbolic information, successfully recognizing characters from ancient Chinese bronze inscriptions [6], Naxi Dongba scripts [5], and detecting seals in painting and calligraphy works [12]. The creation of specialized datasets, for everything from architectural tiles to ancient pottery, has also become a key research activity, forming the foundation for these intelligent systems [13,14]. Despite this progress, a systematic review on archaeological ceramics confirms that challenges of data scarcity and high visual diversity persist [15]. While there is clear success in analyzing murals or inscriptions on relatively flat media, the precise semantic segmentation of artistic motifs on the curved, glazed, and stylistically varied surfaces of artifacts like Chinese export porcelain remains a distinct and underexplored problem. Our work seeks to fill this gap by developing a tailored model that integrates state-of-the-art techniques to address the unique challenges of this domain.

3 Methodology

This section details the formal problem definition, the rigorous protocol for our dataset construction, the architecture of our proposed *Porcelain-SegNet* model, and the metrics used for its evaluation.

3.1 Problem Formulation

The primary objective of this research is formally defined as a binary semantic segmentation task. The goal is to train a deep learning model, f_θ, which learns a mapping from an input image of Chinese export porcelain, I, to a corresponding pixel-wise mask, M. This process can be concisely formulated as:

$$M_{pred} = f_\theta(I) \tag{1}$$

Here, the input $I \in \mathbb{R}^{H \times W \times 3}$ represents a standard color image with height H, width W, and three color channels (RGB). The function f_θ is our deep neural network (e.g., *Porcelain-SegNet*), parameterized by a set of learnable weights θ. The model's output, $M_{pred} \in \{0,1\}^{H \times W}$, is the predicted segmentation mask, which is a 2D map with the same spatial dimensions as the input. Within this mask, each pixel is assigned a class label $c \in \{0,1\}$, where $c = 1$ corresponds to the foreground "Figure" class (including the person and their integral accessories), and $c = 0$ corresponds to the "Background". The entire training process is dedicated to optimizing the parameters θ such that the predicted mask M_{pred} aligns as closely as possible with the manually annotated ground truth mask, M_{gt}.

3.2 Model Architecture: Porcelain-SegNet

To address the unique challenges of segmenting figures from complex porcelain imagery, we designed a novel encoder-decoder network, *Porcelain-SegNet*. Its architecture is detailed below.

Encoder: Swin Transformer Backbone. We employ a Swin Transformer ('Swin-T' variant) [7], pre-trained on ImageNet [16], as our feature encoder. Unlike standard CNNs that use fixed-size kernels, the Swin Transformer's self-attention mechanism, computed within shifted windows, allows it to capture long-range dependencies across the image. This is crucial for understanding the global context of a figure within its intricate decorative surroundings. Furthermore, its hierarchical structure produces multi-scale feature maps, which are essential for recognizing both large overall figures and their minute, fine-grained details.

Decoder with Attention-Guided Skip Connections. The decoder's role is to progressively upsample the high-level feature maps from the encoder to reconstruct the full-resolution segmentation mask. A key innovation in our design lies in the skip connections. Standard additive or concatenation-based skip connections, as in U-Net, can propagate noisy or irrelevant background features, which is particularly problematic for porcelain images with complex patterns. To mitigate this, we embed a Convolutional Block Attention Module (CBAM) [11] into each skip connection. The CBAM module applies both channel and spatial attention, allowing our model to learn to selectively emphasize informative figure-related features from the encoder while suppressing distracting background patterns, leading to cleaner and more accurate feature fusion.

Loss Function: Boundary-Aware Dual Loss. To effectively train our network, we designed a composite loss function that simultaneously optimizes for both regional accuracy and boundary sharpness. This dual-loss strategy ensures the model learns not only *what* to segment but also *how precisely* to segment it, preserving the artistic integrity of the figures' silhouettes. The total loss L_{total} is a weighted sum of a regional segmentation loss and a boundary loss:

$$L_{\text{total}} = L_{\text{seg}} + \lambda \cdot L_{\text{boundary}}$$

where λ is a hyperparameter that balances the two terms. In our experiments, λ was set to 1.0.

Segmentation Loss (L_{seg}). For regional supervision, we employ the Focal Tversky Loss [17], an advanced variant of the Dice Loss. It is particularly effective for scenarios with class imbalance (figure vs. background) and allows for tuning the trade-off between penalizing false positives (FP) and false negatives (FN). First, the Tversky Index (TI) is defined as:

$$\text{TI} = \frac{\text{TP}}{\text{TP} + \alpha\text{FN} + \beta\text{FP}} \quad (2)$$

where TP, FP, and FN are the sums of true positive, false positive, and false negative probabilities over all pixels. The hyperparameters α and β control the penalty magnitude for FNs and FPs, respectively; we use the common setting $\alpha = 0.7$ and $\beta = 0.3$ to place more emphasis on recall. The Focal Tversky Loss is then formulated by applying a focusing parameter γ to the Tversky Loss $(1 - \text{TI})$:

$$L_{\text{seg}} = (1 - \text{TI})^{\gamma} \quad (3)$$

where we set $\gamma = 4/3$. This modification focuses the training on harder-to-classify pixels, which is ideal for capturing small yet significant figural details.

Boundary Loss ($L_{boundary}$). To explicitly enforce crisp outlines, we implement a boundary loss term based on the work of Bokhovkin et al. [10]. This loss leverages the spatial distance of predicted pixels from the ground truth contour.

Let $G \subset \Omega$ be the set of pixels belonging to the ground truth figure region in the image domain Ω. The boundary loss is formulated as a surface integral:

$$L_{\text{boundary}} = \int_{\Omega} \phi_G(p) s_\theta(p) \, dp \tag{4}$$

Here, $s_\theta(p)$ is the model's predicted probability (sigmoid output) for pixel p belonging to the figure class. The term $\phi_G(p)$ represents the signed Euclidean distance from pixel p to the closest point on the boundary of G. Specifically, $\phi_G(p)$ is positive if p is inside G and negative if it is outside. This loss effectively penalizes predictions that are confidently wrong in regions far from the true boundary, directly forcing the network to produce predictions that align precisely with the ground truth contours.

4 Experiments

We constructed a new benchmark dataset, EPF-Seg-1k (Export Porcelain Figures for Segmentation), comprising 1,080 high-resolution images. The images were systematically collected from the archives of major museums and reputable auction house catalogues and were filtered to ensure each contained a clear human figure motif suitable for fine-grained annotation. The final dataset was randomly partitioned into training (60%), validation (20%), and testing (20%) sets.

A comprehensive annotation protocol was established to ensure the reliability of our ground truth masks, with the full process shown in Fig. 1. Using the LabelMe [18] tool, we created precise polygon-based outlines that were converted into binary masks. A cornerstone of this process was a detailed annotation handbook defining the scope of the 'Figure' class. The guiding principle was to annotate *the human figure and its integral accessories* as a single semantic entity. This scope included the human form, all worn items (e.g., hats, robes), handheld objects (e.g., fans, swords), and mounts. Conversely, non-integral background elements like surrounding flora or furniture were strictly excluded.

To guarantee quality and consistency, we implemented a multi-stage control process with two trained annotators. After a calibration phase on a 50-image pilot batch, the main task proceeded with an independent annotation and mandatory peer-review cycle to flag any ambiguities or deviations from the guidelines. This was further validated by a quantitative Inter-Annotator Agreement (IAA) assessment on a randomly selected 100-image subset. The mean Intersection over Union (mIoU) [19] between the annotators' paired masks was 0.92, confirming high reliability. In the final stage, a senior researcher acted as an adjudicator to resolve all disagreements, producing the final, high-quality ground truth for every image and iteratively refining the annotation handbook.

4.1 Implementation and Evaluation

All experiments were conducted on four NVIDIA RTX 4090 GPUs using PyTorch. The Swin Transformer encoder was initialized with ImageNet-1K pre-trained weights via the `timm` library. Images from our EPF-Seg-1k dataset were

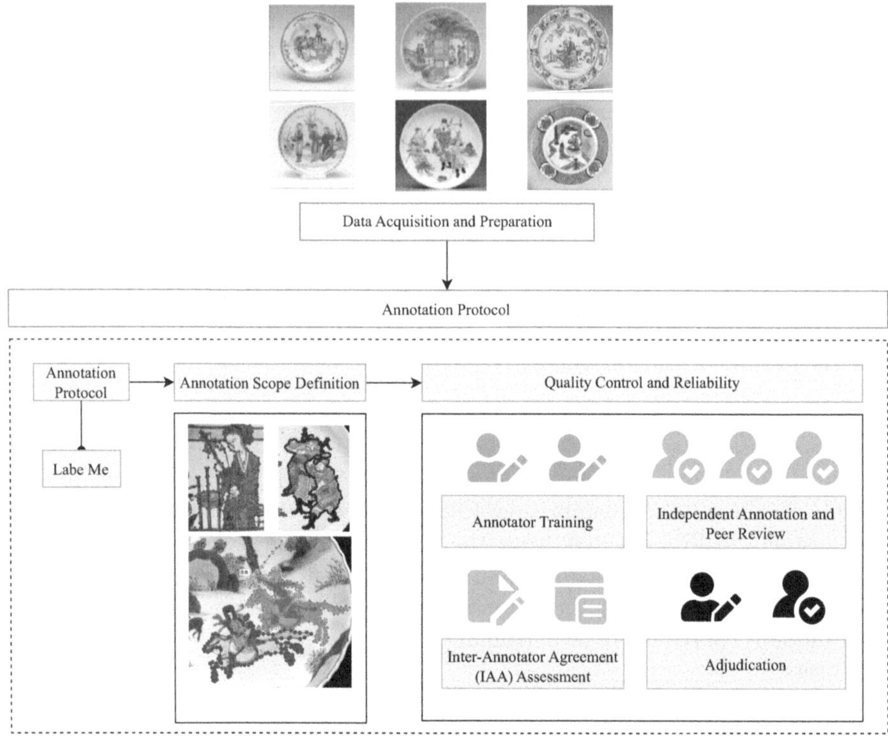

Fig. 1. The complete process of the annotation protocol.

resized to 512×512 pixels. We applied online data augmentation, including random horizontal flipping (50% probability), rotation (within ± 20 degrees), and scaling (from $0.8\times$ to $1.2\times$).

The network was trained for 150 epochs with a batch size of 16 using the AdamW optimizer [20]. We used a cosine annealing learning rate schedule with an initial rate of 5×10^{-5}, a 10-epoch linear warm-up, and a weight decay of 0.01. For our composite loss, $L_{\text{total}} = L_{\text{seg}} + \lambda \cdot L_{\text{boundary}}$, the balancing parameter λ was set to 1.0 to equally weight the segmentation and boundary terms.

We assess model performance using three standard metrics, defined by pixel-level true positives (TP), true negatives (TN), false positives (FP), and false negatives (FN).

Pixel Accuracy (PA): The percentage of correctly classified pixels.

$$\text{PA} = \frac{\text{TP} + \text{TN}}{\text{TP} + \text{TN} + \text{FP} + \text{FN}} \tag{5}$$

Intersection over Union (IoU): The primary metric for segmentation quality. We report IoU for the foreground class and Mean IoU (mIoU) across all classes.

$$\text{IoU} = \frac{\text{TP}}{\text{TP} + \text{FP} + \text{FN}} \tag{6}$$

Dice Coefficient (F1-Score): A common overlap metric, related to IoU.

$$\text{Dice} = \frac{2 \cdot \text{TP}}{2 \cdot \text{TP} + \text{FP} + \text{FN}} \tag{7}$$

4.2 Baselines

To rigorously evaluate Porcelain-SegNet, we benchmark it against established and state-of-the-art segmentation models representing three architectural families.

Baselines include classic encoder-decoders: the pioneering **FCN** [8] for end-to-end prediction and **U-Net** [9], known for its detail-preserving skip connections. As a state-of-the-art CNN, we use **DeepLabv3+** [21] (ResNet-101 backbone), which leverages Atrous Spatial Pyramid Pooling (ASPP) to capture multi-scale context. For a Transformer-based comparison, we employ **SegFormer** [22] (MiT-B2 variant), an efficient architecture with a hierarchical encoder and lightweight MLP decoder.

For a fair comparison, all models were trained from scratch on our EPF-Seg-1k dataset. We used the exact same data splits, augmentation strategies, and training hyperparameters as our method (detailed in Sect. 4.2.1), with backbones initialized from their standard ImageNet pre-trained weights.

4.3 Overall Performance Experiments

In this section, we present a comprehensive quantitative and qualitative analysis of our model's performance on the EPF-Seg-1k test set. We first compare Porcelain-SegNet with the selected baselines, and then conduct a detailed ablation study to validate the effectiveness of its core components.

Table 1 summarizes the primary performance metrics of all models on the held-out test set. As shown, our proposed Porcelain-SegNet achieves state-of-the-art performance, surpassing all baseline models across all three evaluation metrics. Specifically, it obtains a mean Intersection over Union (mIoU) of 90.4%, a Dice Coefficient of 94.9%, and a Pixel Accuracy of 97.2%.

Compared to the classic U-Net, Porcelain-SegNet shows a significant improvement of 4.1% in mIoU, underscoring the benefits of its advanced backbone and attention mechanisms. More notably, it outperforms the powerful DeepLabv3+ and SegFormer baselines by 2.5% and 1.8% in mIoU, respectively. This demonstrates that while general-purpose SOTA models are highly capable, our specialized architecture, which is explicitly designed to handle ambiguous boundaries and preserve fine details, is better suited for the unique challenges presented by Chinese export porcelain imagery. The Swin Transformer's ability to capture global dependencies combined with our attention-guided feature fusion and boundary-aware loss proves critical in achieving this superior performance.

To deconstruct the contributions of our model's key components, we conducted a series of ablation experiments, with the results presented in Table 2.

Table 1. Quantitative comparison with baseline models on the EPF-Seg-1k test set.

Model	mIoU (%)	Dice (%)	PA (%)
FCN [8]	82.7	90.5	95.8
U-Net [9]	86.3	92.6	96.4
DeepLabv3+ [21]	87.9	93.6	96.8
SegFormer [22]	88.6	94.0	97.0
Porcelain-SegNet (Ours)	**90.4**	**94.9**	**97.2**

We started with a baseline model (Base) consisting of the Swin-T encoder and a standard U-Net-like decoder without our proposed modules.

Effect of Attention-Guided Skip Connections (CBAM): By integrating the CBAM module into the skip connections (Base + CBAM), we observed a 1.2% increase in mIoU. This confirms our hypothesis that explicitly guiding the feature fusion process allows the model to suppress irrelevant background patterns and focus on salient figural features, leading to a more refined segmentation.

Effect of Boundary-Aware Dual Loss: Adding only the boundary loss term (L_{boundary}) to the base model (Base + Boundary Loss) resulted in a 1.7% mIoU gain. This demonstrates the critical importance of explicitly penalizing inaccuracies at object contours. This loss component directly forces the network to learn crisp and precise boundaries, which is a primary challenge in this dataset.

Combined Effect: Our full model, Porcelain-SegNet, which synergistically combines both modules, achieves the highest performance. The total gain of 2.7% over the base model is greater than the sum of the individual gains, suggesting a complementary and synergistic relationship between the attention mechanism and the boundary-aware loss function. The CBAM module provides better regional features, which in turn allows the boundary loss to be more effective in refining the final silhouette.

Table 2. Ablation study of Porcelain-SegNet's components on the test set.

Model Configuration	mIoU (%)	Dice (%)	PA (%)
Base (Swin-T + Standard Decoder)	87.7	93.4	96.7
Base + CBAM	88.9	94.1	97.0
Base + Boundary Loss	89.4	94.4	97.1
Porcelain-SegNet (Full Model)	**90.4**	**94.9**	**97.2**

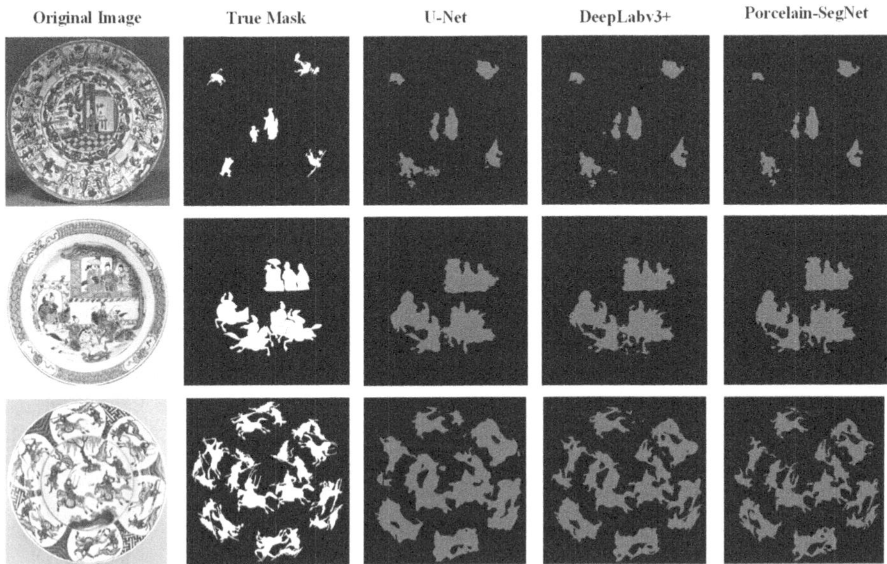

Fig. 2. Qualitative comparison on challenging images from the EPF-Seg-1k test set. From left to right: (a) Input Image, (b) Ground Truth, (c) U-Net, (d) DeepLabv3+, (e) Porcelain-SegNet (Ours). Our model demonstrates superior performance in handling complex backgrounds (top row) and preserving fine details (bottom row), generating significantly cleaner and more accurate segmentation masks.

Qualitative Results. Visual comparisons further illustrate the quantitative advantages of our method. As shown in Fig. 2, we present several challenging examples from the test set. The first row showcases a figure against a complex, similarly-colored floral background. While U-Net and DeepLabv3+ erroneously merge parts of the figure's robes with the background, Porcelain-SegNet successfully delineates a complete and accurate contour. The second row highlights a figure with fine accessories (e.g., a sword and flowing ribbons). Our model excels at preserving these delicate structures, whereas other methods tend to lose or distort these fine-grained details. These qualitative results provide compelling visual evidence that Porcelain-SegNet generates segmentation masks that are not only more accurate but also more faithful to the artistic details of the source material.

5 Conclusion

In this paper, we addressed the challenging task of semantic segmentation for human figures on Chinese export porcelain by proposing a novel, specialized deep learning framework, Porcelain-SegNet. We introduced a new, meticulously annotated benchmark dataset, EPF-Seg-1k, to facilitate research in this domain. Our model, which leverages a Swin Transformer to capture global context, attention-guided skip connections to refine feature fusion, and a boundary-aware dual loss to ensure sharp outlines, has proven to be highly effective. Comprehensive experiments demonstrated that Porcelain-SegNet achieves state-of-the-art performance, significantly outperforming a range of strong baseline models. The ablation studies further validated the contribution of each of our architectural innovations. More importantly, this work provides a foundational tool for a paradigm shift in art-historical analysis, moving from qualitative assessment to quantitative, data-driven investigation. The high-fidelity masks generated by our method can enable large-scale statistical studies of posture, costume, and iconography. Future work will focus on extending the framework to multi-class segmentation for distinguishing finer attributes (e.g., clothing vs. accessories) and applying this technology to cross-corpus stylistic analysis.

Acknowledgement. This work was supported by the Foshan Social Science Planning Project(2025-GJ133), the Guangdong Peizheng College University-Level Research Project(25PZXJQN025) and the Humanity and Social Science Youth Foundation of Ministry of Education of China (21YJCZH202).

References

1. David S.H., John, A.: tChina for the West: Chinese Porcelain and other Decorative Arts for Export. Sotheby Parke Bernet, London & New York, (1974)
2. Krahl, R., Harrison-Hall, J.: Chinese Ceramics: Highlights of the Sir Percival David Collection. British Museum Press, London (2011)
3. Liu, Yu., Ma, Y., Wang, Y., Wang, L., Zhang, Y., Tian, J.: A semantic segmentation method for dunhuang mogao grottoes murals based on dun-net. Heritage Science **12**(1), 183 (2024)
4. Luo, M., Ma, Y., Gao, Y., Wang, L.: Style classification of tang dynasty tomb murals based on an improved residual network. Heritage Sci. **12**(1), 216 (2024)
5. He, Z., Chen, S., Zhu, Z., Wang, R.: Research on the recognition of Chinese naxi minority's dongba script based on deep learning. Heritage Sci. **12**(1), 192 (2024)
6. Wang, Y., Liu, S., Gao, L.: An effective recognition and retrieval model for ancient Chinese bronze inscriptions. Heritage Sci. **12**(1), 132 (2024)
7. Ze, L., et al.: Swin transformer: Hierarchical vision transformer using shifted windows. In: Proceedings of the IEEE/CVF International Conference on Computer Vision (ICCV), pp. 10012–10022, (2021)
8. Jonathan, L., Evan, S., Trevor, D.: Fully convolutional networks for semantic segmentation. In: Proceedings of the IEEE Conference on Computer Vision and Pattern Recognition (CVPR), pp. 3431–3440, (2015)
9. Olaf, R., Philipp, F., Thomas, B.: U-net: Convolutional networks for biomedical image segmentation. In: Medical Image Computing and Computer-Assisted Intervention – MICCAI 2015, pp. 234–241, (2015)
10. Andrey, B., Yury, P., Gleb, K., Ivan, V.: Boundary loss for highly unbalanced segmentation. arXiv preprint arXiv:1908.11307 (2019)
11. Sanghyun, W., Jongchan, P., Joon-Young, L.: In: So, K., (eds.) Cbam: Convolutional block attention module. In: Proceedings of the European Conference on Computer Vision (ECCV), pp. 3–19 (2018)

12. Yuzheng, L., Min, L., Xueqing, Z.: MGCNet: A Multi-scale Grouped Convolution-Based Seal Detection Method for Painting and Calligraphy Works. In: Jin, C., Yang, S., Shang, X., Wang, H., Zhang, Y. (eds.) International Conference on Web Information Systems and Applications pp. 586–594. Springer Nature Singapore, (2024).https://doi.org/10.1007/978-981-97-7707-5_48
13. Springstein, M., Krekeler, F., van der Knaap, E., Othman, A., Ermakova, L., Klinke, H.: Annotile: a lightweight annotation tool for creating datasets of architectural elements on historical tile stoves. ACM J. Comput. Cult. Heritage **16**(3), 1–23 (2023)
14. Shen, C., Fan, J., Jiang, C.: Intelligent classification of ancient pottery from different cultural origins based on deep learning: a case study of longshan and dawenkou cultures. J. Cult. Heritage **63**, 206–214 (2023)
15. Agathos, A., Lympouridis, V., Makantasis, K., Karafotias, G., Chatzichristofis, S.A., Riris, P., Amditis, A.: Findings on machine learning for identification of archaeological ceramics: a systematic literature review. J. Archaeol. Sci. **164**, 105943 (2024)
16. Tsung-Yi, L., et al.: Microsoft coco: Common objects in context. In: Fleet, D., Pajdla, T., Schiele, B., Tuytelaars, T. (eds.) Proceedings of the European Conference on Computer Vision (ECCV), pp. 740–755. Springer, Cham (2014). https://doi.org/10.1007/978-3-319-10602-1_48
17. Nabila, A., Naimul, M.K.: A novel focal tversky loss function with improved attention u-net for lesion segmentation. arXiv preprint arXiv:1910.09424 (2019)
18. Bryan, C.R., Antonio, T., Kevin, P.M., William, T.F.: Labelme: a database and web-based tool for image annotation. Int. J. Comput. Vision, **77**, 157–173, (2008)
19. Crr, L., Kumar, A.: A review on inter-annotator agreement. J. Appl. Sci. Eng. **24**(4), 489–495 (2021)
20. Ilya, L., Frank, H.: Decoupled weight decay regularization. arXiv preprint arXiv:1711.05101 (2017)
21. Liang-Chieh, C., Yukun, Z., George, P., Florian, S., Hartwig, A.: Encoder-decoder with atrous separable convolution for semantic image segmentation. In: Proceedings of the European Conference on Computer Vision (ECCV), pp. 801–818 (2018)
22. Xie, E., Wang, W., Zhiding, Yu., Anandkumar, A., Alvarez, J.M., Luo, P.: Segformer: simple and efficient design for semantic segmentation with transformers. Adv. Neu. Inf. Process. Sys. (NeurIPS) **34**, 12077–12090 (2021)

Federated Learning Adaptive Knowledge Distillation Aggregation for Breast Cancer Assisted Diagnosis

Shan Huang[1,2,3], Jingjie Li[1,2,3], Ruihong Liu[1,2,3], Yongqiang Wu[1,2,3], and Xiaodong Duan[1,2,3](✉)

[1] College of Computer Science and Technology, Dalian Minzu University, Dalian 116650, China
[2] State Ethnic Affairs Commission Key Laboratory of Big Data Applied Technology, Dalian 116650, China
[3] Dalian Key Laboratory of Digital Technology for National Culture, Dalian 116650, China
duanxd@dlnu.edu.cn

Abstract. Breast cancer is one of the common cancers worldwide, and early identification plays a crucial role in its subsequent diagnosis and treatment. However, centralized learning that collects data from distributed institutions may lead to medical privacy breaches, and whether the diagnostic results of the model can be trusted by doctors remains a concern. Therefore, we propose a Federated Learning Adaptive Knowledge Distillation Aggregation (FedAKDA) for breast cancer assisted diagnosis. First, Gaussian Blur Histogram Equalization (GBHE) is used for image preprocessing to address the issue of indistinct tumor regions in mammograms. The FedAKDA method consists of two modules: Adaptive Knowledge Distillation (AKD) and Adaptive Local Aggregation (ALA). These modules are designed to protect patient privacy, compress the model to reduce training time, and maintain the personalization of each client model. Finally, Gradient-weighted Class Activation Mapping (Grad-CAM) is applied to visualize breast cancer lesion regions, enhancing the model's interpretability and aiding doctors in making better judgments. Experimental results show that compared to other methods, the proposed approach achieves superior diagnostic performance for breast cancer.

Keywords: Breast cancer · Personalized Federated Learning · Knowledge Distillation · Class Activation Mapping

1 Introduction

Breast cancer is one of the most common cancers among women worldwide, with the highest incidence rate globally [1]. Early identification plays a crucial role in its subsequent diagnosis and treatment. Medical imaging is commonly used

S. Yang et al. (Eds.): WISA 2025, LNCS 16299, pp. 90–101, 2026.
https://doi.org/10.1007/978-981-95-6426-2_8

in breast cancer diagnosis and treatment, with mammography being the most commonly used screening modality. However, mammography present challenges such as indistinct lesion morphology and ambiguous boundaries, requiring physicians to spend considerable time examining the images, which may compromise diagnostic accuracy.

Recent advances in Artificial Intelligence (AI) have boosted breast cancer diagnosis. Federated Learning (FL) [2] use distributed user data while preserving privacy by iteratively downloading models, training models locally on the clients, uploading models, and aggregating models on the server. A key challenge in FL is heterogeneity environments. Personalized Federated Learning (PFL) addresses this by developing personalized client models, improving performance on Non-Independent and Identically Distributed(non-IID) medical data. Meanwhile, a challenge in medical AI is enhancing model interpretability. FL models often operate as "black boxes" hindering clients' understanding of their decision-making processes. This opacity erodes trust in model predictions, which is particularly problematic in healthcare settings where the reliability of diagnostic outcomes is of utmost importance.

To address these challenges, we propose a Federated Learning Adaptive Knowledge Distillation Aggregation (FedAKDA) framework for breast cancer assisted diagnosis, enabling physicians to diagnose lesion regions more accurately and efficiently while overcoming "data silos" and heterogeneity issues. The key contributions are as follows:

-We propose a Federated Learning Adaptive Knowledge Distillation Aggregation (FedAKDA) algorithm, establishing a dual-module collaborative framework of Adaptive Knowledge Distillation (AKD) and Adaptive Local Aggregation (ALA). The AKD dynamically adjusts the distillation temperature parameter based on the difference value between client accuracy and target accuracy, thereby control the learning intensity from the global teacher model. The ALA dynamically adjusts the weight for each parameter according to local data characteristics to create personalized local models.

-We propose a Gaussian Burl Histogram Equalization (GBHE) preprocessing method. By integrating Gaussian Burl for noise smoothing and Histogram Equalization for contrast enhancement, this method improves the clarity of local details in mammograms and enhances data quality.

-We introduce Gradient-weighted Class Activation Mapping (Grad-CAM) technology, which visualizes breast cancer lesion areas through heatmaps to interpret the model decision basis. This enhancement in model interpretability and trustworthiness aids doctors in making rapid and reliable diagnoses.

2 Related Work

2.1 Personalized Federated Learning

Federated Learning (FL) is a distributed machine learning technology. The widely known traditional FL method FedAvg [2] learns a single global model for all clients by aggregating their local models. However, it often suffers in statistically heterogeneous settings. To address this issue, Li et al. [3] propose FedProx that improves the stability of the FL process through a proximal term.

Recently, personalization has attracted much attention for tackling statistical heterogeneity in FL. In personalized federated learning, Wu et al. [4] employ Knowledge Distillation (KD) to address client heterogeneity, using synthetic training data for KD to train robust federated global models. Liu et al. [5] propose a FedCL that integrates local and global models for contrastive learning, thereby enhancing model generalization. Jin et al. [6] propose personalized edge intelligence via federated self-knowledge distillation, enabling clients to transfer knowledge from past personalized models into current local models and improve the balance between personalization and generalization.

2.2 Knowledge Distillation

Knowledge distillation is a technique to transfer knowledge from a large teacher model to a small student model [7], which is widely used for model compression. For example, Sun et al. [8] propose a BERT-PKD approach that aligns the hidden states of the student model with the teacher using a mean squared error loss. Jiao et al. [9] propose a TinyBERT approach that can additionally transfer useful knowledge from the attention matrix of the teacher model. However, these methods usually require centralized data storage, which may pose privacy issues during data collection.

3 Method

We propose a Federated Learning Adaptive Knowledge Distillation Aggregation (FedAKDA) algorithm for breast cancer assisted diagnosis, as Fig. 1 illustrates. **(1) GBHE preprocessing (introduced in Sect. 3.1) and local training.** Preprocess mammography images using GBHE to enhance tumor regions, then train local models on client datasets. **(2) Local logits generation.** Apply softmax to local model outputs to generate local logits. **(3) Global teacher model aggregation.** Clients upload local logits to the server, which aggregates them to update the global teacher model. **(4) Global model distribution.** Server distributes the updated global teacher model parameters to all clients. **(5) AKD controls the learning intensity (introduced in Sect. 3.2).** Clients dynamically adjusts the distillation temperature based on the difference value between client accuracy and target accuracy, controlling the learning intensity from the global teacher model. **(6) ALA controls the update weights (introduced in Sect. 3.2).** Clients dynamically adjusts the weights of each parameter according to local data characteristics, creating personalized local models. **(7) Model iteration training.** Clients retrain local models using the learned weights and local datasets. Repeat steps (2)-(7) for iterative federated optimization. **(8) Grad-CAM visualization (introduced in Sect. 3.3).** Apply Grad-CAM to the final model to generate heatmaps highlighting breast cancer regions, enhancing interpretability for clinical diagnosis.

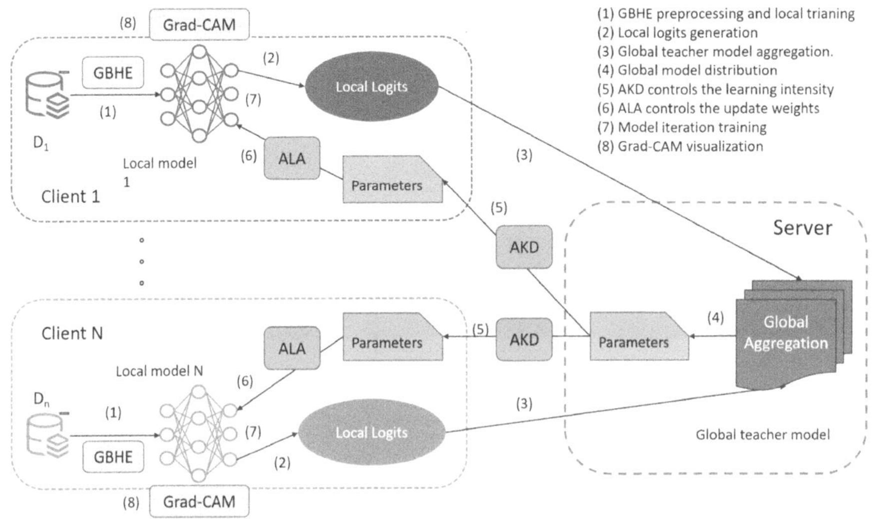

Fig. 1. Federated Learning Adaptive Knowledge Distillation Aggregation (FedAKDA) algorithm for breast cancer assisted diagnosis

3.1 GBHE Preprocessing

Mammography images present challenges for AI model training due to their low contrast, indistinct lump area, and high background noise. These issues obscure diagnostic features. Traditional contrast enhancement techniques amplify noise, while denoising methods blur lesions details and reduce the recognition of key information. Therefore, despite mammography being a common screening modality for breast cancer, developing reliable AI models based on its data remains an under-explored research area.

To address these issues, we propose a GBHE preprocessing, which combines Gaussian Blur and Histogram Equalization (HE) methods. Leveraging Gaussian Blur as a low-pass filter, we first mitigate background noise interference, preserving critical structural details. Subsequent HE enhances image contrast, accentuating subtle features like lump margins and microcalcifications. This dual-stage approach effectively balances noise reduction and feature preservation, generating high-fidelity training data for models to extract discriminative patterns with improved accuracy. The image processing effect is shown in the Fig. 2:

3.2 FedAKDA Algorithm

Problem Statement. Suppose we have N clients with their private training data $D = \{D_1, D_2, \ldots, D_n\}$, which are heterogeneous. D_i is the local dataset of the client i, the samples are (x, y), where x is the input feature, and y is the real label. With the help of a central server, our goal is to collaboratively learn individual local models $\hat{\theta} = \{\hat{\theta}_1, \hat{\theta}_2, \ldots, \hat{\theta}_n\}$, without exchanging the private data.

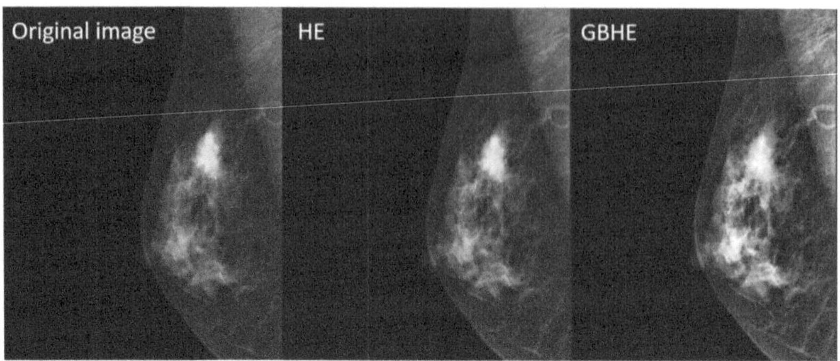

Fig. 2. Comparison chart of GBHE treatment effect

The objective is to minimize the global objective and obtain the reasonable local models, is shown in the formula (1):

$$\{\hat{\theta}_1, \hat{\theta}_2, \ldots, \hat{\theta}_n\} = \arg\min(L_1, L_2, \ldots, L_n) \tag{1}$$

where $L_i = L(\hat{\theta}_i, D_i; \theta), \forall i \in [N]$, and $L(\cdot)$ is the loss function. $theta$ is the global model, which brings external information to client i. The local model serves as the student model and processes the unnormalized output $z_i(x)$ of sample x through softmax function to obtain the predicted probability distribution $p_i(x)$. Concurrently, the global model serves as the teacher model and applying the same processing to $Z(x)$ to derive its predicted probability distribution $q(x)$.

Adaptive Knowledge Distillation Aggregation Algorithm. In FL, the server aggregates the local models trained by the client into a global model, and then updates training parameters to clients via overwriting. This approach often results in poor client model generalization and high communication costs. To address this, we propose a Federated Learning Adaptive Knowledge Distillation Aggregation (FedAKDA) algorithm, establishing a dual-module collaborative framework of AKD and ALA. It enables dynamic adjustment of distillation temperature parameters and local parameter update weights.

In traditional federated training (e.g., FedAvg), the server sends the old global model θ^{t-1} to client i in iteration t, which overwrites its local model to generate a new local model for subsequent local training. As shown in the formula (2):

$$\hat{\theta}_i^t = \theta^{t-1} \tag{2}$$

where θ^{t-1} overwrites the old local model θ_i^{t-1} to obtain the initialized local model $\hat{\theta}_i^t$ for local model training. In FedAKDA, we aggregate global and local models based on knowledge distillation loss and updated parameter weights instead of overwriting, as shown in the formula (3):

$$\hat{\theta}_i^t = \beta_i \cdot \theta_{GT}^{t-1} + (1 - \beta_i) \cdot \theta_{S_i}^{t-1} \tag{3}$$

where θ_{GT}^{t-1} denotes global teacher model parameter, $\theta_{S_i}^{t-1}$ denotes local student model parameter of client i. $\beta_i \in [0, 1]$ is local model updated parameter weights.

AKD Calculates the Parameters. The global teacher model parameters are obtained by aggregating the student model parameters uploaded by clients, as shown in the formula (4):

$$\theta_{GT}^{t} = \frac{1}{N} \sum_{n=1}^{N} \theta_{S_n}^{t} \tag{4}$$

The local student model parameters are optimized by minimizing a combined loss function that includes both the local training loss and the knowledge distillation loss. The local training loss employs the Cross-Entropy Loss function, as shown in the formula (5):

$$L_{local,i} = -\frac{1}{|D_i|} \sum_{(x,y)\in D_i} \sum_{c=1}^{C} y_c \log(P_{i,c}(x)) \tag{5}$$

where C is the number of classes, y_c is the $c-th$ class of the true label, and $P_{i,c}(x)$ is the probability that the local model predicts sample x belongs to the $c-th$ class.

The distillation loss quantifies the discrepancy between the client student model output p_t and the server teacher model output q_t. The distillation loss is typically formulated using the Kullback-Leibler (KL) divergence, as shown in the formula (6):

$$L_{KD,i} = KL(q^T \parallel p_i^T) = \sum_{c=1}^{C} q_c^T \log \frac{q_c^T}{p_{i,c}^T} \tag{6}$$

where $q^T = softmax\left(\frac{Z(x)}{T}\right)$, $p_i^T = softmax\left(\frac{Z_i(x)}{T}\right)$, T denotes the temperature parameter. We introduce a dynamic temperature parameter T that adapts to the accuracy of the local model using linear interpolation. This approach ensures the distillation process balances between knowledge transfer and model specialization, as shown in the formula (7):

$$T_{i,t} = T_{min} + (T_{max} - T_{min}) \times \left(1 - \frac{acc_{i,t}}{acc_{target}}\right) \tag{7}$$

where $T_{i,t}$ denotes the distillation temperature of the client i in the round t of training, and $acc_{i,t}, acc_{target}$ denote the accuracy of the client i in the round t of training and the preset target accuracy. When the $acc_{i,t}$ is lower, the temperature $T_{i,t}$ approaches T_{max}, leading to a smoother output distribution from the teacher model and enabling the student model to learn richer implicit knowledge. As the $acc_{i,t}$ as increases toward acc_{target}, $T_{i,t}$ decreases toward T_{min}, prompting the student model to focus on learning the teacher's decision boundaries. Finally, the total loss function is shown in the formula (8):

$$L_i = \alpha L_{local,i} + (1 - \alpha) L_{KD,i} \tag{8}$$

where α employs to balance the local training loss and distillation training loss.

The local student model updates its parameters by minimizing the combined loss function using Stochastic Gradient Descent (SGD). The parameter update is shown in the formula (9):

$$\theta_{S_i}^t \leftarrow \theta_{S_i}^{t-1} - \eta \nabla_{\theta_{S_i}} L_i \tag{9}$$

where η denotes the learning rate and $\nabla_{W_i} L_i$ denotes the gradient of the total loss with respect to the local model parameters.

ALA Calculates the Aggregation Weights. The adaptive aggregation module is a learnable weight generator. When updating on the client side, the parameters of the downloaded global model and the local model from the previous round are fused based on the calculated aggregation weights. Firstly, randomly sample a portion of local data and train the resulting feature vector. Then input the feature vectors as personalized data into the weight generator, and obtain learnable aggregate weights through a series of functions, which are applied in formula (3). The specific calculation is shown in the formula (10):

$$\beta_i = Sigmoid(W \cdot concat(Z_i, \theta_{GT}^t, \theta_{S_i}^t) + b) \tag{10}$$

where W and b denote the learnable parameters of the weight generator, optimized via backpropagation during training. Z_i denotes the feature vector extracted from the local dataset of client i and $concat(Z_i, \theta_{GT}^t, \theta_{S_i}^t)$ is the feature vectors of the three are concatenated and fed into the weight generator. $Sigmoid(\cdot)$ maps the output to $[0, 1]$, as effective aggregation weights.

3.3 Grad-CAM Visualization

To enhance the reliability of model diagnosis, we employ Grad-CAM technology to visualize the feature areas in mammography images. The lesion areas are highlighted through heat maps, enabling doctors to rapidly identify lesion areas and formulate patient treatment plans based on model predictions. The core principle of Grad-CAM [10] is to extract gradients of specified class. By computing the gradients of the final convolutional layer's feature maps, Class Activation Maps (CAM) are generated. The CAM intuitively highlight the areas that significantly influence the model's decision for that class. Input the image into the trained classifier to obtain the feature map $F \in R^{H \times W \times C}$ from the final convolutional layer. The gradient $\frac{\partial L_c}{\partial F}$ of the target class score c with respect to F is computed, followed by Globally Average Pool (GAP) to derive channel-wise importance weights, is shown in the formula (11):

$$w_c = \frac{1}{H \times W} \sum_{a=1}^{H} \sum_{b=1}^{W} \frac{\partial L_c}{\partial F_{a,b}} \tag{11}$$

where $H \times W$ is the spatial dimension of the feature map. A weighted combination of these weights and the feature maps produces the CAM, is shown in the formula (12):

$$CAM = \sum_{k=1}^{C} w_k F_k \tag{12}$$

Finally, the resulting CAM is then normalized and overlaid onto the original image as a heat map, highlighting the lesion areas to the model's decision for class c. The visualization result of the heatmap is shown in the Fig. 3:

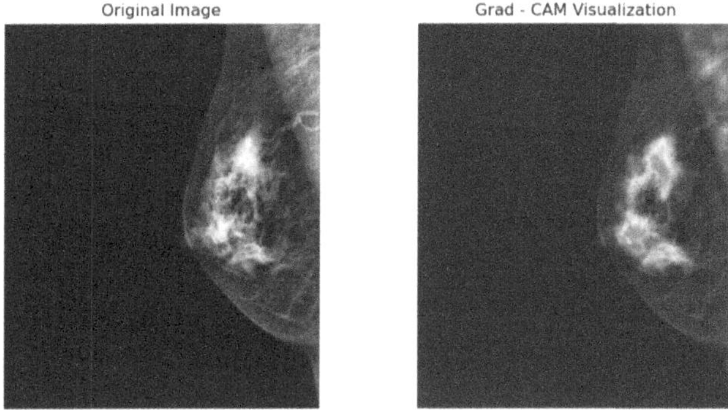

Fig. 3. Grad-CAM visualizes the lesion areas in mammography images

4 Experiment

4.1 Dataset

The Chinese Mammography Medical Database (CMMD) is a benchmark dataset for mammography analysis, comprising 3728 mammography images from 1775 patients with biopsy-confirmed benign or malignant tumors. All image data were acquired using the GE Senographe DS mammography system. In this experiment, 3728 images were partitioned into 10 clients in a non-IID manner to simulate real-world FL training.

4.2 Experimental Settings

The experimental environment is Windows 11 operating system, 11th Gen Intel (R) Core (TM) i9-11900K @ 3.50GHz processor, NVIDIA RTX A4000 GPU graphics card. The computer language is Python 3.10.14, and the experimental framework is based on PyTorch 1.13.1 implementation.

During the experimental, all datasets are partitioned into 10 clients in encrypted form, with each client having different data quantities and distribution ratios of benign and malignant data. Clients store dataset file paths and data labels, which are divided into 80% training and 20% testing set. Clients first use the GBHE method for data preprocessing, and then train local model ResNet50. Client local model employs Cross-Entropy Loss and SGD optimizer with learning rate of 0.005, momentum of 0.9 and batch size of 3, and a total of 100 rounds of training. The dynamic temperature parameter for KD is set within [5, 20] and weight of 0.5. Finally, FedAKDA is employed to update and aggregate the global model.

To comprehensively evaluate the performance of FedAKDA, we use Accuracy, AUC, Training Loss, F1-score and Average Time as the evaluation metrics: **(1) Accuracy.** Reflect the accuracy of the model's classification of mammography images.**(2) AUC.** It's Area Under the Receiver Operating Characteristic (ROC) Curve, quantifying the overall of model discriminative power across all classification thresholds.**(3) Training Loss.** Capturing how well the model fits the training data. **(4) F1-score.** The harmonic mean of precision and recall, servers as a critical metric for evaluating federated learning models on imbalanced medical datasets. **(5) Average Time.** TEvaluate the training efficiency of the AKD module. The timing starts from the initialization of the local model and ends after the final updated of global model. And then calculated the average of each round of running time to reduce accidental errors.

4.3 Comparative Experiments

This experiment compares traditional traditional FL methods, such as FedAvg, and PFL methods, such as FedPAC [11], FedKD, GPFL [12], and FedALA [13]. FedAvg is the typical FL algorithms that updates the global model via weighted averagiing of local model parameters. GPFL balance model commonality and individuality by learning client-specific features while preserving global knowledge, addressing data heterogeneity through a shared-private parameter decomposition. FedPAC employs global semantic knowledge to align local and global feature spaces during training, customizing classifiers for each client to improve personalization. FedKD introduces adaptive mutual knowledge distillation and dynamic gradient compression to reduce communication overhead while enhancing knowledge transfer between global and local models. FedALA incorporates an Adaptive Local Aggregation (ALA) module to dynamically weight client contributions based on data relevance, enabling personalized model aggregation without explicit client modeling.

As shown in Table 1, FedAKDA outperforms other FL methods across all performance metrics, with significant reductions in training time. Compare to FedALA, FedAKDA improves accuracy by 2.3%, AUC by 2.8%, F1-score by 3.8%, and reduced losses by 14.7%. In terms of computational efficiency, FedAKDA is slightly higher average per-round training time than FedAvg with lower communication overhead, but significantly lower than GPFL, FedPAC, and FedALA. Notably, FedAKDA reduces average training time by 3.3 s compared to FedKD. These experimental results prove that FedAKDA significantly reduces communication overhead and outperforms other FL methods in other performance metrics.

Table 1. Performance comparison between FedAKDA and other FL methods

Methods	Accuracy	AUC	Loss	F1-score	Average Time(s)
FedAvg	0.613	0.622	1.586	0.598	41.54
GPFL	0.772	0.783	1.041	0.689	337.28
FedPAC	0.781	0.817	0.854	0.691	182.77
FedKD	0.783	0.799	1.357	0.670	71.45
FedALA	0.785	0.778	0.613	0.687	174.95
FedAKDA	0.791	0.818	0.449	0.707	68.08

4.4 Ablation Experiments

To validate the effectiveness of FedAKDA, we design three sets of ablation experiments: **(1) Centralized training.** Remove the FL framework and adopt centralized training on full dataset. **(2) Single-module training.** Employ the ALA or AKD module individually within the FL framework, to isolate their respective impacts. **(3) Dual-module training.** Employ dual-module collaborative framework of AKD and ALA, to assess their synergistic effects. The experiment is conducted on the CMMD dataset, using accuracy, average training time per round and privacy safety as evaluation metrics.

Table 2. Compare results of three ablation experiments

Methods	Accuracy	Average Time(s)	safety
Resnet50	0.655	-	privacy disclosure
Fed+ALA+Resnet50	0.785	174.95	secure
Fed+AKD+Resnet50	0.783	77.16	secure
Fed+AKDA+Resnet50	0.791	68.08	secure

As shown in Table 2, centralized training with ResNet50 yields suboptimal accuracy due to poor adaptation to distributed non-IID data and poses privacy risks from direct data aggregation. In single-module training, ALA-only training doubles the training time due to its computational complexity, while AKD-only training reduces accuracy due to insufficient local personalization, as the adaptive distillation module alone fails to capture client-specific data characteristics. Dual-module training combines AKD and ALA not only protects data privacy via FL framework but also improves personalized local model accuracy and reduces communication overhead by compared to single-module training, validating the synergistic benefits of AKD and ALA.

4.5 Visualization

To demonstrate the interpretability of FedAKDA, Grad-CAM is employed to generate class activation heatmaps for mammography image predictions. In the heatmap, warmer colors (red, orange) highlight areas with higher contributions to the model's diagnostic decisions, such as lesion areas. And cooler colors (blue, purple) represent areas with less influence. These visualizations accentuate the models' attention to critical features within the mammography images that contribute to the classification decisions.

As shown in Fig. 4, the image classification results of (a) and (b) are malignant tumors, with red areas highlighted in the visualized heatmaps. The focus area concentrated and marked as lesion areas. For (c) and (d), the images are classified as benign, and the focus areas displayed in the visualized heatmap are mostly located at the breast edges with scattered focus areas. These experiments prove the interpretability and reliability of the FedAKDA. By visualizing the feature heatmaps through Grad-CAM technology, doctors can better understand the model's prediction logic and more easily identify lesion ares in mammography images.

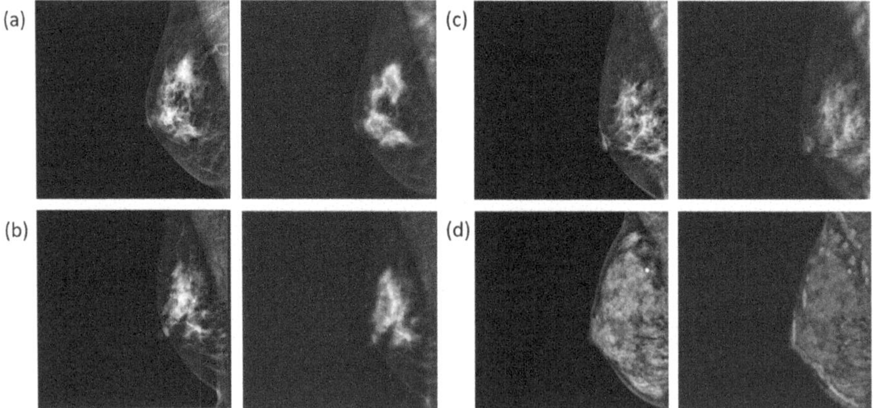

Fig. 4. Crad-CAM visualization of lesion area in mammography images

5 Conclusion

We propose a Federated Learning Adaptive Knowledge Distillation Aggregation (FedAKDA) for breast cancer assisted diagnosis. First, GBHE preprocesses mammography images to enhance lesion areas clarity. The processed dataset is then trained via FedAKDA, which construct a dual-module architecture of AKD and ALA. This approach protects patient privacy, compresses models to reduce training time, and maintains client-specific personalization. Finally, Grad-CAM visualizes lesion areas to enhance model interpretability and assist doctors in diagnosing.

In future research, we plan to incorporate incentive mechanisms to attract more data participants for model training, thereby further enhancing model performance. Additionally, we will explore more advanced neural network architectures to improve classification accuracy and reliability in subsequent work.

References

1. Bray, F., et al.: Global cancer statistics 2022: GLOBOCAN estimates of incidence and mortality worldwide for 36 cancers in 185 countries. CA Cancer J Clin , 229–263 (2024)
2. McMahan, B., Moore, E., Ramage, D.: Communication-Efficient Learning of Deep Networks from Decentralized Data. In: Proceedings of the 20th International Conference on Artificial Intelligence and Statistics (AISTATS), pp. 1273–1282 (2017)
3. Sahu, A.K.: Federated Optimization in Heterogeneous Networks. arXiv: Learning (2018)
4. Wu, C., Wu, F., Liu, R.: FedKD: Communication Efficient Federated Learning via Knowledge Distillation. arXiv2108.13323 (2021)
5. Zhenbing, L., Fengfeng, W., Yumeng, W., Mengyu, Y., Xipeng, P.: FedCL: Federated contrastive learning for multi-center medical image classification. Pattern Recognition, **143**, 0031–3203 (2023)
6. Jin, H.: Personalized edge intelligence via federated self-knowledge distillation. IEEE Trans. on Parallel Distrib. Sys. **34**(2), 567–580 (2023)
7. Geoffrey, H., Oriol, V., Jeff, D.: Distilling the knowledge in a neural network. arXiv preprint arXiv:1503.02531 (2015)
8. Sun, S., Cheng, Y., Gan, Z., Liu, J.: Patient knowledge distillation for bert model compression. In: EMNLP-IJCNLP, pp. 4314–4323 (2019)
9. Jiao, X., et al.: Tinybert: Distilling BERT for natural language understanding. In: EMNLP Findings, pp. 4163–4174 (2020)
10. Selvaraju, R., Cogswell, M., Das, A.: Grad-CAM: visual explanations from deep networks via gradient-based localization. Int. J. Comput. Vision., 336–359 (2020)
11. Xu, J., Tong, X.Y., Huang, S.L.: Personalized Federated Learning with Feature Alignment and Classifier Collaboration. arXiv2306.11867 (2023)
12. Na, S., Liang, Y., Yiu, S.M.: GPFL: A Gradient Projection-Based Client Selection Framework for Efficient Federated Learning (2024)
13. Zhang, J., et al.: FedALA: Adaptive local aggregation for personalized federated learning. In: Proceedings of the AAAI Conference on Artificial Intelligence, pp. 11237–11244 (2023)

LLM4AV: Large Language Model-Driven Autonomous Vehicle Scenario Generation

Henri Patrick Kanimba Ntwali, Youquan Liu(✉), and Junyan Ma

School of Information Engineering, Chang'an University, Xi'an, China
youquan@chd.edu.cn

Abstract. Validation of autonomous driving (AD) functions demands diverse, safety-critical scenarios that are both semantically faithful to natural language requirements and executable in standards-compliant simulators. This paper presents LLM4AV, a text-to-simulation pipeline that converts open-ended descriptions into a domain DSL and renders ASAM OpenSCENARIO 1.x on OpenDRIVE road geometry selected via rule-based map retrieval. A multi-layer validator enforces XML syntax, kinematic feasibility, and behavioral-logic consistency (e.g., maneuver timing, trigger coordination). Closed-loop executions are performed in esmini, and safety is quantified using a segment-scoped protocol that evaluates metrics only during maneuver-valid intervals to avoid dilution from staging or cruising. Across eight scenarios spanning highway maneuvers, urban right-of-way, vulnerable road users, and low-friction braking, LLM4AV consistently produces executable scripts with high semantic fidelity. Observed risk profiles align with domain expectations, including a worst-case TTC = 0.30 s for a bicycle-overtake near an occluder, PET = 0.15 s for an occluded mid-block pedestrian crossing, and peak longitudinal jerk 8.52 m/s^3 under wet-road hard braking; benign cases (e.g., emergency-vehicle yielding) maintained large headways despite high Δv. The contributions are: (i) a standards-aligned NL $\rightarrow$ DSL $\rightarrow$ OpenSCENARIO synthesis chain, (ii) a practical OpenDRIVE retrieval method with feasibility repair, and (iii) a criticality-aware evaluation protocol that concentrates metrics on high-risk windows. The results indicate that LLM4AV improves reproducibility and interpretability of scenario testing while preserving portability across standards-compliant simulators.

Keywords: Autonomous Driving · Scenario Generation · Large Language Models · OpenSCENARIO Simulation-based Testing

1 Introduction

Ensuring AV safety requires exposure to diverse, safety-critical scenarios in controlled simulation, since on-road testing is costly, slow, and potentially hazardous. Scenario generation remains challenging due to the need for semantic richness, behavioral realism, and standards compliance across tools and simulators (e.g., ASAM OpenSCENARIO for dynamics [1] and ASAM OpenDRIVE for road geometry [2]).

S. Yang et al. (Eds.): WISA 2025, LNCS 16299, pp. 102–113, 2026.
https://doi.org/10.1007/978-981-95-6426-2_9

We propose LLM4AV, a modular pipeline that converts open-ended natural-language descriptions into simulation-ready OpenSCENARIO 1.x scripts. LLM4AV (i) generalizes beyond fixed templates and curated logs; (ii) guarantees standards alignment with OpenSCENARIO/OpenDRIVE for cross-simulator portability (esmini [3], CARLA [4], LGSVL [5]); and (iii) introduces criticality-aware evaluation that computes safety metrics only within maneuver-valid windows (inspired by Chat2Scenario).

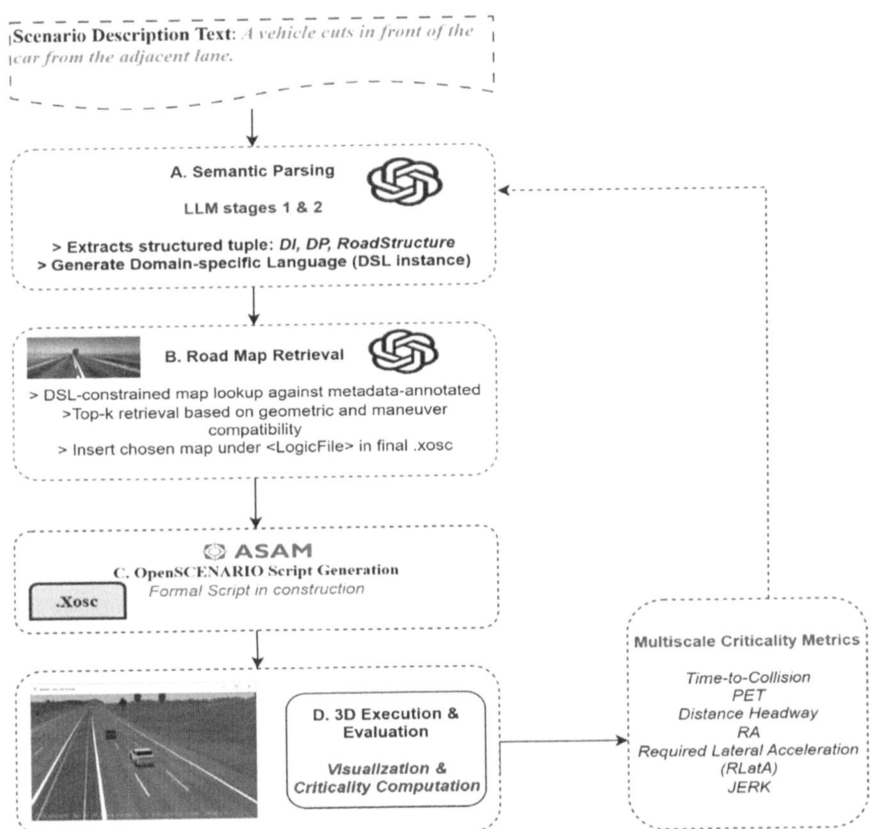

Fig. 1. LLM4AV Framework Architecture. The proposed framework integrates a dual-stage semantic parsing module, DSL-driven OpenDRIVE road map retrieval, standards-compliant OpenSCENARIO script generation, and 3D scenario execution with multiscale criticality-aware evaluation. The closed-loop feedback enables iterative refinement and knowledge accumulation, ensuring semantic fidelity, standards alignment, and cross-simulator interoperability.

This paper makes the following key contributions: (1) End-to-end framework: dual-stage semantic parsing → rule-based OpenDRIVE retrieval → OpenSCENARIO-compliant synthesis → closed-loop execution. (2) Standards-aligned synthesis: DSL-to-XML mappings ensure interoperability across esmini, CARLA, and LGSVL. (3) Criticality-driven assessment: segment-scoped metrics (TTC, PET, DHW, Δv, RLatA, LongJ) computed only during behaviorally valid intervals.

The remainder of this paper is organized as follows. Section 2 reviews related work and situates LLM4AV among emerging frameworks. Section 3 details the proposed methodology and DSL-based pipeline. Section 4 presents experiments, quantitative evaluations.

2 Related Work

Recent advances in Large Language Models (LLMs) have enabled significant progress in autonomous vehicle (AV) scenario generation, supporting natural language-driven synthesis of simulation-ready environments. However, existing frameworks exhibit key limitations in standards compliance, semantic generalization, criticality-driven evaluation, and cross-simulator interoperability, which are addressed by the proposed LLM4AV framework.

Traditional approaches, including TARGET [7] and Driving with Regulation [8], rely on predefined templates, structured crash datasets, and rule-based regulatory inputs. These guarantee deterministic replay but face critical limitations: (1) Limited adaptability—scenarios fail to generalize beyond curated datasets, (2) Restricted semantic flexibility—incapable of interpreting open-ended natural language inputs, and (3) Standards incompatibility—outputs often diverge from ASAM OpenSCENARIO 1.x and OpenDRIVE specifications [1] [2], reducing integration with industrial simulators such as Esmini [3], CARLA [4], and LGSVL [5]. Recent frameworks like VistaScenario [6], ScenarioNet [7], and AutoScenario [8] aim to improve dataset coverage but remain constrained by limited cross-simulator support and lack criticality-focused evaluation.

Several frameworks integrate LLMs to enhance planning and agent-level decision-making but fail to generate OpenSCENARIO-compliant, simulation-ready scenarios. Parking Planning Agent [9] optimizes infrastructure layout but lacks dynamic scenario synthesis.

CoDrivingLLM [10] enables cooperative maneuvers but is limited to policy-level coordination rather than scenario-level execution. LLM-Assist [11] improves closed-loop reasoning but does not output executable XML scripts.

Zero-Shot End-to-End LLM Driving generalizes across tasks but cannot produce standards-compliant outputs [12]. These methods advance multi-agent intelligence but do not deliver end-to-end scenario generation pipelines.

Recent frameworks explore multimodal and generative techniques to improve scene realism and rare-event testing. DriveDreamer-2 [13] and OmniDrive [14] generate high-fidelity driving videos but fail to produce OpenSCENARIO-compatible outputs.

Multimodal LLM Scenario Testing [15] leverages multiple sensor streams for richer datasets but lacks formal XML generation. Traffic Scene Generation (TTSG) [16] improves coverage of complex interactions yet depends on database-driven templates.

LLM-Attacker [17] and SenseRAG [18] specialize in adversarial scenario synthesis but lack cross-simulator adaptability. DriVLMe [19] enhances embodied and social driving agents, contributing improved behavioral realism.

Despite these advances, most multimodal frameworks lack semantic fidelity guarantees and do not integrate criticality-aware evaluation metrics.

Several frameworks aim for standards compliance but face domain constraints. Scenic [20], ScenicNL [21], LimSim+ + [22], and LEADE [23] employ probabilistic modeling and multimodal datasets but rely on domain-specific crash corpora, limiting adaptability to novel driving contexts.

ChatScene [24] and LLM-Nav [25] facilitate scene-level synthesis but lack strict OpenSCENARIO 1.x compatibility.

In contrast, LLM4AV integrates custom DSL-driven map retrieval with a curated OpenDRIVE repository, ensuring full ASAM OpenSCENARIO/OpenDRIVE compliance. It supports deployment across Esmini, CARLA, and LGSVL and uniquely incorporates criticality-aware evaluation for safety-driven AV testing.

Existing frameworks focus on structured dataset replay, agent-level planning, or visual data synthesis, but none achieve standards-compliant, semantically accurate, and cross-simulator-ready scenarios.

In contrast, LLM4AV:

- Translates open-ended natural language descriptions into formal OpenSCENARIO-compliant XML scripts.
- Ensures strict ASAM OpenSCENARIO/OpenDRIVE alignment for interoperability.
- Introduces criticality-aware safety metrics such as TTC, PET, Δv, and RLatA.
- Outperforms prior frameworks in semantic fidelity, runtime efficiency, and scalability.

Significant strides in large language models (LLMs) have recently introduced advanced strategies for improving semantic fidelity and multimodal reasoning, which are directly relevant to enhancing the LLM4AV framework. Wang et al. [26] propose a Transformer-based dual-learning architecture that jointly optimizes bidirectional mappings between natural language and structured representations, leveraging probabilistic alignment constraints to improve cross-task consistency.

Drawing inspiration from this, LLM4AV can benefit from incorporating dual constraints between natural language scenario descriptions and DSL/OpenSCENARIO scripts, thus improving parsing robustness and reducing prompt-repair cycles [27].

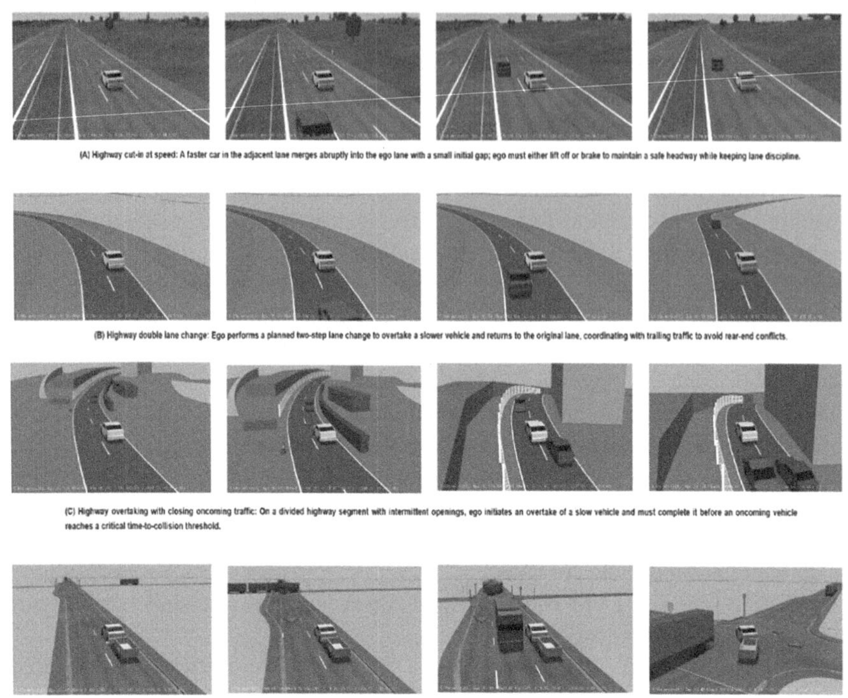

Fig. 2. LLM4AV-generated scenarios in esmini. (A) Highway cut-in at speed; (B) Highway double lane change; (C) Highway overtaking with closing oncoming traffic; (D) Unprotected left at a signalized intersection.

3 Methodology

3.1 Framework

The proposed LLM4AV framework introduces a modular, end-to-end pipeline that converts open-ended natural-language descriptions into simulation-ready scenarios for standards-compliant environments.

As shown in Fig. 1, it comprises: (1) dual-stage semantic parsing—A1 extracts agents, roles, and initial conditions; A2 models pairwise interactions and road constraints—producing a DSL aligned with OpenSCENARIO constructs; (2) rule-based road-map retrieval from a curated OpenDRIVE repository to ensure geometric feasibility; (3) standards-compliant scenario synthesis, generating OpenSCENARIO 1.x XML (agents, maneuvers, triggers, environment) with multi-layer validation for syntax, physics, and behavioral logic; and (4) closed-loop execution with criticality-aware evaluation, executed in esmini for this study.

During execution, segment-scoped safety metrics (TTC, PET, DHW, Δv, RLatA, LongJ) are computed only within maneuver-valid windows. Outcomes are logged to enable iterative refinement, retaining optimal strategies and flagging suboptimal runs for prompt repair, thereby ensuring interoperability, semantic fidelity, and continuous improvement.

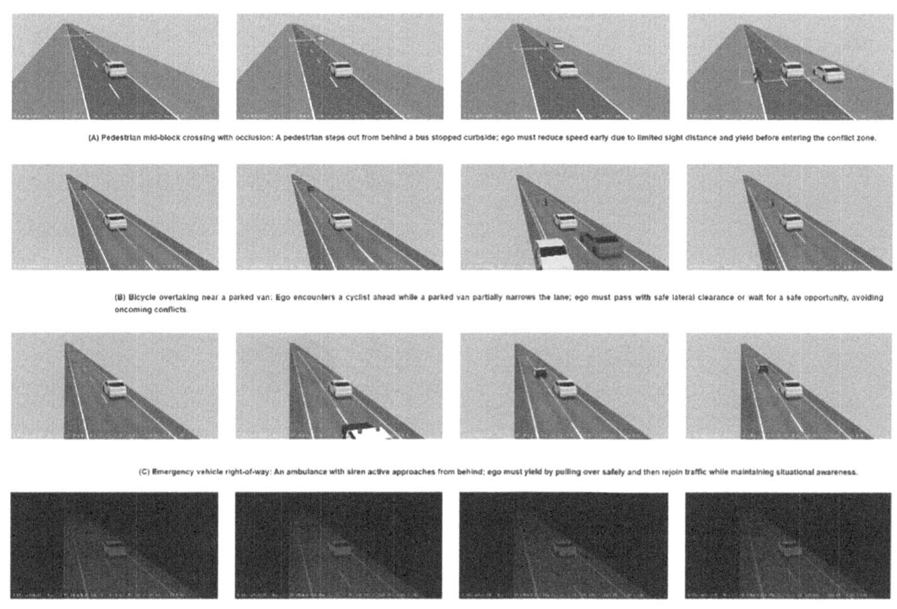

Fig. 3. Urban and adverse-condition scenarios generated by LLM4AV. (A) Pedestrian mid-block crossing with occlusion; (B) Bicycle overtake near a parked van; (C) Emergency-vehicle right-of-way (ego yield); (D) Wet-road long-brake car-following (Esmini replays).

3.2 Dual-Stage Semantic Parsing

LLM4AV converts unstructured natural-language traffic descriptions into a formally constrained DSL aligned with ASAM OpenSCENARIO (see Fig. 4).

Stage A1 performs descriptive initialization—extracting agents, roles, and initial conditions—while Stage A2 infers pairwise interactions and road semantics, enforcing schema constraints.

Unlike rule- or template-based approaches, this two-stage LLM pipeline generalizes beyond curated datasets while preserving semantic fidelity, standards compliance, and cross-simulator interoperability.

1) *Stage A1—Descriptive Initialization (DI):* Decomposes free-form text into static scene attributes: agents and roles (ego/NPC), spatial configuration (lanes, longitudinal positions, headings), and initial kinematics (speed, acceleration). Context (weather, topology, traffic density) is extracted when specified.
2) *Stage A2—Pairwise Interactions and DSL Generation*: Models inter-agent dynamics and converts semantic tuples into an executable DSL aligned with OpenSCENARIO. Identifies maneuver intent, trigger logic, and temporal dependencies; e.g., cut-in is formalized via relative-distance triggers and sequenced actions (see Fig. 4).

3) *Integration with OpenSCENARIO:* Compiles the DSL into OpenSCENARIO 1.x XML for esmini/CARLA/LGSVL. Maneuvers are instantiated under <ManeuverGroup> with synchronized <StartTrigger> / <StopTrigger>; validators enforce syntax, spatial consistency, and dynamic feasibility (no overlap, realistic accelerations).
4) *Advantages of the Dual-Stage Approach:* Separating descriptive initialization (A1) from interaction modeling (A2) improves semantic fidelity and interpretability. The result is a robust, reproducible, standards-compliant representation that supports cross-simulator deployment and criticality-driven safety evaluation.

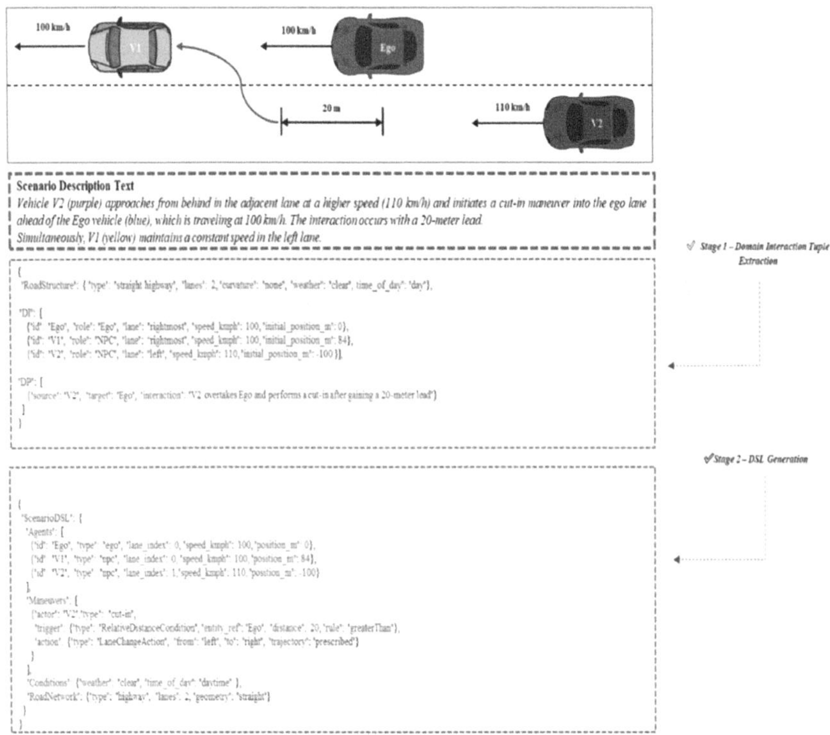

Fig. 4. Dual-Stage Semantic Parsing and DSL Generation in LLM4AV.

3.3 Rule-Based Road Map Retrieval

After semantic parsing, LLM4AV selects an OpenDRIVE map from a curated repository whose geometric, topological, and regulatory properties satisfy the DSL-encoded scenario requirements, thereby bridging high-level semantics and executable simulation context.

Each map is pre-annotated with metadata covering: (i) lane configuration (count, directionality, widths, lateral offsets); (ii) geometry (straights, arcs, spirals, curvature, elevation); (iii) markings/permissions (solid/broken lines, overtaking and lane-change allowances); and (iv) maneuver compatibility flags (e.g., allow_cut_in, allow_lane_change, allow_overtake, allow_junction_turn).

A rule-based matcher compares the DSL road descriptor (e.g., "three-lane curved highway with overtaking") against this metadata via multi-criteria filtering and ranking, returning the best-fit OpenDRIVE file for downstream, standards-compliant scenario synthesis.

The retrieval system parses the DSL's RoadStructure field, extracting attributes such as lane count, curvature, elevation profile, and maneuver flags. These are used to perform structured filtering using a scoring function:

$$score = \lambda_1 Maneuver_{compatibility} + \lambda_2 Lane_{layout} + \lambda_3 Geometry_{match} + \lambda_4 Marking_{support} \tag{1}$$

where each λ_i is a tunable weight. For example, a query requiring "cut-in on curved road" will prioritize maps where at least two adjacent lanes permit lateral transitions, the center lane supports overtaking (i.e., broken lines), and the road contains spiral or arc geometries.

Maps are ranked accordingly, and the top candidate is selected for scenario deployment. This strategy aligns with TTSG's retrieval-and-ranking approach [16] and enhances interpretability and configurability over random map sampling.

3.4 Standards-Compliant Scenario Synthesis

This stage maps the DSL (agents, maneuvers, triggers, constraints, environment) and the selected OpenDRIVE geometry to executable ASAM OpenSCENARIO 1.x XML: agents in <Entities>, maneuvers under <ManeuverGroup>, and activation via synchronized <StartTrigger> / <StopTrigger>.

For cut-ins, the script encodes the NPC lane change and ego braking with explicit temporal coordination. A multi-layer validator enforces syntax conformance, kinematic feasibility, and behavioral-logic consistency, while environmental parameters (friction, weather, lighting, traffic density) ensure realistic configuration.

The scripts are simulator-agnostic; in this study, execution uses esmini. Closed-loop feedback updates the DSL when runs expose unsafe maneuvers or inconsistent triggers, improving semantic fidelity, physical realism, and reproducibility.

3.5 Closed-Loop Execution and Criticality-Aware Evaluation

In the final stage, ASAM OpenSCENARIO 1.x scripts are executed in esmini to assess feasibility, semantic fidelity, and safety. LLM4AV employs segment-scoped evaluation: safety indicators are computed only during maneuver-valid intervals, avoiding dilution from inactive phases.

Agent kinematics (position, velocity, acceleration, steering angle) are tracked and metrics—Time-to-Collision (TTC), Post-Encroachment Time (PET), Dynamic Headway (DHW), relative speed (Δv), required lateral acceleration (RLatA), and longitudinal jerk (LongJ)—are computed; for cut-ins, the window is the active lane-change phase.

Outcomes (trajectories, results, metrics, context) are logged; safe runs become references, while unsafe patterns trigger prompt repair.

Flagged cases feed Stage A2 to refine DSL parameters and trigger constraints, closing the loop and improving semantic precision, physical realism, and safety conformance. Esmini's deterministic playback enables efficient, standards-compliant, and scalable scenario-based validation.

4 Experiment and Result

We validated LLM4AV with eight end-to-end scenarios (Fig. 2 and Fig. 3) generated from natural language via (i) semantic parsing to a constrained DSL, (ii) rule-based OpenDRIVE retrieval, (iii) deterministic ASAM OpenSCENARIO 1.x synthesis, and (iv) closed-loop execution in esmini. The set spans highway, urban, VRU, and low-adhesion conditions: (1) highway cut-in at speed; (2) highway double lane change; (3) highway overtake with closing oncoming traffic; (4) unprotected left at a signalized intersection; (5) occluded mid-block pedestrian crossing; (6) bicycle overtake near a parked van; (7) emergency-vehicle right-of-way; (8) wet-road car-following.

Time-aligned states (trajectories, speeds, lane indices, relative kinematics) were logged, and safety was quantified only over maneuver-valid windows (segment-scoped evaluation). Criticality metrics included TTC, PET, DHW, closing speed Δv, required lateral acceleration (RLatA), and longitudinal jerk (LongJ); PET was computed from time-staggered conflict-zone occupancy, and percentile definitions were used for RLatA (95th) and LongJ (90th). Scenario-wise summaries appear in Table 1.

Table 1. Segment-scoped criticality metrics from esmini logs.

Scenario	TTC (s)	PET (s)	DHW (m)	Δv (km/h)	RLatA (m/s^2)	LongJ (m/s^3)
Highway cut-in at speed	1.28	—	13.95	25.08	0.088	0.001
Highway double lane change	1.55	0.00	6.94	15.00	1.336	0.003
Highway overtaking with closing oncoming traffic	0.35	0.00	0.00	60.67	21.264	0.170
Unprotected left at a signalized intersection	20.00*	0.10	0.00	0.67	10.383	0.000
Pedestrian mid-block crossing with occlusion	0.32	0.15	3.10	15.46	0.000	0.000

(continued)

Table 1. *(continued)*

Scenario	TTC (s)	PET (s)	DHW (m)	Δv (km/h)	RLatA (m/s^2)	LongJ (m/s^3)
Bicycle overtaking near a parked van	0.30	0.00	8.80	11.71	0.000	0.000
Emergency vehicle right-of-way	0.46	0.00	24.77	54.00	0.000	0.000
Wet-road long-brake highway car-following	2.83	—	14.31	16.12	0.000	8.515

Interaction severity is highest for overtaking with closing oncoming traffic and for bicycle/pedestrian cases (TTC $\approx$ 0.30–0.35 s, PET $\approx$ 0), moderate for cut-in (TTC $\approx$ 1.28 s), and lower in wet-road car-following where braking dynamics dominate. Kinematically, Δv peaks in overtaking (60.67 km/h) and emergency right-of-way (54.00 km/h); overtaking also yields the largest lateral demand (RLatA $\approx$ 21.26 m/s2), while wet-road car-following shows the highest longitudinal jerk (LongJ $\approx$ 8.515 m/s^3). Unprotected-left exhibits a capped TTC (20 s*), very low Δv (0.67 km/h), and non-trivial RLatA at commit (10.383 m/s^2). DHW is $\approx$0 m at the conflict apex for overtaking and unprotected-left, but largest when yielding/spacing dominates (emergency right-of-way 24.77 m; wet-road 14.31 m). Non-zero PET occurs only in unprotected-left (0.10 s) and pedestrian (0.15 s), indicating staggered conflict-zone occupancy. Overall, highway maneuvers emphasize longitudinal closing risk (Δv, jerk), whereas VRU interactions show ultra-short TTC with comparatively lower lateral/jerk magnitudes, consistent with cautious yielding.

5 Conclusion and Future Direction

This study presents LLM4AV, a modular, standards-aligned pipeline that maps open-ended natural language to ASAM OpenSCENARIO 1.x via dual-stage semantic parsing, rule-based OpenDRIVE selection, compliant XML generation, and closed-loop evaluation with criticality metrics restricted to maneuver-valid segments.

Esmini experiments over representative highway and urban/VRU interactions demonstrate semantic fidelity, physical feasibility, and cross-simulator portability under strict standards compliance, establishing a reproducible basis for scenario-based AV testing.

Future work: (i) will replicate across CARLA/LGSVL to quantify portability and trigger semantics; (ii) incorporate multimodal prompts (e.g., sketches/HD maps) with simulation-in-the-loop prompt repair; (iii) expand to dense-urban, multi-agent, adverse-weather, and adversarial cases using formal benchmarks and SOTA comparisons; (iv) report end-to-end runtime/cost and release artifacts (DSL schema, .xodr/.xosc corpus, logs) to strengthen reproducibility.

Acknowledgements. This research is supported by National Natural Science Foundation of China (Key Program Grant No. 52131204).

References

1. X. Zhang et al.: A novel scenario-based testing approach for cooperative-automated driving systems. In: Conference Proceedings - IEEE International Conference on Systems, Man and Cybernetics, Institute of Electrical and Electronics Engineers Inc., , pp. 3487–3494 (2023). https://doi.org/10.1109/SMC53992.2023.10394032
2. Diaz-Diaz, A., Ocana, M., Llamazares, A., Gómez-Huélamo, C., Revenga, P., Bergasa, L.M.: Hd maps: Exploiting opendrive potential for path planning and map monitoring. In: 2022 IEEE Intelligent Vehicles Symposium (IV), pp. 1211–1217. IEEE. (2022) https://doi.org/10.1109/IV51971.2022.9827297
3. Baker, P., Mitchell, J., Chodowiec, E., Zhang, X., Khastgir, S., Jennings, P.: Exploring realism in virtual testing: towards a scalable platform using open-source solutions for automated driving systems. In: IEEE Intelligent Vehicles Symposium, Proceedings, Institute of Electrical and Electronics Engineers Inc., pp. 2517–2522 (2024). https://doi.org/10.1109/IV55156.2024.10588761
4. Dosovitskiy, A., Ros, G., Codevilla, F., Lopez, A., Koltun, V.: CARLA: an open urban driving simulator. In: Conference on Robot Learning, pp. 1–16. PMLR (2017)
5. Rong G., et al.: LGSVL Simulator: A High Fidelity Simulator for Autonomous Driving. [Online]. Available: https://www.autoware.auto/
6. Chang, C., et al.: VistaScenario: interaction scenario engineering for vehicles with intelligent systems for transport automation. IEEE Trans. Intell. Veh. (2024). https://doi.org/10.1109/TIV.2024.3400323
7. Li, Q., et al.: Scenarionet: Open-source platform for large-scale traffic scenario simulation and modeling. Adv. Neu. Inf. Process. Sys. **36**, 3894–3920 (2023). https://metadriverse.github.io/scenarionet
8. Lu, Q., Ma, M., Dai, X., Wang, X., Feng, S.: Realistic Corner Case Generation for Autonomous Vehicles with Multimodal Large Language Model (2024). [Online]. Available: http://arxiv.org/abs/2412.00243
9. Jin, Y., Ma, J.: Large language model as parking planning agent in the context of mixed period of autonomous vehicles and human-driven vehicles. Sustain. Cities Soc. **117**, 105940 (2024). https://doi.org/10.1016/j.scs.2024.105940
10. Fang, S., et al.: Towards interactive and learnable cooperative driving automation: a large language model-driven decision-making framework. IEEE Trans. Veh. Technol. (2025). https://doi.org/10.1109/TVT.2025.3552922
11. Sharan, S.P., Pittaluga, F., Chandraker, M.: LLM-Assist: Enhancing Closed-Loop Planning with Language-Based Reasoning (2023). [Online]. Available: http://arxiv.org/abs/2401.00125
12. Deng, Y., Yao, J., Tu, Z., Zheng, X., Zhang, M., Zhang, T.: TARGET: Automated Scenario Generation from Traffic Rules for Testing Autonomous Vehicles via Validated LLM-Guided Knowledge Extraction. (2025). [Online]. Available: http://arxiv.org/abs/2305.06018
13. Zhao, G., et al.: DriveDreamer-2: LLM-Enhanced World Models for Diverse Driving Video Generation (2024). [Online]. Available: http://arxiv.org/abs/2403.06845
14. Wang, S., et al.: OmniDrive: A Holistic Vision-Language Dataset for Autonomous Driving with Counterfactual Reasoning (2025). [Online]. Available: http://arxiv.org/abs/2405.01533
15. Lu, Q., Wang, X., Jiang, Y., Zhao, G., Ma, M., Feng, S.: Multimodal Large Language Model Driven Scenario Testing for Autonomous Vehicles (2024). [Online]. Available: http://arxiv.org/abs/2409.06450

16. Ruan, B.-K., Tsui, H.-T., Li, Y.-H., Shuai, H.-H.: Traffic Scene Generation from Natural Language Description for Autonomous Vehicles with Large Language Model (2025). [Online]. Available: http://arxiv.org/abs/2409.09575
17. Mei, Y., Nie, T., Sun, J., Tian, Y. : LLM-attacker: Enhancing Closed-loop Adversarial Scenario Generation for Autonomous Driving with Large Language Models. (2025). https://doi.org/10.1109/TITS.2025.3578383
18. Luo, X., et al.: Senserag: Constructing environmental knowledge bases with proactive querying for llm-based autonomous driving. In Proceedings of the Winter Conference on Applications of Computer Vision, pp. 989–996 (2025).
19. Huang, Y., Sansom, J., Ma, Z., Gervits, F., Chai, J.: DriVLMe: enhancing llm-based autonomous driving agents with embodied and social experiences. In: IEEE International Conference on Intelligent Robots and Systems, Institute of Electrical and Electronics Engineers Inc., pp. 3153–3160 (2024). https://doi.org/10.1109/IROS58592.2024.10802555
20. Fremont, D.J., et al.: Scenic: a language for scenario specification and data generation. Mach. Learn. **112**(10), 3805–3849 (2023). https://doi.org/10.1007/s10994-021-06120-5
21. Elmaaroufi, K., et al.: ScenicNL: Generating Probabilistic Scenario Programs from Natural Language (2024). [Online]. Available: http://arxiv.org/abs/2405.03709
22. Fu, D., et al.: LimSim++: a closed-loop platform for deploying multimodal llms in autonomous driving. In: IEEE Intelligent Vehicles Symposium, Proceedings, Institute of Electrical and Electronics Engineers Inc., pp. 1084–1090 (2024). https://doi.org/10.1109/IV55156.2024.10588848
23. Tian, H., et al.: LMM-enhanced Safety-Critical Scenario Generation for Autonomous Driving System Testing From Non-Accident Traffic Videos. (2025). [Online]. Available: http://arxiv.org/abs/2406.10857
24. Zhang, J., Xu, C., Li, B.: ChatScene: Knowledge-Enabled Safety-Critical Scenario Generation for Autonomous Vehicles. (2024). [Online]. Available: http://arxiv.org/abs/2405.14062
25. Chang, C., Wang, S., Zhang, J., Ge, J., Li, L.: LLMScenario: large language model driven scenario generation. IEEE Trans Syst Man Cybern Syst **54**(11), 6581–6594 (2024). https://doi.org/10.1109/TSMC.2024.3392930
26. Wang, J., Cao, L., Shan, J., Song, X., Jiang, J. : Dual learning model of code summary and generation based on transformer. In: Lecture Notes in Computer Science (including subseries Lecture Notes in Artificial Intelligence and Lecture Notes in Bioinformatics), Springer Science and Business Media Deutschland GmbH, pp. 41–52 (2024). https://doi.org/10.1007/978-981-97-7707-5_4
27. Li, Z., Pang, N., Zhao, X.: Instruction tuning large language models for multimodal relation extraction using LoRA. In: Lecture Notes in Computer Science (including subseries Lecture Notes in Artificial Intelligence and Lecture Notes in Bioinformatics), Springer Science and Business Media Deutschland GmbH, 2024, pp. 364–376. https://doi.org/10.1007/978-981-97-7707-5_30

Towards More Robust Perceptual Information Collaboration for Multi-agent Decision-Making Systems in Presence of Mixed Real-World Factors

Yunbo Yu, Binyu Zhao, and Wei Zhang(✉)

School of Computer Science and Technology, Harbin Institute of Technology, Harbin, China
weizhang@hit.edu.cn

Abstract. Environmental situation perception and processing serve as critical foundations for multi-agent decision-making systems. Due to the inherent limitations of single-agent perception, recent studies have introduced collaborative perception to enhance perceptual performance by sharing complementary information through communication. However, it is crucial to address mixed real-world factors (e.g., bandwidth, latency) for its applications, which is rarely tackled by most existing works. To break through this bottleneck, we introduce a comprehensive perceptual information collaboration framework *RealCP*, which empowers agents to handle diverse multi-modal inputs and share sparse yet valuable spatial information in a compressed manner, enabling the adaptability to more complex real-world environments. *RealCP* has three distinct advantages: i) it can handle varying number of input sensors by utilizing low-coupling encoders and straightforward fusion strategy. ii) it considers more rational feature collection and compression to achieve higher perceptual performance by distinctive perceptual information preservation and conservatively selection; and iii) it improves the efficiency of cross-agent feature aggregation by more explicitly quantifying the individual contribution of each agent. To evaluate it, we consider 3D object detection in both simulated and real-world scenarios. Experimental results show it consistently outperforms other previous methods, which demonstrates its effectiveness and robustness.

Keywords: Multi-agent decision-making systems · Perceptual information collaboration · Mixed real-world factors

1 Introduction

Multi-agent systems (MAS) attract increasing attention due to the advantages in flexibility, scalability, and robustness. To establish effective decision-making system for MAS, accurate perception and interpretation of environmental information is critical. To enhance the comprehension of the surrounding environment,

S. Yang et al. (Eds.): WISA 2025, LNCS 16299, pp. 114–126, 2026.
https://doi.org/10.1007/978-981-95-6426-2_10

Table 1. Brief summary of recent state-of-the-art works to address real-world factors.

Issues	Methods
Data availability	V2VFormer++ [30], HM-ViT [22], BM2CP [33]
Dynamic input	BM2CP [33]
Pose noise	V2VNet-robust [17], CoAlign [13], Feaco [3]
Communication latency	SyncNet [9], FFNet [32], CoBEVFlow [20]
Safety&Privacy	Adversarial-V2V [16], ROBOSAC [10]

recent studies introduce collaborative perception to enable the exchange of complementary information through communication [5,15], which holds immense potential for diverse real-world applications such as intelligent transportation system [4], unmanned aerial vehicles, and multi-robot systems.

To promote practical applications, recent works have begun to address the inevitable real-world factors, including data availability, bandwidth constraint, communication latency, pose error, etc. However, these works have not comprehensively accounted for the rigorous factors of real-world scenarios. A brief overview of recent state-of-the-art works addressing realistic challenges is presented in Table 1. Although a few efforts [27,29] recognize the complexity of real-world environments and consider multiple challenges concurrently, they rely excessively on historical features, thereby increasing computational overhead.

Based on the above observations, we introduce RealCP, an innovative and robust perceptual information collaboration framework that tackles more real-world factors than previous efforts, encompassing data availability, dynamic input, bandwidth constraint, communication latency, and collaboration heterogeneity. It comprises three essential designs: i) Low-coupling encoders and a diversifying feature collector, which fully discovers perceptual information from a variable number of modalities and integrates them into a semantically-enriched feature map, addressing the issues of data availability and dynamic input; ii) A conservative confidence generator, which generates a spatial confidence map to prioritize potential perceptually critical areas for feature fusion in the near future, addressing the issues of bandwidth constraint, pose error and communication latency; iii) A message aggregator module, which first aligns messages of other agents and then aggregate them by quantifying their contributions in cooperation, addressing the issues of pose error, communication latency, and collaboration heterogeneity.

RealCP has three main contributions. Firstly, it augments the processing prowess for modalities with variable quantities by decoupled modal feature extractions and homogenization fusion strategy. Secondly, it facilitates more logical and efficient feature collection and compression by preserving distinctive perceptual information and conservatively selecting perceptually critical areas from the feature maps. Thirdly, it more explicitly models the cooperation among agents by quantifying their individual contributions. To evaluate RealCP, we have conducted thorough experiments on the collaborative 3D object

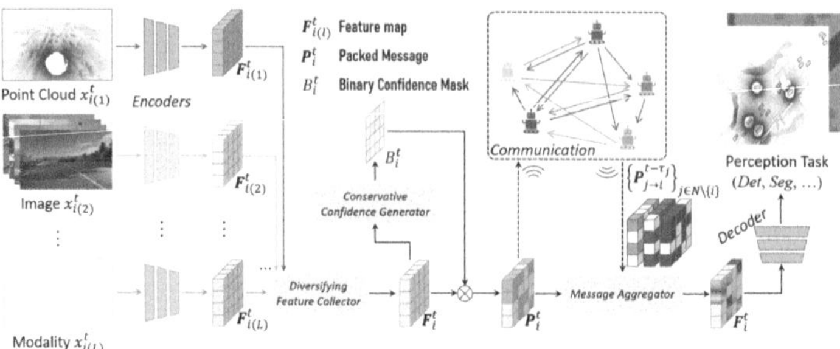

Fig. 1. System overview.

detection task across three datasets: OPV2V [26], V2V4Real [24], and DAIR-V2X [31]. These experiments cover both simulated and real-world scenarios, perfect and noisy conditions, synchronous and asynchronous communications, as well as V2V (vehicle-to-vehicle) and V2X (vehicle-to-everything) interactions. The experimental results demonstrate that RealCP consistently and significantly outperforms prior works across diverse environmental conditions.

2 Related Works

Methods for ideal perceptual information collaboration primarily emphasize enhancing collaborative efficiency and perception performance, bandwidth constraint is often taken into account in the early works. Some studies [17,19,25] employ indiscriminate feature compression across all spatial areas, which dramatically wastes bandwidth. Conversely, other approaches adopt critical feature filtering to reduce communication overhead [3,6,28,29,33].

Methods for collaborating on perceptual information in the real world mainly propose solutions to address real-world factors. First, the issue of pose noise was addressed by V2VNet-robust [17], which proposed a robust framework for feature alignment under inaccurate pose conditions. Then, safety and privacy concerns were addressed by Adversarial-V2V [16] using adversarial training to improve the robustness of the model against potential threats. Subsequently, the communication latency was mitigated by SyncNet [9] through a method designed to minimize delays in vehicle-to-vehicle perception. Following this, data availability challenges were improved by V2VFormer++ [30] with enhanced cross-agent data access. Then, HM-ViT [22] addressed the heterogeneity of the data by introducing a unified framework for collaboration, allowing perception with inputs of a single modality. Most recently, BM2CP [33] handled dynamic input through adaptive processing of varying sensor conditions. However, these approaches tend to focus narrowly on a single issue, limiting their applicability to real-world scenarios that include a diverse array of realistic factors.

3 Methodology

Consider a set of agents $\mathcal{N}$ in the scene and $|\mathcal{N}| = N$. Let $X_i^t = \{x_{i(l)}^t\}_{l=1}^L$ and Y_i^t be the L number of raw modal observations and the task-specific perception supervision of the i-th agent at the current timestamp t, respectively. The objective of perceptual information collaboration is to maximize the perceptual performance

$$\arg\max_{\theta, \boldsymbol{P}_{j\to i}^{t-\tau_j}} \sum_i \varrho(\Psi_\theta(X_i^t, \{\boldsymbol{P}_{j\to i}^{t-\tau_j}\}_{j\in\mathcal{N}\setminus\{i\}}), Y_i^t) \quad (1)$$

where $\varrho(\cdot,\cdot)$ stands for the perceptual evaluation metric. $\Psi_\theta(\cdot,\cdot)$ is the collaborative framework parameterized by θ. $\boldsymbol{P}_{j\to i}^{t-\tau_j}$ is the latest message transmitted from the j-th agent to the i-th agent at timestamp $t-\tau_j$, which have been warped to the i-th agent coordinate according to the 6 DoF pose pairs of the i-th and j-th agents. $\boldsymbol{P}_j^{t-\tau_j} = \phi_\theta(X_j^{t-\tau_j})$, where $\phi_\theta(\cdot)$ is a part of framework $\Psi_\theta(\cdot,\cdot)$ used by the j-th agent.

When agents operate with disparate modal observations or are deprived of some of them, the systems might fail to work functionally. Besides, the temporal synchronization and pose error might significantly degrade the performance due to the erroneous information present in feature maps and spatial misalignment. Therefore, we introduce RealCP to address these issues. The overview is shown in Fig. 1.

Low-Coupling Modal Encoders. In real-world scenarios, agents might be equipped with different sensor devices (data availability), and encounter device changes due to failures or update events (dynamic input). In these cases, the method of previous works may no longer perform effectively or require training from scratch. To address this issue, a flexible solution is urgently required to ensure adaptability and sustainability.

In this sight, we employ a set of low-coupling modal encoders to match all types of observations. For the i-th agent, given its modal observations $X_i^t = \{x_{i(l)}^t\}_{l=1}^L$ at current timestamp t, the final feature map for each modality is derived as $\boldsymbol{F}_{i(l)}^t = \phi_{enc(l)}(x_{i(l)}^t) \in \mathcal{R}^{H\times W\times C}$, where $\phi_{enc(l)}(\cdot)$ represents the encoder specific for the l-th modal observation, and H, W, C are its height, weight and channel, respectively. For all observation-encoder pairs, the encoder extracts feature map from matched observation(s) and finally forms representation(s) of the same size, adapting all feature maps to share the same coordinate system. Low-coupling modal encoders ensures their independent functionality, enabling RealCP to accommodate all modal observations as input. The extracted feature maps are subsequently output to the diversifying feature collector module.

Diversifying Feature Collection. The diversifying feature collector module first discover the similarities and differences in heterogeneous modal feature maps, transforming them into a unified representation and ultimately generating a semantically enriched feature map. Intuitively, it is reasonable to assume that all modal data should exhibit certain shared perceptual characteristics, as they all capture the agent's surrounding environment. Concurrently, each modality also possesses unique information stemming from its distinct perspective. Therefore, it is crucial for the integrated feature map to preserve both the commonalities and the exclusive information of these modalities.

To deal with the inherent heterogeneity among different modalities, we employ a series of learnable embeddings, denoted as $\mathcal{E}_i = \{e_{i(l)}\}_{l=1}^{L} \in \mathcal{R}^{1\times1\times C}$ which transforms all modal feature maps into a homogeneous domain. Specifically, for the modal feature maps $\{\boldsymbol{F}_{i(l)}^t\}_{l=1}^{L}$, each embedding transforms corresponding modal feature map to an unified domain $\overline{\boldsymbol{F}}_{i(l)}^t = e_{i(l)} \times \boldsymbol{F}_{i(l)}^t$, where $l \in \{1, 2, ..., L\}$. The transformed feature maps $\{\overline{\boldsymbol{F}}_{i(l)}^t\}_{l=1}^{L}$ are then concatenated, and a convolutional layer with a kernel size of 1×1 is applied to synthesize a perceptually common feature map $\overline{\boldsymbol{F}}_i^t \in \mathcal{R}^{H\times W\times C}$.

The exclusive feature maps are designed to capture perceptual information from diverse perspectives while maintaining homogeneity with the common feature map. To achieve this, a cross-domain attention (CDA) mechanism $\widetilde{\boldsymbol{F}}_{i(l)}^t = \phi_{CDA}(\boldsymbol{F}_{i(l)}^t, \overline{\boldsymbol{F}}_i^t, \overline{\boldsymbol{F}}_i^t)$, where the modal feature maps $\{\boldsymbol{F}_{i(l)}^t\}_{i=1}^{L}$ serves as queries and the common feature map serves as key and value.

Once the common and exclusive feature maps are collected in the homogeneous domain, the process of feature integration commences $\boldsymbol{F}_i^t = \overline{\boldsymbol{F}}_i^t + \rho(\phi_{FFN}(\sum_{l=1}^{L} \widetilde{\boldsymbol{F}}_{i(l)}^t))$, where $\phi_{FFN}(\cdot)$ denotes the feed-forward network and $\rho(\cdot)$ is the ReLU activation function. In scenarios where a sensor encounters a breakdown or other circumstances that render it unavailable for a period, the corresponding input data $x_{i(l)}^t$ can be conveniently filled with zeros. During the generation of the common feature map, the convolutional operation is only influenced by the non-zero values in the channel dimension. Similarly, when aggregating the exclusive feature maps, the zeros in the input data have minimal impact. And the situation is similar when summing up the exclusive feature maps. This approach ensures that the final feature map is unaffected by unavailable data combining with the utilization of blank sensory data fully filled with zero value.

Conservative Confidence Generator. Due to the existence of communication latency and pose error, agents in the communication network only receive noisy message $\{\boldsymbol{P}_j^{t-\tau_j}\}_{j\in\mathcal{N}\setminus\{i\}}$ from other agents at the current timestamp t, while the packed message $\boldsymbol{P}_i^t$ is destined to be utilized by other agents at future timestamps. This asynchronous and noisy nature leads to spatial information misalignment, seriously undermining the basis of collaboration. We introduce a conservative confidence generator to create a binary confidence mask B_i^t, which highlights the spatially critical features to enable more effective feature aggregation at future timestamps.

Specifically, we derive a normal spatial confidence map $C_i^t = \sigma(\phi_{cls\_head}(\boldsymbol{F}_i^t)) \in [0,1]^{H\times W}$ by passing the feature map through a classification head and subsequently applying a sigmoid activation function $\sigma(\cdot)$. The spatial confidence map C_i^t quantitatively assess the perceptually critical level of all spatial areas. It then employs a refined selection function to generate a normal confidence mask A_i^t, which serves as a decisive indicator in determining the value of sharing the element positioned at $[m,n]$.

$$A_{i[m,n]}^t = \phi_{select}(C_i^t;\epsilon) = \begin{cases} 0 & C_{i[m,n]}^t \leq \epsilon \\ 1 & C_{i[m,n]}^t > \epsilon \end{cases} \quad (2)$$

It is important to note that the normal confidence mask is primarily designed for real-time feature aggregation with accurate pose information at current timestamp t. However, its reliability might be compromised in asynchronous and noisy environments. Considering an agent travelling in the scene, pose errors or other exceptional circumstances may render the agent's actual position uncertain. Despite this uncertainty, we can make two reasonable assumptions: i) The localization and heading errors would remain within acceptable bounds (e.g. not exceeding $3m$ or $2°$); ii) If nearby agents within the communication range receive lower-volume messages at a future timestamp $t+\delta t$, the distance $\delta s = v \times \delta t$ that the agent had travelled would not be excessive.

To account for these potential biases in the feature map, we introduce dilation, a straightforward yet effective technique to collect adjacent information of critical elements, enabling more conservative message packing. We define the conservative binary confidence mask $B_i^t = \phi_{dilate}(A_i^t; d)$, where d is the size of dilated kernel. The packed message of the i-th agent is the obtained by $\boldsymbol{P}_i^t = \boldsymbol{F}_i^t \odot B_i^t$.

Message Aggregator. The message aggregator aims to augment the perceptual information encoded in the final feature map $\hat{\boldsymbol{F}}_i^t \in \mathcal{R}^{H\times W\times C}$ by cooperatively integrating the ego message $\boldsymbol{P}_i^t$ with the latest messages $\{\boldsymbol{P}_{j\to i}^{t-\tau_j}\}_{j\in\mathcal{N}\setminus\{i\}}$ received from other agents.

We first incorporate a Transformer architecture that addresses the spatial misalignment of features due to communication latency and pose error, and collaboration heterogeneity among agents. Inspired by [21], we learn an agent-level embeddings $Q_{j\to i}^{t-\tau_j} = f_{MHA}(\boldsymbol{P}_{j\to i}^{t-\tau_j}, \boldsymbol{P}_{j\to i}^{t-\tau_j}, \boldsymbol{P}_{j\to i}^{t-\tau_j}) + \boldsymbol{P}_{j\to i}^{t-\tau_j}$, then semantic alignment is conducted to derive the activated object-aware feature maps $\ddot{\boldsymbol{P}}_j^{t-\tau_j} = f_{MHA}(\boldsymbol{P}_i^t, Q_{j\to i}^{t-\tau_j}, Q_{j\to i}^{t-\tau_j}) \in \mathcal{R}^{H\times W\times C}$, where $f_{MHA}(\cdot,\cdot,\cdot)$ denotes multi-head attention mechanism.

Then to more explicitly model the cooperation among agents than attention mechanism [3,6,18,22,25,27,29], we introduce Shapley value [1], a well-established solution to quantify contributions in a grand coalition. Specifically, we define the cooperation framework as $G = (\mathcal{N}, v)$. $\mathcal{N} = \{1, 2, ..., n\}$ is the finite agent set. And v is a characteristic function assigning each coalition of agents $\mathcal{S} \subseteq \mathcal{N}$ a value $v(\mathcal{S})$, which reflects the collective contribution of the coalition $\mathcal{S}$ to the cooperative system. With this definition, the Shapley value of the i-th agent D_i^t with respect to the grand coalition $\mathcal{S}$ is defined as

$$D_i^t = \phi_{shapley}(i) = \sum_{\mathcal{S}\subseteq\mathcal{N}\setminus\{i\}} \frac{|\mathcal{S}|!(|\mathcal{N}|-|\mathcal{S}|-1)!}{|\mathcal{N}|!}(v(\mathcal{S}\cup\{i\}) - v(\mathcal{S})) \in \mathcal{R}^{H\times W} \quad (3)$$

where $\varphi(\mathcal{S}) = v(\mathcal{S} \cup \{i\}) - v(\mathcal{S})$ is referred to as the marginal contribution, quantifying the incremental value attributed to the inclusion of the i-th agent in the coalition $\mathcal{S}$. The Shapley value is the average marginal contribution over all possible coalitions. Here we utilize a feed-forward network followed by the ReLU function $\rho(\cdot)$ to serve as the characteristic function to evaluate the marginal contribution of each coalition of messages. This allows us to prioritize messages that positively impact local perception, specifically, those with a positive Shapley value.

After identifying the precise contribution of each agent at locations, the softmax function is applied in agent-wise to normalize the Shapley value $\overline{D}_j^t = \frac{D_j^t}{\sum_{j=1}^{N} D_j^t}$, which makes the sum of Shapley values for each location equals 1. Subsequently, a summation mixer $f_{mixer}(\cdot,\cdot)$ is further employed that monotonically combines these normalized values to facilitate aggregation among agents, assisting attention learning to prefer high-quality and critical features.

$$\hat{\boldsymbol{F}}_i^t = f_{mixer}(\{\ddot{\boldsymbol{P}}_j^{t-\tau_j}\}, \{\overline{D}_j^t\}) = \sum_{i=1}^{N} \ddot{\boldsymbol{P}}_j^{t-\tau_j} \odot \overline{D}_j^t \tag{4}$$

Prediction Head and Objective Function. To conduct 3D object detection task, the regression head $f_{reg\_head}(\cdot)$ is applied to produce the regression output $\hat{Y}_i^{reg} = f_{reg\_head}(\hat{\boldsymbol{F}}_i^t) \in \mathcal{R}^{H \times W \times 7}$, which contains the position, size, and yaw angle of the bounding box at each location. The classification head $f_{cls\_head}(\cdot)$ is applied to distinguish whether each predefined box a target or the background $\hat{Y}_i^{cls} = f_{cls\_head}(\hat{\boldsymbol{F}}_i^t) \in \mathcal{R}^{H \times W \times 2}$. We adopt the smooth L1 loss for both regression $\mathcal{L}_{reg}$ and classification $\mathcal{L}_{cls}$. The overall objective function is $\mathcal{L} = \alpha \cdot \mathcal{L}_{reg} + \beta \cdot \mathcal{L}_{cls}$, where α and β are parameters to balance each loss.

4 Experiments

4.1 Dataset and Implementation Detail

We evaluate RealCP on the 3D object detection task on three public datasets, including OPV2V [26], V2V4Real [24] and DAIR-V2X [31]. OPV2V is the first large-scale vehicle-to-vehicle (V2V) collaborative perception dataset, co-simulated using OpenCDA and CARLA [2]. In contrast to simulated environments, V2V4Real represents the first large-scale real-world multi-modal dataset for V2V perception, collected by two vehicles equipped with multi-modal sensors driving together through diverse scenarios. Extending beyond vehicle-to-vehicle collaboration, DAIR-V2X is the first large-scale real-world vehicle-to-infrastructure perception dataset, containing 100 realistic scenarios and 9K cooperative frames. The training epochs on these datasets are 30, 50 and 30, respectively. The resolution is 0.4 m per pixel in length and width for all datasets. The perception ranges are set as $281.6m \times 76.8m$, $281.6m \times 76.8m$ and $201.6m \times 80m$, respectively.

Table 2. 3D detection quantitative results on OPV2V, V2V4Real and DAIR-V2X datasets."Ayn/Syn" denotes $100ms$ and $0ms$ transmission latencies between the ego agent and collaborators, and "L" is the number of input modalities.

Method	OPV2V(Asy,L=2)		V2V4Real(Asy,L=1)		DAIR-V2X(Syn,L=2)	
	AP@0.5	AP@0.7	AP@0.5	AP@0.7	AP@0.5	AP@0.7
No Fusion	55.24	39.20	49.99	30.42	51.11	44.93
When2com [12]	61.60	41.01	43.69	17.81	47.34	35.47
DiscoNet [11]	78.11	62.46	55.58	30.52	55.68	39.60
V2X-ViT [25]	71.44	53.96	54.97	29.85	53.26	36.91
Where2comm [6]	80.75	60.37	54.83	30.21	60.85	48.34
SCOPE [28]	59.31	51.22	49.16	21.89	55.59	39.85
BM2CP [33]	69.47	43.37	39.36	11.27	64.12	49.16
RealCP	**85.31**	**68.65**	**55.97**	**31.20**	**65.61**	**50.85**

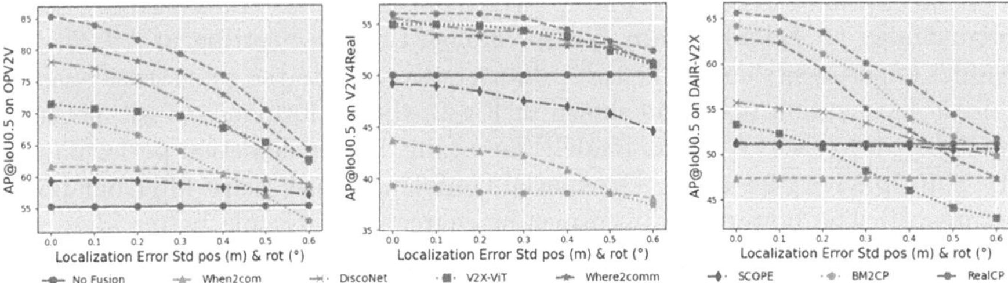

Fig. 2. Detection performance of AP@IoU0.5 under localization error configurations from $0m/^{\circ}$ to $0.6m/^{\circ}$ on OPV2V, V2V4Real and DAIR-V2X datasets.

We compare RealCP with individual observation baseline *No Fusion* and several state-of-the-art (SOTA) methods: *When2com* [12], *DiscoNet* [23], *V2X-ViT* [25], *Where2comm* [6], *SCOPE* [28] and *BM2CP* [33]. All methods are re-implemented based on the pyTorch framework and OpenCOOD [26] codebase, and are trained on one Nvidia RTX 4090 GPU with Adam optimizer and multi-step learning rate scheduler. All methods use the same architecture with no feature compression. PointPillar [8] is used to encode LiDAR data and EfficientNet [7] followed by Lift-Splat-Shoot [14] are used to encode image data. The transmission latencies between the ego agent and collaborators are set to around $\{100ms, 100ms, 0ms\}$ for OPV2V, V2V4Real and DAIR-V2X datasets. The resolution is 0.4 m per pixel in length and width for all datasets. The transmitted BEV intermediate representation has a size of $704 \times 192 \times 64$ for both OPV2V and V2V4Real datasets, and $504 \times 200 \times 64$ for DAIR-V2X dataset.

For hyperparameters, the dilated kernels are set to $\{3, 3, 1\}$ for OPV2V, V2V4Real and DAIR-V2X based on the transmission latencies. The thresholds for selection function $\phi(\cdot;\epsilon)$ are set to $\{0.01, 0.01, 0\}$, respectively. The trade-off parameters in loss function are $\alpha = 2.0$ and $\beta = 1.0$, respectively.

For metrics, we adopt the Average Precision (AP) at Intersection-over-Union (IoU) thresholds of 0.5 and 0.7 to evaluate the 3D object detection performance.

4.2 Quantitative Evaluation

Detection Performance Comparisons. Table 2 compares the object detection performance of RealCP and the SOTA methods. It can be observed that RealCP outperforms previous SOTA methods on all experimental datasets, and achieves significant improvements over other methods on OPV2V dataset. The reasona is that OPV2V additionally provides 4 RGB images per frame. Whereas V2V4Real has no additional modality and DAIR-V2X only provides front-view RGB images, which limits its advantage of exploration and exploit among multiple modalities. Table 2 demonstrate the adaptability of RealCP to both synchronous/asynchronous single/multiple-modality scenarios.

Robustness to Localization Noise. Follow the noise setting in [19,25], we conduct experiments on all the three datasets to validate the robustness against realistic localization noise. As shown in Fig. 2, the performance of all methods inevitably decreases. Whereas RealCP maintains higher detection performance. The plausible reasons are: i) The spatial confidence mask helps filter out noisy features; ii) The powerful transformer in aggregation module better align the collaborative feature maps; iii) The quantification of each agent's contribution further reduces the impact of noisy features in feature fusion. These designs work together to mitigate noise localization distortion effects.

Robustness to Communication Latency. As shown in Table 3, the detection performance of all SOTA methods decreases when the communication latency increases. RealCP can improve the performance under most of time delay levels, which proves that RealCP is more robust to communication latency than previous methods.

The Adaptability to Fit Diverse Inputs. Table 4 shows that RealCP also keeps the strong adaptability against different input settings. It can also be observed that the performance is largely decided by LiDAR sensor. When all agents only use LiDAR for collaboration, the performance on OPV2V and DAIR-V2X datasets are comparable with SOTA methods.

Table 3. Detection performance of AP@IoU0.5 under communication latency configurations on OPV2V dataset.

Method	0	~100	~200	~300
No Fusion	55.24			
When2com [12]	62.35	61.60	57.77	54.27
DiscoNet [11]	78.77	78.11	73.98	69.49
V2X-ViT [25]	72.08	71.44	69.01	66.42
Where2comm [6]	81.74	80.75	76.42	70.33
SCOPE [28]	59.32	59.31	59.03	58.38
BM2CP [33]	70.26	69.47	63.65	57.45
RealCP	86.37	85.31	79.99	72.25

Table 4. The adaptability to dynamic input. "L" denotes the LIDAR and "C" denotes the camera.

	Ego	Adjacent	OPV2V	DAIR-V2X
(a)	C	L	74.59/43.72	19.09/7.28
(b)	L	C	60.28/48.64	51.13/41.71
(c)	L	L	85.21/66.14	65.05/50.45
(d)	C	L, C	74.58/44.29	19.10/7.28
(e)	L	L, C	85.17/66.46	65.15/50.62
(f)	L, C	C	60.53/48.89	51.11/41.71
(g)	L, C	L	85.26/67.54	65.16/50.70
(h)	L, C	L, C	85.31/68.65	65.64/50.85

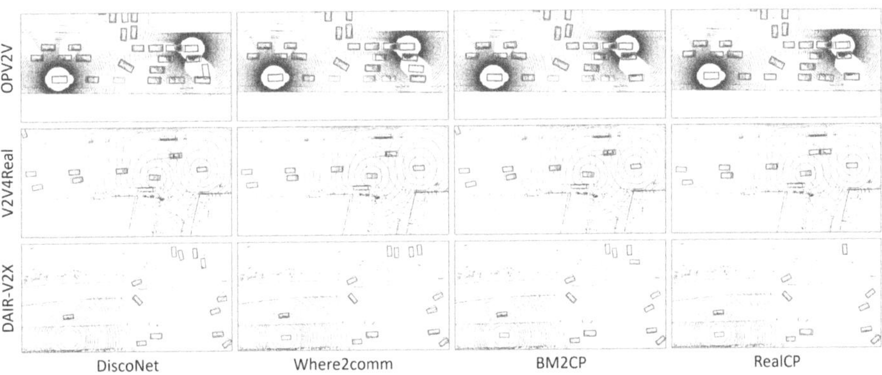

Fig. 3. Detection visual comparison on different datasets. Green and red boxes denote the ground truths and detection results, respectively. (Color figure online)

Table 5. Ablation study (AP@0.5/0.7) of diversifying feature collection (DFC), conservative confidence map (CCM) and shapley value calculation (SVC).

	DFC	CCM	SVC	OPV2V2	V2V4Real	DAIR-V2X
(a)	✓	✓	✓	85.31/68.65	55.97/31.20	65.64/50.85
(b)	✓	✓	×	84.48/66.27	55.80/28.79	62.46/48.10
(c)	✓	×	×	83.49/64.66	51.37/26.41	62.18/47.95
(d)	×	×	×	82.46/64.01	50.25/23.44	60.88/46.21

Table 6. Ablation study of the dilated kernel size in conservative confidence map.

	OPV2V	V2V4Real
1	85.271/67.666	55.965/30.519
3	85.314/68.653	55.965/31.202
5	85.281/67.664	55.948/30.620
7	85.283/67.637	55.931/30.433

4.3 Qualitative Evaluation

Figure 3 provides compared visualizations of the detection results in three scenes. Our model achieves more accurate and less false detection than compared methods. The reason is that RealCP can utilize more semantic information from all input modalities to comprehensively consider the effects of communication latency, pose error on feature maps in spatial and temporal, leading to a more informative fused visual representation and a more holistic perception.

4.4 Ablation Studies

We perform ablation studies to verify the necessity of core components and dilated mechanism. Table 5 assesses the contribution of each component. It can be observed that all components are beneficial to the performance gains. Table 6 compares the choice of the dilated kernel size in conservative confidence generator module. It can be observed that the detection performance is better when dilated kernel size is set to 3 under $\sim 100ms$ communication latency.

5 Conclusion

We present RealCP, a novel framework to make multi-agent decision-making system more robust to mixed real-world factors. It firstly utilizes low-coupling encoders and discover multi-perspective information among different modalities. Then, it introduces a conservative confidence generator to promote pragmatic compression under asynchronous and noisy realistic conditions. Finally, it adopts a messages aggregator to align messages in spatial, and explicitly quantifies the agents' contributions. Extensive experiments on both simulated and real-world datasets show its effectiveness and robustness.

References

1. Chalkiadakis, G., Elkind, E., Wooldridge, M.: Computational aspects of cooperative game theory. Syn. Lect. Artif. Intell. Mach. Learn. **5**(6), 1–168 (2011)
2. Dosovitskiy, A., Ros, G., Codevilla, F., Lopez, A., Koltun, V.: Carla: An open urban driving simulator. In: Conference on Robot Learning, pp. 1–16. PMLR (2017)
3. Gu, J., et al.: Feaco: Reaching robust feature-level consensus in noisy pose conditions. In: Proceedings of the 31st ACM International Conference on Multimedia, pp. 3628–3636 (2023)
4. Han, X., Hu, X., Ma, C., Shang, X.: Troute: Dynamic time-dependent route recommendation on road networks, pp. 573–585 (2024)
5. Han, Y., Zhang, H., Li, H., Jin, Y., Lang, C., Li, Y.: Collaborative perception in autonomous driving: methods, datasets, and challenges. IEEE Intell. Trans. Syst. Mag. (2023)
6. Hu, Y., Fang, S., Lei, Z., Zhong, Y., Chen, S.: Where2comm: communication-efficient collaborative perception via spatial confidence maps. Adv. Neural. Inf. Process. Syst. **35**, 4874–4886 (2022)
7. Koonce, B., Koonce, B.: Efficientnet. Convolutional Neural Networks with Swift for Tensorflow: Image Recognition and Dataset Categorization, pp. 109–123 (2021)
8. Lang, A.H., Vora, S., Caesar, H., Zhou, L., Yang, J., Beijbom, O.: Pointpillars: fast encoders for object detection from point clouds. In: Proceedings of the IEEE/CVF Conference on Computer Vision and Pattern Recognition, pp. 12697–12705 (2019)

9. Lei, Z., Ren, S., Hu, Y., Zhang, W., Chen, S.: Latency-aware collaborative perception. In: European Conference on Computer Vision, pp. 316–332. Springer (2022)
10. Li, Y., Fang, Q., Bai, J., Chen, S., Juefei-Xu, F., Feng, C.: Among us: adversarially robust collaborative perception by consensus. In: Proceedings of the IEEE/CVF International Conference on Computer Vision, pp. 186–195 (2023)
11. Li, Y., Ren, S., Wu, P., Chen, S., Feng, C., Zhang, W.: Learning distilled collaboration graph for multi-agent perception. Adv. Neural. Inf. Process. Syst. **34**, 29541–29552 (2021)
12. Liu, Y.C., Tian, J., Glaser, N., Kira, Z.: When2com: Multi-agent perception via communication graph grouping. In: Proceedings of the IEEE/CVF Conference on computer vision and pattern recognition, pp. 4106–4115 (2020)
13. Lu, Y., et al.: Robust collaborative 3D object detection in presence of pose errors. In: 2023 IEEE International Conference on Robotics and Automation (ICRA), pp. 4812–4818. IEEE (2023)
14. Philion, J., Fidler, S.: Lift, Splat, Shoot: encoding images from arbitrary camera rigs by implicitly unprojecting to 3D. In: Vedaldi, A., Bischof, H., Brox, T., Frahm, J.-M. (eds.) ECCV 2020. LNCS, vol. 12359, pp. 194–210. Springer, Cham (2020). https://doi.org/10.1007/978-3-030-58568-6_12
15. Ren, S., Chen, S., Zhang, W.: Collaborative perception for autonomous driving: current status and future trend. In: Proceedings of 2021 5th Chinese Conference on Swarm Intelligence and Cooperative Control, pp. 682–692. Springer (2022)
16. Tu, J., Wang, T., Wang, J., Manivasagam, S., Ren, M., Urtasun, R.: Adversarial attacks on multi-agent communication. In: Proceedings of the IEEE/CVF International Conference on Computer Vision, pp. 7768–7777 (2021)
17. Vadivelu, N., Ren, M., Tu, J., Wang, J., Urtasun, R.: Learning to communicate and correct pose errors. In: Conference on Robot Learning, pp. 1195–1210. PMLR (2021)
18. Wang, B., Zhang, L., Wang, Z., Zhao, Y., Zhou, T.: Core: Cooperative reconstruction for multi-agent perception. In: Proceedings of the IEEE/CVF International Conference on Computer Vision, pp. 8710–8720 (2023)
19. Wang, T.-H., Manivasagam, S., Liang, M., Yang, B., Zeng, W., Urtasun, R.: V2VNet: vehicle-to-vehicle communication for joint perception and prediction. In: Vedaldi, A., Bischof, H., Brox, T., Frahm, J.-M. (eds.) ECCV 2020. LNCS, vol. 12347, pp. 605–621. Springer, Cham (2020). https://doi.org/10.1007/978-3-030-58536-5_36
20. Wei, S., et al.: Asynchrony-robust collaborative perception via bird's eye view flow. Adv. Neural Inf. Process. Syst. **36** (2024)
21. Wu, D., Dong, X., Shao, L., Shen, J.: Multi-level representation learning with semantic alignment for referring video object segmentation. In: Proceedings of the IEEE/CVF Conference on Computer Vision and Pattern Recognition, pp. 4996–5005 (2022)
22. Xiang, H., Xu, R., Ma, J.: Hm-vit: Hetero-modal vehicle-to-vehicle cooperative perception with vision transformer. In: Proceedings of the IEEE/CVF International Conference on Computer Vision, pp. 284–295 (2023)
23. Xu, R., Tu, Z., Xiang, H., Shao, W., Zhou, B., Ma, J.: Cobevt: Cooperative bird's eye view semantic segmentation with sparse transformers. In: Proceedings of The 6th Annual Conference on Robot Learning, pp. 989–1000. PMLR (2022)
24. Xu, R., et al.: V2v4real: A real-world large-scale dataset for vehicle-to-vehicle cooperative perception. In: Proceedings of the IEEE/CVF Conference on Computer Vision and Pattern Recognition, pp. 13712–13722 (2023)

25. Xu, R., Xiang, H., Tu, Z., Xia, X., Yang, M.H., Ma, J.: V2x-vit: vehicle-to-everything cooperative perception with vision transformer. In: European Conference on Computer Vision, pp. 107–124. Springer (2022)
26. Xu, R., Xiang, H., Xia, X., Han, X., Li, J., Ma, J.: Opv2v: an open benchmark dataset and fusion pipeline for perception with vehicle-to-vehicle communication. In: 2022 International Conference on Robotics and Automation (ICRA), pp. 2583–2589. IEEE (2022)
27. Yang, D., et al.: How2comm: Communication-efficient and collaboration-pragmatic multi-agent perception. Adv. Neural Inf. Process. Syst. **36** (2024)
28. Yang, K., et al.: Spatio-temporal domain awareness for multi-agent collaborative perception. In: Proceedings of the IEEE/CVF International Conference on Computer Vision, pp. 23383–23392 (2023)
29. Yang, K., Yang, D., Zhang, J., Wang, H., Sun, P., Song, L.: What2comm: towards communication-efficient collaborative perception via feature decoupling. In: Proceedings of the 31st ACM International Conference on Multimedia, pp. 7686–7695 (2023)
30. Yin, H., Tiaet al.: V2vformer ++: multi-modal vehicle-to-vehicle cooperative perception via global-local transformer. IEEE Trans. Intell. Trans. Syst. (2023)
31. Yu, H., et al.: Dair-v2x: A large-scale dataset for vehicle-infrastructure cooperative 3D object detection. In: Proceedings of the IEEE/CVF Conference on Computer Vision and Pattern Recognition, pp. 21361–21370 (2022)
32. Yu, H., Tang, Y., Xie, E., Mao, J., Luo, P., Nie, Z.: Flow-based feature fusion for vehicle-infrastructure cooperative 3D object detection. Adv. Neural Inf. Process. Syst. **36** (2024)
33. Zhao, B., Zhang, W., Zou, Z.: Bm2cp: Efficient collaborative perception with lidar-camera modalities. In: Tan, J., Toussaint, M., Darvish, K. (eds.) Proceedings of The 7th Conference on Robot Learning. Proceedings of Machine Learning Research, vol. 229, pp. 1022–1035. PMLR (2023)

ShrimpFormer-X: A Transformer-Based Framework for Counting and Localization of Shrimp Larvae

Jiaxin Lin, Yuefang Gao, Dong Huang, and Zihao Wen(✉)

College of Mathematics and Informatics, South China Agricultural University, Guangzhou, China
jiaxin.lin@stu.scau.edu.cn, gaoyuefang@scau.edu.cn, zihao.wen@hotmail.com

Abstract. Accurate counting of shrimp larvae is essential for feed optimization, density control, and yield prediction in aquaculture. However, larvae counting is challenging due to their small size, high flexibility, and frequent occlusion in dense scenes. Manual methods are time-consuming, error-prone, and potentially harmful to the larvae, while existing automated solutions are often expensive and impractical. In this paper, we propose a Transformer-based framework called ShrimpFormer-X for end-to-end shrimp larvae counting and localization. Unlike conventional methods that rely solely on regression or detection, ShrimpFormer-X integrates both counting and localization tasks. It introduces a hybrid encoder that applies attention selectively to high-level features while leveraging bidirectional fusion to enhance multi-scale representation. A decoupled cross-attention decoder processes contextual and spatial information separately by concatenating image features and positional encodings. Furthermore, a density partition module identifies dense and sparse regions to enable adaptive decoding. Extensive experiments on a large-scale shrimp larvae dataset demonstrate that ShrimpFormer-X achieves state-of-the-art performance in both low and high-density conditions. The model outperforms existing methods in both accuracy and computational efficiency while providing reliable localization.

Keywords: Shrimp larvae counting · Transformer networks · Small object detection · Cross-attention · Aquaculture

1 Introduction

Litopenaeus vannamei, commonly known as whiteleg shrimp, is one of the most farmed shrimp species worldwide and plays a key role in China's aquaculture sector [6]. As shrimp farming continues to scale, the need for precise and automated larvae management is becoming increasingly critical for improving production efficiency and reducing operational costs [23].

A fundamental task in this process is shrimp larvae counting, which supports essential functions such as feed allocation, density control, yield forecasting, and

S. Yang et al. (Eds.): WISA 2025, LNCS 16299, pp. 127–138, 2026.
https://doi.org/10.1007/978-981-95-6426-2_11

trading valuation. Traditional manual counting methods are widely used in practice but are time-consuming, labor-intensive, and error-prone. They often involve sampling followed by extrapolation, and pose risks of physical harm to the larvae [8]. Commercial counting machines have emerged as alternatives, but their high cost and complex setup, which requires specific trays and water levels, restrict their widespread adoption in hatcheries and farms [17].

In recent years, the artificial intelligence technique [3,9–11], especially deep learning [7,12,20,21,27], has shown promise for object counting tasks, including applications in aquaculture [22]. However, existing shrimp larvae counting methods typically fall into two categories: density map regression and detection-based models. While density map methods perform well in dense scenes, they lack precise localization capability. Detection models, on the other hand, struggle with small and overlapping larvae, especially in crowded environments. Both approaches face limitations in balancing accuracy, efficiency, and interpretability.

To address these challenges, we propose a Transformer-based framework called **ShrimpFormer-X** for accurate shrimp larvae counting and localization. The design is tailored to the characteristics of shrimp larvae: small, transparent, and densely distributed. ShrimpFormer-X unifies counting and localization in an end-to-end manner, combining efficiency and accuracy through three core innovations:

- A **hybrid encoder** that selectively applies self-attention to high-level features and enhances spatial detail through multi-scale bidirectional feature fusion.
- A **decoupled cross-attention decoder** that separates contextual and positional information via feature concatenation, improving learning stability and localization accuracy.
- A **density partition module** that adaptively segments feature regions by density, enabling region-specific decoding strategies to handle sparse and dense areas differently.

Notably, ShrimpFormer-X is designed to improve both counting precision and spatial localization, while maintaining a lightweight and scalable architecture suitable for real-world aquaculture applications. We evaluate our method on a challenging open-source dataset with diverse density conditions and show that it outperforms existing solutions in both accuracy and computational efficiency.

2 Related Work

Shrimp larvae counting has evolved from early image processing techniques to modern deep learning-based methods. Existing approaches can be broadly categorized into three groups: image processing-based methods, density map-based methods, and detection-based methods.

2.1 Image Processing-Based Methods

Earlier efforts in shrimp larvae counting relied on traditional image processing algorithms that segment larvae from the background using filtering and edge detection. For example, Khantuwan and Khiripet [13] used Laplacian and Median filters to enhance contrast and applied pattern matching for counting. Similarly, Awalludin et al. [2] employed a combination of improved Canny edge detection and anisotropic diffusion to preserve edge features while reducing noise. These methods are only effective in clean, low-density scenarios with minimal occlusion. Their reliance on hand-crafted features and fixed thresholds limits generalization to real-world dense environments with overlapping larvae.

2.2 Density Map-Based Methods

The density map regression paradigm was introduced by Lempitsky and Zisserman [14], transforming object counting into a density estimation problem where the object count corresponds to the integral over the predicted density map. In aquaculture, Zhang et al. [25] applied this method to shrimp egg counting using a fully convolutional regression network. Qu et al. [19] enhanced robustness to occlusion and background noise by incorporating Bayesian loss functions. Although density-based methods perform well in crowded scenes, they cannot directly localize individual larvae. Localization must be inferred through post-processing such as local peak detection [26], which is not end-to-end trainable and introduces reliance on manually selected thresholds.

2.3 Detection-Based Methods

Detection models such as YOLOv3 and its variants have also been applied to shrimp detection [1]. These models achieve acceptable results in low-density settings where targets are clearly separated. However, they struggle with dense and overlapping larvae due to anchor box limitations and the small size of the targets. Anchor-free methods, such as CAGNet [24], attempt to address this, but performance remains limited.

To reduce labeling costs and improve flexibility, recent crowd counting models use dot annotations and transformer-based detection heads. Models like CLTR [15], PET [16], and APGCC [5] generate point predictions without relying on bounding boxes. While effective in human crowd counting, these models often underperform on shrimp larvae due to the unique visual characteristics: small size, transparency, and limited texture.

3 Proposed Approach

The proposed ShrimpFormer-X is an end-to-end Transformer-based framework for accurate counting and localization of shrimp larvae in both sparse and dense aquaculture environments. It builds upon the Detection Transformer (DETR)

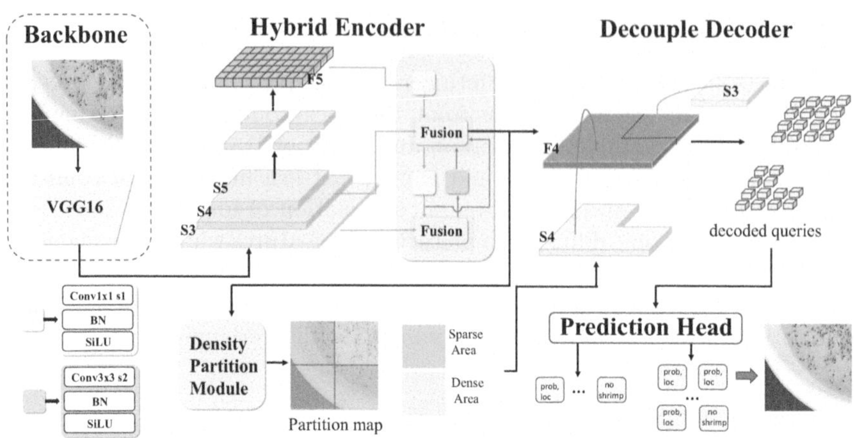

Fig. 1. Overview of the ShrimpFormer-X framework. The input image is processed by a CNN backbone to extract multi-scale features ($S3$, $S4$, $S5$). The hybrid encoder applies window-based self-attention to $S5$ and fuses it with lower-level features using bidirectional fusion, producing feature map $F4$. The density partition module divides the feature map into dense and sparse regions. The decoupled cross-attention decoder processes these regions using context and positional information. Finally, the prediction head outputs point-wise larvae predictions with associated confidence scores.

paradigm and introduces architectural innovations that specifically address the challenges posed by small, overlapping larvae. The overall framework consists of five key components: a CNN backbone, a hybrid encoder, a density partition module, a decoupled cross-attention decoder, and a prediction head.

3.1 Framework Overview

As illustrated in Fig. 1, an input image is first processed by a CNN backbone (VGG-16) to extract hierarchical features from three stages: $S3$, $S4$, and $S5$. These features are refined in the hybrid encoder to enhance contextual and spatial representation. The output, denoted as $F4$, is passed through a density partition module to adaptively divide the image into dense and sparse regions. A decoupled decoder then processes each region accordingly, using contextual cues to guide feature decoding. Finally, a prediction head generates the localization coordinates and presence probabilities of individual shrimp larvae.

3.2 Hybrid Encoder

The encoder enhances semantic features while preserving localization accuracy. To reduce computational overhead, self-attention is applied only to the high-level feature map $S5$, which contains rich semantic information. To retain spatial details from earlier layers, ShrimpFormer-X adopts a bidirectional feature fusion strategy inspired by PANet [18].

This strategy combines the high-resolution spatial cues from $S3$ and $S4$ with the abstracted semantics of $S5$. The fusion process includes 1×1 convolution layers for channel alignment, residual blocks for effective feature mixing, and element-wise addition. The resulting fused feature $F4$ provides a strong basis for accurate small-object detection.

3.3 Density Partition Module

Counting performance can be compromised when applying a fixed decoding strategy across scenes with varying densities. To resolve this, we introduce a density-aware partition module that classifies regions into sparse and dense based on feature responses. This module is adapted from the quadtree-based partitioning approach in PET [16] but simplified for efficiency.

In dense regions, high-resolution features are retained to preserve fine-grained details necessary for detecting overlapping larvae. In contrast, sparse regions use coarser representations to reduce computational load. The density partitioning enables region-specific decoding and facilitates scalable inference.

3.4 Decoupled Cross-Attention Decoder

Conventional cross-attention mechanisms merge content and positional encodings by summation, which can obscure the contribution of each component. ShrimpFormer-X instead adopts a decoupled attention design: content and positional features are concatenated and processed independently. This improves interpretability and allows the model to better learn positional relationships crucial for accurate localization.

Furthermore, the decoder dynamically refines spatial queries by integrating learned positional features from previous decoding stages. This process enhances the spatial discrimination capability and accelerates convergence.

3.5 Prediction Head and Loss Function

The prediction head outputs a set of candidate larvae points, each associated with a confidence score and spatial coordinates. To train the model, we use bipartite matching [4] to assign predicted points to ground truth labels. The total loss consists of three terms, i.e., the classification loss using binary cross-entropy, the localization loss using smooth-ℓ_1 between predicted and true positions, and density partition loss supervising the region classification. Specifically, the overall loss function is defined as

$$\mathcal{L}_{total} = \mathcal{L}_{cls} + \lambda_1 \mathcal{L}_{loc} + \lambda_2 \mathcal{L}_{partition}$$

where λ_1 and λ_2 are weighting coefficients. This design allows ShrimpFormer-X to achieve robust performance across variable densities.

Table 1. Statistics of the Vannamei-983 dataset.

Subset	Number of Images	Number of Labels
Low Density	500	165,307
High Density	483	889,081

4 Experiment

To evaluate the effectiveness of ShrimpFormer-X, we conduct extensive experiments on a large-scale shrimp larvae dataset under varying density conditions.

4.1 Dataset and Evaluation Metrics

We evaluate ShrimpFormer-X on the Vannamei-983 dataset, which consists of images under both low and high-density conditions. As summarized in Table 1, the dataset contains 983 images with more than one million annotated shrimp larvae.

The dataset is split into 80% for training and 20% for testing using stratified sampling. Performance is evaluated using two standard metrics: Mean Absolute Error (MAE) and Mean Squared Error (MSE):

$$MAE = \frac{1}{n}\sum_{i=1}^{n}|Y_i - Q_i|,$$

$$MSE = \frac{1}{n}\sum_{i=1}^{n}(Y_i - Q_i)^2$$

where Y_i and Q_i denote the ground truth and predicted counts for the i-th image.

4.2 Implementation Details

We implement ShrimpFormer-X using PyTorch and train it on a single NVIDIA RTX 3090 GPU. The model is optimized using the Adam optimizer. The learning rate is set to 10^{-5} for the pretrained VGG-16 backbone and 10^{-4} for the encoder-decoder modules. Weight decay is set to 1×10^{-4}. The encoder and decoder consist of 4 and 2 layers, respectively. Hyperparameters are set to $\lambda_1 = 5.0$ and $\lambda_2 = 0.2$ to balance loss components.

During training, we apply basic data augmentation, including random cropping (256×256 patches) and horizontal flipping. High-resolution images are resized while maintaining aspect ratios.

4.3 Results and Analysis

Quantitative Results. Table 2 compares ShrimpFormer-X with several baseline methods on both density levels. Our method achieves the lowest MAE and

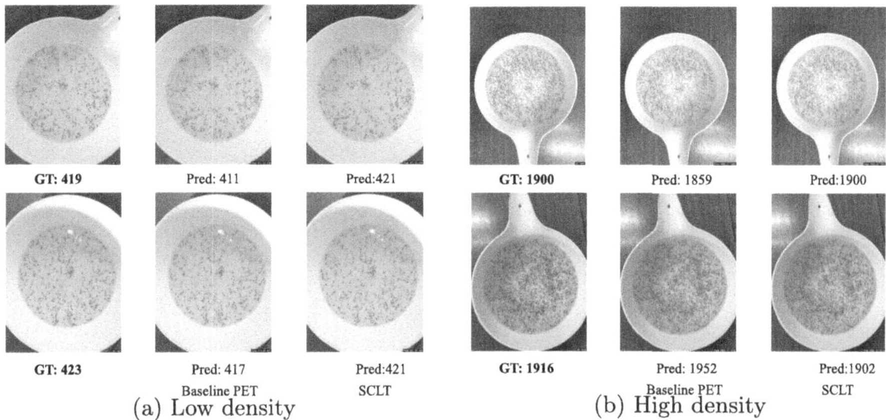

Fig. 2. Examples of the counting results of the ground-truth (left), the baseline PET (middle), and our ShrimpFormer-X (right).

MSE under high-density conditions and maintains competitive accuracy in low-density scenarios. Compared with PET [16], ShrimpFormer-X reduces the high-density MAE from 49.66 to 34.16, demonstrating significant performance gains in crowded scenes.

Table 2. Quantitative comparison of shrimp larvae counting results on the Vannamei-983 dataset. The best performance is in **boldface**, and the second best is underlined.

Venue	Method	Localization	High Density		Low Density	
			MAE	MSE	MAE	MSE
ICCV'2019	BL	✗	295.02	550.45	38.47	42.29
CVPR'2022	MAN	✗	261.14	496.54	119.58	128.41
TMM'2023	FIDTM	✓	50.32	64.31	**3.48**	**5.01**
AAAI'2024	Gramformer	✗	247.21	478.83	24.72	27.48
ECCV'2022	CLTR	✓	192.71	222.47	139.36	150.79
ICCV'2023	PET	✓	49.66	61.95	4.10	<u>5.32</u>
SCIS'2022	TransCrowd	✗	<u>48.38</u>	<u>58.28</u>	4.64	6.24
-	SCLT(Ours)	✓	**34.16**	**46.24**	<u>3.96</u>	5.52

Qualitative Results. Visualization results in Fig. 2 demonstrate the model's ability to localize larvae in both sparse and dense environments. Even under severe occlusion, ShrimpFormer-X provides accurate and consistent localization. Further visualization results are also shown in Figs. 3 and 4.

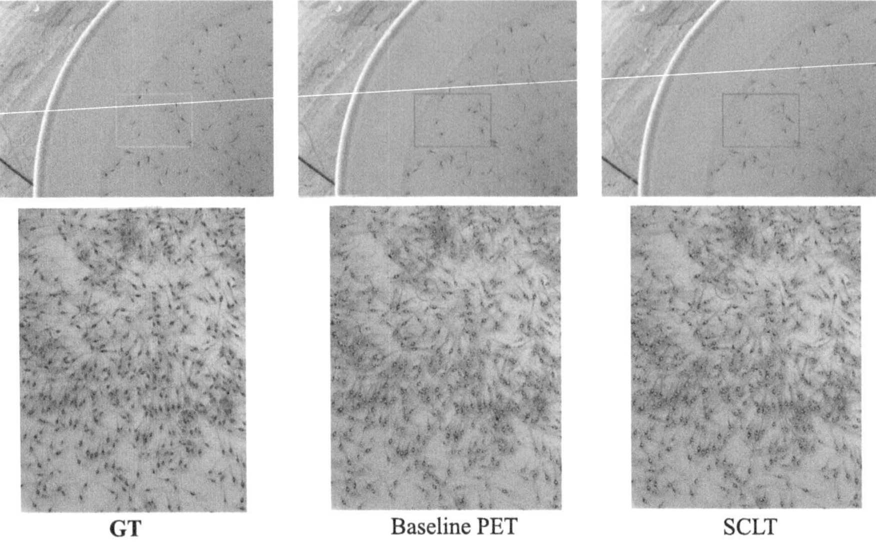

Fig. 3. Visualization of Small Target Detection.

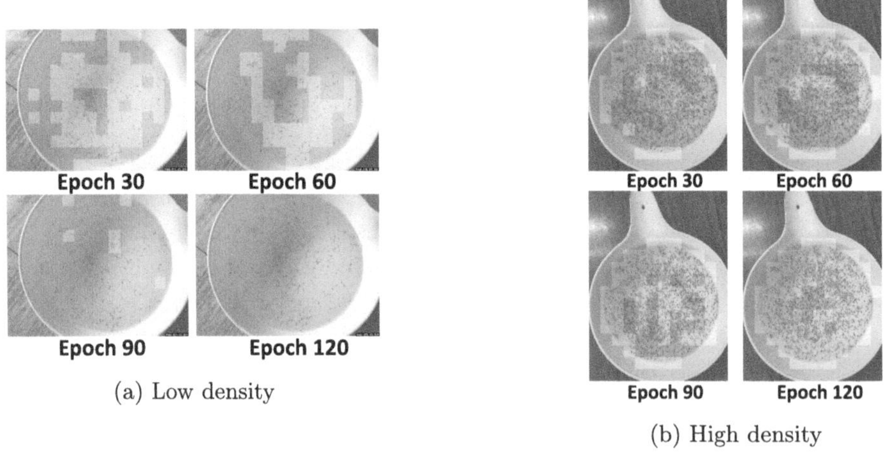

(a) Low density

(b) High density

Fig. 4. Examples of the partition map. The red regions denote the dense regions. (Color figure online)

Ablation Study. We conduct an ablation analysis in Table 3 to evaluate the contributions of the hybrid encoder and decoupled decoder. Each component improves performance independently, and their combination yields the best results. The convergence behavior shown in Fig. 5 also highlights the training stability benefits of the decoupled decoder.

Table 3. Ablation Study for different modules.

Module			High Density		Low Density	
Baseline	Hybrid Encoder	Decouple Decoder	MAE ↓	MSE ↓	MAE ↓	MSE ↓
✓			49.66	61.95	4.10	5.32
	✓		36.45	46.93	3.91	5.17
		✓	35.62	46.18	3.44	4.88
	✓	✓	34.16	46.24	3.96	5.52

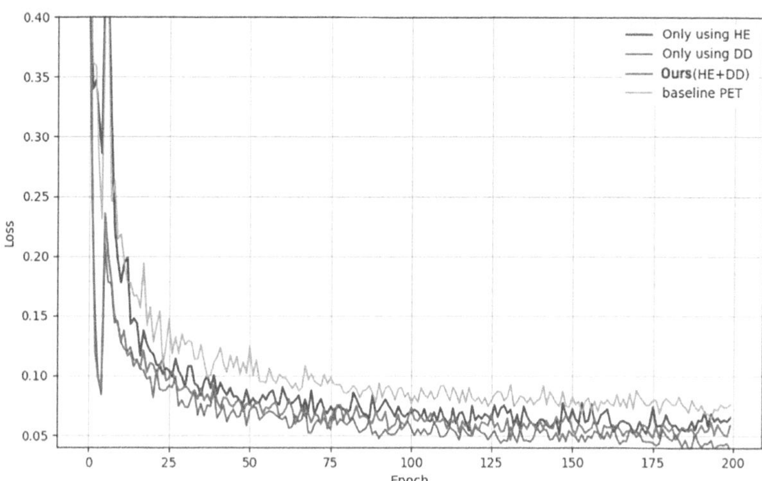

Fig. 5. The loss values of our ShrimpFormer-X during the training process. HE denotes Hybrid Encoder, while DD denotes Decouple Decoder.

Table 4. Influence of different backbones.

Backbone	High Density		Low Density	
	MAE	MSE	MAE	MSE
VGG-16	**34.16**	46.24	**3.96**	**5.52**
ResNet-18	35.20	**45.74**	6.26	7.87
EfficientNet-B0	48.63	65.47	4.66	6.22
DLA-34	37.26	48.88	4.01	5.27

Effectiveness of Density Partitioning. Figure 4 presents the output of the density partition module. It accurately differentiates between sparse and dense regions, enabling efficient region-specific decoding without compromising accuracy.

Backbone Comparison. We compare several backbone architectures in Table 4. VGG-16 achieves the best balance between computational cost and counting accuracy, confirming its suitability for this task.

5 Conclusion

In this work, we propose ShrimpFormer-X, a Transformer-based framework for accurate and efficient shrimp larvae counting and localization. Designed specifically for the challenges of aquaculture environments, ShrimpFormer-X integrates multiple innovations to achieve robust performance under diverse density conditions. The framework incorporates a hybrid encoder that applies self-attention selectively to high-level semantic features and utilizes bidirectional feature fusion to preserve spatial detail. A decoupled cross-attention decoder improves learning stability and enhances localization by independently modeling contextual and spatial cues. Furthermore, a density-aware partition module dynamically separates dense and sparse regions, enabling adaptive decoding and improving computational efficiency. Comprehensive experiments on the Vannamei-983 dataset demonstrate that ShrimpFormer-X consistently outperforms existing state-of-the-art methods in both high- and low-density scenarios.

Acknowledgments. This work was supported by NSFC (62476102) and the Natural Science Foundation of Guangdong Province (2025A1515010245).

References

1. Armalivia, S., Zainuddin, Z., Achmad, A., Wicaksono, M.A.: Automatic counting shrimp larvae based you only look once (YOLO). In: 2021 International Conference on Artificial Intelligence and Mechatronics Systems (AIMS), pp. 1–4. IEEE (2021)
2. Awalludin, E.A., et al.: Combination of canny edge detection and blob processing techniques for shrimp larvae counting. In: 2019 IEEE International Conference on Signal and Image Processing Applications (ICSIPA), pp. 308–313. IEEE (2019)
3. Cai, X., Huang, D., Zhang, G.Y., Wang, C.D.: Seeking commonness and inconsistencies: a jointly smoothed approach to multi-view subspace clustering. Inf. Fusion **91**, 364–375 (2023)
4. Carion, N., Massa, F., Synnaeve, G., Usunier, N., Kirillov, A., Zagoruyko, S.: End-to-end object detection with transformers. In: European Conference on Computer Vision, pp. 213–229. Springer (2020)
5. Chen, I.H., Chen, W.T., Liu, Y.W., Yang, M.H., Kuo, S.Y.: Improving point-based crowd counting and localization based on auxiliary point guidance. In: European Conference on Computer Vision, pp. 428–444. Springer (2024)
6. Chen, Y., et al.: Retrospect of fish meal substitution in pacific white shrimp (litopenaeus vannamei) feed: Alternatives, limitations and future prospects. Rev. Aquac. **16**(1), 382–409 (2024)
7. Deng, X., Huang, D., Chen, D.H., Wang, C.D., Lai, J.H.: Strongly augmented contrastive clustering. Pattern Recogn. **139**, 109470 (2023)

8. Duan, H., et al.: Shrimp larvae counting based on improved yolov5 model with regional segmentation. Sensors **24**(19), 6328 (2024)
9. Fang, S.G., Huang, D., Cai, X.S., Wang, C.D., He, C., Tang, Y.: Efficient multi-view clustering via unified and discrete bipartite graph learning. IEEE Trans. Neural Netw. Learn. Syst. **35**(8), 11436–11447 (2024)
10. Fang, S.G., Huang, D., Wang, C.D., Lai, J.H.: One-step bipartite graph cut: a normalized formulation and its application to scalable subspace clustering. Neural Networks, p. 108003 (2025)
11. Huang, D., Chen, D.H., Chen, X., Wang, C.D., Lai, J.H.: DeepCluE: enhanced deep clustering via multi-layer ensembles in neural networks. IEEE Trans. Emerg. Top. Comput. Intell. **8**(2), 1582–1594 (2024)
12. Huang, D., et al.: Deep clustering with hybrid-grained contrastive and discriminative learning. IEEE Trans. Circuits Syst. Video Technol. **34**(10), 9472–9483 (2024)
13. Khantuwan, W., Khiripet, N.: Live shrimp larvae counting method using co-occurrence color histogram. In: 2012 9th International Conference on Electrical Engineering/Electronics, Computer, Telecommunications and Information Technology, pp. 1–4. IEEE (2012)
14. Lempitsky, V., Zisserman, A.: Learning to count objects in images. Adv. Neural Inf. Process. Syst. **23** (2010)
15. Liang, D., Xu, W., Bai, X.: An end-to-end transformer model for crowd localization. In: European Conference on Computer Vision, pp. 38–54. Springer (2022)
16. Liu, C., Lu, H., Cao, Z., Liu, T.: Point-query quadtree for crowd counting, localization, and more. In: Proceedings of the IEEE/CVF International Conference on Computer Vision, pp. 1676–1685 (2023)
17. Liu, D., et al.: Shrimpseed_net: counting of shrimp seed using deep learning on smartphones for aquaculture. IEEE Access **11**, 85441–85450 (2023)
18. Liu, S., Qi, L., Qin, H., Shi, J., Jia, J.: Path aggregation network for instance segmentation. In: Proceedings of the IEEE Conference on Computer Vision and Pattern Recognition, pp. 8759–8768 (2018)
19. Qu, Y., Jiang, S., Li, D., Zhong, P., Shen, Z.: Slcobnet: Shrimp larvae counting network with overlapping splitting and Bayesian-dm-count loss. Biosys. Eng. **244**, 200–210 (2024)
20. Wang, J., Yang, H., Shan, J., Song, X., Jiang, J.: Attention-based spatial-temporal fusion networks for traffic flow prediction. In: In Proceedings of International Conference on Web Information Systems and Applications (WISA), pp. 500–511 (2024)
21. Xu, Y.K., Huang, D., Wang, C.D., Lai, J.H.: GLAC-GCN: global and local topology-aware contrastive graph clustering network. IEEE Trans. Artif. Intell. **6**(6), 1448–1459 (2025)
22. Yang, X., Zhang, S., Liu, J., Gao, Q., Dong, S., Zhou, C.: Deep learning for smart fish farming: applications, opportunities and challenges. Rev. Aquac. **13**(1), 66–90 (2021)
23. Yu, X., Wang, J., Wen, S., Yang, J., Zhang, F.: A deep learning based feature extraction method on hyperspectral images for nondestructive prediction of tvb-n content in pacific white shrimp (litopenaeus vannamei). Biosys. Eng. **178**, 244–255 (2019)
24. Zhang, G., Shen, Z., Li, D., Zhong, P., Chen, Y.: Cagnet: an improved anchor-free method for shrimp larvae detection in intensive aquaculture. Aquacult. Int. **32**(5), 6153–6175 (2024)
25. Zhang, J., Yang, G., Sun, L., Zhou, C., Zhou, X., Li, Q., Bi, M., Guo, J.: Shrimp egg counting with fully convolutional regression network and generative adversarial network. Aquacult. Eng. **94**, 102175 (2021)

26. Zhou, C., Yang, G., Sun, L., Wang, S., Song, W., Guo, J.: Counting, locating, and sizing of shrimp larvae based on density map regression. Aquacult. Int. **32**(3), 3147–3168 (2024)
27. Zhou, Z.F., Huang, D., Wang, C.D.: Pyramid contrastive learning for clustering. Neural Netw. **185**, 107217 (2025)

BridgeDefectIQ: A Practical AIoT Method for Quantitative Bridge Defect Assessment

Mkhuzo Zulu and Tiexin Wang(✉)

College of Computer Science and Technology, Nanjing University of Aeronautics and Astronautics, Nanjing 210016, China
{mkhuzozulu,tiexin.wang}@nuaa.edu.cn

Abstract. Bridge deterioration poses critical safety risks, with recent collapses resulting in casualties and economic losses. While automated monitoring offers advantages over dangerous and subjective manual inspections, current mainstream deep learning based methods struggle with small defect detection, lack measurement capabilities and are computationally intensive. This study presents an Artificial Intelligence of Things method, named BridgeDefectIQ, that addresses these limitations through a modified YOLOv8 model with a strategically positioned Coordinate Attention and P2 direction path. BridgeDefectIQ not only detects defects but quantifies them, extracting crack dimensions and spallation measurements that provide actionable engineering insights. Validated on an industry-level dataset (1,358 images with 4,024 annotations across five defects), BridgeDefectIQ achieves better performance on balancing accuracy with efficiency, e.g., outperformed all baselines with improvements in precision +3.0%.

Keywords: bridge defects · defect detection · defect quantification

1 Introduction

Bridges are critical infrastructure elements that connect communities across natural and artificial barriers. With over 42,000 U.S. Bridges rated in poor condition[1] maintaining these structures is vital for public safety and economic stability [1]. Recent bridge collapses highlight the devastating consequences of inadequate maintenance. Most notably, the Genoa bridge collapse resulted in 43 deaths and €500 million in damages due to overlooked bridge defects[2] These tragedies emphasize the critical importance of regular and thorough inspections.

Manual visual inspection remains the primary method for bridge maintenance. While inspectors use tools like binoculars and cherry pickers to examine bridge components, this traditional method is inherently subjective, labor-intensive, and potentially dangerous for the personnel involved, especially when

[1] Roads & Bridges, Federal data shows progress in repairing America's bridges, https://www.roadsbridges.com/iija/news/55138223/federal-data-shows-progress-in-repairing-americas-bridges,.

[2] BBC, Genoa bridge disaster: Risk of collapse was known for years https://www.bbc.com/news/world-europe-65684334.

S. Yang et al. (Eds.): WISA 2025, LNCS 16299, pp. 139–151, 2026.
https://doi.org/10.1007/978-981-95-6426-2_12

accessing high or confined spaces [2]. Additionally, traditional manual inspections are infrequent and prone to inconsistency, often failing to detect defects until they become severe [3]. Some of the limitations mentioned have been addressed by using drones equipped with high-resolution cameras enabling remote bridge inspections, reducing safety risks, time, and costs compared to manual methods [4]. However, the vast amounts of data generated requires significant expert interpretation, as automated analysis methods are still limited [5]. While deep learning (DL) has emerged as a promising alternative to manual visual inspections, its application to bridge defect detection faces significant challenges. Current DL models struggle to identify small or complex defects, especially in scenarios with overlapping features or subtle visual cues [6]. Meanwhile, the computational demands of state-of-the-art models, e.g., Faster-CNN, often conflict with the real-time requirements of practical field deployment. Additionally, limited multi-class defect datasets restrict model generalization and real-world applicability.

Traditional bridge inspection methods are inefficient and dangerous. Drones offer safer data collection, the challenge lies in developing an Artificial Intelligence of Things (AIoT) method that can effectively process this data. An ideal solution requires computer vision models that are simultaneously lightweight for edge deployment, accurate in detecting subtle defects and capable of quantifying multiple defects. Our work addresses these challenges by developing CoordYOLO a modified YOLOv8s model optimized for deployment within drone-based bridge inspection systems and extended to quantify specific defects. For instance, critical crack attributes such as width.

Our study makes the following key contributions:

- We present CoordYOLO, integrating CA and a P2 direction path to YOLOv8s for bridge defect detection.
- We propose BridgeDefectIQ, a method for both detection and quantification of bridge defects.
- Comprehensive evaluation against six state-of-the-art models and cross-dataset testing demonstrates superior performance and generalization capability.
- The dataset, along with our open-source implementations of BridgeDefectIQ, is publicly available at http://bit.ly/3GBImAK.

This study is outlined as follows, Sect. 2 describes the related work, the BridgeDefectIQ method is briefly explained in Sect. 3. The experiment evaluation is given in Sect. 4. Finally, we conclude our work in Sect. 5.

2 Related Work

2.1 Automated Detection of Bridge Defects

Object detection models have been used to automate the detection of defects. Hacıefendioğlu et al. [7] applied Faster-RCNN for crack detection with high accuracy but significant computational demands. YOLO-based implementations by

Teng et al. [8] offered improved efficiency but struggled with complex bridge defects. Recent attention enhanced approaches aim to address models struggling with complex bridge defects. Wang et al. [9] developed YOLOv8-CDD integrating convolutional neural networks with transformers. While Wang et al. [10] enhanced YOLOv7 with ACmix attention mechanism for small defect detection.

Although these studies improve model accuracy for challenging defect scenarios, they fail to adequately address computational efficiency and defect quantification, significantly limiting their practical viability in real-world settings.

2.2 Quantification of Defects

While significant progress has been made in detecting defects, fewer studies have focused on accurately quantifying them. Existing quantification methods primarily aim to measure properties such as crack width using image processing, Berrocal et al. [11] employed detailed image analysis techniques to evaluate internal crack morphology in reinforced concrete. Carrasco et al. [12] proposed a crack width quantification method using anisotropic filtering, segmentation, and profile width estimation aligned with the crack curvature. Wang et al. [13] introduced a digital image-based method combining an improved Otsu thresholding technique and perspective correction to identify and quantify damage areas. Although these methods can measure fine cracks and spallation with high precision, they typically focus on a single attribute, which is insufficient for practical use, for example, most bridge standards require both area and depth measurements to accurately assess the severity of a spallation.

Despite significant advancements in detection accuracy, state-of-the-art DL models tend to have substantial computational requirements and slow inference speeds. These factors limit the models practicality. To address these competing demands, we modify YOLOv8s to achieve an optimal balance between detection performance and computational efficiency. We also implement various image processing algorithms to quantify detected defects, namely cracks and spallation.

3 Method

BridgeDefectIQ consists of two main parts, defect detection using a modified YOLOv8s and defect quantification using image processing techniques. As shown in Fig 1, the detection part identifies five types of bridge defects, while the quantification part provides dimensional measurements specifically for cracks and spallation defects.

3.1 Defect Detection

We present CoordYOLO, a modified YOLOv8s comprising the CSPNet backbone, CA module and P2 direction path in Fig 2. The CA module integrates positional information into channel attention, creating direction-aware and position-sensitive feature maps [14], ideal for bridge defects' directional patterns such

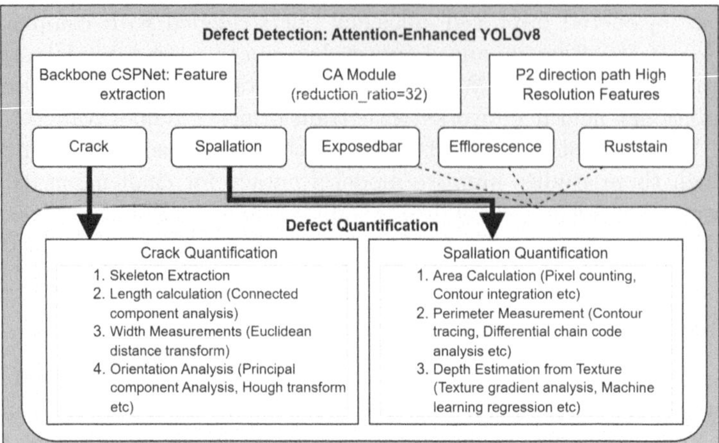

Fig. 1. BridgeDefectIQ: Bridge defect detection and quantification method.

as, elongated cracks, linear exposed bars, and irregular spallation boundaries. This coordinate-specific processing outperforms channel only and/or combined spatial-channel attention by preserving spatial relationships essential for distinguishing visually similar defects.

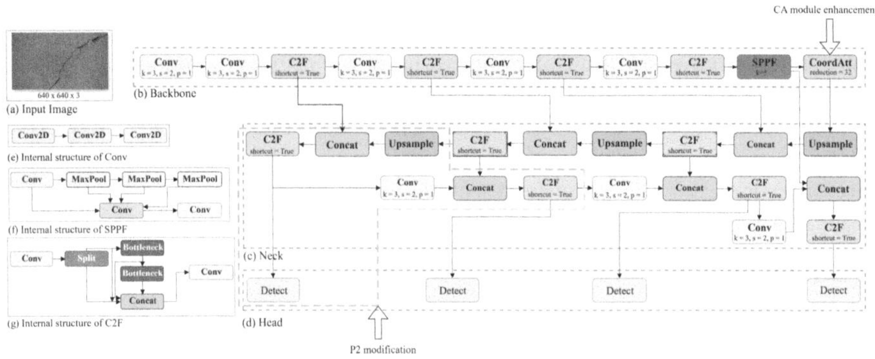

Fig. 2. CoordYOLO: Modified YOLOv8s with CA module and P2 direction path.

CSPNet Backbone. CoordYOLO builds on the YOLOv8s model by incorporating a CA module to enhance spatial feature representation. The model maintains the standard backbone components including Convolutional Layers, Cross-Stage Partial Networks (C2F) and Spatial Pyramid Pooling Fast (SPPF) modules. What distinguishes CoordYOLO is the integration of the CA module after the SPPF module, which improves feature extraction by separately processing spatial information along height and width dimensions, as well as the P2 direction path added in the neck of the model.

CA Module Enhancement. We incorporate an attention module to enhance detection performance, as attention modules have proven effective across various tasks [15]. We integrate a CA module after the SPPF layer in the backbone to enhance defect detection performance, as illustrated in Fig 2. The CA module decomposes spatial attention into separate horizontal and vertical coordinate information processing streams, which proves particularly advantageous for bridge defect detection. CA improves CoordYOLO's ability to distinguish between visually similar bridge defects and accurately localize their boundaries in complex concrete surface textures. The directional encoding process encompasses two adaptive average pooling operations that extract spatial information along both height and width dimensions:

Horizontal pooling:

$$x_h^c(i) = \frac{1}{W} \sum_{j=1}^{W} x^c(i,j) \tag{1}$$

Vertical pooling:

$$x_w^c(j) = \frac{1}{H} \sum_{i=1}^{H} x^c(i,j) \tag{2}$$

where $x_h^c(i)$ and $x_w^c(j)$ represents the horizontal and vertical pooled feature for c, respectively. W is the width feature map and H is the height feature map. $\sum_{j=1}^{W}$ and $\sum_{i=1}^{H}$ indicates summation across all width and height positions from 1 to W and 1 to H respectively. $x^c(i,j)$ denotes the original feature value at spatial position (i,j) in channel c, finally $\frac{1}{W}$ and $\frac{1}{H}$ are the horizontal and vertical normalization factor that averages the summed values.

The concatenated features are processed through a shared bottleneck to reduce the channel dimension. This is followed by direction specific attention where the processed features are split and transformed into separate attention maps:

Horizontal:

$$a_h^c(i) = \sigma \left(\sum_{c'=1}^{C'} W_{c,c'}^h \cdot x_h^{c'}(i) + b_c^h \right) \tag{3}$$

Vertical:

$$a_w^c(j) = \sigma \left(\sum_{c'=1}^{C'} W_{c,c'}^w \cdot x_w^{c'}(j) + b_c^w \right) \tag{4}$$

where $a_h^c(i)$ and $a_w^c(j)$ represent horizontal and vertical attention weight at position i for channel c. σ denotes the sigmoid activation function that normalizes values to the range [0,1]. $\sum_{c'=1}^{C'}$ indicates summation across all intermediate channels from 1 to C'. $W^h_{c,c'}$ and $W^w_{c,c'}$ are the learnable weight parameter connecting the input channel C' to the output channel c for horizontal and vertical attention respectively. $x^{c'}_h(i)$ and $x^{c'}_w(j)$ refers to the horizontally and vertically pooled feature. b^h_c and b^w_c is the learnable bias parameter for channel c in the horizontal and vertical attention mechanism. finally, c' represents the total number of intermediate channels after dimension reduction. The CA module is integrated after SPPF to enhance the highest-level features before they enter the neck, allowing subsequent layers to benefit from spatially-enhanced representations.

The CA module helps CoordYOLO focus on the directional patterns characteristic of bridge defects such as linear cracks and exposed reinforcement, while preserving spatial information critical for distinguishing between visually similar bridge defects.

P2 Modification. The neck integrates a Feature Pyramid Network (FPN) and Path Aggregation Network (PAN) architecture to enhance multi-scale feature representation. Our modified design extends the standard YOLOv8 neck by incorporating an additional P2 feature level with 160 × 160 × 64 resolution, providing finer-grained feature maps critical for small defect detection. The FPN component follows a top-down path, progressively upsampling from P5 (20×20×512) to P4 (40×40×256), P3 (80×80×128), and finally P2 (160×160×64), allowing higher-level semantic information to guide lower-level detailed features. The PAN component then follows a bottom-up path, enhancing localization precision by propagating fine-grained spatial information back through the feature hierarchy.

3.2 Defect Quantification

The defect quantification module converts detection results into dimensional measurements using specialized algorithms for cracks and spallation. These defects have established measurement standards in bridge inspection guidelines, while efflorescence and rust stains are assessed through presence or absence and area coverage already captured by detection bounding boxes. We detail the techniques for measuring crack width and spallation area and depth.

Crack Quantification. After CoordYOLO identifies crack regions, we apply Zhang-Suen skeletonization to extract the medial axis a one-pixel-wide path representing the crack's center while preserving its topological structure. The skeleton S(x,y) is computed as:

$$S(x, y) = \lim_{n \to \infty} T^n(M(x, y)) \tag{5}$$

where $S(x, y)$ is the resulting function after the transformation. $\lim_{n\to\infty}$ represents the limit as n approaches infinity. T^n is the transformation operator T applied n times and $M(x, y)$ is the initial function or matrix being transformed.

For width measurements, we employ the distance transform method. The distance transform $D(x, y)$ computes for each pixel in the skeleton, the distance to the nearest boundary pixel, this is shown as:

$$D(x, y) = \min_{(i,j) \in \Omega} \{d((x, y), (i, j)) \mid M(i, j) = 0\} \quad (6)$$

where Ω is the set of all pixel coordinates in the image, $M(i, j)$ is the binary mask, and $d((x, y), (i, j))$ is the Euclidean distance between points (x, y) and (i, j).

$$d((x, y), (i, j)) = \sqrt{(x - i)^2 + (y - j)^2} \quad (7)$$

For each skeleton pixel at position (x, y), the distance transform $D(x, y)$ computes the minimum Euclidean distance to the nearest boundary pixel.

Spallation Quantification. For spallation quantification, we start by processing the binary mask of the detected spalled region to compute geometric properties. The area A of the spallation is calculated by counting the number of pixels in the mask:

$$A = \sum_{(x,y)} M(x, y) \quad (8)$$

This pixel count is converted to physical area (square centimeters) using the calibration factor:

$$\text{Area (cm}^2\text{)} = \frac{A}{(\text{pixel\_to\_cm\_ratio})^2} \quad (9)$$

Estimating the depth of spallation from 2D images is challenging but essential for severity assessment. We implement a texture-based approach that leverages shadow information and texture gradients to approximate depth. First, we compute the intensity gradient $G(x, y)$ within the spalled region:

$$G(x, y) = \sqrt{G_x(x, y)^2 + G_y(x, y)^2} \quad (10)$$

where Gx and Gy are the Sobel operators in the x and y directions. We then analyze the intensity distribution $I(x, y)$ within the spalled region, focusing on the variation between lighter and darker areas. The depth estimation model is based on the observation that deeper spalls tend to exhibit greater intensity variance, stronger shadow gradients and more complex texture patterns.

We compute a depth estimate $\hat{D}(x, y)$ as:

$$\hat{D}(x, y) = \alpha \times \sigma(I(x, y)) + \beta \times G(x, y) + \gamma \times \text{LBP}(x, y) \quad (11)$$

where $\sigma(I(x, y))$ is the standard deviation of intensities, $\text{LBP}(x, y)$ is the local binary pattern texture complexity, and α, β and γ are weighting coefficients. This estimated depth map is averaged to provide a single depth value for the spalled region, reported in centimeters after appropriate scaling.

The coefficients were calibrated using 10 manually measured spallation samples with ground truth depths obtained via digital depth gauge (± 0.5 mm accuracy). Multiple linear regression with 5-fold cross-validation yielded $\alpha = 0.42$, $\beta = 0.31$ and $\gamma = 0.27$, representing the relative contributions of shadow-based intensity variance, edge gradients and surface texture complexity, respectively. The calibration dataset spanned 5–50 mm depths, covering typical spallation severities.

4 Empirical Analysis and Results

We conduct a series of experiments to address the following research questions (RQs):

- **RQ1:** How does CoordYOLO's detection performance compare to state-of-the-art object detection models for bridge defect detection?
- **RQ2:** How do the model modifications affect computational efficiency?
- **RQ3:** Does BridgeDefectIQ outperform other defect quantification methods?

Through these research questions, we aim to evaluate both the performance gains and the computational trade-offs of our proposed model enhancements.

4.1 Experiment Settings

Implementation Details. All experiments were conducted on Ubuntu 18.04 using PyTorch on an Nvidia GTX 1080Ti GPU with 12 GB memory. using the AdamW optimizer with an initial learning rate of 0.001 and a batch size of 16. We employed standard data augmentation techniques including random horizontal flipping, rotation and mosaic augmentation.

Dataset. CoordYOLO was validated on a self-created dataset. The dataset combines images from China Railway Engineering Group, containing 1358 images with 4024 annotations across five bridge defects: Crack (943), Efflorescence (520), Exposedbar (1123), Ruststain (581) and Spallation (857). Annotations were converted to YOLO and COCO JSON formats. The data was split 75/15/10 for training/validation/testing.

Baselines. We evaluate CoordYOLO against eight models representing key detection paradigms: Faster R-CNN (two-stage), RT-DETR (transformer-based), YOLO variants (YOLOv5, YOLOv8s/m/l) for single-stage detection at different scales, and attention enhanced variants such as, YOLOv8s-SE (channel attention) and YOLOv8s-CBAM (channel and spatial attention). YOLOv8s serves as our direct baseline, enabling comparison of different architectural choices and attention mechanisms.

Evaluation Metrics. We evaluated models using two metric categories, effectiveness and efficiency metrics to assess both model performance and computational demands.

Model Effectiveness Metrics. We evaluated model performance using precision, recall, mean average precision (mAP) and F1-score (F1) metrics. Average precision (AP) is calculated by summing the products of precision and recall changes at each detection threshold, where precision represents the accuracy at each cut-off point and the recall change indicates the improvement in detection coverage between consecutive thresholds. The mAP extends this by averaging AP values across all defect categories, providing a comprehensive measure of detection performance across the entire dataset.

Model Efficiency Metrics. To evaluate model efficiency, we use inference time, Frames Per Second (FPS), number of parameters and Giga Floating-Point Operations Per Second (GFLOPs). FPS measures the processing throughput of a computer vision model, representing the number of images or video frames the model can process in one second. GFLOPs measures the computational complexity by counting floating-point operations (in billions) required per input, with lower values indicating more efficient models.

To validate our quantification method, we developed a synthetic dataset with precise ground truth values. While existing datasets provide mask annotations, we deliberately chose not to use these for validation. The rationale behind this decision was to avoid potential circularity in our evaluation process, any formula we might use to extract measurements from these masks could inadvertently mirror the estimation methods we were attempting to evaluate. Using synthetic data with known values, we established a truly independent benchmark that ensures the objective evaluation of our quantification method. To ensure fair comparison, all methods with calibration settings were set to 1, matching methods without such parameters. We evaluated performance using Mean Absolute Error (MAE), which measures the average absolute difference between predicted and observed values.

4.2 Results and Analysis

Comparative Performance Analysis. Our performance results on the self-created dataset are presented in Table 1, the table demonstrates the comparative model performance of several state-of-the-art object detection models. The models evaluated include Faster R-CNN, RT-DETR, multiple YOLO variants and CoordYOLO. The evaluation metrics include precision, recall, mAP@50 and F1.

Table 1. Comparative performance of object detection models

Model	Backbone	Input	Precision	Recall	mAP@50	F1
Faster R-CNN	ResNet101	1333×800	0.83	0.72	0.82	0.77
RT-DETR	HGNet	640×640	0.84	0.80	0.83	0.82
YOLOv5	CSPDarknet	640×640	0.74	0.78	0.81	0.76
YOLOv8s	CSPNet	640×640	0.87	0.79	0.83	0.83
YOLOv8s-SE	CSPNet	640×640	0.85	0.80	0.85	0.82
YOLOv8s-CBAM	CSPNet	640×640	0.86	0.74	0.81	0.80
YOLOv8m	CSPNet	640×640	0.82	0.79	0.83	0.81
YOLOv8l	CSPNet	640×640	0.82	0.78	0.84	0.80
CoordYOLO	CSPNet	640×640	**0.90**	**0.83**	**0.86**	**0.86**

Each model exhibits distinct performance characteristics. Faster R-CNN's two-stage architecture achieves high precision (0.83) but lower recall (0.72), missing more subtle defects. RT-DETR demonstrates balanced performance (0.82 F1) with a transformer-based design its precision (0.84) suggests some false positives. YOLOv5's lower precision (0.74) indicates difficulty distinguishing defects from surface irregularities. Among single-stage detectors, YOLOv8s achieves the highest baseline precision (0.87). Attention-enhanced variants show varied improvements: YOLOv8s-SE with channel attention achieves respectable precision (0.85), while YOLOv8s-CBAM's channel and spatial attention yields higher precision (0.86). CoordYOLO surpasses all models with 0.90 precision and 0.83 recall through its CA module and P2 path demonstrating the superiority of coordinate-aware processing over channel and/or spatial attention approaches for bridge defect detection, addressing RQ1.

Ablation Study. Table 2 presents our ablation study using YOLOv8s as a baseline achieving a precision of 0.87, recall of 0.79 and mAP@50 of 0.83. Adding CA alone (YOLOv8s+CA) decreased performance. Precision dropped to 0.86 and recall to 0.78, we suspect this is because of the architectural mismatch. The CA module's spatially-enhanced features require appropriate resolution matching. Without P2's high-resolution paths, feature misalignment occurs. However, combining both modifications in CoordYOLO creates synergy, the P2 path's 160×160 feature maps enable CA's directional attention to function optimally, resulting in significant improvements in precision (0.90), recall (0.83) and mAP@50 (0.86). This demonstrates that P2 and CA are complementary components, with P2 providing the resolution necessary for CA to effectively detect small defects, providing sufficient justification for RQ1.

Table 2. Ablation study

Model variant	Precision	Recall	mAP@50	F1
Baseline-YOLOv8s	0.87	0.79	0.83	0.83
YOLOv8s+CA	0.86	0.78	0.84	0.83
YOLOv8s+P2	0.81	0.79	0.84	0.80
CoordYOLO	**0.90**	**0.83**	**0.86**	**0.86**

Comparative Computational Analysis. Our comparative computational results are presented in Table 3, The evaluation metrics include inference time, FPS, and model complexity in terms of parameters and GFLOPs. Table 3 shows Faster-RCNN, RT-DETR, and YOLOv8l require over 100 GFLOPs, 30M parameters, and inference times of more than 40 ms, limiting the model's practicality. Addressing RQ2, CoordYOLO achieves superior efficiency (13 ms, 74 FPS) through optimization techniques including feature map pruning to reduce memory bandwidth and mixed precision FP16 inference. These optimizations enable real-time performance despite additional architectural components, demonstrating that enhanced detection capability need not compromise computational efficiency.

Table 3. Comparative computational efficiency of object detection models

Model	Infer(ms)	FPS	Params(M)	GFLOPs
Faster R-CNN	70	14	46	458
RT-DETR	61	16	32	104
YOLOv5	26	39	25	65
YOLOv8s	14	73	**11**	28
YOLOv8s-SE	17	63	**11**	37
YOLOv8s-CBAM	17	62	**11**	37
YOLOv8m	23	44	26	79
YOLOv8l	63	16	44	165
CoordYOLO	**13**	**74**	**11**	**26**

Quantification Validation. Our quantification method was validated on a synthetic datasets, we designed the dataset to include varying lighting conditions and shadows, multiple texture patterns mimicking concrete surfaces, and noise and blur effects simulating field conditions. The results are presented in Table 4, the tables demonstrate the comparative performance of three other quantification methods from Carrasco [12], Berrocal, [11] and Wang [13]. The methods were evaluated against our proposed quantification method. MAE was used as the evaluation metric.

Table 4. Comparison of defect quantification methods

Method	Crack width	Spallation area	Spallation depth
Carrasco	3.2701	-	-
Berrocal	3.2701	-	-
Wang	-	9.0195	-
Our method	**3.2640**	**9.0121**	**2.2688**

In Table 4, dashes indicate methods that don't output certain attributes. Carrasco and Berrocal methods only measure crack width, both achieving MAE of 3.2701 using the same formula: $width = 2 * dist_{transform}[y, x]$. Our method improves to 3.2640 MAE. For spallation, Wang's method only estimates area with a MAE of 9.0195, while our method quantifies both area with a MAE of 9.0121 and depth MAE of 2.2688, Our quantification method simultaneously outputs critical attributes required by bridge inspection standards. These results demonstrate superior accuracy and practical utility, supporting RQ3.

5 Conclusion

This study combines an enhanced object detection model with specialized quantification algorithms to address critical gaps in current inspection practices. By modifying YOLOv8s with a CA module and a P2 direction path, CoordYOLO effectively captures the directional patterns characteristic of bridge defects while maintaining computational efficiency suitable for field deployment. The quantification algorithms transform visual detections into engineering metrics that provide more value for informed maintenance decisions.

References

1. McLaughlin, E., Charron, N.: Automated defect quantification in concrete bridges using robotics and deep learning. J. Comput. Civil Eng. (2020)
2. PLafiosca, N., Fan, I.-S.: Review of non-contact methods for automated aircraft inspections. Insight (2020)
3. Yin, T., Shen, G., Yin, L., Shi, G.: Bridge surface defect localization based on panoramic image generation and deep learning-assisted detection method. Buildings (2024)
4. Olaszek, P., Maciejewski, E., Rakoczy, A., Cabral, R., Santos, R.A., Ribeiro, D.: Remote inspection of bridges with the integration of scanning total station and unmanned aerial vehicle data. Rem. sens. (2024)
5. Zampokas, G., et al.: A UAV-captured dataset for semantic segmentation of energy infrastructure. In: 2022 International Conference on Unmanned Aircraft Systems (ICUAS) (2022)
6. Chen, C.-W., Tserng, H.P., Chiang, E.: Preliminary study on enhancing detection of concrete bridge surface spalling by infrared thermography. Proc ISARC (2024)

7. Kemal Hacıefendioğlu and Hasan Baçağa. Concrete road crack detection using deep learning-based faster R-CNN method. Iranian J. Sci. Technol. Trans. Civil Eng. (2021)
8. Teng, S., Liu, Z., Chen, G., Cheng, L.: Concrete crack detection based on well-known feature extractor model and the $yolo_v2$ network. Appl. Sci. (2021)
9. Wang, C., et al.: Yolov8-CDD: an improved concrete defect detection method combined CNN with transformer. Measure. Sci. Technol. (2024)
10. Wang, Z., Chen, Z., Abba, A.A.: An improved yolov7 based concrete defect detection method. In: 2024 7th International Conference on Pattern Recognition and Artificial Intelligence (2024)
11. Carlos, G.: Berrocal, Ingemar Löfgren, Karin Lundgren, Niclas Görander, and Christopher Halldén. Characterisation of bending cracks in r/frc using image analysis, Cement and Concrete Research (2016)
12. Carrasco, M., Araya-Letelier, G., Velázquez, R., Visconti, P.: Image-based automated width measurement of surface cracking. Sensors (2021)
13. Wang, J., Liu, H., Han, Z., Wang, Y.: Automatic identification method of bridge structure damage area based on digital image. Sci. Rep. (2023)
14. Hou, Q., Zhou, D., Feng, J.: Coordinate attention for efficient mobile network design. In: 2021 IEEE/CVF Conference on Computer Vision and Pattern Recognition (CVPR) (2021)
15. Chen, J., Meng, D., Gao, X., Zhang, L., Kong, C.: Agne: attentional graph convolutional network embedding for knowledge concept recommendation in MOOCS. In: Web Information Systems and Applications (2024)

DCBF: Deep Convolutional Boosted Forest for PM2.5 Concentration Inversion with Multi-source Data

Jiaqi Wu[1,2], Kewen Xia[1,2], and Shurui Fan[1,2](✉)

[1] Innovation and Research Institute, Hebei University of Technology, Shijiazhuang 050299, China
fansr@hebut.edu.cn

[2] School of Electronic and Information Engineering, Hebei University of Technology, Tianjin 300401, China

Abstract. $PM_{2.5}$ is a major airborne pollutant affecting air quality and public health and exhibits complex spatiotemporal patterns driven by multi-factor interactions. We propose Deep Convolutional Boosted Forest (DCBF), an enhanced deep learning framework integrating multi-source data including satellite observations, meteorological records, and auxiliary data for seamless day-night hourly $PM_{2.5}$ inversion. The method replaces conventional multi-grained scanning with an Efficient Channel Attention-enhanced 1D convolutional neural network (1DCNN-ECA) to improve local feature extraction, while employing CatBoost to reconstruct cascade forest structures for enhanced global feature interactions and generalization capability. Experimental results demonstrate that DCBF shows significant improvement over baseline models. When applied to analyze a heavy pollution episode in Shijiazhuang, DCBF successfully captured the spatiotemporal evolution patterns of $PM_{2.5}$, validating its practical utility in environmental monitoring.

Keywords: PM2.5 concentration inversion · One dimensional convolutional neural network · Efficient channel attention · Multi grained cascade forest · Multi source data

1 Introduction

Fine-grained, continuous mapping of $PM_{2.5}$ concentration is critical for air quality monitoring, yet existing ground stations are sparse [1]. Traditional retrieval methods rely on polar-orbit Aerosol Optical Depth (AOD) data and geographic proxies, which suffer from spatial gaps and low temporal resolution. We leverage geostationary FY-4A Top-of-Atmosphere reflectance [2] and integrate nighttime VIIRS light data with high-frequency MERRA-2 AOD to achieve seamless, all-time $PM_{2.5}$ inversion [3].

Deep learning techniques have made significant progress in multi-scale representation, multi-source data fusion, and efficient feature extraction, driving advances in $PM_{2.5}$ inversion and complex spatiotemporal prediction tasks. In the domain of 1D

S. Yang et al. (Eds.): WISA 2025, LNCS 16299, pp. 152–160, 2026.
https://doi.org/10.1007/978-981-95-6426-2_13

convolutional networks and physics-informed modeling, Meng et al. [4] developed the ST-KMRN model, which integrates temporal convolutional networks, graph convolution encoders, and Koopman operator-based forecasting. Their approach employs a two-layer 1D CNN gate to effectively combine physical priors with deep representations, achieving superior accuracy and generalizability in long-sequence predictions for multi-resolution spatiotemporal data. In attention mechanism research, Wang et al. [5] introduced the Efficient Channel Attention (ECA) module, a lightweight yet powerful solution that uses adaptive 1D convolutions to capture channel-wise dependencies without dimensionality reduction, directly inspiring the design of the 1DCNN-ECA component in DCBF. For multi-source data fusion, Yu et al. [6] proposed a cascaded architecture with cross-modal attention and temporal adaptive layers for precipitation forecasting, while Bai et al. [7] designed Dust-Mamba, a multimodal network leveraging Mamba-style architectures and selective attention to effectively integrate satellite and ground-based meteorological data. In ensemble learning, Grotmol et al. [8] presented an enhanced CatBoost framework incorporating feature engineering and non-negative least squares for output fusion, offering valuable insights for $PM_{2.5}$ modeling optimization. Deep forest models are widely used for regression tasks in ensemble learning. However, their cascade random forests are sensitive to outliers, leading to reduced accuracy [9, 10]. Collectively, these advancements provide critical methodological foundations for the DCBF model's development across its 1D CNN architecture, attention mechanisms, multi-source integration strategies, and ensemble learning approaches. Shap value analysis method is widely used in model verification.

2 Multi-source Dataset Construction

We conduct hourly seamless inversion of $PM_{2.5}$ using Shijiazhuang City as the study area. The required data include ground-level $PM_{2.5}$ concentration data, satellite data, meteorological data, and auxiliary data, all of which need to be unified to the same spatiotemporal resolution. The details of each dataset are shown in Table 1.

The ground-level $PM_{2.5}$ data were obtained from hourly concentration measurements at 50 national and provincial monitoring stations in the Shijiazhuang area. For daytime $PM_{2.5}$ inversion, this study utilized Bands 1, 2, and 6 of the FY-4A satellite along with four angular datasets. For nighttime inversion, the Day/Night Band (DNB) of NPP/VIIRS nighttime light remote sensing data was employed. Hourly AOD data were sourced from the MERRA-2 reanalysis dataset. Meteorological data were derived from the ERA5 reanalysis dataset, with 10 selected variables used in this study. Auxiliary data included the Normalized Difference Vegetation Index (NDVI), ground elevation data, and population density data. The construction process of multi-source dataset is shown in Fig. 1.

Table 1. Details of the data.

Dataset	Variable	Temporal resolution	Spatial resolution
$PM_{2.5}$	$PM_{2.5}$	Hourly	Site
Satellite data (daytime)	TOA	Hourly	0.02° × 0.02°
	SAZ,SAA,SOZ,SOA		0.04° × 0.04°
Satellite data (nighttime)	DNB	Daily	500 m
AOD	MERRA-2 AOD	Hourly	0.5° × 0.625°
Meteorological factors	U/V10M, BLH, SP, TCW, TCO, T2M,RH,LH,LL	Hourly	0.25 × 0.25°
Auxiliary data	NDVI	5-Day	10 m
	DEM	Annually	30 m
	Population density	Annually	1 km

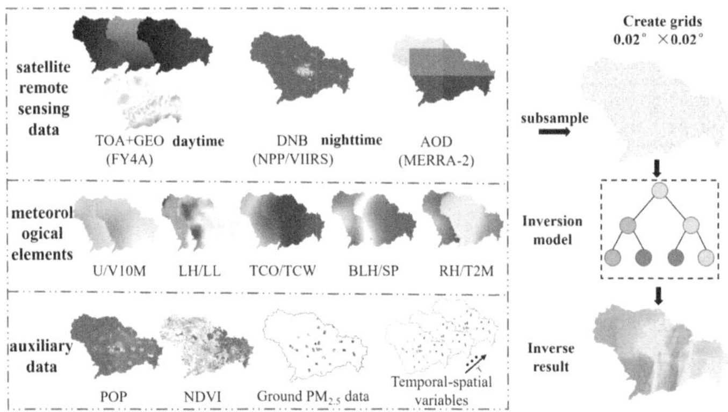

Fig. 1. The flowchart for constructing a multi-source dataset

3 Proposed Methodology

This section provides a detailed explanation of the model's approach. The overall framework of the model is illustrated in Fig. 2.

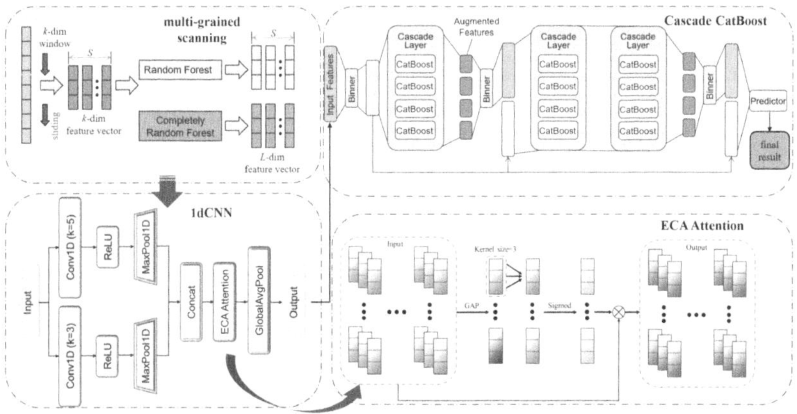

Fig. 2. The overall framework of DCBF

3.1 Multi-grained Scanning

Deep Forest is a tree-based ensemble algorithm where decision trees are grown by splitting nodes using optimal features, thereby eliminating the need for data normalization. The algorithm first employs multi-grained scanning to perform feature learning on the training set, followed by a cascade forest structure to learn from the processed data.

In the multi-grained scanning phase: Input data with p-dimensional features are processed using a sliding window of length k to extract features. With a stride of n, this generates S k-dimensional feature segments and the calculation method of S is as follows:

$$S = (p - k)/n + 1 \tag{1}$$

These segments are then fed into both random forest and completely-random forest models, producing S L-dimensional probability vectors. The cascade forest consists of multiple cascade layers, where each layer contains two Random Forests and two Completely-Random Tree Forests.

3.2 1DCNN Feature Extraction

To effectively capture local temporal patterns and multi-scale correlations among environmental features, we design a 1D Convolutional Neural Network module as a deep feature extractor. This component processes sequential tabular inputs using multiple parallel convolution branches with different kernel sizes, thereby enabling adaptive representation of patterns at various temporal scales. The extracted features are subsequently fused with original input features to support downstream modeling task.

(1) Input Representation

Given a sample with L features (or time steps), the input vector is denoted by: $\mathbf{x} \in \mathbb{R}^L$. For a batch of N samples, the input matrix is: $\mathbf{X} \in \mathbb{R}^{N\times L}$. This 1D sequence serves as the raw input for the convolutional layers.

(2) First Convolutional Layer: Multi-scale Encoding

The first layer employs multiple parallel 1D convolutional filters with different kernel sizes, such as $k \in \{3, 5\}$, to extract short-range and mid-range local patterns. For each kernel size k, the convolution operation followed by non-linear activation and max-pooling is given by:

$$\mathbf{h}_k^{(1)} = \text{MaxPool}\left(\sigma\left(\mathbf{W}_k^{(1)} * \mathbf{X} + \mathbf{b}_k^{(1)}\right)\right) \tag{2}$$

where: $*$ denotes the 1D convolution operation, $\mathbf{W}_k^{(1)}$ and $\mathbf{b}_k^{(1)}$ are the weights and bias of the convolution filter with kernel size k, $\sigma(\cdot)$ denotes the ReLU activation function, $\text{MaxPool}(\cdot)$ performs temporal downsampling.

(3) Second Convolutional Layer: Deeper Abstraction

Each convolutional branch continues with a second 1D convolution layer to further abstract higher-order features:

$$\mathbf{h}_k^{(2)} = \text{MaxPool}\left(\sigma\left(\mathbf{W}_k^{(2)} * \mathbf{h}_k^{(1)} + \mathbf{b}_k^{(2)}\right)\right) \tag{3}$$

The same kernel size k is maintained in both layers for consistency within each branch.

(4) Multi-Branch Feature Fusion

After the two convolutional layers, feature maps from all branches are concatenated along the channel dimension. Let C be the number of output channels from each branch, then:

$$\mathbf{H}_{\text{cat}} = \text{Concat}\left(\mathbf{h}_3^{(2)}, \mathbf{h}_5^{(2)}\right) \in \mathbb{R}^{N \times 2C \times T'} \tag{4}$$

where T' is the length of the sequence after two levels of down sampling. This concatenation integrates multi-scale features and strengthens temporal feature diversity.

(5) Global Average Pooling

To obtain a fixed-length vector for each sample, we apply Global Average Pooling (GAP) over the temporal dimension:

$$\mathbf{f} = \frac{1}{T'} \sum_{t=1}^{T'} \mathbf{H}_{\text{cat}}[:, t] \in \mathbb{R}^{2C} \tag{5}$$

3.3 ECA Attention

To enhance the model's ability to focus on informative feature channels, we incorporate the Efficient Channel Attention (ECA) module, a lightweight and parameter-efficient mechanism that selectively recalibrates channel-wise feature responses.

(1) Channel-wise Feature Aggregation

Let the feature map extracted from the 1D-CNN module be $\mathbf{H} \in \mathbb{R}^{C \times T}$, where C is the number of channels and T is the temporal length. ECA begins by applying GAP across the temporal dimension to obtain a compact channel descriptor: $\mathbf{z} \in \mathbb{R}^{C}$, where $z_c = \frac{1}{T}\sum_{t=1}^{T} H_{c,t}$.

(2) Local Cross-Channel Interaction via 1D Convolution

Instead of using fully connected layers, ECA employs a 1D convolution to capture local cross-channel dependencies:

$$\mathbf{s} = \sigma(\text{Conv1D}(\mathbf{z}, k)) \tag{6}$$

Here, $\text{Conv1D}(\cdot, k)$ denotes a 1D convolution with kernel size k, and $\sigma(\cdot)$ is the Sigmoid activation function, producing channel attention weights $\mathbf{s} \in \mathbb{R}^{C}$. The kernel size k is dynamically determined as a function of the number of channels:

$$k = \psi(C) = \left| \frac{\log_2 C}{\gamma} + b \right|_{\text{odd}} \tag{7}$$

where γ and b are constants, and $|\cdot|_{\text{odd}}$ rounds to the nearest odd integer to maintain symmetry.

(3) Feature Recalibration

The attention weights are applied multiplicatively to the original feature map across channels:

$$\widetilde{\mathbf{H}}_c = s_c \cdot \mathbf{H}_c, \quad c = 1, 2, \ldots, C \tag{8}$$

The output $\widetilde{\mathbf{H}} \in \mathbb{R}^{C \times T}$ represents the recalibrated feature map, which enhances informative channels and suppresses less relevant ones.

3.4 Improved GcForest

CatBoost is another robust machine learning algorithm based on Gradient Boosted Decision Tree (GBDT) that requires minimal preprocessing. It employs specialized tree structure adjustments to minimize prediction bias during training. The core idea is to partition the data during training, ensuring that the gradient of the current sample is not influenced by future samples. This can be mathematically represented as follows:

$$g_i = \frac{1}{|S_i|} \sum_{j \in S_i} L'(y_j, \hat{y}_j) \tag{9}$$

where g_i is the gradient estimate of the i-th sample, S_i is the set of samples preceding sample i in the data permutation, and L' is the derivative of the loss function with respect to the predicted value.

The improved cascade forest structure processes features through a Binner module for discretization, fusion, and selection, then passes them through sequential CatBoost layers that dynamically enhance features step-by-step. This design replaces traditional random forests with CatBoost to reduce noise and overfitting, supports GPU acceleration for faster training, and incorporates adaptive depth adjustment that automatically stops training when performance plateaus—optimizing efficiency and avoiding overcomputation.

4 Experiments and Analysis

4.1 Performance Comparison of Different Models

To evaluate the proposed model, we compare it with mainstream machine learning methods for satellite-based $PM_{2.5}$ retrieval. All models are trained on the same dataset. As shown in Table 2, a clear accuracy gap exists between deep neural networks (DNN) and ensemble models. Among ensemble methods, Random Forest reduces overfitting through dual randomization and outperforms CatBoost with default parameters. Deep Forest further lowers prediction errors and improves generalization, but its cascade layers propagate only class probabilities, leading to significant information loss. In contrast, Gradient Boosted Decision Trees (GBDT) learn nonlinear transformations between layers and retain richer features. The proposed DCBF model improves upon the baseline Deep Forest and achieves the best overall performance. Most models also show slightly higher accuracy at night, suggesting stronger regularity in nocturnal $PM_{2.5}$ concentrations.

Table 2. Performance metrics of different machine learning models for day-night inversion.

Model	R^2		RMSE($\mu g \cdot m^{-3}$)		MAE($\mu g \cdot m^{-3}$)	
	day	night	day	night	day	night
DNN	0.794	0.730	18.91	21.63	13.09	15.10
CatBoost	0.914	0.907	12.23	13.00	8.51	9.13
RF	0.940	0.944	10.23	10.03	6.97	6.84
GcForest	0.945	0.956	9.81	8.96	6.70	6.10
DF	0.951	0.959	9.20	8.57	6.31	5.84
CNNGcf	0.954	0.956	8.95	8.48	6.25	5.82
DCBF	**0.958**	**0.961**	**8.49**	**8.43**	**5.81**	**5.73**

4.2 Analysis of a Heavy Pollution of $PM_{2.5}$

By integrating spatiotemporally matched feature data from grid points across Shijiazhuang, the model achieves seamless day-night $PM_{2.5}$ inversion with full coverage. As shown in Fig. 3, the 4-interval concentration maps from 12:00 Oct. 28 to 08:00 Oct. 30,

2023 clearly track pollution evolution: accumulation in the east on the 28th, westward spread to the urban center, peak downtown concentrations on the evening of the 29th, and centralized clustering with peripheral improvement by the 30th.

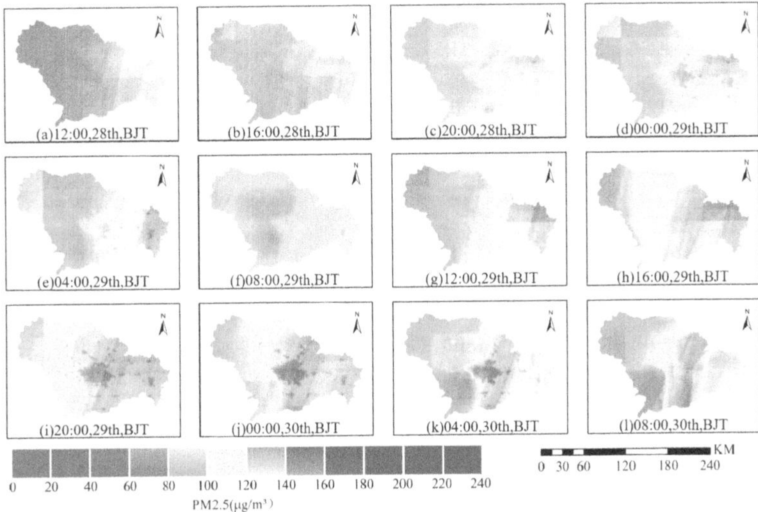

Fig. 3. Estimation of $PM_{2.5}$ concentration

5 Conclusion

In this study, we proposed DCBF, a Deep Convolutional Boosted Forest framework, for high-resolution, seamless day-night $PM_{2.5}$ concentration inversion using multi-source data. By replacing the traditional multi-grained scanning module with an Efficient Channel Attention-enhanced 1D convolutional neural network (1DCNN-ECA), the model improves local feature representation. The integration of CatBoost in the cascade structure further enhances global feature interactions and generalization performance. Extensive experiments confirm that DCBF outperforms existing baseline models in both accuracy and stability. Furthermore, a case study of a severe haze event in Shijiazhuang demonstrates DCBF's capability in capturing the spatiotemporal dynamics of $PM_{2.5}$, indicating its strong applicability for real-world environmental monitoring and air quality management.

Acknowledgments. This study was funded by Science and Technology Cooperation Special Project of Shijiazhuang (grant number SJZZXA24006).

Disclosure of Interests. The authors have no competing interests to declare that are relevant to the content of this article.

References

1. Yang, Q., et al.: Optimization of hourly PM 2.5 inversion model integrating upper-air meteorological elements. IEEE Access **11**, 35421–35428 (2023)

2. Shen, H., Li, T., Yuan, Q., Zhang, L.: Estimating regional ground-level PM2. 5 directly from satellite top-of-atmosphere reflectance using deep belief networks. J. Geophy. Res.: Atmos., **123**(24), 13–875 (2018)
3. Choo, G.H., Lee, K., Kim, G.: Estimation of nighttime PM2. 5 concentrations over Seoul using Suomi NPP/VIIRS day/night band. Atmos. Environ. **338**, 120861 (2024)
4. Meng, C., Wen, H., Wang, D., et al.: Physics-Informed long-sequence forecasting from multi-resolution spatiotemporal data. In: IJCAI 2022, pp. 406–414 (2022)
5. Wang, Q., Wu, B., Zhu, P., Li, P., Zuo, W., Hu, Q.: ECA-Net: efficient channel attention for deep convolutional neural networks. In: Proceedings of the IEEE/CVF Conference On Computer Vision And Pattern Recognition, pp. 11534–11542 (2020)
6. Yu, D., et al.: Integrating multi-source data for long sequence precipitation forecasting. In: Proceedings of the AAAI Conference on Artificial Intelligence, Vol. 39, No. 27, pp. 28539–28547 (2025)
7. Bai, C., Zhang, Y., Wang, W., et al.: Dust-Mamba: An efficient dust storm detection network with multiple data sources. In: AAAI 2025, pp. 1221–1229 (2025)
8. Grotmol, K., Pedersen, M., Langseth, H.: A robust and scalable stacked ensemble for day-ahead forecasting. In: AAAI 2023, pp. 1273–1280 (2023)
9. Wang, Y., Hu, J., Fan, X., et al.: Prediction Method of Type 2 Diabetes Mellitus Based on a Combination of Hybrid Feature Selection and Random Forest. In: Yuan, L., Yang, S., Li, R., Kanoulas, E., Zhao, X. (eds.) WISA 2024. LNCS, vol. 14883, pp. 439–450. Springer, Cham (2024)
10. Zhang, Y., Li, X., Wang, J., et al.: Improved Multi-Grained Cascade Forest Model for Transformer Fault Diagnosis. CSEE J. Power Energy Sys. **11**, 468–476 (2025)

Towards Real-World Event-Guided Motion Deblurring

Shuang Ma[1], Zhanwen Liu[1(✉)], Yang Wang[1], Shangyu Xie[1], Huanna Song[1], and Xing Fan[2]

[1] School of Information Engineering, Chang'an University, Xi'an 710018, China
{2023124091,zwliu}@chd.edu.cn

[2] School of Electronic and Control Engineering, Chang'an University, Xi'an 710018, China

Abstract. Image-based deblurring methods encounter difficulties in handling complex motion blur under real-world conditions, primarily due to significant information loss. Event-based methods alleviate this issue by using events to provide detailed motion information. However, discrepancies in spatial resolution between images and events pose a major challenge for event-based deblurring. To tackle this problem, we propose a Cross-Modal Scale-Aware Fusion Network (CSFNet), inspired by the fact that events predominantly capture high-contrast edge information. The network includes a Spatial Scale-Aware Super-Resolution Module (SASR), which adaptively aligns event resolution with image resolution by incorporating refined image features. Furthermore, a Global Interaction Fusion Module (GIFM) is introduced to improve cross-modal feature integration through global spatial aggregation and channel-wise correlation. Experimental evaluations on both synthetic and real-world datasets validate the effectiveness of our approach, which achieves a PSNR gain of 0.12 dB on the Ev-REDS dataset.

Keywords: Image deblurring · Event camera · Super resolution · Cross-modal fusion

1 Introduction

Event cameras [4], bio-inspired visual sensors, exhibit a high sensitivity to changes in illumination within a scene. They respond asynchronously to pixel-wise relative brightness changes and generate events with microsecond temporal resolution. These characteristics enable event cameras to capture fine-grained motion information, making them particularly suitable for motion deblurring.

Several studies [3,9,12,13] have utilized events to recover motion information lost in blurred images, thereby improving detail restoration. However, the asynchronous and sparse nature of events poses challenges to extract effective

Supported by organization x.

S. Yang et al. (Eds.): WISA 2025, LNCS 16299, pp. 161–169, 2026.
https://doi.org/10.1007/978-981-95-6426-2_14

motion cues directly. To address this challenge, existing methods [8–10] transform spatially sparse events into dense representations and employ advanced network architectures to extract high-quality motion features for cross-modal fusion. Some methods [9,13] further design specialized fusion strategies to integrate structural details provided by events with the color information of the images. Despite these efforts, most methods overlook the problem of inconsistent spatial resolution between event cameras and frame-based cameras in hybrid imaging systems.

To address this challenge, we propose a Cross-Modal Scale-Aware Fusion Network (CSFNet) that performs cross-scale and cross-modal fusion by combining scale-aware super-resolution with global interaction fusion. Leveraging the spatial consistency between event-triggered regions and high-contrast image edges, we design a Scale-Aware Super-Resolution (SASR) module to generate high-resolution event features. After aligning the resolution of event features with image features, we introduce a Global Interaction Fusion Module (GIFM) that integrates both modalities through spatial aggregation and channel interaction, utilizing the global modeling capabilities of Mamba and Transformer. The contributions are summarized as follows:

- We propose a Scale-Aware Super-Resolution (SASR) module to align the resolutions of events and images with uncertain spatial resolution discrepancy by leveraging the spatial correlation between event locations and high-contrast image edges.
- We introduce a Global Interaction Fusion Module (GIFM) that enables global feature integration by performing spatial aggregation within each modality and channel interaction across modalities.
- We validate the effectiveness of the proposed Cross-Modal Scale-Aware Fusion Network (CSFNet) on the Ev-REDS dataset and the HS-ERGB dataset with different spatial scales, where it achieves state-of-the-art performance.

2 Related Work

2.1 Image-Based Motion Deblurring

Image-based motion deblurring methods have achieved remarkable progress in recent years. Traditional approaches [1,17] employ manually designed constraints to explicitly estimate motion and address the inverse problem of restoring sharp images from blurred inputs. However, their effectiveness is constrained by the accuracy of motion estimation, which limits their applicability. With the advent of deep learning, research has shifted toward data-driven methods [7,11] that learn the mapping between blurred and sharp images using carefully designed networks. Despite significant advancements, most of these methods rely on paired training data and often fail in real-world scenarios under extreme motion.

2.2 Event-Based Motion Deblurring

Research on event-based motion deblurring has gained considerable attention owing to the ability of event cameras to capture fine-grained motion information. Pan *et al.* introduce the Event-Based Double Integral (EDI) model [8], which provides a theoretical model for reconstructing sharp images from blurry inputs and events. Building on this foundation, several approaches [3,9,12,13] have been proposed to design more effective cross-modal fusion strategies. Sun *et al.* [9] apply cross-modal attention to fuse event and image features. EIFNet [12] employs dual cross-attention to capture long-range interactions. TRMD [3] introduces a dual cross-modal fusion module to improve residual estimation. STCNet [13] adopts a modality-differential collaboration strategy to reduce redundancy. Although these methods have achieved notable progress, they generally assume consistent spatial resolution between events and images. To address spatial resolution discrepancies in real-world scenarios, Zhang *et al.* [16] propose a multi-scale feature fusion module combined with a spatial generalization loss. Despite these efforts, existing methods do not fully consider the distribution characteristics of event data.

3 Method

3.1 Problem Statement

We formulate event-based motion deblurring as the reconstruction of a latent sharp image $L(t)$ from a blurred image B_T and the corresponding event stream $E(\mathcal{T})$ with unknown spatial resolution discrepancies. The spatial scale $R_s(\cdot,\cdot)$ is defined as the spatial resolution ratio between the blurred image and the events, which strongly affects the quality of restoration. Larger values of R_s make it more challenging to obtain high-resolution event representations for cross-modal fusion. Unlike prior methods that typically assume $R_s = 1$, we generalize the task to unknown spatial scales as:

$$L(t) = Deblur(t; B(T); E_{R_s}(\mathcal{T})), \quad \forall\, t \in \mathcal{T},\ R_s = \frac{S_B}{S_E}, \tag{1}$$

where $E_{R_s}(\mathcal{T})$ refers to the event representation at the scale R_s, and S_B and S_E denote the spatial resolutions of blurry image and events, respectively.

3.2 Proposed Framework

The framework of the proposed Cross-Modal Scale-Aware Fusion Network (CSFNet) is illustrated in Fig. 1. During the feature encoding stage, a blurry image B_T and an event voxel grid E_G are processed by separate encoders. A two-step fusion strategy is then employed in the cross-modal fusion stage: low-resolution event features F_{EL} are first aligned in the Scale-Aware Super-Resolution Module (SASR), and subsequently fused with high-resolution image features F_B through the Global Interaction Fusion Module (GIFM). Finally, a sharp image is reconstructed in the feature decoding stage.

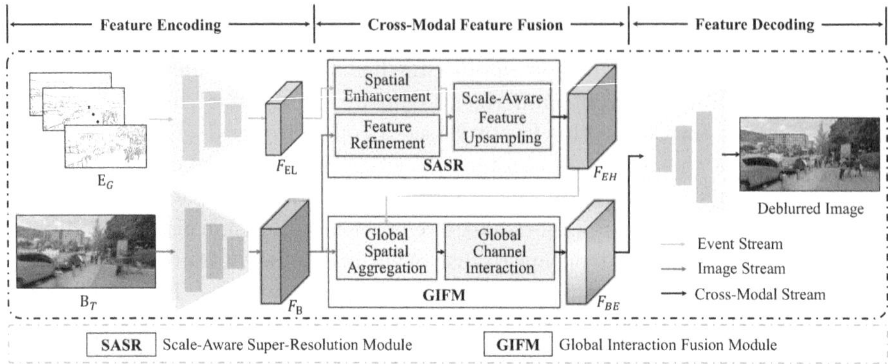

Fig. 1. The framework of Cross-Modal Scale-Aware Fusion Network (CSFNet).

3.3 Scale-Aware Super-Resolution Module

The disparity in spatial resolution between events and images hinders direct cross-modal fusion, necessitating the alignment of event resolution with images through arbitrary-scale super-resolution. As illustrated in Fig. 2, the Scale-Aware Super-Resolution Module (SASR) improves the resolution of sparse event features by incorporating refined image features.

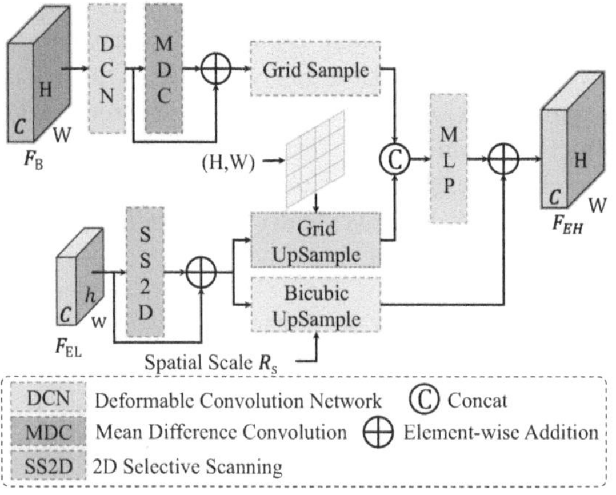

Fig. 2. The structure of the Scale-Aware Super-Solution Module (SASR).

Spatial Enhancement. Since events are locally dense but globally sparse in spatial distribution, we adopt a Mamba-based 2D Selective Scanning (SS2D) mechanism [6] to capture global spatial correlations and suppress noise.

Feature Refinement. Motivated by the observation that event-triggered regions correspond primarily to high-contrast edges in the image, we extract high-frequency image features using a Deformable Convolution Network (DCN) and Mean Difference Convolution (MDC) to preserve continuous edge information in high-resolution event features.

Scale-Aware Feature Upsampling. Leveraging the strengths of implicit neural representation for arbitrary-scale super-resolution, we employ a Multilayer Perceptron (MLP) to integrate globally enhanced event features with refined image features. While grid upsampling preserves local spatial continuity, it lacks sufficient contextual perception. To address this limitation, we additionally apply bicubic upsampling to low-resolution event features to capture spatial correlations within sparse neighborhoods according to the spatial scale R_s. Finally, pixel-wise summation aggregates the outputs of the feature-level super-resolution process.

3.4 Global Interaction Fusion Module

To integrate detailed information from events with contextual information from images, we design the Global Interaction Fusion Module (GIFM) to enable cross-modal fusion by modeling global spatial correlations within each modality and global channel interactions across modalities, as illustrated in Fig. 3.

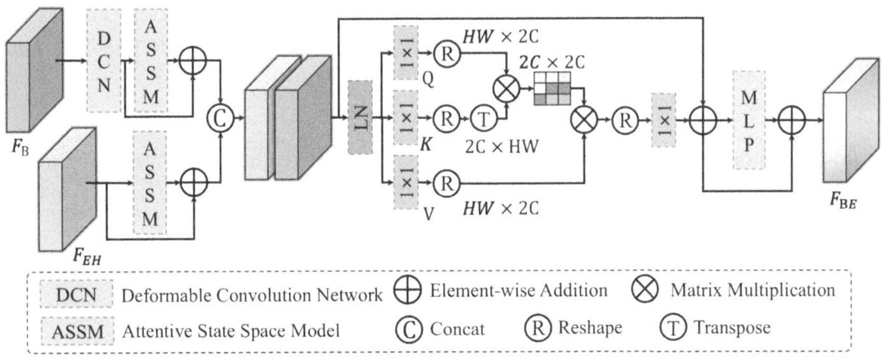

Fig. 3. The structure of the Global Interaction Fusion Module (GIFM).

Global Spatial Aggregation. To reduce the impact of noise on fusion, we introduce the Attentive State Space Model (ASSM) [5], which models global spatial correlations of event and image features independently before fusion.

Global Channel Interaction. Leveraging the complementary and redundant characteristics of different feature channels, we apply self-attention to compute cross-channel correlations after concatenating high-resolution event features with image features along the channel dimension. This process enables comprehensive interaction among channels and facilitates the selection of informative features within each modality for fusion.

4 Experiment

4.1 Experimental Settings

Our Cross-Modal Scale-Aware Fusion Network (CSFNet) is implemented in PyTorch and trained on a NVIDIA GeForce RTX 3090 GPU. The training process uses a batch size of 4 with a patch size of 64. We utilize the Adam optimizer, with an initial learning rate that follows a cosine decay schedule from 10^{-4} to 10^{-5}.

4.2 Quantitative and Qualitative Results

To evaluate the robustness of our model across different spatial scales, we conduct a quantitative comparison with state-of-the-art image-based deblurring approaches (DeblurINR [14], MotionETR [17], LEVS [7]) and event-based methods (EVDI [15], SSDM [2], GEM [16], EIFNet [12], TRMD [3], STCNet [13]) on the Ev-REDS dataset ($R_s = 4$) and the HS-ERGB dataset ($R_s = 1$). The results in Table 1 show that our model outperforms the second-best method by 0.12 dB on the Ev-REDS dataset and 0.07 dB on the HS-ERGB dataset, demonstrating both the adaptability of the Scale-Aware Super-Resolution Module (SASR) and the effectiveness of the Global Interaction Fusion Module (GIFM).

Table 1. Quantitative comparison results of our method and other state-of-the-art methods on the Ev-REDS dataset ($R_s = 4$) and HS-ERGB dataset ($R_s = 1$), where SL denotes supervised learning and SSL denotes self-supervised learning. The best and second-best results are indicated in **bold** and underlined, respectively.

Method	Events	Training	Ev-REDS		HS-ERGB	
			PSNR↑	SSIM↑	PSNR↑	SSIM↑
DeblurINR	✗	SSL	15.86	0.2609	–	–
Motion-ETR	✗	SL	18.23	0.4292	23.79	0.6276
LEVS	✗	SL	18.62	0.4612	22.13	0.5548
EVDI	✓	SSL	17.83	0.4356	25.13	0.7027
S-SDM	✓	SSL	19.90	0.5081	–	–
EIFNet	✓	SL	23.88	0.6647	26.19	0.7328
GEM	✓	SSL	23.95	0.6647	26.22	0.7292
TRMD	✓	SL	24.15	0.6762	25.81	0.7105
STCNet	✓	SL	24.26	**0.6887**	26.17	**0.7378**
Ours	✓	SSL	**24.38**	0.6828	**26.24**	0.7314

As illustrated in Fig. 4, qualitative results on the Ev-REDS dataset with $R_s = 4$ demonstrate that our method yields sharper edges with fewer artifacts compared to DeblurINR, EVDI, and S-SDM.

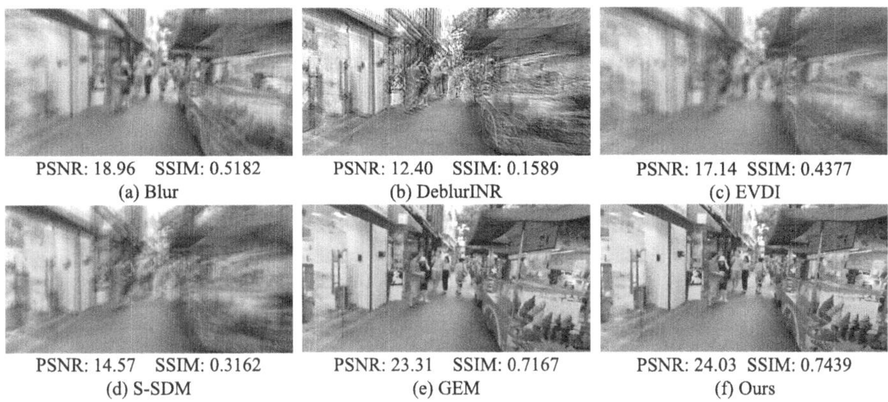

Fig. 4. The Qualitative comparison results of our method and other state-of-the-art methods on the Ev-REDS dataset.

4.3 Ablation Study

As reported in Table 2, we evaluate the contribution of SASR and GIFM on the EvREDS dataset with $R_s = 4$. When incorporating SASR alone improves PSNR by 0.21 dB (model #1 vs. #2), while GIFM alone increases PSNR by 0.22 dB (model #1 vs. #3). When SASR and GIFM are integrated, PSNR is further enhanced by 0.43 dB (model #1 vs. #4). These results demonstrate that the proposed event super-resolution module effectively addresses the spatial resolution inconsistency between events and images, and that the cross-modal feature fusion module strengthens the fusion process.

Table 2. The ablation of the individual components.

ID	SASR	GIFM	PSNR↑	SSIM↑
#1	✗	✗	23.95	0.6647
#2	✓	✗	24.16	0.6714
#3	✗	✓	24.17	0.6743
#4	✓	✓	**24.38**	**0.6828**

5 Conclusion

In this paper, we present the Cross-Modal Scale-Aware Fusion Network (CSFNet) to tackle the issue of spatial resolution discrepancies in event-based motion deblurring under real-world conditions. To alleviate spatial scale inconsistencies between events and images, we introduce a Scale-Aware Super-Resolution Module (SASR) that aligns event features with image features at arbitrary spatial resolutions. Furthermore, we propose a Global Interaction Fusion Module (GIFM) to facilitate thorough integration of cross-modal information through joint spatial and channel modeling, thereby improving complementary feature representation. Extensive experiments on both synthetic and real-world datasets across multiple scales validate the efficacy of the proposed approach.

Acknowledgements. This work is supported by the Key Research and Development Program of Shaanxi Province (2023-LL-QY-24) and the National Natural Science Foundation of China (U24B20127).

References

1. Bahat, Y., Efrat, N., Irani, M.: Non-uniform blind deblurring by reblurring. In: Proceedings of the IEEE International Conference on Computer Vision, pp. 3286–3294 (2017)
2. Chen, K., Chen, S., Zhang, J., Zhang, B., Zheng, Y., Huang, T., Yu, Z.: Spikereveal: Unlocking temporal sequences from real blurry inputs with spike streams. Adv. Neural. Inf. Process. Syst. **37**, 62673–62696 (2024)
3. Chen, K., Yu, L.: Motion deblur by learning residual from events. IEEE Trans. Multimedia **26**, 6632–6647 (2024)
4. Gallego, G., et al.: Event-based vision: a survey. IEEE Trans. Pattern Anal. Mach. Intell. **44**(1), 154–180 (2020)
5. Guo, H., et al.: Mambairv2: attentive state space restoration. In: Proceedings of the Computer Vision and Pattern Recognition Conference, pp. 28124–28133 (2025)
6. Guo, H., Li, J., Dai, T., Ouyang, Z., Ren, X., Xia, S.T.: Mambair: a simple baseline for image restoration with state-space model. In: European Conference on Computer Vision, pp. 222–241. Springer (2024)
7. Jin, M., Meishvili, G., Favaro, P.: Learning to extract a video sequence from a single motion-blurred image. In: Proceedings of the IEEE Conference on Computer Vision and Pattern Recognition, pp. 6334–6342 (2018)
8. Pan, L., Scheerlinck, C., Yu, X., Hartley, R., Liu, M., Dai, Y.: Bringing a blurry frame alive at high frame-rate with an event camera. In: Proceedings of the IEEE/CVF Conference on Computer Vision and Pattern Recognition, pp. 6820–6829 (2019)
9. Sun, L., et al.: Event-based fusion for motion deblurring with cross-modal attention. In: European Conference on Computer Vision, pp. 412–428. Springer (2022)
10. Sun, Z., Fu, X., Huang, L., Liu, A., Zha, Z.J.: Motion aware event representation-driven image deblurring. In: European Conference on Computer Vision, pp. 418–435. Springer (2024)

11. Wang, J., Yang, H., Shan, J., Song, X., Jiang, J.: Attention-based spatial-temporal fusion networks for traffic flow prediction. In: International Conference on Web Information Systems and Applications, pp. 500–511. Springer (2024)
12. Yang, W., Wu, J., Li, L., Dong, W., Shi, G.: Event-based motion deblurring with modality-aware decomposition and recomposition. In: Proceedings of the 31st ACM International Conference on Multimedia, pp. 8327–8335 (2023)
13. Yang, W., Wu, J., Ma, J., Li, L., Shi, G.: Motion deblurring via spatial-temporal collaboration of frames and events. In: Proceedings of the AAAI Conference on Artificial Intelligence. vol. 38, pp. 6531–6539 (2024)
14. Zhang, T., Quan, Y., Ji, H.: Cross-scale self-supervised blind image deblurring via implicit neural representation. In: The Thirty-eighth Annual Conference on Neural Information Processing Systems (2024)
15. Zhang, X., Yu, L.: Unifying motion deblurring and frame interpolation with events. In: Proceedings of the IEEE/CVF Conference on Computer Vision and Pattern Recognition, pp. 17765–17774 (2022)
16. Zhang, X., Yu, L., Yang, W., Liu, J., Xia, G.S.: Generalizing event-based motion deblurring in real-world scenarios. In: Proceedings of the IEEE/CVF International Conference on Computer Vision, pp. 10734–10744 (2023)
17. Zhang, Y., Wang, C., Maybank, S.J., Tao, D.: Exposure trajectory recovery from motion blur. IEEE Trans. Pattern Anal. Mach. Intell. **44**(11), 7490–7504 (2021)

The Improved TS-APO Path Planning Method for UAV Antenna Near-Field Measurement

Wenqi Lv[1,2], Shurui Fan[1,2](✉), Li Wang[1,2], Xiaoyu An[1,2], Yujie Wang[1,2], Dongnan Zhang[1,2], and Xiaoliang Wang[3]

[1] Innovation and Research Institute, Hebei University of Technology, Shijiazhuang 050299, China
fansr@hebut.edu.cn

[2] School of Electronic and Information Engineering, Hebei University of Technology, Tianjin 300401, China

[3] Hebei WeisaiTe Technology Co., Ltd., Shijiazhuang 050299, China

Abstract. To address the issues of cumbersome preparation, high costs, and low efficiency in antenna near-field measurements, this paper proposes a method for waypoint selection and path planning using unmanned aerial vehicles (UAVs) for the measurement area, aiming to improve the efficiency of antenna near-field measurements. First, an Electromagnetic-induction-based Adaptive Sampling strategy (EAS) is employed to select waypoints on the measurement plane. The electromagnetic field propagation principle is used to construct an electromagnetic distribution matrix, followed by a dynamic threshold adjustment strategy to filter the waypoints that meet the required conditions. Secondly, this paper introduces an improved Arctic Puffin Optimization algorithm based on tabu search, called the Tabu Search-Arctic Puffin Optimization algorithm (TS-APO), for path planning of the selected waypoints. Compared to traditional antenna near-field measurement methods, the proposed approach effectively reduces the number of measurement points and the scanning path length. Experimental results comparing with similar path planning algorithms show that the TS-APO algorithm shortens planning time while maintaining strong optimization capability. To meet the UAV's flight dynamics constraints, a piecewise uniform B-spline curve method is applied to smooth the generated path. UAV software-in-the-loop simulations using AirSim validate the feasibility of the algorithm. Data validation with a 3-m parabolic antenna operating at 1 GHz demonstrates that the proposed method can significantly improve measurement efficiency while ensuring the accuracy of antenna near-field measurements.

Keywords: UAV · Antenna Measurement · Near-Field · Path Planning · TSP

1 Introduction

With technological advancements, UAVs have become a key enabler of the low-altitude economy, demonstrating significant potential in antenna measurement [1]. Antenna, as a critical electromagnetic wave device, plays a crucial role in ensuring system stability. Traditional UAV-based antenna measurements often use uniform sampling combined with

S. Yang et al. (Eds.): WISA 2025, LNCS 16299, pp. 170–178, 2026.
https://doi.org/10.1007/978-981-95-6426-2_15

bow-shaped or spiral scanning paths [2]. While effective in basic tasks, these methods face challenges in efficiency and resource use, especially in large-aperture, wide-range measurements. UAVs offer advantages in mobility and flexibility, providing an efficient, flexible solution for antenna measurement.

To utilize UAVs for measurement tasks, path planning is a key element. The core of this task lies in how to quickly plan the UAV's flight trajectory to efficiently cover the antenna measurement area while ensuring the accuracy of the measurement data.

This work introduces an efficient and accurate UAV-based near-field antenna measurement approach to effectively address the aforementioned challenges.

1. An Electromagnetic-induction-based Adaptive Sampling strategy (EAS) is used to perform key selection of numerous measurement points, with the selected key measurement points serving as UAV waypoints.
2. The Tabu Search-Arctic Puffin Optimization algorithm (TS-APO) is employed to quickly traverse the key waypoints, solving the UAV Traveling Salesman Problem (TSP).
3. The method is verified for feasibility using the Robot Operating System (ROS) and the AirSim simulation platform. The accuracy of the near-field measurement is validated through specific antenna data, confirming that the measurement accuracy meets the expected results.

2 Related Work

A large number of feasible methods have been proposed by researchers to address the Traveling Salesman Problem (TSP) [3]. Tung and Liu et al. [4] combined genetic algorithms with improved operators to solve the integrated TSP and coverage path planning problem (TSP-CPP). Dewantoro et al. [5] proposed a hybrid method (ACO-TS) combining ant colony optimization (ACO) and tabu search (TS) algorithms to solve the TSP. This method uses TS as a local search to improve the performance of ACO.

Integrating algorithmic research with practical application contexts enables research outcomes to address real engineering problems, facilitating the transformation of theoretical results into practical productivity and avoiding the issue of traditional TSP solutions being optimal yet impractical. Krishnan et al. [6] focused on energy efficiency in UAV-assisted IoT data collection and innovatively proposed a TSP-based solution space reduction strategy. Hu et al. [7] proposed an improved path planning algorithm through real-time monitoring and analysis of wind farm environments, providing technical support for the intelligent operation and maintenance of wind farms. Yu [8] studied the path planning of an atmospheric environment monitoring system using a simulated annealing-enhanced whale optimization algorithm (SA-WOA), improving data collection efficiency.

Existing research has laid a solid foundation for this field's development. However, real-world applications are diverse, and a single algorithm often lacks the versatility to solve all problems. While exact algorithms provide optimal solutions, their exponential complexity necessitates improvements in efficiency [9]. We aim to offer a new approach to solving the TSP, expanding its applicability in complex real-world scenarios.

3 Problem Modeling

We adopt a vertical-plane configuration for near-field antenna measurement, where the antenna aperture faces upward and the measurement region is defined as a square area above the antenna. Key points are extracted within this region based on predefined criteria and treated as UAV waypoints, with their path optimized by solving the TSP. The UAV antenna near-field measurement problem is modeled as shown in Fig. 1.

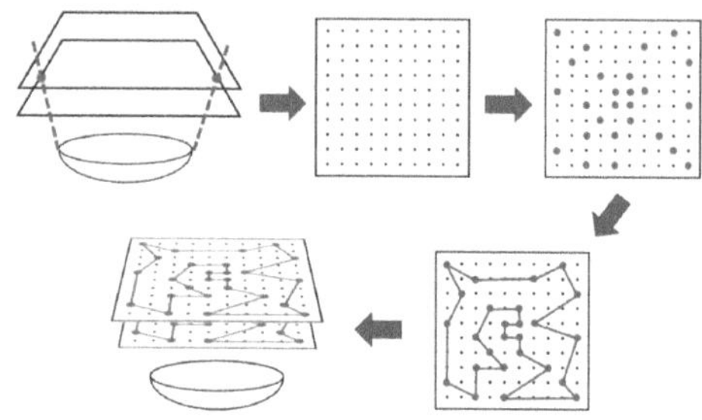

Fig. 1. Modeling of the UAV Antenna Near-Field Measurement Problem.

4 Proposed Method

4.1 Sampling of the Measurement Area

The Electromagnetic-induction-based Adaptive Sampling strategy (EAS) constructs an electromagnetic propagation matrix to extract a series of measurement points that contribute most to the electromagnetic field, thereby ensuring that they effectively reflect the overall performance of the antenna within the measurement area.

The initial stage of EAS is characterized by the configuration of global parameters, where the antenna aperture is set to 3 m, the operating frequency to 1 GHz, the measurement region to a side length of 20 m, and the measurement height to 3 m.

Subsequently, source points and measurement points are generated. The Euclidean distance between each measurement point p_m and the source point p_s is calculated as follows:

$$d_{m,s} = \sqrt{(x_m - x_s)^2 + (y_m - y_s)^2 + (z_m - z_s)^2} \tag{1}$$

The electromagnetic propagation matrix B is constructed as follows:

$$B_{m,s} = \frac{e^{-jk_0 d_{m,s}}}{d_{m,s}} \tag{2}$$

where j is the imaginary unit. The quantity k_0 is the wave number.

Perform a QR decomposition on matrix B. The QR decomposition decomposes the matrix B^T into an orthogonal matrix Q and an upper triangular matrix R:

$$B^T = QR \tag{3}$$

The diagonal elements $R_{i,i}$ indicate each measurement point's contribution to the electromagnetic field, and QR decomposition, unlike LU decomposition, is less sensitive to ill-conditioning, ensuring more reliable component extraction [10].

A threshold τ is defined as the criterion for identifying measurement points. Points satisfying the condition for R_{nn} must meet the following requirements:

$$\frac{|R_{nn}|}{|R_{11}|} \geq \tau \tag{4}$$

If the number of points satisfying the selection criteria is insufficient for practical near-field antenna measurement, a dynamic threshold adjustment strategy is employed. In this approach, the threshold is appropriately reduced to extract a sufficient set of waypoint indices.

The selected measurement points are treated as UAV waypoints, allowing the UAV to quickly traverse with the test probe and transmit the corresponding antenna characteristic data for each point.

4.2 Improved TS-APO Path Planning Algorithm

This paper presents an improved TS-APO path planning algorithm that achieves shorter planning time and stronger optimization capability than other methods, with clear advantages when handling large numbers of measurement points. The flowchart of the TS-APO algorithm is shown in Fig. 2.

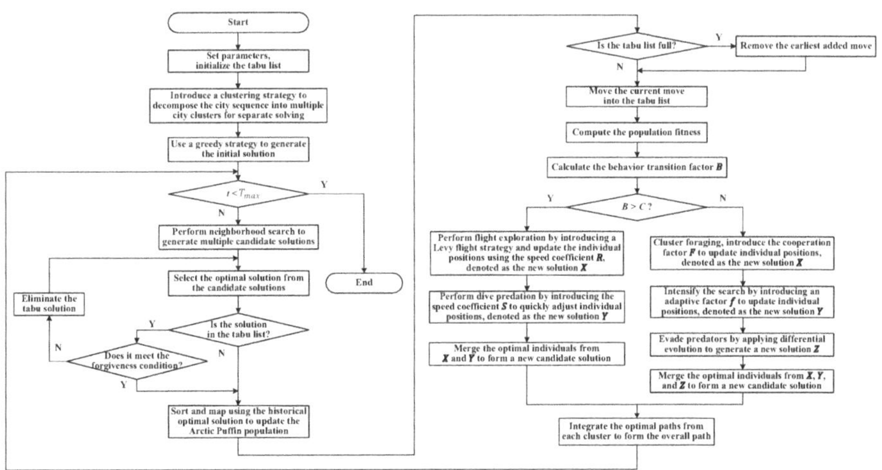

Fig. 2. Algorithm flowchart of TS-APO.

Arctic Puffin Optimization (APO) is designed for continuous optimization [11], while path planning involves discrete nodes. A sorting-based mapping bridges these spaces, enabling feasible neighborhoods and effective solution generation.

In solving large-scale TSP, a clustering strategy can be employed to decompose the problem and reduce its complexity [12]. Using the K-Means clustering algorithm, all cities are partitioned into several clusters based on the Euclidean distances between cities. The K-Means algorithm iteratively optimizes the cluster centers and their members [13], minimizing the sum of distances within each cluster. Mathematically, the objective function is defined as:

$$J = \sum_{j=1}^{K} \sum_{i \in C_j} \|x_i - \mu_j\|^2 \tag{5}$$

Here, K denotes the number of clusters, C_j represents the j-th cluster, μ_j is the centroid of cluster C_j, x_i is the coordinate of the i-th city, and $\|x_i - \mu_j\|$ is the Euclidean distance between the city point x_i and the centroid μ_j of the cluster.

Given that the performance of the APO algorithm is highly sensitive to the initial solution, the Tabu Search (TS) algorithm combined with a greedy initialization strategy is employed to generate high-quality initial solutions within a relatively short time [14].

The Lévy flight strategy, an advanced stochastic search mechanism, is cleverly integrated into the global exploration and local exploitation phases of the APO algorithm. Its purpose is to balance the algorithm's global search capability and local exploitation ability [15].

The formula for generating the Lévy flight step size is as follows:

$$\text{step} = \frac{u}{|v|^{\frac{1}{\beta}}} \tag{6}$$

where u and v are random variables that follow a Gaussian distribution. β is the characteristic exponent of the stable distribution, typically taking values in the range $0 < \beta \leq 2$ The characteristic exponent β determines the shape of the step size distribution. σ is the scaling factor for the Lévy flight, and its calculation formula is as follows:

$$\sigma = \left[\frac{\Gamma(1+\beta) * \sin\left(\pi * \frac{\beta}{2}\right)}{\beta * \Gamma\left(\frac{1+\beta}{2}\right) * 2^{\frac{\beta-1}{2}}} \right]^{\frac{1}{\beta}} \tag{7}$$

where $\Gamma(\cdot)$ is the Gamma function, which ensures that the mathematical form of the step size distribution aligns with the physical characteristics of Lévy flights, enabling the algorithm to exhibit reasonable search behavior in practical applications.

4.3 Algorithm Comparison Experiments

The experiments were conducted on a 2024 ROG Zephyrus G14 Air equipped with an AMD Ryzen 9 8945HS processor and 16 GB of RAM. The software environment is PyCharm Professional 2024.3.4, and the configured virtual environment uses Python 3.7 as the processor version.

Typical path planning algorithms, such as IMGA [4], DSMO [16], PSO, SA [17], and ACO-TS [5], were selected for comparative analysis. Each algorithm was tested using the eil76 and gil262 datasets, with 50 repetitions performed for each algorithm. The optimal results, average results, and average convergence times for each algorithm were recorded. As shown in Table 1.

Table 1. Comparison of Path Planning Algorithm Results.

Dataset	Performance index	IMGA [4]	DSMO [16]	PSO	SA [17]	ACO-TS [5]	TS-APO
eil76	Optimal Result (m)	569.54	558.68	582.35	562.79	568.30	548.84
	Average Result (m)	584.24	572.70	598.77	581.06	572.51	552.54
	Average Convergence Time (s)	4.88	7.56	12.62	1.44	47.19	3.13
gil262	Optimal Result (m)	2744.17	2543.15	2832.89	2731.81	2720.15	2522.75
	Average Result (m)	2753.75	2627.87	2861.44	2772.62	2733.52	2550.61
	Average Convergence Time (s)	57.45	42.39	49.30	40.15	263.17	36.90

As shown in the table, TS-APO outperforms other algorithms in various metrics to different extents. Specifically, in the eil76 dataset, its performance is only 2.70% worse than the optimal path of 538. Compared with other algorithms, TS-APO achieves a 3.49%–7.72% improvement in average results with shorter convergence time, highlighting its superiority in path planning tasks.

To avoid flight instability or loss of control caused by frequent path variations of the UAV and to improve energy efficiency, the path is smoothed using piecewise uniform B-spline curves. The waypoint distribution and the comparison between the paths before and after smoothing are shown in Fig. 3.

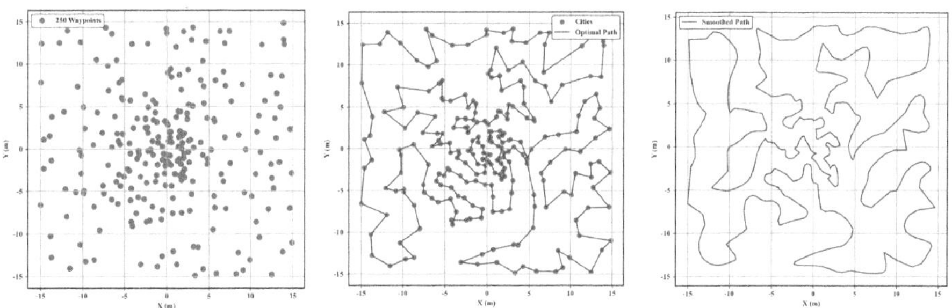

Fig. 3. Comparison of the Path Before and After Smoothing.

5 Simulation Experiment

Considering cost and safety constraints, this work employs software-in-the-loop (SIL) simulation for algorithm validation and system testing using AirSim on Unreal Engine 4.27.2, with plugins compiled in Visual Studio 2022, as illustrated in Fig. 4.

Fig. 4. Flight Simulation Experiment using Airsim.

The near-field data are transformed into far-field radiation patterns via discrete Fourier transform (DFT) and compared with theoretical results from simulation. As shown in Fig. 5, the transformed patterns show good agreement with the theoretical ones within the $\pm$ 45° main-lobe region, with particularly high consistency observed around the 7° half-power beamwidth.

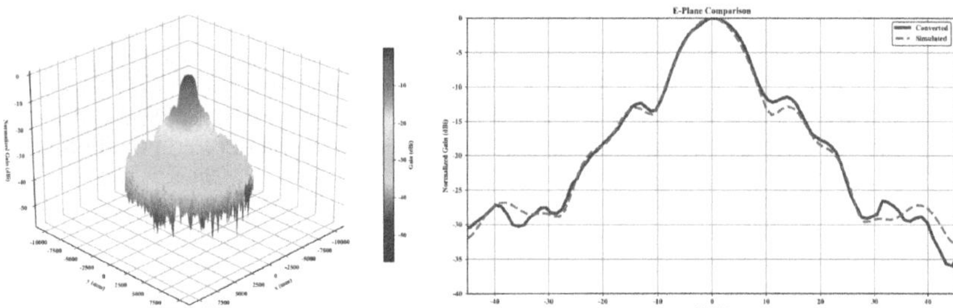

Fig. 5. (a) 3D Radiation Pattern (b) E-plane Comparison Pattern.

6 Conclusions and Prospect

In this work, a UAV-based method for waypoint selection and path planning was proposed, where waypoints were selected using EAS and optimized through the TS-APO algorithm. Simulation experiments in AirSim, combined with verification using parabolic antenna data, confirmed that the proposed approach achieves the expected accuracy and enhances the efficiency of near-field antenna measurements. Future work will extend the validation to real-world UAV experiments under outdoor conditions, incorporating factors such as wind disturbances and complex terrain to further assess the robustness and applicability of the algorithm.

Acknowledgments. This study was funded by Science and Technology Cooperation Special Project of Shijiazhuang (grant number SJZZXB24009).

Disclosure of Interests. The authors have no competing interests to declare that are relevant to the content of this article.

References

1. Umeyama, A., Salazar-Cerreno, J., Fulton, C.: UAV-based far-field antenna pattern measurement method for polarimetric weather radars: simulation and error analysis. IEEE Access **8**, 191124–191137 (2020)
2. Ciorba, L., Paonessa, F., Righero, M., et al.: Electromagnetics-aware path for UAV-based near-field antenna measurements implemented with a high dynamic range SDR. IEEE Trans. Instrument. Meas. (2024)
3. Zhang, C., Cai, Y., Hu, P., et al.: Logistics distribution route optimization using hybrid ant colony optimization algorithm. In: International Conference on Web Information Systems and Applications, pp. 510–517. Springer, Cham (2022)
4. Tung, W.-C., Liu, G.-S.: Solution of an integrated traveling salesman and coverage path planning problem by using a genetic algorithm with modified operators. IADIS Int. J. Comput. Sci. Inf. Syst. **14**(2), 95–114 (2019)
5. Dewantoro, R.W., Sihombing, P.: The combination of ant colony optimization (ACO) and tabu search (TS) algorithm to solve the traveling salesman problem (TSP). In: 3rd International Conference on Electrical, Telecommunication and Computer Engineering (ELTICOM), pp. 160–164. IEEE, New York (2019)
6. Krishnan, S., et al.: Energy-efficient UAV-assisted IoT data collection via TSP-based solution space reduction. In: IEEE Global Communications Conference (GLOBECOM), 2023
7. Xiaobing, H.U., Ze, L.U., Hang, L.I., et al.: Unmanned aerial vehicle wind farm inspection path planning based on real-time potential landing safety constraints. China Saf. Sci. J. **35**(2), 28 (2025)
8. Yu, B., Fan, S., Cui, W., et al.: A Multi-UAV cooperative mission planning method based on SA-WOA algorithm for three-dimensional space atmospheric environment detection. Robotica **42**(7), 2243–2280 (2024)
9. Laporte, G.: The traveling salesman problem: an overview of exact and approximate algorithms. Eur. J. Oper. Res. **59**(2), 231–247 (1992)
10. Zhuang, F., He, H., Ye, A., et al.: Towards faster and robust solution for dynamic LR and QR factorization. Sci. Rep. **14**(1), 27923 (2024)
11. Wang, W., Tian, W., Xu, D., et al.: Arctic puffin optimization: a bio-inspired metaheuristic algorithm for solving engineering design optimization. Adv. Eng. Softw. **195**, 103694 (2024)
12. Zhang, K., Sun, C., Shen, D., et al.: High-dimensional nearest neighbour search-based blocking in entity resolution. In: Jin, C., Yang, S., Shang, X., Wang, H., Zhang, Y. (eds.) WISA 2024. LNCS, vol. 14883, pp. 215–226. Springer, Singapore (2024)
13. Yu, H., Du, Y., Zhang, X., et al.: A dynamic convergence criterion for fast K-means computations. In: Jin, C., Yang, S., Shang, X., Wang, H., Zhang, Y. (eds.) WISA 2024. LNCS, vol. 14883, pp. 191–202. Springer, Singapore (2024)
14. Sun, S., Tong, Y., Qi, B., et al.: Hybrid particle swarm optimization algorithm for traveling salesman problem based on ternary optical computer. Procedia Comput. Sci. **243**, 1280–1287 (2024)
15. Li, J., An, Q., Lei, H., et al.: Survey of levy flight-based metaheuristics for optimization. Mathematics **10**(15) (2022)

16. Akhand, M.A.H., et al.: Discrete spider monkey optimization for travelling salesman problem. Appl. Soft Comput. **86**, 105887 (2020)
17. Wang, L.J., Cai, R.Y., Lin, M., et al.: Enhanced list-based simulated annealing algorithm for large-scale traveling salesman problem. IEEE Access **7**, 144366–144380 (2019)

Direction-Aware Completion of Sparse SfM Point Clouds and Its Application in UAV Imagery-Based Road Reconstruction

Jialu Liu[1(✉)], Gaoli Cheng[2], Guozhan Zhang[1], and Ronggui Ma[1]

[1] Chang'an University, Xi'an, China
2023224113@chd.edu.cn
[2] Shaanxi Highway Mechanization Engineering Co., Ltd., Xi'an, China

Abstract. Low-altitude UAV photogrammetry is increasingly used for 3D reconstruction of road environments. However, sparse point clouds produced by Structure-from-Motion (SfM) often suffer from occlusions and feature matching failures, leading to incomplete surfaces.

We propose Direction-Aware Void Completion (DAVC), a geometric method that improves point cloud continuity by leveraging local directional cues. DAVC identifies deficient regions, extracts dominant directions via Principal Component Analysis (PCA), and extrapolates new points along structural lines, followed by density-aware filtering to avoid redundancy.

Experiments on real-world UAV datasets show that DAVC achieves a mean completion error of 3.8 cm, outperforming Moving Least Squares, Point Completion Network, and GRNet. It improves surface recovery by 39%, reduces discontinuities by 34%, and runs 2.1 × faster than deep learning-based methods. Without requiring image data or model training, DAVC offers an efficient and structure-consistent solution for large-scale UAV-based road reconstruction.

Keywords: Sparse Point Cloud Completion · Directional Geometry Prior · UAV Road Reconstruction · Structure-from-Motion (SfM)

1 Introduction

1.1 Background and Motivation

In recent years, the application of Unmanned Aerial Vehicles (UAVs) in low-altitude photogrammetry has become a cornerstone for 3D modeling of road environments. This technology is central to smart transportation, road surface monitoring, and digital twin city construction. With the advancement of computer vision algorithms, especially Structure-from-Motion (SfM) and Multi-View Stereo (MVS), it is now feasible to reconstruct sparse and dense point clouds directly from aerial imagery, without relying on expensive LiDAR sensors or prior maps.

However, in practical UAV flight missions, especially in urban or traffic-heavy road scenes, the reconstructed sparse point clouds often suffer from severe structural incompleteness. Factors such as occlusions (e.g., vehicles, pedestrians, trees), repeated textures,

S. Yang et al. (Eds.): WISA 2025, LNCS 16299, pp. 179–190, 2026.
https://doi.org/10.1007/978-981-95-6426-2_16

and weak visual features can lead to point cloud voids, broken surfaces, and missing lane or edge lines. These deficiencies not only degrade visual quality but significantly affect downstream tasks, including lane detection, road segmentation, and HD map construction.

1.2 Related Work

Early point cloud completion methods primarily leveraged geometric priors and surface reconstruction. Hoppe et al. [1] used signed distance functions for surface recovery from unorganized point sets. Alexa et al. [2] introduced the Moving Least Squares (MLS) projection to define smooth surfaces from noisy samples, while Dey and Sun [3] enhanced normal and feature estimation via local covariance analysis.

Recent learning-based methods, such as PCN [4] and GRNet [5], have achieved strong performance by learning geometric priors or combining voxel and point representations. However, these models are typically trained on synthetic CAD datasets like ShapeNet, limiting their effectiveness in real-world UAV scenes with sparse, irregular road data.

Traditional SfM/MVS pipelines (e.g., COLMAP) remain prevalent for UAV-based modeling but often suffer from incompleteness in occluded or oblique regions. To address this, Kim et al. [7] and Zhang and Huang [8] explored occlusion-aware UAV path planning to improve scene coverage. Xu and Stilla [6] emphasized semantic guidance in large-scale urban reconstructions, and Calì and Ambu [9] demonstrated UAV photogrammetry's potential for broad-scope surface modeling. Recent WISA work like Han et al. [10] 's TRoute models direction- and condition-dependent road dynamics to improve route prediction. Similarly, we leverage directional priors to resolve sparse or occluded UAV observations for road geometry completion.

Despite these efforts, structure-consistent completion under occlusion and sparse sampling in UAV road scenes remains unresolved, motivating our proposed Direction-Aware Void Completion (DAVC) approach.

1.3 Paper Organization

The remainder of this paper is organized as follows: Sect. 2 presents the overall pipeline and key steps of the proposed Direction-Aware Void Completion (DAVC) method. Section 3 describes the experimental setup, datasets, both quantitative and qualitative comparisons with existing approaches, and discusses the completion results, performance analysis, and application scenarios. Finally, Sect. 4 concludes the paper and outlines future research directions.

2 Method

To address the structural incompleteness in road regions reconstructed from UAV imagery, we propose Direction-Aware Void Completion (DAVC), a geometric approach designed to restore missing structures in sparse point clouds. DAVC leverages directional cues derived from Principal Component Analysis (PCA) and incorporates density-aware control to enhance continuity without introducing redundancy. The entire process consists of four steps, as illustrated in Fig. 1.

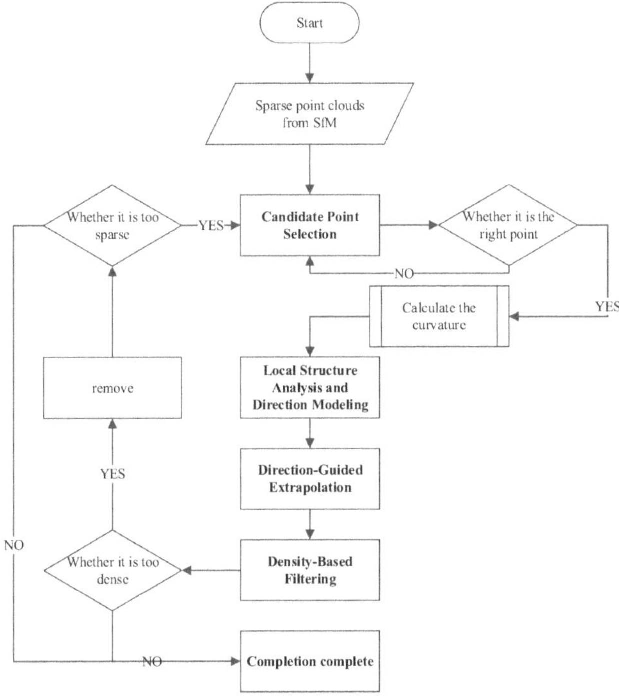

Fig. 1. Algorithm flowchart

2.1 Candidate Point Selection

To selectively target structurally incomplete areas in the sparse point cloud, DAVC begins the completion process by selecting seed points that are geometrically or observationally deficient. These candidate points guide subsequent neighborhood modeling and extrapolation while avoiding unnecessary augmentation in well-sampled regions.

Given a sparse input point cloud P, defined as:

$$P = \{p_i \in R^3 | i = 1, \ldots, N\} \tag{1}$$

where each point pi is associated with the following attributes:

$o_i \in N$: observation multiplicity, i.e., the number of distinct views from which p_i is triangulated, A small o_i suggests unreliable reconstruction due to occlusion, oblique viewing, or texture-deficient regions;

$\rho i \in R^+$: estimated local point density.

The neighborhood $N_r(p_i)$ is defined within a Euclidean ball of radius r as:

$$N_r(p_i) = \{p_j \in P \big| \|p_j - p_i\|2 \le r\} \tag{2}$$

The pointwise local density is defined as:

$$\rho_i = \frac{1}{V_r} \sum_{p_j \in N_r(p_i)} exp\left(-\frac{\|p_j - p_i\|_2^2}{2\sigma^2}\right) \tag{3}$$

where $V_r = \frac{4}{3}\pi r^3$ is the neighborhood volume, and σ is a bandwidth parameter for Gaussian smoothing. This density estimate captures both the number and proximity of neighboring points.

A point p_i is labeled as a candidate for void completion if it satisfies the compound condition:

$$p_i \in C \Leftrightarrow (o_i < \tau_o) \vee \left(\rho_i < \tau_\rho\right) \tag{4}$$

where $\tau_o \in N$ and $\tau_\rho \in R^+$ are empirically determined thresholds. In this work, we empirically set the visibility threshold to $\tau_o = 3$, consistent with default SfM filtering heuristics in COLMAP. The density threshold $\tau_\rho = 0.08$ was selected based on observed neighborhood distributions in our UAV dataset. Although not extensively tuned, these values were found to provide a robust trade-off between coverage and redundancy.

To ensure that completion only targets meaningful road-related areas, we apply a semantic filtering step using a road region indicator function $X_{road}(p_i) \in \{0,1\}$, which determines whether a 3D point lies within a drivable surface.

We use DeepLabv3 + with a ResNet-101 backbone, pretrained on the Cityscapes dataset, to segment road regions from UAV images. The resulting 2D binary road masks are back-projected into 3D using camera intrinsics and extrinsics estimated by Structure-from-Motion. The final filtered set becomes:

$$C := \left\{p_i \in P | X_{road}(p_i) = 1 \wedge (o_i < \tau_o) \vee \left(\rho_i < \tau_\rho\right)\right\} \tag{5}$$

To further characterize structural boundaries, we estimate the local curvature κ_i from the eigenvalues $\lambda_1 \geq \lambda_2 \geq \lambda_3$, of the neighborhood covariance matrix. The curvature is defined as:

$$\kappa_i = \frac{\lambda_2 + \lambda_3}{\lambda_1} \tag{6}$$

A higher value of κ_i indicates a more isotropic or abrupt surface, suggesting potential edges or feature boundaries. We use this as an auxiliary signal to prioritize candidate points near structural voids.

In practice, we apply a curvature threshold τ_c to select candidate points satisfying $\kappa_i < \tau_c$ for reliable extrapolation.

This refined candidate set $C \subset P$ becomes the basis for directional modeling in Sect. 2.2.

2.2 Local Structure Analysis and Direction Modeling

For each candidate point$p_i \in C$, , DAVC performs a local structure analysis to estimate the dominant geometric direction along which extrapolation should occur. This is achieved via Principal Component Analysis (PCA) on the spatial distribution of its neighboring points. We have defined a neighborhood with a fixed radius, as shown in formula (2).

The covariance matrix of the neighborhood is computed as:

$$C_i = \frac{1}{|N_r(p_i)|} \sum_{p_j \in N_r(p_i)} (p_j - \overline{p})(p_j - \overline{p})^T \quad (7)$$

where $\overline{P}$ is the centroid of the neighborhood. By performing eigenvalue decomposition on C_i, we obtain eigenvalues $\lambda_1 \geq \lambda_2 \geq \lambda_3$ and corresponding eigenvectors v_1, v_2, v_3. The dominant direction is defined as:

$$d_i = v_1 \quad (8)$$

To ensure directional reliability, we impose an eigenvalue ratio constraint:

$$\frac{\lambda_1}{\lambda_2} > \tau_{dir} \quad (9)$$

Only when this condition holds is the direction d_i considered valid for extrapolation. Otherwise, p_i is skipped to avoid unreliable estimation in isotropic or noisy regions.

To assist in identifying structurally meaningful candidates, we also utilize the local curvature κ_i as defined in Sect. 2.1. While it is not used directly for direction estimation, curvature serves as an auxiliary indicator for edge transitions, and may be used to prioritize extrapolation in high-curvature regions.

2.3 Direction-Guided Extrapolation

Once the dominant direction d_i for each candidate point $p_i \in C$ is estimated (as described in Sect. 2.2), DAVC performs geometry-guided extrapolation along that direction to generate new 3D points that fill local voids.

Given a unit direction vector $d_i \in R^3$ and a fixed step size Δ, the extrapolated points are computed as:

$$p_i^{(k)} = p_i + k \cdot \Delta \cdot d_i, p_i^{(-k)} = p_i - k \cdot \Delta \cdot d_i, k = 1,2, \ldots, K \quad (10)$$

where K is the maximum number of extrapolation steps. These points are generated symmetrically along the estimated local structure direction.

To handle curving structures such as road edges or turning lanes, we implement a sliding-window PCA refinement during extrapolation. After each step k, the newly generated point $p_i^{(k)}$ is appended to a local history buffer and then perform PCA on W_k to obtain an updated tangent direction, which is used to generate the next point as:

$$W_k = \left\{p_i, p_i^{(1)}, \ldots, p_i^{(k)}\right\}, p_i^{(k+1)} = p_i^{(k)} + \Delta \bullet d_i^{(k)} \quad (11)$$

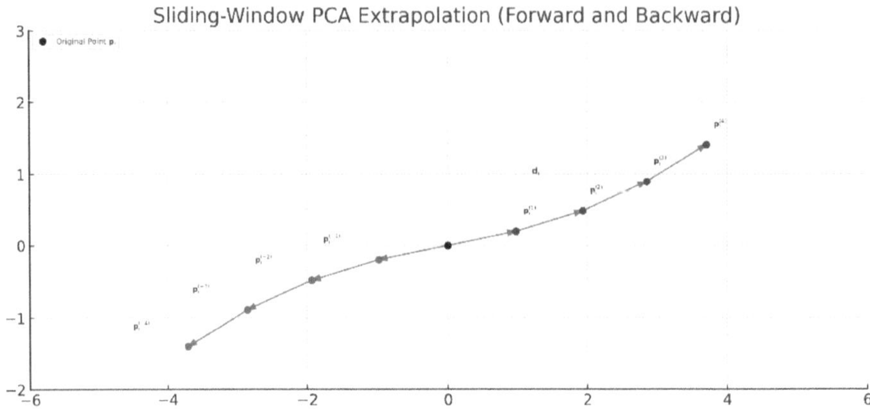

Fig. 2. Sliding-Window PCA Extrapolation

Figure 2 illustrates the sliding-window PCA refinement process:

To ensure the reliability of extrapolated points, several filtering conditions are applied. A newly generated point $p_i^{(k)}$ is accepted only if it satisfies:

i. The point must lie within the 3D bounding box of the original point cloud and, if available, within the road mask region.
ii. The point is discarded if it lies too close to any existing point:

$$\min_{p_j \in P} \|p_i^{(k)} - p_j\|_2 \leq \tau_r \tag{12}$$

iii. The extrapolation direction should not deviate significantly from the local tangent direction:

$$\theta = \arccos\left(\frac{d_i^{(k)} \bullet d_{local}}{\|d_i^{(k)}\| \bullet \|d_{local}\|}\right) < \tau_\theta \tag{13}$$

In our implementation, we set $\tau_\theta = 15^\circ$ to allow moderate curvature while preserving directional smoothness. The reference direction d_{local} is defined as the vector between two consecutive extrapolated points.

2.4 Density-Based Filtering

After extrapolation, the generated point set S_{gen} may contain redundant points in already dense regions. To remove unnecessary clutter while preserving structural continuity, we perform a density-based filtering step using a soft, kernel-weighted density estimator.

For each generated point $q_i \in S_{gen}$, the local density is computed as:

$$\rho_i = \frac{1}{V_r} \sum_{p_j \in N_r(q_i)} \exp\left(-\frac{\|q_i - p_j\|_2^2}{2\sigma^2}\right) \tag{14}$$

where q_i is extrapolated point under evaluation and σ is Gaussian kernel width (typically $0.5r$), controlling spatial influence.

A point is retained only if its estimated density is below a threshold:

$$q_i \in S_{final} \iff \rho_i < \tau_\rho \tag{15}$$

This ensures that new points are added only in sparse regions. In our setting, $r = 0.2 meters$, $\sigma = 0.1$, and $\tau_\rho = 0.08$.

Figure 3 demonstrates the results of directional extrapolation and density-based filtering. In the middle, new points are extrapolated along the local dominant direction, indicated by a arrow.

After density evaluation, Points identified with cross-marks were removed during density filtering, while those circled in blue were retained in the final output.

This process improves road surface completeness while preserving structural continuity.

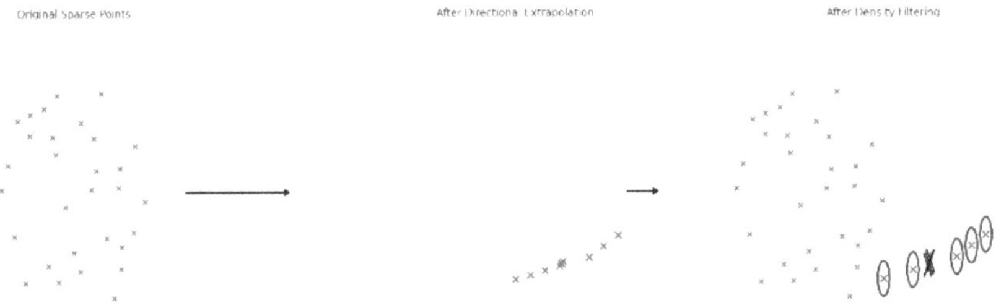

Fig. 3. Directional extrapolation and density filtering in DAVC.

3 Experiments

3.1 Experimental Setup

We conducted experiments on real-world UAV road scene datasets to evaluate the effectiveness and efficiency of DAVC compared to existing completion methods, experiments are conducted on UAV-based road scene datasets. The datasets consist of real-world aerial images captured over national highways in Lantian County, Xi'an, Shaanxi Province, China, and reconstructed into sparse point clouds using a Structure-from-Motion (SfM) pipeline. As shown in Fig. 4.

Fig. 4. The real UAV imagery

The camera intrinsic parameters are obtained through bundle adjustment, and semantic road masks are generated using a pre-trained DeepLabv3 + model with a ResNet-101 backbone.

The experiments were conducted on a workstation with an Intel Core i7-8750H processor, 16 GB of RAM, and an NVIDIA GTX 1050 Ti GPU.

The SfM reconstruction is carried out using COLMAP, and semantic segmentation is processed via PyTorch.

Key DAVC hyperparameters are set empirically as follows: local neighborhood radius $r = 0.2$ meters, density threshold $\tau_\rho = 0.08$, extrapolation step size $\Delta = 0.1$ meters, and angular consistency threshold $\tau_\theta = 15°$.

For baseline comparison, both classical geometric methods (e.g., Moving Least Squares [1]) and recent learning-based point cloud completion models (e.g., PCN [4], GRNet) are considered.

Quantitative and qualitative evaluations are performed to demonstrate the improvements brought by DAVC.

3.2 Qualitative Experiments

In addition to quantitative evaluation, qualitative comparisons are conducted to visually assess the completion performance of DAVC and baseline methods.

Figure 5 illustrates the sparse input point clouds and the completion results produced by Moving Least Squares (MLS), Point Completion Network (PCN), GRNet, and the proposed DAVC method.

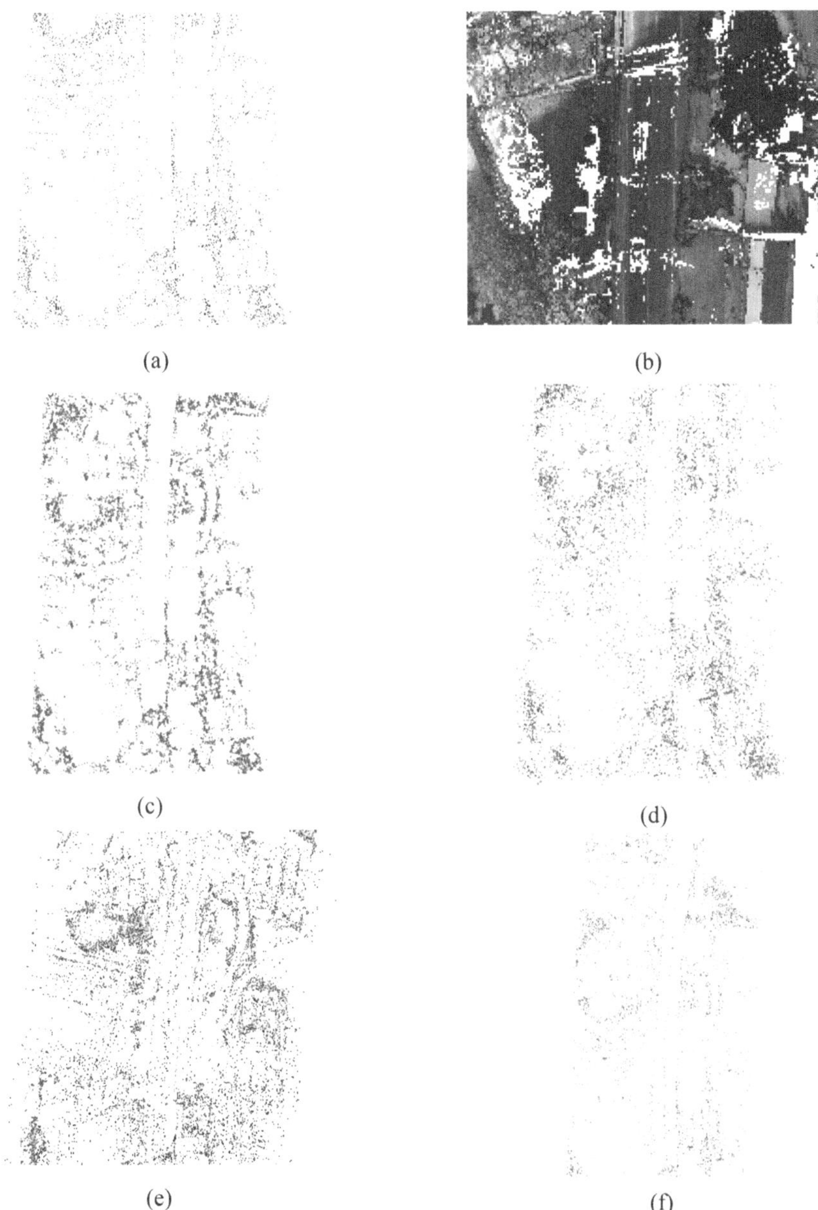

Fig. 5. (a) Sparse Point Cloud (b) Dense Point Cloud (c) MLS-Completed Point Cloud (d) PCN-Completed Point Cloud (e) GRNet-Completed Point Cloud (f) DAVC- Completed Point Cloud

As shown in Fig. 5, MLS often over-smooths the point cloud, leading to the loss of detailed structures along road edges.

PCN and GRNet generate relatively dense completions but often introduce artifacts and irregular clusters in occluded areas, due to insufficient adaptation to outdoor road environments. GRNet performs well in completing specific objects such as houses, but its effectiveness is limited in road scenes, where the large-scale outdoor environment hampers more accurate completion.

In contrast, DAVC produces road surfaces that are notably more continuous and structurally consistent compared to the baselines.

It successfully recovers missing lane markings and reduces surface discontinuities without introducing excessive artifacts.

These qualitative results visually demonstrate the effectiveness of DAVC in enhancing road surface completeness while preserving geometric continuity.

3.3 Comparison Experiments

To validate the effectiveness of the proposed DAVC To address the structural incompleteness, we conduct comparison experiments against representative classical and learning-based point cloud completion approaches.

We select Moving Least Squares (MLS) [1], a classical geometric method that fits smooth surfaces locally to sparse or noisy points, and learning-based models including Point Completion Network (PCN) [4] and GRNet, which have demonstrated strong performance on object-centric datasets. However, these learning-based methods often struggle to generalize to large-scale, irregular UAV road environments.

The evaluation focuses on completion accuracy, surface recovery, structural continuity, and runtime efficiency.

Table 1. Quantitative Comparison of Completion Performance

Method	Completion Error (cm) ↓	Surface Recovery Rate (%)↑	Discontinuity Reduction (%) ↑	Runtime Efficiency (×) ↑
MLS	7.4	18.2	20.5	1.0
PCN	6.2	25.7	28.1	0.6
GRNet	5.8	28.3	30.4	0.5
DAVC (Ours)	3.8	39.3	34.2	1.2

Quantitative results are summarized in Table 1, DAVC achieves the lowest mean completion error of 3.8 cm, outperforming both geometric and learning-based baselines.

It improves the surface recovery rate by 39.3% and reduces structural discontinuities by 34.2%, significantly enhancing road surface continuity.

Moreover, DAVC achieves higher computational efficiency without requiring model training. It is particularly effective in road environments or scenarios with continuous, structured surfaces, where local geometric directionality can be reliably exploited. In highly unstructured or irregular object reconstructions, learning-based methods may still offer advantages due to their global shape priors.

3.4 Curvature Ablation Study

To assess the impact of local curvature filtering on completion performance, we conduct an ablation study by disabling the curvature constraint in candidate selection.

As defined in Sect. 2.1, according to Eq. (6), the local curvature κ_i characterizes structural boundaries based on neighborhood eigenvalues. In this ablation study, we evaluate the effect of applying or removing curvature filtering during candidate point selection.

We compare two configurations:

Table 2. Effect of Curvature Filtering on Completion Performance

Method	Completion Error (cm) ↓	Surface Recovery Rate (%)↑	Discontinuity Reduction (%) ↑
DAVC (with curvature filtering)	3.8	39.3	34.2
DAVC (without curvature filtering)	5.1	42.5	21.7

The effect of curvature filtering is evaluated in Table 2, incorporating curvature filtering reduces the completion error and improves structural continuity by avoiding extrapolation near discontinuous boundaries.

Although disabling curvature filtering slightly increases surface recovery rate, it introduces more artifacts and irregular completions, leading to degraded geometric consistency.

4 Conclusion

In this work, we proposed Direction-Aware Void Completion (DAVC) to address structural incompleteness in sparse UAV road point clouds.By guiding extrapolation with local geometric directionality and controlling completion through curvature and density filtering, DAVC effectively improves road surface continuity without requiring model training.Experimental results show that DAVC outperforms classical interpolation and learning-based methods on real UAV datasets.

DAVC demonstrates strong performance in road structure completion; however, large-scale scene completion still heavily relies on learning-based approaches.Future work will focus on integrating global learning priors to further enhance completion in highly irregular outdoor environments.

Acknowledgments. This work was supported in part by the Science and Technology Project of Shaanxi Highway Mechanization Engineering Co., Ltd. (Project No. KY01-24), the 2023 Transportation Science and Technology Project of Shaanxi Provincial Department of Transportation (Project No. 23-80X), and the Key Technologies of Pavement Digital Twin and Disease Simulation Based on Multi-source Information Fusion Project supported by Chang'an University (Project No. 25-31X).

Disclosure of Interests. The authors have no competing interests to declare that are relevant to the content of this article.

References

1. Hoppe, H., DeRose, T., Duchamp, T., McDonald, J., Stuetzle, W.: Surface reconstruction from unorganized points. In: Proceedings of the 19th Annual Conference on Computer Graphics and Interactive Techniques (SIGGRAPH), pp. 71–78. ACM (1992)
2. Alexa, M., Behr, J., Cohen-Or, D., Fleishman, S., Levin, D., Silva, C.T.: Computing and rendering point set surfaces. IEEE Trans. Vis. Comput. Graph. **9**(1), 3–15 (2003)
3. Dey, T.K., Sun, J.: Normal and feature estimation from noisy point clouds. In: Botsch, M., Chen, B. (eds.) Proceedings of the 2006 Eurographics Symposium on Point-Based Graphics, pp. 21–30. Eurographics Association (2006)
4. Yuan, W., Khot, T., Held, D., Mertz, C., Hebert, M.: PCN: point completion network. In: Proceedings of the 2018 International Conference on 3D Vision (3DV), pp. 728–737. IEEE (2018)
5. Xie, H., Yao, H., Zhou, S., Mao, J., Zhang, S., Sun, W.: GRNet: gridding residual network for dense point cloud completion. In: Vedaldi, A., Bischof, H., Brox, T., Frahm, J.M. (eds.) Computer Vision – ECCV 2020, LNCS, vol. 12354, pp. 365–381. Springer, Cham (2020)
6. Xu, Y., Stilla, U.: Towards building and civil infrastructure reconstruction from point clouds: a review on data and key techniques. IEEE J. Sel. Top. Appl. Earth Obs. Remote Sens. **14**, 2857–2885 (2021)
7. Kim, S., Lee, D., Park, C.: Path planning for UAVs considering obstacles using visibility graph and dubins path. In: Proceedings of the 2018 International Conference on Wireless Information and Signal Processing (WISA), pp. 207–218. Springer, Cham (2018)
8. Zhang, J., Huang, H.: Occlusion-aware UAV path planning for reconnaissance and surveillance. Drones **5**(3), 98 (2021)
9. Calì, M., Ambu, R.: Advanced 3D photogrammetric surface reconstruction of extensive objects by UAV camera image acquisition. Sensors **18**(8), 2815 (2018)
10. Han, X., Hu, X., Ma, C., Shang, X.: TRoute: dynamic time-dependent route recommendation on road networks. In: Web Information Systems and Applications (WISA 2024). Lecture Notes in Computer Science, vol. 14883, pp. 573–585. Springer, Singapore (2024)

Intelligent Computing

MSTRLG: Multi-Scale Trajectory Recovery via Local-Global Similarity Fusion

Jiafan Liu[1], Yixiao Tong[1], Wenyu Wu[1], and Jiali Mao[1,2](✉)

[1] East China Normal University, Shanghai, China
{51275903116,52265903007,52275903019}@stu.ecnu.edu.cn
[2] Shanghai Engineering Research Center of Big Data Management, Shanghai, China
jlmao@dase.ecnu.edu.cn

Abstract. Accurate trajectory recovery is crucial for location-based services like traffic prediction and route optimization. However, raw trajectory data contain missing or noisy positions due to environmental interference and sensor limitations. Existing methods perform poorly in complex urban environments, such as irregular overpasses and parallel roads with varying movement patterns. To overcome these limitations, we propose MSTRLG, a *Multi-Scale Trajectory Recovery via Local-Global Similarity Fusion* framework. Our solution introduces (i) an adaptive multi-scale grid representation that dynamically weights optimal spatial scales to capture geometric features in overpass trajectories, and (ii) a novel fusion mechanism that combines local geometric patterns from similar trajectories (identified by learnable Fourier features) with global motion continuity constraints. Experiments on real-world datasets show that MSTRLG outperforms state-of-the-art methods, reducing *RMSE* by 16.01% for trajectories with ten consecutive missing points.

Keywords: trajectory recovery · multi-scale grid representation · local-global similarity

1 Introduction

With the advancement of positioning technologies, vehicle trajectories—defined as sequential records of positional points—are extensively collected on urban road networks. These data enable various transportation applications, including map inference [1], travel time estimation [2], and route planning [3]. However, positioning errors and environmental constraints often cause continuous data gaps, which can significantly compromise the performance of downstream applications. Although existing methods can recover geographic coordinates for incomplete trajectories, their accuracy proves particularly limited when dealing with complex road network structures characterized by irregular geometries and close-proximity neighboring relationships.

S. Yang et al. (Eds.): WISA 2025, LNCS 16299, pp. 193–206, 2026.
https://doi.org/10.1007/978-981-95-6426-2_17

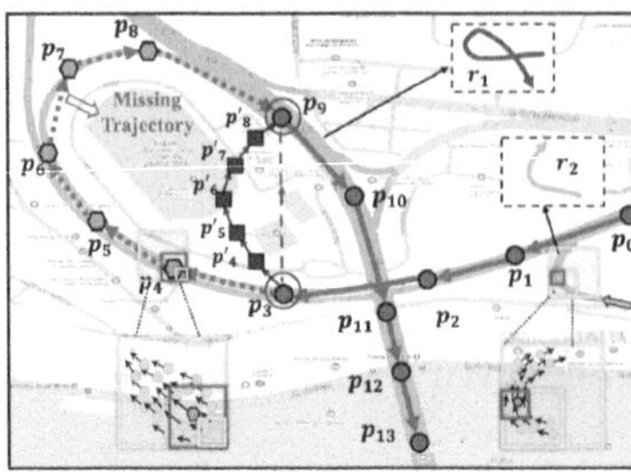

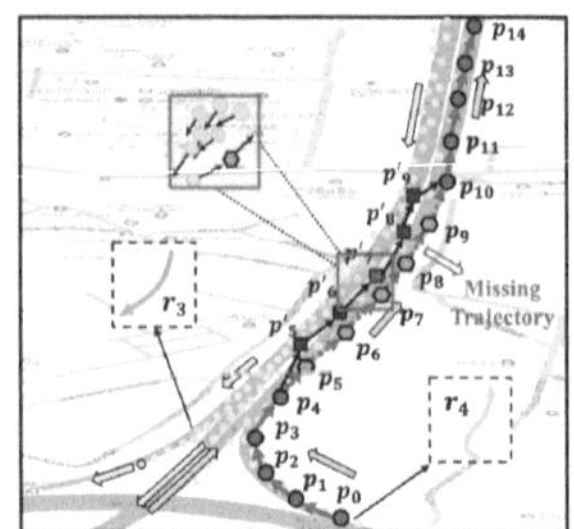

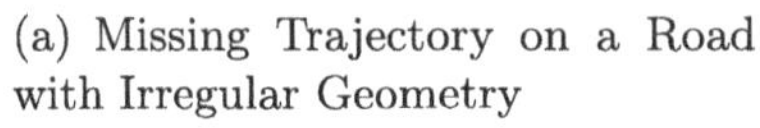

(a) Missing Trajectory on a Road with Irregular Geometry

(b) Missing Trajectory on a Complex Road with Ambiguous Adjacency

Fig. 1. Challenges in Trajectory Recovery Due to Complex Road Network Structures.

Challenge 1: The road where the trajectory is missing exhibits an irregular morphology. The proliferation of complex road infrastructures (e.g., overpasses) in urban networks results from growing traffic demands and spatial limitations. These structures often exhibit irregular geometric configurations, as illustrated by ring road r_1 in Fig. 1 a, creating substantial challenges for trajectory recovery. While existing methods [9,13,14] achieve satisfactory performance on regular road geometries, their accuracy significantly deteriorates when reconstructing missing trajectories on irregularly shaped roads. For example, in Fig. 1 a, the inferred positions between $p4$ and $p8$ are incorrectly estimated as p'_4 and p'_8. This limitation stems from either inappropriate spatial modeling scales or reliance on morphologically dissimilar reference trajectories (e.g., segments $p_0 - p_2$ or $p_{10} - p_{13}$). These observations underscore the critical need for developing robust trajectory inference methods capable of handling such complex road topologies.

Challenge 2: The road containing missing trajectory data is in immediate proximity to adjacent roadways. The emergence of closely spaced parallel roads in urban networks—typically necessitated by terrain constraints or historical preservation requirements—introduces significant challenges for trajectory recovery. As illustrated in Fig. 1 b, road r_4 (containing missing trajectory points between p_5 and p_9) runs parallel to counter-directional road r_3 with a separation distance of merely 9 meters. This spacing is substantially narrower than the standard 60-meter separation typical of urban roads[1]. The existing trajectory recovery methods [7,8,10] employ heuristic approaches that identify neighboring trajectory points within fixed distance thresholds. However, their dependence on localized observations renders them inadequate for such constrained configurations. Specifically, the selection of inappropriate distance thresholds in these methods often leads to erroneous reconstructions, as demonstrated by the incorrect mapping of points p'_5 and p'_9 to r_3 rather than the true trajectory on r_4.

[1] https://nacto.org/publication/urban-street-design-guide/.

These limitations highlight the necessity of developing approaches that extend beyond purely local analysis to account for complex spatial configurations.

To address the challenge of irregular road morphology, we propose a *Multi-Scale Contextual Grid Representation* module based on *Graph Attention Networks* [19]. This module associates each trajectory point with three fixed-scale spatial grids (*coarse*, *medium*, and *fine*) to capture geometric features at different resolutions. Through an attention mechanism, the model adaptively learns neighborhood interactions in each scale and dynamically assigns weights to optimal scales based on local road characteristics. As shown in Fig. 1 a(the most important grid scale is highlighted by a red square), road r_2 with sharp turns primarily utilizes fine-grid features to capture detailed directional changes, while smoother roads like r_1 are effectively represented by medium-scale grids. For the challenge where local neighbors near p_7 are predominantly from adjacent roads (Fig. 1 b), we propose a *Local-Global Adaptive Fusion Network-based Trajectory Recovery* module. This Network combines two complementary information sources: local morphologically similar segments from parallel roads and global motion continuity cues from known trajectory segments ($p_0 - p_4$ and $p_{10} - p_{14}$). The module employs learnable Fourier features [20] to identify similar trajectories, then integrates both the geometric patterns from similar trajectories and the global motion continuity from existing segments into the trajectory recovery task. Such approach enables accurate trajectory recovery while avoiding misleading local neighbors with divergent movement patterns.

To summarize, we propose a *Multi-Scale Trajectory Recovery via Local-Global Similarity Fusion* framework (called MSTRLG) that integrates three key components: (i)a *Spatial Temporal Dual Feature Extraction* module that captures behavioral patterns of trajectory points and multi-scale grid features through hierarchical geographical partitioning; (ii)a *Multi-Scale Contextual Grid Representation* module that generates comprehensive contextual embeddings by aggregating features from both the current grid and its neighboring grids at each scale level, while dynamically assigns dominant weights to optimal scales; (iii) a *Local-Global Adaptive Fusion Network-based Trajectory Recovery* module that fuses local geometric patterns from similar trajectories via spectral analysis, global motion continuity from known segments, and multi-scale contextual cues for autoregressive trajectory recovery. Overall, our contributions are 4 folds:

- To enhance trajectory recovery accuracy, we propose a *Multi-Scale Trajectory Completion via Local-Global Similarity Fusion* framework, comprising spatial-temporal dual feature extraction, multi-scale contextual grid representation, similarity-aware global trajectory encoding, and a local-global adaptive fusion network for trajectory reconstruction.
- To address erroneous reconstructions on roads with irregular geometries, we design a *multi-scale contextual grid representation* module that hierarchically integrates trajectory features from neighboring points across different scales, enabling the model to dynamically prioritize the most geometry-appropriate scale for diverse road morphologies.

- To mitigate negative impact of close-proximity neighbors on missing trajectory recovery, we design a *Local-Global Adaptive Fusion Network-based Trajectory Recovery* module that integrates morphologically similar trajectories identified through spectral analysis, and the current trajectory's intrinsic motion continuity, reducing errors by avoiding misleading local patterns.
- Experimental results on two real-world trajectory datasets demonstrate that the proposed method reduces *RMSE* by 16.01% in cases with 10 consecutive missing points, validating its effectiveness.

2 Related Work

Early trajectory completion methods primarily relied on *graph-based heuristics*. Chen et al. introduced a *Maximum Probability Product* algorithm that reconstructs missing segments by identifying optimal paths in transfer networks [4]. Luo et al. developed a temporal path query method to extract frequent routes from historical trajectories within specified time windows [5]. Subsequent approaches incorporated more sophisticated graph construction techniques: Wei et al. proposed a collaborative learning framework to build routable graphs from uncertain trajectories [6], while Elshrif et al. [7] and Isufaj et al. [8] employed shortest-path algorithms on connectivity graphs for trajectory recovery.

Recent advances leverage *deep learning* to model sequential dependencies in trajectory data. Wang et al. pioneered an *RNN-based Seq2Seq* framework with *Kalman filter* calibration to reduce predictive uncertainty [9]. Subsequent works introduced *attention mechanisms*: Zhang et al. combined rule extractors with attention for similarity-enhanced reconstruction [10], while Deng et al. employed cross-attention to model inter-trajectory relationships [11]. Xia et al. extended this through dual self-attention across *intra-* and *inter-trajectory* contexts [12]. The field then evolved to incorporate end-to-end approach. Ren et al. developed *GRU-based* networks with road-network alignment [13]. Chen et al. proposed a *multi-task transformer variant* for network-constrained recovery [14]. Most recently, Wei et al. introduced a graph encoder-decoder that jointly models *micro-* and *macro-semantics* for sparse trajectory completion [15].

The aforementioned trajectory recovery methods demonstrate satisfactory performance in standard scenarios, but exhibit two critical limitations: (1) strong dependence on precise road network data, which restricts their generalizability to road-agnostic environments, and (2) significantly degraded accuracy when handling irregular road geometries (e.g., complex overpasses) or ambiguous spatial adjacency (e.g., parallel roads with <10 m spacing).

3 Problem Definition

In this section, we first present definitions of the basic concepts. Then, we give a formal problem definition.

Definition 1 (Trajectory). *A trajectory tr refers to a sequence of position points generated by a moving object, denoted* $tr = p_0, p_1, \ldots, p_n$, *where each* $p_i = (lng_i, lat_i, t_i)$, $1 \leq i \leq n$, *represents spatial coordinates* (x_i, y_i) *at timestamp* t_i. *The sampling interval* $\varepsilon = t_i - t_{i-1}$ *remains constant for all* $2 \leq i \leq n$.

Definition 2 (Missing Trajectory). *Given a trajectory tr, a missing trajectory* $tr_c = \{p_{i+1}, p_{i+2}, \ldots, p_{j-1}\}$ *is a consecutive subsequence bounded by observed points* p_i *and* p_j, *where* $1 \leq i < j \leq n$ *and* $j - i > 1$.

Problem Definition. Given a trajectory tr with uniform sampling interval ε and a missing segment tr_c bounded by observed points p_i and p_j, the objective is to recover the consecutive point sequence of the missing trajectory tr_c in tr.

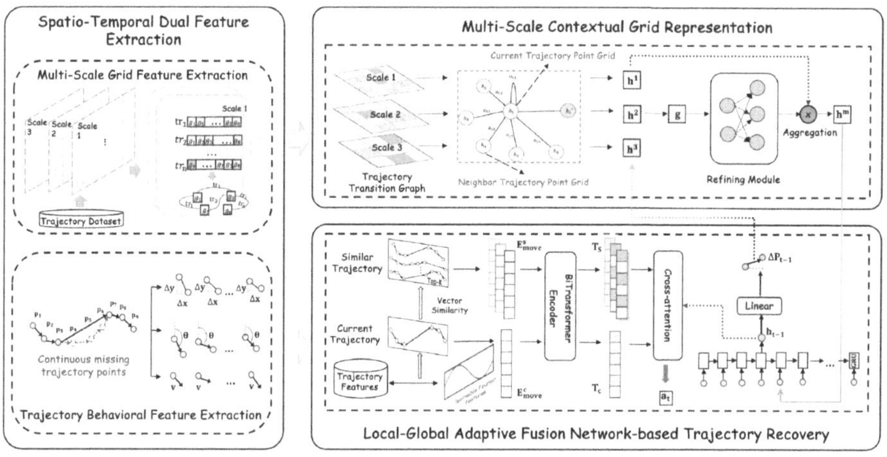

Fig. 2. Overall Framework of MSTRLG.

4 Proposed Model

4.1 Framework

We present a Multi-Scale Trajectory Recovery via Local-Global Similarity Fusion framework MSTRLG. As shown in Fig. 2, MSTRLG involves *Spatial Temporal Dual Feature Extraction*, *Multi-Scale Contextual Grid Representation*, and *Local-Global Adaptive Fusion Network-based Trajectory Recovery*.

4.2 Spatio-Temporal Dual Feature Extraction

Considering that trajectory behavior patterns vary across different spatial scales in complex road regions with diverse structures, and both the current observation and its neighboring points contain critical information for accurately recovering missing trajectory, we extract dual features for each trajectory point: (i) trajectory behavior features, and (ii) multi-scale grid features.

Trajectory Behavioral Feature Extraction. Trajectory behavioral features capture the motion patterns between consecutive points, including: (i) *Point Displacement*, which calculates the relative position of each trajectory point to the previous time-step, denoted as $\Delta p_i = (p_i.lon - p_{i-1}.lon,\ p_i.lat - p_{i-1}.lat)$ (ii) *Point Speed*, v_i, which is calculated by dividing the distance between it and its subsequent one by their time interval; (iii) *Point Direction*, θ_i, which is the direction of the line segment connecting it to its subsequent one. After that, we obtain a trajectory behavioral feature vector $\mathbf{f}_{\text{move}} \in \mathbb{R}^4$ for each point.

Multi-Scale Grid Feature Extraction. To enhance the spatial expressiveness of behavior features under various road shapes and scales, we divide the whole geospatial region into square grids at different scales: 50 m, 100 m, and 200 m, and extract behavioral features for each grid based on the trajectories passing through them. As shown in Fig. 3, for each scale, grid features include: (i) *Point Density Distribution*, which counts of trajectory points within each grid, (ii) *Point Direction Distribution*, which denotes the direction distribution of trajectory points within each grid. Here the point's direction is consistent with the trajectory behavioral feature extraction, and is further discretized into 6 bins(60° each) to form a normalized histogram. Finally, for each scale, we build $\mathbf{f}_{\text{grid}} \in \mathbb{R}^7$ for each grid, forming a feature matrix $\mathbf{H} \in \mathbb{R}^{7\times M\times N}$. Note that M and N vary across scales due to differences in grid scale.

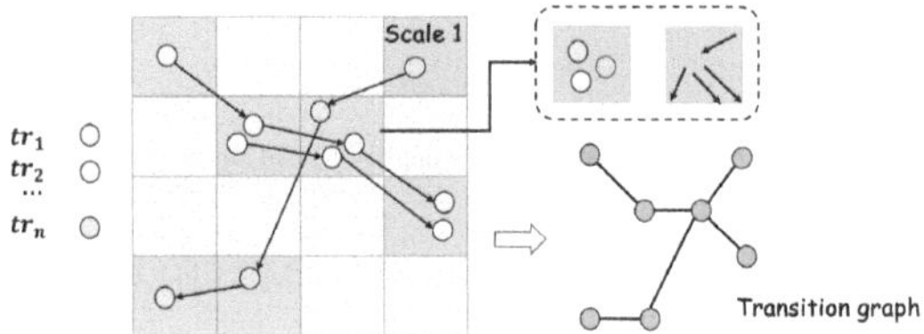

Fig. 3. Extraction of grid features and construction of trajectory transition graph.

4.3 Multi-Scale Contextual Grid Representation

To better reconstruct missing trajectories' shape on roads with irregular geometries, considering the behavior of missing trajectories are influenced by surrounding points, we design a *Multi-Scale Contextual Grid Representation* module. Specially, we first construct a trajectory transition graph to represent the linking between grids at different scales. Then, we employ a GAT to aggregate features from the current point and its neighboring points into grid representations.

Trajectory Transition Graph Construction. We propose to build a trajectory transition graph based on trajectories to represent the linking relationship of different grids, with each grid as a vertex. For each pair of adjacent trajectory points, construct an edge for the grids they belong to. As shown in Fig. 3, the graph reflects road constraints through motion patterns when the actual road network is unavailable.

Neighborhood Feature Encoding. Considering the robustness and remarkable performance of GAT in learning vertex embeddings, we employ a GAT to generate the vertex representations [19,21,22]. For each scale, we first apply a 1×1 convolution to project the feature matrix $\mathbf{H} \in \mathbb{R}^{7 \times M \times N}$ into initial vertex representations $\mathbf{H}^0 \in \mathbb{R}^{F \times M \times N}$, where F is the vertex dimension. GAT introduced the attention mechanism to dynamically assign attention scores to neighboring vertex. The attention score between vertex i and its neighbor j is computed as:

$$e_{ij} = \text{LeakyReLU}\left(\mathbf{W}_2 \left[\mathbf{W}_1 \mathbf{h}_i \,\|\, \mathbf{W}_1 \mathbf{h}_j\right]\right), \tag{1}$$

$$a_{ij} = \frac{\exp(e_{ij})}{\sum_{k \in \mathcal{N}_i} \exp(e_{ik})}, \tag{2}$$

where $\mathbf{W}_1 \in \mathbb{R}^{F \times F}$ and $\mathbf{W}_2 \in \mathbb{R}^{2F \times 1}$ are learnable parameters, $\mathbf{h}_i, \mathbf{h}_j \in \mathbb{R}^F$ are the vertex representations of vertex i and j, and $\mathcal{N}_i$ denotes the neighbors of vertex i in the trajectory transition graph. Then we update the representation $\mathbf{h}_i$ of the grid i through integrating the features of its neighborhood vertices :

$$\mathbf{h}_i = \sigma\Big(\sum_{j \in \mathcal{N}_i} a_{ij} \cdot \mathbf{W}_3 \mathbf{h}_j\Big), \tag{3}$$

where $\mathbf{W}_3 \in \mathbb{R}^{F \times F}$ denotes the learnable parameters, σ represents a non-linear function. After N layers GAT, the final vertex representation is $\mathbf{h} \in \mathbb{R}^F$. For each trajectory point, we have three grids representation: $\mathbf{h}^1$, $\mathbf{h}^2$, and $\mathbf{h}^3$. To dynamically assigns dominant weights to optimal scales, we introduce a learnable fusion mechanism. Specifically, we first compute a contextual vector $\mathbf{g}$ through average pooling across three scales, and then fed into a refining module consisting of two fully connected layers to generate three projection matrices $\mathbf{W}_1, \mathbf{W}_2, \mathbf{W}_3$. Finally, the fused grid representation is obtained as:

$$\mathbf{h}^m = \text{ReLU}\left(\mathbf{W}_1 \mathbf{h}^1 + \mathbf{W}_2 \mathbf{h}^2 + \mathbf{W}_3 \mathbf{h}^3\right), \tag{4}$$

4.4 Local-Global Fusion Network-Based Trajectory Recovery

Local neighboring trajectory points on closely spaced parallel roads often exhibit varying movement patterns, which can mislead existing methods to recover trajectories on incorrect roads. However, morphologically similar trajectory segments from parallel roads provide useful local geometric cues, while the known portions of the current trajectory impose global motion continuity constraints. Inspired by these, we propose a novel fusion mechanism that leverages learnable Fourier features to first identify morphologically similar trajectories for local geometric modeling, and then combines them with current trajectory's global motion continuity and multi-scale contextual information to enable robust trajectory recovery.

Similar Trajectory Retrieval. Motivated by the success of learnable Fourier features in trajectory modeling [20], we represent trajectories using Fourier feature encoding to facilitate robust similarity retrieval.

Formally, we aim to learn a mapping function $\Phi(\mathbf{f}_{\text{move}}) : \mathbb{R}^4 \rightarrow \mathbb{R}^d$, which transforms $\mathbf{f}_{\text{move}}$ into $\mathbf{e}_{\text{move}}$:

$$\Phi(\mathbf{f}_{\text{move}}) = \frac{1}{\sqrt{d}} \left[\cos(\mathbf{f}_{\text{move}} \mathbf{W}_{\mathbf{r}}) \parallel \sin(\mathbf{f}_{\text{move}} \mathbf{W}_{\mathbf{r}})\right], \tag{5}$$

where $\mathbf{W}_{\mathbf{r}} \in \mathbb{R}^{4 \times d/2}$ is a random projection matrix sampled from a Gaussian distribution. For each trajectory, this process generates representation $\mathbf{E}^c_{\text{move}} \in \mathbb{R}^{n \times d}$, where n denotes the number of points in the trajectory. The similarity between two features $\Phi(x)$ and $\Phi(y)$ can be computed via their dot product:

$$S(x, y) = \Phi(x) \cdot \Phi(y) = \frac{1}{d} \sum \cos((x - y)\mathbf{W}_{\mathbf{r}}), \tag{6}$$

which corresponds to a translation-invariant kernel function parameterized by W_r and ensures similarity depends only on the relative offset $(x-y)$, and is robust to local inconsistencies. By computing the point-wise dot product between trajectories and averaging the results, we retrieve the top-K most similar trajectories representation $\mathbf{E}^s_{\text{move}} = \left\{\mathbf{E}^i_{\text{move}} \in \mathbb{R}^{n \times d}\right\}_{i=1}^K$.

Bidirectional Trajectory Encoding. To jointly leverage contextual information from both preceding and succeeding points, we design a BiTransformer architecture composed of two Transformer encoders, which capture temporal dependencies in two directions. Multi-head attention is the core operation of Transformers, allowing the model to capture long-range dependencies across the entire input. Formally, the current trajectory representation $\mathbf{E}^c_{\text{move}}$ and its similar trajectories representation $\mathbf{E}^s_{\text{move}}$ are fed into the BiTransformer to generate new representations$\mathbf{T}_c \in \mathbb{R}^{n \times d}$ and $\mathbf{T}_S = \left\{\mathbf{T}^i_s \in \mathbb{R}^{n \times d}\right\}_{i=1}^K$

Trajectory Recovery via Local-Global Fusion. After constructing grid and trajectory representations, we recover missing segments in an autoregressive manner using a GRU, whose initial hidden state $\mathbf{H}_0$ is derived from $\mathbf{T}_c$ through averaging and linear projection. Considering the varying contributions of different points in similar trajectories and current trajectory to the recovery of missing trajectory points, we employ a cross-attention mechanism that adaptively weights each trajectory point based on its relevance to the current hidden state. Specifically, the weighted trajectory representation $\mathbf{a}^s_t$ is computed as:

$$\mathbf{a}_t = \sum_{i=1}^{K} \sum_{j=1}^{n} \alpha_t^{ij} \mathbf{T}^{ij} \tag{7}$$

$$\alpha_t^{ij} = \frac{\exp(\mathbf{u}_t^{ij})}{\sum_{j'=1}^{n} \exp(\mathbf{u}_t^{ij'})} \tag{8}$$

$$\mathbf{u}_t^{ij} = \mathbf{W}_6 \tanh\left(\mathbf{W}_4\mathbf{H}_{t-1} + \mathbf{W}_5\mathbf{T}^{ij}\right) \tag{9}$$

where $\mathbf{W}_4, \mathbf{W}_5 \in \mathbb{R}^{d\times d}$ and $\mathbf{W}_6 \in \mathbb{R}^{1\times d}$ are learnable parameters. Here, $\mathbf{T}^{ij}$ denotes the j-th point of the i-th trajectory, with $i = 0$ referring to the current trajectory $\mathbf{T}_c$ and $i \in 1,\ldots,K$ to the similar trajectories $\mathbf{T}_S$. At each timestep t, the GRU takes four inputs: (1) the previous hidden state $\mathbf{H}_{t-1}$; (2) the previous predicted displacement Δp_{t-1}; (3) the fused grid representation $\mathbf{h}_t^m$ for $\hat{p_{t-1}}$; (4) the weighted trajectory representation $\mathbf{a}_t$. The hidden state is updated as:

$$\mathbf{H}_t = \mathrm{GRU}\left(\mathbf{H}_{t-1}, \Delta p_{t-1}, \mathbf{h}_t^m, \mathbf{a}_t\right) \tag{10}$$

$\mathbf{H}_t$ is then passed through a linear transformation to estimate $\Delta\mathbf{p}_t$, which is added to the previous coordinate to obtain the current point: $\hat{p_t} = \hat{p_{t-1}} + \Delta p_t$.

4.5 Model Traning

We simply use mean square error between the locations of true trajectory $p_t \in \mathbb{R}^{2\times n}$ and the locations of predicted trajectory $\hat{p_t} \in \mathbb{R}^{2\times n}$ as the loss function:

$$\mathcal{L} = \frac{1}{n}\sum_{t=1}^{n} ||p_t - \hat{p_t}||_2^2 \tag{11}$$

5 Experiments

5.1 Experimental Settings

Datasets. To evaluate the recovery performance of MSTRLG, we use the publicly available Porto taxi trajectory dataset (**Porto2013**[2]) and the private Shanghai cycling trajectory dataset collected near the Huangpu River (**Shanghai2020**). The original sampling period of the Porto dataset is 15 s. For easier comparison with previous studies [9,10], we resample it to one minute during preprocessing. The detailed statistics of them are given in Table 1.

Table 1. Description of experimental datasets

Dataset	Shanghai2020	Porto2013
Time Range	30/06/2020–30/06/2021	01/07/2013–30/06/2014
Trajectory	46,870	280,525
Latitude range	(31.204398,31.24151)	(41.111975,41.177462)
Longitude range	(121.465479,121.51698)	(−8.667057,−8.585305)
Sampling Interval	10 s	60 s

[2] https://www.kaggle.com/c/pkdd-15-predict-taxi-service-trajectory-i.

Evaluation Metrics. To evaluate the recovery performance, we use four metrics:(1) **RMSE**: Root Mean Square Error between the true and predicted positions of missing points;(2) **NDTW**: Normalized Dynamic Time Warping distance, which divides Dynamic Time Warping distance (DTW) [18] by the trajectory length;(3) **Hausdorff Distance** [16]: Measures the maximum distance from a point on one trajectory to the closest point on the other;(4) **Frĉchet Distance** [17]: Adds sequential constraints to Hausdorff and better captures overall curve similarity.

Baselines. We compare MSTRLG with the following five baselines: (1) **LI** uses linear interpolation to recover missing points; (2) **DHTR** [9] first uses a subseq2seq model to predict grid sequences, then calibrate points by Kalman Filter; (3) **SimiDTR** [10] extracts similar trajectories using rule-based extractors and uses them to recover missing coordinates; (4) **TrImpute** [7] is a heuristic search algorithm which selects candidate trajectory points based on distance and angle thresholds, then finds the shortest path between segment endpoints. (5) **GTI** [8] is a heuristic search algorithm which constructs a directed graph from missing segments and finds shortest paths on the graph.

Experimental Settings. All experiments are conducted on a GPU-CPU platform with Tesla V100. The program and baselines are implemented in PyTorch 1.7.1 and Python 3.8. Based on observations, the lengths of missing trajectory typically range from 5 to 10. To simulate missing data, we randomly remove m consecutive points ($m = 5, 10$) from each trajectory. We set the embedding dimensionality d to 256. The number of GAT-layers is set to $M = 2$, and the dimensionality of node features is $F = 16$. The Transformer module contains 6 layers and uses 4 attention heads. For the Shanghai2020 dataset, the number of similar trajectories K is set to 4, and for Porto2013, K is set to 6. We train MSTRLG using the AdamW optimizer and set the learning rate to 1e−4.

Table 2. Comparison of Trajectory Recovery Performance on Different Datasets

Methods	*Shanghai2020*								*Porto2013*							
	$m = 5$				$m = 10$				$m = 5$				$m = 10$			
	RMSE	*NDTW*	*Hausdorff*	*Fréchet*	*RMSE*	*NDTW*	*Hausdorff*	*Fréchet*	*RMSE*	*NDTW*	*Hausdorff*	*Fréchet*	*RMSE*	*NDTW*	*Hausdorff*	*Fréchet*
Li	0.0610	0.0262	0.1130	0.0659	0.0891	0.287	0.0907	0.0928	0.2610	0.1210	0.2992	0.4402	0.4893	0.2867	0.5924	0.6091
TrImpute	0.0443	0.0229	0.0578	0.0787	0.0784	0.0248	0.0732	0.0824	0.2733	0.1252	0.2852	0.4521	0.4352	0.2651	0.6293	0.6064
GTI	0.0426	0.0204	0.0539	0.0495	0.0762	0.0234	0.0695	0.0804	0.2628	0.0639	0.2752	0.3382	0.4242	0.2412	0.6040	0.5568
DHTR	0.0560	0.0257	0.0556	0.0893	0.0866	0.0303	0.0887	0.1291	0.2402	0.0616	0.2133	0.3468	0.4524	0.1735	0.4186	0.5665
SimiDTR	0.0434	0.0235	0.0744	0.0707	0.0767	0.0249	0.0775	0.0870	0.1565	0.0589	0.3349	0.3075	0.3585	0.1438	0.5769	0.5297
MSTRLG	**0.0409**	**0.0184**	**0.0427**	**0.0452**	**0.0735**	**0.0206**	**0.0549**	**0.0712**	**0.1498**	**0.0546**	**0.2099**	**0.3013**	**0.3011**	**0.1403**	**0.4097**	**0.5125**

5.2 Performance Comparison

Quantitative Comparison. Table 2 reports the quantitative results of diffrent methods for trajectory recovery results for different lengths of missing trajectory on two datasets. We have the following observations:

- On the Shanghai2020 dataset (dense bike trajectories with short intervals), heuristic methods (TrImpute, GTI) outperform deep learning approaches (DHTR, SimiDTR) due to regular spatial continuity enabling effective local graph-based search. However, their reliance on computationally expensive graph construction limits scalability for large datasets.
- Conversely, Porto2013's taxi trajectories feature larger sampling intervals and complex road geometries where deep learning excels through modeling intricate spatiotemporal correlations. However, no single baseline model can consistently deliver the best performance.
- Finally, our model achieves promising performance compared to all the baselines. We attribute the above phenomena to two reasons:(1)it extracts neighboring points features at multiple scales;(2)it combines local geometric patterns from similar trajectories with own global motion continuity constraints.

Visualization Comparison. We visualize two cases to compare the recovery results of different methods. As shown in Fig. 4, the ground truth trajectory is represented by the green line, the recovered trajectory by the red line, and the missing trajectory by the blue dashed line. In Case 1, we choose a trajectory on a ring road to validate the effectiveness of the multi-scale contextual grid representation module. MSTRLG leverages the behavioral features of neighboring trajectory points at multiple scales, leading to recovered trajectories that better align with the true trajectory. In Case 2, a missing trajectory on the closely spaced parallel roads is chosen to test the local-gobal fusion network-based trajectory recovery module. MSTRLG uses local geometric patterns from similar trajectories and global motion continuity from known segments, resulting in recovered trajectories that are more consistent with the true trajectory.

Table 3. MSTRLG Ablation Analysis Results

Methods	*Shanghai2020*								*Porto2013*							
	$m = 5$				$m = 10$				$m = 5$				$m = 10$			
	RMSE	*NDTW*	*Hausdorff*	*Fréchet*	*RMSE*	*NDTW*	*Hausdorff*	*Fréchet*	*RMSE*	*NDTW*	*Hausdorff*	*Fréchet*	*RMSE*	*NDTW*	*Hausdorff*	*Fréchet*
w/o GR	0.0442	0.0214	0.0493	0.0523	0.0766	0.0231	0.0637	0.0813	0.1582	0.0588	0.2121	0.3109	0.3321	0.1485	0.4182	0.5275
w/o SR	0.0426	0.0201	0.0473	0.0493	0.0749	0.0219	0.0644	0.0781	0.1541	0.0572	0.2104	0.3182	0.3176	0.1452	0.4145	0.5217
MSTRLG	**0.0409**	**0.0184**	**0.0427**	**0.0452**	**0.0735**	**0.0206**	**0.0549**	**0.0712**	**0.1498**	**0.0546**	**0.2099**	**0.3013**	**0.3011**	**0.1403**	**0.4097**	**0.5125**

5.3 Model Analysis

Ablation Study. We design two model variants for ablation: 1) w/o GR removes the Multi-Scale Contextual Grid Representation module; 2) w/o SR removes the Similar Trajectory Retrieval module, encoding only the current trajectory. As shown in Table 3, removing either module leads to performance degradation, confirming that integrating multi-scale contextual aggregation and local-global adaptive fusion is essential for improving trajectory recovery accuracy of irregular overpasses and parallel roads with varying movement patterns.

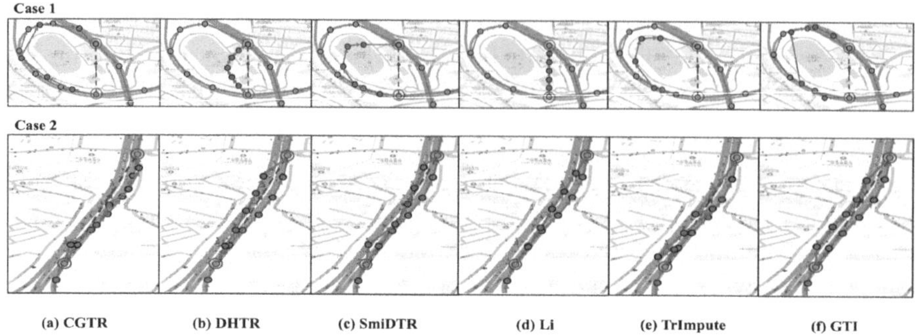

Fig. 4. Visualization of missing trajectory.

Parameter Sensitivity. We further conduct a parameter study to analyze the impacts of the parameter K in *Similar Trajectory Retrieval.* As shown in Fig. 5, we observe that MSTRLG performs the best under K = 4,6 in Shanghai2020, Porto2013. On the one hand, setting K too small limits spatiotemporal information, leading to a high RMSE. On the other hand, setting K too large increases computational complexity and may introduce noise or irrelevant trajectories, which is harmful to recovery trajectory.

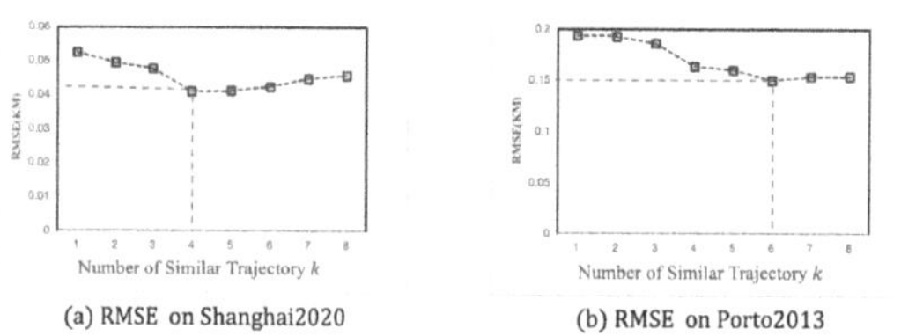

Fig. 5. Impact of K on two datasets.

6 Conclusion

This paper presents *MSTRLG*, an innovative trajectory recovery framework that effectively addresses two critical challenges: the accurate recovery of irregularly shaped trajectories and trajectories in close proximity to adjacent roads. The proposed framework achieves this through a novel integration of *multi-scale contextual aggregation* and *adaptive local-global similarity fusion*, which enhance trajectory recovery accuracy. Extensive experiments conducted on two real-world datasets demonstrate that *MSTRLG* achieves superior performance compared to state-of-the-art methods. Looking forward, we plan to extend this work by investigating multi-modal data fusion techniques, to further improve trajectory recovery in particularly challenging road environments such as tunnels.

References

1. Ahmed, M., Wenk, C.: Constructing street networks from GPS trajectories. In: ESA(2012)
2. Li, X., Cong, G., Sun, A., Cheng, Y.: Learning travel time distributions with deep generative model. In: WWW, pp. 1017–1027 (2019)
3. Zhao, W. X., Zhou, N., Sun, A., Wen, J. R., Han, J., Chang, E. Y.: A time-aware trajectory embedding model for next-location recommendation. Knowl. Inf. Syst., 559–579(2018)
4. Chen, Z., Shen, H., Zhou, X.: Discovering popular routes from trajectories. In: ICDE, pp. 900–911. IEEE (2011)
5. Luo, W., Tan, H., Chen, L., Ni, L.M.: Finding time period-based most frequent path in big trajectory data. In: SIGMOD, pp. 713–724 (2013)
6. Wei, L.Y., Zheng, Y., Peng, W.C.: Constructing popular routes from uncertain trajectories. In: KDD, pp. 195–203 (2012)
7. Elshrif, M.M., Isufaj, K., Mokbel, M.F.: Network-less trajectory imputation. In: GIS, pp. 1–10 (2022)
8. Isufaj, K., Elshrif, M.M., Abbar, S., Mokbel, M.: GTI: a scalable graph-based trajectory imputation. In: GIS, pp. 1–10 (2023)
9. Wang, J., Wu, N., Lu, X., Zhao, W.X., Feng, K.: Deep trajectory recovery with fine-grained calibration using kalman filter. IEEE Trans. Knowl. Data Eng. **33**(3), 921–934 (2019)
10. Zhang, Y., Deng, L., Zhao, Y., Chen, J., Xie, J., Zheng, K.: Simidtr: deep trajectory recovery with enhanced trajectory similarity. In: DASFAA, pp. 431–447 (2023)
11. Deng, L., Zhao, Y., Sun, H., Yang, C., Xie, J., Zheng, K.: Fusing local and global mobility patterns for trajectory recovery. In: DASFAA, pp. 448–463 (2023)
12. Xia, T., Qi, Y., Feng, J., Xu, F., Sun, F., Guo, D., Li, Y.: Attnmove: history enhanced trajectory recovery via attentional network. In: AAAI, pp. 4494–4502 (2021)
13. Ren, H., Ruan, S., Li, Y., Bao, J., Meng, C., Li, R., Zheng, Y.: Mtrajrec: map-constrained trajectory recovery via seq2seq multi-task learning. In: SIGKDD (2021)
14. Chen, Y., Zhang, H., Sun, W., Zheng, B.: Rntrajrec: road network enhanced trajectory recovery with spatial-temporal transformer. In: ICDE (2023)
15. Wei, T., Lin, Y., Lin, Y., Guo, S., Zhang, L., Wan, H.: Micro-macro spatial-temporal graph-based encoder-decoder for map-constrained trajectory recovery. IEEE Trans. Knowl. Data Eng. (2024)
16. Huttenlocher, D.P., Kedem, K., Kleinberg, J.M.: On dynamic voronoi diagrams and the minimum Hausdorff distance for point sets under Euclidean motion in the plane. In: SCG, pp. 110–119 (1992)
17. FRÉCHET, M.: Sur quelques points du calcul fonctionnel. In: Rendiconti del Circolo Matematico di Palermo, vol. 22, no. 1, pp. 1–72 (1906)
18. Agrawal, R., Faloutsos, C., Swami, A.: Efficient similarity search in sequence databases. In: FODO, pp. 69–84 (1993)
19. P. Velĭckovĭc, G. Cucurull, A. Casanova, A. Romero, P. Lió, and Y. Bengio.: Graph attention networks. In: ICLR (2018)
20. Li, Y., Si, S., Li, G., Hsieh, C. J., Bengio, S.: Learnable fourier features for multi-dimensional spatial positional encoding. In: NeurIPS, pp. 15816–15829 (2021)
21. Chen, J., Meng, D., Gao, X., Zhang, L., Kong, C.: AGNE: attentional graph convolutional network embedding for knowledge concept recommendation in MOOCs. In: WISA, pp. 463–475 (2024)

22. Cao, S., Lu, W., Xu, Q.: Grarep: learning graph representations with global structural information. In: CIKM, pp. 891–900 (2015)

Streaming Rank-Happiness Maximization Queries Under Group Fairness Constraints

Conghao Liu[1] and Jiping Zheng[1,2](✉)

[1] College of Computer Science and Technology, Nanjing University of Aeronautics and Astronautics, Nanjing, China
{liuconghao,jzh}@nuaa.edu.cn

[2] State Key Laboratory for Novel Software Technology, Nanjing University, Nanjing, China

Abstract. In multi-criteria decision-making, the happiness maximization query, a.k.a. the regret minimization query, has been an important tool besides the top-k and skyline queries, and its objective is to select a small subset of tuples that retains the best score across multiple criteria. As a variant of happiness-based queries, the rank-happiness maximization query integrates the concept of *rank* into the happiness ratio, preventing the ratio from being perceived as an arbitrary numerical value which frequently confuses users. Meanwhile, in many real-world scenarios, data arrives in a streaming manner while fairness concerns remain persistently relevant. However, existing algorithms for multi-criteria decision-making neglect the fairness concerns when processing streaming data. In this paper, we study the streaming rank-happiness maximization queries under group fairness constraints. We employ the average rank-happiness ratio as a metric to measure users' satisfaction, thereby avoiding bias toward a few dissatisfied users. As the average rank-happiness function is monotone and submodular, we formulate the group fairness constraints as matroid constraints. Thus, we convert our problem into a streaming submodular maximization problem under a matroid constraint and propose the FairStream algorithm with provable theoretical guarantees to solve the problem. Extensive experiments on real-world and synthetic datasets confirm the effectiveness and efficiency of our proposed algorithm.

Keywords: Rank-happiness maximization query · Data stream · Group fairness · Submodular optimization · Top-k queries · Skyline queries

1 Introduction

Selecting a few representative tuples from a large dataset is an important functionality in many applications, e.g., multi-criteria decision-making, recommendation and Web search. The most widely used tools to fulfill this task are the top-k query [9,13], the skyline query [3] and the happiness maximization query (HMQ) a.k.a. the regret minimization query (RMQ) [14,16]. HMQ or RMQ [14]

S. Yang et al. (Eds.): WISA 2025, LNCS 16299, pp. 207–218, 2026.
https://doi.org/10.1007/978-981-95-6426-2_18

is proposed as a balanced alternative to address the uncontrollable output sizes of skyline queries and the limitation of top-k queries which require users to explicitly provide utility functions. The minimization of regret ratio in RMQ is often equivalent to the maximization of happiness ratio, but the happiness ratio is more favorable to conduct theoretical analysis [12,16]. Since the happiness ratio or the regret ratio is usually an artificial number with no direct significance, the rank-regret representative (RRR) problem [1] is introduced which employs the concept of rank-regret to quantify user dissatisfaction based on ranking positions. Correspondingly, the query with rank-happiness is referred as the rank-happiness maximization query (RHMQ).

In many real-world applications, on the one hand, data points are not static and often arrive in the streaming manner. On the other hand, fairness is receiving more attention. For example, consider an educational institution that is hiring teachers. When processing applications that arrive sequentially, the institution must assess not only the candidates' qualifications but also ensure balanced representation across subjects and genders to meet diverse teaching tasks. Although existing studies [23,25] expand their focus to dynamic environments, and investigate the group fairness constraints in happiness maximization queries [24], the integration of fairness constraints into dynamic or streaming settings remains unexplored. Motivated by this, we study the problem of streaming rank-happiness maximization queries under group fairness constraints (Streaming-FairRHMQ). Specifically, we introduce the *group fairness* constraints, which is a widely used form of fairness in many real-world problems [17]. Moreover, to avoid bias towards a few dissatisfied users, we introduce the average form of rank-happiness ratio which incorporates the probability distribution of user utility functions to maximize users' satisfactions.

To solve the problem above, we propose an algorithm called FairStream. In the streaming setting, the algorithm can only learn from each point upon its arrival. For every arriving point, we incorporate a random subsampling technique [8] to improve the efficiency. Moreover, we sacrifice partial spatial resources to cache points that are probabilistically discarded through subsampling, thereby enhancing the effectiveness and robustness of the algorithm. To sum up, our main contributions are listed as follows:

- We formulate a new problem of Streaming-FairRHMQ and show the submodularity and monotonicity of the average rank-happiness ratio function.
- We propose an algorithm called FairStream for this problem with approximation guarantee, i.e., $1/4$.
- Extensive experiments on both synthetic and real-world datasets are conducted to verify the effectiveness, efficiency, and scalability of our proposed FairStream algorithm.

2 Related Work

The happiness maximization query (HMQ) initially known as the k-regret query was first proposed by Nanongkai et al. [14]. HMQ is identical to the regret minimization query (RMQ) where the happiness ratio is defined as one minus the regret ratio. Several efficient methods have been developed to the problem of HMQ [22]. However, RMQ/HMQ fails to deal with the probability distribution of utility functions, which may lead to sub-optimal satisfaction in certain scenarios. Zeighami et al. [20] first proposed the average regret minimizing sets problem and provided an approximation algorithm based on the supermodularity and monotonicity of the average regret ratio. Subsequently, Luenam et al. [12] proposed and solved the problem of average happiness maximization queries. Moreover, as an important variant of RMQ/HMQ, the rank-regret minimization query was introduced [1,18] to evaluate users' regret by rank rather than numerical values, thus mitigating the inadequacy of the regret/happiness ratio in precisely quantifying users' happiness levels.

Since the results of HMQ may show discrimination or bias against specific groups, some studies [24,26] have integrated fairness with RMQ/HMQ. However, all aforementioned works focus on the static setting. For the streaming and dynamic settings, some efficient algorithms for HMQ have been proposed [23,25]. As the average rank-happiness function shows submodularity and monotonicity, another related work is submodular maximization. In the streaming setting, typical algorithms such as IndependentSetImprovement [5], Sieve-Streaming [2], Sieve-Streaming++ [10] and ThreeSieves [4] have been proposed. Furthermore, Halabi et al. [6] investigate the relationship between fairness and matroids in the streaming model; however, their proposed two-pass algorithm achieves only $\frac{1}{11.656}$-approximation ratio and fails to strictly satisfy the fairness constraints. They also introduce the GreedyFairStreaming algorithm, which demonstrates strong empirical performance; however, it generally lacks worst-case guarantees. In a nutshell, there still lack efficient and effective algorithms for the incorporation of fairness constraints within streaming contexts in the field of data querying.

3 Preliminaries

3.1 Problem Definition

Let $\mathcal{D}$ be a d-dimensional dataset containing n data points. For each point $p \in \mathcal{D}$, $p = \langle p[1], p[2], \ldots, p[d] \rangle \in \mathcal{D}$ is described by d numerical attributes which are normalized to the range $[0, 1]$.

Definition 1. (Utility Function). *A utility function f is a mapping $f : \mathbb{R}_+^d \rightarrow \mathbb{R}_+$. Given a utility function f, the utility of a data point p is denoted as $f(p)$, which shows how satisfied the user is with p.*

Following the conventions in the literature [14,19,24], we assume the form of the utility functions to be linear, i.e., f is represented as a d-dimensional vector $v = \langle v[1], v[2], \ldots, v[d] \rangle \in \mathbb{R}_+^d$, where $v[i]$ denotes the importance of the i-th dimension in users' happiness. W.l.o.g., we assume that v is normalized such that $||v||_1 = \sum_{i=1}^{d} v[i] = 1$. Thus, the utility of a data point p w.r.t. f is expressed as $f(p) = v \cdot p = \sum_{i=1}^{d} v[i] \cdot p[i]$.

Definition 2. (Rank-happiness ratio). *Given a set $S \subseteq \mathcal{D}$ and a utility function f, the rank-happiness ratio function is defined as,*

$$rhr_f(S) = \frac{n - \nabla_f(S)}{n} = \frac{\max_{p \in S}(n - \nabla_f(p))}{n} = \frac{n - \min_{p \in S} \nabla_f(p)}{n}. \tag{1}$$

$\nabla_f(p)$ denotes the rank of p in the sorted list of $\mathcal{D}$ in descending order as to f. Given that users prefer a larger value in each dimension of p, the smaller $\nabla_f(p)$, the better p is.

Definition 3. (Average Rank-happiness Ratio). *Given a subset $S \subseteq \mathcal{D}$ and a utility function class $\mathcal{FC}$ with the probability density function $\eta(\cdot)$ associated with a probability distribution Θ, the average rank-happiness ratio (arhr) of S is defined as*

$$arhr_{\mathcal{FC}}(S) = \int_{f \in \mathcal{FC}} rhr_f(S) \cdot \eta(f) df. \tag{2}$$

Since the class of all possible linear utility functions $\mathcal{FC}$ is uncountable, we utilize a sampling technique from [21] to compute the average rank-happiness ratio with a theoretical bound. Specifically, we sample m utility functions according to the distribution Θ and obtain a set of sampled utility functions $\mathcal{F} = \{f_1, f_2, \cdots, f_m\}$. Using $\mathcal{F}$, we compute the estimated average rank-happiness ratio by averaging the rank-happiness ratio of the m sampled utility functions, as follows,

$$arhr_{\mathcal{F}}(S) = \frac{1}{m} \sum_{f \in \mathcal{F}} rhr_f(S). \tag{3}$$

Fairness Model. We adopt a well-established model for the group fairness in the literature [24]. Besides the numerical attributes in the data model, the points in dataset $\mathcal{D}$ are associated with categorical attributes representing protected characteristics such as *gender* or *race*. Suppose that $\mathcal{A}$ is the domain of an attribute A with $C = |\mathcal{A}|$. The dataset $\mathcal{D}$ is partitioned into C disjoint groups $\{\mathcal{D}_1, \mathcal{D}_2, \ldots, \mathcal{D}_C\}$ by attribute A with $\mathcal{D} = \bigcup_{c=1}^{C} \mathcal{D}_c$.

Given a dataset $\mathcal{D}$ with C groups $\{\mathcal{D}_1, \mathcal{D}_2, \ldots, \mathcal{D}_C\}$, the group fairness constraints is defined by restricting the number of tuples selected from each group $\mathcal{D}_c$ to lie between a lower bound l_c and an upper bound u_c in $\mathbb{Z}_+$ with $l_c \leq u_c$. Formally, a subset $S \subseteq \mathcal{D}$ is considered fair if and only if $l_c \leq |S \cap \mathcal{D}_c| \leq u_c$ for each $c \in [C]$. Furthermore, we limit the total number of data points in the result set S to a positive integer $k \in \mathbb{Z}_+$.

Next, we formally define the streaming rank-happiness maximization query under fairness constraints (Streaming-FairRHMQ) based on the above notions.

Definition 4. (Streaming-FairRHMQ.) *Given a data stream $\mathcal{D}$ with C disjoint groups $\{\mathcal{D}_1, \ldots, \mathcal{D}_C\}$, a user-specified positive integer k, and a set of $2C$ positive integers $l_1, \ldots, l_C$ and $u_1, \ldots, u_C$, the Streaming-FairRHMQ problem asks for a subset $S \subseteq \mathcal{D}$ of size k such that $arhr_{\mathcal{F}}(S)$ is maximized and $l_c \leq |S \cap \mathcal{D}_c| \leq h_c$ for each group $\mathcal{D}_c$, i.e., returns a subset S^*,*

$$S^* = \underset{S \subseteq \mathcal{D}: |S| = k}{\arg\max}\ arhr_{\mathcal{F}}(S) \quad s.t.\ l_c \leq |S \cap \mathcal{D}_c| \leq u_c, \forall c \in [C]. \tag{4}$$

In the streaming setting we assume that the data points in $\mathcal{D}$ arrive sequentially in some adversarially chosen order, and the algorithm learns about each point only when it arrives. From the analysis in [18], we obtain that the RHMQ problem is NP-hard for any $d \geq 3$. Consequently, Streaming-FairRHMQ is also NP-hard for $d \geq 3$.

3.2 Properties

We first introduce the monotonicity and submodularity properties of the average rank-happiness ratio function as they are useful for designing our approximation algorithm.

Given a non-negative set function g: $2^{\mathcal{D}} \rightarrow \mathbb{R}_+$ and a subset $S \subseteq \mathcal{D}$, we use $g(p|S) := g(S \cup \{p\}) - g(S)$. A set function g is **monotone** iff for any $S_1 \subseteq S_2 \subseteq \mathcal{D}$, it holds that $g(S_2) \geq g(S_1)$. And g is **submodular** iff for any $S_1 \subseteq S_2 \subseteq \mathcal{D}$ and $p \in \mathcal{D} \backslash S_2$, it holds that $g(p|S_1) \geq g(p|S_2)$.

Lemma 1. *The average rank-happiness ratio function $arhr(\mathcal{F}, S)$ is a monotone non-decreasing submodular function.*

Lemma 1 can be easily obtained by the definitions of submodularity and monotonicity.

A concept closely related to the fairness constraints is matroid. A matroid $\mathcal{M}$ is defined on a finite ground set $\mathcal{D}$ and a family $\mathcal{I}$ of subsets of $\mathcal{D}$ called *independent sets*. For a comprehensive introduction to matroids, see [11]. We define a matroid $\mathcal{M} = (\mathcal{D}, \mathcal{I})$ [7] based on the group fairness constraints, where the collection of independent sets is characterized as follows:

$$\mathcal{I} = \Big\{ S \subseteq \mathcal{D} : \sum_{c=1}^{C} \max\{|S \cap \mathcal{D}_c|, l_c\} \leq k \wedge |S \cap \mathcal{D}_c| \leq h_c, \forall c \in [C] \Big\}. \quad (5)$$

Hence, the group fairness constraints will be treated as a special case of a matroid constraint in our proposed algorithm. Therefore, our Streaming-FairRHMQ problem can be reformulated as a streaming submodular maximization problem under a matroid constraint.

4 The Algorithm for Streaming-FairRHMQ

In this section, we provide the FairStream algorithm which employs a probabilistic subsampling strategy, whereby each data point in a stream is accepted or disregarded with an appropriate probability. The random acceptance or disregard of points undoubtedly enhances the algorithm's efficiency by reduce the processing overhead. However, the subset potentially comprises points contributing significantly to the solution's value. Moreover, for the groups with small size, the subsampling approach may lead to no points of the entire group with non-zero probability are selected, thereby violating the group fairness constraints. To avoid these limitations, our algorithm caches a limited number of disregarded points in the backup sets $B = \{B_1, B_2, \cdots, B_C\}$, ensuring that high-value points are retained with controllable probability and the minimal representation from all protected groups is maintained. The points that have not been disregarded constitute the main component of the solution set, denoted as set A. Finally, the algorithm optimizes the main set A using backup sets B to form the final solution set S.

In the following sections, we use $f_a(\cdot)$ to denote the average rank-happiness ratio function $arhr(\mathcal{F}, S)$. We denote by $p_1, p_2, \cdots, p_n$ the points of $\mathcal{D}$ in the order when they arrive. For simplicity, we use $A+p$, $A-p$ to denote $A \cup \{p\}$ and $A \backslash \{p\}$, respectively. For a point $p_i \in \mathcal{D}$ and a set $A \subseteq \mathcal{D}$, we use the shorthand $f_a(p_i : A) = f_a(p_i \mid A \cap \{p_1, p_2, \ldots, p_{i-1}\})$.

Algorithm 1: The FairStream algorithm

Input: A matroid $(\mathcal{D}, \mathcal{I})$, low bounds $l_1, l_2, \cdots, l_C$
Output: Solution set S

1 $A \leftarrow \emptyset, B_c \leftarrow \emptyset$ for $c \in [C]$
2 **for** *every arriving point* p_i *of group* c **do**
3   **if** *with probability* x **then**
4   // Update the Main Set
5   **if** $A + p_i \in \mathcal{I}$ **then**
6       $A \leftarrow A + p_i$
7   **else if** $f_a(p_i|A) > 0$ **then**
8       $U \leftarrow \{e \in A : A + p_i - e \in \mathcal{I}\}$
9       $q = \arg\min_{e \in U} f_a(e : A)$
10   **if** $f_a(p_i|A) > (1+b) \cdot f_a(q : A)$ **then**
11         $A \leftarrow A - q + p_i$
12   **else**
13   // Update the Backup Sets
14   **if** $B_c < l_c$ **then**
15       $B_c \leftarrow B_c + p_i$
16   **else**
17       $p' = \arg\min_{e \in B_c} f_a(e)$
18   **if** $f_a(p_i) > f_a(p')$ **then**
19         $B_c \leftarrow B_c - p' + p_i$
20 $S = $ UpdateWithBackup$(A, B_1, \cdots, B_C)$
21 **return** S

As shown in Algorithm 1, for an arriving point p_i, the algorithm first subsamples it with probability x. Then p_i is added to the main set A only under two conditions. If adding p_i to A preserves independence, p_i is accepted. Otherwise, the algorithm evaluates the marginal gain of p_i. This step does not affect the final output, as replacements occur only with positive marginal gains, thus enhancing efficiency as points with zero marginal gain are immediately discarded without further processing. If the marginal gain is positive, a set U is initialized with points e from A such that replacing e with p_i maintains independence. A replacement of the point $q \in U$ with minimal $f_a(q : A)$ is triggered when $f_a(p_i|A) > (1 + b) \cdot f_a(q : A)$ (Lines 5–11).

If p_i is disregarded, it may be added to the backup set B_c based on its quality. In summary, for the disregarded points, we cache l_c points with the highest $f_a(\cdot)$ values from group c (Lines 14–19). After a certain number of points have arrived, the main set A is updated with backup sets (Line 20). As shown in Algorithm 2, points from B_c supplement A if the number of group c is below the lower bound l_c. Replacements are made if they increase the objective function value of A while maintaining independence. To conclude, Algorithm 2 serves two key purposes. (1) Fairness guarantee: making the solution set meet lower-bound constraints for each group $c \in [C]$. (2) Quality preservation: preventing the loss of high-quality disregarded points to some extent.

Algorithm 2: UPDATEWITHBACKUP($A, B_1, \cdots, B_c$)

```
1  for c ∈ 1, ..., C do
2      for p ∈ B_c do
3          if |A ∩ D_c| < l_c then
4              A = A + p
5          else
6              U ← {e ∈ A : A + p − e ∈ I}
7              for p' ∈ U in order of increasing f_a(p') do
8                  if f_a(A + p − p') > f_a(A) then
9                      A ← A − p' + p
10                     break
11 return A
```

Theoretical Analysis. The following theorem shows the approximation ratio and the space complexity of our algorithm.

Theorem 1. *The* FAIRSTREAM *algorithm is a* $\frac{b}{(1+b)^2}$*-approximation algorithm and its space complexity is* $O(k)$.

Proof. Let $\mathtt{OPT}$ be the optimal solution set and V denote the set of points that were added to the main set A at some timestamp. From [8], we know that the main set A, selected from the remaining points, satisfying, $\mathbb{E}\left[f_a\left(A_n\right)\right] \geq \frac{b}{(1+b)^2} \cdot \mathbb{E}[f_a(V \cup \mathtt{OPT})]$, where $x = \frac{1}{b+2}$. And it is evident from Algorithm 2 that $f_a(S) \geq f_a(A)$. By the monotonicity of f_a we have:

$$\mathbb{E}\left[f_a\left(S\right)\right] \geq \mathbb{E}\left[f_a\left(A\right)\right] \geq \frac{b}{(1+b)^2} \cdot \mathbb{E}[f_a(V \cup \texttt{OPT})] \geq \frac{b}{(1+b)^2} \cdot f_a(\texttt{OPT}). \quad (6)$$

Hence, the FairStream algorithm is with $\frac{b}{(1+b)^2}$-approximation. The approximation ratio is at most $1/4$ when $b = 1$. For the backup sets Bs and the solution set S are both with $O(k)$ complexity, the space complexity of the FairStream algorithm is $O(k)$.

5 Experiments

5.1 Setup

All the algorithms were implemented in C++ on a PC running Ubuntu 20.04 LTS with a 3.00GHz processor and 32GB memory. We used the CPU time of each algorithm and the average rank happiness ratios (ARHR) of their solutions as metrics for the efficiency and effectiveness. Furthermore, we used the number of fairness violations defined in [7], i.e., for a given S, $err(S) = \sum_{c \in [C]} \max\{|S \cap \mathcal{D}_c| - h_c, l_c - |S \cap \mathcal{D}_c|, 0\}$ to measure how unfair a solution is compared to the one satisfying the group fairness constraints.

Datasets. We conducted our experiments on both synthetic and real datasets. The synthetic dataset includes Anti-Correlated with $d \in \{2, \ldots, 7\}$ and $n \in \{10^2, \ldots, 10^6\}$ created by the generator in [3]. Each dataset is divided into $C \in \{2, \ldots, 10\}$ groups. The real datasets include Adult[1], Lawschs[2],Credit[3] and Compas[4]. Adult is a $5d$ dataset with 32,561 tuples divided into groups by *gender* and *race*. Lawschs includes 65,494 students' information which is divided by *gender* and *race*. Credit is a $7d$ dataset of 1,000 tuples divided by *job* and housing condition. Compas is a $9d$ dataset of size 4,743 partitioned by *gender* and *whether the insured is recidivous*.

Algorithms. Since there are currently no efficient algorithms that can address our problem, we first compare our algorithm with Greedy which is a classical algorithm provided in [15] and was used in [20] for the ARMQ problem. Although Greedy is a static algorithm without fairness constraints, we used Greedy as the benchmark algorithm providing typically the best solution quality, which allows us to compare the relative performance of other algorithms. Moreover, we compare with three classical streaming algorithms ThreeSieves [4], IndependentSetImprovement [5] and SieveStreaming++ [10] for streaming submodular maximization problem. The last one is GreedyFairStreaming [6], which performs quite well for streaming submodular maximization under fairness constraints, however it does not provide any worst-case guarantee. For simplicity, we refer to each algorithm by its initial letter except for our FairStream algorithm, e.g., G for the Greedy, TS for the ThreeSieves.

[1] https://archive.ics.uci.edu/ml/datasets/adult.
[2] http://www.seaphe.org/databases.php.
[3] https://archive.ics.uci.edu/dataset/144/statlog+german+credit+data.
[4] https://github.com/propublica/compas-analysis.

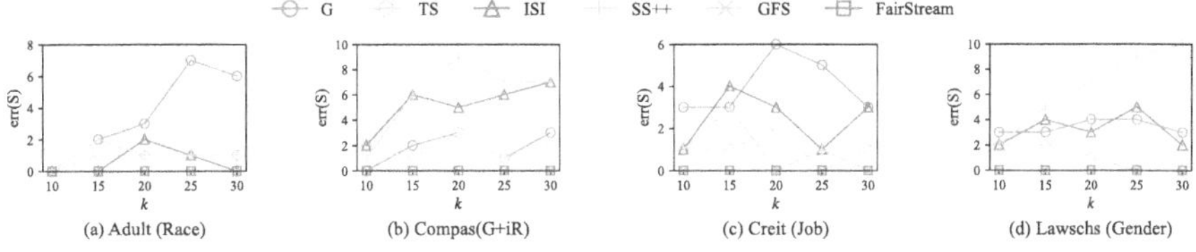

Fig. 1. Numbers of fairness violations of different algorithms.

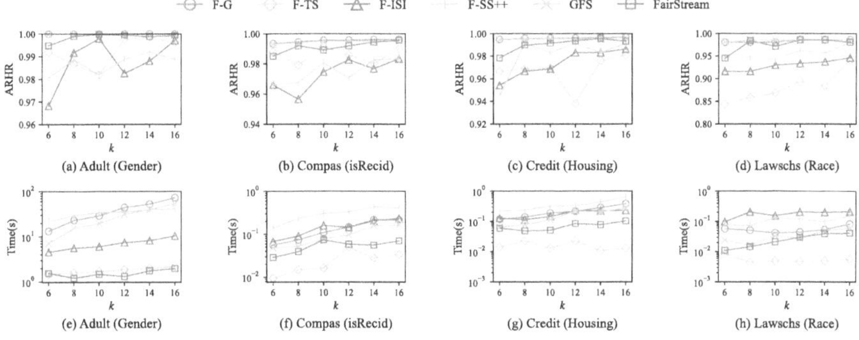

Fig. 2. The ARHR and the CPU time of each algorithm on real datasets by varying k.

Preparation. We sample a set of $m = 500$ linear utility functions uniformly from the utility function space. We employ the average value derived from 50 simulations as the final output of our algorithm. We set the parameter $T = 1000$ in the TS algorithm. Moreover, to evaluate the average rank happiness ratio of all the algorithms and obtain a fair comparison, for the algorithms above which don't consider group fairness, we divide the given dataset into groups while ensuring that the points selected in each group meet the upper and lower bounds. Then we run them separately on each group and take the union of each group solutions as the final solution. This fairness-adapted version of the above algorithm has a prefix "F-" before its name, e.g., F-TS for the adapted TS. We normalize the numerical attributes of all the data points within each dataset to the range $[0, 1]$. The skyline sets derived from these datasets serve as the input for each algorithm. For group fairness constraints, we set l_c to $\left\lfloor (1-\alpha)k \cdot \frac{|\mathcal{D}_c|}{|\mathcal{D}|} \right\rfloor$ and h_c to $\left\lceil (1+\alpha)k \cdot \frac{|\mathcal{D}_c|}{|\mathcal{D}|} \right\rceil$, where $\alpha = 0.1$.

5.2 Experimental Results

Fairness Violations: In Fig. 1, we report the number of fairness violations for the solution S generated by each algorithm across varying solution sizes k. Specifically, the baseline algorithms (except GFS) consistently violate the group fairness constraints, as their original designs lack fairness concerns, leading to biased solutions that favor advantaged groups and under-represent disadvan-

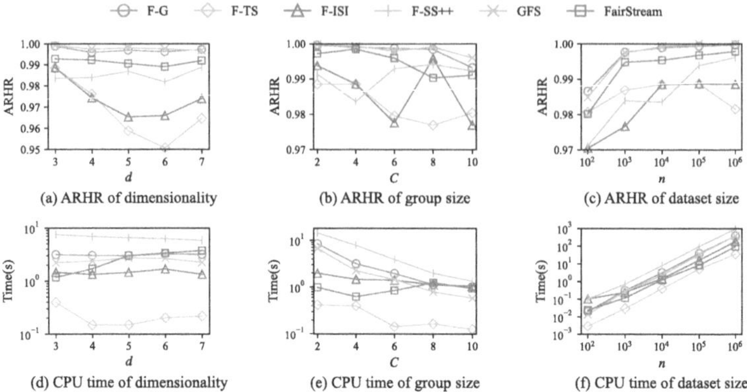

Fig. 3. Performance of different algorithms on the anti-correlated datasets for solution size $k = 15$ by varying dimensionality d, number of groups C, and dataset size n.

taged ones. This highlights the need for integrating fairness-aware mechanisms. Our proposed FairStream algorithm always obtains a solution with $err(S) = 0$ because it strictly follows the group fairness constraints.

Results on Real Datasets: In Fig. 2, we show the efficiency by comparing our FairStream algorithm with the baseline algorithms on real datasets. The experiments are performed to show the running time of different algorithms by varying the output size k to get the average rank happiness ratios. Regarding the quality of solutions, generally, the ARHRs of all the algorithms exhibit an increasing trend as k increases. The GFS generally achieves the best performance compared to other algorithms. However, it lacks worst-case approximation guarantee, which may lead to suboptimal results in extreme scenarios. And compared with our FairStream algorithm, it takes more time. In most cases, our proposed algorithm FairStream outperforms the three classical streaming algorithms e.g., F-TS, F-ISI and F-SS++, which demonstrates the effectiveness of our algorithm. In terms of efficiency, we observe that the F-TS algorithm achieves the best performance on almost all the datasets. This is because it takes $O(1)$ time to process each arriving point and does not need to cache any additional points. Our FairStream algorithm exhibits comparable efficiency to F-TS. Meanwhile, FairStream consistently surpasses F-TS in solution quality under all the situations.

Overall, FairStream outperforms the baseline algorithms both in solution quality and efficiency on most of the datasets.

Scalability: To evaluate the scalability of different algorithms with respect to *dimensionality* d, *number of groups* C and *dataset size* n, we performed extensive experiments on the generated anti-correlated datasets with d ranging from 3 to 7 (with $n = 10^4$ and $C = 3$), C ranging from 2 to 10 (with $d = 3$ and $n = 10^4$), and n ranging from 10^2 to 10^6 (with $d = 3$ and $C = 4$). We fix the solution size k to 15 in all the experiments above. The results are shown in Fig. 3.

As d increases, the ARHR generally exhibits a declining trend. The runtime of most of the algorithms remains relatively stable, except for our FairStream algorithm. The runtime of FairStream grows with d because the number of filtered points decreases as d increases (see Line 7 of Algorithm 1). When C increases, the ARHRs of baseline algorithms generally decline. This is because larger C imposes stricter fairness constraints. Regarding runtime, the algorithms that enforce fairness by processing each group separately show a decrease in runtime, while others remain unchanged. As n increases, the ARHRs of the baseline algorithms typically increase. In terms of runtime, all the algorithms exhibit a near-linear increase in running time as n increase. In summary, compared with the baseline algorithms, our algorithm has better scalability, higher solution quality and higher efficiency.

6 Conclusion

In this paper, we study the problem of streaming rank-happiness maximization query under group fairness constraints (Streaming-FairRHMQ). We first formally define the Streaming-FairRHMQ problem and show its NP-hardness. Then, we propose the FairStream algorithm with 1/4 approximation ratio based on the submodularity and monotonicity properties of the rank-happiness ratio function. Extensive experiments on real and synthetic datasets confirm the efficiency, effectiveness, and scalability of our proposed algorithm.

Acknowledgments. This work is partially supported by the Basic Research Program of Jiangsu under grant No. BK20252032, the Fundamental Research Funds for the Central Universities under grant No. NS2024056 and the Postgraduate Research & Practice Innovation Program of NUAA under grant No. XCXJH20241601.

References

1. Asudeh, A., Nazi, A., Zhang, N., Das, G., Jagadish, H.V.: RRR: rank-regret representative. In: SIGMOD, pp. 263–280 (2019)
2. Badanidiyuru, A., Mirzasoleiman, B., Karbasi, A., Krause, A.: Streaming submodular maximization: massive data summarization on the fly. In: SIGKDD, pp. 671–680 (2014)
3. Borzsony, S., Kossmann, D., Stocker, K.: The skyline operator. In: ICDE, pp. 421–430 (2001)
4. Buschjäger, S., Honysz, P.J., Pfahler, L., Morik, K.: Very fast streaming submodular function maximization. In: ECML-PKDD, pp. 151–166 (2021)
5. Chakrabarti, A., Kale, S.: Submodular maximization meets streaming: matchings, matroids, and more. Math. Program. **154**, 225–247 (2015)
6. El Halabi, M., Fusco, F., Norouzi-Fard, A., Tardos, J., Tarnawski, J.: Fairness in streaming submodular maximization over a matroid constraint. In: ICML, pp. 9150–9171 (2023)

7. El Halabi, M., Mitrović, S., Norouzi-Fard, A., Tardos, J., Tarnawski, J.M.: Fairness in streaming submodular maximization: algorithms and hardness. In: NIPS, pp. 13609–13622 (2020)
8. Feldman, M., Karbasi, A., Kazemi, E.: Do less, get more: streaming submodular maximization with subsampling. In: NIPS, pp. 732–742 (2018)
9. Ilyas, I.F., Beskales, G., Soliman, M.A.: A survey of top-k query processing techniques in relational database systems. CSUR **40**(4), 1–58 (2008)
10. Kazemi, E., Mitrovic, M., Zadimoghaddam, M., Lattanzi, S., Karbasi, A.: Submodular streaming in all its glory: tight approximation, minimum memory and low adaptive complexity. In: ICML, pp. 3311–3320 (2019)
11. Korte, B., Vygen, J.: Combinatorial Optimization: Theory and Algorithms. Springer, Heidelberg (2012)
12. Luenam, P., Chen, Y.P., Wong, R.: Approximating happiness maximizing set problems. arXiv preprint arXiv:2102.03578 (2021)
13. Meng, X., Zhang, Z., Cui, S., Huo, H.: Top-k collective spatial keyword approximate query. In: WISA, pp. 227–238 (2024)
14. Nanongkai, D., Sarma, A.D., Lall, A., Lipton, R.J., Xu, J.: Regret-minimizing representative databases. In: VLDB, pp. 1114–1124 (2010)
15. Nemhauser, G.L., Wolsey, L.A., Fisher, M.L.: An analysis of approximations for maximizing submodular set functions—i. Math. Program. **14**, 265–294 (1978)
16. Qiu, X., Zheng, J., Dong, Q., Huang, X.: Speed-up algorithms for happiness-maximizing representative databases. In: APWebWAIM DS Workshop, pp. 321–335 (2018)
17. Wang, Y., Fabbri, F., Mathioudakis, M.: Fair and representative subset selection from data streams. In: WWW, pp. 1340–1350 (2021)
18. Xiao, X., Li, J.: Rank-regret minimization. In: ICDE, pp. 1848–1860 (2022)
19. Xie, M., Wong, R.C.W., Lall, A.: An experimental survey of regret minimization query and variants: bridging the best worlds between top-k query and skyline query. VLDB J. **29**(1), 147–175 (2020)
20. Zeighami, S., Wong, R.C.W.: Minimizing average regret ratio in database. In: SIGMOD, pp. 2265–2266 (2016)
21. Zeighami, S., Wong, R.C.W.: Finding average regret ratio minimizing set in database. In: ICDE, pp. 1722–1725 (2019)
22. Zheng, J., Dong, Q., Qiu, X., Huang, X.: Accelerated algorithms for maximizing average happiness ratio in databases. Front. Comp. Sci. **15**(6), 1–3 (2021). https://doi.org/10.1007/s11704-020-0178-7
23. Zheng, J., Ma, W., Wang, Y., Wang, X.: Continuous k-regret minimization queries: a dynamic coreset approach. TKDE **35**(6), 5680–5694 (2022)
24. Zheng, J., Ma, Y., Ma, W., Wang, Y., Wang, X.: Happiness maximizing sets under group fairness constraints. In: VLDB, pp. 291–303 (2022)
25. Zheng, J., et al.: Hybrid regret minimization: a submodular approach. TKDE **36**(7), 3151–3165 (2024)
26. Zhu, K., Zheng, J., Yang, Z., Dong, J.: Identifying rank-happiness maximizing sets under group fairness constraints. In: APWeb-WAIM, pp. 325–341 (2024)

A Storage-Compute Separation Blockchain Architecture Based on Distributed File System for Storage Scalability

Guoshuo Guan, Yingying Zhou, and Jie Song(✉)

Software College, Northeastern University, Shenyang 110819, China
songjie@mail.neu.edu.cn

Abstract. The rapid growth of blockchain systems such as Bitcoin and Ethereum has led to an escalating storage burden on nodes. The storage scalability of blockchain has become an urgent problem. On-chain storage approaches introduce cross-shard transaction coordination and index maintenance management problems. Hybrid on-chain/off-chain storage approaches execute transactions off-chain, which face reliance on third-party and data consistency problems. To solve the above problems, we propose a novel storage-compute separation blockchain architecture. Computations related to transactions and consensus are still processed on-chain by blockchain nodes. While the block data is stored in distributed file systems shared by blockchain nodes. Meanwhile, to ensure data consistency, integrity and read/write efficiency, we design a new read and write workflow based on four modules for the proposed architecture. We implement the proposed architecture based on Hyperledger Fabric. Experimental results for three scenarios (transfers, world-state queries, and historical data queries) show that compared to Hyperledger Fabric, our architecture reduces storage by 49.9% with only TPS loss of 3.29% and latency loss of 9.20% on average.

Keywords: Blockchain Storage Scalability · Distributed File System · Storage-Compute Separation

1 Introduction

In recent years, blockchain systems represented by Bitcoin [1] and Ethereum [2] have attracted increasing attention, with a corresponding growth in storage requirements. In blockchain networks, full nodes store all transaction data, while light nodes store all block headers to verify transactions that have occurred. This storage mechanism ensures that blockchain data cannot be easily altered. However, with the continuous generation of large volumes of data, the storage burden on full nodes is becoming increasingly heavy, and in the future, even light nodes may find it difficult to join the blockchain [3]. As of June 2025, the size of the Bitcoin blockchain has exceeded 633 GB chain community[1], while Ethereum's

[1] https://etherscan.io/chart/tx.

S. Yang et al. (Eds.): WISA 2025, LNCS 16299, pp. 219–231, 2026.
https://doi.org/10.1007/978-981-95-6426-2_19

blockchain size has surpassed 1322 GB[2]. The storage scalability of blockchain has become an urgent problem that needs to be addressed in blockchain development.

Many studies have attempted to address this issue. The sharding approaches [4] partition the blockchain horizontally into multiple parallel chains, each of which is maintained by a subset of nodes. However, sharding introduces additional challenges such as cross-shard transaction handling and potential shard attacks. In addition, the semi-full node approach [5] involves each node storing a subset of blocks, while all nodes collectively maintain the complete blockchain. Nevertheless, these approaches require blockchain nodes to maintain block distribution index tables, which is an additional storage burden beyond the blockchain data. The above two approaches are on-chain storage approaches and, in essence, do not reduce the overall size of the blockchain. Other studies explore hybrid on-chain/off-chain storage [6]. Transactions are executed off-chain, and only key information, such as file hashes and metadata, is preserved on-chain, while original data, such as documents and transaction records, is maintained off-chain. However, hybrid on-chain/off-chain storage relies on third-party guarantors to verify transaction correctness and encounters challenges regarding data consistency and synchronization between the on-chain and off-chain layers [7].

To address the storage scalability of blockchain, we propose a storage-compute separation blockchain architecture based on a shared distributed file system (shared DFS). In this architecture, computations such as transaction execution, block packaging, and consensus are still performed on-chain by blockchain nodes (BNs), while block data is stored in shared DFS. BNs belonging to the same community jointly access and maintain the same shared DFS, which stores complete replicas of the blockchain. The main contributions of this paper are as follows.

- We propose a storage-compute separation blockchain architecture based on shared DFS, which not only significantly reduces the burden of local storage and index maintenance for nodes, but also preserves the advantages of on-chain computation and governance.
- We propose new read and write workflows based on our architecture, including four modules: virtual write manager, access control, block mapping, and metadata table. These modules support collaborative access and storage for the consistency and integrity of blockchain data.
- We implement our architecture based on Hyperledger Fabric and compare it with Hyperledger Fabric in terms of TPS, latency, and storage. Experimental results show that our approach achieves significant storage savings with low losses in TPS and latency.

2 Related Work

Storage scalability has long hindered the development of blockchain technology, primarily because the nodes in traditional blockchain store the entire ledger.

[2] https://etherscan.io/chartsync/chainarchive.

Numerous studies have sought to reduce the storage burden on individual nodes, which can be categorized into on-chain and hybrid on-chain/off-chain approaches.

2.1 On-Chain Approaches

Sharding is a widely adopted on-chain scalability technique for blockchains [8]. Xu et al. [9] proposed a method that groups nodes into multiple shards based on geographic location, with each shard independently handling local transactions and consensus, thereby enabling hierarchical management of distributed ledgers. Deep reinforcement learning [10] was used to automatically optimize the combination and adjustment of sharding, enabling the sharding structure to dynamically adapt to the network changes. Although processing transactions within each shard separately improves system performance, transactions involving multiple shards require coordination to ensure data consistency and atomicity across shards. Furthermore, attackers may target and gain control over a single shard, compromising the shard's data or launching double-spending attacks with malicious nodes.

In addition to sharding, the semi-full node approach alleviates storage pressure by allowing nodes to store only a subset of blocks [11]. Feng et al. [12] applied content-centric networking addressing and distributed caching mechanisms to blockchain storage, enabling nodes to efficiently share and retrieve block data on demand. Zhao et al. [11] designed a semi-full node that sits between full and light nodes. Semi-full nodes reduce storage requirements by selectively deleting blocks using an MCMC-based random algorithm. To facilitate querying of blocks not stored locally, these studies require maintaining block distribution index tables. When nodes frequently join or leave the network and the blockchain grows rapidly, the index tables may become outdated, potentially leading to inaccurate query results.

2.2 Hybrid On-Chain/Off-Chain Approaches

Beyond on-chain approaches, hybrid on-chain/off-chain storage approaches [13] divide data into two parts: on-chain and off-chain. Important, sensitive, or frequently verified data are stored in the chain (on-chain), while large volumes of raw data that do not require frequent verification are kept off the chain (off-chain), such as in distributed file systems [14] and cloud storage [15]. SlimChain [16] stores ledger state and transaction execution off-chain for parallel processing, while only maintaining state digests and essential proofs on-chain. Spatio-temporal data management [17] can also leverage hybrid storage, with raw spatio-temporal data preserved in off-chain cloud storage. However, hybrid on-chain/off-chain approaches handle transactions off the blockchain and require third-party guarantors to verify the correctness of transactions, and the effectiveness of this verification cannot be ensured.

To this end, we devise a shared distributed storage method. Transactions are still executed on-chain; however, unlike the original blockchain, multiple nodes share the block data in a distributed file system. This not only avoids the shortcomings of off-chain transaction execution but also reduces the storage burden on nodes.

3 Architecture Design

In this section, we describe our proposed architecture in detail. First, we present the corresponding system architecture (see Sect. 3.1). Then, we explain the read and write workflow in detail (see Sect. 3.2). Finally, we analyze the decentralization degree of our architecture (see Sect. 3.3).

3.1 Architecture Overview

As illustrated in Fig. 1, we implement a storage-compute separation blockchain architecture, which contains three distinct layers: the Blockchain Layer, the Control Layer, and the Storage Layer. The blockchain layer performs computation tasks related to transactions and consensus, the control layer processes read and write requests initiated by the blockchain layer, and the storage layer utilizes the shared DFS to store and manage block data. Storage-compute separation and block data sharing enhance the storage scalability of blockchain.

Blockchain Layer. In traditional blockchain architecture, full nodes store complete copies of all blocks. Our architecture separates node computation from storage by sharing a DFS. This approach enables Blockchain Nodes (BNs) to focus exclusively on computational tasks, such as validating transactions, establishing consensus, and packaging new blocks, while offloading storage-related functions to the shared DFS. In our architecture, BNs form multiple communities, with each community sharing the same DFS. For example, in consortium blockchains, BNs belonging to the same organization typically constitute a community and share a single DFS to store block data. BNs only engage in data exchange with the shared DFS when they need to read and write historical transactions.

Control Layer. The control layer acts as an intermediary communication interface that coordinates between the blockchain layer and the storage layer. It contains four key modules. The Access Control module processes read and write requests, validating permissions and hashes to ensure the consistency of the storage layer's contents. The Virtual Write Management module analyzes request volumes. It determines the number of block replicas and dynamically selects the appropriate blockchain nodes for writing based on the current load and write frequency. The Block Mapping module utilizes a consistent hashing algorithm to efficiently map requests to specific storage nodes. Meanwhile, BNs determine which shared DFS to access based on the smart routing algorithm implemented in this module. The Metadata Table module records writer ID, replica location, Checksum, and other relevant information for rapid indexing during read operations.

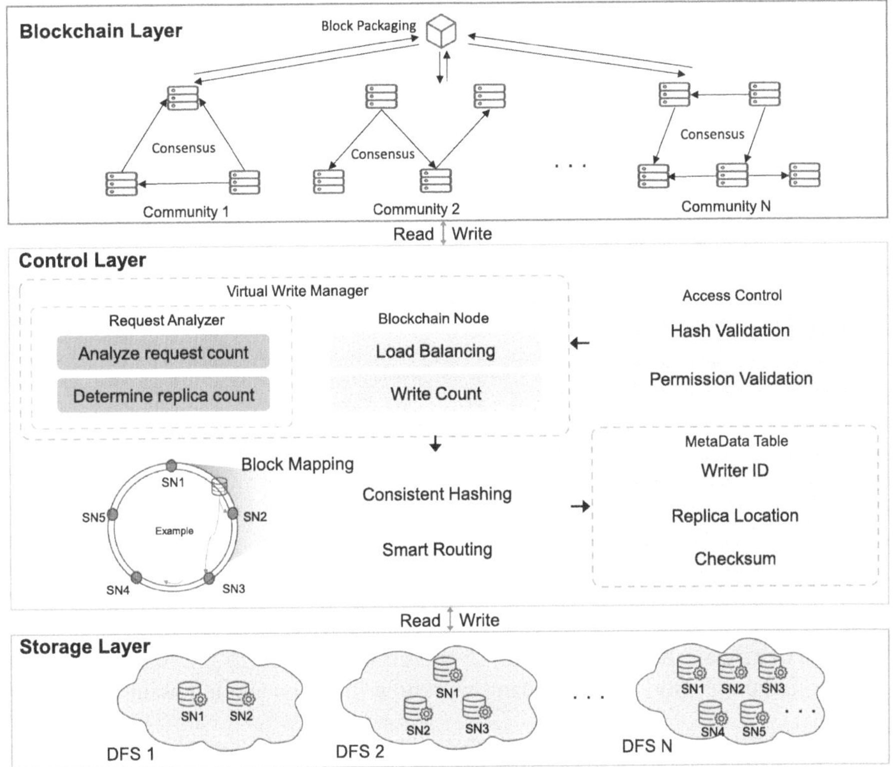

Fig. 1. System architecture.

Storage Layer. The Storage layer offers an external storage service for block data. This layer consists of multiple shared DFSs for BNs in communities. Each DFS contains multiple Storage Nodes (SNs) that collectively store block data in a distributed fashion. Furthermore, various DFSs need to communicate periodically to ensure data integrity and consistency, using mechanisms such as gossip protocols for data synchronization.

3.2 Read And Write Workflow

In the architecture proposed above, due to the separation of storage and computation, BNs do not store block data on the local file system. Instead, they access a shared DFS for read/write operations. The shared DFS raises concerns regarding data consistency and fault tolerance. To address these potential challenges and ensure read/write efficiency under complex requests, we manage the read and write processes through the control layer. The workflow of reading and writing is illustrated in Fig. 2.

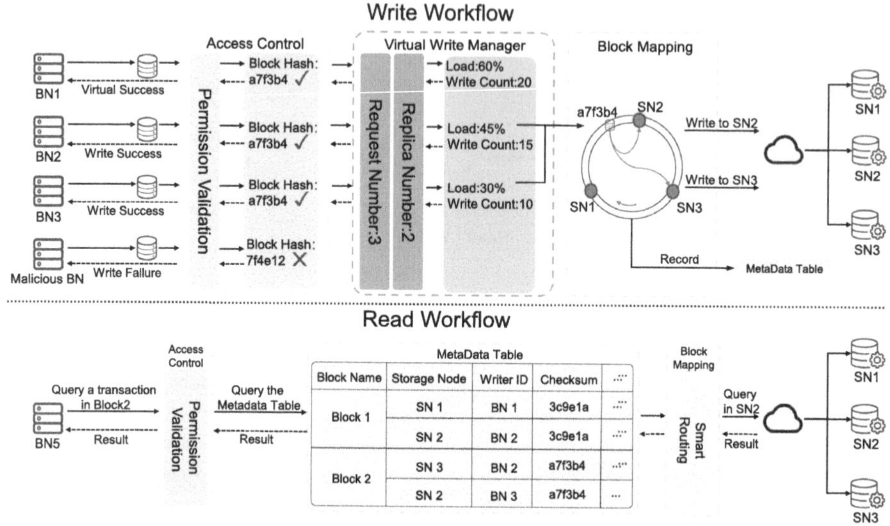

Fig. 2. Read and write workflow of proposed architecture.

Write Workflow. As shown in Fig. 2, we simulate four BNs: BN 1, BN 2, BN 3, and a malicious Node. The writing workflow includes the following steps:

- Step 1: After completing the consensus process, all BNs send write requests for the new block to the control layer.
- Step 2: Upon receiving the new block write request, the control layer first performs authentication to verify whether the requesting BN is registered in the blockchain network. This step ensures that only trusted BNs are allowed to participate in data writing.
- Step 3: Once the request validation has been completed, the hash verification module performs a hash check on the block data uploaded by different BNs. The write request will only be forwarded if the identical hash value reaches a predetermined threshold. For identified malicious blocks, the hash verification module will return a write failure response.
- Step 4: Write requests from BNs that pass the hash verification are forwarded to the virtual write management module. This module dynamically evaluates and determines the actual writing nodes based on the number of active nodes and the set hyperparameters for the number of replicas. The selection process is based on a scoring system, where the score for each BN n is defined as: $\text{Score}(n) = w_F \cdot F_n + w_L \cdot (1 - L_n)$. Where F_n represents the write frequency of BN n, and L_n indicates the load status of BN n, normalized to the range of $[0, 1]$. The weights w_F and w_L are respectively assigned to the write frequency and load status. In this example scenario, BN2 and BN3 are ultimately selected as the actual writing BNs. BN1 directly returns a response indicating a virtual write success and releases the associated handles and resources.

- Step 5: For the selected actual writing nodes, the requests continue to the block mapping module. In the architecture proposed above, different BNs, following consensus and hash verification at the control layer, submit block files with identical filenames and contents. This characteristic can lead traditional filename-hashed storage systems to store identical block files on the same SN, creating load imbalance under high concurrency. This module employs a sequential consistent hashing algorithm for load balancing, as illustrated in Algorithm 1, which maps block data sequentially across multiple SNs. Unlike the traditional consistent hashing algorithm, which maps nodes and data files to a hash ring using hash functions to determine file placement by their hash values, our method distributes identical block files across distinct nodes by sequentially selecting targets on the hash ring.
- Step 6: Finally, upon successful writing, the system records the information of this write operation in the metadata table to ensure the traceability of all operations and fast indexing for read workflow.

Algorithm 1: Sequential consistent hashing

Input: $blocks$: the block data; $storageNodes$: the set of available storage nodes;R:the number of replica copies;V:the number of virtual nodes per storage node
Output: $targetNodes$: the set of selected storage nodes
$virtualNodes \leftarrow [\,]$;
foreach $node \in storageNodes$ **do**
  for $j \leftarrow 1$ **to** V **do**
    $virtualNode \leftarrow node + \_VN + j$;
    $virtualNodes$.append($virtualNode$);
    $nodeHash[virtualNode] \leftarrow$ HashFunction($virtualNode$);
$blockHash \leftarrow$ HashFunction($block$);
$initialTargetNode \leftarrow$ FindNext($blockHash, nodeHash$);
$targetNodes$.append($initialTargetNode$);
for $i \leftarrow 0$ **to** $R - 1$ **do**
  $index \leftarrow$ indexOf($initialTargetNode$);
  $nextIndex \leftarrow (index + i)$ mod length($nodeHash$);
  $nextTargetNode \leftarrow$ FindNext($nodeHash[nextIndex], nodeHash$);
  $targetNodes$.append($nextTargetNode$);

Read Workflow. As shown in Fig. 2, we simulate a BN: BN5. The read workflow includes the following steps:

- Step 1: A BN initiates a query for transaction information within a specific block. In this example scenario, BN5 sends a query request for Block 2.
- Step 2: In the control layer, the access control module performs permission verification to ensure that the request comes from a BN with the appropriate permissions.

- Step 3: Based on the metadata table, the control layer identifies the actual SN for the corresponding block. In this sample scenario, Block 2 resides on SN2 and SN3.
- Step 4: Based on the smart routing algorithm, the control layer routes the query request to the SN with a comparatively lower average load to achieve inter-node load balancing. In this scenario, the routing algorithm evaluates both SN2 and SN3, and ultimately directs the query request to SN2.
- Step 5:Upon receiving the request, the SN transmits the block content to the control layer, which then relays the result back to the requesting BN.

3.3 Decentralization Degree Analysis

In the public blockchain, using our architecture requires BNs to trust a shared DFS, which can lead to additional trust issues and reduce the degree of decentralization. However, in a consortium blockchain, the shared DFS is managed by a specific organization for its internal BNs. Since BNs within the organization inherently trust each other, our architecture introduces almost no additional trust issues. The degree of decentralization can be measured by node equivalence and the level of transaction verification. The node equivalence measures the consistency of functions and processing capabilities among different BNs in the network. The level of transaction verification is classified into four levels: L1, unable to verify payments; L2, verify all the payments; L3, verify partial transactions; and L4, verify all the transactions [18]. Based on the above definitions, we conduct the Comparison of decentralization among Blockchain Platforms (Table 1)

Table 1. Comparison of decentralization among blockchain platforms

Platform	Node Type	Trust Requirement	Degree of Decentralization	
			Node equivalence	Level of tran. verification
Ethereum	Light Node	Full Node	Fair	L2
	Full Node	Null		L4
Ethereum+our architecture	Node	SDFS	High	L2
Fabric	Peer Node	Orderer Node	Low	L3
	Orderer Node	Null		L1
Fabric+our architecture	Peer Node	Orderer Node+SDFS	Low	L3
	Orderer Node	Null		L1

In Ethereum, light nodes store block headers and can verify all payments, but they rely on full nodes to verify all transactions(L2). Full nodes can verify all transactions (L4). Light nodes and full nodes have similar functions; therefore, their node equivalence is fair. Our architecture requires nodes to utilize a shared DFS to verify all transactions (L2). Additionally, our architecture ensures uniform functions across all nodes, enhancing node equivalence. In Fabric, peer nodes depend on endorsement outcomes from orderer nodes. Constrained by channel configurations and endorsement policies, they can only verify partial transactions (L3). Orderer nodes cannot verify payments(L1). Since peer node and orderer node have entirely different functions, the Fabric exhibits low node equivalence. Our architecture requires peer nodes to utilize a shared DFS. Still, due to the existence of the world-state database, this dependency does not affect peers' ability to verify transactions independently(L3). Therefore, in consortium blockchain platforms, such as Hyperledger Fabric, adopting our architecture introduces no additional trust issues and does not reduce decentralization. This makes our architecture particularly suitable for consortium blockchain implementations.

4 Experiments

In this section, we conduct a rigorous comparison to systematically assess the performance of our architecture. We present the experimental setup (see Sect. 4.1) and the evaluation of performance(see Sect. 4.2).

4.1 Experimental Setup

Considering that the architecture proposed above is particularly suitable for consortium blockchain platforms, the blockchain layer utilizes the consortium blockchain platform Hyperledger Fabric. The control layer employs File System in User Space (FUSE) to intercept and forward I/O operations from the blockchain layer. The storage layer adopts GlusterFS as the shared storage file system. The experimental server is equipped with an 8-core, 16-thread CPU (AMD Ryzen R9 6900HS) and 32 GB of RAM, and Fabric V2.4.9 is installed, operating in the Ubuntu 20.04 LTS environment. The BNs are deployed through Docker containers.

Hyperledger Caliper is a key benchmarking framework in the blockchain community[3]. We customized its testing scenarios by modifying the chaincode logic and configuration files. All experiments were conducted using the customized Caliper tool, with results averaged over ten runs to ensure objectivity and reproducibility. The schematic figure of the experimental scenario is presented in Fig. 3.

The experimental scenario in Fig. 3 involves two organizations: Organization One consists of three peer nodes utilizing a shared DFS, while Organization Two has one peer node and one orderer node. The consensus mechanism adopts the SOLO consensus algorithm and employs LevelDB as the world-state database. The experimental setup employs the single-replica strategy across all scenarios except for the storage scenario.

[3] https://hyperledger-caliper.github.io/caliper/0.6.0/.

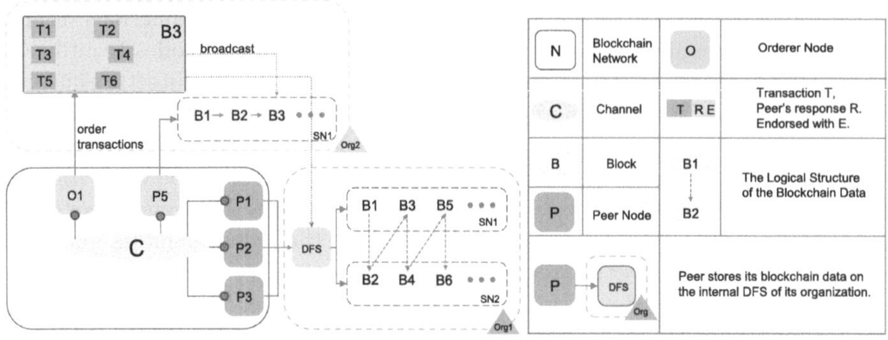

Fig. 3. Experimental setup with two organizations.

4.2 Evaluation of Performance

Figure 4 compares the Hyperledger Fabric with our architecture in a transfer scenario. We evaluated the impact of varying transaction sending rates on transactions per second (TPS) and latency. Both schemes exhibit a linear increase in TPS with increasing sending rates. However, due to fixed processing overhead, latency remains elevated under low transaction send rates. As the send rate increases, latency decreases gradually. Compared to the baseline, our architecture resulted in a 3.86% decrease in average TPS and an 8.69% increase in average latency due to the introduction of external storage.

Figure 5 presents the TPS and latency comparison for the Caliper built in a world-state query scenario, which only involves querying world-state data. Since our architecture does not modify the world-state database, the performance of both schemes is comparable, with a peak TPS of around 1400 and an average latency of 0.01 s.

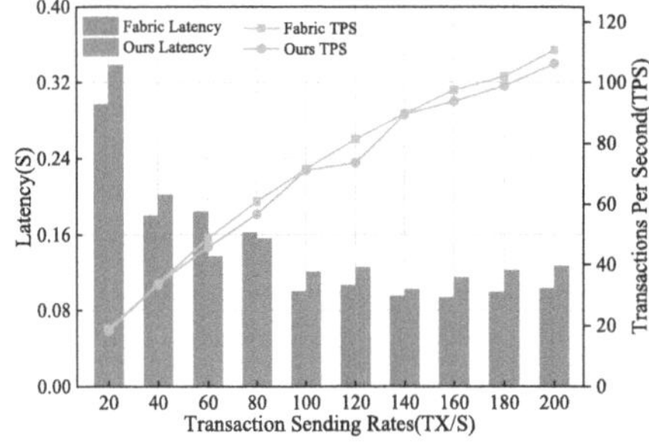

Fig. 4. Transfer scenario.

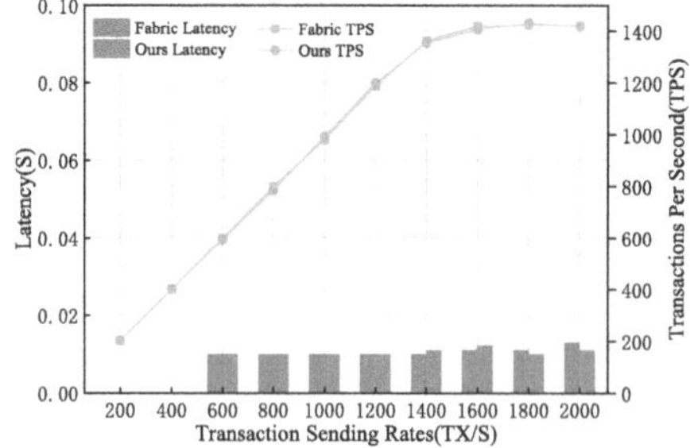

Fig. 5. World-State query scenario.

Figure 6 shows the impact of varying transaction numbers on TPS and latency in our customized block data query scenario. Because block data queries involve traversing block files, performance is significantly affected by the number of transactions. Due to communication overhead from external storage and traversal costs inherent to DFS, our solution exhibits marginally degraded performance in this scenario. The results indicate that the performance of both schemes decreases linearly as the number of transactions increases. Compared to the baseline, our architecture resulted in a 6% decrease in TPS and an 18.9% increase in latency.

Figure 7 compares the storage performance of the Hyperledger Fabric to our architecture with one replica and our architecture with two replicas. The results show that, compared to the baseline, the two-replica storage space is reduced by 24.8%, and the one-replica storage space is reduced by 49.9%. The performance of our scheme is strongly correlated with the number of BNs using Shared DFS and the number of replicas. Thus, in a real consortium blockchain scenario, if an organization allows a large number of peer nodes to use a single shared DFS, the storage space savings will be more considerable.

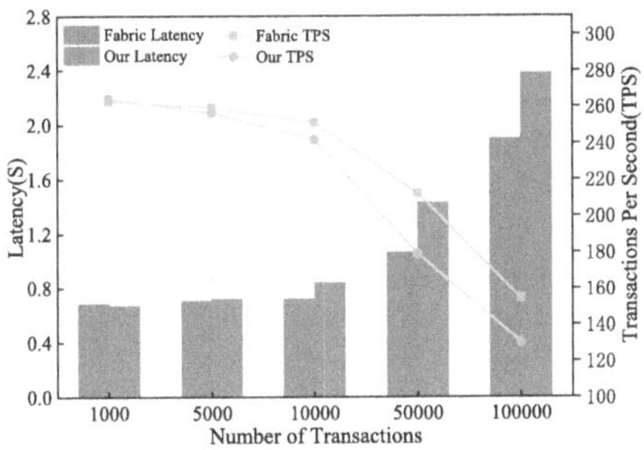

Fig. 6. Block data query scenario.

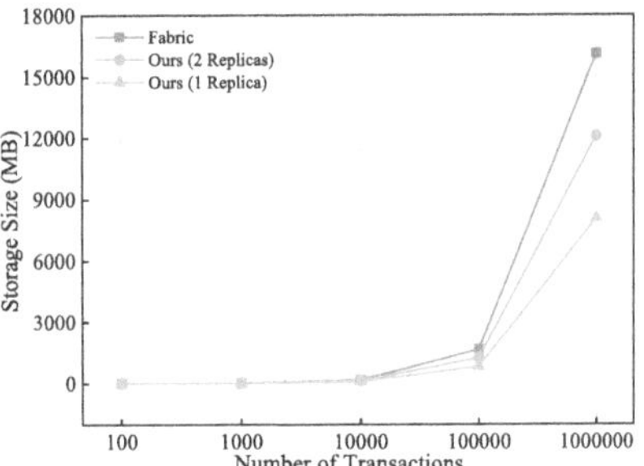

Fig. 7. Storage scenario.

5 Conclusion

In this paper, we propose a novel storage-compute separation architecture based on shared DFS to improve the storage scalability of blockchain systems. This architecture enables BNs to focus on computational tasks related to transactions and consensus while offloading data storage to the shared DFS. In addition, we design a new read and write workflow comprising four modules: virtual write manager, access control, block mapping, and metadata table. These modules facilitate collaborative access and maintain the integrity of blockchain data. Our experimental implementation using Hyperledger Fabric demonstrates substantial improvements in storage. Specifically, our architecture achieved a 49.9% reduction in storage requirements compared to Hyperledger Fabric, with only a 3.29% loss in TPS and a 9.20% loss in latency on average.

Acknowledgments. This paper is supported by the National Natural Science Foundation of China under Grant No.62472077.

References

1. Nakamoto, S.: Bitcoin: A peer-to-peer electronic cash system. (2008)
2. Buterin, V.: Ethereum white paper. (2013)
3. Wakinaka, F., Ogawa, T.: Block group distribution methods that ensure persistence in blockchain. In: 2024 IEEE 29th Asia Pacific Conference on Communications (APCC), pp. 515–518. IEEE (2024)
4. Ren, J., Li, J., Liu, H., et al.: Task offloading strategy with emergency handling and blockchain security in SDN-empowered and fog-assisted healthcare IoT. Tsinghua Sci. Technol. **27**(4), 760–776 (2021)
5. Huang, Y., Zhang, J., Duan, J. et al.: Resource allocation and consensus on edge blockchain in pervasive edge computing environments. In: 2019 IEEE 39th International Conference on Distributed Computing Systems (ICDCS), pp. 1476–1486. IEEE (2019)
6. Miyachi, K., Mackey, T.K.: hOCBS: a privacy-preserving blockchain framework for healthcare data leveraging an on-chain and off-chain system design. Inf. Process. Manage. **58**(3), 102535 (2021)
7. Ding, Y., Wu, Z., Miao, Y., et al.: Genuine on-chain and off-chain collaboration: achieving secure and non-repudiable file sharing in blockchain applications. IEEE Trans. Netw. Serv. Manage. **21**(2), 1802–1816 (2023)
8. Wang, G., Shi, Z.J., Nixon, M. et al.: Sok: Sharding on blockchain. In: Proceedings of the 1st ACM Conference on Advances in Financial Technologies, pp. 41–61. ACM (2019)
9. Xu, G., Zhou, Z., Song, X. et al.: Research on transaction allocation strategy in blockchain state sharding. Future Gener. Comput. Syst. 107756 (2025)
10. Yang, X., Xu, T., Zan, F., et al.: An overlapping self-organizing sharding scheme based on DRL for large-scale IIoT blockchain. IEEE Internet Things J. **11**(4), 5681–5695 (2023)
11. Zhao, P., Cheng, H., Fang, Y., et al.: A secure storage strategy for blockchain based on MCMC algorithm. IEEE Access **8**, 160815–160824 (2020)
12. Feng, H., Wang, J., Li, Y.: A blockchain storage architecture based on information-centric networking. Electronics **11**(17), 2661 (2022)
13. Wang, J., Li, L., Tan, A. et al.: Multi-User on-chain and off-chain collaborative query optimization based on consortium blockchain. In: Web Information Systems and Applications - 20th International Conference, WISA, pp. 476–487 (2023)
14. Yu, J., Zhang, X., Wang, J. et al.: Robust and Trustworthy data sharing framework leveraging on-chain and off-chain collaboration. Comput. Mater. Continua **78**(2) (2024)
15. Liu, C., Hou, C., Jiang, T. et al.: FACOS: Enabling privacy protection through fine-grained access control with on-chain and off-chain system. IEEE Trans. Inf. Forensics Sec. (2024)
16. Xu, C., Zhang, C., Xu, J., et al.: SlimChain: scaling blockchain transactions through off-chain storage and parallel processing. Proc. VLDB Endowment **14**(11), 2314–2326 (2021)

17. Ren, Y., Huang, D., Wang, W., et al.: BSMD: a blockchain-based secure storage mechanism for big spatio-temporal data. Futur. Gener. Comput. Syst. **138**, 328–338 (2023)
18. Fan, X., Niu, B., Liu, Z.: Scalable blockchain storage systems: research progress and models. Computing , 1–28 (2022). https://doi.org/10.1007/s00607-022-01063-8

Hardware Collaborated Vector Similarity Search

Yuxiang Xu[1], Ye Yuan[2(✉)], Yishu Wang[1], and Dong Jiang[1]

[1] School of Computer Science and Engineering, Northeastern University, Shenyang 110169, China

[2] School of Computer Science and Technology, Beijing Institute of Technology, Beijing 100081, China

yuan-ye@bit.edu.cn

Abstract. With the rapid advancement of artificial intelligence and big data technologies, high-dimensional vector retrieval has become a fundamental technology in areas such as recommender systems and information retrieval. When facing billion-scale datasets, traditional graph-based approximate nearest neighbor search (ANNS) methods struggle to balance index construction efficiency, query performance, and system scalability due to single hardware limitations. This paper proposes a CPU-GPU collaborative pipelined vector retrieval framework targeting large-scale datasets. The framework addresses GPU memory constraints through collaborative clustering with dynamic load balancing and asynchronous double buffering. During graph construction, we employ a two-phase pipelined approach where GPU performs parallel k-NN computation while CPU simultaneously optimizes graph structures. For query processing, the system implements collaborative graph navigation with intelligent prefetching and adaptive workload distribution. Experimental results on billion-scale datasets demonstrate significant improvements: our method achieves sub-400 ms query latency with over 90,000 QPS, representing 150–200 times improvement over traditional approaches and 2–3 times better than BANG. On deep learning features, query latency is reduced by 50–70% while maintaining high recall accuracy.

Keywords: Vector retrieval · Approximate nearest neighbor · CPU-GPU collaboration · Pipelined acceleration · Large-scale data

1 Introduction

With the rapid development of deep learning technologies, high-dimensional vector representations have become the foundation for numerous artificial intelligence applications, including recommender systems, information retrieval [1], and large language models. In these scenarios, efficient ANNS techniques are crucial for handling massive datasets. Currently, ANNS algorithms can be categorized into four major types: tree-based, hash-based, quantization-based [2], and graph-based approaches. Among them, graph-based ANNS methods have

S. Yang et al. (Eds.): WISA 2025, LNCS 16299, pp. 232–244, 2026.
https://doi.org/10.1007/978-981-95-6426-2_20

gained widespread adoption in practice due to their outstanding retrieval efficiency and accuracy.

However, as data volumes grow to the billion-scale, traditional graph-based ANNS methods face significant challenges. On the one hand, constructing high-quality graph indices has a time complexity close to $O(n^2 \cdot d)$, making the graph building process extremely time-consuming. On the other hand, the immense storage demands and computational load during graph search make it difficult for a single hardware platform to process efficiently. To address these issues, researchers propose various optimizations. Solutions such as SONG [3] and GGNN [4] attempt to leverage GPU parallelism to accelerate graph search, but are limited by GPU memory size. This forces them to employ extreme compression ratios for GPU data processing, significantly degrading accuracy. DiskANN [5] and BANG [6] explore hybrid solutions combining memory and external storage, but suffer from high I/O latency and irregular memory access patterns that prevent them from achieving the high query response.

In this paper, we propose a CPU-GPU collaborative computing framework for billion-scale datasets, aiming to fully utilize the complementary strengths of both hardware platforms for efficient vector similarity search. The core innovations of this framework include: (1) CPU-GPU cooperative clustering method with asynchronous double buffering, which addresses GPU memory limitations through batch processing while implementing intelligent two-level clustering and maximizing the computational capabilities of both platforms; (2)efficient centroid-based k-NN graph construction and optimization, adopting *build-while-transfer* and *optimize-while-transfer* modes to reduce device idle time; (3) collaborative graph navigation and intelligent prefetching strategy, where the CPU pre-extracts cluster data corresponding to high in-degree centroids during GPU-based graph search, reducing query data transfer latency; and (4) dynamic load distribution and collaborative computation mechanism, combining adaptive workload allocation with asynchronous parallel processing for efficient query result generation.

Experimental results show that our framework achieves significant performance improvements on multiple standard datasets: compared to traditional approaches such as DiskANN and SPANN [7], our method is nearly 200 times faster. Furthermore, compared to the latest hardware collaborative solution BANG, our framework achieves a 2–3× speedup with 50–70% lower query latency while maintaining high recall accuracy.

2 Background

2.1 CPU-GPU Collaborative Computing

CPU-GPU Collaborative Computing is a computational paradigm that fully leverages the complementary advantages of heterogeneous hardware platforms. CPUs offer strong single-thread performance, complex control logic, and large memory capacity, making them well-suited for tasks involving intricate control flows, irregular memory access patterns, and large caching requirements.

GPUs, on the other hand, consist of thousands of lightweight compute cores and high memory bandwidth. With their Single Instruction Multiple Thread (SIMT) architecture, they excel at large-scale, regular, and parallel computations.

2.2 Clustering and K-NN Graph

Clustering is an unsupervised learning technique aimed at partitioning a dataset into several subsets with similar characteristics. K-means++ is one of the most classic clustering algorithms, which partitions the data into Kclusters by iterative optimization, with each cluster represented by its centroid. The objective of the algorithm is to minimize the within-cluster sum of squared errors.

To enable efficient navigation over these centroids, we construct a k-nearest neighbor (k-NN) graph where each centroid serves as a node connected to its k most similar neighboring centroids.

In a k-NN graph, each data point x in a dataset $X = \{x_1, x_2, \ldots, x_n\}$ is represented as a node in the graph G. We define N_x as the local neighborhood of x consisting of its k nearest neighbors. The edge set E of the graph is defined as (x, y) where $y \in N_x$. Specifically, the resulting graph is a directed graph $G = (X, E)$, where $E = \{(x, y) \mid x \in X, y \in N_x\}$ and generally $(x, y) \in E \nRightarrow (y, x) \in E$.

The search process generally employs a greedy strategy, starting from an entry point and moving to the neighbor that is closest to the query point.

2.3 Limitations of Existing Methods

Despite significant progress in vector similarity search, existing methods face two critical challenges when applied to billion-scale datasets:
Inefficient Index Construction. Traditional approaches suffer from sequential bottlenecks in both clustering and graph building phases. CPU-only clustering is computationally slow for massive datasets, while GPU-only processing is constrained by memory limitations. Moreover, existing graph construction methods follow a *construct-then-optimize* paradigm with significant device idle time, failing to leverage pipelined execution for efficiency.
Suboptimal Search-Time Collaboration. Current search methods lack effective CPU-GPU coordination during query processing, often resulting in one device waiting while the other works. The absence of intelligent data prefetching and dynamic workload balancing leads to communication bottlenecks and underutilized hardware resources.

3 k-NN Graph Construction

3.1 CPU-GPU Collaborative Construction Framework

To address the problem that billion-scale data cannot be fully stored in GPU memory, this paper proposes a centroid-based k-NN graph construction method that adopts a CPU-GPU collaborative architecture, dividing the entire construction process into two core stages: collaborative clustering and pipeline graph construction. The framework comprises three key modules: (i)*asynchronous double buffering prefetch mechanism*, divides the dataset into mini-batches and implements ping-pong buffer. (ii)*CPU-GPU parallel processing synchronization mechanism*, allowing both platforms to work simultaneously. (iii)*intelligent two-level clustering*, automatically detects and decomposes imbalanced clusters; the graph construction optimization module employs a staged pipeline mode to ensure maximum utilization of each hardware type for tasks it excels at.

3.2 CPU-GPU Collaborative Clustering Algorithm Design

The collaborative clustering algorithm addresses billion-scale data processing challenges through three key technical innovations. First, the **asynchronous double buffering prefetch mechanism** divides the dataset into mini-batches and implements ping-pong buffer switching between Buffer1 and Buffer2, enabling computation-transmission overlap while the CPU asynchronously prefetches the next batch during current batch processing. Second, the **CPU-GPU parallel processing synchronization mechanism** achieves true hardware collaboration by allowing both platforms to work simultaneously on their assigned data subsets, with OpenMP parallelization on the CPU side and multi-CUDA stream processing on the GPU side, followed by weighted averaging to merge results. Third, the **intelligent two-level clustering strategy** automatically detects imbalanced clusters through coefficient of variation analysis and performs adaptive decomposition when cluster size ratios exceed predefined thresholds, ensuring uniform data distribution across the entire dataset.

These technical innovations are systematically integrated in Algorithm 1, which presents the complete collaborative clustering workflow. The algorithm begins with K-means++ initialization(Line 1) and iteratively processes the dataset through mini-batch collaborative computation until convergence(Lines 2–15). Specifically, the core prefetch mechanism implements asynchronous double buffering: while processing the current mini-batch, the CPU asynchronously prefetches the next batch to the backup buffer using *async* operations, ensuring continuous data flow and effectively hiding I/O latency (Lines 4–5). The synchronization mechanism demonstrates true hardware collaboration where CPU threads handle distance computation for their assigned data subsets using OpenMP parallelization, while GPU processes its assigned subsets through asynchronous memory transfer and multi-CUDA streams, followed by weighted averaging to merge results (Lines 7–9):

$$C_j^{new} = \frac{|S_j^{CPU}| \cdot C_j^{CPU} + |S_j^{GPU}| \cdot C_j^{GPU}}{|S_j^{CPU}| + |S_j^{GPU}|} \tag{1}$$

where C_j^{CPU} and C_j^{GPU} represent the positions of centroid j computed by the CPU and GPU respectively, with $|S_j^{\mathrm{CPU}}|$ and $|S_j^{\mathrm{GPU}}|$ denoting the corresponding number of data points.Additionally, the two-level strategy automatically determines second-level clustering necessity through multi-dimensional evaluation, with the number of sub-clusters determined when cluster imbalance is detected, ensuring balanced decomposition (Lines 16–18):

$$n_{sub} = \max(2, \min(\frac{|C_{large}|}{size_{target}}, \sqrt{|C_{large}|})) \quad (2)$$

where $|C_{large}|$ is the size of the large cluster and $size_{target}$ is the target cluster size. Finally, the algorithm returns the computed centroids and assignments as the clustering results (Line 19).

Algorithm 1. CPU-GPU Collaborative K-means Clustering

Require: Dataset D, vectors n, dimension d, clusters k, batch size B
Ensure: Centroids C, assignments A
1: Initialize centroids C using K-means++
2: **for** $iter = 1$ **to** $MaxIterations$ **do**
3:   **for** each batch $\in D$ **do**
4:     Asynchronously prefetch next batch
5:     $\alpha = \alpha_{base} + \min(\alpha_{max}, \frac{T_{transfer}}{T_{compute}} \times \beta)$ {Dynamic load balancing}
6:     $batch_{cpu}, batch_{gpu} \leftarrow$ SplitBatch(batch, α)
7:     Launch CPU_thread: ProcessBatch($batch_{cpu}$) $\rightarrow result_{cpu}$
8:     Launch GPU_thread: ProcessBatch($batch_{gpu}$) $\rightarrow result_{gpu}$
9:     Synchronize and merge
10:   **end for**
11:   Update centroids: $c_j = \frac{\sum_{x \in C_j} x}{|C_j|}$ for all clusters j
12:   **if** converged **then**
13:     **break**
14:   **end if**
15: **end for**
16: **for** each large cluster c where $|c| > threshold$ **do**
17:   SubCluster(c, n_{sub}) and update assignments
18: **end for**
19: **return** C, A

3.3 Pipelined Centroid-Based k-NN Graph Construction

After obtaining centroids through clustering, the system adopts a CPU-GPU collaborative pipeline approach to construct the k-NN graph between centroids. The core innovation lies in dividing the construction process into two strictly separated phases:

Graph Construction Phase. The GPU focuses on parallel computation of k-NN relationships between centroids using a *blocked pipeline strategy* and *multi-CUDA stream parallel processing*, while asynchronously transferring completed graph blocks to the CPU. During this phase, the CPU solely handles data reception and storage without performing any optimization.

Graph Refinement Phase. Upon receiving the complete k-NN graph, the CPU partitions it into multiple refinement regions. Each region undergoes parallel optimizations including *reverse edge addition*, *long-range connection optimization*, and *connectivity enhancement*. The CPU asynchronously transfers optimized results back to the GPU immediately after processing each region, realizing an efficient *optimize-and-stream* paradigm.

This phased design is based on the key insight that graph refinement requires global structural information. Performing local optimizations during construction would not only yield suboptimal results but also introduce unnecessary computational overhead.

4 k-NN Graph Search Process

4.1 Collaborative Search Framework and Process Design

The centroid-based k-NN graph search process builds upon the k-NN graph structure constructed in the previous stage, where query points navigate through the graph to locate their nearest neighbors by following edges to progressively closer centroids. To accelerate the search of k-NN graphs, this work proposes a CPU-GPU collaborative search framework that fully exploits the complementary advantages of heterogeneous hardware platforms.

The collaborative search framework employs a two-stage pipeline architecture: (i) *collaborative graph navigation* and *intelligent prefetching*, where GPU executes beam search while CPU simultaneously performs predictive data loading; (ii) *dynamic load distribution* and *collaborative computation*, where workloads are adaptively allocated between CPU and GPU based on real-time performance feedback, followed by asynchronous result consolidation.

4.2 Collaborative Graph Navigation and Intelligent Prefetching

The **collaborative graph navigation** employs true hardware parallelism in which the GPU and the CPU execute complementary tasks simultaneously within the same time period. The GPU leverages its massive parallel computing capability for batch beam search, utilizing thread-block collaborative processing where each query is allocated an independent thread block that maintains candidate centroid arrays, distance arrays, and access marker arrays in shared memory. Meanwhile, the CPU utilizes its large memory capacity for **intelligent data prefetching** based on four fusion strategies: high-in-degree prefetching prioritizes loading central node data, neighbor-relationship prefetching preloads related data based on spatial locality, query-pattern prefetching learns transition probabilities through historical access sequences, and adaptive hybrid strategy dynamically adjusts weights based on real-time hit rates. The CPU ensures true parallelism with graph search through asynchronous I/O and multithreading architecture, avoiding traditional sequential processing bottlenecks.

These technical innovations are systematically implemented in Algorithm 2 through parallel execution design. The algorithm begins with establishing initial cache for high-degree centroids (Lines 1–2) and iteratively processes query batches through simultaneous GPU navigation and CPU prefetching operations until all batches are completed (Lines 3–17). Specifically, for each query batch, `BatchBeamSearch(`B_i`, `G`, `p`, `w`)` implements thread-block collaborative parallel beam search on the GPU side (Line 4). Each query uses a deterministic random strategy to select starting centroids, finds optimal candidates through warp-level parallel reduction, expands graph neighbors, and updates the beam with key optimizations including coalesced memory access, squared distance comparison, and early stopping mechanisms, ultimately returning a batch-level candidate centroid matrix $P_i \in \mathbb{R}^{b \times p}$. Simultaneously, `AsyncPrefetch(`$P_i$`, `$M$`)` launches multithreaded prefetching tasks in parallel on the CPU side (Line 5). The CPU predicts data access patterns using the four fusion strategies and updates the cache state M through asynchronous I/O operations, ensuring true parallelism with GPU graph search through event synchronization mechanisms.

4.3 Dynamic Load Distribution and Collaborative Computation

The collaborative computation stage maintains CPU-GPU cooperation through cross-batch adaptive performance optimization and asynchronous parallel processing. The system implements cross-batch adaptive mechanisms where GPU workload ratio ρ is calculated based on historical performance statistics and adaptively adjusted across different batches. The **dynamic load distribution mechanism** prioritizes complex tasks for GPU processing and assigns lightweight tasks to CPU, automatically partitioning workloads based on real-time performance feedback and task characteristics. The **asynchronous parallel collaborative computation** enables GPU workers to batch process work units through CUDA kernels for large-scale parallel distance computation, while CPU workers simultaneously process tasks using cached data from the prefetching stage and OpenMP multithreading techniques. Both hardware types work asynchronously through independent work queues, avoiding mutual waiting, followed by efficient result consolidation that combines CPU and GPU computation results through global top-k sorting.

Algorithm 2 implements these mechanisms through dynamic workload partitioning, parallel execution, and intelligent result merging. After obtaining candidate centroids from the navigation stage, the system iteratively processes work partitioning and collaborative computation until all candidate centroids are processed (Lines 6–17). To be specific, `PartitionWork(`P_i`, `ρ`)` dynamically allocates work units to W_{GPU} and W_{CPU} queues based on the calculated GPU workload ratio ρ (Lines 6–7). Subsequently, the system launches true parallel execution where GPU workers process W_{GPU} through set union operations and CUDA kernel distance computation (Lines 8–12), while CPU workers simultaneously handle W_{CPU} using cached data and OpenMP parallelization (Lines 13–15). Subsequently, `UpdateRatio(`T_{GPU}`, `T_{CPU}`)` provides real-time feedback by updating

Algorithm 2. CPU-GPU Collaborative KNN Graph Search

Require: Query set $Q = \{q_1, q_2, ..., q_n\}$, centroid-based k-NN graph G, data D, parameters k, p
Ensure: Top-k nearest neighbors for each query
1: $C_{high} \leftarrow$ TopDegree(G, $\alpha \cdot |V|$) {High-degree centroids}
2: $\mathcal{M} \leftarrow$ WarmupCache(C_{high}) {Initialize prefetch cache}
3: **for** batch $B_i \subseteq Q$ with $|B_i| = b$ **do**
4: $P_i \leftarrow$ BatchBeamSearch(B_i, G, p, w) **on GPU** $\{P_i \in \mathbb{R}^{b \times p}\}$
5: $\mathcal{M} \leftarrow$ AsyncPrefetch(P_i, $\mathcal{M}$) **on CPU** {Update cache}
6: $\rho \leftarrow \frac{T_{GPU}}{T_{GPU}+T_{CPU}}$ {Adaptive GPU ratio}
7: $W_{GPU}, W_{CPU} \leftarrow$ PartitionWork(P_i, ρ)
8: **for all** $w \in W_{GPU} \cup W_{CPU}$ **in parallel do**
9: **if** $w \in W_{GPU}$ **then**
10: $R_w \leftarrow$ TopK(ExactDistance(q_w, $D[\bigcup_{c \in w} R[c]]$), k) **on GPU**
11: **else**
12: $R_w \leftarrow$ TopK(ExactDistance(q_w, $\mathcal{M}[\bigcup_{c \in w} R[c]]$), k) **on CPU**
13: **end if**
14: **end for**
15: $Results_i \leftarrow$ Merge($\{R_w\}$)
16: $\rho \leftarrow$ UpdateRatio(T_{GPU}, T_{CPU}) {Load balance feedback}
17: **end for**
18: **return** $Results = \bigcup_i Results_i$

load distribution parameters based on actual execution times, enabling continuous optimization across batches (Line 16). Finally, the algorithm implements result consolidation by merging CPU and GPU computation results through global top-k sorting to generate final results (Line 18).

4.4 Complexity Analysis

Let N be the number of data vectors, d the dimensionality, K the number of clusters, k_g the average degree of the centroid graph, b the GPU mini-batch size, and W the maximum number of prefetched raw vectors in GPU during query.

For index construction, cooperative clustering and centroid graph building require $O(Nd + Kk_g d + K \log K)$ time. For query processing, each search incurs $O(k_g d + \log K + k \log k)$ operations. In terms of space, the CPU stores the full dataset and index with $O(Nd + Kd + Kk_g)$, while the GPU only maintains $O(Kd + Kk_g + Wd)$, much smaller than $O(Nd)$. This reduction explains the scalability of our framework on memory-limited GPUs.

5 Experimental Environment

5.1 Environment Configuration

All experiments were conducted on a server running Ubuntu 18.04.1 LTS. The server is equipped with two Intel Xeon Silver 4210 CPUs (2.2 GHz, 20 cores per

CPU, totaling 40 threads), 130 GB of DDR4 main memory, and two NVIDIA Tesla T4 GPUs (each with 15 GB of GDDR6 memory). The storage subsystem includes a 7.3 TB MR9361-8i hard disk drive (HDD) and a 960 GB SDLF1DAR NVMe solid-state drive (SSD). All benchmarks were executed on this platform.

The C++ source code was compiled using `g++` version 12.2.0 with the `-O3`, `-std=c++14`, and `-fopenmp` flags to enable optimization and OpenMP-based multithreading. CUDA kernels were compiled using `nvcc` version 12.3 with the `-O3` optimization flag to ensure high performance on the GPU.

5.2 Datasets and Benchmarks

Table 1. Statistics of representative large-scale datasets.

Dataset	Dimension	Raw Data Size	Data Type	Domain
SIFT1B	128	119 GB	uint8	Image
SPACEV1B	100	93 GB	int8	Web Search
DEEP1B	96	358 GB	float32	Image

In this study, we employ three classic billion-scale datasets widely used in the ANNS domain in Table 1. The SIFT1B [8] dataset contains 1 billion 128-dimensional SIFT feature vectors, serving as a widely adopted benchmark in the computer vision domain. The Deep1B [9] dataset comprises 1 billion 96-dimensional deep learning feature vectors extracted from the final hidden layer outputs of neural networks. For experimental requirements, we converted the Deep dataset from float format to int8 format to meet our testing specifications. The Spacev1B [10–14] dataset consists of 1 billion 100-dimensional randomly generated vectors, designed to evaluate algorithm performance on uniformly distributed data.

Additionally, we conduct experiments using 100M-scale versions of these three datasets, obtained by extracting the first 100 million data points from their respective 1B counterparts. For experimental comparison, we select the following three state-of-the-art algorithms:

SPANN: A large-scale approximate nearest neighbor indexing method based on multi-level clustering and sparse graph structures, balancing high retrieval accuracy with computational efficiency.

DiskANN: A high-performance approximate nearest neighbor search system designed for billion-scale datasets with external memory support (SSD/HDD), enabling efficient ANN queries under limited memory constraints.

BANG: Utilizes quantization-compressed datasets on a single GPU to perform distance computations while efficiently accessing graph indices stored in host memory.

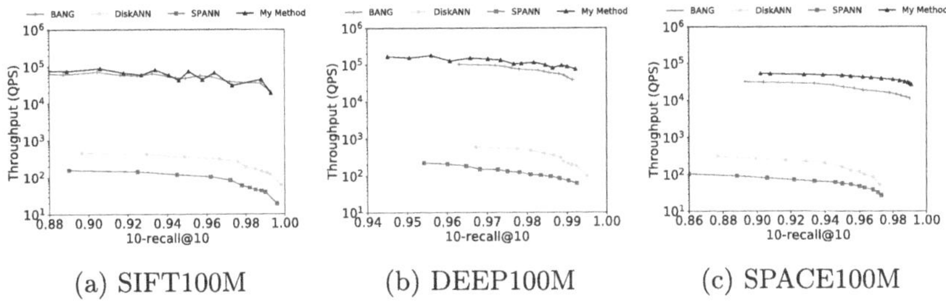

(a) SIFT100M (b) DEEP100M (c) SPACE100M

Fig. 1. QPS-recall curves on 100 M datasets.

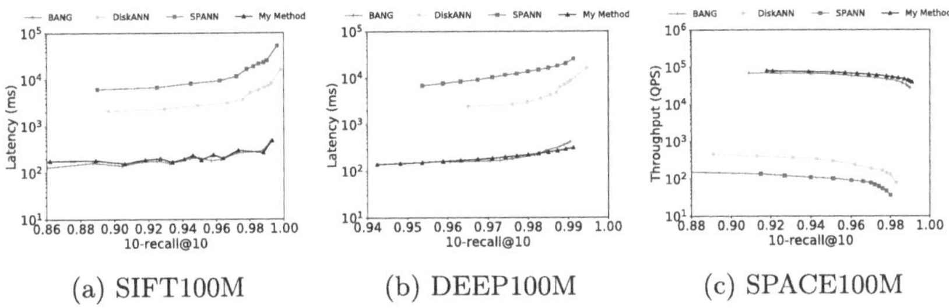

(a) SIFT100M (b) DEEP100M (c) SPACE100M

Fig. 2. Latency-recall curves on 100 M datasets.

5.3 Performance on 100 M Datasets

On 100M-scale datasets, we evaluated the performance of our proposed CPU-GPU collaborative pipeline indexing method against mainstream approaches including SPANN, DiskANN, and BANG. The test datasets included SIFT100M, Deep100M, and Space100M, representing different data distributions and feature dimensions. The experiments focused on key metrics including query throughput (QPS) and recall rate, with results shown in Fig. 1.

In terms of query performance, our method achieved significant improvements in query throughput while maintaining high recall rates (Recall@10 exceeding 90%). On SIFT100M, with a query batch size of 10,000, the QPS reached 80,000, which is tens to hundreds of times higher than DiskANN and SPANN, and comparable to BANG's performance. From a latency perspective, our method maintains latency within 135–487 ms on SIFT100M, essentially matching BANG's 130–465 ms, while DiskANN exhibits latency as high as 2,104–16,038 ms and SPANN reaches 6,235–52,432 ms, which shown in Fig. 2.

Similar trends were observed on Deep100M and Space100M datasets. Notably, our method demonstrates the most outstanding performance on Deep100M, with latency only 140–315 ms, outperforming BANG's 156–419 ms. This thoroughly validates the superiority of the collaborative pipeline architecture in medium-scale big data scenarios.

5.4 Performance on 1B Datasets

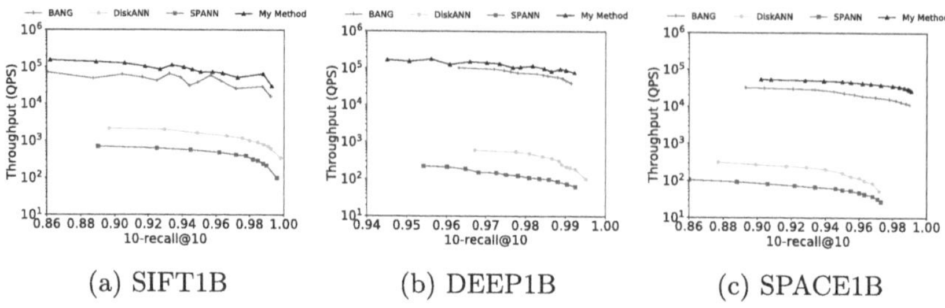

(a) SIFT1B (b) DEEP1B (c) SPACE1B

Fig. 3. QPS-recall curves on 1B datasets.

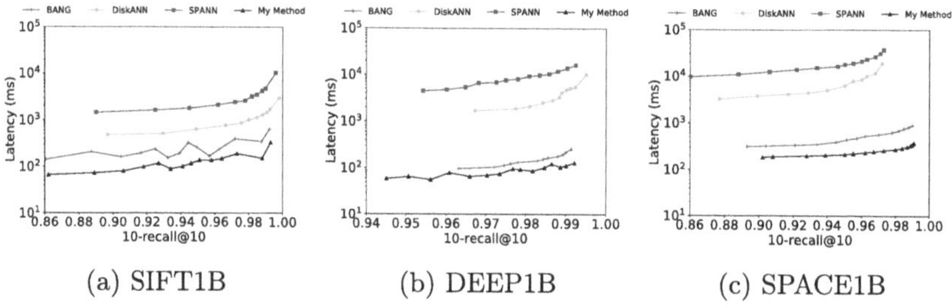

(a) SIFT1B (b) DEEP1B (c) SPACE1B

Fig. 4. Latency-recall curves on 1B datasets.

Experimental results demonstrate that as data scale expands to billion-level, our CPU-GPU collaborative method exhibits more significant performance advantages. On the SIFT1B dataset, in Fig. 3, our collaborative framework achieves over 90,000 QPS at approximately 0.95 Recall@10, significantly outperforming BANG's 60,000 QPS and exceeding DiskANN and SPANN by approximately 200 times.

Figure 4 shows that the latency analysis reveals consistent superiority across all billion-scale datasets. Our method consistently delivers sub-400 ms response times, with particularly impressive performance on Deep1B where latency is reduced by over 50% compared to BANG. This improvement becomes more pronounced as recall requirements increase, demonstrating the robustness of our collaborative architecture under demanding accuracy constraints. The performance gains are attributed to effective load balancing between CPU and GPU components, where intelligent prefetching and asynchronous processing eliminate traditional bottlenecks that plague single-platform solutions.

6 Related Work

Approximate Nearest Neighbor Search (ANNS) has been widely studied, with methods evolving from tree structures, hashing, and quantization to more recent graph-based approaches such as NN-Descent, EFANNA [15], and HNSW [16], which improve recall and efficiency but face high index construction cost and memory overhead on large datasets. Leveraging GPUs, works like CAGRA [17] and GPU-accelerated NN-Descent exploit parallelism, optimized data structures, and prefetching to accelerate graph construction and querying, while hybrid storage systems (e.g., DiskANN, SPANN, BANG) and emerging hardware solutions (e.g., SmartSSD [18], FusionANNS [19]) further enhance scalability under resource constraints. Despite these advances, challenges remain in resource collaboration, load balancing, and adaptability in extreme-scale vector retrieval scenarios.

7 Conclusion

This paper addresses the computational and storage bottlenecks in large-scale vector retrieval by proposing a CPU-GPU collaborative pipeline heterogeneous retrieval framework for billion-scale datasets. The framework innovatively integrates mechanisms including mini-batch collaborative clustering, pipeline asynchronous k-NN graph construction and optimization, intelligent prefetching, and dynamic load allocation, achieving efficient CPU-GPU resource collaboration and parallel task scheduling. This significantly enhances the overall efficiency and scalability of large-scale vector retrieval systems.

Acknowledgments. This work is supported by the National Key R&D Program of China (2024YFE0209000), the NSFC (Grant Nos.61932004, 62225203, U21A20516, 62302084, U2441237) and the Beijing Municipal Science and Technology Commission and Zhongguancun Science Park Management Committee (Z231100007423003).

References

1. Zhang, K., Sun, C., Shen, D., Nie, T., Kou, Y.: High-dimensional nearest neighbor search-based blocking in entity resolution. In: International Conference on Web Information Systems and Applications, pp. 215–226. Springer (2024)
2. Johnson, J., Douze, M., Jégou, H.: Billion-scale similarity search with GPUs. IEEE Trans. Big Data **7**(3), 535–547 (2019)
3. Zhao, W., Tan, S., Li, P.: SONG: approximate nearest neighbor search on GPU. In: 2020 IEEE 36th International Conference on Data Engineering (ICDE), pp. 1033–1044. IEEE (2020)
4. Groh, F., Ruppert, L., Wieschollek, P., Lensch, H.P.: GGNN: graph-based GPU nearest neighbor search. IEEE Trans. Big Data **9**(1), 267–279 (2022)
5. Jayaram Subramanya, S., Devvrit, F., Simhadri, H.V., Krishnawamy, R., Kadekodi, R.: DiskANN: fast accurate billion-point nearest neighbor search on a single node. In: Advances in neural information processing Systems, vol. 32 (2019)

6. Khan, S., Singh, S., Simhadri, H.V., Vedurada, J., et al.: BANG: billion-scale approximate nearest neighbor search using a single GPU. arXiv preprint: arXiv:2401.11324 (2024)
7. Chen, Q., et al.: Spann: Highly-efficient billion-scale approximate nearest neighborhood search. In: Advances in Neural Information Processing Systems, vol. 34, pp. 5199–5212 (2021)
8. Jégou, H., Tavenard, R., Douze, M., Amsaleg, L.: Searching in one billion vectors: re-rank with source coding. In: 2011 IEEE International Conference on Acoustics, Speech and Signal Processing (ICASSP), pp. 861–864. IEEE (2011)
9. Babenko, A., Lempitsky, V.: Efficient indexing of billion-scale datasets of deep descriptors. In: Proceedings of the IEEE Conference on Computer Vision and Pattern Recognition, pp. 2055–2063 (2016)
10. Xu, Y., et al.: SPFresh: incremental in-place update for billion-scale vector search. In: Proceedings of the 29th Symposium on Operating Systems Principles, pp. 545–561 (2023)
11. Zhang, Q., et al.: {VBASE}: Unifying online vector similarity search and relational queries via relaxed monotonicity. In: 17th USENIX Symposium on Operating Systems Design and Implementation (OSDI 23), pp. 377–395 (2023)
12. Chen, Q., et al.: SPTAG: a library for fast approximate nearest neighbor search (2018)
13. Wang, J., Li, S.: Query-driven iterated neighborhood graph search for large scale indexing. In: Proceedings of the 20th ACM International Conference on Multimedia, pp. 179–188 (2012)
14. Wang, J., Wang, J., Zeng, G., Tu, Z., Gan, R., Li, S.: Scalable k-NN graph construction for visual descriptors. In: 2012 IEEE Conference on Computer Vision and Pattern Recognition, pp. 1106–1113. IEEE (2012)
15. Fu, C., Cai, D.: EFANNA: an extremely fast approximate nearest neighbor search algorithm based on KNN graph. arXiv preprint: arXiv:1609.07228 (2016)
16. Malkov, Y.A., Yashunin, D.A.: Efficient and robust approximate nearest neighbor search using hierarchical navigable small world graphs. IEEE Trans. Pattern Anal. Mach. Intell. **42**(4), 824–836 (2018)
17. Ootomo, H., Naruse, A., Nolet, C., Wang, R., Feher, T., Wang, Y.: CAGRA: highly parallel graph construction and approximate nearest neighbor search for GPUs. In: 2024 IEEE 40th International Conference on Data Engineering (ICDE), pp. 4236–4247. IEEE (2024)
18. Tian, B., Liu, H., Duan, Z., Liao, X., Jin, H., Zhang, Y.: Scalable billion-point approximate nearest neighbor search using {SmartSSDs}. In: 2024 USENIX Annual Technical Conference (USENIX ATC 24), pp. 1135–1150 (2024)
19. Tian, B., et al.: FusionANNs: an efficient CPU/GPU cooperative processing architecture for billion-scale approximate nearest neighbor search. arXiv preprint: arXiv:2409.16576 (2024)

eGLSM: A High Performance GPU LSM with Low Memory Usage

Jianyuan Wang[1](✉), Zhaokun Zhang[1], Yong Zhang[2], and Chunxiao Xing[2]

[1] Department of Computer Science and Technology, Tsinghua University, Beijing, China
{wangjian24,zhangzk24}@mails.tsinghua.edu.cn
[2] BNRIST, Tsinghua University, Beijing, China
{zhangyong05,xingcx}@tsinghua.edu.cn

Abstract. Log-Structured Merge-tree (LSM tree) enables fast write throughput and is a key component in many database systems, especially key-value stores. While there has been a considerable amount of research work on LSM optimization techniques, recent work has begun exploring the potential to leveraging modern GPUs to implement data structures. GPU is well-known for its immense computing power with high degrees of parallelism, which also brings challenges to developers. However, existing GPU-based LSM trees still suffer from high memory usage and sub-optimal performance. In this paper, we first do a root cause analysis and identify the inefficiencies of existing work. We then propose eGLSM, a novel LSM tree structure on GPU. eGLSM utilizes the GPU memory by mixing the use of buffer area and data area and saves almost 33% of the GPU memory space compared to previous work. We then improve the performance by proposing a coalesced batch insertion algorithm and an optimized lookup algorithm. Our experimental results demonstrate eGLSM's superior performance with fine stability and scalability, improving the insertion and lookup performance by at least 30% and 10% respectively. The batch insertion algorithm significantly reduces the CPU-GPU synchronization overhead by orders of magnitude. This work takes an important step towards high-performance LSM tree on GPU.

Keywords: Log Structured Merge tree · GPU · Memory Utilization · Batch Insertion

1 Introduction

Nowadays, enterprises heavily rely on database systems, especially key-value stores, to support their services. Well-known key-value stores include LevelDB and BigTable [3] at Google, Voldemort at LinkedIn, RocksDB [4] at Facebook. Log Structured Merge tree (LSM) [9], a predominant data structure in modern key-value storage engines, is pivotal for ensuring efficient data handling in real-world production environments.

With the proliferation of large-scale Web applications—such as social networks, content recommendation platforms, and online retail systems—there is a growing need to handle massive volumes of concurrent writes and real-time queries [5]. These Web-scale

S. Yang et al. (Eds.): WISA 2025, LNCS 16299, pp. 245–257, 2026.
https://doi.org/10.1007/978-981-95-6426-2_21

workloads often involve high ingestion rates of user-generated data, clickstreams, session logs, and time-series metrics, imposing considerable pressure on memory, throughput, and latency. LSM-tree-based key-value stores are widely adopted to cope with such challenges due to their high write throughput and compaction-friendly design.

Database technologies have rapidly evolved, with a major trend being the use of GPUs to accelerate data processing. By leveraging the massive parallelism and high memory bandwidth of modern GPUs, researchers have enhanced core operations such as high-precision computation, relational query execution, storage engine optimization, and heterogeneous offloading in real-time IoT scenarios [10, 11]. These GPU-based approaches significantly improve throughput and reduce latency. Representative GPU-accelerated database systems include Milvus, optimized for large-scale unstructured data and AI applications; HeavyDB, which supports high-performance SQL and geospatial queries; and CrystalDB, tailored for SaaS with multi-tenant and multi-cloud features. However, traditional data structures, when ported to GPUs, are often hampered by inherent GPU challenges such as branching and synchronization overheads.

Due to the nature of the LSM operation executed in batches [7], our focus gravitates towards exploiting the capability of GPU to optimize LSM batch operations. Yet, previous LSM structures implemented on GPU [1] have not fully exploited GPU's computing capabilities. More specifically, the inefficiency of traditional tree-like structure insertion algorithms on GPUs arises from the root-to-leaf traversal. Our studies highlight the unnecessary GPU synchronization overhead in this process, which is a critical performance bottleneck and can be largely reduced [8]. We also identify the optimizing opportunity in lookup operation. Meanwhile, a huge amount of temporary GPU memory space is neglected and wasted.

Addressing these gaps, we introduce **eGLSM**, an **e**fficient **G**PU **LSM** data structure. Our contributions are threefold:

- A memory-efficient design that leverages a state-switching mechanism, potentially saving up to a third of GPU memory space.
- We propose a coalesced batch insertion algorithm and an optimized lookupalgorithm, which gain at least 30% and 10% speedup respectively.
- We conduct comprehensive experiments on both insertion and lookup operations, showing the effectiveness of our proposed method.

The rest of this paper is organized as follows: Sect. 2 provides background on Log-Structured Merge-trees (LSM) and GPU hardware to set the foundation for our work. Section 3 details the architecture of the proposed eGLSM system and its main components. Section 4 describes its implementation, including key design decisions and optimizations. Section 5 presents performance evaluations of eGLSM under various workloads. We conclude the paper with a summary in Sect. 6.

2 Background and Motivation

2.1 LSM Trees

LSM trees, fundamental to modern key-value stores like RocksDB, LevelDB, BigTable, prioritize efficient, high-speed write operations. A typical LSM tree consists of multiple components. The memory component (C_0), or Memtable, temporarily holds incoming write operations. Subsequent disk-based components (C_1, C_2, ..., C_N) store data in the Sorted Sequence Table (SST) format, which is immutable and in key-value form sorted by key. The primary operation of LSM trees is the write operation. Data is initially written to C_0. When full, its contents are sorted and transferred to disk components, with smaller files periodically merged for optimal query performance. For lookup operation, Searches start from C_0 and proceed through disk components using binary search within sorted SSTs.

2.2 GPU Hardware

Graphics Processing Units (GPUs) excel at parallel processing of large data blocks, surpassing general-purpose CPUs. A GPU consists of multiple Streaming Multiprocessors (SMs) and off-chip global (device) memory. Each SM can run hundreds of threads in parallel and provides low-latency on-chip shared memory compared to global memory. We implement eGLSM on NVIDIA GPUs using CUDA, a C++ -based parallel computing platform. GPU programs center on kernels—functions executed on the GPU—though kernel launches incur higher overhead than CPU calls. Data transfer between CPU and GPU typically occurs over the PCIe bus: CPU memory → GPU memory → computation → results back to CPU.

2.3 Motivation

Due to the nature of LSM organization in SSTs, which can be created and merged in batches, it comes to exploit the massive parallelism provided by GPU. We conducted a simple test that compares the parallel sort operation on CPU and GPU. The sorting speed on the GPU is an order of magnitude higher than that of the CPU. Furthermore, if considering the time required for data transfer, GPU sorting still holds a significant advantage.

A previous work [6] proposed that CPU-GPU communication is split into two categories, residual report and kickoff latency. The former is the latency for transferring the data from the device to the host, and the latter is introduced when launching a kernel. The union of the above is called the sync overhead.

Previous work [1] typically follows a traditional paradigm where a loop is used, and each iteration launches a GPU kernel to process a single layer. The CPU then waits and decides whether to continue, introducing significant synchronization overhead. While this approach is simpler to implement, synchronization blocks the CPU thread until all prior GPU kernels finish execution. In earlier studies, this synchronization was found to consume up to 20% of total runtime. To improve efficiency, we eliminate this overhead by pre-computing the final state of the LSM and launching a single GPU kernel, significantly boosting performance. Additionally, lookup efficiency is improved by skipping empty levels.

The limited capacity of device memory is another important factor to be considered in GPU data structure design. The unit price of GPU memory is higher than CPU main memory, thus low memory utilization in GPU data structure is unacceptable and we have designed a novel structure that takes the large wasted buffer space into account to fully leverage the GPU device memory space.

3 Methodology

3.1 Overview

Where eGLSM differs from previous work comes from two major parts: its space efficient structure utilizing a mixed storage scheme, and its optimized operations, highlighting the coalesced batch insertion and lookup algorithm.

LSM structures traditionally use an out-of-place compaction algorithm for performance, but eGLSM optimizes for the valuable GPU space by employing a mixed storage scheme that dynamically uses free data space as temporary space for compaction, reducing space needs significantly. If space runs low, it switches to in-place sorting, which minimally impacts the performance according to our experiments.

In the coalesced batch insertion algorithm, we almost eliminate the synchronization overhead between CPU and GPU by abandoning the looping. Instead, we launch the whole compaction process in a single GPU kernel call by precomputing the final status of eGLSM. Through reducing the synchronization overhead, eGLSM achieves significant insertion performance gain by at least 30%.

3.2 Structure

Figure 1 illustrates the eGLSM's structure, which leverages both CPU and GPU. The CPU manages eGLSM's metadata such as data_ptr which is the the GPU data starting address and batch_num for the current batch account. Data manipulation functions on eGLSM are accessed from the CPU.

Data in eGLSM is manipulated in batches and stored in consecutive memory space, represented as fixed-length block memories or entries in Fig. 1. The size of each entry depends on configuration by specifying how many keys to be packed in a single batch, i.e., batch size.

Entries are categorized by their status: NULL (for uninitialized storage), DATA (for actual data storage), or MERGE-BUF (a temporary compaction buffer using an out-of-place merging algorithm).

In eGLSM, data is logically organized into a series of levels. The mixed storage scheme is a key feature of eGLSM, as the compaction buffer area and data area can be dynamically transformed into each other, while in GPU LSM implementation, the huge waste of buffer area is neglected. When the data area is not being actively used, it can be temporarily transferred to the buffer area.

The out-of-place algorithm for sorting and merging requires O(n) space, which is especially large when performing the last layer compaction. The mixed storage improves the device memory utilization by more than 33%, compared to the design of allocating a large specialized area for compaction.

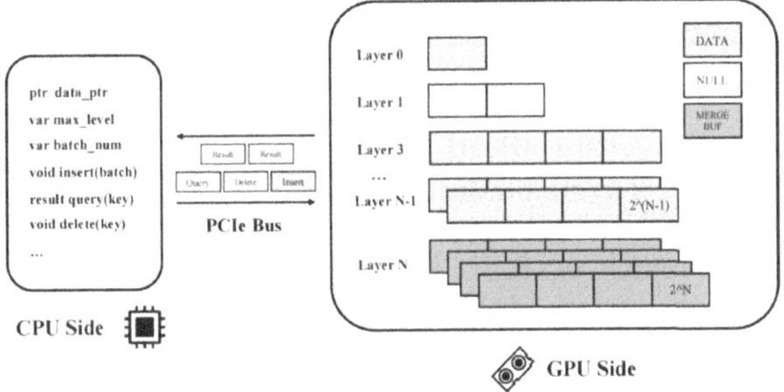

Fig. 1. The overview structure of eGLSM.

3.3 Efficient LSM Operations for GPU

1) Efficient Coalesced Batch Insertion

Core Idea of eGLSM Insertion. The mixed storage scheme introduces a trade-off between space and time. When there is sufficient buffer space, eGLSM can use the out-of-place compaction algorithm. However, to hold more data when other levels are full, MERGE-BUF status has to be converted to DATA status. At that time, no larger room can be used as MERGE-BUF, then eGLSM needs to perform an in-place sorting to ensure data is in order. This trade-off is particularly important in the context of GPU devices. It is important to note that switching to in-place sorting happens only when previous layers are full, meaning the frequency of invoking in-place sorting matches the maximum level count, which minimally impacts performance.

In the classic looping style of leveled compaction procedure, CPU side will check the current visiting level at the beginning of each loop. If full, CPU will launch a kernel to the GPU to merge the data with the data in the buffer to its lower level. Loop process will continue until an empty level is found.

The batch insertion procedure of eGLSM is listed in Algorithm 1: eGLSM eliminates looping, which requires control to frequently return to the CPU to decide when to end the loop, creating substantial sync overhead. As a result, eGLSM achieves considerable performance gains by avoiding this inefficiency. This is the core idea of eGLSM efficient insertion algorithm.

```
Algorithm 1: eGLSM_BatchInsert (bk)
   Input: bk: a batch of keys to be inserted
1:   begin
2:     dest ← eGLSM. getInsertDestLevel()
3:     buf ← eGLSM. getAvailableBufLevel()
4:     if dest == buf then
5:       memcpy(level buf, bk, BATCH_SIZE)
6:       inplaceSort(level buf, levelSize(level buf))
7:     else
8:       cpy_size ← levelSize(level dest) -BATCH_SIZE
9:       memcpy(level buf, root, cpy_size)
10:      offset ← levelSize(level dest) - BATCH_SIZE
11:      memcpy(level buf, input, BATCH_SIZE)
12:      cub::DeviceRadixSort(level buf, level dest, getLevelSize(dest))
```

```
Algorithm 2: getInsertDestLevel (bn)
   Input: bn: current number of stored batches
   Output: l: destination level
1:   begin
2:     count-trailing-zero
3:     operation to bn
4:     return CTZ(~bn)
```

```
Algorithm 3: getAvailableBufLevel (bn)
   Input: bn: current number of stored batches
   Output: bl: current available buffer
1:   begin
2:     ret_level ← 32 - CTZ(bn) -1
3:     while bn & (1 << ret_level) do
4:       ret_level = ret_level -1
5:     return ret_level
```

This is the core idea of eGLSM efficient insertion algorithm. eGLSM will switch to in-place compaction at a certain point. We will introduce this coalesced batch insertion algorithm first. The timing and choice of in-place algorithm are discussed later.

In Algorithm 1: line 1 and 2, the eGLSM first pre-computes two essential values dest and buf. eGLSM will use them to decide if to use an out-of-place insertion or not. Two functions are defined in Algorithm 2: and Algorithm 3: respectively. The efficient out-of-place insertion algorithm designed in eGLSM utilizes the binary level capacity scheme and involves some bitwise operations.

Algorithm 2: computes the final position that the batches are to be merged into. Firstly we apply a bitwise NOT on bn and CTZ is an assembly language instruction and stands for count trailing zeros.

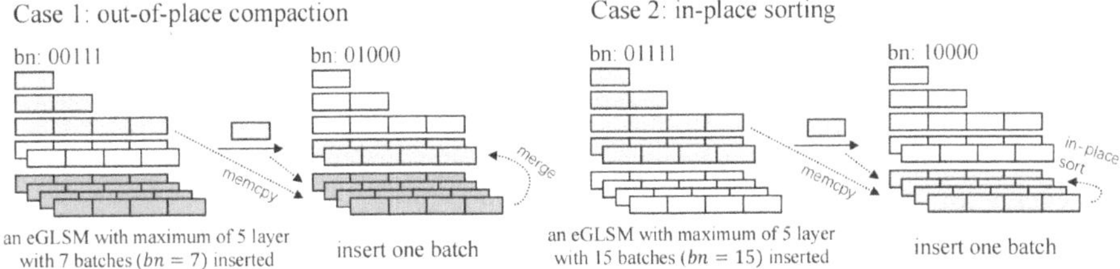

Fig. 2. eGLSM insertion examples

An Example of Insertion. Case 1 in Fig. 2 shows an example of how eGLSM insertion performs. The variable bn indicates the current number of data batches in eGLSM which can be regarded as a bitmap of current eGLSM storage.

Timing to Switch to In-Place Algorithm. Figure 2 shows two cases. If dest equals buf, which means that now layer buf cannot act as a buffer for compaction, then an in-place sort algorithm should be done to satisfy the ordering constraint for each layer. Fortunately, there exists previous work on efficient GPU in-place sorting algorithms. We use the implementation from this in our prototype implementation.

2) **Lookup Operation**

Algorithm 4: eGLSM_Lookup (*bk*)

Input: *bk*: keys to be queried
Output: *res*: query results

1: **for** *q in bk* **do**
2: *result* ← -1; // Assume result is -1 when not found
3: **for** *level* ← 0 **to** max_level - 1 **do**
4: **if** (*batch_num* & (1 ≪ *level*)) = 0 **then**
5: continue; // skip if current level do not have data
6: *levelKeys* ← **getLevelKeys();**
7: *levelValues* ←***getLevelValues*();**
8: **if** *found* ≠−1 **then**
9: *result* ←*levelV alues[found];*
10: **break;**
11: *res*[*q*] ←*result*; // Stores the result

eGLSM runs the lookup operation given a batch of keys, leveraging GPU parallel processing where each thread is independently tasked with resolving a specific query. Each thread begins at the top level. Since elements within each level are sorted, a binary search is performed at each level until the key is located, otherwise the search result will return -1. Algorithm 4: shows the procedure of eGLSM lookup operation.

It is worth noting that in Algorithm 4:, lines 4 and 5, eGLSM intelligently utilizes the information from batch_num, which follows the binary scheme, to efficiently skip the empty levels and thereby avoid needless search operations that would otherwise waste computational resources and increase query latency. Extensive experimental results consistently demonstrate a significant performance improvement in both speed and accuracy compared to previous state-of-the-art approaches in similar large-scale search scenarios.

4 Analysis

In this section, we analyze the space efficiency and time usage of both eGLSM and GPU LSM, which explains the effectiveness of our design and optimization.

4.1 Getting Optimal Batch Size Configuration

We propose two notions of memory utility in Definition 1 and 2.

Definition 1. For a GPU data structure DS, Structure Utility is defined as:

$$StructureUtility_{DS} = \frac{DatasructureSize}{DataCapacity} \times 100\% \tag{1}$$

Definition 2. Given a GPU data structure, DS and a type of GPU D, Device Space Utility is defined as:

$$DeviceUtility_{DS,D} = \frac{DataCapacity}{GPU\_MemCapacity(D)} \times 100\% \tag{2}$$

The first definition describes the proportion of data capacity that occupies the data structure size. The latter equals the data capacity plus the buffer area size, and the second definition describes the proportion of data capacity that occupies the GPU memory capacity under a specific GPU model. Two definitions measure the memory utility from both theoretical and practical perspectives.

Suppose the GPU LSM implementation has n layer of data batches. One additional layer is a must for the compaction buffer. The compaction buffer will be as large as the last data layer, which in the extreme case will hold all the data batches from 1 to n -1 layer plus the newly inserted batch and get compacted into the last n-th layer.

The $\text{StructureUtility}_{\text{eGLSM}}$ is 100% due to its elimination of the buffer area. However the structure utility is only a theoretical value, the real device memory usage cannot be neglected. For eGLSM, we argue that $(3 \cdot 2^n - 1) \times$ batchsize should be as close to GPU Device Capacity as possible. Since we operate eGLSM in the binary form, batchsize in the power of 2 is more desirable. For simplicity, we consider int-int key-value storage, and the real storage would be items count times eight bytes. Then the optimal configuration can be pre-computed by calculating from some typical GPU memory capacity configuration. For example, for an NVIDIA RTX-4090 GPU with 24GB of device memory, when configuring batch size 4096, $\text{DeviceUtility}_{\text{eGLSM, 2080Ti}}$ equals 100%.

4.2 Time Analysis

The total computational time, for the insertion procedure in GPU LSM or eGLSM within the CUDA programming paradigm can be divided into four parts, including time for data transferring, host-device synchronization, device data copying, and calculating. In the context of data structures on GPU, calculating is associated with the computational processes during the merging of pre-existing data with the newly inserted batch into the subsequent layer.

Fortunately, the proportion of time consumption of each part can be directly observed by the NVIDIA Nsight System, which is a performance analysis tool that helps developers optimize and profile the performance of GPU accelerated applications with visualization aids. Table 1 compares GPU LSM and eGLSM CUDA summary on API and Kernel calls on inserting 15, 625 batches with the batch size of 2^{15}.

The major difference between the GPU LSM and eGLSM is that eGLSM has almost eliminated the proportion of cudaDeviceSynchronize time due to the coalesced compaction scheme. Meanwhile, the time spent on cub::DeviceRadixSort also decreases. It is worth noting that cudaMemcpy refers to the timespan of calling CUDA APIs on the CPU side, while CUDA memcpy DtoD/HtoD refers to the actual memcpy operation process on the GPU side. Here a slight increase in cudaMemcpy time is observed for eGLSM, because eGLSM packs the merging layers into one call. However, the time spent on the GPU side (CUDA memcpy DtoD) decreases, indicating that eGLSM reduces the amount of data copying. Overall, the optimizations in eGLSM reduce synchronization overheads and sorting time, leading to faster insertions.

Table 1. Comparison of time and proportion of different CUDA API/Kernel calls

API/Kernel	GPU LSM		eGLSM	
	Time	Ratio	Time	Ratio
cudaLaunchKernel	405.0 ms	5.10%	387.5 ms	6.70%
cudaMemcpy	1.7 s	20.80%	1.9 s	32.60%
CUDA memcpy DtoD	284.3 ms	3.60%	108.6 ms	1.90%
CUDA memcpy HtoD	719.7 ms	9.10%	723.8 ms	12.50%
cudaDeviceSynchronize	1.7 s	21.20%	72.6 us	**0.00%**
cub::DeviceRadixSort	2.1 s	26.30%	1.8 s	30.80%
cudaMalloc	356.3 ms	4.50%	321.9 ms	5.60%
cudaMemset	510.1 ms	6.40%	395.1 ms	6.80%
CUDA memset	243.3 ms	3.10%	185.1 ms	3.20%
cudaStreamCreate	12.4 us	0.00%	12.5 us	0.00%

5 Experiments and Results

This section shows both the insertion and lookup performance of eGLSM compared to two previous works in various aspects.

5.1 Experiment Settings

In our evaluation, we use the dataset LGN, which is randomly generated based on the log-normal distribution. We create 1024M random key-value pairs.

We compare eGLSM with two GPU data structures as baselines:

- **GPU LSM** [1], the LSM structure implemented on GPU supporting insertions and deletions with retrieval operations of lookup, count, and range query operations. We implemented the GPU LSM according to the pseudocode provided in their paper.
- **GPU B-Tree** [2], designed by Awad et al. with a warp cooperative worksharing strategy. We use their open-sourced code.

We implement eGLSM atop CUDA and NVIDIA's CUB library, using NVIDIA's NVCC compiler (version 11.0.211) with the -O3 optimization flag. All experiments are run on a server with an Intel(R) Xeon(R) Gold 6248 CPU@2.50GHz and an 11GB NVIDIA 2080Ti GPU.

5.2 Experiment Results

Find Optimal Batch Configuration. First, we investigate the optimal batch configuration for the eGLSM data structure, as well as the GPU LSM and GPU B-Tree. We perform experiments using various batch configurations ranging from 2^{10} to 2^{18}, with each configuration being a power of 2. Our goal is to identify the optimal batch size for unified experiment settings that yield the best performance while keeping the flexibility for data manipulation.

Table 2. Total insertion time (Unit: Seconds) with different batch sizes under Lgn dataset

Batch Size	eGLSM	GPU LSM	GPU B-Tree
210	23.98	33.73(+41%)	77.67%(+223%)
211	15.30	22.71(+48%)	40.38(+164%)
212	10.91	17.58(+61%)	20.97(+92%)
213	8.74	11.53(+32%)	11.78(+35%)
214	5.34	7.35(+38%)	7.13(+34%)
215	3.79	5.65(+49%)	4.63(+22%)
216	3.29	4.72(+43%)	3.92(+9%)
217	3.12	4.37(+40%)	3.59(+15%)
218	3.08	4.26(+38%)	3.42(+10%)

Table 2 shows that eGLSM, GPU LSM, and GPU B-Tree exhibited a significant decrease in insertion time as the batch size increased from 2^{10} to 2^{14}, which can be attributed to the fact that batch size less than 2^{14} fails to fully leverage the parallelism brought by CUDA cores. The performance gains from increasing the batch size became much smaller after the size of 2^{15}, which is close to 8 times larger than the number of CUDA cores (i.e., 4352 cores). It is worth noting that a larger batch size may reduce the flexibility of data manipulation due to the larger batch granularity. Considering the GPU utilization and the flexibility of data manipulation, we choose 2^{15} as the setting for later experiments.

The experimental results clearly demonstrate that eGLSM significantly outperforms the GPU LSM in terms of insertion performance, achieving improvements ranging from 32% to 61% across various datasets and workload configurations. Furthermore, when compared with the GPU B-Tree, eGLSM exhibits even greater performance gains, with insertion throughput improvements ranging from 10% up to 223%, depending on the data distribution and batch size.

Stability of eGLSM Insertion Performance. To evaluate the stability of eGLSM and baselines' insertion performance, we continuously inserted 15625 batches of elements and measured the insertion time of each batch. The results show that eGLSM can sustain stable insertion rates with relatively low tail latency.

The statistics in Table 3 demonstrate that eGLSM has significantly lower latency than GPU LSM and GPU B-Tree. Specifically, eGLSM has a p99 latency of 0.692 ms with a maximum of 68.031 ms, while GPU LSM has a higher p99 latency of 1.610ms and a maximum of 133.281 ms, showing that our coalesced batch insertion makes substantial improvement. Though GPU B-Tree has better performance on p99 and maximum latency, it exhibits lower performance in other metrics.

Overall, the results indicate that eGLSM provides higher and more stable insertion performance with fewer spikes, making it suitable for workloads demanding both speed and consistency.

Lookup performance of eGLSM. To evaluate the lookup performance, we conduct the experiments under 3 different workloads at different proportions of LGN dataset. Query size is 1024×32, where 1024 is the maximum threads within a block on NVIDIA 2080Ti, i.e., 32678 keys are queried simultaneously. We calculate the average time of 1000 different trials. Workloads including RANDOM, HALF, and FULL. RANDOM refers to generating queries randomly and most of them are non-existing in the dataset. HALF means half of the query keys are sampled from the dataset and FULL is generated in the same manner. Figure 3 shows that eGLSM outperforms the other two baselines. Due to eGLSM skipping needless binary search on empty levels, eGLSM performance is much more stable compared to GPU LSM.

Table 3. Latency statistics for 15265 insertions on eGLSM, GPU LSM and GPU B-Tree in microseconds (bz = 215).

Metrics/Methods	eGLSM	GPU LSM	GPU B-Tree
Average	0. 202	0. 257(+27%)	0. 323(+59%)
P25	0. 132	0. 082(-38%)	0. 313(+137%)
Median	0. 194	0. 185(-4%)	0. 329(+69%)
P99	0. 692	1. 610(+132%)	0. 414(-41%)
Max	68. 031	133.281(+96%)	30. 872(-45%)

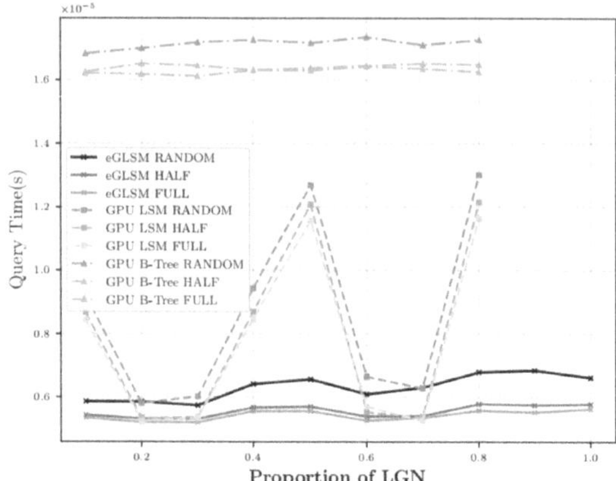

Fig. 3. Lookup performance of eGLSM, GPU LSM and GPU B-Tree under 3 different workloads (bz = 215).

6 Conclusions

In conclusion, this paper addresses the inefficiencies of existing GPU-based LSM tree methods, including high memory usage and sub-optimal performance. We propose eGLSM, a novel LSM tree with a mixed storage scheme. Data area and merge buffer can be dynamically converted. Switching between in-place and outout-place insertion algorithms reduces memory footprint by a third compared to previous work. We optimize operations in eGLSM including batch insertion and lookup, improving insertion throughput by at least 30%, read performance by 10% with more stability. Our coalesced batch insertion algorithm significantly reduces the synchronization overhead between CPU and GPU. Overall, the eGLSM approach presents a promising solution for leveraging the computing power of modern GPU to optimize LSM tree implementations in key-value stores and other database systems.

References

1. Ashkiani, S., Li, S., Farach-Colton, M., Amenta, N., Owens, J.D.:GPU LSM: a dynamic dictionary data structure for the GPU. In: IPDPS, pp. 430–440 (2018)
2. Awad, M.A., Ashkiani, S., Johnson, R., Farach-Colton, M., Owens, J.D.: Engineering a high-performance GPU B-Tree. In PPoPP, pp. 145–157 (2019)
3. Fay, C., et al.: BigTable: a distributed storage system for structured data. ACM Trans. Comput. Syst. (TOCS) **26**(2), 1–26 (2008)
4. Dong, S., Callaghan, M., Galanis, L., Borthakur, D., Savor, T., Strum, M.: Optimizing Space Amplification in RocksDB. In: CIDR, vol. 3, p. 3, (2017)
5. Wu, Y., et al.: MemQ: A graph-based query memory prediction framework for effective workload scheduling. In: ICDE, pp. 3876–3889 (2025)
6. Lee, S., Ryu, D., Ko, H.S.: Reduction of CPU-GPU synchronization overhead for accelerating implicit clothing simulator. In: Eurographics, pp. 9–12 (2018)

7. Jeon, G., Jeon, S., Seo, J.T.: A survey on attack cases with VBS malware in windows. In: WISA, pp. 284–295. Springer, Singapore (2024)
8. Shie, C.H., Fang, P.H., Fan, C.I.: Fuzzing JavaScript engines with diversified mutation strategies. In: WISA, pp. 243–254. Springer, Singapore (2024)
9. O'Neil, P., Cheng, E., Gawlick, D., O'Neil, E.: The log-structured merge-tree (LSM-tree)Acta Informatica **33**, 351–385 (1996)
10. Sheng, M., et al.: A multi-source heterogeneous medical data enhancement framework based on lakehouse. Health Inf. Sci. Syst. **12**, 37 (2024)
11. Wang, S., Zhang, Y., Wang, K., Wang, J., Luo, Y., Hao, R.: MQRLD: a multimodal data retrieval platform with query-aware feature representation and learned index based on data lake. Inf. Process. Manag. **62**(4), 104101 (2025)

HQIA: An Index Advisor for Hybrid Query Workloads

Kun Chao[1], Kaijun Wen[1(✉)], Kaige Wang[2(✉)], Jiahao Li[1], Peng Ren[3], and Chunxiao Xing[3]

[1] School of Software, Henan University, Kaifeng 475004, China
{chaokun,wenkaijun,lijiaohao}@henu.edu.cn
[2] School of Artificial Intelligence, Henan University, Zhengzhou 450046, China
wkg@henu.edu.cn
[3] BNRist, DCST, RIIT, Tsinghua University, Beijing 100084, China
{renpeng,xingcx}@tsinghua.edu.cn

Abstract. With the advent of the era of LLMs, the demand for exploring multimodal data continues to grow. Queries over such data often involve hybrid workloads that combine structured attribute filtering with high-dimensional semantic matching. Efficient index advising for such hybrid query workloads is crucial for improving query performance. However, existing index advisors typically optimize individual types of queries in isolation and lack the capability to recommend scalar and vector indexes simultaneously for hybrid workloads. To address this limitation, this paper proposes HQIA, a hybrid query index advisor based on RIBE. Within this framework, we design a feature representation method tailored for hybrid queries, which integrates query cost and attribute distribution information. By incorporating physical query plans with histogram features, the method gains the ability to represent the predicate coverage of queries, thereby improving the distinguishability among different queries and ultimately enhancing workload compression and clustering. Building upon this foundation, we extend RIBE's index advising mechanism to support hybrid queries. Experimental results show that HQIA achieves superior tuning effectiveness and significant performance improvements over existing methods in hybrid query scenarios.

Keywords: Hybrid query · Index advisor · Pgvector

1 Introduction

In recent years, with the rapid development of large language models (LLMs), the demand for managing and retrieving massive amounts of heterogeneous data has grown significantly. In modern application scenarios, query workloads often involve both structured attribute filtering and high-dimensional vector similarity matching, forming hybrid query workloads [1]. This query pattern is prevalent in domains such as recommendation systems, intelligent search, e-commerce,

S. Yang et al. (Eds.): WISA 2025, LNCS 16299, pp. 258–269, 2026.
https://doi.org/10.1007/978-981-95-6426-2_22

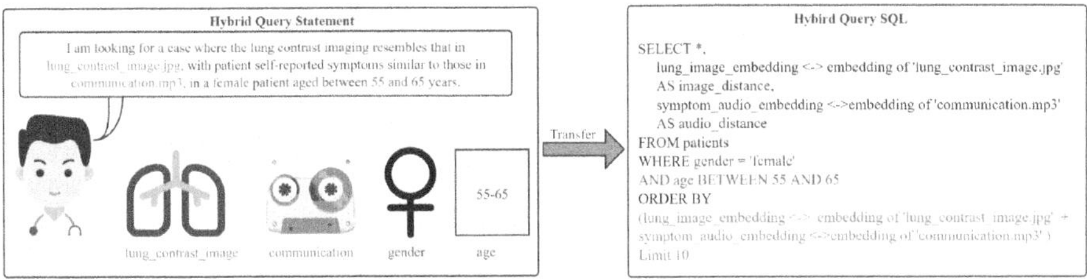

Fig. 1. Hybrid query example in a medical multimodal scenario.

and healthcare, placing higher demands on the query processing capabilities of database systems.

For example, in the healthcare domain, a physician may wish to retrieve historical cases that match specific patient demographics (e.g., female patients aged between 55 and 65) and exhibit similar imaging features and self-reported symptoms. As illustrated in Fig. 1, such workloads involve both filtering based on structured attributes (e.g., gender, age) and matching based on vector similarity (e.g., medical imaging, voice descriptions). Such hybrid workloads present new challenges for database systems, particularly in supporting efficient query processing across both structured and unstructured data modalities.

Index advising is one of the commonly used techniques for database optimization. However, existing index recommendation methods primarily target scalar query workloads, typically focusing on index selection over scalar attributes, and struggle to effectively address the needs of vector queries in hybrid workloads [2]. Meanwhile, although vector databases have made significant progress in supporting high-dimensional data retrieval, most existing work focuses on accelerating index structures themselves, such as HNSW [3] and IVF [4], and lacks a systematic framework for index recommendation. Furthermore, current mainstream what-if analysis tools (e.g., HypoPG [5]) do not support hypothetical evaluation of vector indexes, leaving a technical gap in optimization strategies for vector queries during hybrid index configuration.

To address the above challenges, this paper proposes HQIA (Hybrid Query Index Advisor), a framework designed to provide efficient and accurate index recommendations under hybrid query workloads. Building on the RIBE framework [6], HQIA introduces systematic extensions tailored to the characteristics of hybrid queries. The main contributions are as follows:

- We design a feature representation method for hybrid queries, which jointly considers query predicates, data distribution, and vector similarity information. This enhances the distinguishability of query features, thereby improving query clustering and workload compression effectiveness.
- We extend RIBE's index recommendation workflow [6] by introducing an index evaluation mechanism that jointly optimizes scalar and vector queries in hybrid workloads. In particular, to address the lack of what-if simulation support for vector indexes, HQIA estimates the benefits of vector indexes by

actually creating the indexes and caching the benefit data, enabling effective benefit estimation for vector indexes and enhancing overall index recommendation quality and query performance.
- We conduct comprehensive experimental evaluations on an extended TPC-H [7] hybrid query workload, demonstrating that HQIA outperforms existing methods in terms of both recommendation quality and query performance.

This work provides a new approach to database optimization for hybrid query scenarios and offers a potential direction for future research on index tuning in multimodal data management systems.

2 Related Work

Recent index recommendation methods aim to reduce tuning complexity by compressing query workloads and leveraging machine learning techniques. For example, GSUM [8] and ISUM [9] reduce the tuning workload through workload compression, although this strategy inevitably leads to some information loss, potentially impacting the quality of index recommendations. DISTILL [10] further introduces machine learning models to replace parts of the "what-if" calls, but it still faces challenges regarding compatibility with workload compression and model maintenance. RIBE [6] effectively overcomes the limitations of traditional methods by introducing a workload matrix and deep learning techniques. By compressing the query workload, eliminating redundant queries, and leveraging fast benefit estimation based on the workload matrix, RIBE significantly reduces tuning complexity and computational overhead. Unlike GSUM and ISUM, RIBE focuses on retaining queries that have a significant impact on tuning outcomes, thus avoiding critical information loss and improving tuning quality. In addition, RIBE employs the deep learning model ChangeFormer [6], which uses a self-attention mechanism to predict whether an index configuration will alter the query plan structure, thereby reducing reliance on the optimizer and improving prediction accuracy. However, despite its strong performance on scalar queries, RIBE is not well-suited to handle the demands of high-dimensional vector retrieval in hybrid query workloads.

In the context of vector database management systems (VDMSs) such as pgvector [11], Milvus [12] and Qdrant [13], various efficient approximate nearest neighbor search algorithms have been proposed, significantly improving retrieval efficiency for large-scale data. Nevertheless, these systems largely depend on manual selection of appropriate index types and parameters, lacking automated index recommendation mechanisms. VDTuner [14], as an automated tuning framework, leverages Bayesian optimization to enhance query performance. DKTune [15] also uses Bayesian optimization to tune system knobs. However, VDTuner's tuning process for vector indexes remains independent of traditional scalar index tuning and does not achieve coordinated optimization for hybrid workloads.

Recently, optimization for hybrid query workloads has attracted increasing attention. Platforms like MQRLD [1] attempt to combine traditional indexing techniques with vector databases to optimize query performance in complex queries, especially for multimodal data retrieval. Although this method supports the joint optimization of structured and vector data, it still faces challenges such as the lack of a unified and automated index recommendation mechanism under large-scale workloads.

By introducing hybrid query feature representation, benefit estimation, and workload compression techniques, HQIA demonstrates significant advantages in improving both index recommendation quality and query performance, making progress toward bridging the gap in hybrid query index recommendation in current research.

3 Architecture

Figure 2 illustrates the overall architecture of HQIA, which aims to generate an index set capable of simultaneously accelerating scalar and vector queries within hybrid workloads.

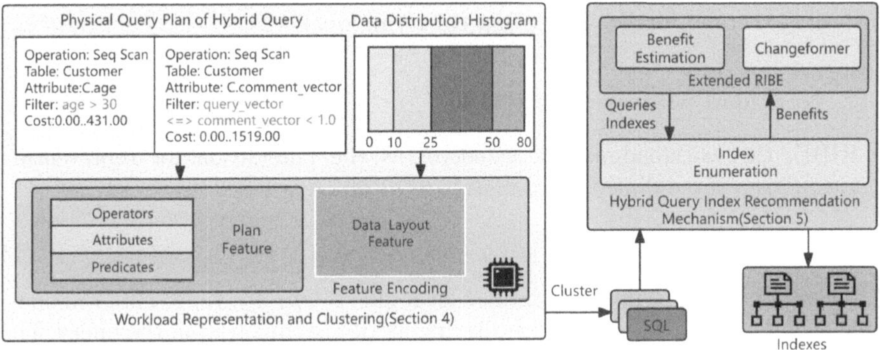

Fig. 2. The architecture of HQIA.

HQIA takes a hybrid query workload as input and processes each query by extracting its physical execution plan from PostgreSQL [16] and attribute distribution features from equi-depth histograms. These are combined to construct a multidimensional feature representation, which is then encoded—integrating operators, attributes, predicate conditions, and data layout information—into unified feature vectors. Finally, clustering is applied to generate a compressed workload.

Building on the extended RIBE framework [6], HQIA performs index benefit estimation over the compressed hybrid workload. For scalar queries, it leverages RIBE's cost modeling approach to rapidly estimate index benefits. For vector queries, it evaluates benefits by physically creating indexes and measuring their effects, while also maintaining cached benefit information to avoid redundant index creations. To further reduce reliance on the optimizer, HQIA employs an adapted ChangeFormer module [6] to predict whether a given index configuration will alter the query plan structure.

Further details on workload representation and clustering (Sect. 4), as well as the hybrid query index recommendation mechanism (Sect. 5), will elaborate on how feature representation, query clustering, and hybrid-aware index recommendation mechanism are employed to optimize hybrid query workloads.

4 Hybrid Query Workload Representation and Clustering

Accurate workload representation is critical for improving the effectiveness of index tuning in hybrid query optimization. Traditional methods such as RIBE represent queries using a workload matrix C [6], but this approach has certain limitations, particularly in capturing the distribution range of query predicates. To enhance the precision of workload representation, we propose an augmented method that incorporates histogram-based features, aiming to more accurately depict the predicate coverage of queries. Furthermore, we introduce a weighting adjustment mechanism to control the influence of different features during query clustering, thereby further improving clustering quality.

4.1 Workload Matrix C in RIBE

In RIBE, the workload matrix C serves as the foundation for representing queries. C is a three-dimensional matrix, typically expressed as:

$$C[q, OpType, Attrs] \tag{1}$$

where q denotes a unique identifier for a query, $OpType$ represents the operation type, and $Attrs$ refers to the filtering attributes involved in the query. The values in the C matrix are typically the execution cost of the query or relevant information from the query plan [6].

This representation can capture the basic operations and cost information of queries; however, it has certain limitations when representing the predicate range of a query. In particular, for queries with the same predicates but covering different data ranges, the C matrix cannot effectively distinguish between them.

Since it only includes query operations and attributes, it fails to reflect the impact of data distribution on query execution, which may cause similar queries to be incorrectly clustered into different groups and ultimately affect the accuracy of index recommendations.

4.2 Improvement of Representation Accuracy

To address the limitations of RIBE's [6] workload matrix representation, we incorporate equi-frequency histogram features to more precisely capture predicate coverage—particularly in scalar and vector queries where data distribution varies significantly. This enhances query distinguishability and provides more accurate feature representations for downstream clustering and index recommendation.

For scalar data, we adopt equi-frequency binning [16] to represent the data distribution. Given an attribute A with range $[A_{\min}, A_{\max}]$, the range is divided into n bins, each containing approximately the same number of data points. Unlike equal-width binning, this method is robust to data skew and thus accounts for the actual data distribution more accurately, allowing for a better representation of query coverage.

Let B be the filtering condition of a query predicate, $[B_{\min}, B_{\max}]$. We map this predicate onto the equi-frequency histogram and represent it as a binary vector. For example, if the range of attribute A is divided into 10 bins, the corresponding binary vector for B may be:

$$\text{Hist}_B = \begin{bmatrix} 0\ 1\ 1\ 1\ 1\ 1\ 0\ 0\ 0\ 0 \end{bmatrix} \tag{2}$$

where each 1 indicates that the predicate overlaps with the corresponding bin. This allows for a precise representation of the query's access pattern as part of its feature vector.

For vector data, since PostgreSQL's built-in `pg_stats` [16] does not support histograms, we estimate distribution via binning based on distances to the geometric center of the vector space. Specifically, we compute the distance of each vector to the center, then apply equi-frequency binning. This enables a similar binary representation of vector predicate coverage:

$$\text{Hist}_V = \begin{bmatrix} 0\ 0\ 0\ 0\ 1\ 0\ 0\ 0\ 0\ 0 \end{bmatrix} \tag{3}$$

This representation captures distribution in high-dimensional space and improves the discriminative power for downstream clustering and index recommendation.

4.3 Weighted Adjustment and Regularization in Clustering

During query clustering, the workload matrix C and the histogram vector Hist_q provide complementary feature representations. However, due to the typically higher dimensionality of histograms, they may dominate the similarity computation and bias clustering results.

Let the matrix-based representation of query q be $C_q = C[q, OpType, Attrs]$, and its histogram vector be Hist_q. To balance the contributions of these representations, the weighting coefficient α is dynamically calculated based on their respective dimensionalities. Let d_C be the dimensionality of C_q and d_H be the dimensionality of Hist_q. The coefficient α is defined as:

$$\alpha = \frac{d_H}{d_C + d_H} \tag{4}$$

This calculation gives a higher weight to the similarity score of the lower-dimensional feature, preventing the higher-dimensional feature from dominating the result. The similarity between queries q and p is computed as:

$$\mathrm{Sim}(q,p) = \alpha \cdot \mathrm{Sim}_C(C_q, C_p) + (1-\alpha) \cdot \mathrm{Sim}_{Hist}(\mathrm{Hist}_q, \mathrm{Hist}_p) \tag{5}$$

where Sim_C and Sim_{Hist} denote the similarity measures based on the matrix and histogram representations, respectively.

To prevent the high-dimensional histogram vectors from skewing the similarity calculation, we further apply normalization to the histogram features:

$$\hat{\mathrm{Hist}}_q = \frac{\mathrm{Hist}_q}{||\mathrm{Hist}_q||} \tag{6}$$

This regularization ensures that histogram features contribute proportionally, regardless of their dimensionality.

5 Hybrid Query Index Recommendation Mechanism

Building on prior sections, we now describe the design and implementation of the hybrid query index recommendation mechanism in HQIA. Specifically, we discuss how benefit estimation and the ChangeFormer model [6] are leveraged to optimize index configuration for hybrid queries.

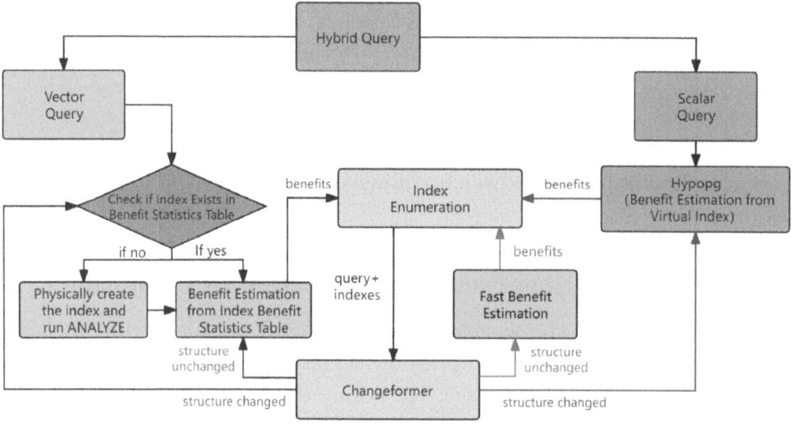

Fig. 3. The Workflow of Index Recommendation for Hybrid Queries in HQIA.

5.1 Index Benefit Estimation

HQIA extends the RIBE-based index recommendation mechanism [6] to effectively support hybrid query workloads, which typically involve both structured attribute filtering (e.g., range conditions in SQL) and high-dimensional vector similarity search (e.g., pgvector-based retrieval). To accommodate such complex queries, HQIA introduces a unified approach to estimate the benefits of both scalar and vector indexes.

For scalar queries, HQIA uses HypoPG [5] to simulate virtual indexes and estimate their impact on the query plan. However, since tools like HypoPG do not support "what-if" simulation for vector indexes, HQIA physically creates the index and runs `ANALYZE` to gather statistics for benefit estimation.

To reduce overhead, HQIA maintains an Index Benefit Statistic Table (IBST) to cache previously computed benefits. As shown in Fig. 3, when a vector query is received, HQIA first checks whether the required index exists. If it does, the benefit is retrieved from IBST; otherwise, the index is created and analyzed, and the results are stored in IBST for future reuse. This avoids repeated index creation and improves overall efficiency.

The IBST table records various query features, such as query attributes, query record count, predicates, selectivity, and index benefit. Each row of data represents a query configuration and its corresponding index benefit. This information helps HQIA quickly assess the potential benefits of building vector indexes, thus reducing the significant overhead caused by repeatedly creating indexes.

5.2 Extended Use of the ChangeFormer Model

RIBE points out that structural changes in a query plan can significantly affect execution cost. Therefore, when such changes are predicted, a full cost re-evaluation is necessary; otherwise, fast estimation based on existing statistics can be used to avoid unnecessary computation [6]. To further enhance index recommendation for hybrid queries, HQIA extends the use of the ChangeFormer model from RIBE. ChangeFormer is based on the Transformer architecture and predicts whether a newly added index will structurally alter the query plan.

In HQIA, ChangeFormer's functionality for scalar queries remains unchanged and is not elaborated here. For vector queries, HQIA customizes ChangeFormer's behavior based on the predicted structural impact:

Structural change predicted: The system checks whether the target index exists in the IBST. If it exists, benefit estimation is performed directly. Otherwise, the system physically creates the index and runs `ANALYZE` to gather statistics.

No structural change predicted: The system bypasses index creation and performs lightweight benefit estimation using cached statistics.

This design allows HQIA to efficiently balance estimation accuracy and computational overhead, accelerating the index recommendation process for hybrid workloads. The overall workflow is illustrated in Fig. 3.

6 Experiment

6.1 Experimental Setup

- **Hardware Configuration:** The experiments were conducted on a virtual machine configured with 16 CPU cores and 16GB of memory. The virtual

machine was hosted on a physical machine running Windows 11, equipped with an Intel Core i5-13490F processor (16 threads) and 32GB of memory.
- **Virtual Machine Operating System:** The virtual machine operated on CentOS 7, with necessary dependencies and tools installed to support the PostgreSQL [16] database system.
- **Database System:** PostgreSQL 13 was used as the database management system, with the following extensions installed:
  - `pgvector` [11]: Enables storage and querying of vector data, particularly for high-dimensional vector representations.
  - `hypopg` [5]: Supports the simulation of virtual indexes, which is crucial for evaluating index recommendation algorithms.

6.2 Dataset and Workloads

- **Dataset:** The experiments used the TPC-H [7] dataset with a scale factor (SF) of 1, extended to accommodate the requirements of hybrid queries. Specifically, textual columns (such as address and comment) in each table were transformed into 384-dimensional vectors using the all-MiniLM-L6-v2 model [17], and these vectors were stored in corresponding columns. As a result, the dataset contains not only traditional scalar data but also vector data, enabling optimization for mixed queries that combine scalar and vector-based conditions.
- **Scalar Query Workload:** The scalar query workload in this experiment is generated based on the standard TPC-H query templates. For each template in TPC-H, we randomly generated 30 batches of queries, excluding templates 2, 17, and 20 to avoid the dominant effect of excessively long execution times on the experimental results [18]. To increase the complexity of index recommendation, we also incorporated a set of synthetic queries into the workload. These queries were generated using a context-free SQL grammar that supports advanced features such as joins, nested queries, and vector-based queries. Constants in the queries were randomly selected from actual attribute values to ensure non-empty results and enhance query diversity. This workload is referred to as TPCH(Scalar).
- **Hybrid Query Workload:** Building on the scalar query workload, we introduced several modifications to the queries. For example, in the `WHERE` clause, in addition to conventional attribute-based filtering conditions, we incorporated vector similarity matching queries. This allows for more precise matching based on vector data while filtering scalar data. We also added `ORDER BY` clauses to sort results by vector similarity. These enhancements expand the query set to include both traditional scalar queries (e.g., range queries on integers and floating-point numbers) and vector similarity queries. In this way, we simulate more complex workloads that are representative of real-world applications, enabling coordinated optimization of scalar and vector data. This workload is referred to as TPCH(Hybrid).

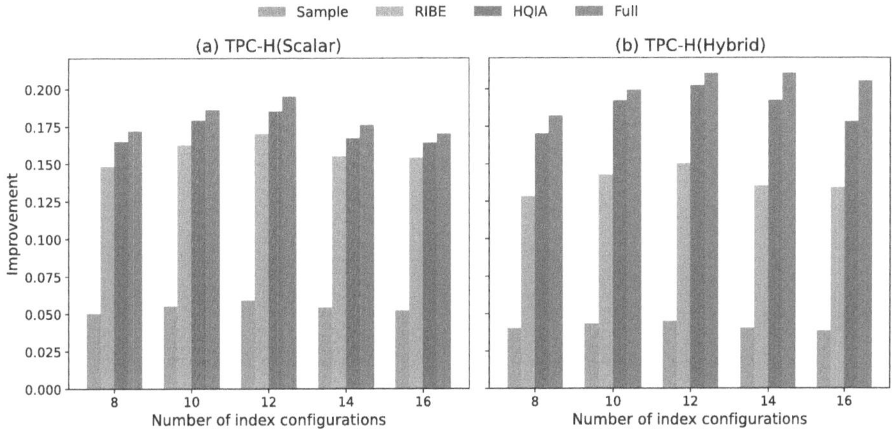

Fig. 4. Cost improvements achieved by different index tuning methods.

6.3 Experimental Results and Analysis

This experiment compares the performance improvement of four optimization strategies—**Sample**, **Full**, **RIBE** [6], and **HQIA**—under both hybrid query workloads and scalar query workloads. **Full** disables workload compression, leading to the highest execution time and higher computational overhead. **Sample** compresses the workload by randomly sampling a specified number of queries, resulting in the lowest execution time but potentially missing the optimal index configurations. Therefore, we focus the comparison on **RIBE** and **HQIA** below.

As shown in Fig. 4, HQIA demonstrates a performance advantage in both types of query workloads, particularly in the hybrid query workload scenario, where HQIA shows a more significant improvement compared to RIBE [6], which only supports scalar index recommendations. The main reason for this improvement is that HQIA, in the index recommendation process, introduces vector similarity matching queries in addition to traditional structured filtering conditions. By incorporating vector indexing, HQIA performs excellently in high-dimensional feature matching queries, effectively accelerating queries and significantly reducing overall query latency. This result validates HQIA's effective extension of traditional index recommendation methods, especially in handling multi-modal and complex workloads, showcasing stronger adaptability and optimization capabilities.

However, Fig. 5 shows that the optimization time for HQIA is slightly higher compared to RIBE [6]. Although HQIA offers significant functional optimization, the increase in optimization time is mainly due to the richer query predicate distribution information introduced during the feature acquisition phase of the optimization method, which leads to the processing of more complex feature vectors during compression and index recommendation, thus increasing computational overhead. Particularly in the hybrid query workload optimization process, HQIA needs to evaluate vector indexing, adding additional time

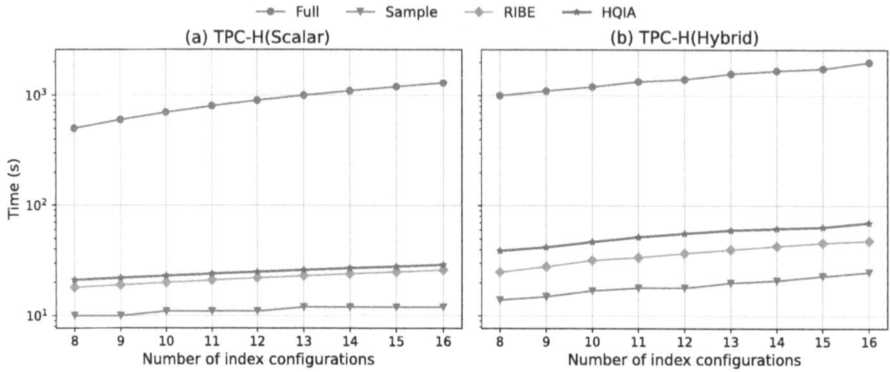

Fig. 5. Running time of different index tuning methods.

overhead. Nevertheless, overall, despite the slight increase in optimization time, HQIA shows a noticeable improvement in execution cost for hybrid query workloads, indicating that HQIA still provides strong optimization effects in complex query workload scenarios.

In conclusion, HQIA enhances query performance compared to RIBE [6], with a more pronounced advantage in hybrid query scenarios, demonstrating its adaptability in multi-modal and complex workloads. Although there is a slight increase in optimization time, it is justified by the significant improvement in query performance achieved. HQIA's index recommendation for hybrid query workloads is highly meaningful, especially when dealing with large-scale and high-complexity query workloads.

7 Conclusion and Future Work

This paper proposes the HQIA (Hybrid Query Index Advisor) framework for hybrid query workloads combining structured attribute filtering and high-dimensional vector matching, aiming to improve the index recommendation capability in hybrid query scenarios. HQIA enhances the distinguishability of query workloads by integrating query cost information and attribute distribution features, thereby improving workload compression and index optimization performance. Based on this, the paper extends the RIBE index recommendation process to achieve joint optimization of scalar queries and vector queries, further enhancing query performance under hybrid query workloads. Experimental results show that HQIA significantly improves query performance under hybrid query workloads. Compared to RIBE, which only supports scalar index recommendations, HQIA demonstrates clear advantages in execution efficiency, especially in multi-modal query scenarios, where it shows stronger adaptability. Although the introduction of additional feature information leads to a slight increase in optimization time, HQIA still demonstrates significant performance improvements on hybrid query workloads, proving its optimization potential in complex query workloads.

Future work includes automatic vector index selection and tuning, online adaptation for dynamic workloads, and evaluating HQIA's generalizability on real-world datasets.

References

1. Sheng, M., et al.: MQRLD: a multimodal data retrieval platform with query-aware feature representation and learned index based on data lake. Inf. Process. Manag. **62**(4), 104101 (2025)
2. Wu, Y., Zhou, X., Zhang, Y., Li, G.: Automatic database index tuning: a survey. IEEE Trans. Knowl. Data Eng. (2024)
3. Malkov, Y.A., Yashunin, D.A.: Efficient and robust approximate nearest neighbor search using hierarchical navigable small world graphs. IEEE Trans. Pattern Anal. Mach. Intell. **42**(4), 824–836 (2018)
4. Johnson, J., Douze, M., Jégou, H.: Billion-scale similarity search with GPUs. IEEE Trans. Big Data **7**(3), 535–547 (2019)
5. Rouhaud, J.: HypoPG (2024). https://github.com/HypoPG/hypopg. version 1.3.0. Accessed June 2025
6. Yu, T., Zou, Z., Sun, W., Yan, Y.: Refactoring index tuning process with benefit estimation. Proc. VLDB Endowment **17**(7), 1528–1541 (2024)
7. Transaction Processing Performance Council: TPC-H (2024). http://www.tpc.org/tpch/. Accessed June 2025
8. Deep, S., Gruenheid, A., Koutris, P., Naughton, J., Viglas, S.: Comprehensive and efficient workload compression. arXiv preprint: arXiv:2011.05549 (2020)
9. Siddiqui, T., Jo, S., Wu, W., Wang, C., Narasayya, V., Chaudhuri, S.: ISUM: efficiently compressing large and complex workloads for scalable index tuning. In: Proceedings of the 2022 International Conference on Management of Data, pp. 660–673 (2022)
10. Siddiqui, T., Wu, W., Narasayya, V., Chaudhuri, S.: DISTILL: low-overhead data-driven techniques for filtering and costing indexes for scalable index tuning. Proc. VLDB Endowment **15**(10), 2019–2031 (2022)
11. Kane, A.: pgvector (2024). https://github.com/pgvector/pgvector. version 0.5.1. Accessed June 2025
12. Zilliz: Milvus (2024). https://milvus.io. version 2.3. Accessed June 2025
13. Team, Q.: Qdrant (2024). https://qdrant.tech. version 1.7. Accessed June 2025
14. Yang, T., et al.: VDTuner: automated performance tuning for vector data management systems. In: 2024 IEEE 40th International Conference on Data Engineering (ICDE) (2024)
15. Yue, Z., Cai, P.: Database parameters tuning via Bayesian optimization with domain knowledge. In: International Conference on Web Information Systems and Applications, pp. 277–289. Springer (2024)
16. PostgreSQL Global Development Group: PostgreSQL (2024). https://www.postgresql.org. version 13. Accessed June 2025
17. Reimers, N., Gurevych, I.: all-MiniLM-L6-v2 (2021). https://huggingface.co/sentence-transformers/all-MiniLM-L6-v2. Accessed June 2025
18. Kossmann, J., Halfpap, S., Jankrift, M., Schlosser, R.: Magic mirror in my hand, which is the best in the land? An experimental evaluation of index selection algorithms. Proc. VLDB Endowment **13**(12), 2382–2395 (2020)

Quantification-Driven Coordinate Scheduling for Heterogeneous Resources

Mingdong He(✉), Zhaoyi Yang, Lei Xu, Xiaodong Zhang, Bo Li, Xuewu Li, and Hongyu Liao

Information Center, Guangdong Power Grid Corporation, Guangzhou, China
147369785@qq.com, zhangxiaodong@csg.cn, libo@gdxx.csg.cn, liaohongyu@xxzx.gd.csg.cn

Abstract. Modern power-grid information systems require efficient and intelligent use of heterogeneous computing resources, including CPU, GPU, memory, and network bandwidth. To address this, we propose a method for quantitative evaluation and coordinated scheduling of heterogeneous computing resources. The framework consists of three modules: First, a resource sensing and abstraction module continuously collects multidimensional resource usage data and maps them into normalized indices to reflect CPU load, memory pressure, GPU intensity, and network demand, forming a unified resource status view. Second, a scheduling and allocation module evaluates each task's resource demand and priority level, and calculates a feasibility index to determine schedulability. A multi-objective strategy generates allocation plans balancing task urgency, system load, and resource efficiency. Third, a monitoring and feedback module evaluates execution using indicators such as task completion rate and latency, enabling dynamic policy adjustment. This approach enables fine-grained, adaptive scheduling of heterogeneous resources in power-grid computing scenarios, improving resource utilization and service responsiveness.

Keywords: Coordinated scheduling · Heterogeneous computing · Quantitative computing-network evaluation · Dynamic feedback

1 Introduction

Large-scale power-grid information centres are rapidly evolving from monolithic data silos into cyber-physical hubs where computing, storage, and high-speed networks must be orchestrated in real time. Modern workloads, ranging from SCADA [19] stream analytics and contingency assessment to AI-based market forecasting, demand heterogeneous resources [16] (CPU, GPU, memory, and bandwidth) whose utilization changes with load curves and fault events. Conventional schedulers, however, still optimize a single dimension or rely on static quotas, resulting in resource imbalance, low energy efficiency, and missed service-level objectives [4, 14]. A root cause is the absence of a unified, quantitative view of heterogeneous resources.

S. Yang et al. (Eds.): WISA 2025, LNCS 16299, pp. 270–283, 2026.
https://doi.org/10.1007/978-981-95-6426-2_23

Raw counters such as CPU cycles, GPU kernel occupancy, page-fault rate and link throughput are measured on incomparable scales, making global reasoning impossible [12]. Without normalisation, even sophisticated algorithms degenerate into ad-hoc rules that fail to balance timeliness, utilisation and energy.

To bridge this gap we propose a Quantitative Evaluation and Heterogeneous Resource Scheduling (QE-HRS) framework. First, a resource sensing and abstraction module continuously maps multi-source telemetry into four normalised indices-CPU load, GPU intensity, memory pressure and network demand—and fuses them into a composite resource view. Second, a scheduling and allocation module evaluates each task's resource demand and priority, derives a feasibility index, and employs a multi-objective strategy to generate allocations that jointly optimise deadline satisfaction, system load and energy consumption. Third, a monitoring and feedback module computes execution indicators such as task-completion rate and latency; whenever the composite scheduling-quality metric degrades, the policy is fine-tuned and re-issued.

Our method is consistent with, and extends, the classical Five-Minute Rule. Whereas the original rule weighs the cost of keeping a data block in memory versus retrieving it from disk [9], we generalise the cost-benefit trade-off to a multi-resource plane and embed it inside a feedback controller that recalculates benefits with live indices, enabling sub-minute policy adjustment.

In summary, this paper makes the following contributions:

1. We introduce a quantitative evaluation framework for computing-network resources, providing unified metrics for CPU, GPU, memory, and bandwidth utilization across power grid computing environments.
2. We propose a coordinated heterogeneous resource scheduling method that jointly considers task urgency, priority, and system constraints, supporting both global and local optimization strategies.
3. We develop a feedback-driven optimization mechanism that dynamically adjusts scheduling strategies based on multi-dimensional execution metrics, improving adaptability and stability in practical deployments.

2 Related Work

Existing research on resource scheduling in cloud and data center environments often focuses on optimizing a single resource type, such as CPU or GPU utilization [6,13]. These methods typically apply static load balancing or heuristic optimization to improve throughput [2,15], but fail to consider dynamic interactions across other critical resources such as memory, storage, and network [3]. As a result, these approaches are limited in adapting to real-time workload variations or ensuring overall system efficiency.

Several energy-aware scheduling strategies [1,23,24] focus primarily on minimizing power consumption or maximizing throughput, without jointly optimizing for latency, bandwidth, or resource contention. While they improve cost-effectiveness for specific tasks, these methods often lack fine-grained coordination across diverse heterogeneous resources. For example, greedy scheduling for

MapReduce jobs [24] or data center task energy heuristics [3] may neglect broader system-level bottlenecks.

Some works propose static or application-specific load balancing methods [11,21,22], which fail to adapt to dynamic or unpredictable workloads. These techniques are often designed for specific patterns and cannot be generalized to modern elastic infrastructures.

Efforts have also been made toward multi-objective or workflow-based scheduling [10,20], seeking to balance multiple goals such as energy, delay, and cost. However, many rely heavily on metaheuristics, without incorporating real-time system feedback or quantitative modeling of subsystem interactions. For instance, virtual-machine-based strategies improve flexibility [20], but often ignore bandwidth contention or task urgency.

Survey-based studies [18] offer broad overviews of cloud scheduling techniques, but lack practical evaluation frameworks or dynamic coordination strategies. Some task schedulers targeting heterogeneous clouds [8,16] attempt to incorporate diverse resource types, but often remain limited to loosely coupled metrics or lack support for system-wide optimization. Similarly, works focused on deep learning inference [7] propose reinforcement learning for QoS-aware scheduling, but struggle to generalize across varied infrastructure demands.

3 Module Architecture

3.1 Resource Sensing and Abstraction Module

To support fine-grained and adaptive scheduling across heterogeneous computing environments, this module (see Fig. 1) unifies the representation of computing-network resources—including CPU, GPU, memory, bandwidth, and I/O subsystems—through a standardized quantitative indexing system. It collects real-time and event-driven telemetry from multiple resource dimensions and filters raw data such as CPU usage, memory occupancy, network bandwidth, and I/O pressure.

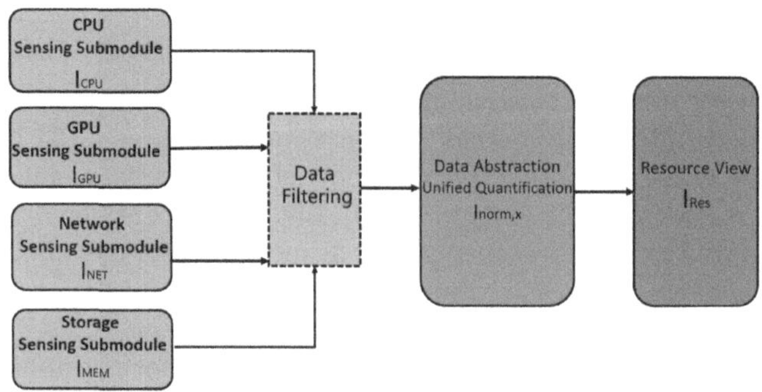

Fig. 1. Architecture of Resource Sensing and Abstraction Module

CPU Sensing Submodule. This submodule captures the real-time utilization status of CPU resources, including core occupancy and instruction execution rate. The CPU Utilization Index is defined as:

$$I_{CPU} = \frac{Active\_Cycles}{Total\_Cycles} \times \left(1 + \alpha \times \frac{IPC}{IPC_{ref}}\right) \times 100\% \tag{1}$$

GPU Sensing Submodule. This submodule monitors GPU core occupancy and memory usage to quantify the GPU load.

$$I_{GPU} = \frac{GPU\_ActiveTime}{GPU\_TotalTime} \times \left(\frac{MemUsed}{MemTotal}\right)^{\beta} \times 100\% \tag{2}$$

Network Sensing Submodule. This submodule captures network bandwidth usage, round-trip delay, and packet loss.

$$I_{NET} = \left(\frac{DataRate}{MaxBand}\right) \times \left(1 + \gamma \times \frac{RTT_{avg}}{RTT_{ref}}\right) \times 100\% \tag{3}$$

Memory Sensing Submodule. This submodule monitors memory usage and access pressure.

$$I_{MEM} = \left(\frac{MemUsed}{MemTotal}\right) \times \left(1 + \delta \times \frac{PageFaultRate}{PFR_{ref}}\right) \times 100\% \tag{4}$$

3.2 Coordinated Scheduling and Allocation Module

In the Coordinated Scheduling and Allocation Module(see Fig. 2), a task-driven scheduling framework is established to address multi-objective constraints in heterogeneous resource scheduling within power grid operations, covering resources such as CPU, GPU, network bandwidth, and memory. With "task demand" as the core, the module relies on the unified abstract indices output by the Resource Sensing and Abstraction Module. By incorporating each task's priority level, deadline requirement, and resource intensity information, the module supports functions such as resource feasibility judgment, priority-based task filtering, initial task allocation, and local optimization, ultimately generating a structured scheduling strategy.

Task Analysis Submodule. This submodule receives and analyzes task requests submitted by upper-level services or users, and transforms them into quantifiable task attributes. It analyzes task characteristicsto estimate the execution time and resource demand intensity of each task, serving as the basis for subsequent scheduling optimization.

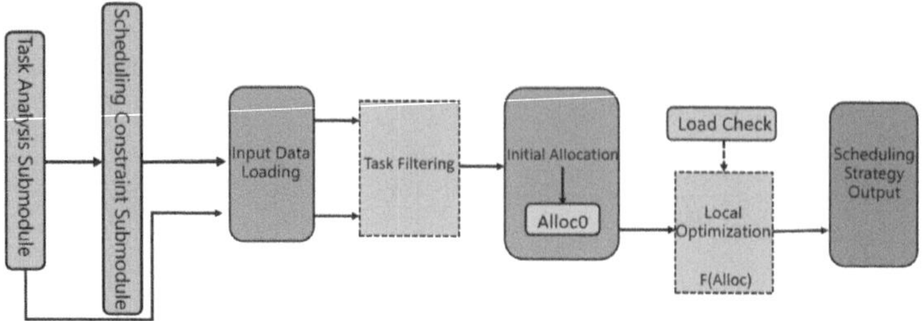

Fig. 2. Architecture of Coordinated Scheduling and Allocation Module

Task Resource Demand Index ($I_{\text{TaskReq,i}}$)

$$I_{\text{TaskReq,i}} = \alpha_{\text{CPU}} \cdot \text{CPU\_Req}_i + \alpha_{\text{GPU}} \cdot \text{GPU\_Req}_i + \alpha_{\text{MEM}} \cdot \text{Mem\_Req}_i + \alpha_{\text{NET}} \cdot \text{Net\_Req}_i \quad (5)$$

where:

- `CPU_Req`, `GPU_Req`, `Mem_Req`, `Net_Req`: Resource demands of task i for CPU, GPU, memory, and network bandwidth.
- $\alpha_{\text{CPU}}, \alpha_{\text{GPU}}, \alpha_{\text{MEM}}, \alpha_{\text{NET}}$: Weights representing the importance of CPU, GPU, memory, and network bandwidth dimensions.

Task Priority Adjustment Index ($I_{\text{prio},i}$)

$$I_{\text{prio},i} = \lambda_1 \cdot \frac{1}{Deadline_i - t_{now}} + \lambda_2 \cdot PriorityLevel_i \quad (6)$$

Where:

- $\texttt{Deadline}_i$: The deadline of task i.
- t_{now}: The current time.
- $\texttt{PriorityLevel}_i$: The priority level of task i (can be a discrete value like 1, 2, 3, ..., where a higher number indicates a higher priority).
- λ_1, λ_2: Weight coefficients.

Scheduling Constraint Submodule. Based on task requirements and the current system resource status, the task resource demand index $I_{\text{TaskReq},i}$ and the priority adjustment index $I_{\text{Prio},i}$ are used to define scheduling objectives and system resource constraints.

Scheduling Feasibility Index (I_{feas})

$$I_{\text{feas}} = \frac{\sum_{i \in \mathcal{T}} I_{\text{TaskReq},i} \times f(I_{\text{Prio},i})}{(1 - I_{\text{Res}} + \varepsilon)} \quad (7)$$

Where:

- $I_{\text{TaskReq},i}$: Task i's resource demand index.

- $I_{\text{Prio},i}$: Task i's priority adjustment index.
- I_{Res}: Composite resource index representing current overall system resource utilization.
- ε: A smoothing term (e.g., 0.01) to prevent denominator from approaching 0.

Scheduling Strategy Generation Submodule. Upon receiving the indices computed from the preceding submodules, this submodule generates the final scheduling strategy table based on multi-objective optimization principles. Under the constraints of resource and timing requirements, it determines the concrete resource allocation for each task. The strategy ensures that tasks with higher priority and those approaching their deadlines receive preferential treatment, while the overall system load remains within an acceptable resource usage range.

Data Input Loading. This step retrieves all pending scheduling tasks $\{T_i\}$. For each task T_i, the system reads the task resource demand index $I_{TaskReq,i}$ and the priority adjustment index $I_{Prio,i}$.

The system also reads the scheduling feasibility index I_{feas} from the constraint submodule, the composite resource index I_{Res}, the safety margin δ (typically 5%âĂŞ10%), and the optimization weights α, β, and γ for subsequent objective functions. At this stage, the system only loads and stores the data in memory without performing filtering or computations.

Task Screening. The system determines scheduling feasibility as follows: if $I_{feas} > 1$, the total task resource demand exceeds $I_{Res} \times (1 - \delta)$, indicating insufficient resources. Tasks with lower priorities are removed. When $I_{feas} > 1$, tasks with smaller $I_{Prio,i}$ values are sequentially removed until the feasibility index I_{feas} drops below or equals 1. If $I_{feas} \leq 1$, no removal is performed.

Initial Allocation. Comprehensive priority value is calculated: For each remaining task T_i, the following is computed:

$$V_i = \frac{I_{feas}}{I_{TaskReq,i}} \tag{8}$$

- Tasks are sorted in descending order of V_i, resulting in the sequence $\{T_0, T_2, \ldots\}$.
- Greedy allocation: Initialize the available resource set as $ResFree = TotalRes$. For each task $T(k)$, if $I_{TaskReq,k} \leq ResFree$, update $ResFree = ResFree - I_{TaskReq,k}$.
- The resulting allocation is recorded in *Alloc*0, storing for each task the resource quantity and node information. Tasks that exceed resource limits are placed into a pending queue.

Local Optimization. The objective function is defined as:

$$F(Alloc) = \alpha \times T_{max}(Alloc) + \beta \times E_{use}(Alloc) + \gamma \times L_{delay}(Alloc) \tag{9}$$

where:

- $T_{max}(Alloc)$: Maximum completion time across all scheduled tasks.
- $E_{use}(Alloc)$: Resource usage efficiency or energy estimate.
- $L_{delay}(Alloc)$: Penalty for tasks exceeding deadline (set to 0 if no deadline).

Exchange improvement: Select two tasks T_x and T_y from *Alloc*0 to attempt reassignment. If swapping or micro-adjusting their CPU/GPU allocation yields improved performance such that $F(Alloc') < F(Alloc0)$, retain the adjustment. Repeat iteratively until no further improvement can be made. Load balancing: Recompute load ratio for all nodes in *Alloc*0. If the power threshold is exceeded, migrate low-priority tasks to reduce usage and reattempt reallocation until constraints are satisfied.

Scheduling Strategy Output. The records in `Alloc2` are converted into execution instructions containing {Task ID, Node ID, allocated CPU/GPU/memory quantity, and estimated execution interval (with optional sequencing control)}. Each record corresponds to the execution assignment of a task on a specific resource node.

3.3 Dynamic Monitoring and Feedback Optimization Module

The dynamic monitoring and feedback optimization module (see Fig. 3) evaluates the effectiveness of system scheduling strategies based on real-time monitoring results and multi-dimensional indicators. By continuously collecting data on resource usage and task execution, the module assesses the effectiveness of the current scheduling strategy and triggers global or local optimization adjustments as needed. This module provides a feedback mechanism for the scheduling system to ensure that, under the power grid scheduling scenario, the system can automatically adapt the resource allocation strategy based on actual operating conditions, thereby achieving optimal performance.

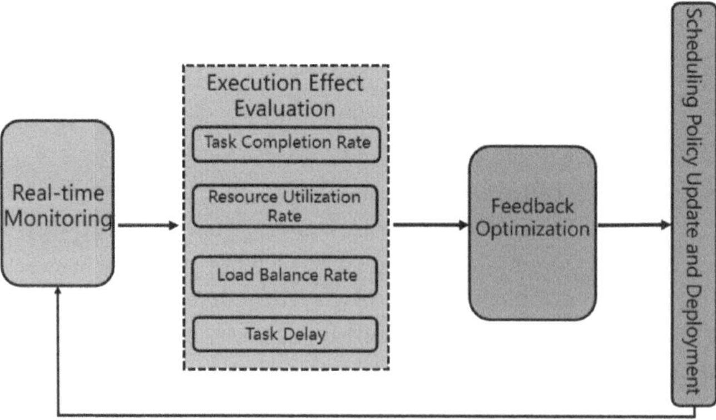

Fig. 3. Architecture of Dynamic Monitoring and Feedback Optimization Module

Real-Time Monitoring Submodule. This submodule collects real-time information on the usage, load status, and task execution of various resources (CPU, GPU, network, and storage). It obtains resource usage data during task execution, such as CPU/GPU utilization at each node, execution duration, and task delays.

Execution Effect Evaluation Submodule. Based on the data collected in real time, this submodule calculates and evaluates the effectiveness of scheduling strategies. It assesses key indicators such as task completion rate, resource utilization, and load balancing to assist in analyzing the effectiveness of current strategies.

Task Completion Rate (I_{comp})

$$I_{comp} = \frac{Completed\ Tasks}{Total\ Tasks} \tag{10}$$

Resource Utilization (I_{util})

$$I_{util} = \frac{I_{CPU} + I_{GPU}}{Total\ Resources} \tag{11}$$

Load Balancing ($I_{LoadBal}$)

$$I_{LoadBal} = \frac{|\sum I_{CPU} - \sum I_{GPU}|}{\sum I_{CPU} + \sum I_{GPU}} \tag{12}$$

Task Latency ($I_{latency}$)

$$I_{latency} = \frac{\sum_{Tasks} Delay\ Time}{Total\ Tasks} \tag{13}$$

Feedback Optimization Submodule. Based on the indicators output by the execution effect evaluation submodule, this component determines whether the current scheduling strategy requires optimization. If so, it adjusts task allocation and resource utilization through optimization methods, ensuring the system achieves optimal performance during task execution.

Scheduling Effectiveness Evaluation Index ($I_{SchedOpt}$)

$$I_{SchedOpt} = \alpha \cdot I_{comp} + \beta \cdot I_{util} + \gamma \cdot I_{LoadBal} + \delta \cdot I_{latency} \tag{14}$$

- By adjusting weights α, β, γ, and δ, the system prioritizes increasing the task completion rate I_{comp} and resource utilization rate I_{util}, while minimizing load imbalance $I_{LoadBal}$ and task latency $I_{latency}$.

Strategy Update and Deployment Submodule. Based on the optimization results from the feedback optimization module, the system updates the current scheduling strategy and deploys it to the execution system for adjustment. Meanwhile, the real-time monitoring module continues operating, enabling periodic or event-driven evaluation to ensure that the updated strategy can be effectively executed within the system.

Instruction Sequence Generation. According to the optimized scheduling strategy, the system updates the records in `Alloc2` and transforms them into concrete execution instructions, which include task ID, node ID, resource allocation (e.g., CPU/GPU/memory), and estimated execution time. Each instruction also specifies the scheduling order and the quantity of resources required to ensure proper task execution at the specified resource node.

Instruction Deployment. The generated scheduling instructions are dispatched to the resource nodes for actual execution. These instructions specify how each task should be executed, including trigger conditions and resource usage.

Return to Real-Time Monitoring. After the strategy update, the execution module actively reports execution status back to the real-time monitoring and feedback optimization modules. These modules continue monitoring the task execution. The feedback mechanism ensures that, if resource contention or delay occurs, adjustments can be made promptly based on feedback to re-optimize the scheduling strategy.

4 Experimental Evaluation

To evaluate the effectiveness of our proposed QE-HRS scheduling framework, we conduct a series of simulations using a customized cloud computing simulation environment based on CloudSim Plus [5]. Simulation-based evaluation enables us to flexibly model heterogeneous infrastructure and workload characteristics without being constrained by physical deployment. We define a data center environment composed of multiple computing nodes with varied configurations in CPU cores, memory, bandwidth, and optional GPU units, reflecting the heterogeneity commonly observed in electric power grid computing environments.

Each computing node is represented as a physical host equipped with configurable computing resources, allowing us to simulate varying degrees of heterogeneity. Virtual machines are instantiated on top of each host to execute the submitted tasks. Tasks are synthetically generated with different priorities, resource demands, and deadline constraints to emulate a mixture of real-world electricity grid applications, including data aggregation, control algorithms, and delay-sensitive workloads [17].

The simulated environment supports periodic monitoring and real-time feedback collection. The normalized resource indices and the overall resource index proposed in this paper are used to guide the scheduling decisions during simulation. We compare our method against several baseline scheduling algorithms to verify improvements in resource utilization, task latency, and scheduling effectiveness under dynamic conditions.

4.1 Ablation Study on Local and Feedback Optimization

To evaluate the effectiveness of each optimization module in the proposed QE-HRS framework, we conduct an ablation study involving four scheduling strategies: (1) a base version with no optimization; (2) a version with only local optimization; (3) a version with only feedback optimization; and (4) the complete QE-HRS system.

We simulate workloads consisting of 10, 100, and 1000 tasks, each with randomized CPU, GPU, memory, and network requirements, as well as predefined deadlines. For each strategy, we measure the average task completion time, average resource utilization, and task success rate (percentage of tasks meeting deadlines).

As illustrated (see Fig. 4), both local and feedback optimizations contribute to improved scheduling performance. The full QE-HRS system demonstrates the best performance, particularly under high workload conditions, where the synergy between local and feedback mechanisms enables more balanced and adaptive scheduling.

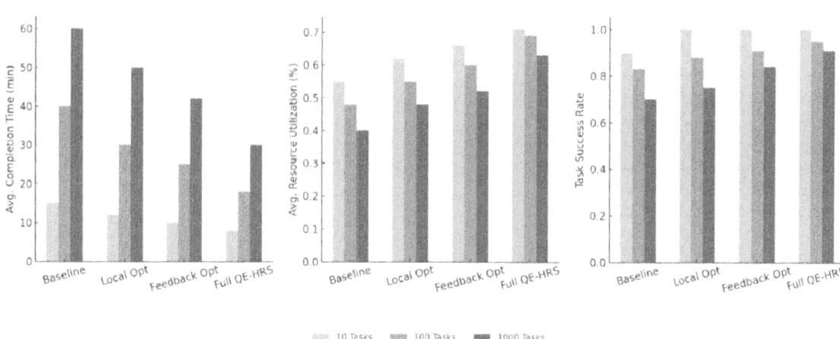

Fig. 4. Normalized resource indices and composite index for four heterogeneous servers

4.2 Evaluation of Scheduling Responsiveness and Scalability

To further assess the responsiveness and scalability of the proposed heterogeneous resource scheduling framework, we conduct experiments under varying workload intensities. The goal is to investigate how the system reacts as the number of submitted tasks increases, focusing on two key performance indicators: average response time (from task submission to the start of execution) and average completion time (from task submission to task finish).

In this experiment, we simulate lightweight real-time tasks such as service queries, data scanning, and low-latency control logic. These tasks require rapid scheduling but minimal computing resources, which reflects many real-world edge and online applications. We vary the number of tasks from 10 to 1000 to emulate different system loads.

As the Fig. 5 shows, both the response time and completion time grow with the task volume. However, the increase remains smooth and sublinear, indicating that the proposed strategy maintains scheduling efficiency under increasing pressure. The relatively narrow gap between response and completion time further reflects the scheduler's ability to minimize task queuing delays and ensure timely processing.

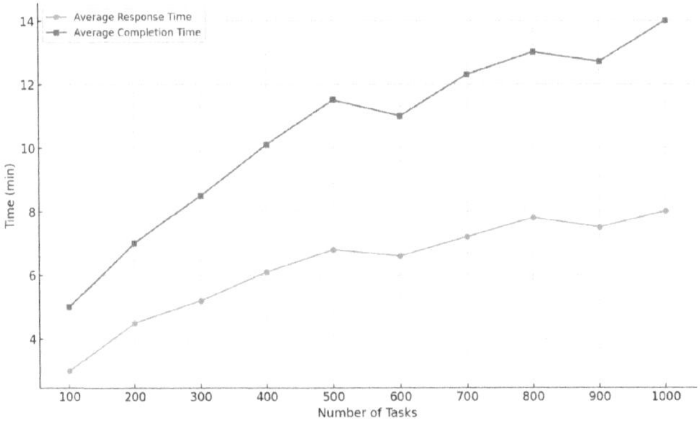

Fig. 5. Average response time and task completion time under increasing task loads

4.3 Comparison with Baseline Scheduling Strategies

To further evaluate the effectiveness and robustness of the proposed resource scheduling framework, we compare it against three widely adopted baseline strategies: static quota allocation, round-robin scheduling, and least-loaded dispatching. These strategies are commonly used in traditional resource management systems, but each exhibits inherent limitations, particularly under heterogeneous and dynamic workload conditions.

Static quota allocation distributes resources evenly among tasks without considering their real-time requirements, leading to inefficient utilization and potential bottlenecks when task demands vary significantly. Round-robin scheduling assigns tasks cyclically across available computing nodes, disregarding differences in resource availability and workload characteristics, often resulting in resource contention and load imbalance under fluctuating demands. The least-loaded dispatching strategy, while more dynamic by directing tasks to the least-utilized resources, still overlooks detailed task-specific resource requirements such as

CPU intensity, memory pressure, or bandwidth sensitivity, which may cause suboptimal scheduling decisions and degraded system performance under high heterogeneity.

In contrast, the proposed QE-HRS framework quantitatively evaluates multidimensional resource demands and adopts dynamic feedback mechanisms, enabling more adaptive and efficient task scheduling across diverse workloads. This experiment (see Fig. 6) aims to highlight the advantages of QE-HRS over baseline methods in terms of average task completion time and task success rate under increasing task loads.

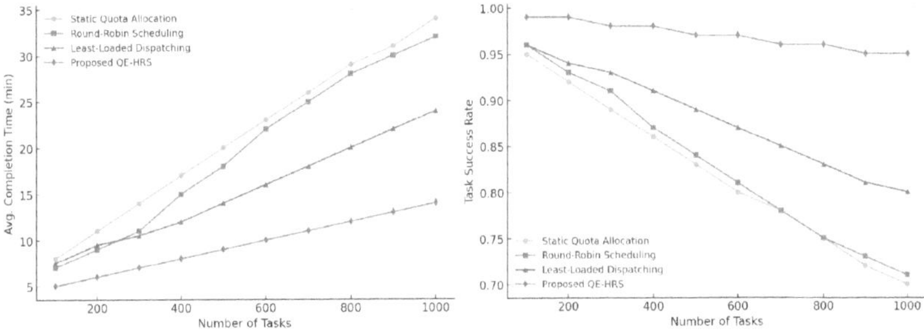

Fig. 6. Comparison of Task Completion Time and Success Rate under Different Scheduling Strategies

5 Conclusion

This paper proposes a multi-objective scheduling framework for heterogeneous computing resources under power grid coordination scenarios. By quantifying resource demands and constraints in terms of CPU, GPU, memory, and network, the model enables fine-grained evaluation and dynamic adjustment of scheduling policies. It achieves joint optimization across key metrics such as performance, energy consumption, and latency. The framework adapts to real-time workload demands and system status, effectively improving resource utilization efficiency and ensuring reliable and responsive operation of power grid information services.

References

1. Bampis, E., Chau, V., Letsios, D., Lucarelli, G., Milis, I., Zois, G.: Energy efficient scheduling of MapReduce jobs. In: Silva, F., Dutra, I., Santos Costa, V. (eds.) Euro-Par 2014. LNCS, vol. 8632, pp. 198–209. Springer, Cham (2014). https://doi.org/10.1007/978-3-319-09873-9_17
2. Behera, I., Sobhanayak, S.: Task scheduling optimization in heterogeneous cloud computing environments: a hybrid GA-GWO approach. J. Parallel Distrib. Comput. **183**, 104766 (2024)

3. Beloglazov, A., Abawajy, J., Buyya, R.: Energy-aware resource allocation heuristics for efficient management of data centers for cloud computing. Futur. Gener. Comput. Syst. **28**(5), 755–768 (2012)
4. Beloglazov, A., Buyya, R.: Energy efficient resource management in virtualized cloud data centers. In: 2010 10th IEEE/ACM International Conference on Cluster, Cloud and Grid Computing, pp. 826–831. IEEE (2010)
5. Calheiros, R.N., Ranjan, R., Beloglazov, A., De Rose, C.A., Buyya, R.: CloudSim: a toolkit for modeling and simulation of cloud computing environments and evaluation of resource provisioning algorithms. Softw.: Pract. Experience **41**(1), 23–50 (2011)
6. Chhabra, S., Singh, A.K.: Dynamic resource allocation method for load balance scheduling over cloud data center networks. J. Web Eng. **20**(8), 2269–2284 (2021). https://doi.org/10.13052/jwe1540-9589.2083
7. Fang, Z., Yu, T., Mengshoel, O.J., Gupta, R.K.: QoS-aware scheduling of heterogeneous servers for inference in deep neural networks. In: Proceedings of the 2017 ACM on Conference on Information and Knowledge Management, pp. 2067–2070 (2017)
8. Gill, S.S., Buyya, R.: Resource provisioning based scheduling framework for execution of heterogeneous and clustered workloads in clouds: from fundamental to autonomic offering. J. Grid Comput. **17**, 385–417 (2019)
9. Gray, J., Putzolu, F.: The 5 minute rule for trading memory for disk accesses. Tandem Corporation Tech. Rep. **86** (1985)
10. Hosseinzadeh, M., Ghafour, M.Y., Hama, H.K., Vo, B., Khoshnevis, A.: Multi-objective task and workflow scheduling approaches in cloud computing: a comprehensive review. J. Grid Comput. **18**(3), 327–356 (2020)
11. Li, K., Xu, G., Zhao, G., Dong, Y., Wang, D.: Cloud task scheduling based on load balancing ant colony optimization. In: 2011 Sixth Annual Chinagrid Conference, pp. 3–9 (2011). https://api.semanticscholar.org/CorpusID:21285250
12. Panda, S.K., Jana, P.K.: Normalization-based task scheduling algorithms for heterogeneous multi-cloud environment. Inf. Syst. Front. **20**, 373–399 (2018)
13. Ravi, V.T., Becchi, M., Jiang, W., Agrawal, G., Chakradhar, S.: Scheduling concurrent applications on a cluster of CPU-GPU nodes. In: 2012 12th IEEE/ACM International Symposium on Cluster, Cloud and Grid Computing (CCGrid 2012), pp. 140–147 (2012). https://doi.org/10.1109/CCGrid.2012.78
14. Sun, C., Guo, H., Shen, D., Nie, T., Hou, Z.: Temporal convolution and multi-attention jointly enhanced electricity load forecasting. In: International Conference on Web Information Systems and Applications, pp. 39–51. Springer (2023)
15. Tsai, J.T., Fang, J.C., Chou, J.H.: Optimized task scheduling and resource allocation on cloud computing environment using improved differential evolution algorithm. Comput. Oper. Res. **40**(12), 3045–3055 (2013)
16. Tuli, S., et al.: HUNTER: AI based holistic resource management for sustainable cloud computing. J. Syst. Softw. **184**, 111124 (2022)
17. Vijayakumar, S., Zhu, Q., Agrawal, G.: Dynamic resource provisioning for data streaming applications in a cloud environment. In: 2010 IEEE Second International Conference on Cloud Computing Technology and Science, pp. 441–448 (2010). https://doi.org/10.1109/CloudCom.2010.95
18. Vinothina, V.V., Sridaran, R., Ganapathi, P.: A survey on resource allocation strategies in cloud computing. Int. J. Adv. Comput. Sci. Appl. **3**(6) (2012)
19. Wikipedia contributors: Scada — Wikipedia, the free encyclopedia (2025). https://en.wikipedia.org/w/index.php?title=SCADA&oldid=1282835103. Accessed 28 Apr 2025

20. Xiao, Z., Song, W., Chen, Q.: Dynamic resource allocation using virtual machines for cloud computing environment. IEEE Trans. Parallel Distrib. Syst. **24**(6), 1107–1117 (2013). https://doi.org/10.1109/TPDS.2012.283
21. Xu, G., Pang, J., Fu, X.: A load balancing model based on cloud partitioning for the public cloud. Tsinghua Sci. Technol. **18**(1), 34–39 (2013). https://doi.org/10.1109/TST.2013.6449405
22. Ye, L., Xia, Y., Yang, L.: A cost-aware scheduling algorithm for reliable workflow in IaaS clouds. In: 2021 33rd Chinese Control and Decision Conference (CCDC), pp. 275–280 (2021). https://doi.org/10.1109/CCDC52312.2021.9601373
23. Ying, C.T., Yu, J., Yang, X.Y.: Energy-aware task scheduling algorithms in cloud computing. Microelectron. Comput. **29**(5), 188–192 (2012)
24. Yousefi, M.H.N., Goudarzi, M.: A task-based greedy scheduling algorithm for minimizing energy of MapReduce jobs. J. Grid Comput. **16**, 535–551 (2018)

Balancing Fairness Among User Groups in Happiness Maximization Queries

Jie Dong(✉), Chaoyi Jiang, Conghao Liu, and Teng Teng

College of Computer Science and Technology, Nanjing University of Aeronautics and Astronautics, Nanjing, China
{djnuaa,cyjiang,liuconghao,tengteng}@nuaa.edu.cn

Abstract. As a pivotal technique of multi-criteria decision making, the happiness maximization query is used to select a compact subset from a large dataset to meet the requirements of various users. Most existing researches on the happiness maximization query merely concentrate on maximizing overall users' satisfaction, but overlook the potential dissatisfaction that can be aroused among users. However, this situation likely fails to capture the diverse needs and preferences of different user groups, leading to negative consequences. In this paper, we propose and study the problem of fairness among user groups in the happiness maximization query. By introducing two classical fairness notions, i.e., envy-freeness and proportionality into the problem, we turn to study solutions for the two forms of this problem namely EF1HMQ and MMSHMQ. Then we propose the algorithm framework FairHMQ using two fairness strategies Round-Robin and Envy-Cycle Elimination respectively. Extensive experiments on real-world and synthetic datasets confirm the effectiveness and efficiency of our proposed algorithms.

Keywords: Happiness Maximization Queries · User Group Fairness · Envy-freeness · Maximin Share Fairness

1 Introduction

Extracting a small representative subset from a large database that supports decisions is a crucial task in many data-driven applications [8]. Since the data typically involves multiple attributes, this task is known as multi-criteria decision making (MCDM). The happiness maximization query (a.k.a. regret minimization query) [7,9,11] has been identified as a key technology in MCDM, which combines the advantages of top-k queries [5] and skyline queries [3], and already been studied for over a decade [1]. It measures the overall satisfaction of users in the worst case and tends to maximize it. However, the query results may exist potential "envy" relationships among different user groups which exhibit various preferences, thus resulting in some user groups' disappointment [10]. Fairness among user groups in the happiness maximization query is ignored.

In this paper, we study the fairness issues of the happiness maximization query from the perspective of different user groups. The core challenge lies in

S. Yang et al. (Eds.): WISA 2025, LNCS 16299, pp. 284–292, 2026.
https://doi.org/10.1007/978-981-95-6426-2_24

balancing diverse preferences, which essentially translates to obtaining a feasible set that satisfies all user groups. Further, we propose and study this problem combined with two predominant fairness notions *envy-freeness* and *proportionality* [2] in fair allocation field, and then formulate the EF1HMQ and MMSHMQ problems respectively. Specifically, EF1HMQ is rooted in the principle of *envy-freeness up to one item*, and guarantees that no user group envies the set of another group after the removal of at most one item, which property provides a practical relaxation of strict *envy-freeness*. Relatively, MMSHMQ bases on the concept of *maximin share fairness*, which seeks to ensure that each user group receives at least their proportional "fair share". Then, we propose a FAIRHMQ algorithm framework. By adopting fairness strategies, i.e., ROUND-ROBIN and ENVY-CYCLE ELIMINATION, we explore the feasibility and effectiveness of the framework.

To sum up, the main contributions of this paper are listed as follows:

- We formulate the problem of fairness among user groups in the happiness maximization query and study two forms namely EF1HMQ and MMSHMQ.
- We propose an algorithm framework FAIRHMQ based on the two fairness strategies ROUND-ROBIN and ENVY-CYCLE ELIMINATION, then discuss the theoretical results respectively.
- Comprehensive experiments on real-world and synthetic datasets show the effectiveness and efficiency of the proposed algorithms.

2 Preliminaries

2.1 Basic Concepts

Let D be a set of n d-dimensional points with positive real values. For each point $p \in D$, the value on the i-th dimension is represented as $p[i]$ i.e., $p = \{p[1], p[2], ..., p[d]\} \in \mathbb{R}^d_+$.

Linear Utility Function [1]. The utility function $u = \{u[1], u[2], \cdots, u[d]\}$ composed with d nonnegative real values, represents user's preference over d attributes. A linear utility of p can be expressed as the inner product of a utility vector u and a point p, i.e., $u(p) = u \cdot p = \sum_{i=1}^{d} u[i] \cdot p[i]$, which shows how satisfied the user is with p. Since the scale invariance of happiness ratio [7], the utility vector is rescaled to be a unit vector based on the l_1-norm, i.e., $\sum_{i=1}^{d} u[i] = 1$. All user groups can be defined as a set of linear utility functions $U = \{U_1, U_2, ..., U_t\}$, where U_i denotes the preference of user group i.

Definition 1 (Happiness Ratio). *Given a dataset D, a subset $S \subseteq D$ and a utility function u, the happiness ratio of S, represented as $hr_{u,D}(S)$, is defined as*

$$hr_{u,D}(S) = \frac{\max_{p \in S} u(p)}{\max_{p \in D} u(p)}.$$

The hr function's range is $[0, 1]$, and obviously user with utility function u will be happier if the hr approaches 1.

Definition 2 (Minimum Happiness Ratio). *Given a dataset D, a subset $S \subseteq D$ and a class of utility functions U. The minimum happiness ratio of S, represented as $mhr_{U,D}(S)$, is defined as*

$$mhr_{U,D}(S) = \min_{u \in U} \frac{\max_{p \in S} u(p)}{\max_{p \in D} u(p)}.$$

2.2 Fairness Model

We focus on the additive valuation function [2] for each user group i, i.e., the utility $v_i(S_i) = \sum_{g \in S_i} hr_{U_i,D}(S_i)$, where g denotes an item (data point) and the dataset D, S_i corresponds to U_i. Then, we present some concepts of the envy-freeness fairness model and the proportionality fairness model.

Definition 3 (Envy-freeness (EF)). *A set S is EF if:*

$$v_i(S_i) \geq v_i(S_j), \forall i, j \in [t], i \neq j.$$

The existence of envy-freeness is not always guaranteed, then Lipton et al. [6] and Budish et al. [3] propose a relaxed form.

Definition 4 (Envy-freeness up to one item (EF1)). *A set S is EF1 if:*

$$v_i(S_i) \geq v_i(S_j \setminus \{g\}), \forall i, j \in [t], i \neq j, \exists g \in S_j.$$

Envy-freeness fairness model relies on above two concepts.

Definition 5 (Proportionality (PROP)). *A set S is PROP if:*

$$v_i(S_i) \geq \frac{v_i(S)}{n}, \forall i \in [t].$$

The existence of proportionality also cannot be guaranteed, a relaxation of proportionality is maximin share fairness, defined below.

Definition 6 (Maximin Share fairness (MMS)). *A set S is MMS if:*

$$v_i(S_i) \geq MMS_i = \mu_i^t = \max_{S \in \mu_i^t} \min_{i \in [t]} v(S_i), \forall i \in [t],$$

where μ_i^t is the maximum value ensuring that user group i to keep the worst of the t disjoint bundles divided from all the items of a set S.

Although MMS significantly weakens the requirement of proportionality, it is still an NP-hard problem to compute the maximin share of each user group using a reduction from partition. Therefore, the majority of the literature has focused on approximate MMS, then we present the notion of *α-MMS*.

Definition 7 (α-Maximin Share fairness (α-MMS)). *Let $\alpha \in (0, 1]$, S is α-MMS if:*

$$v_i(S_i) \geq \alpha \cdot MMS_i = \alpha \cdot \mu_i^t, \forall i \in [t].$$

Our proposed proportionality fairness model is based on the above concepts.

2.3 Problem Formulation

The problem of the fairness among user groups in the happiness maximization query, following the ralaxed fairness notions of *envy-freeness/proportionality*, we give the relaxation fairness problem forms.

Definition 8 (EF1HMQ). *Given t user groups $U = \{U_1, U_2, \ldots, U_t\}$, a dataset D, a positive integer k, the goal is to select a size-k subset $S \subseteq D$ ensuring EF1 fairness. Formally,*

$$S^* = \underset{S \subseteq D:|S|=k}{\arg\max}\ mhr_{U,D}(S)\ s.t.\ v_i(S_i) \geq v_i(S_j \setminus \{g\}),$$
$$\exists g \in S_j, \forall i, j \in [t], i \neq j,\ \cup_{i \in t} S_i = S.$$

Definition 9 (MMSHMQ). *Given t user groups $U = \{U_1, U_2, \ldots, U_t\}$, a dataset D, a positive integer k, the goal is to select a size-k subset $S \subseteq D$ ensuring MMS fairness. Formally,*

$$S^* = \underset{S \subseteq D:|S|=k}{\arg\max}\ mhr_{U,D}(S)\ s.t.\ v_i(S_i) \geq MMS_i,$$
$$\forall i, j \in [t], i \neq j,\ \cup_{i \in t} S_i = S.$$

3 The FAIRHMQ Algorithm Framework

To ensure fairness among user groups in happiness maximization queries, we first select candidate sets of k items that maximize overall satisfaction for each group. These sets are then evaluated using fairness strategies tied to the *happiness ratio*. Our algorithm framework FAIRHMQ, is illustrated in Fig. 1.

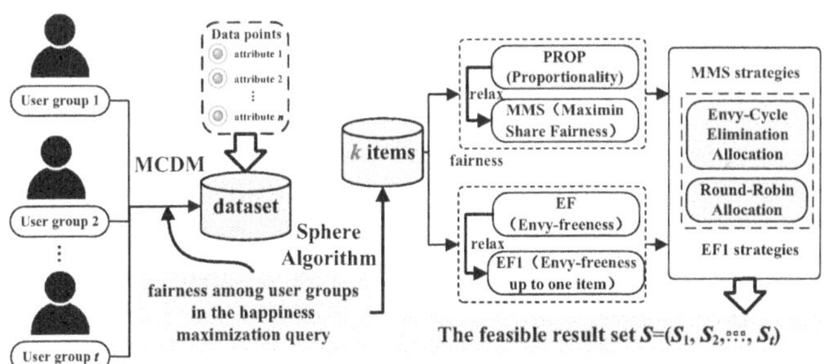

Fig. 1. The Algorithm Framework.

Initially, the FAIRHMQ algorithm framework shown in Algorithm 1 uses a classic happiness maximization queries' algorithm to compute a candidate set C_i of size k based on every user group's preferences U_i as a backup (Lines 1–4). This algorithm obtains a representative set of utility functions and then selects

Algorithm 1: FAIRHMQ

Input: Dataset D, t user groups $U = \{U_1, U_2, ..., U_t\}$, the interger k.
Output: A feasible solution S.

```
1  for i ← 1,...t do
2      while |C_i| < k do
3          q ← the point that realizes mhr for user group i; C_i ← C_i ∪ {q};
4  candidate set C ← ∪_{i=1}^t C_i;
5  while |S| = 0 or exists unfairness do
6      S ← FAIR STRATEGIES(U, C, k);
7  return S;
8  Function ROUND-ROBIN STRATEGY (U, C, k)
9      for each user group i ∈ [1, t] do
10         S_i ← ∅;
11     for ℓ = 1 to k do
12         Let i ← ℓ mod t, g* ∈ argmax_{g∈C} v_i(g);
13         S_i ← S_i ∪ {g*}, C ← C \ {g*};
14     return S ← (S_1, ..., S_t);
15 Function ENVY-CYCLE ELIMINATION STRATEGY (U, C, k)
16     for each user group i ∈ [1, t] do
17         S_i ← ∅;
18     for l ← 1 to k do
19         Create the envy graph G depending on the U;
20         if there exists a node i with in-degree 0 in G then
21             Let i be an unenvied user group,
               g* ∈ argmax_{g∈C} v_i(g); S_i ← S_i ∪ {g*}, C ← C \ {g*};
22         else
23             Find an envy-cycle EC = (i_1, ..., i_d) and resolve it; Update G;
24     return S ← (S_1, ..., S_t);
```

the relatively best points in these functions to achieve satisfactory results. We use the fair strategies to adjust the result set $S = \{S_1, ..., S_t\}$ for each user group until S isn't empty and there does not exist unfairness (Lines 5–6). There are several simple, polynomial-time algorithms for computing *fair results*.

Round-Robin Strategy. The ROUND-ROBIN strategy assigns items to user groups over multiple rounds following a fixed order. In each round, each user group picks their most valuable item from the remaining ones according to the given sequence. First, initialize the set S_i for each user group i as an empty set. Then each iteration corresponds to assigning one item: let $i \leftarrow \ell \mod t$, which determines the current user group based on the round number ℓ (Line 12). The most preferred item g^* is selected for the current user group i and g^* is assigned to the user group i while the removal of g^* is also carried out on the set of available items (Line 13).

Envy-Cycle Elimination Strategy. Unlike ROUND-ROBIN, the ENVY-CYCLE ELIMINATION strategy dynamically allocates items to disadvantaged user groups rather than following a fixed order. It uses an envy graph where nodes represent user groups and edges indicate envy between them (Line 19). In each round, the algorithm checks for nodes with an in-degree of zero, and these unenvied groups receive their most preferred item from the remaining pool (Lines 20–22). If no such node exists, an envy cycle is identified and resolved through item redistribution among the user groups in the cycle (Lines 23–24).

Theorem 1. ROUND-ROBIN *and* ENVY-CYCLE ELIMINATION *strategies compute an EF1HMQ [4,6].*

Proof. ROUND-ROBIN ensures EF1 by having groups choose items in a fixed order, granting priority to each in turn. ENVY-CYCLE ELIMINATION guarantees EF1 by only allocating items to a group if no other group envies its current set in that round.

Theorem 2. *The* ROUND-ROBIN *and* ENVY-CYCLE ELIMINATION *strategies compute 1/2-MMSHMQ.*

Proof. The ROUND-ROBIN's approximate ratio can be computed by $n \cdot v_i(A_i) \geq v_i(S) - \sum_{j \in t} v_i(g_j) \geq n \cdot \mu_i^t - \frac{n}{2} \cdot \mu_i^t = \frac{n}{2} \cdot \mu_i^t \quad \Rightarrow v_i(A_i) \geq \frac{1}{2} \cdot \mu_i^t$;

The ENVY-CYCLE ELIMINATION's approximate ratio can be computed by $v_i(A_i) \geq \frac{1}{2} \cdot \frac{v_i\left(\bigcup_{j \in t_2} A_j\right)}{|t_2|} \geq \frac{1}{2} \cdot \mu_i^{t-t_1}\left(M \setminus \bigcup_{j \in t_1} A_j\right) \geq \frac{1}{2} \cdot \mu_i^t$.

4 Experiments

4.1 Setup

All algorithms are implemented in C++, and all experiments are conducted on a PC running Ubuntu 20.04 LTS with a 3.00 GHz processor and 32 GB memory.

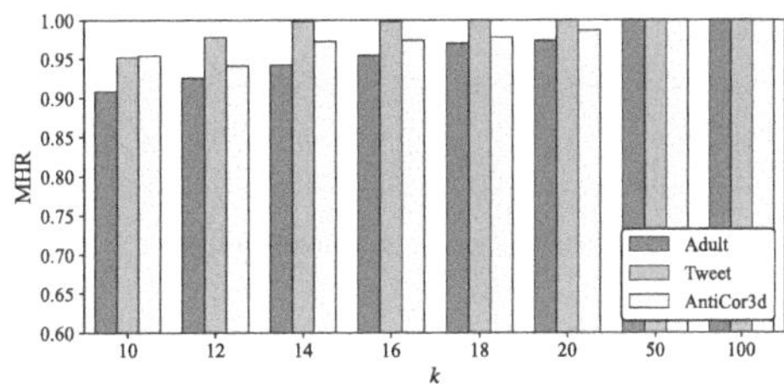

Fig. 2. Results for the total user groups' *MHR*s of different datasets.

Datasets. We run the experiments on synthetic datasets and two real-world datasets as follows: (i) **AntiCor** are synthetic datasets created by the generator [3]. (ii) **Adult** contains 32,561 tuples with five attributes. (iii) **Tweet** contains 12,197 tuples with five attributes.

Baselines. We compare our algorithm utilizing two strategies i.e., Round-Robin, Envy-Cycle Elimination with two baselines: (i) NaiveGreedy simply assigns each item to the user group that holds the highest value. (ii) RandHMQ randomly assigns HMQ results to each user group.

Metrics. The algorithms are evaluated using four metrics: *Runtime* for efficiency and MHR, *mms value*, $EFerr$ for solution quality. Larger values of MHR and *mms value* are better, while smaller $EFerr$ is preferred. The number of user groups is set to $d+1$, where d is the dataset's dimensionality.

4.2 Experimental Results

MHRs. Figure 2 shows the total MHRs across user groups. Without fairness considerations, the algorithm already achieves high MHRs, ensuring overall satisfaction and generally increases with k. Figure 3 shows per-group MHRs for varying k (only $k = \{10, 16, 20, 100\}$ shown due to space limits). Our algorithms maintain high performance (MHRs$>$ 0.94) across all datasets, despite a slight decline from fairness constraints. In contrast, NaiveGreedy and RandHMQ show unstable MHRs distributions due to strict greedy strategies and randomness.

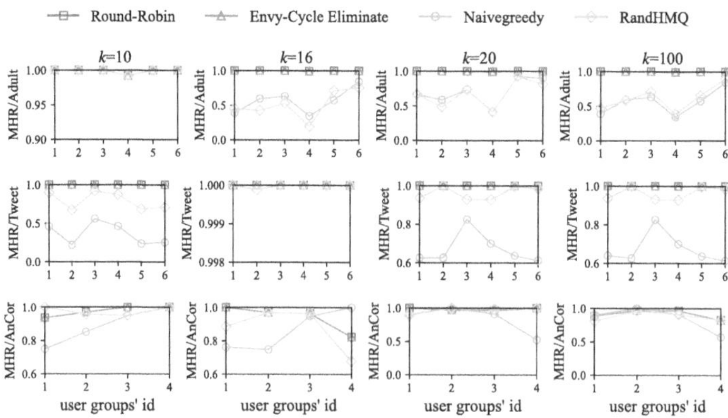

Fig. 3. Results for the MHRs of different user groups on different datasets.

Figure 4. simultaneously shows results of $EFerr$, *mms value* and *Runtime*. In terms of time efficiency, Round-Robin algorithm involves only simple loops,

ENVY-CYCLE ELIMINATION algorithm accelerates execution by using disjoint set union technique instead of DFS search, and RANDHMQ algorithm consistently runs quickly due to its randomness. As a result, we observe that these algorithms generally remain at a fast pace while the NAIVEGREEDY algorithm involves nested sorting loops for user preferences leading to the low efficiency.

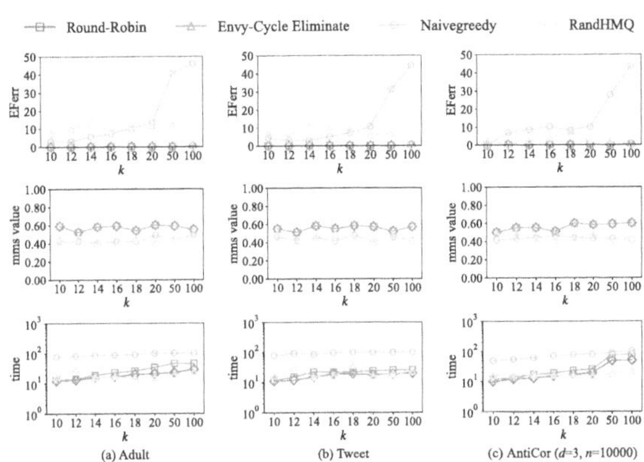

Fig. 4. Results for the $EFerr, mms\ value, time$ of different datasets.

Fairness Performance. By varying the solution size k, it is clearly evident that the ROUND-ROBIN and ENVY-CYCLE ELIMINATION achieve *zero unfairness*. Specifically, $EFerr = 0$ and *mms values* for four algorithms (using the minimum value achieved within each user group) align with the theoretical results of our algorithms. These results confirm the fairness and effectiveness of our algorithms, but other baselines cannot follow the fairness constraints all the time.

5 Conclusion

In this paper, we investigate the problem of the fairness among user groups in the happiness maximization query and propose two forms namely EF1HMQ and MMSHMQ based on the fairness notions in fair allocation field. By incorporating classical fair strategies ROUND-ROBIN and ENVY-CYCLE ELIMINATION, the FAIRHMQ framework successfully delivers fair solutions for user groups with different preferences and demonstrates strong empirical performance.

Acknowledgments. This work is partially supported by the Fundamental Research Funds for the Central Universities under grant No. NS2024056 and the Postgraduate Research & Practice Innovation Program of NUAA under grant No. XCXJH20241601.

References

1. Agarwal, P.K., Kumar, N., Sintos, S., Suri, S.: Efficient algorithms for k-regret minimizing sets. In: SEA, pp. 7:1–7:23 (2017)
2. Amanatidis, G., et al.: Fair division of indivisible goods: recent progress and open questions. Artif. Intell. **322**(C), 103965 (2023)
3. Borzsony, S., Kossmann, D., Stocker, K.: The skyline operator. In: ICDE, pp. 421–430 (2001)
4. Caragiannis, I., Kurokawa, D., Moulin, H., Procaccia, A.D., Shah, N., Wang, J.: The unreasonable fairness of maximum nash welfare. TEAC **7**(3), 1–32 (2019)
5. Ilyas, I.F., Beskales, G., Soliman, M.A.: A survey of top-k query processing techniques in relational database systems. ACM Comput. Surv. (CSUR) **40**(4), 1–58 (2008)
6. Lipton, R.J., Markakis, E., Mossel, E., Saberi, A.: On approximately fair allocations of indivisible goods. In: EC, pp. 125–131 (2004)
7. Nanongkai, D., Sarma, A.D., Lall, A., Lipton, R.J., Xu, J.: Regret-minimizing representative databases. In: PVLDB, vol. 3, pp. 1114–1124 (2010)
8. Zhang, Y., Meng, D., Zhang, L., Kong, C.: DINE: dynamic information network embedding for social recommendation. In: WISA, vol. 14094, pp. 76–87 (2023)
9. Zheng, J., Ma, W., Wang, Y., Wang, X.: Continuous k-regret minimization queries: a dynamic coreset approach. TKDE **35**(6), 5680–5694 (2023)
10. Zheng, J., Ma, Y., Ma, W., Wang, Y., Wang, X.: Happiness maximizing sets under group fairness constraints. In: PVLDB, vol. 16, pp. 291–303 (2022)
11. Zheng, J., et al.: Hybrid regret minimization: a submodular approach. TKDE **36**(7), 3151–3165 (2024)

Accelerating Subgraph Matching Using Isolated Vertices and BFS Data Structures

Guiyang Zhang, Bo Ning(✉), and Xin Zhou

Dalian Maritime University, Dalian, China
{ningbo,zhouxin314159}@dlmu.edu.cn

Abstract. Subgraph matching is a fundamental problem in graph analysis, with widespread applications in social networks, bioinformatics, and knowledge engineering. Traditional subgraph matching algorithms typically rely on backtracking search and recursive computation, which often lead to high computational complexity and an exponential growth of the search space when dealing with large scale graph data. This paper proposes ADR-IVE(AuxiliaryDualReorder-IVE) subgraph matching algorithm based on a backtracking enumeration framework, aiming to enhance the efficiency of subgraph matching and optimize query processing performance. The proposed method leverages the attributes of both the data graph and the query graph to optimize the generation order of query graph vertices, thereby effectively reduce the candidate space of query vertices. Experimental results demonstrate that the proposed method achieves an overall performance improvement of approximately 60% compared to state-of-the-art algorithms when handling large-scale and structurally complex subgraph matching tasks. This significant advantage highlights the method's potential for widespread application and practical value in large-scale subgraph matching tasks.

Keywords: Subgraph Isomorphism · Subgraph Matching · Subgraph Query Processing

1 Introduction

1.1 Background

Subgraph matching locating all isomorphic embeddings of a query graph within a data graph is pivotal for frequent-pattern mining, social-network analytics, cheminformatics and knowledge-graph completion. Existing solutions fall into two paradigms: join-based methods [1] that decompose and reunite sub-patterns, and backtracking schemes that recursively enumerate vertex mappings under a predefined order. Owing to the NP-completeness of the problem, we enhance the IVE [2] framework with a novel candidate generation strategy, yielding substantial gains in both efficiency and scalability.

S. Yang et al. (Eds.): WISA 2025, LNCS 16299, pp. 293–300, 2026.
https://doi.org/10.1007/978-981-95-6426-2_25

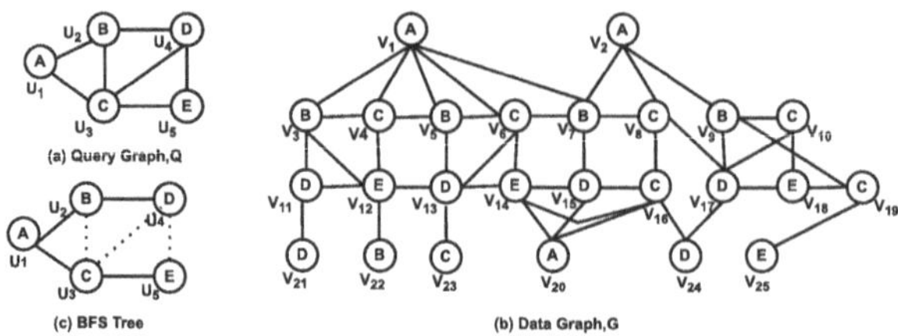

Fig. 1. (a) Query graph, (b) Data graph, and (c) Breadth-First Search (BFS) spanning tree of the query graph.

1.2 Related Work

Ullmann's [3] neighborhood-driven candidate generation, subsequent techniques progressively shrink the search space: QuickSI [4] favors rare vertices/edges, TurboIso [5] leverages vertex similarity, CFLMatch [6] prioritizes dense cores, DAF [7] employs DAG-based filtering with candidate spaces and failure sets, GuP [8] guards candidates for early pruning, and VEQs [9] exploits static and dynamic vertex equivalence plus neighbor-safe compression to refine ordering and storage.

Definition 1. (Graph Isomorphism)Let $G = (V, E, L)$ be a labeled graph where $L : V \rightarrow \Sigma$ assigns labels. Two graphs G_A and G_B are isomorphic ($G_A \equiv G_B$) if there exists a bijective function $f : V_A \rightarrow V_B$ such that $L_A(v) = L_B(f(v))$ and $(v_i, v_j) \in E_A \Rightarrow (f(v_i), f(v_j)) \in E_B$.

Definition 2. (Subgraph Matching Query) Given a data graph G , and a query graph Q, a subgraph matching query retrieves all the subgraphs g of the data graph G that are isomorphic to the query graph q(i.e.,g≡q).

Extending the isolated-vertex exploration (IVE) [2]framework, we present ADR-IVE, which refines candidate generation and vertex ordering. The DMDE reordering strategy exploits dual-graph properties to minimize search cost. Comprehensive experiments confirm superior runtime and scalability over state-of-the-art methods.

2 ADR-IVE (AuxiliaryDualReorder-IVE)

2.1 Overview

This paper provides a schematic illustration and detailed elaboration of the proposed method using the query graph and data graph shown in Fig. 1 as examples. ADR-IVE adopts an algorithmic structure based on the filtering-sorting-enumeration framework. Algorithm 1 presents the pseudocode implementation of the overall framework.

Algorithm 1. Subgraph Matching

```
1: Input: Query graph Q, Data graph G
2: Output: All matched subgraphs isomorphic to Q
3: C ← FilterVertices(Q, G)
4: φ ← DMDE(Q, G, C)
5: Enumerate(Q, G, C, φ, {}, 0)
6: function ENUMERATE(Q, G, C, φ, M, i)
7:     if termination condition met then
8:         Output TerminateFunc(Q, G, C, φ, {}, M)
9:     else if u is a non-isolated vertex then
10:        NonIsoEnumerate(Q, G, C, φ, M, i+1)
11:    else
12:        IsoEnumerate(Q, G, C, φ, M, i+1)
13:    end if
14: end function
```

2.2 Filtering: BFS-Based Filtering

In the filtering phase, candidate sets for query vertices are generated by imposing constraints based on label matching, degree threshold, and neighbor label count. An initial root vertex is selected according to the minimum ratio of label frequency to vertex degree(e.g., u_1 in Fig 1a–1b). Subsequently, candidate sets are generated sequentially in a breadth-first search (BFS) order starting from this root vertex. This process facilitates the subsequent generation of the vertex matching order. The corresponding BFS tree is illustrated in Fig 1c.

2.3 Sorting: DMDE

We propose a reordering strategy to maximize isolated vertices in the query graph, minimizing non-isolated ones that necessitate costly backtracking. Since matching isolated vertices is solvable in polynomial time via bipartite matching, overall efficiency is significantly improved. The resulting complexity is $O(mn' \cdot n^2 \cdot m)$, as analyzed in [2], where m, n, and n' denote the number of data vertices, non-isolated, and isolated query vertices, respectively. While brute-force search guarantees optimality, its exponential complexity $O(2^{|V_q|} \cdot |V_q|^2)$ is impractical. Instead, we propose the DMDE algorithm, which generates near-optimal matching orders in $O(|V_q|^3)$ time by leveraging structural properties of the query graph. Algorithm 3 outlines DMDE. It first selects a start vertex with the smallest candidate-to-degree ratio, adds it to the order, and initializes an available set with its neighbors. A graph copy q' tracks edge deletions. Iteratively, the vertex with the smallest candidate-degree ratio (with ties broken by smallest ID) is selected from the available set and added to the order; its neighbors are then added to the available set. This process continues until all vertices are ordered. By maximizing edge deletions per iteration, DMDE increases residual isolated vertices, reducing backtracking. Although not globally optimal, DMDE achieves a practical trade-off between quality and efficiency. After ordering, vertices are classified as isolated (if all backward neighbors are matched) or non-isolated, enabling differentiated embedding strategies.

Algorithm 2. Sorting: DMDE(Q, $\mathcal{A}$)

1: **Input:** Query graph Q, AuxilaryData $\mathcal{A}$
2: **Output:** o, the matching order
3: **function** DMDE($Q, \mathcal{A}$)
4: let $u_0 \in Q$ be a start vertex with minimal (candidate vertices) / degree
5: $o \leftarrow [u_0]$
6: $Available \leftarrow N_q(u_0)$
7: $q' \leftarrow q$
8: Remove edges $(u_0, u' \mid u' \in N_q(u_0))$ in q'
9: **while** $|Available| > 0$ **do**
10: $u_{\max} \leftarrow$ NULL
11: **for** u in $Available$ **do**
12: **if** $u_{\max} =$ NULL or $|N_{q'}(u_{\max})|$ **then**
13: $u_{\max} \leftarrow u$
14: **end if**
15: Remove edges $(u_{\max}, u' \mid u' \in N_{q'}(u_{\max}))$ in q'
16: $Available \leftarrow Available \cup N_{q'}(u_{\max})$
17: o.add($u_{\max}$)
18: **end for**
19: **end while**
20: **return** o
21: **end function**
22: **function** CLASSIFY(Q, o)
23: $IsoV \leftarrow \{\}$
24: $NonIsoV \leftarrow \{\}$
25: **for** $i = 0, 1, \ldots, |V_q| - 1$ **do**
26: $u_i \leftarrow o[i]$
27: **if** all u_i . backward neighbor already visited **then**
28: $IsoV \leftarrow u_i \cup IsoV$
29: **else**
30: $NonIsoV \leftarrow u_i \cup NonIsoV$
31: **end if**
32: **end for**
33: **return** $IsoV, NonIsoV$
34: **end function**

2.4 Enumeration

This subsection outlines the subgraph matching process. Following the framework in [2], we apply distinct strategies for non-isolated and isolated vertices. Non-isolated vertices are matched via backtracking enumeration, while isolated vertices are matched by forming a bipartite graph [10] between themselves and their candidate sets—a process solvable in polynomial time. The final embeddings are obtained by combining matches from both groups.

3 Experiment

3.1 Experimental Environment

All code was implemented in C++20 and compiled with CMake 3.26.0 and GCC 9.4.0 on Ubuntu 20.04.5 LTS. Experiments were run on a server equipped with an Intel Xeon Platinum 8352V CPU (32 cores, 2.10 GHz) and 60 GB of RAM.

Table 1. Properties of Data Graph

Dataset	$\|V\|$	$\|E\|$	$\|\Sigma\|$	d
DBLP	317080	1049866	15	6.6
Human	4674	86282	44	36.9
US Patents	3774768	16518947	20	8.8
Youtube	1134890	2987624	25	5.3

3.2 Dataset Description

This study uses four standard SI datasets: DBLP, Human, Patents, and YouTube (see Table 1). In Human, edges denote protein-protein interactions and labels are from Gene Ontology; for the others, labels are randomly assigned. Query graphs are connected random walk-induced subgraphs, all satisfiable. For each dataset, we generate 200 queries per size: $|V(q)| = 4$–20 for Human, 4–32 for others. Each size includes four dense ($d(q) \geq 3$) and four sparse ($d(q) < 3$) sets.

3.3 Comparison Algorithms

This study compares the proposed method against four state-of-the-art subgraph isomorphism algorithms: DPiso [7], GQL [11], CFL [6], and IVE [2]. Implementations of DPiso, GQL, and CFL are publicly available via [12], and IVE via [2].

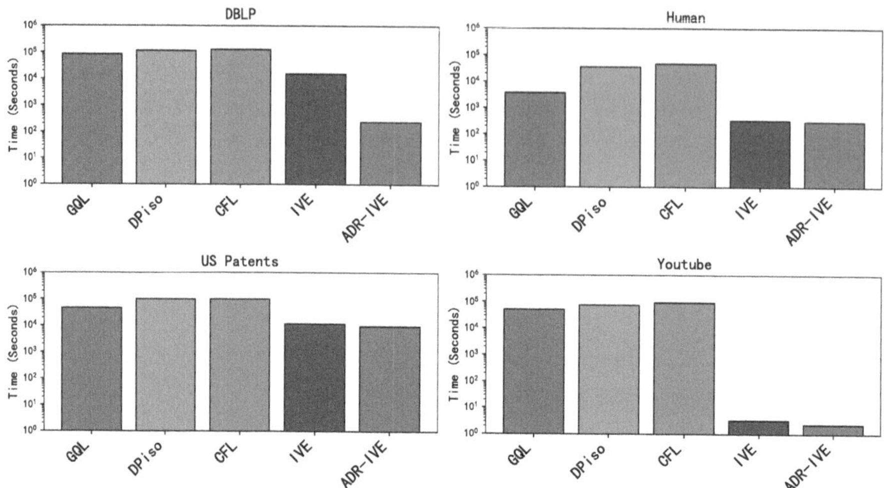

Fig. 2. Figure 2 presents the total runtime performance of various algorithms on different datasets.

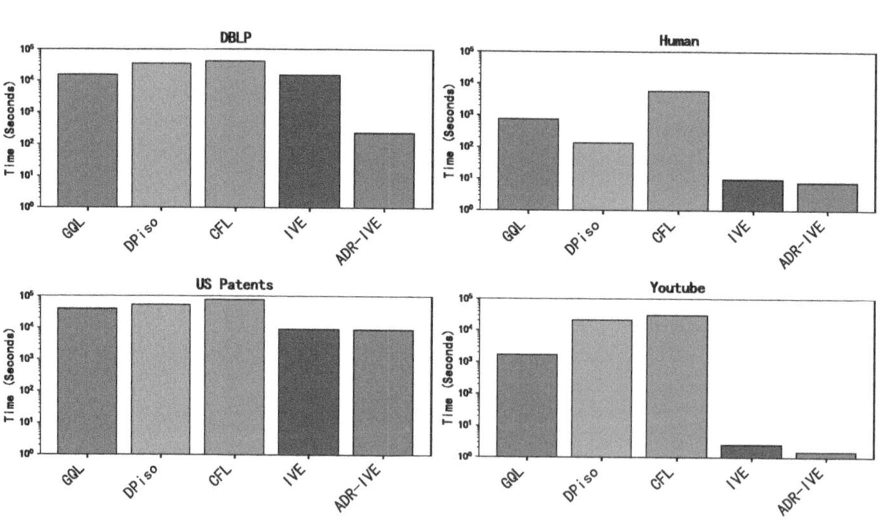

Fig. 3. Figure 3 demonstrates the runtime of various algorithms on dense query graphs across different datasets.

3.4 Experimental Analysis

We evaluate the proposed ADR-IVE algorithm on total matching time and dense queries. As shown in Fig. 2, ADR-IVE outperforms all baselines across datasets.

The improvement stems from structure-aware enumeration and efficient candidate generation, which prune the search space effectively. Figure 3 show that ADR-IVE performs robustly under dense queries.

4 Conclusion

This paper exploits the structural properties of both the query and data graphs to optimize the enumeration order, substantially improving enumeration efficiency. Furthermore, BFS-based auxiliary structures are introduced to efficiently generate candidate sets. These structures directly identify the minimal candidate set for each query vertex, reducing candidate sets and enhancing enumeration efficiency.

References

1. Lai, L., Qin, L., Lin, X., Chang, L.: Scalable subgraph enumeration in MapReduce. Proc. VLDB Endow. **8**(10), 974–985 (2015). https://doi.org/10.14778/2794367.2794368
2. Jiang, Z., Zhang, S., Hou, X., Yuan, M., You, H.: IVE: accelerating enumeration-based subgraph matching via exploring isolated vertices. In: 2024 IEEE 40th International Conference on Data Engineering (ICDE), Utrecht, Netherlands, pp. 4208–4221 (2024). https://doi.org/10.1109/ICDE60146.2024.00321.
3. Ullmann, J.R.: An algorithm for subgraph isomorphism. J. ACM **23**(1), 31–42 (1976). https://doi.org/10.1145/321921.321925
4. Shang, H., Zhang, Y., Lin, X., Yu, J.X.: Taming verification hardness: an efficient algorithm for testing subgraph isomorphism. Proc. VLDB Endow. **1**(1), 364–375 (2008). https://doi.org/10.14778/1453856.1453899
5. Han, W.S., Lee, J., Lee, J.H.: Turboiso: towards ultrafast and robust subgraph isomorphism search in large graph databases. In: Proceedings of the 2013 ACM SIGMOD International Conference on Management of Data (SIGMOD '13), pp. 337–348. Association for Computing Machinery, New York (2013). https://doi.org/10.1145/2463676.2465300
6. Bi, F., Chang, L., Lin, X., Qin, L., Zhang, W.: Efficient subgraph matching by postponing cartesian products. In: Proceedings of the 2016 International Conference on Management of Data (SIGMOD '16), pp. 1199–1214. Association for Computing Machinery, New York (2016). https://doi.org/10.1145/2882903.2915236
7. Han, M., Kim, H., Gu, G., Park, K., Han, W.S.: Efficient subgraph matching: harmonizing dynamic programming, adaptive matching order, and failing set together. In: Proceedings of the 2019 International Conference on Management of Data (SIGMOD '19), pp. 1429–1446. Association for Computing Machinery, New York (2019). https://doi.org/10.1145/3299869.3319880
8. Arai, J., Fujiwara, Y., Onizuka, M.: GuP: fast subgraph matching by guard-based pruning. Proc. ACM Manag. Data **1**(2), Article 167, p. 26 (2023). https://doi.org/10.1145/3589312
9. Liu, X., Li, H., Xu, G., Chen, Z., Huang, X., Lu, R.: Privacy-enhanced federated learning against poisoning adversaries. IEEE Trans. Inf. Forensics Secur. **16**, 4574–4588 (2021). https://doi.org/10.1109/TIFS.2021.3108434

10. Gao, Z., Chu, D., Zhang, F., Wang, K., Yuan, L.: Efficient p-biclique query on large bipartite networks. WISA 2024. Lecture Notes in Computer Science, vol. 14883. Springer, Singapore (2024). https://doi.org/10.1007/978-981-97-7707-5_18
11. He, H., Singh, A.K.: Graphs-at-a-time: query language and access methods for graph databases. In: Proceedings of the 2008 ACM SIGMOD International Conference on Management of Data (SIGMOD '08), pp. 405–418. Association for Computing Machinery, New York (2008). https://doi.org/10.1145/1376616.1376660
12. Sun, S., Luo, Q.: In-memory subgraph matching: an in-depth study. In Proceedings of the 2020 ACM SIGMOD International Conference on Management of Data (SIGMOD '20), pp. 1083–1098. Association for Computing Machinery, New York (2020). https://doi.org/10.1145/3318464.3380581

MAHI: Graph Index for Multi-attribute Constrained Vector Search

Qian Kang[1], Yuan Ye[2(✉)], Wang Yishu[1], and Jiang Dong[1]

[1] School of Computer Science and Engineering, Northeastern University, Shenyang 110169, China

[2] School of Computer Science and Technology, Beijing Institute of Technology, Beijing 100081, China

yuan-ye@bit.edu.cn

Abstract. Approximate nearest neighbors search (ANNS) combined with attribute filtering is crucial for practical vector search tasks involving vector similarity and multi-attribute constraints. Existing methods focus only on discrete or continuous attributes, ignoring their joint constraints. Meanwhile, existing solutions for continuous attribute filtering directly combined with discrete attribute methods can lead to poor segment balance, which affects construction and query efficiency. To address these issues, we propose MAHI (multi-attribute hybrid index), an index framework that efficiently supports ANNS under hybrid attribute constraints. We use prefix trees to partition the original dataset based on discrete attributes. Within each generated block, we combine the KMeans algorithm with two different indexing methods to mitigate the data distribution unevenness caused by discrete attribute partitioning, thereby restoring the structure of continuous attributes. Experiments on large-scale datasets demonstrate that MAHI significantly outperforms existing state-of-the-art baseline algorithms in terms of recall, throughput, and indexing efficiency.

Keywords: Vector search · Attribute filtering · Range attribute filtering · Graph index

1 Introduction

In recent years, with the rapid development of big data and artificial intelligence technologies, vector representation learning has become the mainstream method for modelling the semantics of complex entities. Against this backdrop, approximate nearest neighbour search (ANNS [1]) has emerged as a core technology for high-dimensional vector retrieval, finding widespread application in practical scenarios such as recommendation systems [2–5], search engines [6,7], and retrieval-augmented generation (RAG) for large language models [8–10].

In many practical scenarios, user queries often involve constraints on both discrete and continuous attributes [11]. For example, in an e-commerce scenario, a user may want to search for shoes from brand X priced between 100 and 200,

S. Yang et al. (Eds.): WISA 2025, LNCS 16299, pp. 301–311, 2026.
https://doi.org/10.1007/978-981-95-6426-2_26

where shoes serve as the search vector, brand is a discrete attribute constraint, and the continuous attribute constraint is the range [100, 200]; in academic literature retrieval, researchers may need to retrieve papers published since 2022 from the NeurIPS conference on the topic of graph neural networks. Where the papers serve as the search vector, NeurIPS and graph neural networks are the discrete attributes, and the time range [2022, 2025] acts as the continuous attribute constraint. Such queries that combine vector similarity with attribute constraints are collectively referred to as attribute filtering approximate nearest neighbors search (Attribute Filtering ANNS, abbreviated as FANNS).

Although FANNS has become an important capability for vector retrieval systems [11–15], the existing main methods [16–20] still face significant challenges in supporting 'multi-attribute hybrid queries'. Most existing methods focus on supporting filtering for a single attribute type.

To address these issues, this paper proposes a vector indexing framework supporting multi-attribute hybrid filtering search—MAHI (multi-attribute hybrid hierarchical index). This method is designed for query scenarios involving both discrete and continuous attribute constraints, featuring a unified multi-layer index structure with systematic design across data organisation, segmentation strategies, graph index construction, and query optimisation. The main contributions of this paper are as follows:

- **Hybrid Attribute Query Definition:** We formalize the problem of Approximate Nearest Neighbor Search (ANNS) under hybrid attribute constraints, explicitly defining query scenarios involving both discrete and continuous attributes.
- **Multi-attribute hybrid hierarchical index(MAHI)** We propose a novel indexing framework that integrates prefix tree partitioning for discrete attributes and density-aware segmentation for continuous attributes. This design addresses the challenge of disrupted continuous attribute distributions after discrete partitioning, enabling efficient and accurate hybrid queries.
- **Query Optimization via Segment Coverage Estimation:** We introduce a lightweight estimation mechanism to prune segments with low relevance to the query's continuous attribute range, thereby avoiding excessive or irrelevant comparisons and improving overall search precision.

2 Problem Definition

Let $I(A)$ be a set of discrete attribute values and $I(B) = [l, r]$ be a closed interval of a continuous attribute B. Let $I(AB)$ denote the combination of $I(A)$ and $I(B)$, and define the filtered subset of the dataset D as:

$$D[I_{AB}] = \{v \in D \mid v(A) \in I(A) \wedge v(B) \in [l, r]\} \tag{1}$$

Let $d > 0$ be an integer and δ be a distance metric over the d-dimensional vector space $\mathbb{R}^d$ (e.g., Euclidean distance). Let D be a dataset of n points in $\mathbb{R}^d$, where $n > 0$. Then we define:

Definition 1 (Multi-attribute Hybrid kANNS). *Let $D = \{v_1, v_2, \ldots, v_n\}$ be a d-dimensional vector dataset. A multi-attribute hybrid query is defined as $Q = (q, I(A), [l, r], k)$, where:*

- *$q \in \mathbb{R}^d$ is the query vector,*
- *$I(A)$ is a set of discrete attribute values,*
- *$[l, r]$ is an interval of the continuous attribute B, with $r > l$,*
- *k is the number of nearest neighbors to retrieve.*

The query returns a subset $S \subseteq D$ of size k such that:

- *$\forall v_x \in S$, it holds that $v_x(A) \in I(A)$ and $v_x(B) \in [l, r]$,*
- *$\forall v_x \in S$, $\forall v_y \in D \setminus S$, it holds that either:*

$$v_y(A) \notin I(A) \quad or \quad v_y(B) \notin [l, r] \quad or \quad \delta(v_x, q) \leq \delta(v_y, q)$$

3 MAHI: Multi-attribute Hybrid Hierarchical Index

In order to support efficient vector search that simultaneously includes discrete and continuous attributes, and to solve the problem of uneven data distribution caused by the direct combination of discrete and continuous attribute filtering methods, we propose MAHI (Multi-attribute Hybrid Hierarchical Index), a two-stage indexing framework. MAHI first decomposes the search space according to discrete attribute combinations and then applies range-aware indexing over continuous attributes within each discrete group. This hierarchical structure enables fine-grained filtering, reduced search space, and improved query performance.

MAHI breaks down the query processing pipeline into three closely coordinated stages:

Stage 1: Discrete attribute partitioning. We partition the dataset based on combinations of discrete attributes. Each unique combination defines a block, and all such blocks are organised in a prefix tree (Trie [21]) to enable efficient constraint filtering and pruning.

Stage 2: Continuous attribute clustering. Within each block (i.e., each data node in the prefix tree), we apply the K-means clustering algorithm to continuous attribute dimensions to mitigate the impact of uneven distribution exacerbated by discrete attribute partitioning.

Stage 3: Block Graph Index Construction within Segments. Finally, based on the size of the clusters after clustering, we constructed two different nearest neighbors graph indexes for each cluster.

3.1 Discrete Attribute Partitioning

We begin by partitioning the dataset using discrete attributes. First, extract all discrete attribute combinations satisfied by the points in the original dataset, with each discrete attribute combination corresponding to a data block. Each block contains vectors sharing identical values for a fixed set of discrete attributes (e.g., brand = "Nike", category = "Trainers"). To organize these blocks efficiently, we construct the Trie where: Each node represents a specific attribute value; Red nodes (data nodes) represent complete discrete attributes combinations and are linked to the corresponding data blocks. This hierarchical structure transforms the multi-dimensional matching problem into a path traversal process. First, we initialise an empty root node as the starting point of the prefix tree. Then, for each discrete attribute block $G_C \in G$, we extract its corresponding attribute value sequence A_d. These attribute values are arranged in a predefined order to form the unique attribute path of the block. Next, starting from the root node, we iterate through each attribute value attr_value in A_d. At the current node being traversed, if there is no existing child node labelled with that attribute value, we create a new path node and connect it to the current node. Then, we move the current node pointer to that child node and continue processing the next attribute value. Once the attribute value sequence has been fully traversed, the current node corresponds to the end point of the attribute combination A_d. We mark this node as a red node and attach the corresponding data block G_C to it. We repeat this process until all discrete attribute blocks have been inserted into the prefix tree.

For example, the discrete attribute combinations satisfied by dataset D are $\{(1,2),(1,6),(2,3,5),(2,5),(3,4),(3,5,6)\}$. We can use the above algorithm to construct a trie as shown in Fig. 1. The red nodes correspond to datasets that satisfy the attribute combinations, while the black nodes, which do not have any attribute combinations satisfied in the dataset, are marked as path nodes. The path combinations from the root node to any red node in Fig. 1 represent an attribute combination, and the red nodes correspond to datasets that satisfy this attribute combination and store their entry point lists. By structuring discrete attribute filtering as prefix tree traversal, we reduce the complexity of block identification from a linear scan over combinations to a logarithmic-time path search, laying the foundation for efficient hybrid indexing in the next stages.

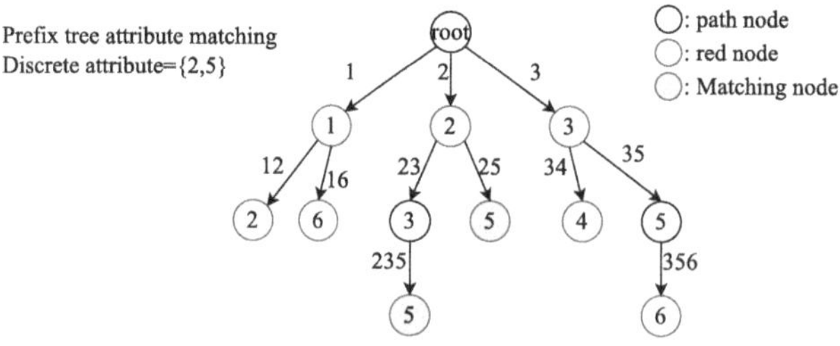

Fig. 1. Prefix tree for discrete attribute filtering

3.2 Continuous Attribute Clustering

Although discrete partitioning reduces the candidate space, each block may still contain tens of thousands of vectors, and these vectors with unevenly distributed continuous attributes become even more unevenly distributed after being partitioned by the prefix tree based on discrete attributes. Basic full-range index or fixed-width interval range indexes can lead to poor segment balance, thereby affecting construction and query efficiency.

To address this issue, we apply the K-Means clustering algorithm to continuous attributes within each data node of the prefix tree. Our approach dynamically partitions blocks into cohesive, density-aware intervals, enabling the robust use of the median (instead of the mean) to handle outliers while storing each cluster as a segment with explicit attribute range boundaries for early pruning during queries. After processing with the prefix tree, we cluster the data subsets represented by each data node in the prefix tree. For example, for the subset satisfying the attributes (2, 3, 5), after attribute processing, we obtain five data clusters. The continuous attribute ranges covered and the number of data points in each cluster are shown in the figure.

3.3 Block Graph Index Construction Within Segments

Once the data is segmented, we build graph indexes over each segment based on its size:

Complete Block Graph (small segments): For segments with size below a threshold δ, the first data point is the entry point, and the rest are normal points. We construct a shallow 0-layer HNSW [22] graph index where The neighbors of an entry point include all points, while the neighbors of a normal point only include the nearest neighbors. This flat structure is fast and efficient for small segments.

Standard Block Graph (large segments): for larger segments, we use a binary-range-based HNSW hierarchy, inspired by SeRF [19]. Each level corresponds to a nested range (e.g., $[1, 64], [1, 32], [17, 32]$), and each sub-segment stores neighbors as shown in Fig. 1, but at this point, the entry point is not connected to all points in the segment. This design supports efficient range-aware search and recursive refinement.

3.4 Query Processing with Optimization

Although the widely used greedy search method can correctly execute attribute filtering queries, it has obvious shortcomings. In particular, when the query range $[l, r]$ overlaps only slightly with the continuous attribute range of the candidate paragraphs, the algorithm still performs indiscriminate search operations on these paragraphs. This 'out-of-range search' introduces a large number of invalid data points, reducing query efficiency and diluting the proportion of valid results, thereby significantly affecting the recall rate.

To address the above issues, we introduced a coverage-aware pruning strategy during the query process to optimise filtering at the segment level. The main idea is that after finding all blocks that satisfy the discrete attribute constraints in the prefix tree, we further utilise the continuous attribute range information of each segment to determine whether it effectively overlaps with the query interval

$[l, r]$. Specifically, as shown in Algorithm 1, for each candidate segment S, we check whether its attribute range B_S satisfies the coverage condition: it is fully covered by $[l, r]$, or; the coverage range exceeds $1/16$ of the length of $[l, r]$, and the coverage range exceeds $1/32$ of B_S. If the coverage condition is satisfied, S is directly added to the candidate set; otherwise, S is skipped. The coverage condition effectively avoids redundant computations and false positives caused by accessing irrelevant segments during the query process.

Algorithm 1. Attribute-Filtering Query Processing

Require: Query vector q with attributes $I_q(A)$ and $I_q(B) \in [l, r]$; Trie index structure `Trie`; Block graph index `I`; Search width L; Top-K neighbors count K; Entry points per graph `num_entry_points`
Ensure: Top-K nearest neighbors R satisfying attribute filters
1: Use `Trie` to find matching attribute blocks G_{match} where $I_A(G_C) \supseteq I_q(A)$
2: Initialize candidate set $R \leftarrow \emptyset$
3: **for all** attribute block $G_C \in G_{\text{match}}$ **do**
4: Initialize segment set $\mathcal{S}_{\text{cand}} \leftarrow \emptyset$
5: **for all** segment $S \in G_C$ **do**
6: **if** attribute range of S is fully covered by $[l, r]$ **then**
7: Add S to $\mathcal{S}_{\text{cand}}$
8: **else if** S does not satisfy range filter conditions **then**
9: Skip S
10: **else**
11: Recursively check the left and right children of S
12: **for all** segment $S \in \mathcal{S}_{\text{cand}}$ **do**
13: Perform graph index search on S using entry points and width L
14: Add results to candidate set R
15: Merge candidates from all segments into R
16: Resort R by distance to q
17: **return** Top-K results from R

3.5 Complexity Analysis

Constructing the prefix tree over $|L|$ label sets with $|P|$ attributes costs $O(|L| \cdot |P|)$ time and $O(|L| \cdot |P|)$ space. For the HNSW-based block graphs, the indexing cost is $O(n \log n)$ with $O(knm)$ space, where m is the max out-degree and k is the number of layers. Overall, the time complexity of MAHI is $O(|L| \cdot |P| + n \log n)$ and the space complexity of MAHI is $O(|L| \cdot |P| + knm)$.

4 Experiments

Table 1. Dataset statistics.

Dataset	Dimension	Number of points	Source	Number of queries
SIFT1M	128	1,000,000	Image	1,000
Deep10M	96	10,000,000	Image	10,000
SIFT100M	128	100,000,000	Image	10,000

To evaluate the efficiency and effectiveness of MAHI under multi-attribute constraints, we conducted experiments on three widely used vector datasets. As shown in Table 1, the three datasets contain 1M, 10M, and 100M data points, respectively. SIFT1M and SIFT100M are 128-dimensional image vectors, while DEEP10M is a 96-dimensional image vector. The query set for SIFT1M contains 1k data points, while the query sets for SIFT100M and DEEP10M contain 10k data points.

4.1 Settings

As shown in Fig. 2a, the discrete attribute descriptions are generated using the Zipf method described in [25], involving label combinations ranging from 1 to 12, whose frequency distribution follows a truncated long-tail pattern. Additionally, a continuous attribute is introduced for evaluation purposes. As shown in Fig. 2b, the continuous attribute values range from 0 to 10,000 and follow a bimodal distribution combined with a long-tail distribution, where the bimodal peaks have means of 300 and 3500 (standard deviations 150 and 1000 respectively), accounting for 95% of data, and the remaining 5% follows a long-tail distribution. Query attribute constraints are generated following the method in [25], with discrete labels randomly selected and continuous attributes sampled with range boundaries aligned to their distribution patterns. The query results are obtained through a brute-force search method and pre-stored.

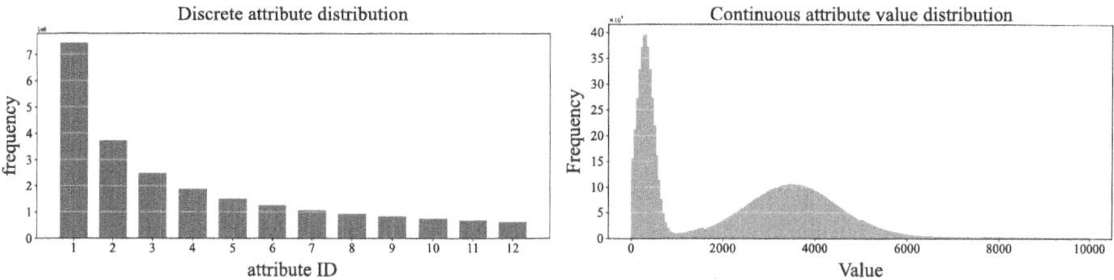

(a) The frequency of discrete attributes (b) The frequency of continuous attributes

Fig. 2. attribute distribution

4.2 Algorithms and Parameters

We compare our proposed MAHI with the state-of-the-art UNG [23], which is used as the baseline due to its strong performance. As UNG only supports discrete attributes, we extend it with post-filtering to handle continuous attributes. Evaluation metrics include queries per second (QPS), and recall. We also compare index construction time/space and search time, using L_{Build} and L_{Search} as tunable parameters. UNG uses the Vamana [24] graph indexing algorithm with $\alpha = 1.2$ and max outdegree is 32. Following its original settings, we set the cross-group edge parameter $\delta = 6$. MAHI adopts HNSW as the underlying index for block graphs, with a maximum layer height of 8 and max degree is 32. Continuous attributes are clustered into 8 centers.

4.3 Index Construction

For each algorithm and different parameters in the experiment, we used 32 threads to build the index. We varied the candidate neighbor queue size L during index construction to evaluate its impact on performance.

As shown in Figs. 3a–3c, MAHI achieves a 5% to 10% speedup in index building time compared to UNG under identical multi-threaded settings. Experiments have shown that UNG takes a long time to construct cross-group edges, due to the overhead of constructing cross-group edges despite their fewer numbers.

As shown in Figs. 3d–3f, on the SIFT100M large dataset, the MAHI index size is approximately 15% larger than the UNG in the SIFT100M dataset. On DEEP10M, MAHI's space occupancy is approximately 10% higher than UNG's. In the SIFT1M small dataset, the occupancy of the space is approximately 8%. In general, despite an approximate 10% increase in index size, MAHI achieves an approximate 10% faster index construction compared to UNG.

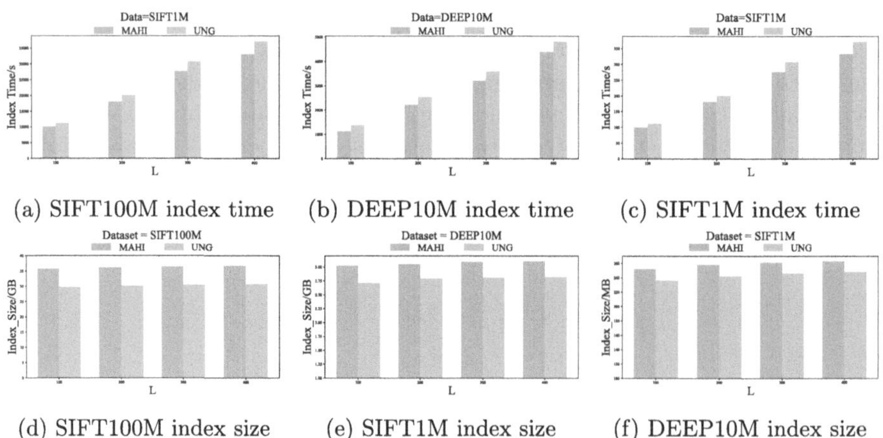

(a) SIFT100M index time (b) DEEP10M index time (c) SIFT1M index time

(d) SIFT100M index size (e) SIFT1M index size (f) DEEP10M index size

Fig. 3. Index time varying L for construction processing

4.4 Overall Query Performance

Figures 4a–4c illustrate that MAHI consistently outperforms UNG in terms of query throughput (QPS), particularly when the search queue length is small. On larger datasets, MAHI achieves significant improvements in both recall and efficiency. Under a 90% recall target, it delivers up to 1.3× higher QPS compared to UNG. As shown in Figs. 4d–4f, as the query queue L increases, both MAHI and UNG query efficiency gradually increase.

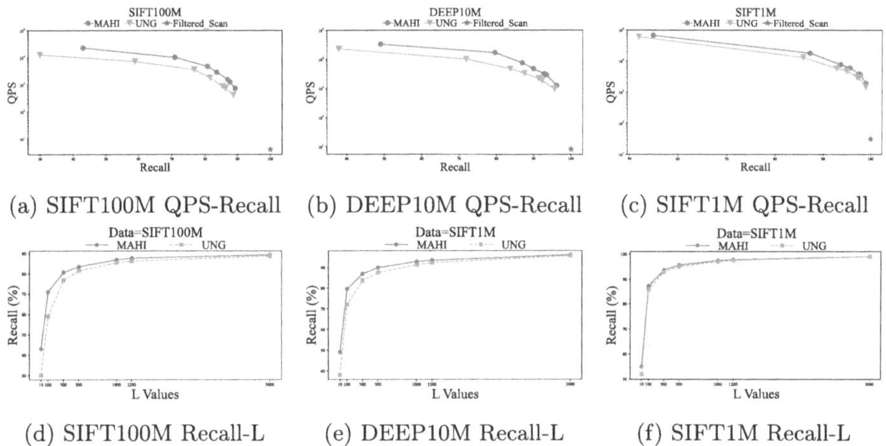

(a) SIFT100M QPS-Recall (b) DEEP10M QPS-Recall (c) SIFT1M QPS-Recall

(d) SIFT100M Recall-L (e) DEEP10M Recall-L (f) SIFT1M Recall-L

Fig. 4. Query performance in multiple attribute constraints

5 Conclusion

This paper presents MAHI, a hybrid indexing framework for efficient approximate nearest neighbor search under both discrete and continuous attribute constraints. By combining prefix-tree-based discrete filtering, density-aware segmentation, and hierarchical graph indexing, MAHI enables accurate and scalable multi-attribute retrieval. Experimental results on large-scale datasets demonstrate that MAHI consistently outperforms existing methods in recall, query efficiency, and indexing speed, highlighting its practical value and strong potential for real-world applications.

Acknowledgements. This work is supported by the National Key R&D Program of China (2024YFE0209000), the NSFC (Grant Nos.61932004, 62225203, U21A20516, 62302084, U2441237) and the Beijing Municipal Science and Technology Commission and Zhongguancun Science Park Management Committee (Z231100007423003).

References

1. Indyk, P., Motwani, R.: Approximate nearest neighbors: towards removing the curse of dimensionality. In: Proceedings of STOC, pp. 604–613 (1998)
2. Lakshman, V., et al.: Embracing structure in data for billion-scale semantic product search, n.d. Unpublished manuscript
3. Li, S., et al.: Embedding-based product retrieval in taobao search. In: Proceedings of the 27th ACM SIGKDD Conference on Knowledge Discovery and Data Mining (2021)
4. Nigam, P., et al.: Semantic product search. In: Proceedings of the 25th ACM SIGKDD International Conference on Knowledge Discovery and Data Mining (2019)
5. Wu, Z., Liu, Y., Shi, X., Zhao, X., Wang, Y., Zhang, G.: GTGNN: global graph and taxonomy tree for graph neural network session-based recommendation. In: WISA 2024, pp. 29–40 (2024)
6. Huang, P.-S., He, X., Gao, J., Deng, L., Acero, A., Heck, L.: Learning deep structured semantic models for web search using clickthrough data. In: Proceedings of the 22nd ACM International Conference on Information and Knowledge Management (2013)
7. Liu, Y., et al.: Pre-trained language model for web-scale retrieval in baidu search. In: Proceedings of the 27th ACM SIGKDD Conference on Knowledge Discovery and Data Mining (2021)
8. Gao, Y., et al.: Retrieval-augmented generation for large language models: a survey (2023)
9. Zhao, P., et al.: Retrieval-augmented generation for ai-generated content: a survey (2024)
10. Xu, D., Li, X., Zhang, Z., et al.: Harnessing large language models for knowledge graph question answering via adaptive multi-aspect retrieval-augmentation. In: Proceedings of the AAAI Conference on Artificial Intelligence, vol. 39, pp. 25570–25578 (2025)
11. Zhang, Q., Xu, S., Chen, Q., et al.: Vbase: unifying online vector similarity search and relational queries via relaxed monotonicity. In: 17th USENIX Symposium on Operating Systems Design and Implementation (OSDI 23), pp. 377–395 (2023)
12. Pinecone. Pinecone vector database. https://www.pinecone.io/. Accessed 12 June 2025
13. Wang, J., et al.: A purpose-built vector data management system. In: Proceedings of SIGMOD 2021, pp. 2614–2627 (2021)
14. Weaviate. Weaviate vector database. https://weaviate.io/. Accessed 12 June 2025
15. Douze, M., Guzhya, A., Deng, C., et al.: The faiss library (2024)
16. Gupta, G., Yi, J., Coleman, B., et al.: Caps: a practical partition index for filtered similarity search (2023)
17. Gollapodi, S., Karia, N., Sivashankar, V., et al.: Filtered-diskann: graph algorithms for approximate nearest neighbor search with filters. In: Proceedings of the ACM Web Conference 2023, pp. 3406–3416 (2023)
18. Wang, M., Lv, L., Xu, X., et al.: An efficient and robust framework for approximate nearest neighbor search with attribute constraint. Adv. Neural Inf. Process. Syst. **36** (2024)
19. Zuo, C., Qiao, M., Zhou, W., et al.: Serf: Segment graph for range-filtering approximate nearest neighbor search. In: Proceedings of the ACM on Management of Data, vol. 2, no. 1, pp. 1–26 (2024)

20. Patel, L., Kraft, P., Guestrin, C., et al.: Acorn: performant and predicate-agnostic search over vector embeddings and structured data. In: Proceedings of the ACM on Management of Data, vol. 2, no. 3, pp. 1–27 (2024)
21. Morrison, D.R.: Patricia-practical algorithm to retrieve information coded in alphanumeric. J. ACM (JACM) **15**(4), 514–534 (1968)
22. Malkov, Y.A., Yashunin, D.A.: Efficient and robust approximate nearest neighbor search using hierarchical navigable small world graphs. IEEE Trans. Pattern Anal. Mach. Intell. **42**(4), 824–836 (2020)
23. Cai, Y., et al.: Navigating labels and vectors: a unified approach to filtered approximate nearest neighbor search. In: Proceedings of the ACM on Management of Data, vol. 2, no. 6, pp. 1–27 (2024)
24. Subramanya, S.J., Devvrit, E., Simhadri, H.V., Krishnawamy, R., Kadekodi, R.: Diskann: fast accurate billion-point nearest neighbor search on a single node. Adv. Neural Inf. Process. Syst. **32** (2019)

Efficient Dense Diverse Subgraph Search in Attributed Graphs

Luyao Gao, Sijin Wang, Yikai Zhang(✉), and Can Lu

Guangzhou University, Guangzhou, China
{gaoluyao,wsjin}@e.gzhu.edu.cn, {ykzhang,lucan}@gzhu.edu.cn

Abstract. Given a user-specified query vertex, the task of finding a cohesive subgraph containing the query, known as subgraph search, is fundamental in graph analytics and has numerous applications. In many real-world graphs, vertices are associated with attributes. Though subgraph search in such attributed graphs has been extensively studied, most existing approaches emphasize structural and attribute cohesiveness, often overlooking the representation of minority groups. When vertices carry sensitive attributes such as race, gender, or political opinion, this can lead to biased and unrepresentative subgraphs. Therefore, we investigate the problem of diverse subgraph search, which aims to identify a dense subgraph containing a query vertex while ensuring attribute diversity. Specifically, we formulate the problem of *densest at-least-$\boldsymbol{k}$ subgraph search* ($\mathsf{Dal}\vec{k}\mathsf{SS}$). Despite its NP-hardness, we develop two efficient approximation algorithms to address the problem. Experimental results on real-world datasets demonstrate the effectiveness and efficiency of our algorithms, underscoring the practical value of our study.

Keywords: Attributed graph · Diversity · Subgraph search

1 Introduction

With the widespread availability of graph data, graph analytics has attracted considerable attention from both research and industry communities. A fundamental problem in graph analytics is *subgraph search*, which aims to identify a cohesive subgraph containing a given query vertex. This problem has been extensively studied, particularly in the context of *personalized* graph analysis. However, existing subgraph search methods on *attributed graphs* [6,8,13] primarily emphasize attribute cohesiveness, often overlooking other important aspects of subgraph quality, such as attribute diversity.

In light of this, we propose to study the problem of *densest at-least-$\boldsymbol{k}$ subgraph search*, abbreviated as $\mathsf{Dal}\vec{k}\mathsf{SS}$, with the goal of understanding its computational hardness and designing practically efficient algorithms. Specifically, given an attributed graph G and a query vertex q, the $\mathsf{Dal}\vec{k}\mathsf{SS}$ problem seeks to identify the densest subgraph S^* that contains q and satisfies the constraint that, for each attribute value, the number of vertices with that attribute value is no less than the corresponding lower bound specified by the integer vector $\boldsymbol{k}$.

S. Yang et al. (Eds.): WISA 2025, LNCS 16299, pp. 312–319, 2026.
https://doi.org/10.1007/978-981-95-6426-2_27

Dal$\overrightarrow{k}$SS has the potential to support a variety of applications, including organizational psychology [4] and mitigation of polarization [2]. For instance, in a social network where each user is labeled with a particular skill, Dal$\overrightarrow{k}$SS can assist a user q in forming a cohesive work team centered around them. The resulting team not only maintains structural cohesiveness, but also ensures that a sufficient number of members possess the required skills, an aspect that can positively impact overall team performance.

Our main contributions are as follows:

(1) Unlike prior works that focus on either *attribute-cohesive* subgraph search or diverse subgraph *discovery*, we study the problem of dense diverse subgraph search, which jointly emphasizes attribute diversity and query-based subgraph search. We introduce a formulation called *densest at-least-**k** subgraph search* (Dal$\overrightarrow{k}$SS), which supports a variety of real-world applications. We also establish the NP-hardness of the problem.

(2) In view of the NP-hardness of Dal$\overrightarrow{k}$SS, we develop two approximation algorithms to solve the problem in polynomial time. Specifically, the first algorithm is based on linear programming and provides a $\frac{1}{3}$-approximation guarantee; that is, the density of the subgraph it returns is at least one third that of the optimal solution. The second algorithm adopts a greedy peeling strategy to significantly reduce computational overhead, though it sacrifices a provable guarantee on solution quality.

(3) We conducted experiments to verify the efficiency and effectiveness of our algorithms on real-world attributed graphs. The results show that both our approximation algorithms can achieve an approximation ratio higher than 0.8, indicating their effectiveness. In addition, compared with an integer-programming-based baseline, our second algorithm can achieve a two-orders-of-magnitude speedup, demonstrating its efficiency.

2 Related Work

The first line of research closely related to our work is attribute-cohesive subgraph *search*. Liu *et al.* [8] propose the *vertex-centric attributed community* problem on vertex-attributed graphs, which identifies k-truss-based subgraphs centered around query vertices and enforces attribute cohesiveness within the subgraph. Li *et al.* [6] study the *edge-attributed community search* problem, which aims to find a subgraph in an edge-attributed graph that contains the given query vertex and maximizes similarity among edge attributes. Wang *et al.* [13] develop algorithms for approximate community search based on an attributed subgraph model that incorporates a query-centric attribute cohesiveness metric within the k-core framework. It is important to note that these approaches prioritize attribute cohesiveness (*i.e.*, similarity), whereas our work centers on attribute diversity, a fundamentally different objective.

Another line of research related to our work is cohesive diverse subgraph *discovery* which aims to identify qualified subgraphs *globally* without any query constraints. Anagnostopoulos *et al.* [1] devise a spectral algorithm to address the *fair densest subgraph* problem, which is to identify a densest subgraph in which the vertices are evenly distributed across the two values of a binary attribute. Miyauchi *et al.* [9] propose two notions of dense diverse subgraph: the *densest diverse subgraph*, which places an upper bound on the maximum fraction of monochromatic vertices, and the *densest at-least-$\boldsymbol{k}$ subgraph* ($\mathsf{Dal}\vec{k}\mathsf{S}$), which uses absolute count lower bounds to ensure that no attribute value is overlooked. In addition, they develop efficient algorithms to find these subgraphs with provable guarantees. Our formulation of dense diverse subgraph follows the $\mathsf{Dal}\vec{k}\mathsf{S}$ model. Chowdhary *et al.* [3] incorporate *k-core* structural constraints and model attribute diversity using the Jaccard similarity metric. They further develop an algorithm within a branch-and-bound framework, enhanced by effective pruning strategies. Note that all these works are fundamentally different from ours in that they focus on discovering dense diverse subgraphs globally within the graph, whereas our work aims to identify a local dense diverse subgraph centered around a given query vertex, *i.e.*, subgraph discovery versus subgraph search.

3 The Preliminaries

Let $\mathbb{C}$ be a set of colors, where each color represents an abstraction of certain attributes, such as gender (*e.g.*, male/female) or sentiment (*e.g.*, positive/negative). Let $G = (V, E, \ell)$ be an undirected vertex-attributed graph, where (1) V and E denote the sets of vertices and edges in G, respectively; and (2) $\ell : V \to \mathbb{C}$ is a function that assigns a color from $\mathbb{C}$ to each vertex in V. Given a subset $S \subseteq V$, we use $\mathsf{deg}_S(v)$ to denote the degree of vertex v within S, *i.e.*, the number of neighbors of v that belong to S. In addition, for a color $c \in \mathbb{C}$, we use the notation S_c to denote the set of vertices in S associated with color c, that is, $S_c = \{v \in S \mid \ell(v) = c\}$. The subgraph induced by S, denoted $G(S)$, is defined as $G(S) = (S, E(S))$, where $E(S) = \{(u, v) \in E \mid u \in S, v \in S\}$ is the set of all edges with both endpoints in S. The *density* of the induced subgraph, denoted $d(S)$, is defined as $d(S) = \frac{|E(S)|}{|S|}$, which reflects the cohesiveness of the vertex set S. We use boldface letters to denote vectors (*e.g.*, $\boldsymbol{x}$, $\boldsymbol{y}$, $\boldsymbol{k}$) and use regular font to denote individual elements of a vector (*e.g.*, x_e, y_v, k_c).

Following [9], we define a diverse cohesive subgraph using an integer vector $\boldsymbol{k} = (k_c)_{c\in\mathbb{C}}$, which specifies a lower bound on the number of vertices required for each color. Specifically, for each color $c \in \mathbb{C}$, k_c denotes the minimum number of vertices of color c that a valid subgraph must contain. To quantify the cohesiveness of a subgraph $G(S)$ induced by a vertex subset $S \subseteq V$, we use the density measure $d(S)$. Formally, we focus on the following *densest at-least-$\boldsymbol{k}$ subgraph search* problem, referred to as $\mathsf{Dal}\vec{k}\mathsf{SS}$.

Definition 1 ($\mathsf{Dal}\vec{k}\mathsf{SS}$). *Given an attributed graph $G = (V, E, \ell)$, a query vertex $q \in V$, and a color-wise lower bound vector $\boldsymbol{k} = (k_c)_{c\in\mathbb{C}}$, find a subset of vertices $S \subseteq V$ such that (1) $q \in S$; (2) $|S_c| \geq k_c$ for any $c \in \mathbb{C}$, where $S_c = \{v \in S \mid \ell(v) = c\}$ denotes the set of vertices in S with color c; and (3) the density $d(S) = \frac{|E(S)|}{|S|}$ of S is maximized.*

Algorithm 1: LP-Based Algorithm

Input: $G = (V, E, \ell)$, q, $\boldsymbol{k} = (k_c)_{c\in\mathbb{C}}$
Output: an approximate solution $S \subseteq V$

1 **for** *each* $\boldsymbol{h}$ *such that* $\boldsymbol{k} \leq \boldsymbol{h} \leq (|V_c|)_{c\in\mathbb{C}}$ **do**
2 Solve $\mathsf{LP}(\boldsymbol{h})$ and let $(\boldsymbol{x}^*, \boldsymbol{y}^*)$ be its optimal solution;
3 Compute candidates $\mathsf{Cand}(\boldsymbol{h}) \leftarrow \{\,\mathsf{Feasible}(S(r)) \mid r \in \{y_v^* \mid v \in V\} \cup \{0\}\}$;
4 Choose the best candidate $\mathsf{Best}(\boldsymbol{h}) \leftarrow \mathrm{argmax}_{S\in\mathsf{Cand}(\boldsymbol{h})} d(S)$;
5 **return** the one among $\mathsf{Best}(\boldsymbol{h})$'s that maximizes density;

6 **Procedure** $\mathsf{Feasible}(S)$**:**
7 **for** *each* $c \in \mathbb{C}$ **do**
8 **while** $|S_c| < k_c$ **do**
9 Select $v \in V_c \setminus S_c$ that has the maximum number of neighbors in S;
10 $S \leftarrow S \cup \{v\}$;
11 **return** S;

The following theorem characterizes the hardness of $\mathsf{Dal}\vec{k}\mathsf{SS}$. Its proof, along with the proofs of other lemmas and theorems, can be found in the full version.

Theorem 1. *The* $\mathsf{Dal}\vec{k}\mathsf{SS}$ *problem is NP-hard.*

4 Approximation Algorithms

Since $\mathsf{Dal}\vec{k}\mathsf{SS}$ is NP-hard, we focus on approximate solutions. Specifically, we first design a linear programmingâĂŞbased algorithm that offers an approximation guarantee but incurs high computational cost. To improve efficiency, we then propose a faster heuristic algorithm, albeit without a formal approximation bound.

4.1 A Linear-Programming-Based Algorithm

Let $\boldsymbol{h} = (h_c)_{c\in\mathbb{C}}$ be a vector satisfying $\boldsymbol{h} \geq \boldsymbol{k}$, *i.e.*, $h_c \geq k_c$ for $\forall c \in \mathbb{C}$. Consider the following linear program (LP) which introduces a variable x_e for each edge $e \in E$ and a variable y_v for each vertex $v \in V$.

$$\begin{aligned}
\mathsf{LP}(\boldsymbol{h}): \quad & \text{maximize} && \sum_{e\in E} x_e \\
& \text{subject to} && x_e \leq min\{y_u, y_v\} && \forall e = (u, v) \in E, \\
& && \sum_{v\in V_c} y_v = \frac{h_c}{\|\boldsymbol{h}\|_1} && \forall c \in \mathbb{C}, \\
& && y_q = \frac{1}{\|\boldsymbol{h}\|_1}, \quad y_v \leq \frac{1}{\|\boldsymbol{h}\|_1} && \forall v \in V, \\
& && x_e \geq 0, \; y_v \geq 0 && \forall e \in E, \; \forall v \in V
\end{aligned}$$

It is worth noting that the above LP formulation is fundamentally different from those for $\mathsf{Dal}\overrightarrow{k}\mathsf{S}$ [9] and $\mathsf{Dal}k\mathsf{S}$ [5], as it additionally enforces the inclusion of the query vertex q (to be explained later), which is an essential component of the solution. To derive the final algorithm, we next analyze the relationship between the solution of $\mathsf{LP}(\boldsymbol{h})$ and the optimal solution S^* of $\mathsf{Dal}\overrightarrow{k}\mathsf{SS}$.

We start with some notations. Let $(\boldsymbol{x}^*, \boldsymbol{y}^*)$ be an optimal solution to $\mathsf{LP}(\boldsymbol{h})$. For each $r \in \{y_v^* \mid v \in V\} \cup \{0\}$, define $S(r) = \{v \in V \mid y_v^* \geq r\}$. It is easy to verify the following properties of $S(r)$: (1) the number of such $S(r)$'s is finite given that $\{y_v^* \mid v \in V\} \cup \{0\}$ is a finite set; (2) $S(r_1) \subseteq S(r_2) \subseteq V$ holds whenever $r_1 \geq r_2$; and (3) $q \in S(r)$ always holds, as the LP formulation ensures $y_q^* \geq y_v^*$ for all v. We consider such sets $S(r)$ as candidate solutions to $\mathsf{Dal}\overrightarrow{k}\mathsf{SS}$. However, it is noteworthy that some $S(r)$ may not satisfy the cardinality constraints imposed by $\boldsymbol{k}$. To address this, we define a procedure $\mathsf{Feasible}(S(r))$, which augments $S(r)$ by adding vertices with underrepresented colors (see lines 6–11 of Algorithm 1 for more details). The output of $\mathsf{Feasible}(S(r))$ is a superset of $S(r)$ that fulfills all the cardinality requirements specified by $\boldsymbol{k}$.

Next, we establish a lemma that is key to the design of our final algorithm.

Lemma 1. *Let S^* be the optimal solution to $\mathsf{Dal}\overrightarrow{k}\mathsf{SS}$ and define the vector $\boldsymbol{h}^*$ such that $h_c^* = |S_c^*|$ for $\forall c \in \mathbb{C}$. Let $(\boldsymbol{x}^*, \boldsymbol{y}^*)$ be an optimal solution to $\mathsf{LP}(\boldsymbol{h}^*)$ and let λ^* denote the corresponding value of the objective function. Then there exists $S \in \{S(r) \mid r \in \{y_v^* \mid v \in V\} \cup \{0\}\}$ such that $q \in \mathsf{Feasible}(S)$ and $d(\mathsf{Feasible}(S)) \geq \frac{1}{3} \cdot d(S^*)$, where $d(\cdot)$ denotes the density function.*

We now present the complete algorithm. According to Lemma 1, if the optimal solution S^* to $\mathsf{Dal}\overrightarrow{k}\mathsf{SS}$ were known, we could invoke the above linear program using the corresponding h^* (*i.e.*, $\mathsf{LP}(\boldsymbol{h}^*)$) to obtain an approximate solution with provable guarantee. However, since S^* is not available a priori, we instead enumerate all integer vector $\boldsymbol{h}$ such that $\boldsymbol{k} \leq \boldsymbol{h} \leq (|V_c|)_{c \in \mathbb{C}}$. For each such $\boldsymbol{h}$, we compute candidate solutions via $\mathsf{Feasible}(S(r))$ and evaluate their densities. Among all candidates, the one with the optimal density is selected as the final result. See Algorithm 1 for implementation details.

We have the following theorem regarding Algorithm 1 for $\mathsf{Dal}\overrightarrow{k}\mathsf{SS}$.

Theorem 2. *Given the constraint vector $\boldsymbol{k}$, the time complexity of Algorithm 1 is $\mathcal{O}\left(T_{LP} \cdot \prod_{c \in \mathbb{C}}(|V_c| - k_c)\right)$, where T_{LP} denotes the time required to solve the linear program $\mathsf{LP}(\boldsymbol{h})$. Furthermore, Algorithm 1 guarantees a $\frac{1}{3}$-approximation.*

4.2 A Hybrid Algorithm

One major drawback of Algorithm 1 is its requirement to solve a polynomial number of linear programming (LP) instances, incurring prohibitive computational cost. To address this issue, we adopt a simple yet effective algorithm based on a greedy peeling strategy. Specifically, starting with $S = V$, the algorithm iteratively removes the vertex (excluding the query vertex q) with the

Algorithm 2: Hybrid Algorithm

Input: $G = (V, E, \ell)$, q, $\boldsymbol{k} = (k_c)_{c \in \mathbb{C}}$
Output: an approximate solution $S \subseteq V$

```
  // invoke the peeling process to get the first candidate solution
1 S_1 ← Peeling(G, q, k);
  // invoke the LP process to compute another candidate solution
2 S_2 ← LP(k);
3 return argmax_{S∈{S_1,S_2}} d(S);
4 Procedure Peeling(G, q, k):
5     let S^(n) ← V and i ← n;
6     while |S_c^(i)| > k_c for ∀c ∈ C with k_c ≥ 1 do
7         v* ← arg min_{v∈S^(i)\{q}} deg_{S^(i)}(v);
8         update S^(i−1) ← S^(i) \ {v*} and set i ← i − 1;
9     return argmax_{S∈{S^(n),...,S^(i)}} d(S);
```

lowest degree $deg_S(v)$ from the current subgraph S. This process continues until the cardinality constraints for each color may no longer be satisfied. Among all intermediate subgraphs encountered during this peeling process, the one with the highest density, denoted S^+, is selected as the final output. See lines 4–9 of Algorithm 2 for details. We have the following theoretical guarantee for S^+.

Lemma 2. *Given the constraint vector* $\boldsymbol{k}$, *let* S^* *be the optimal solution to* Dal$\vec{k}$SS. *If* $|S_c^*| > k_c$ *for* $\forall c \in \mathbb{C}$ *with* $k_c \geq 1$, *then the final result* S^+ *of the peeling process satisfies* $d(S^+) \geq \frac{1}{2} \cdot d(S^*)$, *where* $d(\cdot)$ *is the density function.*

Algorithm 2 presents our second algorithm. It starts by invoking the peeling procedure to generate a candidate solution S_1. Note that S_1 is not guaranteed to be a $\frac{1}{2}$-approximation unless the optimal solution S^* satisfies $|S_c^*| > k_c$ for $\forall c \in \mathbb{C}$ with $k_c \geq 1$, as shown in Lemma 2. To improve the quality of the result, the algorithm additionally solves LP($\boldsymbol{k}$) to obtain another candidate solution S_2. The algorithm then returns the better of S_1 and S_2 as the final output. We emphasize that the existence of a formal approximation guarantee for this algorithm remains an open question.

Theorem 3. *Given the constraint vector* $\boldsymbol{k}$, *Algorithm 2 runs in* $\mathcal{O}(T_{LP})$ *time where* T_{LP} *is the time complexity to solve linear program* LP($\boldsymbol{k}$).

5 Experiments

We conducted experiments to evaluate our algorithms. The experiments were carried out on a server with Intel i9-13900K CPU and 128GB of main memory. Linear programs and integer programs are solved with SciPy and HiGHS. We used four real-world graph datasets: Amazon [10], Deezer [12], GitHub [11],

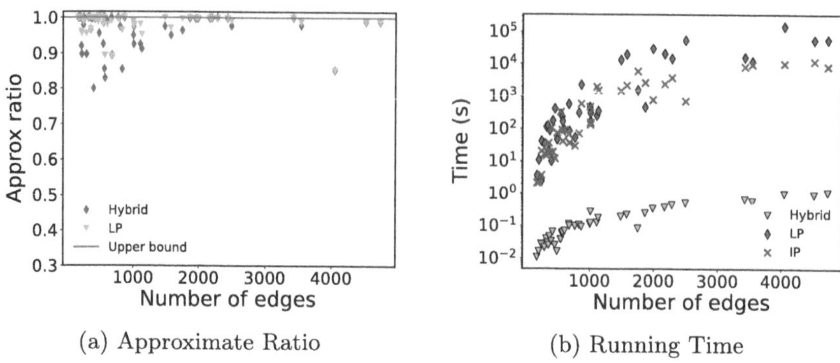

(a) Approximate Ratio

(b) Running Time

Fig. 1. Comparison of the Quality of Subgraphs Returned by IP, LP, and Hybrid

and DBLP. Due to space limitations, we primarily present the results on Amazon, which comprises 299 graphs. In each graph, vertices represent Amazon products, and edges indicate co-purchasing relationships between them. Each graph contains exactly two product categories, and each vertex is associated with exactly one of these categories. Details of the remaining datasets and additional results are provided in the full version of the paper.

We compared the following algorithms in our experiments: (1) IP: an exact integer-programming-based algorithm (see the full version for details), (2) LP: the linear-programming-based algorithm (Sect. 4.1), and (3) Hybrid: the hybrid algorithm that capitalizes on the greedy peeling strategy (Sect. 4.2). By default, the cardinality threshold for each color is set to $k_c = 5$.

We compared the densities of the subgraphs returned by IP, LP, and Hybrid. To allow IP to complete within a reasonable time and serve as a ground-truth baseline for optimal density, we selected small graphs from the Amazon dataset. Specifically, following [9], we used graphs containing no more than 1,000 vertices and 5,000 edges. Figure 1a shows the approximation ratios of LP and Hybrid for each tested graph. The approximation ratio of IP is always 1 and thus is omitted. As observed, LP consistently outperforms Hybrid, owing to its stronger theoretical guarantee. Nevertheless, Hybrid still achieves an empirical approximation ratio exceeding 0.8 across all tested instances, indicating its practical effectiveness despite the absence of a formal approximation bound.

Figure 1b reports the running time of IP, LP and Hybrid on graphs from the Amazon dataset containing no more than 1,000 vertices and 5,000 edges. As observed, Hybrid consistently achieves the best performance among all algorithms. In particular, it is approximately two orders of magnitude faster than both LP and IP in all instances tested.

6 Conclusion

We formulate the problem of *densest at-least-$\boldsymbol{k}$ subgraph search* (Dal$\overrightarrow{k}$SS) and investigate its computational complexity, establishing its NP-hardness. To address $\mathsf{Dal}\overrightarrow{k}\mathsf{SS}$ efficiently, we develop two approximation algorithms, including a linear-programming-based method with provable guarantees and a hybrid algorithm Hybrid that balances efficiency and effectiveness. Experimental results on real-world datasets demonstrate that Hybrid achieves strong empirical performance in terms of both solution quality and running time, making it a practical solution for diverse subgraph search in attributed graphs.

Acknowledgements. This research was funded by the National Natural Science Foundation of China (Grant Nos. 62002236 and 62402127) and the Science and Technology Projects of Guangzhou, China (Grant Nos. 2024A04J4301 and 2025A04J4166).

References

1. Anagnostopoulos, A., Becchetti, L., Fazzone, A., Menghini, C., Schwiegelshohn, C.: Spectral relaxations and fair densest subgraphs. In: CIKM (2020)
2. Boxell, L., Gentzkow, M., Shapiro, J.M.: Is the internet causing political polarization? Evidence from demographics. Technical report, National Bureau of Economic Research (2017)
3. Chowdhary, A.A., Liu, C., Chen, L., Zhou, R., Yang, Y.: Finding attribute diversified communities in complex networks. In: DASFAA (2020)
4. Horwitz, S.K.: The compositional impact of team diversity on performance: theoretical considerations. Human Res. Dev. Rev. **4**(2) (2005)
5. Khuller, S., Saha, B.: On finding dense subgraphs. In: ICALP (2009)
6. Li, L., Zhao, Y., Luo, S., Wang, G., Wang, Z.: Efficient community search in edge-attributed graphs. TKDE **35**(10) (2023)
7. Lin, L., Ouyang, D., He, Z., Li, C.: Parallelize accelerated triangle counting using bit-wise on GPU. In: WISA (2023)
8. Liu, Q., Zhu, Y., Zhao, M., Huang, X., Xu, J., Gao, Y.: Vac: vertex-centric attributed community search. In: ICDE (2020)
9. Miyauchi, A., Chen, T., Sotiropoulos, K., Tsourakakis, C.E.: Densest diverse subgraphs: how to plan a successful cocktail party with diversity. In: SIGKDD (2023)
10. Ni, J., Li, J., McAuley, J.: Justifying recommendations using distantly-labeled reviews and fine-grained aspects. In: EMNLP-IJCNLP (2019)
11. Rozemberczki, B., Allen, C., Sarkar, R.: Multi-scale attributed node embedding. J. Complex Netw. **9**(2) (2021)
12. Rozemberczki, B., Sarkar, R.: Characteristic functions on graphs: birds of a feather, from statistical descriptors to parametric models. In: CIKM (2020)
13. Wang, Y., et al.: Scalable community search with accuracy guarantee on attributed graphs. In: ICDE (2024)

Compression Scope: An Interactive Demonstration for Visualizing Performance Trade-Offs of Compression Algorithms

Benzhao Tang[1], Shiyu Yang[1](✉), and Zhitao Shen[2]

[1] Guangzhou University, Guangzhou 510006, China
benzhaotang@e.gzhu.edu.cn, syyang@gzhu.edu.cn
[2] Ant Group, Hangzhou 310013, China
zhitao.szt@antgroup.com

Abstract. The rapid expansion of big data demands compression algorithms that balance efficiency and effectiveness. However, selecting the appropriate algorithm is challenging due to inherent trade-offs between speed and compression ratio, further complicated by varying hardware environments. This paper presents Compression Scope, an interactive system for exploring these trade-offs through benchmarking, visualization, and analysis. The system can run predefined compression tasks on heterogeneous servers, demonstrate results through intuitive charts and highlight Pareto-optimal algorithms for data-driven decision-making, supporting aggregation, trend exploration, and exportable plots. By integrating automated benchmarking with interactive trade-off analysis, it empowers users to identify compression algorithms suited to their specific performance and resource constraints.

Keywords: Data Compression · Performance Trade-offs · Visualization

1 Introduction

In the era of big data, vast amounts of data are generated at an astonishing rate [1]. The massive scale of these data introduces significant computational, transmission, and storage overheads for their collection and storage. Data compression algorithms are regarded as a crucial method for efficiently organizing and managing large-scale data while reducing resource expenditure, and thus are widely applied. For instance, Zstd [2] is utilized by AWS Redshift and Facebook's RocksDB for field compression, and LZ4 [3] is widely used in Apache Hadoop. To enhance compression performance on specific data types, numerous specialized compression algorithms have been proposed, such as algorithms for floating-point numbers [4], logs [5], and machine-generated data [6].

Data compression algorithms reduce substantial redundancy within the data by identifying and exploiting its internal characteristics, representing the original data in a more compact format. When the original data are needed,

S. Yang et al. (Eds.): WISA 2025, LNCS 16299, pp. 320–323, 2026.
https://doi.org/10.1007/978-981-95-6426-2_28

they are recovered through a decompression process. However, compression and decompression processes inevitably introduce additional computational overhead and latency. Generally, as a compression algorithm expends more resources to uncover more optimal data features, it achieves a higher compression ratio but at the cost of lower compression and decompression speeds. Concurrently, compression and decompression speeds are influenced by hardware performance and the operating environment. Therefore, a significant challenge arises in how to make trade-offs among the three most critical metrics: compression ratio, compression speed, and decompression speed, to select an algorithm that suits the specific needs of enterprises and users.

This demonstration introduces Compression Scope, a system designed to assist enterprises and users in selecting suitable compression algorithms by enabling them to conveniently benchmark, visualize, and analyze the performance of multiple algorithms under various hardware conditions, thereby facilitating an efficient evaluation of performance trade-offs to meet their specific requirements. Compression Scope is composed of three principal modules: Execute Compressions (EC), Performance Visualization (PV), and Pareto-Optimality (PO).

EC facilitates the benchmarking of user-uploaded datasets by executing a series of predefined compression tasks across heterogeneous hardware environments and subsequently storing the resulting performance data in a database. **PV** enables users to query stored compression results and renders these data as intuitive, interactive visualizations that include features for data aggregation, toggling between detailed and average views, trend analysis, and export of chart. **PO** plots the performance trade-offs for each algorithm in a chart of speed versus compression ratio, highlighting the algorithms that lie on the Pareto frontier.

2 System Design

2.1 Overview

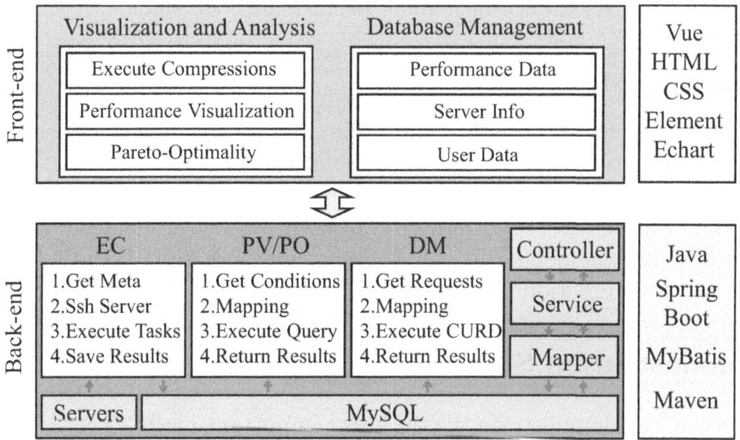

Fig. 1. System Overview.

Compression Scope is a user-friendly and comprehensive framework designed to evaluate the performance of various compression algorithms across multiple dimensions and under different hardware conditions. As shown in Fig. 1, the system adopts a decoupled front-end/back-end architecture. The front-end, implemented with Vue, HTML, and CSS, employs Element UI and ECharts for visualization and interaction, handling user requests and rendering back-end responses. The back-end, developed with Spring Boot in a three-tier structure (Controller, Service, Mapper), processes requests, executes services, and interfaces with the database. Compression Scope comprises one core component, Visualization and Analysis (VA), and one foundational component, Database Management (DM). Next, we elaborate on each component below.

2.2 Visualization and Analysis

The VA component, central to Compression Scope, comprises three core modules: Execute Compressions (EC), Performance Visualization (PV), and Pareto-Optimality (PO).

The EC module parses uploaded datasets, extracts metadata (e.g., size, row count, samples), and stores them in the database. It combines metadata with predefined compression algorithms to generate tasks (e.g., applying multiple algorithms on different hardware) and dispatches them via SSH to remote servers. Upon completion, results are stored in the `Compression_Results` table.

The PV module translates user-defined conditions into database queries, retrieves matching performance data, and renders bar charts for compression ratio and speed, with switchable axes between algorithms and datasets. It supports aggregation into average-value charts, trend filtering via sliders, and chart data downloads.

The PO module queries relevant performance data to plot an "Average Compression Performance Trade-offs" chart, where each algorithm is a circle positioned by compression speed (x-axis) and ratio (y-axis), sized by decompression speed. Algorithms on the Pareto frontier are automatically highlighted, and sliders allow filtering based on user-defined performance thresholds.

2.3 Database Management

The DM component encapsulates database operations and offers user-friendly management functions via the interface, enabling efficient querying and management of tables in the Compression Scope database (e.g., dataset metadata, compression results, user information). It translates user requests into corresponding CRUD statements through Mapper interfaces and MyBatis XML files, ensuring secure interaction with the MySQL database.

3 Demonstration Scenario

In this section, we demonstrate the two most important modules of Compression Scope: Performance Visualization (PV) and Pareto-Optimality (PO). We use the

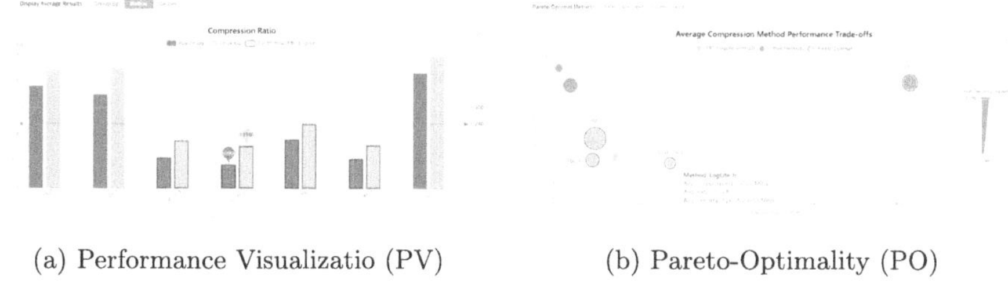

(a) Performance Visualizatio (PV) (b) Pareto-Optimality (PO)

Fig. 2. Visualization and Analysis of Compression Scope.

EC module to execute multiple stream compression algorithms on the Apache.log and Linux.log datasets on Server 1 and Server 2, and the compression results are stored in the database.

As shown in Fig. 2 a, the PV module plots the compression ratio results as a bar chart. This chart allows quick identification of the algorithm with the optimal compression ratio, which is LogLite-b. The average compression ratios for the two datasets are 0.300 and 0.240, respectively, and the specialized algorithms all perform better than the average. The toggle switch in the upper-left corner can be used to calculate the average compression ratio for each algorithm. The "Group By" option allows for switching the x-axis. The bar chart for speed is similar, with the distinction that it can differentiate speeds on different servers.

As depicted in Fig. 2 b, the PO module highlights Pareto-optimal algorithms, namely PBC, PBC-F, LogLite-b and FSST, with red circles. It is evident that PBC emphasizes decompression speed, LogLite-b prioritizes compression ratio, and FSST focuses on compression speed. Using the dropdown menu at the top, users can select the combination of metrics to consider for the Pareto frontier analysis. The three sliders enable users to set ranges to filter the algorithms displayed in the chart.

References

1. Worldwide IDC Global DataSphere Forecast Update, 2025–2029: Regional Analysis. https://my.idc.com/getdoc.jsp?containerId=US53383425, Accessed 01 Aug 2025
2. Yann, C., Murray, K.: Zstandard compression and the application/zstd media type (2018)
3. Collet, Y.: LZ4: Fast Compression Algorithm (2011). https://github.com/lz4/lz4, Accessed 01 Aug 2025
4. Li, R., Li, Z., Wu, Y., Chen, C., Zheng, Y.: Elf: erasing-based lossless floating-point compression. In: Proceedings of the VLDB Endowment, vol. 16, no. 7, pp. 1763–1776. VLDB Endowment (2023)
5. Tang, B., Yang, S., Shen, Z., Zhang, W., Lin, X., Tian, Z.: LogLite: lightweight plug-and-play streaming log compression. In: Proceedings of the VLDB Endowment, vol. 18, no. 11, pp. 3757–3770. VLDB Endowment (2025)
6. Zhang, J., et al.: High-ratio compression for machine-generated data. In: Proceedings of the ACM on Management of Data, vol. 1, no. 4, pp. 1–27. ACM (2023)

Demonstration of Memory Database Hot Loading System

ZiJing Wang[1], Yibo Zhao[1], Zeng Chun[1(✉)], WenKui Zheng[1], and Zhang Yong[2]

[1] School of Software, Henan University, Kaifeng 475004, Henan, China
zengchun@henu.edu.cn

[2] National Research Center for Information Science and Technology, Tsinghua University, Beijing 100084, China

Abstract. Loading data files into the in-memory database has become one of the bottlenecks affecting service continuity. Our demonstration showcases a hot-loading technique for in-memory databases: upon start-up, the system only creates the memory-mapped address space for the in-memory table space files, bypassing the traditional step of loading the entire data files from disk into memory. The database becomes ready for service within seconds while background threads continue to load the actual data pages in parallel.The demo will replay the complete hot-loading workflow and present a side-by-side comparison of startup latency and transaction throughput with and without hot-loading enabled.

Keywords: In-memory Database · Hot Loading · Data Dumping

1 Introduction

In today's era of Web, big data and real-time computing, the time of loading data files into in-memory databases has become a critical bottleneck for service continuity [1].

To mitigate this bottleneck, the industry typically relies on techniques such as hot-standby failover or incremental loading [2, 3], yet these approaches still suffer from high resource redundancy and complex data-consistency maintenance. In large-scale data scenarios, the fundamental issue of cold-start latency remains unresolved.

This paper presents and demonstrates an innovative hot-loading technique that overturns the conventional startup paradigm. Under a specific parameter mode, the database merely creates the memory-mapped address space for each table space file, bypassing the conventional step of loading data from disk into memory; the data files are then populated in parallel by background threads while the system is already serving clients.The breakthrough lies in transforming the startup behavior from an "active full load" to a "passive on-demand loading," enabling sub-second startup times. This advance is revolutionary for fault recovery, rolling upgrades, and disaster-recovery failovers alike [4].

Our demonstration shows that with hot-loading enabled, the in-memory database starts up in a fraction of the time on hundreds of gigabytes of data, while TPC-C transaction throughput rises significantly.

S. Yang et al. (Eds.): WISA 2025, LNCS 16299, pp. 324–327, 2026.
https://doi.org/10.1007/978-981-95-6426-2_29

2 System Overview

The architecture diagram for hot loading is shown in Fig. 1.When the database is running normally, the background dump process writes the current PCH (Page Control Header) information, including page number and data file number, into a dump file on disk at regular intervals according to a priority policy. When the database restarts, the dump file will be read. In memory mode, the physical pages corresponding to the dump file will be loaded into the buffer in priority, and then the remaining pages will be fully loaded. In disk mode, the physical pages corresponding to the dump file will be loaded into the buffer accordingly.

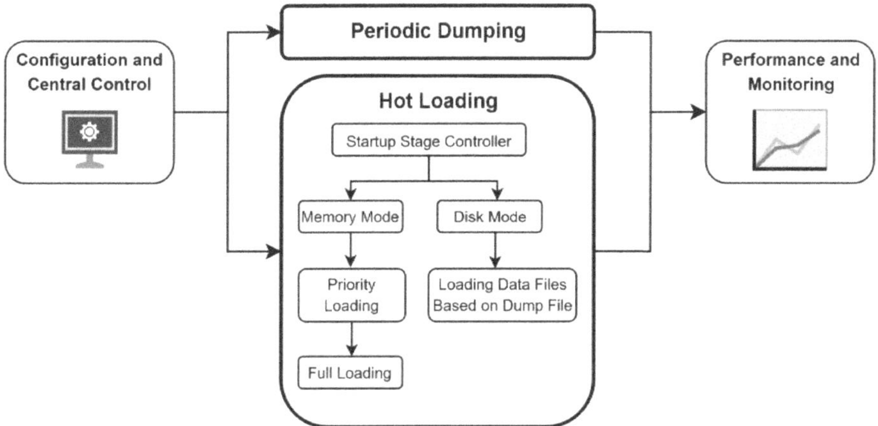

Fig. 1. Hot-loading Architecture

Our hot loading technology only creates the mapping space for the data files of the memory table space in a specific parameter mode during the original startup process, skipping the process of loading the data files from disk to memory, achieving the goal of rapid database startup. Due to the fact that the data files of the memory table space were not actually loaded into memory, but the space for the memory mapping area and PCH array has already been allocated and created. Therefore, in order to avoid changing the usage logic of the memory table space, it is still a normal process to allocate page IDs and initialize the structure and information of PCH in the mapping area. Moreover, it is not necessary to assign PCH information during the initialization process. After the page is actively or passively loaded into memory, the information of PCH is then filled in.

3 Demonstration Overview

To demonstrate the complete hot-loading process, we used the TPC-C stress testing tool and designed a demonstration comparing the database with and without hot- loading mode enabled. The specific steps include: (i) Start the database in normal mode and record the startup time. (ii) Enable the hot-loading feature for in-memory tables via database configuration parameters, start the database, and record the startup time. (iii) Run a TPC-C benchmark and capture the resulting graphs, backend dump process will generate dump files periodically. (iv) Restart the database and the physical pages corresponding to the dump file will be loaded into the buffer in priority. (v) Run the TPC-C benchmark again and capture the resulting graphs to compare transaction throughput. Our demo tool, as shown in Fig. 2, will be used to demonstrate the above steps.

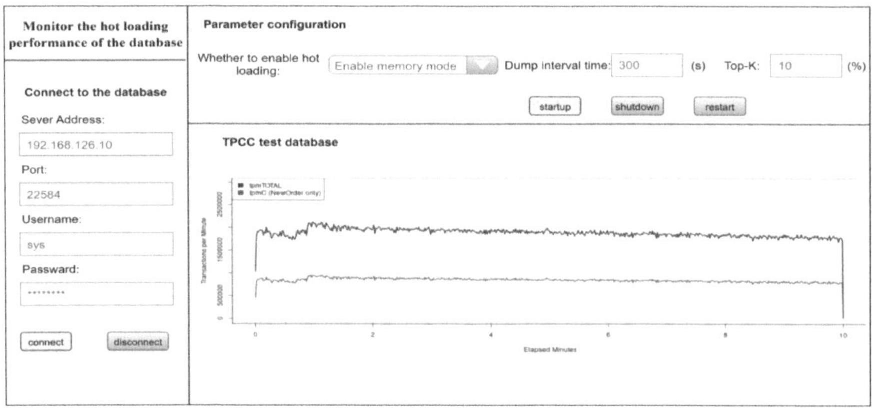

Fig. 2. Demo Tool

Startup Time: We measured the database startup time with and without hot-loading enabled, using 360 GB of data files. The average startup time without hot-loading is 73 s, and the average startup time with hot loading enabled is 4 s.

Transaction Throughput: Our testing shows that enabling hot loading during TPC-C runs can increase tpmC (NewOrder only) and tpm TOTAL by 2.5 times.

4 Technical Details

4.1 Periodic Dumping

When the database is running normally, the background process writes the current control header information (PCH for memory mode), including page number and data file number, into the dump file at intervals determined by the user. The database reads the dump file upon restart and schedules the corresponding physical page to load it into the buffer. This process can achieve a certain degree of performance improvement during the cold start phase.

During the dumping, the TopK quick select algorithm is used to calculate the Kth largest value of the user specified ratio K based on the number of page accesses. All page information larger than this value is dumped. At the same time, the user can also determine a minimum allowed access value. If the Kth largest value calculated during this dumping is smaller than this minimum access value, the dump will be canceled.

4.2 Backend Loading

When the database startup is completed, the background hot-loading process loads the hot data pages based on the dump files generated during the dump phase. Hot-loading processing can improve system performance during the initial run phase after database restart, enabling the database system to provide highly available services faster.

We change their original memory loading mode and load the memory tables that were originally located during the database startup process after the database startup. This solution can significantly accelerate the database startup speed and use a priority loading method to load the most frequently accessed data pages in the memory tables, allowing the database system to improve available services faster. After the priority loading is completed, we perform a full disk memory table loading to improve subsequent services.

After a quick startup, in order to quickly restore the database to its normal state, the database will synchronously create a thread for automatic loading of memory pages. At the same time, in order to prioritize the performance of the database, the design hopes to load more commonly used pages into memory first. Therefore, the entire automatic loading process is divided into two stages. The first stage is the priority loading of hot data, and the second stage is the sequential loading of all data.

During the process of loading a page into memory due to the need to use certain page information for database transaction execution, the page may not have been actually loaded into memory yet, so it is necessary to determine the status of the page. If the page is not yet loaded in memory, it needs to be loaded first. The loading process adopts multi-threaded parallel loading, and to prevent duplicate loading and content overwriting, the page status is checked and locked before loading.

References

1. Lake, P., Crowther, P.: In-Memory Databases (2013).https://doi.org/10.1007/978-1-4471-5601-7_8
2. Oriani, A., Garcia, I.C.: From backup to hot standby: high availability for HDFS. In: Reliable Distributed Systems. IEEE (2013).https://doi.org/10.1109/SRDS.2012.33
3. Ahn, M., Willhalm, T., May, N., et al.: An examination of cxl memory use cases for in-memory database management systems using sap hana. Proc. VLDB Endowment **17**(12), 3827–3840 (2024)
4. Yang, X., Wang, B., Li, C., Wang, J., Xie, X.: Efficient direct search on compressed genomic data. In: 2013 IEEE 29th International Conference on Data Engineering (ICDE), Brisbane, QLD, Australia, pp. 961–972 (2013). https://doi.org/10.1109/ICDE.2013.6544889

TaaSDB: A Transaction Management Service Platform Supporting Cross-Model Transactions

Yuan Teng, Weixing Zhou, Zeshun Peng, Jiacheng Li, Yiming Xu, and Yanfeng Zhang(✉)

Northeastern University, Shenyang 110100, Liaoning, China
zhangyf@mail.neu.edu.cn

Abstract. Modern applications increasingly require ACID transactions across multiple data models, but existing systems struggle with this due to the heterogeneity of data stores. Traditional approaches like XA and 2PC are too rigid for diverse models (relational, graph, key-value, etc.), forcing developers to implement complex consistency logic. This paper presents TaaSDB, a novel data platform that offers cross-model ACID transactions through a Transaction-as-a-Service abstraction. Key innovations include: 1) a scalable three-tier architecture, 2) intelligent conflict detection for atomicity, and 3) dynamic model mapping for unified operations. TaaSDB simplifies multistore usage as easily as a single database.

Keywords: Cloud-native Database · Transaction-as-a-Service · Transaction Processing · Heterogeneous data management · Cross-model transactions

1 Introduction

In modern enterprise systems, the growing complexity of business scenarios and data diversity has made hybrid data models standard practice. A typical e-commerce order workflow may combine relational databases for transactions, KV stores for inventory, and graph databases for user relationships [1]. While polyglot persistence optimizes individual component performance, it creates systemic challenges:(1) In cloud-native architectures, execution-storage decoupling fails to prevent quadratic growth of transaction coordination overhead with node scaling, while storage-layer computational constraints limit elastic scaling; (2) The lack of cross-engine atomic commit protocols compromises multi-model transactional consistency.

TaaSDB decouples transaction processing as an independent service, forming a three-layer architecture of execution, transaction, and storage. Unlike traditional cloud native databases that tightly integrate transaction processing with computation or storage [2], TaaSDB's decoupled transaction service prevents storage engine bottlenecks from computational resource constraints while minimizing coordination overhead during execution engine scaling. Each layer can scale independently, enhancing scalability.

S. Yang et al. (Eds.): WISA 2025, LNCS 16299, pp. 328–332, 2026.
https://doi.org/10.1007/978-981-95-6426-2_30

TaaSDB seamlessly integrates with diverse storage engines while guaranteeing atomicity in hybrid transactions through an optimized conflict detection mechanism. The system provides unified APIs that allow users to submit operations across different data engines within a single transaction, supporting multiple query languages. Through a dynamic model mapping mechanism, TaaSDB abstracts all data as <key, operation, value> triples, transforming heterogeneous operations into unified read-write sets for consistent processing.

By addressing scalability, atomicity, and operational challenges, TaaSDB enables efficient multi-model data management for demanding scenarios like financial transactions and real-time logistics.

2 System Overview

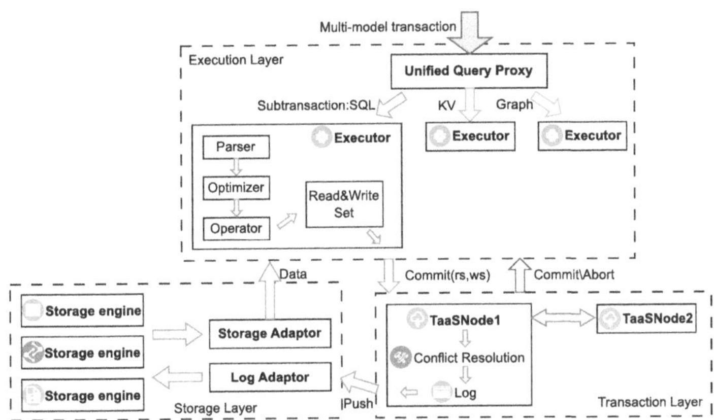

Fig. 1. TaaSDB architecture and overflow.

To support cross-model transactions, TaaSDB builds a three-layer decoupled architecture of execution, transaction, and storage, as shown in Fig. 1.

Execution Layer. The execution layer includes a Unified Query Proxy (UQP) and multiple execution engines. The UQP maintains cross-model transactions, distinguishes the commands sent by the user, and sends them to the appropriate execution engine for execution. After executing a statement, the execution engine converts it into a unified read-write set. When a transaction is committed, the UQP and the execution engines send the unified sets to the transaction layer for conflict resolution.

Transaction Layer. The Transaction Layer adopts the multi-master architecture and employs an Epoch-based Multi-master OCC algorithm [3,4]. It centrally resolves transaction conflicts while supporting parallel transaction processing across multi-master nodes. After conflict resolution, it sends logs to the Storage Layer for data updating.

Storage Layer. The Storage Layer integrates diverse storage nodes through adapter patterns. Storage adapters abstract structural differences, enabling transparent data access, while log record adapters convert generic operations into engine-specific formats for persistence.

3 Demonstration Description

In e-commerce scenarios, TaaSDB offers flexible configuration capabilities by utilizing Elastic Computing Service (ECS) through cloud APIs. As shown in Fig. 2, the three-layer architecture enables fine-grained layer-specific configurations. In the following, we demonstrate how to build TaaSDB for users:

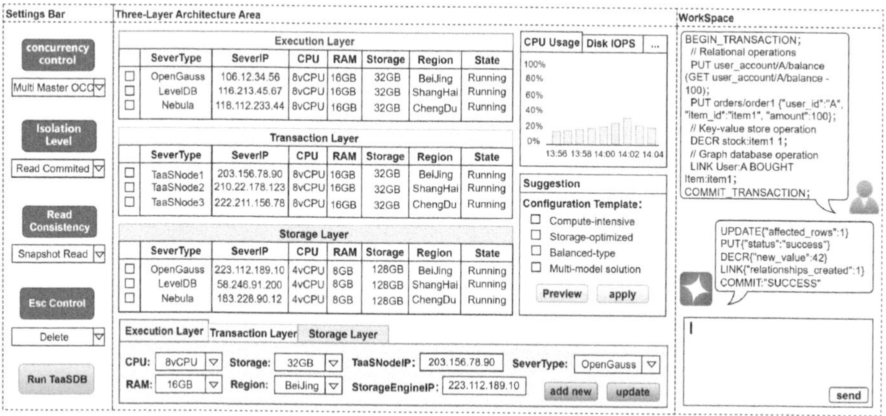

Fig. 2. TaaSDB SystemDemonstration.

Step 1: System Initialization and Architecture Configuration. During the initial configuration process, users select an appropriate combination of execution and storage engines (e.g., KV, SQL, Graph) in the Three-Layer Architecture Area based on the data characteristics of the business scenario. Each data engine supports independent resource configuration, allowing users to fine-tune parameters such as CPU cores, memory capacity, and storage space. This enables an optimal balance between resource utilization and cost efficiency.

Step 2: Transaction Policy Configuration. In the transaction layer, TaaS nodes require both resource configuration (e.g., CPU and memory) and critical transaction-related parameter settings via the left Setting Bar. TaaSDB offers a comprehensive range of concurrency control mechanisms, including optimistic and pessimistic algorithms for read/write-intensive workloads. It also supports configurable transaction isolation levels, from read committed to serializable, and read consistency modes, allowing flexible selection between strong and eventual consistency to meet diverse business requirements for real-time data performance.

Step 3: Cross-model Transaction Execution. By integrating multiple execution and storage engines supporting different data models into TaaSDB, a comprehensive cross-model database system is established. After that, Users can launch TaaSDB via the Setting Bar and submit cross-model transactions in the WorkSpace area. When TaaSDB is running, performance monitoring graphs (e.g., CPU usage, disk I/O) of the instance can be viewed on the right side.

The workspace panel in Fig. 2 demonstrates an example of executing a hybrid multi-model transaction. The user initiates a cross-model transaction by entering "BEGIN_TRANSACTION;" in the dialog box below and sending the command. Subsequently, the user can sequentially input and submit multiple transaction execution statements involving different data models. This process continues until the user sends a "COMMIT;" command to finalize the cross-model transaction. (For illustrative purposes, Fig. 2 consolidates the entire cross-model transaction within a single dialog bubble, though this is not a technical requirement.) Upon processing the cross-model transaction, TaaSDB returns the execution results in the dialog box.

Optional: Intelligent Template Application and Deployment. To improve deployment efficiency, the Suggestion area offers pre-validated preset configuration templates. Upon selecting a template for a specific business scenario, the system automatically maps and configures parameters across all layers. Users can compare the template recommendations with their current settings through a difference view and make custom adjustments if necessary. The deployment wizard validates the configurationâĂŹs effectiveness and constructs the network topology among nodes in the execution, transaction, and storage layers, significantly reducing deployment complexity.

4 Summary and Future Work

This paper presents TaaSDB, a novel Transaction-as-a-Service platform that enables ACID transactions across heterogeneous data models (relational, graph, key-value, etc.) through its innovative three-layer architecture. By decoupling transaction processing as an independent service layer and employing intelligent conflict detection with dynamic model mapping, TaaSDB simplifies multi-model data management while maintaining consistency and scalability. Future work will focus on optimizing performance for latency-sensitive applications and expanding support for emerging data models in complex enterprise environments.

References

1. Yu, X., et al.: Epoxy: ACID transactions across diverse data stores. Proc. VLDB Endow. **16**(11), 2742–2754 (2023)
2. Verbitski, A., et al.: Amazon aurora: design considerations for high throughput cloud-native relational databases. In: Proceedings of the 2017 ACM International Conference on Management of Data, pp. 1041–1052. ACM, New York (2017)
3. Zhou, W., et al.: GeoGauss: strongly consistent and light-coordinated OLTP for geo-replicated SQL database. Proc. ACM Manag. Data **1**(1), 1–27 (2023)
4. Zhou, W., et al.: Concurrency control as a service. Proc. VLDB Endow. **18**(9), 2761–2774 (2025)

Intelligent Service

MVDE-MSA: A Multi-view Disentangled Enhancement Framework with Matching Scenario Awareness for Robust Address Matching

Yifan Zhu[1], Yuqi Gao[1], Jiali Mao[1(✉)], Shuangli Wu[2], Pengcheng Ding[2], and Feifei Wang[2]

[1] East China Normal University, Shanghai, China
{51265903116,51275903062}@stu.ecnu.edu.cn, jlmao@dase.ecnu.edu.cn
[2] The Urban Operation Management Center of Putuo District, Shanghai, China
{wusl,dpc,wangfeif}@shpt.gov.cn

Abstract. Standardized address data plays a pivotal role in urban governance applications such as emergency response systems, yet existing address matching methods face significant challenges due to inconsistent data collection standards across departments. These challenges manifest as colloquial expressions, missing elements, and high textual similarity among geographically proximate addresses, collectively degrading matching performance. To address these limitations, we propose MVDE-MSA, a robust address matching framework incorporating *Multi-View Disentangled Enhancement* and *Matching Scenario Awareness*. The framework features two key innovations: a *Multi-View Disentanglement-Enhanced Encoding module* that explicitly preserves complementary information across different address representations, and a *Matching Scenario-Aware Dynamic Weight Allocation network* that adaptively adjusts view-specific similarity weights based on scenario characteristics. Comprehensive evaluations on real-world datasets reveal that *MVDE-MSA* achieves a 6.0% improvement in *Recall@1* over state-of-the-art methods, proving particularly effective in handling ambiguous and incomplete address data.

Keywords: Address matching · Multi-view disentanglement · Matching scenario awareness

1 Introduction

Chinese government has implemented "block data" technology to enhance the precision of urban governance by establishing a novel framework based on geospatial standardization. This framework relies on a unified geospatial coding system that divides urban areas into standardized grid units, facilitating systematic integration and cross-departmental sharing of government data. By doing so,

S. Yang et al. (Eds.): WISA 2025, LNCS 16299, pp. 335–348, 2026.
https://doi.org/10.1007/978-981-95-6426-2_31

it enables the accurate correlation and synthesis of originally fragmented governance elements-such as population, enterprise, housing, and incident records-that are typically dispersed across different government agencies [1].

Address standardization serves as the foundational component of this precision governance system. However, multi-source address data often suffers from inconsistencies, including non-standardized formats, ambiguous expressions, and missing elements. To address these issues, the prevailing methodology employs a standardized address database, which ensures normalized alignment of address elements through similarity matching between non-standard (query) addresses and a reference database.

The core computational task in this process is address similarity calculation, wherein the reference address with the highest similarity score to the query address is selected as the matching result. Current state-of-the-art approaches typically encode both query and standardized addresses into vector representations by extracting textual, geographic, and functional features [2–9]. These representations are then compared using various similarity metrics to determine the optimal match. Despite advancements in address similarity matching technologies, two critical challenges persist in practice.

Challenge 1: Incomplete or colloquial query addresses. As demonstrated in Table 1, query addresses frequently exhibit structural incompleteness (containing only 1–2 administrative-level elements) or incorporate colloquial spatial descriptors (e.g., "market beneath X complex"), while standardized addresses maintain complete hierarchical structures (city, district,..., apartment number). Existing approaches attempt to address textual discrepancies through deep semantic modeling of address elements [2–4]. However, their reliance on inter-element contextual relationships becomes ineffective when query addresses either lack sufficient elements or contain non-standard spatial references. This limitation stems from the models' inability to generate discriminative representations from sparse or unconventional inputs, resulting in significant low accuracy.

Table 1. Query addresses with missing address elements or colloquial descriptions

Standardized Address	Query addresses with missing address elements	Query addresses with colloquial descriptions
Ganquan Farmers' Market, No. 500 Ganquan Road, Mingdu Residents' Committee, Ganquan Road Subdistrict, Putuo District, Shanghai (上海市普陀区甘泉路街道名都社区 居民委员会甘泉路500号甘泉集贸市场)	No. 500 Ganquan Road (甘泉路500号)	10 m northeast of the intersection of Ganquan Road and Huangling Road (甘泉路黄陵路路口东北角10米)
	Ganquan Market (甘泉集市)	Vegetable Market beneath Xinhuang Residential Complex (新黄小区楼下菜市场)

Challenge 2: High textual similarity among proximate standardized addresses. Figure 1 illustrates how geographically adjacent standardized addresses (e.g., orange/yellow and blue/green pairs) often share identical administrative elements while differing only in *POI* names. This phenomenon causes

text-based methods [2,4] to frequently misalign query addresses to neighboring standardized addresses. For instance, the query "Bookstore at the east gate of East China Normal University" might incorrectly match to "East China Normal University Putuo Campus" (blue) rather than the correct "Daxia Bookstore" (green) due to textual proximity. Recent solutions incorporate geographic coordinates [6,8,9] or functional semantics [5] as auxiliary features. However, these approaches adopt static weighting schemes for different feature views (text, location, function), failing to adapt to scenario-dependent variations in feature reliability. We define a matching scenario as the unique combination of textual discrepancy, spatial relationship, and semantic relevance between address pairs. The fixed-weight paradigm prevents models from dynamically prioritizing the most reliable features in each scenario, ultimately constraining matching performance.

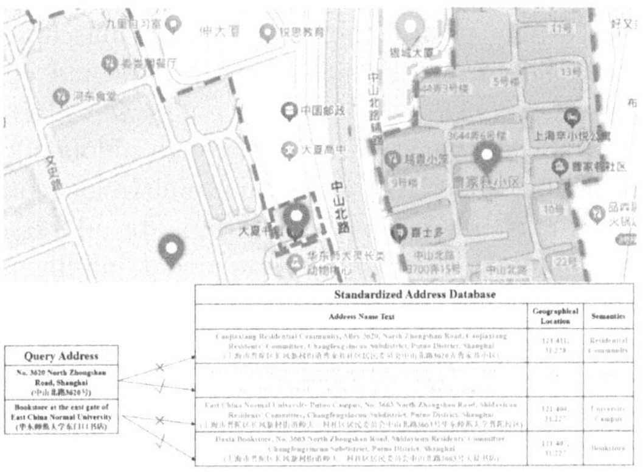

Fig. 1. Query addresses and spatially proximate standardized addresses.

To address Challenge 1, we propose a *multi-view disentanglement-enhanced encoding* strategy that independently encodes each view (textual, geographic, functional semantic) and computes separate similarity scores, preserving complementary information more effectively than existing fusion approaches [5,8,9]. We further enhance this through *adversarial perturbation-based hard negative sampling*, which generates view-specific challenging negatives to strengthen discriminative feature learning. For Challenge 2, we develop a *matching scenario-aware dynamic weighting network* that combines gated fusion networks [10] and multi-head attention mechanisms. This system first integrates multi-view representations to characterize the matching scenario, then uses self-attention to encode scenario differences, and finally dynamically allocates weights to each view's similarity score through softmax normalization, enabling adaptive importance assignment based on contextual relevance.

Overall, We propose a novel address matching framework integrating *Multi-View Disentanglement Enhancement* and *Matching Scenario Awareness*, which operates through three coordinated components: (i) an *Adversarial Perturbations Enhanced Hard Negative Sampling module* that generates challenging triplets to enforce discriminative representation learning; (ii) a *Multi-View Disentanglement-Enhanced Encoding system* utilizing three specialized encoders to produce *view-specific representations* (textual, geographic, and functional) while maintaining separate similarity computations for each view; and (iii) a *Matching Scenario-Aware Dynamic Weight Allocation mechanism* that first synthesizes multi-view representations through *gated fusion*, then employs *multi-head self-attention* to capture scenario-specific variations, ultimately determining optimal view-weighting via *softmax normalization*. Our contributions are 4 folds:

- We propose a *Multi-View Disentangled Enhancement Framework with Matching Scenario Awareness for Robust Address Matching*, called ***MVDE-MSA***, which systematically integrates hard negative sampling, multi-view encoding, and dynamic weight allocation to notably improve matching accuracy.
- To overcome challenges from incomplete addresses and colloquial descriptions, we develop *adversarial triplet-trained multi-view disentangled encoders* that preserve view-specific characteristics through independent encoding and similarity computation, effectively enhancing multi-view complementarity.
- To solve erroneous matches to spatially proximate addresses, we introduce a *scenario-aware dynamic weighting network* that intelligently adjusts view importance based on matching context through adaptive score weighting.
- Comprehensive experiments on two real-world datasets demonstrate MVDE-MSA's superiority, particularly achieving a 6.0% improvement in *Recall@1* on *Putuo District*'s standardized address dataset compared to state-of-the-art baselines.

2 Related Work

Research on address matching has evolved through three distinct phases. The initial phase employed rule-based methods and statistical models, exemplified by Song's Hidden Markov Model for address element recognition [11] and Wang et al.'s hierarchical matching approach starting from granular elements [12]. These methods, while pioneering, suffered from maintenance burdens due to their reliance on manually crafted rules.

The second phase leveraged *deep learning* for semantic understanding. Lin et al. pioneered this direction using *word2vec embeddings* with *Enhanced Sequential Inference Models* [2], while Huang et al. addressed rare addresses through *heterogeneous graph networks* [3]. Subsequent innovations included Chen et al.'s *BiLSTM-CNN* hybrid for multi-scale feature extraction [4], Shan et al.'s *web-augmented context* [5], Zhao et al.'s *coordinate embeddings* [6], and Qiang et al.'s *contrastive alignment* of textual-spatial features [13]. Although these methods demonstrated the value of multi-view information, their early fusion strategies inadvertently suppressed complementary inter-view relationships, particularly detrimental when handling addresses with substantial textual disparities.

The current phase incorporates *geographically-enhanced pretrained models* to overcome the spatial knowledge limitations of conventional language models [14–16]. Notable advancements include Huang et al.'s *query-click correlation learning* [17], Ding et al.'s *map-modality encoding* [8], Wu et al.'s *graph-based topology learning* [9], and Huang et al.'s *geographic tool integration* [18]. Despite these innovations, a fundamental challenge persists: the methods' limited discriminative power when processing proximate addresses with near-identical standardized texts frequently leads to erroneous neighbor matches. Our work addresses this critical limitation through scenario-aware dynamic view weighting, explicitly accounting for matching context variability to achieve more robust address discrimination.

3 Problem Definition

Definition 1 (Address). *A geographic address A is a structured representation of a physical location, formally defined as $A = (A_{text}, A_{geo}, A_{sem})$, where A_{text} denotes the textual description comprising hierarchical administrative elements and optional colloquial references, A_{geo} specifies the geospatial coordinates, and A_{sem} captures functional semantics including but not limited to POI categories.*

Definition 2 (Standardized Address). *A Standardized Address $SA = (E, G, S)$ is a canonical representation of a geographic entity, where $E = \{e_1, ..., e_k\}$ denotes the complete hierarchy of administrative elements (e.g., province, city, district, street, building number) ordered from highest to lowest administrative level, $G = (lat, lon)$ specifies the geospatial coordinates (latitude, longitude), and $S = (t_1, ..., t_j)$ captures semantic attributes including functional type t_i (e.g., residential, commercial, school,...).*

Definition 3 (Standardized Address Database). *A Standardized Address Database P_Δ is a collection of standardized addresses corresponding to all geographic entities within a given region Δ. A standardized address database containing n standardized addresses can be denoted as: $P_\Delta = \{SA_1, ..., SA_n\}$.*

Problem Definition (Address Standardized Matching). Given a query address $A = (A_{text}, A_{geo}, A_{sem})$ and a standardized address database $P_\Delta = \{SA_1, ..., SA_n\}$ for region Δ, where each $SA_i = (E_i, G_i, S_i)$ is a complete standardized address, determine the target standardized address $SA^* \in P_\Delta$, s.t. : $SA^* = \arg\max_{SA_i \in P_\Delta} Sim(A, SA_i)$, where $Sim(\cdot,\cdot)$ is a multi-view similarity metric considering:(i)Textual similarity between A_{text} and E_i,(ii)Spatial proximity between A_{geo} and G_i,(iii)Semantic consistency between A_{sem} and S_i.

4 Framework

We present a *Multi-View Disentangled Enhancement Framework with Matching Scenario Awareness* for robust address matching (Fig. 2), which systematically addresses the challenges through three integrated components: (i) *Adversarial Perturbations Enhanced Hard Negative Sampling*, (ii) *Multi-View Disentanglement-Enhanced Encoding*, and (iii) *Matching Scenario-Aware Dynamic Weight Allocation*. It operates by first generating discriminative training samples, then learning view-specific representations, and finally performing context-aware matching.

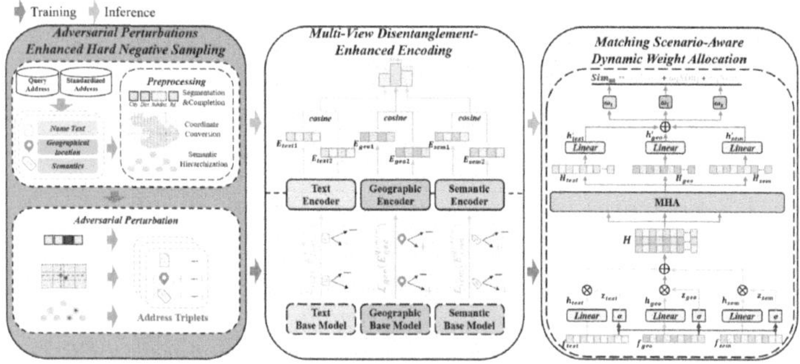

Fig. 2. Overall Framework of MVDE-MSA.

4.1 Adversarial Perturbations Enhanced Hard Negative Sampling

Preprocessing. We perform comprehensive data preprocessing to enable effective multi-view encoding through three key operations. First, for *text standardization*, we employ the *MGeo community model*[1] to segment raw address name texts into hierarchical administrative elements, removing non-address components and reconstructing them in standardized order. Missing higher-level elements (province/city/district) are manually completed based on data provenance to reduce ambiguity. Second, for *spatial representation*, we adopt the grid-based method from [8], transforming raw (lat, lon) coordinates into discrete grid cell indices by partitioning the map space into an $N * N$ grid and computing each point's absolute grid position (Fig. 3). Third, to address semantic hierarchy challenges, we manually construct an *address semantic hierarchy tree* (Fig. 4) that organizes functional attributes from general to specific (e.g., "Public Places"-"Market"-"Farmers' Market"). Each address's semantics are encoded as an ordered path from root to leaf node, preventing cross-level mismatches (e.g., distinguishing "university" from other "school" subtypes). This tripartite preprocessing ensures consistent representation across three views.

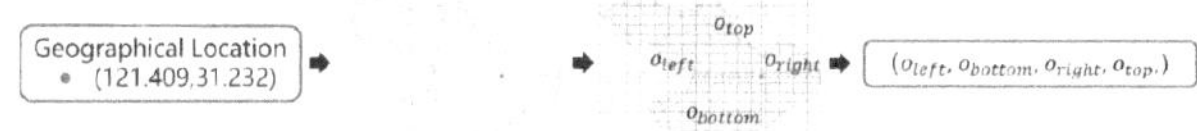

Fig. 3. Coordinate Conversion.

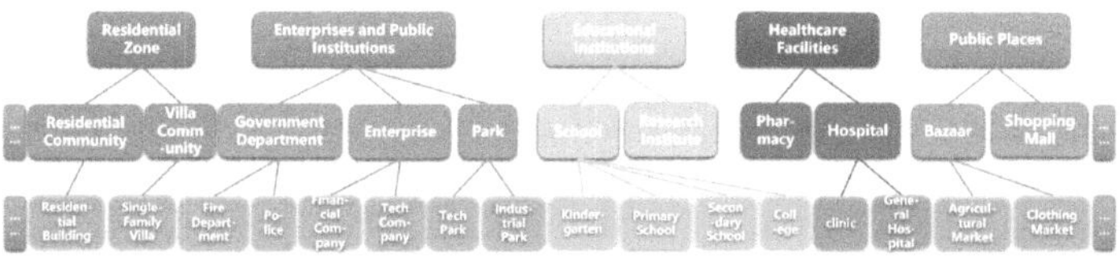

Fig. 4. Address Semantic Hierarchy Tree.

Adversarial Perturbations Enhanced Sampling. This step generates training data for the subsequent stages by applying adversarial perturbations to the preprocessed multi-view information to produce address triplets containing hard negatives. Each query address functions as an anchor, with its corresponding standardized address in the standardized address database acting as the positive sample. Adversarial perturbations are applied to three views of the address(*name text*, *geographical location*, and *address semantics*), yielding a total of nine hard negatives per anchor (three from each view). For the address name text, road names, road numbers, and *POI* names in the positive sample are randomly replaced with alternative names of the same administrative level. For the geographical location, three random coordinates are generated within a *500-meter* radius of the positive sample. For address semantics, three semantic labels are randomly selected from the same hierarchical level in the semantic tree as the positive sample.

4.2 Multi-view Disentanglement-Enhanced Encoding

To encode the address name text and address semantics, we employ *BERT-based* models as the base encoders, using the [CLS] token's output as the representation. For geographical location encoding, we utilize a *multilayer perceptron* (*MLP*) to project the coordinates into a high-dimensional vector space. For a triplet sample $< q, +, - >$, we obtain its encoded vectors as follows:

$$E_x^{q/+/-} = Encoder_x(q/+/-), x \in \{text, geo, sem\} \tag{1}$$

Three encoders are trained based on their respective triplet loss $\mathcal{L}_x$:

$$\mathcal{L}_x = max\{f(E_x^q, E_x^+) - f(E_x^q, E_x^-) + \delta, 0\}, x \in \{text, geo, sem\} \tag{2}$$

[1] https://modelscope.cn/models/iic/mgeo_geographic_elements_tagging_chinese_base/summary.

where f is an $L2$ norm distance function and δ is the margin hyperparameter. As a result, we obtain three encoders, each representing one of the three views of an address:

$$Encoder_x = f(\cdot;\theta_x^*), \theta_x^* = \arg\min_{\theta} \mathcal{L}_x(\theta), \theta^{(0)} = \theta_{base}, x \in \{text, geo, sem\} \quad (3)$$

For a query address $A = (A_{text}, A_{geo}, A_{sem})$ and a standardized address $B = (B_{text}, B_{geo}, B_{sem})$, we calculate the similarity in each of the three views: name text similarity $Sim_t(A, B)$, geographical location similarity $Sim_l(A, B)$, and address semantics similarity $Sim_s(A, B)$ as follows:

(i) Calculate the cosine similarity between the name texts embedding vectors:

$$E_{text1} = Encoder_{text}(A_{text}), E_{text2} = Encoder_{text}(B_{text}) \quad (4)$$

$$Sim_t(A, B) = cosine\_similarity(E_{text1}, E_{text2}) \quad (5)$$

(ii) Since two addresses that are geographically far apart are unlikely to refer to the same entity, we apply a distance-based correction factor $\mu(A, B)$ to the cosine similarity of the geographic embeddings:

$$E_{geo1} = Encoder_{geo}(A_{geo}), E_{geo2} = Encoder_{geo}(B_{geo}) \quad (6)$$

$$Sim_l(A, B) = \mu(A, B) \cdot cosine\_similarity(E_{geo1}, E_{geo2}) \quad (7)$$

where $\mu(A, B) = e^{-\frac{distance(A,B)}{\lambda}}$, $distance(A, B)$ is the distance between A and B (in kilometers), λ is a decay parameter that controls how quickly μ decreases with $distance$.

(iii) Calculate the cosine similarity between the semantic embedding vectors:

$$E_{sem1} = Encoder_{sem}(A_{sem}), E_{sem2} = Encoder_{sem}(B_{sem}) \quad (8)$$

$$Sim_s(A, B) = cosine\_similarity(E_{sem1}, E_{sem2}) \quad (9)$$

4.3 Matching Scenario-Aware Dynamic Weight Allocation

The dynamic weight allocation module, illustrated in Fig. 2, processes the representations and similarity scores of two addresses across three distinct views (text, geography, and semantics) to compute view-specific weighting factors. The module begins by concatenating the encoded representations and their corresponding similarity score for each view, yielding view-specific feature vectors f (e.g., $f_{text} = Concat[E_{text1}, E_{text2}, Sim_t]$). These vectors are then fed into a *dynamic gated fusion network*, which learns inter-view dependencies and produces a unified matching representation H, capturing the composite matching scenario across all views:

$$H = Stack[z_{text}h_{text}, z_{geo}h_{geo}, z_{sem}h_{sem}] \quad (10)$$

where $h_{text} = W_{text} f_{text} + b_{text}$, $z_{text} = \sigma(W_1 Concat[f_{text}, f_{geo}, f_{sem}])$ and so on. Then, multi-head self-attention is employed to capture cross-view interactions, encoding the differences in matching scenarios into enhanced representations of each view $H_{text}, H_{geo}, H_{sem}$:

$$Q_i = QW_i^Q, K_i = KW_i^K, V_i = VW_i^V, i = 1, 2, ..., m \tag{11}$$

$$head_i = Attention(Q_i, K_i, V_i) = softmax(\frac{Q_i K_i^T}{\sqrt{d_k}})V_i, i = 1, 2, ..., m \tag{12}$$

$$MHA(Q, K, V) = Concat[head_1, head_2, ..., head_m]W^O \tag{13}$$

$$H_{attn} = MHA(H, H, H) \tag{14}$$

$$H_{text}, H_{geo}, H_{sem} = H_{attn}[:, 0], H_{attn}[:, 1], H_{attn}[:, 2] \tag{15}$$

Finally, the vector representation of each view is projected into a scalar via a linear: $h'_{text} = W'_{text} H_{text} + b'_{text}$, and so on, followed by a softmax operation to calculate $\omega_t, \omega_l, \omega_s$:

$$\omega = softmax(Concat[h'_{text}, h'_{geo}, h'_{sem}]) \tag{16}$$

$$\omega_t, \omega_l, \omega_s = \omega[:, 0], \omega[:, 1], \omega[:, 2] \tag{17}$$

We calculate the weighted sum $Sim_m(A, B)$ and select the standardized address with the highest $Sim_m(A, B)$ as the final matching result:

$$Sim_m(A, B) = \omega_t Sim_t(A, B) + \omega_l Sim_l(A, B) + \omega_s Sim_s(A, B) \tag{18}$$

During the training phase, the label is assigned a value of 1 if A and B constitute a matching pair, and 0 otherwise. The module is optimized by minimizing the *mean squared error* (MSE) between the predicted similarity score Sim_m and the label as the loss function.

5 Experiments

5.1 Experimental Settings

Datasets. Our experiments utilize two distinct datasets to evaluate the proposed method. The first dataset, **Putuo**, comprises 5,230 standardized addresses from *Shanghai's Putuo District*, each containing name text, $WGS84$ coordinates, and functional type information. We supplement this with 1743 query addresses obtained from the *Shanghai Open Data Platform*[2], where geographic coordinates were acquired via the *Gaode Map API*[3] and ground truth matches were manually annotated. The dataset follows a 1,143:300:300 split for training, validation, and testing respectively. The second dataset, **GeoTES-rerank**, represents a large-scale benchmark for *query-POI* matching, containing approximately 100,000 samples distributed as 50,000 for training, 20,000 for validation, and 30,002 for testing. Each query is paired with t candidate $POIs$ ($t = 20$ for training, $t = 40$ for validation/testing), with exactly one is correct.

[2] https://data.sh.gov.cn/.

[3] https://lbs.amap.com/api/webservice/guide/api/georegeo.

Settings. All experiments are conducted on a GPU-CPU platform equipped with a *Tesla V100*. Both our model and baselines are developed using *Python 3.8*. Grid partitioning parameter $N = 200$, the margin hyperparameter $\delta = 1$, the decay parameter $\lambda = 8$, and the number of heads $m = 4$. The base model for the *Text Encoder* is *MGeo* [8], for the *Geographic Encoder* is a 4-layer *MLP* with layer sizes (4, 32, 32, 16), and for the *Semantic Encoder* is *SBERT* [16].

Metrics. The evaluation of matching accuracy employs two metrics: ***Recall@k***, which quantifies the percentage of queries where the correct standardized address appears within *top-k* results, and ***MRR@k*** (*Mean Reciprocal Rank*), which computes the average reciprocal of the rank at which the correct standardized address is positioned among *top-k* results.

Baselines. To evaluate the performance of MVDE-MSA, we conduct comparative experiments with following address matching methods: (1)**Geocoding**[4] employs *address normalization* and *IK Analyzer-based tokenization* followed by *cosine similarity matching*, (2)**Rule-Bottom-Up** adopts a hierarchical matching strategy that first identifies fine-grained address elements and then performs bottom-up matching, (3)**ESIM** [2] utilizes an *Enhanced Sequential Inference Model* to extract address representations for similarity computation, (4)**PALM** [6] incorporates spatial correlations through coordinate embeddings and fuses them with semantic representations using *MLP* before similarity calculation, (5)**PLMs**, we fine-tuned *BERT* [14], *RoBERTa* [15], and *SBERT* [16] on address name texts, using the [CLS] token representation for similarity measurement, (6)**MGeo-base** [8] represents the first cross-modal approach integrating map and address information, where we used its released community version[5], (7)**G2PTL** [9] models spatial topological relationships between addresses, implemented using its official pre-trained model[6].

Ablation. To evaluate the contribution of key components in our model, we design three variants for ablation: (1) **w/o DynamicWeight**, where we remove the dynamic weight allocation module and instead determine optimal fixed weights through grid search on the validation set; (2) **w/o GeographicEncoder**, which eliminates the geographic encoding component; and (3) **w/o SemanticEncoder**, which eliminates the semantic encoding component.

5.2 Results and Analysis

The experimental results presented in Table 2 reveal several key findings. *Rule-based* methods demonstrate limited performance, primarily due to their vulnerability to errors in address element recognition. Although *ESIM* and *PALM*

[4] https://github.com/fjklqq/Geocoding.
[5] https://modelscope.cn/models/iic/mgeo_geographic_entity_alignment_chinese_base/summary.
[6] https://huggingface.co/Cainiao-AI/G2PTL.

Table 2. Results of comparison approaches

	Putuo				GeoTES-rerank			
	Recall@1	Recall@5	Recall@10	MRR@5	Recall@1	Recall@5	Recall@10	MRR@5
Geocoding	0.347	0.407	0.503	0.37	0.216	0.444	0.603	0.338
Rule-Bottom-Up	0.387	0.467	0.61	0.423	0.313	0.53	0.627	0.427
ESIM	0.487	0.65	0.74	0.506	0.498	0.587	0.618	0.539
PALM	0.583	0.69	0.807	0.643	0.562	0.621	0.686	0.587
PLMs BERT	0.75	0.84	0.863	0.788	0.587	0.633	0.707	0.603
RoBERTa	0.72	0.85	0.867	0.774	0.573	0.619	0.723	0.599
SBERT	0.747	0.84	0.863	0.786	0.596	0.627	0.716	0.612
MGeo-base	0.803	0.91	0.91	0.852	0.763	0.857	**0.936**	0.808
G2PTL	0.763	0.87	0.913	0.819	0.71	0.853	0.917	0.793
MVDE-MSA	**0.863**	**0.92**	**0.923**	**0.889**	**0.787**	**0.866**	0.906	**0.821**
-w/o DynamicWeight	0.81	0.903	0.91	0.849	0.756	0.848	0.891	0.801
-w/o GeographicEncoder	0.783	0.89	0.913	0.843	0.751	0.844	0.887	0.799
-w/o SemanticEncoder	0.823	0.91	0.92	0.874	0.776	0.853	0.898	0.813

capture certain latent semantic patterns in addresses, their performance remains constrained by insufficient consideration of complementary relationships among multi-view information. Fine-tuned general pre-trained models achieve moderate performance, while *MGeo* and *G2PTL* show superior results by incorporating geographic knowledge during training. Our proposed method surpasses all baseline approaches in the address standardization task, achieving a 6.0% improvement in *Recall@1* on the *Putuo* dataset compared to previous state-of-the-art methods. This significant enhancement stems from our model's ability to effectively preserve complementary relationships across multiple views while dynamically adjusting similarity weights based on matching scenario differences.

The ablation study yields important insights into our model's components. Eliminating the *dynamic weight allocation module* and replacing it with optimal fixed weights (determined through validation set grid search) leads to substantial performance degradation, with *Recall@1* decreasing by 5.3% on *Putuo* and 3.1% on *GeoTES*. This confirms the module's crucial role in capturing cross-view interactions and adapting to varying matching scenarios. Further analysis of view-specific contributions shows that removing the geographic encoder results in more severe performance drops (8.0% on *Putuo* and 3.6% on *GeoTES*) compared to removing the semantic encoder (4.0% and 1.1% respectively). These findings validate our method's effective utilization of multi-view information and their complementary relationships for accurate matching, and suggest that geographic information provides more discriminative power for address matching, likely due to its inherent precision and intuitive relevance in spatial contexts.

Parameter Sensitivity. To thoroughly investigate parameter sensitivity, we conducted an empirical analysis of how the grid partitioning parameter N and decay parameter λ of the distance-based correction factor μ affect *Recall@1* performance on the *Putuo* dataset. As illustrated in Fig. 5, optimal performance occurs at $N = 200$ (yielding approximately 45-meter grid resolution in *Putuo*) and $\mu = 8$. The grid parameter N demonstrates a clear trade-off: smaller values produce excessively large grids that fail to distinguish geographically distant addresses, while larger values not only increase computational overhead but may artificially separate standardized addresses from the same *POI* (such as adjacent buildings in a residential complex) into discrete grids, thereby distorting their spatial relationships. Regarding the decay parameter λ, we observe a critical balance in its setting. Higher λ values lead to overly gradual decay of μ, diminishing the correction effect, whereas lower values cause excessively rapid decay that significantly amplifies spatial differences, even geographically close addresses (e.g., within 1 km) may exhibit low geographic similarity. These findings demonstrate the importance of careful parameter calibration to maintain appropriate geographic discrimination while avoiding either excessive generalization or overspecification of spatial relationships.

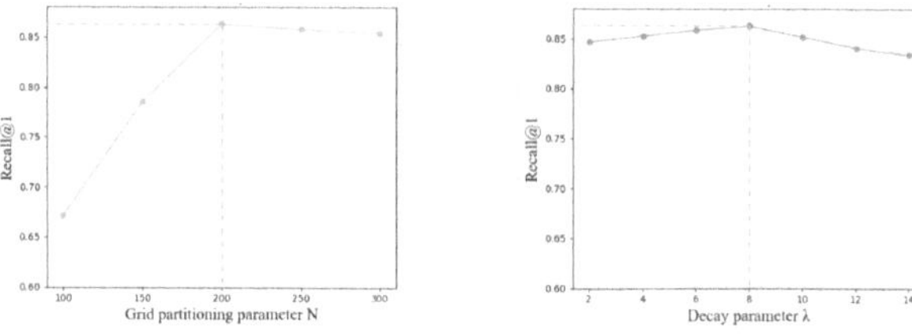

Fig. 5. Impact of N and λ on Putuo.

Case Study. Table 3 presents a comparative case study between MVDE-MSA and the previous state-of-the-art approach *MGeo*, using three representative query addresses from *Putuo* dataset that exemplify distinct matching challenges. The first case demonstrates MVDE-MSA's robustness to significant address element omissions, the second showcases its ability to handle colloquial address descriptions, and the third illustrates its effectiveness in resolving ambiguous matches caused by high textual similarity with incorrect standardized addresses. The bold-highlighted differences in matching results clearly indicate MVDE-MSA's superior accuracy, which stems from its comprehensive preservation of multi-view information complementarity and adaptive capability across varying matching scenarios. These empirical observations provide concrete validation for the effectiveness of MVDE-MSA in real-world address matching situations.

Table 3. Comparative Analysis of Matching Results Between MVDE-MSA and MGeo

Query Address	Matching Result	
	MGeo-base [8]	MVDE-MSA
Building 84, Chezhan Xincun Residential (车站新村85号楼)	Building 84, Chezhan Xincun Residential Compound, Taopu Road, Chezhan Xincun Community Residents' Committee, Zhenru Subdistrict, Putuo District, Shanghai (上海市普陀区真如镇街道车站新村 社区居民委员会桃浦路**车站新村**小区**84号楼**)	Building 85, North Zone of Caoyangyuan, Zhichuan Community Residents' Committee, Zhenru Subdistrict, Putuo District, Shanghai (上海市普陀区真如镇街道芝川社区 居民委员会**曹杨苑**北区**85号楼**)
Bank at Northeast Corner of the Intersection of Anyuan Road and Yejiazhai Road (安远路叶家宅路口东北角银行)	No. 84 Anyuan Road, Jinxiuli Community Residents' Committee, Changshou Road Subdistrict, Putuo District, Shanghai (上海市普陀区长寿路街道锦绣里 社区居民委员会**安远路84**号)	China Construction Bank Shanghai Anyuan Road Branch, No. 742 Anyuan Road, Meishanyuan Community Residents' Committee, Changshou Road Subdistrict, Putuo District, Shanghai (上海市普陀区长寿路街道梅山苑社区居民 委员会**安远路742**号中国**建设**银行上**海**安远**路**支行)
Building A, 18F(B), No. 1118 Changshou Road (长寿路1118号A幢18F(B))	Room 18F, Building A, Yueda International Tower, No. 1118 Changshou Road, Zhiyinyuan Community Residents Committee, Changshou Road Subdistrict, Putuo District, Shanghai (上海市普陀区长寿路街道知音苑社区 居民委员会长寿路1118号悦达国际大厦**A**座**18F**室)	Room 18B, Building A, Yueda International Tower, No. 1118 Changshou Road, Zhiyinyuan Community Residents Committee, Changshou Road Subdistrict, Putuo District, Shanghai (上海市普陀区长寿路街道知音苑社区 居民委员会长寿路1118号悦达国际大厦**A**座**18B**室)

6 Conclusion

This paper presents *MVDE-MSA*, a novel multi-view disentangled enhancement framework that addresses two fundamental challenges in address matching: handling incomplete or colloquial address descriptions and distinguishing geographically proximate addresses with highly similar textual patterns. The framework achieves robust matching performance through two key innovations: (i) *enhanced multi-view information disentanglement* that preserves complementary address representations, and (ii) *scenario-aware dynamic weighting* that adaptively adjusts view importance based on matching context. Experimental results on real-world datasets demonstrate *MVDE-MSA*'s superiority over state-of-the-art methods. Future work will explore incorporating additional semantic dimensions, such as event-related features, to further enrich address representations and improve matching accuracy.

References

1. Li, Y., et al.: Spatio-temporal data fusion techniques for modeling digital twin city. Geo-Spatial Inf. Sci., 1–24 (2024)
2. Lin, Y., Kang, M., Wu, Y., Du, Q., Liu, T.: A deep learning architecture for semantic address matching. Int. J. Geogr. Inf. Sci. **34**(3), 559–576 (2020)
3. Huang, J., et al.: Hgamn: heterogeneous graph attention matching network for multilingual poi retrieval at baidu maps. In: SIGKDD, pp. 3032–3040 (2021)
4. Chen, J., Chen, J., She, X., Mao, J., Chen, G.: Deep contrast learning approach for address semantic matching. Appl. Sci. **11**(16), 7608 (2021)
5. Shan, S., Li, Z., Qiang, Y., Liu, A., Xu, J., Chen, Z.: Deepam: deep semantic address representation for address matching. In: APWeb-WAIM, pp. 45–60 (2019)

6. Zhao, J., et al.: Incorporating semantic similarity with geographic correlation for query-poi relevance learning. In: AAAI. vol. 33, pp. 1270–1277 (2019)
7. Yan, X., Luo, W., Wang, J., Shen, X.: A study on context-matching-based joint training for chinese coreference resolution. In: International Conference on Web Information Systems and Applications, pp. 126–137. Springer (2024)
8. Ding, R., et al.: MGEO: multi-modal geographic language model pre-training. In: SIGIR, pp. 185–194 (2023)
9. Wu, L., et al.: G2ptl: a geography-graph pre-trained model. In: CIKM, pp. 4991–4999 (2024)
10. Arevalo, J., Solorio, T., Montes-y Gómez, M., González, F.A.: Gated multimodal units for information fusion. arXiv preprint arXiv:1702.01992 (2017)
11. Song, Z.: Address matching algorithm based on chinese natural language understanding. J. Remote Sens **17**(4), 788–801 (2013)
12. Yong, W., Jiping, L., Qingsheng, G., An, L.: The standardization method of address information for pois from internet based on positional relation. Acta Geodaetica et Cartographica Sinica **45**(5), 623 (2016)
13. Qiang, Y., Zheng, J., Wu, L., Wen, H., Lou, J., Deng, M.: A momentum contrastive learning framework for query-poi matching. In: ICDM, pp. 833–838 (2024)
14. Devlin, J., Chang, M.W., Lee, K., Toutanova, K.: Bert: pre-training of deep bidirectional transformers for language understanding. In: NAACL-HLT, pp. 4171–4186 (2019)
15. Liu, Y., et al.: Roberta: a robustly optimized bert pretraining approach. arXiv preprint arXiv:1907.11692 (2019)
16. Reimers, N., Gurevych, I.: Sentence-Bert: sentence embeddings using siamese Bert-networks. In: EMNLP-IJCNLP, pp. 3982–3992 (2019)
17. Huang, J., et al.: Ernie-Geol: a geography-and-language pre-trained model and its applications in baidu maps. In: SIGKDD, pp. 3029–3039 (2022)
18. Huang, C., et al.: Geoagent: To empower LLMS using geospatial tools for address standardization. In: ACL, pp. 6048–6063 (2024)

MSAE-SQL: A Multi-layer Semantic-Aware Enhanced NL2SQL Model

Minxuan Li, Kexin Ding, Derong Shen(✉), Tiezheng Nie, Yue Kou, and Minghe Yu

Computer Science and Technology, Northeastern University, Shenyang, China
{2401857,shendr,nietiezheng,kouyue,yuminghe}@mail.neu.edu.cn

Abstract. Building cross-domain NL2SQL parsers relies heavily on accurate schema linking—the ability to align question tokens with database schema elements. However, existing methods based on n-gram or static similarity often fail when natural language expressions differ from database terms. To address this, we propose MSAE-SQL, a multi-layer semantic-aware enhanced NL2SQL model. Our approach leverages pre-trained language models (PLMs) to capture semantic correlations, constructs a schema linking graph by combining hyperbolic space distances and cosine similarity, and introduces relation-guided graph encoding to enhance the contextual representation of question-schema pairs. Finally, we apply irrelevant table pruning in decoding to improve SQL generation accuracy. Experiments on Spider, Spider-SYN, and Spider-DK benchmarks demonstrate the superior performance and robustness of our model.

Keywords: NL2SQL · Semantic-Aware · Schema Linking · Encoder · Decoder

1 Introduction

Natural Language to SQL (NL2SQL) has become a pivotal research area, enabling users to interact with relational databases through everyday language. By translating natural language questions into executable SQL queries, this technology significantly lowers the barrier to data access, making it a prominent semantic parsing task. A fundamental challenge constraining its domain generalization is **schema linking**: accurately bridging the lexical and semantic gap between informal questions and formal database schemas.

Existing methods often struggle with the inherent ambiguity and variation of natural language, making their schema linking capabilities brittle. Approaches ranging from simple fuzzy matching to sophisticated neural parsers can fail to align entities with non-trivial lexical gaps (e.g., mapping "new mexico" to a 'state' column, as shown in Fig. 1) [6,18,21,26] or misinterpret polysemous words due to context-insensitive embeddings (e.g., distinguishing "millet" as a technology brand versus a food item) [8,22]. These inaccuracies in the initial schema linking stage are a primary source of error, which inevitably propagate and lead to the generation of incorrect SQL queries.

S. Yang et al. (Eds.): WISA 2025, LNCS 16299, pp. 349–361, 2026.
https://doi.org/10.1007/978-981-95-6426-2_32

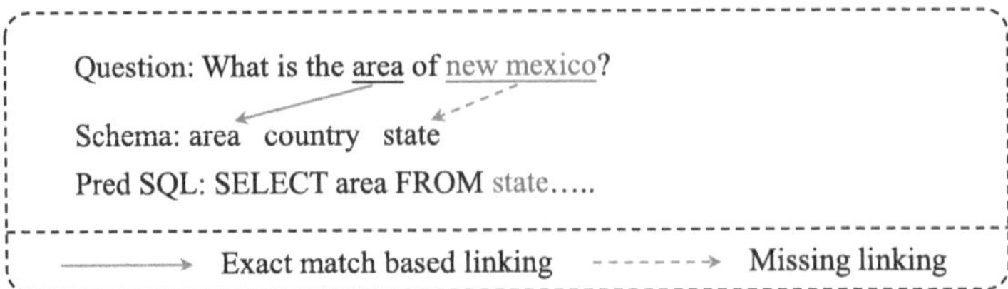

Fig. 1. An example of a common schema linking failure, where "new mexico" is incorrectly linked.

To overcome these challenges, this paper introduces **MSAE-SQL**, a novel framework for text-to-SQL generation. At the core of our approach is a new schema linking method that constructs a robust graph representation to more accurately capture the complex relationships between question words and schema items. This enriched representation is then processed by an enhanced encoding-decoding architecture to generate the final query with high fidelity. The efficacy and generalization capabilities of MSAE-SQL are demonstrated through extensive experiments on the Spider, Spider-syn, and Spider-DK benchmarks.

The main contributions of this paper can be summarized as follows:

1. We propose a novel schema linking method that couples hyperbolic space with cosine similarity. This synergistic approach constructs an enhanced graph representation that improves the linking accuracy between natural language entities and database schema elements.
2. We introduce a relationship-guided graph attention encoder (E-RGAT). This module leverages similarity relations to guide the SQL encoding process, further boosting model accuracy.
3. We design a decoder that incorporates a table pruning mechanism. By filtering out irrelevant tables during generation, this method significantly improves the precise matching rate of the final SQL query.
4. We conduct extensive evaluations and analyses on multiple challenging benchmarks to demonstrate the effectiveness and robustness of our proposed MSAE-SQL model.

2 Related Work

2.1 Evolution of NL2SQL Methods

The field of NL2SQL has rapidly evolved from early sequence-to-sequence [23,29] and grammar-based approaches [10] to the current paradigm dominated by large Pre-trained Language Models (PLMs) [1,18]. Modern methods leverage the deep semantic understanding of PLMs to map natural language questions and database schemas to executable SQL. This modern technical framework, as illustrated in Fig. 2, generally involves encoding the question and schema, performing schema linking to bridge the lexical gap, and finally decoding the query. Despite this progress, accurately bridging this gap remains a central challenge.

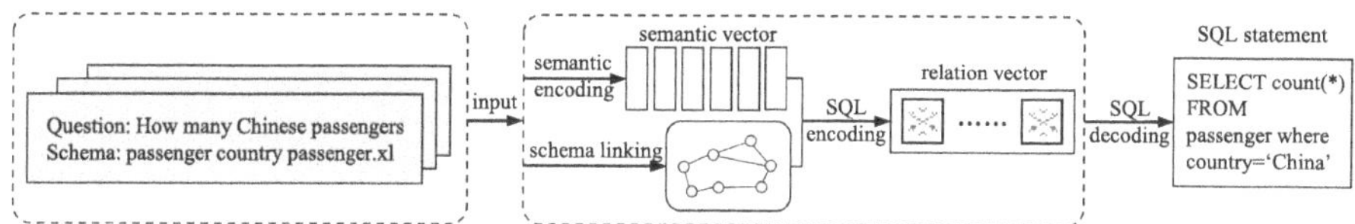

Fig. 2. NL2SQL technology framework diagram.

2.2 Schema Linking and Graph-Based Representation

Schema linking, which identifies correspondences between question entities and schema elements, is a critical step for model accuracy and generalization. To move beyond simple text matching [14], a predominant strategy has been to represent the question and schema as a unified graph. Pioneering works like RAT-SQL [26] demonstrated the effectiveness of this approach, where schema links are explicitly modeled as relations in the graph.

Subsequent research has focused on refining these graph-based representations. One major direction is to enhance the edge information by designing specialized relational embeddings or improving contextual encodings to achieve better alignment [5,7,17]. Another direction focuses on better leveraging the intrinsic knowledge of PLMs to directly improve the linking process itself [20].

2.3 Graph-Based Encoding and Decoding

The de-facto standard for processing these structured inputs is the **graph neural network (GNN)** encoder [4]. These encoders, often based on the powerful Transformer architecture [15,26], are adept at learning relation-aware representations that fuse the semantics of the question with the structure of the schema. These rich representations are then passed to a decoder. Modern decoders have themselves evolved from rule-based [2] or statistical methods [27] to neural, structure-aware generators [13]. A common technique is to generate the query's Abstract Syntax Tree (AST), which ensures the syntactic validity of the output. Our work operates within this graph-based paradigm, but argues that the quality of the initial graph construction via schema linking remains a key bottleneck, which we aim to address directly.

3 Methodology

3.1 Problem Formulation

The task of Natural Language to SQL (NL2SQL) aims to translate a user's natural language question Q into an executable SQL query Y, conditioned on a given database schema S. Formally, the input is a pair $X = (Q, S)$, where the question is a sequence of tokens $Q = (q_1, q_2, \ldots, q_{|Q|})$, and the schema $S = \{t_1, t_2, \ldots, t_{|T|}\}$ is a set of tables. Each table t_i contains a set of columns $C_i = \{c_{i,1}, c_{i,2}, \ldots, c_{i,|C_i|}\}$, with table and column names being natural language phrases. The objective is to learn a model f that approximates the mapping $f : X \to Y$ and maximizes the likelihood of generating the correct query.

3.2 Model Architecture Overview

To address this task, we propose MSAE-SQL, a novel framework based on an enhanced encoder-decoder architecture, as illustrated in Fig. 3. The encoder is designed to generate a rich and contextualized representation of the input by constructing and embedding a semantic-aware schema linking graph. This graph explicitly models the relationships between the question tokens and the elements of the database schema. Subsequently, the decoder utilizes this representation to generate the target SQL query by constructing its Abstract Syntax Tree (AST) in a depth-first traversal order. The following subsections will elaborate on each component of our model.

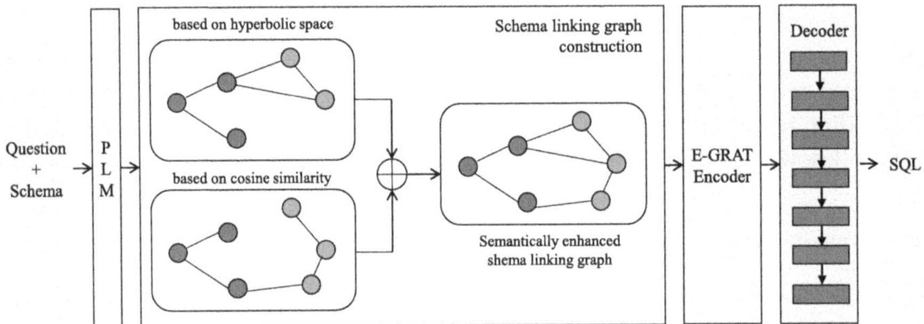

Fig. 3. Overview of MSAE-SQL.

3.3 Semantic-Aware Schema Linking Graph Construction

To effectively capture the intricate relationships between question words $Q = (q_1, \dots, q_{|Q|})$ and schema items $S = (s_1, \dots, s_{|S|})$, we construct a hybrid schema linking graph. This graph is built upon two complementary signals: a hierarchical relationship score derived from contextual perturbations in hyperbolic space, and a direct semantic similarity score based on learnable cosine similarity.

Hierarchical Relations via Hyperbolic Perturbation. First, to model hierarchical dependencies, we measure the importance of a question word q_i to a schema item s_j by quantifying the embedding shift of s_j when q_i is masked, as illustrated in Fig. 4. Let c_j be the contextual embedding of s_j from a PLM, and $c_j^{q_i}$ be its embedding when q_i is replaced by a `[MASK]` token [9]. We measure the difference between these two embeddings using the Poincaré distance in hyperbolic space, which is adept at capturing hierarchical structures with low distortion [3,30].

Both vectors are first mapped into the Poincaré ball model via $M(v) = \tanh(\frac{\mu\|v\|}{2})\frac{v}{\|v\|}$ to obtain c_j' and $c_j^{q_i'}$. The relationship score is then computed as:

$$h(q_i, s_j) = 2\tanh^{-1}\left(\left\|-c_j' \oplus c_j^{q_i'}\right\|\right) \tag{1}$$

where $\oplus$ denotes Möbius addition. We binarize these scores with a predefined threshold to produce a sparse adjacency matrix HS.

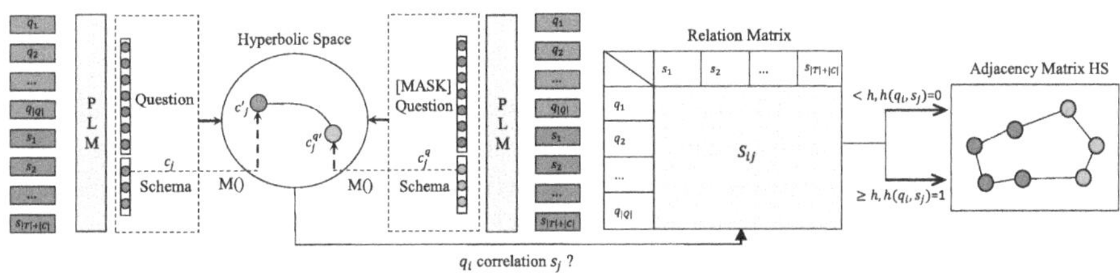

Fig. 4. Construction process based on hyperbolic space schema linking graph.

Semantic Similarity via Learnable Cosine. Second, to capture direct semantic correspondence, we compute a learnable weighted cosine similarity between question words and schema items, as depicted in Fig. 5. This allows the model to learn fine-grained alignments beyond static embeddings. The similarity is calculated as:

$$S_{ij} = \text{ReLU}\left(\frac{\langle q_i W_1, s_j W_2\rangle}{\|q_i W_1\| \cdot \|s_j W_2\|}\right) \tag{2}$$

where W_1 and W_2 are learnable weight matrices. To ensure a cleaner mapping, for each schema item s_j, we only retain the link to the question word q_i with the highest similarity score, producing a refined adjacency matrix CS.

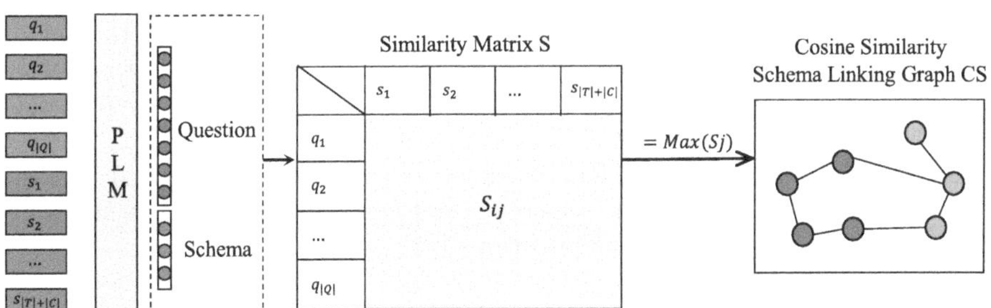

Fig. 5. Schema linking graph constructed based on cosine similarity.

Hybrid Graph Fusion. Finally, the two complementary matrices—HS, which captures hierarchical dependencies, and CS, which captures direct semantic similarity—are fused to form the final semantically enhanced schema linking graph. The fusion is performed using Möbius addition, a native operation in hyperbolic geometry. The resulting graph adjacency matrix serves as the input for our graph encoder.

3.4 Graph Encoding and Decoding Architecture

Our architecture consists of a novel graph encoder, E-RGAT, which incorporates schema linking signals, followed by a decoder equipped with a table pruning mechanism.

Semantic Relation-Guided Graph Encoder (E-RGAT). To effectively capture the rich relational information in the schema graph, we build our encoder upon the Relational Graph Attention Network (RGAT) [28], an extension of GAT [25] that models different edge types. The key innovation of our encoder, which we term **E-RGAT (Enhanced RGAT)**, is to directly inject the semantic similarity scores computed during schema linking into the attention mechanism. This allows the model to prioritize relations that are more semantically relevant to the input question.

Specifically, let Q_{ji} be the pre-computed similarity score between question node x_i and schema node x_j. We modify the standard scaled dot-product attention of RGAT by integrating Q_{ji} as a relation-aware modulation factor. The attention weights α'_{ji} and the resulting node representations $x_i^{(l+1)}$ at layer l are computed as follows:

$$\alpha'_{ji} = \left((x_i W_q)\left(x_j W_k + Q_{ji} \cdot F(E_{ji})\right)\right)^T \tag{3}$$

$$\hat{\alpha}'_{ji} = \text{softmax}_j\left(\frac{\alpha'_{ji}}{\sqrt{d}}\right) \tag{4}$$

$$\hat{x}'_i = \sum_{j \in \mathcal{N}(i)} \hat{\alpha}'_{ji}\left(x_j W_v + Q_{ji} \cdot F(E_{ji})\right) \tag{5}$$

where W_q, W_k, W_v are trainable projection matrices, $F(E_{ji})$ is the feature vector for the explicit relation type between nodes j and i, and $\mathcal{N}(i)$ is the set of neighbors of node i. As shown in Eq. (3) and (5), the similarity score Q_{ji} directly enhances the relational features, guiding the attention to focus on more plausible links.

Finally, the output of the attention layer is passed through a feed-forward network (FFN) with a residual connection and layer normalization, a standard practice in Transformer-based architectures [24]:

$$x_i^{(l+1)} = \text{LayerNorm}\left(x_i^{(l)} + \text{FFN}(\hat{x}'_i)\right) \tag{6}$$

The overall architecture of our E-RGAT module is illustrated in Fig. 6.

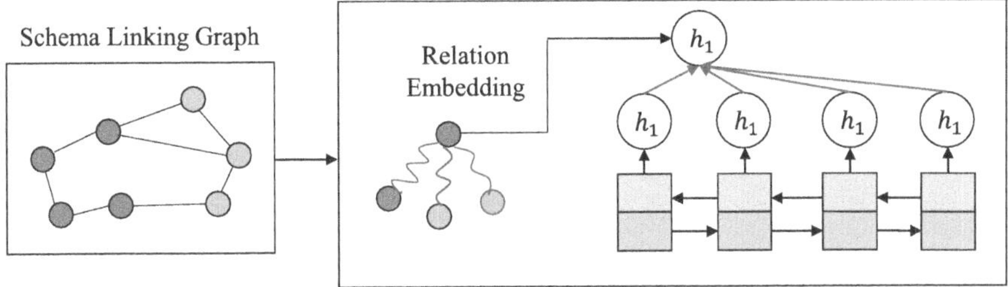

Fig. 6. The E-RGAT architecture, illustrating how the similarity score Q, computed from hyperbolic and cosine similarity, is injected as an enhanced relationship to modulate the attention calculation.

Decoder with Irrelevant Table Pruning. Our decoder generates SQL by constructing an Abstract Syntax Tree (AST) [31] in a depth-first order. An LSTM [16] predicts a sequence of actions from three categories: APPLYRULE(r), SELECTTABLE(t), and SELECTCOLUMN(c).

To improve accuracy and efficiency, we introduce a **table pruning mechanism** that reverses the traditional selection order. The decoder first uses a pointer network to predict a target column, leveraging a memory module to handle co-references. This selection immediately prunes the search space for the subsequent SELECTTABLE action to only the tables containing that column, effectively eliminating irrelevant candidates. If a column is unique, its corresponding table is selected deterministically. As illustrated in Fig. 7, this column-first approach leverages the strong structural correspondence between columns and tables to significantly reduce generation errors.

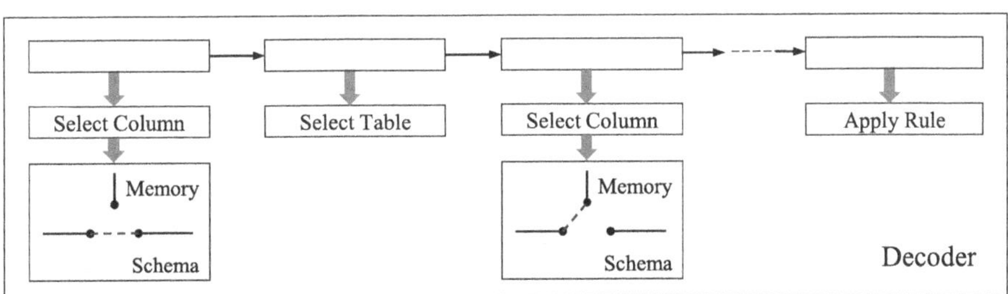

Fig. 7. The decoding process with table pruning. By first selecting a column, the decoder dynamically filters out irrelevant tables, thereby simplifying the subsequent table selection step.

4 Experiments

4.1 Experimental Setup

Datasets. We conducted extensive experiments with three publicly available benchmark datasets, as shown in Table 1.

Table 1. Overview of the benchmark dataset.

Dataset	Overview
Spider	Spider [32] is a large-scale cross-domain Text-to-SQL benchmark dataset. It contains 8659 training samples from 146 databases and 1034 evaluation samples from 20 databases.
Spider-DK	Spider-DK [12] is a manually curated dataset based on Spider. It extracts 535 Question-SQL pairs from 10 databases of the Spider development set and modifies them to integrate domain knowledge. Schema information is implicit in Spider-DK, so complex inference is required.
Spider-SYN	Spider-SYN [11] is another challenging variant of the Spider dataset. Spider-SYN is built by manually modifying natural language questions using synonym substitution. A total of 1034 samples are available in Spider-SYN's 20 databases.

Evaluation Metric. Following the official standard [32], we use Execution Accuracy (EM) as the primary evaluation metric. A predicted SQL query is considered correct if its execution result on the given database is identical to the result of the ground-truth query. The EM score is calculated as Eq. (7), where Nex is the number of correctly predicted queries and N is the total number of samples.

$$EM = Nex/N \tag{7}$$

Baselines. We compare our model against several strong baselines: **(1) IRNet** [14]: Utilizes a grammar-based approach with an intermediate representation (SemQL). **(2) RATSQL** [26]: Employs a relation-aware self-attention mechanism to encode the schema structure. **(3) ETA** [20]: A weakly supervised model using a prune-then-refine strategy for schema linking. **(4) LGESQL** [5]: Leverages line graphs to explicitly model relations within the database schema. **(5) RYANSQL** [7]: Adopts a sketch-based approach that recursively constructs queries using Statement Position Codes (SPC). **(6) S2SQL** [17]: Injects syntactic dependency information from the question into its graph encoder.

4.2 Main Results and Analysis

As shown in Table 2, our proposed MSAE-SQL model consistently outperforms all baseline methods across the Spider, Spider-DK, and Spider-SYN datasets.

Table 2. Exact matching accuracy (%) in three different datasets.

Model	Spider-DK	Spider-SYN	Spider
IRNet	38.6	36.7	63.9
RYANSQL	40.1	47.8	70.6
ETA	41.8	50.6	70.8
LGESQL	44.7	53.1	71.7
RATSQL	40.9	48.9	69.7
S2SQL	45.2	55.1	72.1
Our	**46.3 (↑1.1)**	**56.5(↑1.4)**	**72.6(↑0.5)**

Notably, the performance gains are most significant on the more challenging datasets. On Spider-SYN, our model achieves an execution accuracy of 56.4%, surpassing the state-of-the-art S^2SQL by 1.3%. This result highlights the effectiveness of our enhanced schema linking graph. It successfully helps the decoder identify correct schema items even when question terms (e.g. synonyms) are inconsistent with the schema, thereby improving the model's domain generalization capability.

Furthermore, on the Spider-DK dataset, which features implicit domain knowledge, our model achieves a 0.9% improvement over S^2SQL. This demonstrates that our schema linking approach effectively captures latent relationships between the question and the database schema, showcasing the model's robustness. The improvements on the standard Spider benchmark further validate the overall effectiveness of our method.

4.3 Ablation Study

To validate each component of MSAE-SQL, we conducted an ablation study (Table 3). Removing the full schema linking mechanism (*w/o Schema Linking*) caused the most significant performance drop of 1.06%. Disabling its constituent parts, hyperbolic space (*w/o Hyperbolic Space*) and cosine similarity (*w/o Cosine Similarity*), decreased accuracy by 1.03% and 0.77% respectively, confirming their complementary effect. Finally, removing relation embedding (*w/o Relation Embedding*) and table pruning (*w/o Table Pruning*) led to performance drops of 0.43% and 0.33%, respectively.

Table 3. Ablation experiments of different modules.

Approach	Spider	Spider-SYN	Spider-DK
w/o hyperbolic space	72.0	55.0	45.3
w/o cosine similarity	72.1	55.5	45.5
w/o schema linking	70.7	55.7	45.8
w/o relation embedding	72.2	55.8	46.1
w/o irrelevant table pruning	72.3	56.0	46.1
Our	**72.6**	**56.5**	**46.3**

4.4 Case Study

To illustrate our model's advantages, Table 4 presents four cases where MSAE-SQL succeeds while the baseline S^2SQL fails, highlighting its superior schema linking and intent understanding.

Our model excels at **accurate schema linking**. In the first case, S^2SQL fails to map the synonym "institution" to the correct table, `organization`, instead choosing `group`. Our model makes the correct link, a success visualized in Fig. 8, where our correlation matrix shows a significantly stronger connection. In the next case, our model also correctly identifies the necessary linking table `ENROLLED_IN`, whereas S^2SQL selects the wrong source table, `course`.

Furthermore, MSAE-SQL better **captures user intent**. In the third case, it correctly interprets the ambiguous phrase "total passengers" as the `main_services` column, while S^2SQL defaults to an incorrect `COUNT(*)`. In the final case, our model avoids hallucinating an unrequested column (`PetType`), adhering strictly to the user's query.

These cases demonstrate that MSAE-SQL overcomes key Text-to-SQL challenges—including synonymy, relation selection, and semantic ambiguity—to generate more accurate and reliable SQL.

Table 4. Case Study Results

Question	Return me the number of the institution.
S^2SQL	`SELECT COUNT(DISTINCT name) FROM group`
Our	`SELECT COUNT(DISTINCT name) FROM organization;`
Gold	`SELECT COUNT(DISTINCT name) FROM organization;`
Question	Find the first name of students not enrolled in any course.
S^2SQL	`SELECT Fname FROM STUDENT WHERE StuID NOT IN (SELECT StuID FROM course)`
Our	`SELECT Fname FROM STUDENT WHERE StuID NOT IN (SELECT StuID FROM ENROLLED_IN)`
Gold	`SELECT Fname FROM STUDENT WHERE StuID NOT IN (SELECT StuID FROM ENROLLED_IN)`
Question	Show the names and total passengers for all train stations not in London.
S^2SQL	`SELECT name, count(*) FROM station ORDER BY total_passengers DESC LIMIT 3;`
Our	`SELECT name, main_services FROM station ORDER BY total_passengers DESC LIMIT 3;`
Gold	`SELECT name, main_services FROM station ORDER BY total_passengers DESC LIMIT 3;`
Question	Find pets born after 2020-01-01 and their weight.
S^2SQL	`SELECT Pets.PetType, Pets.weight FROM Pets WHERE Pets.birthdate > "2020-01-01";`
Our	`SELECT petid, weight FROM pets WHERE birthdate > '2020-01-01';`
Gold	`SELECT petid, weight FROM pets WHERE birthdate > '2020-01-01';`

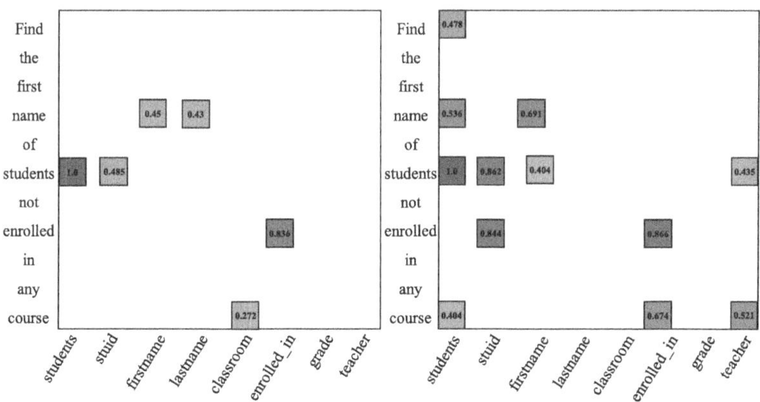

Fig. 8. Correlation between question words and schema items via cosine similarity (left) and hyperbolic space (right); darker colors denote stronger links.

5 Conclusion

This paper introduced **MSAE-SQL**,a Text-to-SQL model that improves cross-domain generalization by tackling the schema linking challenge. Our approach fuses semantic and hierarchical information in a schema linking graph, which, combined with a relation-aware encoder and table pruning, demonstrably outperforms existing methods on challenging benchmarks involving synonyms and implicit domain knowledge. Future work will extend this approach to support interactive feedback, resolve query ambiguity, and query complex hierarchical data by integrating table structure explanation techniques [19].

Acknowledgments. This work was supported by the National Key Research and Development Program of China (2023YFB4503601); the National Natural Science Foundation of China (62172082, 62072084, 62072086).

References

1. Affolter, K., Stockinger, K., Bernstein, A.: A comparative survey of recent natural language interfaces for databases. VLDB J. **28**(5), 793–819 (2019). https://doi.org/10.1007/s00778-019-00567-8
2. Awasthi, A., Chakrabarti, S., Sarawagi, S.: Structured case-based reasoning for inference-time adaptation of text-to-SQL parsers. In: Proc. AAAI, pp. 12536–12544 (2023)
3. Balažević, I., Allen, C., Hospedales, T.: Multi-relational poincaré graph embeddings. NeurIPS **32**, 4465–4475 (2019)
4. Bogin, B., Berant, J., Gardner, M.: Representing schema structure with graph neural networks for text-to-sql parsing. In: Proc. ACL, pp. 4560–4565 (2019)
5. Cao, R., Chen, L., Chen, Z., Zhao, Y., Zhu, S., Yu, K.: LGESQL: line graph enhanced text-to-SQL model with mixed local and non-local relations. In: Proc. ACL-IJCNLP, pp. 2541–2555 (2021)
6. Chen, Z., et al.: Shadowgnn: graph projection neural network for text-to-sql parser. In: Proc. NAACL-HLT, pp. 5567–5577 (2021)
7. Choi, D., Shin, M.C., Kim, E., Shin, D.R.: RYANSQL: recursively applying sketch-based slot fillings for complex text-to-SQL in cross-domain databases. CL **47**(2), 309–332 (2021)
8. Deng, X., Awadallah, A.H., Meek, C., Polozov, O., Sun, H., Richardson, M.: Structure-grounded pretraining for text-to-sql. In: Pro. NAACL-HLT, pp. 1333–1346 (2021)
9. Devlin, J., Chang, M.W., Lee, K., Toutanova, K.: Bert: pre-training of deep bidirectional transformers for language understanding. In: Proc. NAACL-HLT. vol. 1, pp. 4171–4186 (2019)
10. Dong, L., Lapata, M.: Language to logical form with neural attention. In: Proc. ACL, pp. 33–43 (2016)
11. Gan, Y., Chen, X., Huang, Q., Purver, M., Woodward, J.R., Xie, J., Huang, P.: Towards robustness of text-to-SQL models against synonym substitution. In: Proc. ACL-IJCNLP, pp. 2505–2515 (2021)
12. Gan, Y., Chen, X., Purver, M.: Exploring underexplored limitations of cross-domain text-to-SQL generalization. In: Proc. EMNLP, pp. 8926–8931 (2021)
13. Guo, C., Tian, Z., Tang, J., Wang, P., Wen, Z., Yang, K., Wang, T.: A case-based reasoning framework for adaptive prompting in cross-domain text-to-sql. CoRR (2023)

14. Guo, J., et al.: Towards complex text-to-sql in cross-domain database with intermediate representation. In: Proc. ACL, pp. 4524–4535 (2019)
15. He, P., Mao, Y., Chakrabarti, K., Chen, W.: X-sql: reinforce schema representation with context. arXiv preprint arXiv:1908.08113 (2019)
16. Hochreiter, S., Schmidhuber, J.: Long short-term memory. Neural Comput. **9**(8), 1735–1780 (1997)
17. Hui, B., et al.: S^2SQL: injecting syntax to question-schema interaction graph encoder for text-to-SQL parsers. In: Muresan, S., Nakov, P., Villavicencio, A. (eds.) Findings of ACL, pp. 1254–1262 (2022)
18. Lin, X.V., Socher, R., Xiong, C.: Bridging textual and tabular data for cross-domain text-to-sql semantic parsing. In: Findings of EMNLP, pp. 4870–4888 (2020)
19. Liu, F., Tong, S., Shen, D., Nie, T., Kou, Y.: A hierarchical structure explanation method for complex tables. In: Proc. WISA, pp. 308–316 (2024)
20. Liu, Q., Yang, D., Zhang, J., Guo, J., Zhou, B., Lou, J.G.: Awakening latent grounding from pretrained language models for semantic parsing. In: Findings of ACL-IJCNLP, pp. 1174–1189 (2021)
21. Qi, J., et al.: Rasat: integrating relational structures into pretrained seq2seq model for text-to-sql. In: Proc. EMNLP, pp. 3215–3229 (2022)
22. Shi, P., et al.: Learning contextual representations for semantic parsing with generation-augmented pre-training. In: Proc. AAAI, pp. 13806–13814 (2021)
23. Sutskever, I., Vinyals, O., Le, Q.V.: Sequence to sequence learning with neural networks. In: NeurIPS. NIPS'14, vol. 2, pp. 3104–3112 (2014)
24. Vaswani, A., et al.: Attention is all you need. NeurIPS **30**, 5998–6008 (2017)
25. Veličković, P., Cucurull, G., Casanova, A., Romero, A., Liò, P., Bengio, Y.: Graph attention networks. In: ICLR (2018)
26. Wang, B., Shin, R., Liu, X., Polozov, O., Richardson, M.: RAT-SQL: relation-aware schema encoding and linking for text-to-sql parsers. In: Proc. ACL (2020)
27. Wang, B., Gao, Y., Li, Z., Lou, J.G.: Know what i don't know: Handling ambiguous and unknown questions for text-to-sql. In: Findings of ACL, pp. 5701–5714 (2023)
28. Wang, K., Shen, W., Yang, Y., Quan, X., Wang, R.: Relational graph attention network for aspect-based sentiment analysis. In: Proc. ACL, pp. 3229–3238 (2020)
29. Yan, Z., Ma, J., Zhang, Y., Shen, J.: SQL generation via machine reading comprehension. In: Proc. COLING, pp. 350–356 (2020)
30. Yang, M., et al.: Hyperbolic graph neural networks: a review of methods and applications. arXiv preprint arXiv:2202.13852 (2022)
31. Yu, T., Yasunaga, M., Yang, K., Zhang, R., Wang, D., Li, Z., Radev, D.: Syntaxsqlnet: syntax tree networks for complex and cross-domain text-to-sql task. In: Proc. EMNLP, pp. 1653–1663 (2018)
32. Yu, T., et al.: SPIDER: a large-scale human-labeled dataset for complex and cross-domain semantic parsing and text-to-SQL task. In: Proc. EMNLP, pp. 3911–3921 (2018)

SSCP-HGC: Structural and Semantic Commonality Perception in Heterogeneous Graph Contrastive Learning for Recommendation

Shiquan Luo[1], Shaojie Ji[1], Guohua Chen[2,5](✉), Feiyi Tang[3,5], Ronghua Lin[1], Weisheng Li[1], and Yong Tang[1,4](✉)

[1] School of Computer Science, South China Normal University, Guangzhou, China
[2] Network and Information Center, South China Normal University, Guangzhou, China
[3] School of Information Engineering, Guangzhou Panyu Polytechnic, Guangzhou, China
[4] Institute of Data Intelligence, Guangdong University of Science and Technology, Guangzhou, China
ytang4@qq.com
[5] Pazhou Lab, Guangzhou, China
chengh@m.scnu.edu.cn

Abstract. In heterogeneous graph recommendation tasks, the interaction data between users and items contains heterogeneous characteristics and high-order semantic information, providing abundant data support for the optimization of recommendation systems. Most existing mainstream methods construct multi-graph representations based on metapath mechanisms. However, the subjectivity and limitations in meta-path selection often lead to the introduction of noisy data or the omission of critical information, which in turn constrains the improvement of recommendation performance. To address this issue, this paper proposes a structural and semantic commonality perception heterogeneous graph contrastive learning model (SSCP-HGC) for recommendation tasks, aiming to reduce the dependence on the quality of meta-paths by leveraging both structural and semantic commonalities in heterogeneous graphs. Specifically, SSCP-HGC introduces a structural commonality enhancement module, which reinforces the structural commonality across graphs by capturing stable and implicit connections in multiple graphs. Meanwhile, a semantic commonality generation module is designed to extract core semantics from the high-order semantic information encoded by meta-paths, revealing the semantic consistency of nodes of the same type across multiple graphs. These modules jointly optimize the graph representations within a contrastive learning framework. Experimental results on multiple public datasets show that the proposed method outperforms existing methods, validating its effectiveness in reducing meta-path quality dependency and enhancing recommendation performance.

S. Yang et al. (Eds.): WISA 2025, LNCS 16299, pp. 362–373, 2026.
https://doi.org/10.1007/978-981-95-6426-2_33

Keywords: Recommendation System · Heterogeneous Graph · Contrastive Learning · Structural Commonality · Semantic Commonality

1 Introduction

In the research of heterogeneous graph recommendation systems, the interaction data between users and items exhibits significant heterogeneity and contains rich semantic information, such as latent preferences of users [1] and intent perception [2]. Current mainstream methods typically rely on meta-path mechanisms to reconstruct the user-item interaction graph into multiple meta-path graphs [3–5], aiming to extract high-order semantic, correlations through these graphs. However, this approach faces several challenges. Firstly, although constructing multiple views using predefined meta-paths can encode specific semantics, low-quality meta-paths often introduce noise or omit critical information [6], and this issue is further amplified in independently optimized contrastive learning frameworks, ultimately degrading recommendation performance [7]. Second, existing methods often overlooks the collaborative modeling of global semantic structures across graphs, making it difficult to comprehensively capture the complex semantic associations in user-item interactions, thereby limiting the improvement of recommendation performance.

To address the above challenges, we proposed solutions from the dimensions of structural commonality and semantic commonality. Specifically, structural commonality refers to node pairs that maintain stable connections across most meta-path graphs (e.g., the connections between s1 and s2, and between s2 and s3 in Fig. 1 (b)), as well as node pairs with implicit connections (e.g., the potential link between s1 and s3). Such structural commonality constitutes an essential part of the high-order semantic information in meta-paths [8]. In addition, we introduce the concept of semantic commonality, which essentially refers to the consistent part in the semantic spaces of different graphs and can effectively suppress invalid associations introduced by meta-paths [9]. Overall, the main contributions of this paper are as follows:

- We propose an innovative heterogeneous graph contrastive learning model, SSCP-HGC, for recommendation tasks. The structural commonality enhancement module within this model is designed to strengthen stable and implicit connections across multiple meta-path graphs, effectively reducing the dependence on meta-path quality.
- We propose a semantic commonality generation module by employing multi-head attention to aggregate global graphs and extract semantic commonality, thereby enhancing the consistency and generalization ability of node representations.
- Extensive experiments conducted on four public datasets demonstrate that the proposed model consistently outperforms existing baseline methods.

2 Related Work

2.1 Heterogeneous Graph Recommendation System

Frontier research on heterogeneous graph recommendation systems focuses on exploring the synergistic mechanisms between meta-paths and graph neural networks (GNN). For instance, the HGE-DED [10] model identifies and aggregates multi-type neighbors of target nodes via defined meta-paths and adopts a bilateral differentiation strategy to account for edge-type differences in GNN message passing, thereby enhancing its ability to capture the complex structures of heterogeneous graphs. The M2GNN [11] model constructs a cross-domain heterogeneous information network based on predefined meta-path patterns and employs a hierarchical GNN aggregation framework to perform aggregation within meta paths for filtering irrelevant labels and extracting user interests. With their strong capabilities in semantic mining and structural exploitation, heterogeneous graph recommendation systems have been widely applied in social recommendation [3], e-commerce [12], and media recommendation [13], demonstrating broad prospects for development.

2.2 Graph Contrastive Learning

Graph Contrastive Learning (GCL) introduces the principles of contrastive learning into GNN [14], optimizing through the construction of positive and negative sample pairs for contrastive optimization [15]. GCL methods are generally categorized into graph-enhancement-based and structure-aware approaches. In graph-enhancement-based methods, the AdaGCL framework [16] employs a graph generator and denoising mechanism to produce personalized augmented graphs, effectively addressing data sparsity and noise in recommendation tasks. In structure-aware methods, LS-GCL [17] constructs semantic-guided subgraphs and multi-level neighbor graphs, introducing a local-global structural contrast strategy to enhance the structural fidelity of node representations.

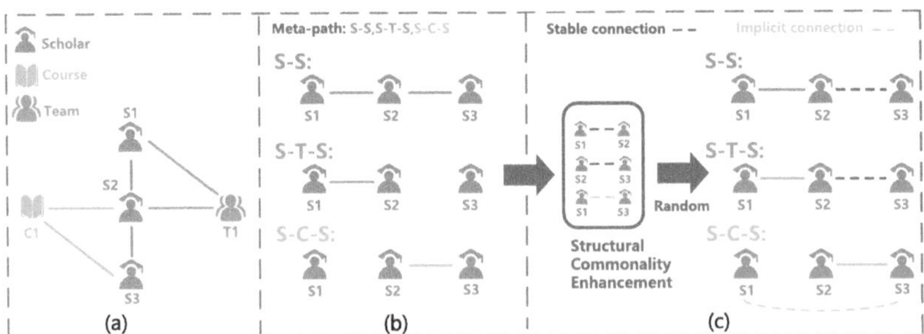

Fig. 1. (a) The heterogeneous graph of the Scholat dataset includes three types of nodes: scholars, courses, and teams. (b) Meta-path graphs constructed based on different meta-paths. (c) Meta-path graphs after structural commonality enhancement.

3 Preliminaries

3.1 Meta-Path Graph

In a heterogeneous graph $\mathcal{G}$, a meta-path ρ is defined as a sequence composed of node types and edge types [18], typically represented as: $\mathcal{T}_{v_i} \xrightarrow{\mathcal{T}_{e_1}} \mathcal{T}_{v_2} \xrightarrow{\mathcal{T}_{e_2}} \cdots \xrightarrow{\mathcal{T}_{e_k}} \mathcal{T}_{v_{k+1}}$, where $\mathcal{T}_{v_i} \in \mathcal{T}_v$ and $\mathcal{T}_{e_j} \in \mathcal{T}_e$, where $\mathcal{T}_v$ denotes the set of node types and $\mathcal{T}_e$ denotes the set of edge types. Based on the concept of meta-path, the meta-path graph $\mathcal{G}_\rho$ is constructed by connecting all pairs of nodes in $\mathcal{G}$ that match the start and end node types of the meta-path ρ. This process involves removing all edges in the original graph $\mathcal{G}$ that are irrelevant to the meta-path ρ, as well as the intermediate nodes of the meta-path instances. Only the nodes at the ends of the meta-path instances are retained, and new edges are established between these node pairs.

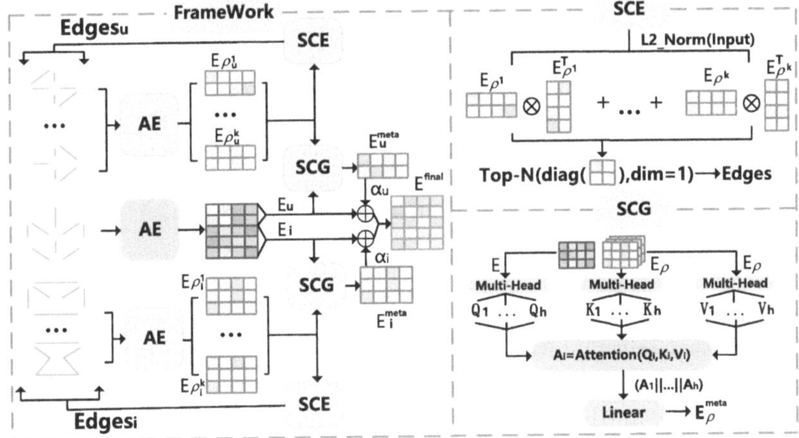

Fig. 2. The complete framework of our proposed SSCP-HGC. The model input is first transformed into embedding representations through an autoencoder (AE). The structure commonality enhancement (SCE) module then optimizes the meta-path graphs based on these embeddings, while the semantic commonality generation (SCG) module extracts semantic commonality features from meta-path graphs of the same category.

4 Methodology

Figure 2 shows the overall framework of the proposed SSPC-HGC model. The SSPC-HGC model consists of four core components: embedding generation, structural commonality enhancement module, semantic commonality generation module, and optimization objective.

4.1 Embedding Generation

The initial input of the model includes a core user-item graph and multiple meta-path graphs. To distinguish between user and item groups during recommendation, we enforce that the start and end node types of a meta-path must be identical, categorizing meta-paths into user and item categories [19].

To balance the distribution of initial embeddings and enhance the robustness of the model in noisy environments and scenarios with missing features, this study adopts the xavier initialization strategy [20] for the initialization of node embeddings. Simultaneously, a random masking operation is applied to the node feature set to simulate data incompleteness [21]. On this basis, an independent autoencoder [22] is constructed for each graph, where both the encoder and decoder modules are implemented based on lightGCN [23]. Specifically, the neighborhood aggregation of lightGCN follows the following paradigm:

$$\mathbf{E}^{(l+1)} = \left(\mathbf{D}^{-\frac{1}{2}}\mathbf{A}\mathbf{D}^{-\frac{1}{2}}\right)\mathbf{E}^{(l)}; \quad \mathbf{E} = \sum_{l=0}^{L} \alpha_l \mathbf{E}^{(l)} \tag{1}$$

where $\mathbf{E}^{(l)}$ is the feature embedding after the l-th layer of lightGCN convolution, and $\mathbf{D}^{-\frac{1}{2}}\mathbf{A}\mathbf{D}^{-\frac{1}{2}}$ is the normalized adjacency matrix. The final encoder output $\mathbf{E}$ is obtained by taking a weighted sum of the embeddings $\mathbf{E}^{(l)}$ from different layers, where each embedding is multiplied by a corresponding weight coefficient α_l. Typically, these weights are set to be uniformly distributed. For different graphs, we define the output of the autoencoder as:

$$\textbf{User} - \textbf{Item}: E_u,\ E_i \quad \textbf{User}: E_{\rho_u^k} \quad \textbf{Item}: E_{\rho_i^k} \tag{2}$$

where E_u and E_i denote the outputs of the autoencoder for user nodes and item nodes, respectively, after the overall graph convolution in the core user-item graph. Meanwhile, $E_{\rho_u^k}$ is defined as the autoencoder output extracted from meta-path ρ_u^k in the user category. Correspondingly, $E_{\rho_i^k}$ is defined as the encoder output generated from meta-path ρ_i^k in the item category.

4.2 Structural Commonality Enhancement

In this section, we design structural commonality enhancement(SCE) module to reduce the reliance on metapath quality. In each iteration of the model, we compute the similarity matrix between nodes of the same type within the same graph. Then, we fuse the mean similarity across same category graphs. Taking the user category meta-path graph as an example:

$$\textbf{SimM}_u = \frac{1}{K_u}\sum_{k=0}^{K_u} \textbf{diag}(f_{sim}(\mathbf{E}_{\rho_u^k})) \tag{3}$$

where $f_{sim}(\cdot)$ is used to compute the cosine similarity between any two embeddings within a given matrix. The diag$(\cdot)$ sets the diagonal elements of the similarity matrix to zero, eliminating self-similarity between nodes. K_u denote the number of meta-paths in the user category. $SimM_u$ represents the average node similarity across meta-path graphs in the user category. Similarly, an average similarity matrix $SimM_i$ between item node pairs can also be obtained

We focus on establishing connections for each node with the top n most relevant nodes, which may either already exist in the original graph or be implicit connections:

$$\mathbf{Edges}_u = \{(i,j) \mid j \in \mathrm{TopN}_{k=1}^{n}\left(\mathrm{SimM}_u(i,k)\right),\ i = 1,2,\ldots,n\} \tag{4}$$

The operator TopN indicates the indices of the top-N highest similarity values. Edges_u represents the set of top-N relevant connections for users. Finally, we replace or insert these structural commonality into the corresponding meta-path graphs of the same type with a certain probability(as illustrated in Fig. 1 (c)). The enhanced meta-path graphs will be used in the next iteration of the model. The structural commonality of item category, Edges_i, is similarly injected into the meta-path graphs of each item type.

4.3 Semantic Commonality Generation

In addition, inspired by the idea of semantic fusion [24], we introduce a semantic commonality generation module to extract the semantics commonality of homogeneous nodes from different graph structures. We employ a multi-head attention mechanism [25] in this module and assign separate multi-head attention networks for users and item categories. Taking user category as an example:

$$Q = E_u W^Q, \quad K = E_{\rho_u^k} W^K, \quad V = E_{\rho_u^k} W^V \tag{5}$$

Q, K, and V are the query matrix, key matrix, and value matrix, respectively, each with a dimension of d. The corresponding learnable weight matrices are W^Q, W^K, and W^V. Owing to the utilization of the multi-head attention mechanism, we decompose the original matrices along the feature dimension. Consequently, each head is associated with its distinct Q_i, K_i, and V_i, and the feature dimension for each head is $\frac{d}{h}$. The attention scores for each head are computed as:

$$\mathrm{Attention}_i = \mathrm{softmax}\left(\frac{Q_i K_i^\top}{\sqrt{\frac{d}{h}}}\right) V_i \tag{6}$$

where $\mathrm{Attention}_i$ is the normalized attention score for head i. Subsequently, the outputs of all attention heads are concatenated and projected back to the original feature dimension through a new weight matrix, resulting in the semantic commonality embedding $E_{\rho_u^k}^{meta}$ for user nodes. Similarly, the semantic commonality embedding $E_{\rho_i^k}^{meta}$ for item nodes is obtained.

Finally, we obtain the final embeddings for user and item category nodes, respectively:

$$E_u^{\mathrm{final}} = E_u + \frac{\alpha_u}{K_u} \sum_{k=1}^{K_u} E_{\rho_u^k}^{\mathrm{meta}}; \quad E_i^{\mathrm{final}} = E_i + \frac{\alpha_i}{K_i} \sum_{k=1}^{K_i} E_{\rho_i^k}^{\mathrm{meta}} \tag{7}$$

The hyperparameters α_u and α_i control the weights of the semantic commonality embeddings for user and item nodes, respectively. K_u and K_i denote the number of meta-paths for user and item categories, respectively. E_u^{final} and E_i^{final} represent the final embedding representations for users and items, respectively.

4.4 Optimization Objective

In this model, after processing the core user-item graph through autoencoder, the user and item embeddings may suffer from potential distribution bias due to the independence of feature extraction, which could negatively impact the accuracy of recommendation results. Inspired by the uniformity loss studies in the NFGCL [26] model, this research introduces a uniformity loss function. This loss function independently computes the uniformity loss of user embeddings in the latent semantic space, thereby optimizing the performance of the model. Taking the uniformity loss of users as an example:

$$\mathcal{L}_{uf}^{user} = \frac{2}{n(n-1)} \sum_{1 \le i < j \le n} \exp(-\|e_{u_i} - e_{u_j}\|^2) \tag{8}$$

The function calculates the loss by computing the average Euclidean distance between embeddings of nodes of the same type. e_{u_i} and e_{u_j} are the embeddings of the i and j users in the core user-item graph, respectively. Similarly, the item uniformity loss $\mathcal{L}_{uf}^{item}$ can be derived accordingly. Next, we introduce the pairwise Bayesian Personalized Ranking (BPR) loss [27], which is commonly used in recommendation tasks. The optimization objective is achieved by maximizing the preference score difference of users between positive and negative samples:

$$\mathcal{L}_{\text{bpr}} = -\frac{1}{|D|} \sum_{(u,i,j) \in D} \ln \sigma \left(E_u^{\text{final}}(u,:) \cdot \left(E_i^{\text{final}}(i,:) - E_i^{\text{final}}(j,:) \right) \right) \tag{9}$$

where D is set of triples obtained through random sampling, where each triplet consists of user u, positive item i, and negative item j. σ is the sigmoid function. Given the effective application of the InfoNCE [28] loss function in graph contrastive learning, we design an InfoNCE loss computation function based on multi-path graph contrastive learning. This loss function calculates the InfoNCE loss between any two graphs of the same type and takes the average as the InfoNCE loss for that category of graphs. Taking the user category as an example:

$$\mathcal{L}_{\text{in}}^{user} = \frac{2}{K_u(K_u-1)} \sum_{1 \le a < b \le K_u} \sum_{i \in \mathcal{D}_{\text{user}}} -\log \frac{\exp(s(e_i^{\rho_a}, e_i^{\rho_b})/\tau)}{\sum_{j \in \mathcal{D}_{\text{user}}} \exp(s(e_i^{\rho_a}, e_j^{\rho_b})/\tau)} \tag{10}$$

where $\mathcal{D}_{\text{user}}$ denotes the set of users in the sampled batch, $e_i^{\rho_a}$ represents the embedding vector of user i under the metapath ρ_a graph, s is the similarity function (e.g., cosine similarity), τ is the temperature coefficient. Similarly, we can also obtain the $\mathcal{L}_{\text{in}}^{item}$ based on multiple item graphs. The final loss function of SSPC-HGCL is $\mathcal{L} = \mathcal{L}_{\text{bpr}} + (\lambda_{uf}^1 \cdot \mathcal{L}_{\text{uf}}^{user} + \lambda_{uf}^2 \cdot \mathcal{L}_{\text{uf}}^{item}) + (\lambda_{in}^1 \cdot \mathcal{L}_{\text{in}}^{user} + \lambda_{in}^2 \cdot \mathcal{L}_{\text{in}}^{item})$. These four λ are hyperparameters used to adjust the weights of the uniformity loss and the contrastive learning loss.

5 Experiments

In this section, we conduct a series of experiments to validate the effectiveness of the SSPC-HGC model.

Table 1. The statistical information of experimented datasets.

Datasets	Users	Items	Interactions	Meta-paths
Amazon	6170	2753	195791	UIU, UIBIU, IUI, IBI, ICI
Yelp	16239	14284	198397	UU, UBU, BCiB, BCaB
Douban Book	13024	22347	792062	UU, UGU, UBU, BUB, BAB, BYB
Scholat	2302	800	32790	UU, UTU, UCU, TUT

5.1 Experimental Settings

Datasets. The effectiveness of the SSCP-HGC model is empirically validated through experiments conducted on four real-world datasets: Amazon [29], Yelp [30], Douban Book [19], and Scholat[1]. Table 1 presents the statistical details of each dataset. For the Scholat dataset (formally referred to as the SCHOLAT Multiplex Network), we adapted it to fit the heterogeneous graph scenario. Additionally, to mitigate the impact of the long-tail distribution, users with fewer than three interactions were removed from the selected datasets.

Baselines. To evaluate SSPC-HGC's performance, this study compares it with two mainstream types of baseline methods. The first includes graph neural network-based approaches like LightGCN [23], SimGCL [31], and HGCL [20]. The second covers intent-based methods such as BIGCF [32] and IHGCL [19]. To ensure fair experimental comparisons, each baseline model is independently tuned to ensure it reaches the theoretically optimal performance.

5.2 Overall Performance

Table 2 summarizes the recommendation performance evaluation results of various baseline models on four public datasets. Statistical analysis shows that the proposed SSCP-HGC model demonstrates significant advantages on four of the benchmark datasets, outperforming the SOTA baseline models based on heterogeneous graph neural networks. Specifically, compared to existing SOTA methods, the SSCP-HGC model achieves an average improvement of 5.88% in normalized discounted cumulative gain and 4.29% in recall, validating the effectiveness of its architectural design. Further analysis shows intent-aware methods outperform traditional heterogeneous graph models. The SSCP-HGC model enhances recommendation stability by strengthening stable node connections and simulating intention perception through latent relationship mining.

[1] https://www.scholat.com/research/opendata/#multiplex_network.

Table 2. Recommendation performance comparison between the SSPC-HGC model and baseline methods. R@20 denotes Recall@20, and N@20 denotes NDCG@20.

Dataset	Amazon		Yelp		Douban Book		Scholat	
Metric	R@20	N@20	R@20	N@20	R@20	N@20	R@20	N@20
LightGCN	0.1694	0.1228	0.0974	0.0687	0.1671	0.1553	0.1091	0.0860
SimGCL	0.1875	0.1407	0.1107	0.0776	0.1936	0.1789	0.1228	0.1074
HGCL	0.1787	0.1327	0.1064	0.0757	0.1850	0.1753	0.1139	0.0953
BIGCF	0.1880	0.1408	0.1142	0.0781	0.1908	0.1807	0.1271	0.1057
IHGCL	0.1934	0.1435	0.1166	0.0817	0.1963	0.1828	0.1282	0.1085
SSPC-HGC	**0.1982**	**0.1526**	**0.1241**	**0.0872**	**0.2085**	**0.1933**	**0.1308**	**0.1136**

Table 3. Ablation study of SSPC-HGC.

Dataset	Amazon		Yelp		Douban Book		Scholat	
Metric	R@20	N@20	R@20	N@20	R@20	N@20	R@20	N@20
w/o-SCE	0.1793	0.1385	0.1032	0.0694	0.1723	0.1672	0.1102	0.0972
w/o-SCG	0.1859	0.1427	0.1025	0.0684	0.1898	0.1829	0.1163	0.1066
SSPC-HGC	**0.1982**	**0.1526**	**0.1241**	**0.0872**	**0.2085**	**0.2036**	**0.1308**	**0.1136**

5.3 Ablation Study

In the ablation study, we design two variants of SSCP-HGC for analysis. Specifically, w/o-SCE denotes the removal of the structural commonality enhancement module, while w/o-SCG indicates the removal of the semantic commonality generation module. The experimental results are presented in Table 3.

In the w/o-SCE variant, the model performance drops significantly, indicating that the SCE module plays a crucial role in enhancing the discriminative power of embeddings by capturing structural commonality in the heterogeneous graph. The performance degradation of the w/o-SCG variant is slightly lower than that of w/o-SCE, which can be attributed to the absence of the SCG module weakening the semantic alignment among homogeneous nodes, thereby reducing the quality of the final node embeddings.

5.4 Performance on Low Quality Meta-Paths

In this experiment, we simulate the presence of noise or missing information caused by low-quality meta-paths by randomly adding or removing a certain proportion of edges in the meta-path graphs. We then systematically evaluate the dependency of baseline models and SSCP-HGC variants on meta-path quality. The experimental results are presented in Fig. 3.

In the experiments conducted on baseline models under varying levels of edge perturbation, SSCP-HGC exhibits the smallest fluctuations in recommendation

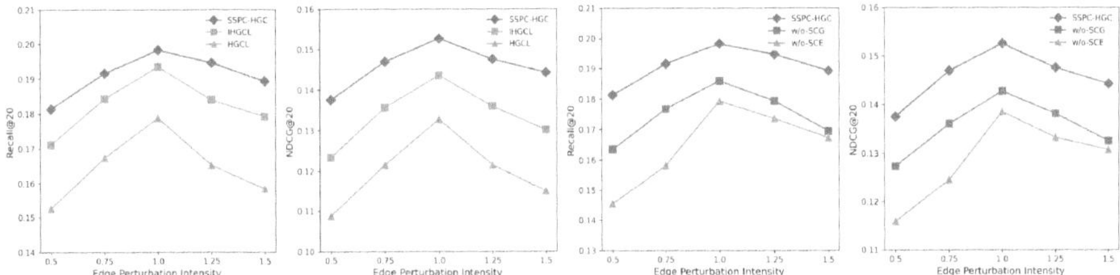

Fig. 3. Comparison of recommendation performance of models under different edge perturbation levels on Amazon. The x-axis represents perturbation intensity, defined as the ratio of edges in the perturbed graph to the original graph.

performance, demonstrating its effectiveness in mitigating the reliance on meta-path quality. In the experiments on the SSPC-HGC variant, when additional invalid connections are introduced (edge perturbation intensity greater than 1.0), the recommendation performance of the w/o-SCG variant degrades more significantly. This phenomenon underscores the critical role of the SCG module in filtering out irrelevant connections by generating semantic commonality embeddings. In contrast, under experimental conditions where effective connections are removed (edge perturbation intensity less than 1.0), the recommendation performance of the w/o-SCE variant experiences a sharp decline. This result strongly demonstrates that the SCE module effectively enhances meaningful connections between nodes by reinforcing structural commonality features.

6 Conclusion

In this paper, we propose a novel heterogeneous graph contrastive learning recommendation model, SSCP-HGC. This model addresses the strong reliance of heterogeneous graph recommendation systems on meta-path quality from the dual perspectives of structural and semantic commonalities. Specifically, the structural commonality enhancement module effectively addresses the issues of low-quality meta-paths introducing irrelevant associations or missing critical information by reinforcing stable and implicit connections across multiple graphs of the same category. The semantic commonality generation module leverages a multi-head attention mechanism to extract semantic consistency across graphs, thereby significantly enhancing the model's generalization ability to complex interaction patterns. Extensive experimental results demonstrate that SSCP-HGC significantly outperforms existing baseline models in recommendation performance while exhibiting lower sensitivity to meta-path quality.

References

1. Wu, J., Fan, W., Chen, J., Liu, S., Li, Q., Tang, K.: Disentangled contrastive learning for social recommendation. In: Proceedings of the 31st ACM International Conference on Information & Knowledge Management, pp. 4570–4574 (2022)

2. Fan, S., et al.: Metapath-guided heterogeneous graph neural network for intent recommendation. In: Proceedings of the 25th ACM SIGKDD International Conference on Knowledge Discovery & Data Mining, pp. 2478–2486 (2019)
3. Jiang, Y., Ma, H., Zhang, X., Li, Z., Chang, L.: Incorporating metapath interaction on heterogeneous information network for social recommendation. Front. Comp. Sci. **18**(1), 181302 (2024)
4. Tan, L., Gong, D., Jinmao, X., Li, Z., Liu, F.: Meta-path fusion based neural recommendation in heterogeneous information networks. Neurocomputing **529**, 236–248 (2023)
5. Zhang, Y., Liao, W., Wang, Y., Zhu, J., Chen, R., Zhang, Y.: Meta-path automatically extracted from heterogeneous information network for recommendation. World Wide Web **27**(3), 26 (2024)
6. Yang, C., Gong, X., Shi, C., Philip, Y.: A post-training framework for improving heterogeneous graph neural networks. In: Proceedings of the ACM Web Conference, vol. 2023, pp. 251–262 (2023)
7. Yu, J., Ge, Q., Li, X., Zhou, A.: Heterogeneous graph contrastive learning with meta-path contexts and adaptively weighted negative samples. IEEE Trans. Knowl. Data Eng. **36**(10), 5181–5193 (2024)
8. Jiang, J., Xu, L.: Jointly learning structure-augmented semantic representation and logical rules for knowledge graph completion. In: International Conference on Web Information Systems and Applications, pp. 52–63. Springer, Heidelberg (2023). https://doi.org/10.1007/978-981-99-6222-8_5
9. Zhang, D., Yin, J., Zhu, X., Zhang, C.: MetaGraph2Vec: complex semantic path augmented heterogeneous network embedding. In: Phung, D., Tseng, V.S., Webb, G.I., Ho, B., Ganji, M., Rashidi, L. (eds.) PAKDD 2018. LNCS (LNAI), vol. 10938, pp. 196–208. Springer, Cham (2018). https://doi.org/10.1007/978-3-319-93037-4_16
10. Chen, Y., et al.: Heterogeneous graph embedding with dual edge differentiation. Neural Netw. **183**, 106965 (2025)
11. Huai, Z., Yang, Y., Zhang, M., Zhang, Z., Li, Y., Wu, W.: M2gnn: metapath and multi-interest aggregated graph neural network for tag-based cross-domain recommendation. In: Proceedings of the 46th International ACM SIGIR Conference on Research and Development in Information Retrieval, pp. 1468–1477 (2023)
12. Zheng, J., Li, Q., Liao, J.: Heterogeneous type-specific entity representation learning for recommendations in e-commerce network. Inf. Process. Manag. **58**(5), 102629 (2021)
13. Cai, D., Qian, S., Fang, Q., Jun, H., Ding, W., Changsheng, X.: Heterogeneous graph contrastive learning network for personalized micro-video recommendation. IEEE Trans. Multimedia **25**, 2761–2773 (2022)
14. You, Y., Chen, T., Sui, Y., Chen, T., Wang, Z., Shen, Y.: Graph contrastive learning with augmentations. Adv. Neural. Inf. Process. Syst. **33**, 5812–5823 (2020)
15. Xue, X., Xu, B., Dong, X., Cai, Q., Gao, K.: Contrastive learning-based cross-domain data augmentation for aspect-based sentiment analysis. In: International Conference on Web Information Systems and Applications, pp. 180–188. Springer, Heidelberg (2024). https://doi.org/10.1007/978-981-97-7707-5_16
16. Jiang, Y., Huang, C., Huang, L.: Adaptive graph contrastive learning for recommendation. In: Proceedings of the 29th ACM SIGKDD Conference on Knowledge Discovery and Data Mining, pp. 4252–4261 (2023)
17. Yang, K., Liu, Y., Zhao, Z., Ding, P., Zhao, W.: Local structure-aware graph contrastive representation learning. Neural Netw. **172**, 106083 (2024)

18. Dong, Y., Chawla, N.V., Swami, A.: metapath2vec: scalable representation learning for heterogeneous networks. In: Proceedings of the 23rd ACM SIGKDD International Conference on Knowledge Discovery and Data Mining, pp. 135–144 (2017)
19. Sang, L., Wang, Y., Zhang, Y., Zhang, Y., Wu, X.: Intent-guided heterogeneous graph contrastive learning for recommendation. IEEE Trans. Knowl. Data Eng. (2025)
20. Chen, M., Huang, C., Xia, L., Wei, W., Xu, Y., Luo, R.: Heterogeneous graph contrastive learning for recommendation. In: Proceedings of the Sixteenth ACM International Conference on Web Search and Data Mining, pp. 544–552 (2023)
21. Tian, Y., Dong, K., Zhang, C., Zhang, C., Chawla, N.V.: Heterogeneous graph masked autoencoders. In: Proceedings of the AAAI Conference on Artificial Intelligence, vol. 37, pp. 9997–10005 (2023)
22. Wang, W., et al.: Hgate: heterogeneous graph attention auto-encoders. IEEE Trans. Knowl. Data Eng. **35**(4), 3938–3951 (2021)
23. He, X., Deng, K., Wang, X., Li, Y., Zhang, Y., Wang, M.: Lightgcn: simplifying and powering graph convolution network for recommendation. In: Proceedings of the 43rd International ACM SIGIR conference on research and development in Information Retrieval, pp. 639–648 (2020)
24. Wang, X., et al.: Heterogeneous graph attention network. In: The World Wide Web Conference, pp. 2022–2032 (2019)
25. Ziniu, H., Dong, Y., Wang, K., Sun, Y.: Heterogeneous graph transformer. In: Proceedings of the Web Conference 2020, pp. 2704–2710 (2020)
26. Xiao, Y., Ma, R., Sang, J.: NFGCL: a negative-sampling-free graph contrastive learning framework for recommendation. Inf. Sci. **695**, 121732 (2025)
27. Rendle, S., Freudenthaler, C., Gantner, Z., Schmidt-Thieme, L.: BPR: bayesian personalized ranking from implicit feedback. arXiv preprint arXiv:1205.2618 (2012)
28. Chen, T., Kornblith, S., Norouzi, M., Hinton, G.: A simple framework for contrastive learning of visual representations. In: International Conference on Machine Learning, pp. 1597–1607. PmLR (2020)
29. Shi, C., Hu, B., Zhao, W.X., Yu, P.S.: Heterogeneous information network embedding for recommendation. IEEE Trans. Knowl. Data Eng. **31**(2), 357–370 (2018)
30. Wang, H., Zhou, K., Zhao, X., Wang, J., Wen, J.-R.: Curriculum pre-training heterogeneous subgraph transformer for top-n recommendation. ACM Trans. Inf. Syst. **41**(1), 1–28 (2023)
31. Yu, J., Yin, H., Xia, X., Chen, T., Cui, L., Nguyen, Q.V.H.: Are graph augmentations necessary? Simple graph contrastive learning for recommendation. In: Proceedings of the 45th International ACM SIGIR Conference on Research and Development in Information Retrieval, pp. 1294–1303 (2022)
32. Zhang, Y., Sang, L., Zhang, Y.: Exploring the individuality and collectivity of intents behind interactions for graph collaborative filtering. In: Proceedings of the 47th International ACM SIGIR Conference on Research and Development in Information Retrieval, pp. 1253–1262 (2024)

Automated Modularization of Kernel Components Based on Type Dependencies and Interface Semantics

Linghao Xu and Wenjia Zhao(✉)

Xi'an Jiaotong University, Xi'an 710049, Shannxi, China
2201412018@stu.xjtu.edu.cn, zhaowenjia@xjtu.edu.cn

Abstract. With the development of software technology, the internal structure of programs is becoming more and more complex. Typically, programs lack effective internal isolation, making them vulnerable to attackers who exploit security flaws to compromise the entire system. Program modularization is a security mechanism to solve this problem. In this paper, we introduce an automated modularization technique for operating system kernels, leveraging type-based dependency analysis and kernel interface features. The proposed approach initially performs a dependency analysis based on data types to generate a dependency relationship network. Then, existing modular information within this network is extracted using community detection, employing GN-MQ as a metric for evaluating module quality and an improved Louvain algorithm to optimize the partitioning. The features of the operating system kernel interface are also extracted and analyzed for the operating system kernel interface, which is used as enlightening information to guide the modularization process and make the algorithm converge faster. We evaluated the effectiveness of the proposed approach using two distinct versions of the Linux kernel source code. The results demonstrate that several modules identified by our technique in the older kernel version have been subsequently refactored or removed in the newer version, confirming the practical relevance of our approach. The accuracy of the experimental results is further validated through comparative grouping using the MoJoFM algorithm.

Keywords: Dependence Analysis · Program Modularization · Community Detection · Operating System Kernel

1 Introduction

With the development of modern software technology, software has become more and more feature-rich, and consequently the complexity of software code continues to grow. In the traditional software design process, since there is usually no internal isolation in the program, once there is a security vulnerability in the

S. Yang et al. (Eds.): WISA 2025, LNCS 16299, pp. 374–385, 2026.
https://doi.org/10.1007/978-981-95-6426-2_34

program, a malicious attacker can take advantage of the vulnerability to damage the actual memory and then control the control or data flow of the whole program. Operating system as the core component in a computer system, the security of the operating system is often related to the security of the whole computer system.

Program modularization has emerged as an effective solution to this challenge. Modularization divides complex programs into smaller, self-contained units (modules) with clearly defined functionalities and isolated memory spaces. By limiting memory access between modules, this approach reduces the propagation of vulnerabilities and mitigates their impact on the entire system. Beyond enhancing security, modularization also improves software maintainability and facilitates program refactoring and optimization.

A prerequisite for effective modularization is dependency analysis, which identifies potential control and data dependencies among program elements. The standard procedure involves determining control dependencies and then conducting data flow analysis over a reachable control flow graph. However, performing accurate dependency analysis in large-scale C/C++ programs remains challenging due to factors such as indirect call targets, pointer aliasing, and path explosion. In these contexts, determining the precise scope of memory access or influence becomes computationally infeasible.

To address these limitations, Lu et al. [8] proposed a type-based modular dependency analysis technique (TyPM), which provides an efficient means of identifying inter-module dependencies by leveraging type information. Building upon this idea, modular structures in dependency networks can be further explored using community detection techniques. In graph theory, community detection aims to partition a network into subgroups (communities) with dense internal connections, and this concept can be effectively applied to extract modular information from program dependency graphs [2].

In this paper, we propose a modularization framework for operating system kernels that integrates type-based dependency analysis with kernel interface semantics. First, we construct a dependency network using type relationships and apply community detection using an improved Louvain algorithm guided by the GN-MQ modular quality metric. To further accelerate convergence and improve modular granularity, we incorporate semantic features of kernel interfaces–particularly function names and parameter lists–to provide heuristic guidance. Since functionally related interfaces tend to reside in the same module, analyzing these semantic cues improves modular partitioning accuracy.

To summarize, the main contributions of this paper are as follows:

1. We develop an automated tool for performing modular partitioning on operating system kernels by analyzing dependencies extracted from binary-level program representations.
2. We adapt and enhance a community detection algorithm to operate effectively on dependency call graph networks, enabling accurate modular segmentation.
3. We incorporate semantic similarity analysis based on interface features to guide and refine the modularization process.

4. We conduct extensive evaluations to validate the effectiveness and accuracy of the proposed modularization method.

2 Related Work

Over the years, a variety of techniques have been developed for program modularization. Bunch [9] constructs Module Dependency Graphs (MDGs) using tools such as CIA and Acacia, and formulates modularization as an optimization problem aimed at maximizing the Modularity Quality (MQ) metric. Clustering is performed using strategies such as hill-climbing and genetic algorithms. Bunch also supports user-defined clustering constraints, reduces the scope of automatic clustering, accelerates the clustering process, enables incremental structural maintenance, and fine-tunes subsystem structures. However, its MDG model is unweighted and does not capture inter-module dependency strength, limiting its precision. Bavota et al. [1] proposed the R3 tool, which incorporates semantic analysis to enhance modularization. Their method extracts semantic terms from the source code and evaluates their importance using a TF-IDF weighting scheme. Static analysis is employed to derive dependencies, which are then encoded into a term-document matrix. Topic modeling techniques (e.g., RTM) are used to generate topic distributions, and class relocation is guided by semantic similarity to reduce coupling and improve modular cohesion. BCD [5] adopts a community detection approach for modularization. It constructs a weighted, directed decomposition graph by combining code locality (Sequence Graph, SG), data references (Data Reference Graph, DRG), and function calls (Call Graph, CG). The Newman algorithm is then applied to identify communities without requiring a predefined number of clusters. By optimizing a modularity function, the method achieves effective decomposition based on structural and semantic relationships. Li et al. [7] proposed using a unified code property graph (UCPG) as an intermediate representation of source code, introducing FCG and NCS based on CPG. They used graph neural networks to implement cross-function vulnerability detection, further verifying the importance of weighted dependencies and semantic information in program structure analysis. Modularization of C/C++ binary programs remains challenging due to the loss of structural information during compilation [11]. Nevertheless, recent studies [4] have demonstrated that binary-level modularization can still provide benefits, such as detecting small, functionally cohesive code segments. In contrast to prior works, this paper focuses on modularizing operating system kernel components by combining type-based dependency analysis and semantic interface extraction. We leverage community detection techniques to partition the dependency graph and incorporate kernel interface semantics to guide the modularization process, improving both modular granularity and algorithmic convergence.

3 Design

3.1 Overview

Figure 1 illustrates the overall workflow of our approach. The input to our system is a binary representation of the program in LLVM IR format. Two core processes are executed in parallel: type-based dependency analysis to construct an inter-module dependency network, and interface feature extraction to capture semantic relationships among modules. The semantic similarity is computed based on edit distance (Levenshtein distance) applied to function names and parameter lists, thereby reducing the search space and improving efficiency. Once both the dependency network and semantic similarity information are obtained, the community structure among modules is determined using the Louvain algorithm, which operates on the dependency graph to identify cohesive subgraphs. The resulting community structure serves as the final modularization output.

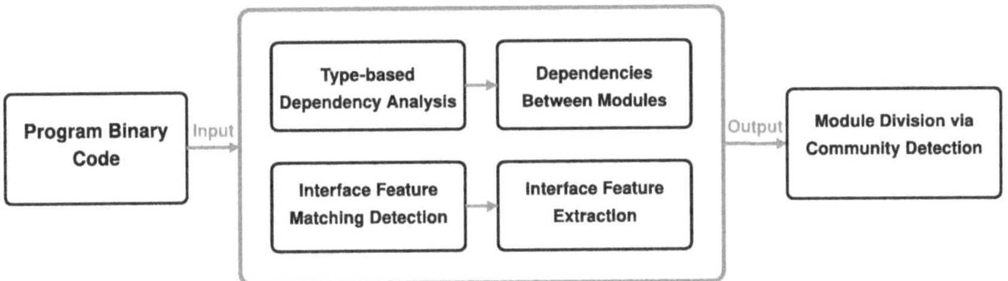

Fig. 1. The overview of our technique.

3.2 Type-Based Module Dependency Graph Construction

Dependencies between program modules can be modeled as directed graphs, where nodes represent modules and edges represent inter-module dependencies. Let $\mathcal{M} = \{M_1, M_2, \ldots, M_n\}$ denote the set of program modules, where each module M_i corresponds to a source-level file compiled to LLVM IR (i.e., a `.bc` file). Let $\mathcal{T}$ be the set of all data types used across the modules, including struct types, pointers, arrays, and function types.

We aim to construct a **directed dependency graph** $G = (\mathcal{M}, E)$, where:

- Each node $M_i \in \mathcal{M}$ represents a module.
- Each directed edge $(M_i \rightarrow M_j) \in E$ indicates that module M_i may influence memory-access behavior in module M_j via type-level data transfer.

The construction is based on the TyPM [8] principle, which states:

A module M_i is dependent on module M_j if M_i may pass an object of type $T \in \mathcal{T}$ (or a reference to it) that can be accessed by a memory-related instruction in M_j.

This forms a relatively accurate, type and direction labeled dependency network graph using global variables and function parameters was obtained for use in subsequent community detection algorithms.

3.3 Community Detection Algorithm

In the constructed graph G, functionally related modules often exhibit strong interdependencies, forming densely connected subgraphs known as *communities*. By drawing an analogy between modularization and community detection, we can leverage network science methods to improve software modularity, thereby enhancing maintainability, scalability, and development efficiency.

The quality of community detection is commonly evaluated using the **modularity** metric, which measures the density of intra-community connections relative to inter-community connections. A commonly used metric is the Girvan-Newman Modularity Quality (GN-MQ) [10], formally defined as:

$$Q = \frac{1}{2m}\sum_{i,j}\left[A_{ij} - \frac{k_i k_j}{2m}\right]\delta(c_i, c_j), \quad m = \frac{1}{2}\sum_{i,j} A_{ij} \tag{1}$$

Here, A_{ij} represents the adjacency matrix, k_i, k_j are the degrees of nodes i and j, m is the total number of edges, c_i is the community to which i belongs to, and $\delta(c_i, c_j)$ equals 1 if nodes i and j are in the same community, and 0 otherwise.

To reflect the directional nature of program dependencies, we modify the modularity function by replacing the undirected term $-\frac{k_i k_j}{2m}$ with a directed form $-\frac{k_i^{\text{out}} k_j^{\text{in}}}{2m}$. The resulting modularity is:

$$Q = \frac{1}{2m}\sum_{i,j}\left[\omega_{ij} - \frac{k_i^{out} k_j^{in}}{2m}\right]\delta(c_i, c_j), \quad m = \frac{1}{2}\sum_{i,j} \omega_{ij} \tag{2}$$

Here, ω_{ij} is the weight of the edge between node i to node j, k_i^{out} and k_j^{in} are the weighted out-degree and in-degree of nodes i and j, respectively.

Given that exact optimization of modularity is NP-hard [3], approximation algorithms are essential for large-scale graphs. To maximize this modularity, we adopt the **Louvain algorithm** [2], a widely used greedy optimization algorithm known for its efficiency on large and sparse graphs. Figure 2 show the schematic diagram of the Louvain algorithm. The Louvain algorithm is also capable of generating hierarchical modular structures, offering further insights into program architecture. Louvain proceeds in two iterative phases:

- **Phase 1 (Local Optimization):** Initially, each node is assigned to its own community. Then, each node i is visited in random order. For each neighbor j of i, the algorithm evaluates the modularity gain from moving i into j's community and chooses the move that yields the highest positive gain (if any). This process repeats until no further improvement is possible.

- **Phase 2 (Community Aggregation):** The current community structure is compressed into a new graph, where each community becomes a super-node. Edges between communities are aggregated accordingly, and Phase 1 is reapplied to the new graph. The procedure continues iteratively until convergence.

The modularity gain ΔQ when moving node i into community C is computed as:

$$\Delta Q = \left[\frac{\sum_{\text{in}} + k_{i,\text{in}}}{2m} - \left(\frac{\sum_{\text{tot}} + k_i}{2m}\right)^2\right] - \left[\frac{\sum_{\text{in}}}{2m} - \left(\frac{\sum_{\text{tot}}}{2m}\right)^2 - \left(\frac{k_i}{2m}\right)^2\right] \tag{3}$$

Here, $\sum_{\text{in}}$ is the total weight of intra-community edges, $\sum_{\text{tot}}$ is the total degree of the community, $k_{i,\text{in}}$ is the weight of edges from node i into the community, and k_i is the total degree of node i. A simplified version used for efficiency is:

$$\Delta Q = \frac{1}{2m}\left(k_{i,\text{in}} - \frac{k_i \sum_{\text{tot}}}{m}\right) \tag{4}$$

Fig. 2. Schematic diagram of the Louvain algorithm

The fast unfolding Louvain algorithm is efficient and fast, mainly spending time in the first iteration. Reduce traversal nodes by compressing the graph during community coalescence, accelerating subsequent iterations. However, it has resolution limitations for small communities. Louvain provides meaningful intermediate results through hierarchical clustering, and specific level results can be obtained by setting the maximum number of iterations. The size of the community typically decreases with cohesion and the results are significant when the number of communities has an inflection point, which can be used to set the maximum iterations and terminate early.

In practice, node traversal order can affect results, causing local optima and repeated node movements, slowing detection. Although node order has minimal impact on modularity, it affects the calculation time [2]. This paper uses random access for each layer and traversal to reduce errors, despite increased time, as it positively impacts local optima optimization. Some nodes have unclear community tendencies and swing between similar communities. The algorithm is improved to record node moves and set a maximum to prevent this. Nodes may have the same modularization gain among multiple communities, with minor changes causing attribute shifts but small gains. The backward phase mostly involves such walks. Louvain has low time complexity, quantitative evaluation, and hierarchical results, but also has defects like slow convergence for large communities and a tendency to fall into local optima as a greedy algorithm. This can lead to adding peripheral points to compact communities, creating locally advantageous but globally inferior solutions.

To mitigate these issues, we introduce two enhancements:

- We incorporate semantic similarity scores between functions, derived from interface features (see Subsect. 3.4), as bias terms in modularity gain.
- We improve robustness by tracking node move history and introducing early stopping criteria when fewer than 0.1% of nodes change communities per iteration [15].

Since the main research object is the operating system kernel, oriented to the kernel interface of the operating system, the interface features are analyzed and extracted to obtain the similarity, and the bias is introduced to ΔQ. The modified modularity gain becomes:

$$\Delta Q = \Delta Q' \times B_l \times B_e, \tag{5}$$

Here, $\Delta Q'$ is from Eq. (4), and B_l, B_e are similarity-based bias factors derived from lexical and semantic features, respectively. These enhancements help avoid overfitting to locally optimal structures and reduce the inclusion of semantically unrelated nodes in dense clusters.

3.4 Interface Feature Similarity

In modern software systems, it is common to expose multiple entry points to provide services or interfaces for higher-level applications. Functions that serve similar purposes typically share syntactic and semantic characteristics. Extracting such interface-level features to guide community detection can provide valuable heuristic information and significantly improve algorithmic efficiency. We utilize both string literals and function parameter characteristics as syntactic features. Among these, string literals play a particularly important role because they often contain distinctive values that help uniquely identify specific functionality. If two functions across different modules share the same string literal, there is a high probability that they provide equivalent or closely related functionality [14].

However, exact matches of literal values are relatively rare. To address this, we introduce a similarity metric based on the Levenshtein edit distance [6], which computes the minimum number of operations (insertion, deletion, or substitution) required to transform one string into another. Since function names in kernel code are typically short, using dynamic programming to calculate Levenshtein distance has minimal performance impact. The recurrence relation is as follows:

$$D_{ij} = \begin{cases} \max(i,j) & \min(i,j) = 0 \\ \min \begin{cases} D_{i-1,j-1} + w_a \\ D_{i-1,j} + w_b \\ D_{i,j-1} + w_c \end{cases} & else \end{cases} \tag{6}$$

Here, w_a, w_b, w_c represent the weights for replacement, deletion, and insertion operations, respectively. In our implementation, all weights are set to 1. Based on the computed distance, the normalized similarity score B_l is defined as:

$$B_l = 1 - \frac{D_{mn}}{max\,(m,n)} \tag{7}$$

Here, D_{mn} denotes the edit distance, and m and n are the lengths of the two strings.

In the Linux kernel, function names often follow a "noun_verb" format, with components separated by underscores ("_"). To better capture structural semantics, we first split function names and semantically segment the components. Verbs are excluded from similarity analysis using a stop-word list, under the rationale that identical nouns often indicate belonging to the same module, while identical verbs might reflect similar functionalities across modules. By removing verbs and comparing only the noun components, we aim to enhance the module-level granularity of similarity measurement. However, relying solely on function names may lead to incomplete information. To mitigate this limitation, we incorporate additional semantic features extracted via the LLVM framework, including parameter lists, argument types, and return types. Each matching feature contributes to an accumulated similarity score, with customizable weights for fine-tuning.

The final similarity score B_e is defined as:

$$B_e = \begin{cases} 1 & similarity = 1 \\ similarity \times coefficient & else \end{cases} \tag{8}$$

This score reflects the aggregated similarity derived from multiple semantic and syntactic dimensions. The similarity coefficient can be adjusted depending on how strongly interface features are intended to influence the overall modularization process.

4 Implementation

We use C++ to implement our system. It is developed on the basis of LLVM version 15.0.0, as the overall analysis level of the program is based on LLVM IR. There are three main steps consisting of: type-based dependency analysis, feature extraction from the interface, and community detection in the dependency graph. The type-based dependency analysis tool uses open-source tool, TyPM [8], and the features of the interface are mainly based on string literals and information provided by LLVM.

5 Evaluation

5.1 Setup

To evaluate the effectiveness of our proposed modularization approach, we conducted comparative experiments using the Linux kernel source code. The Linux kernel is a well-established, large-scale software system developed by highly experienced programmers. As the kernel evolves, new versions often include extensive refactoring, which improves modular separation and optimizes system structure. Given these characteristics, we selected two stable releases for our evaluation: Linux kernel versions 5.15.0 and 6.5.0. The experimental environment was set up using QEMU (version 2.5), simulating a virtualized x86_64 CPU with 8 cores at 2000 MHz and 32 GB of allocated RAM. The experiments were conducted on Ubuntu 22.04.2 LTS. Our implementation is written in C++ and built on LLVM version 15.0.0. For type-based dependency analysis, we used the TyPM tool [8], while interface features were extracted using string literal analysis and metadata available in LLVM IR.

5.2 Modularization

We applied our modularization method independently to two versions of the Linux kernel. For version 5.15.0, a total of 2,542 modules were analyzed, yielding 150 aggregated communities. In contrast, version 6.5.0 involved 2,807 modules and produced 92 aggregated communities, compared to version 5.15.0, version 6.5.0 shows evidence of structural refinement. Specifically, we identified 309 newly added modules, among which 27 appeared in newly formed aggregated communities, suggesting an increase in cohesion and modular clarity. One clear example is the *drivers/pci/setup-irq.bc* module from version 5.15.0. In version 6.5.0, this functionality has been refactored and redistributed into *drivers/pci/irq.bc*, demonstrating logical separation of IRQ setup from the broader PCI setup code.

This structural evolution validates the effectiveness of our modularization technique in capturing refactoring behaviors and aligning with improved software architecture. Table 1 presents a comparison of several key modules across the two versions, highlighting those that were reorganized or merged based on community detection results.

5.3 Similarity Comparison

To assess the rationality and consistency of the modularization results, we performed similarity comparisons between the generated module partitions and reference structures. In general, the evaluation of clustering quality benefits from expert-defined gold standards. However, in the case of mature software systems such as the Linux kernel, historical version comparisons serve as a valuable proxy for ground truth.

Table 1. Modules that have been optimized as a result of dividing communities in 5.15.0 vs. 6.5.0

Modules	5.15.0 vs. 6.5.0
drivers/pci/pcie/portdrv_pci.bc arch/x86/kernel/crash_core_64.bc mm/vmacache.bc drivers/pci/setup-irq.bc drivers/video/fbdev/core/fb_cmdline.bc drivers/firmware/sysfb.bc drivers/gpu/drm/drm_dsc.bc drivers/gpu/drm/drm_hashtab.bc drivers/gpu/drm/i915/dma_resv_utils.bc fs/efivarfs/efivarfs.mod.bc kernel/sched/wait_bit.bc kernel/crash_dump.bc	Optimization is recommended in 5.15.0 and has done in 6.5.0
drivers/video/console/dummycon.bc drivers/tty/tty_ioctl.bc drivers/acpi/acpica/tbfind.bc drivers/tty/vt/vt.bc drivers/acpi/acpica/tbxfload.bc kernel/cpu.bc	Optimization is recommended in two versions and in the same community
drivers/input/input-compat.bc drivers/acpi/acpica/exmutex.bc mm/mmzone.bc ...	Optimization is recommended in two versions but in the different community

To measure similarity between partitioning results, we adopted the MoJoFM metric [13], a normalized enhancement of the original MoJo metric [12]. The MoJoFM metric quantifies the minimum number of "Move" and "Join" operations required to transform one modularization into another. It is formally defined as:

$$MoJoFM(A,B) = \left(1 - \frac{mmo(A,B)}{\max(mmo(\forall A,B))}\right) \times 100\% \tag{9}$$

Here, mno(A,B) denotes the total number of minimum "move" and "merge" steps required to transform division A into division B, $\max(mmo(\forall A,B))$ denotes the the maximum such value among all possible divisions.

We utilize comparative evaluations across different stable versions of the Linux kernel and design both positive and negative references for validation. To perform a dual-perspective evaluation, we define the following configurations: $A(v)$, representing the modular structure of kernel version v as defined by its source code directory hierarchy; $R(v)$, denoting the automated modularization result produced by our method for version v; and $Ran(v)$ a randomly generated clustering based on version v, which serves as a negative control for assessing structural coherence. Based on modular consistency, we hypothesize the following inequality should hold:

$$\begin{aligned} &\text{MoJoFM}(\text{A}(6.5.0), \text{R}\ (6.5.0)) > \\ &\text{MoJoFM}(\text{A}(5.15.0), \text{R}(5.15.0)) > \\ &\text{MoJoFM}(\text{Ran}(5.15.0), \text{R}(5.15.0)) \end{aligned} \tag{10}$$

Table 2 confirmed this hypothesis. The modularization produced by our method in version 6.5.0 shows a higher similarity to its reference than that in 5.15.0, reflecting the refinement and structural improvements in the newer kernel version. Furthermore, the stark contrast between the automated result and the randomized baseline further validates the structural integrity and semantic relevance of our approach.

Table 2. MoJoFM similarity comparison results.

MoJoFM	results
MoJoFM(A(6.5.0), R(6.5.0))	96.29
MoJoFM(A(5.15.0), R(5.15.0))	91.19
MoJoFM(Ran(5.15.0), R(5.15.0))	13.61

6 Conclusions

In this paper, we presented a modularization tool tailored for operating system kernel code, implemented on top of the LLVM framework. Our method performs type-based dependency analysis to construct an inter-module dependency network and applies an improved Louvain algorithm for community detection. To enhance the effectiveness of modular partitioning, we further incorporate semantic similarity measures derived from kernel interface features, including Levenshtein distance and string-based similarity metrics. We evaluated our approach on two stable versions of the Linux kernel, and demonstrated its practicality and accuracy.

Acknowledgment. This work was supported in part by the National Natural Science Foundation of China (No. 62302388), in part by China Postdoctoral Science Foundation (No. 2023M742793).

References

1. Bavota, G., Gethers, M., Oliveto, R., Poshyvanyk, D., Lucia, A.: Improving software modularization via automated analysis of latent topics and dependencies. ACM Trans. Softw. Eng. Methodol.(TOSEM) **23**(1), 1–33 (2014)
2. Blondel, V.D., Guillaume, J.L., Lambiotte, R., Lefebvre, E.: Fast unfolding of communities in large networks. J. Stat. Mech: Theory Exp. **2008**(10), P10008 (2008)
3. Brandes, U., et al.: Maximizing modularity is hard. arXiv preprint physics/0608255 (2006)
4. Haq, I.U., Caballero, J.: A survey of binary code similarity. ACM Comput. Sur. (CSUR) **54**(3), 1–38 (2021)
5. Karande, V., Chandra, S., Lin, Z., Caballero, J., Khan, L., Hamlen, K.: BCD: decomposing binary code into components using graph-based clustering. In: Proceedings of the 2018 on Asia Conference on Computer and Communications Security. pp. 393–398 (2018)
6. Levenshtein, V.I., et al.: Binary codes capable of correcting deletions, insertions, and reversals. In: Soviet physics doklady. vol. 10, pp. 707–710. Soviet Union (1966)
7. Li, W., Li, X., Feng, W., Jin, G., Liu, Z., Jia, J.: Vulnerability detection based on unified code property graph. In: Yuan, L., Yang, S., Li, R., Kanoulas, E., Zhao, X. (eds.) Web Information Systems and Applications, pp. 359–370. Springer Nature Singapore, Singapore (2023)
8. Lu, K.: Practical program modularization with type-based dependence analysis. In: 2023 IEEE Symposium on Security and Privacy (SP). pp. 1256–1270. IEEE (2023)
9. Mancoridis, S., Mitchell, B.S., Chen, Y., Gansner, E.R.: Bunch: A clustering tool for the recovery and maintenance of software system structures. In: Proceedings IEEE International Conference on Software Maintenance-1999 (ICSM'99). Software Maintenance for Business Change (Cat. No. 99CB36360). pp. 50–59. IEEE (1999)
10. Newman, M.E., Girvan, M.: Finding and evaluating community structure in networks. Phys. Rev. E **69**(2), 026113 (2004)
11. Sarhan, Q.I., Ahmed, B.S., Bures, M., Zamli, K.Z.: Software module clustering: An in-depth literature analysis. IEEE Trans. Software Eng. **48**(6), 1905–1928 (2020)
12. Tzerpos, V., Holt, R.C.: Mojo: A distance metric for software clusterings. In: Sixth working conference on reverse engineering (cat. no. PR00303). pp. 187–193. IEEE (1999)
13. Wen, Z., Tzerpos, V.: An effectiveness measure for software clustering algorithms. In: Proceedings. 12th IEEE International Workshop on Program Comprehension, 2004. pp. 194–203. IEEE (2004)
14. Yang, C., Xu, Z., Chen, H., Liu, Y., Gong, X., Liu, B.: Modx: binary level partially imported third-party library detection via program modularization and semantic matching. In: Proceedings of the 44th International Conference on Software Engineering. pp. 1393–1405 (2022)
15. Zhang, J., Fei, J., Song, X., Feng, J.: An improved louvain algorithm for community detection. Math. Probl. Eng. **2021**(1), 1485592 (2021)

Boundary-Aware Multi-scale Spiking Neural Network for Image Semantic Segmentation

Yumo Li, Sunan Ge(✉), Xueqing Zhao, and Xin Shi

School of Computer Science, Xi'an Polytechnic University, Shaanxi Key Laboratory of Clothing Intelligence, Xi'an, China
230721100@stu.xpu.edu.cn, {gesunan,zhaoxueqing,shixin}@xpu.edu.cn

Abstract. Semantic segmentation of imagery is critical for remote sensing feature extraction and image generation, but faces challenges like microscopic structure recognition in high-resolution images and ambiguous semantic boundaries in complex scenes—leading to suboptimal precision, detail loss, and insufficient frequency-spatial feature integration. To address these, we propose a multiscale spatio-temporal boundary-aware semantic segmentation framework inspired by biological neural responses, consisting of three core modules: dual-path feature extraction (using spiking convolution kernels, DCT multi-band enhancement, and cross-domain RGB concatenation for info complementarity), multi-scale feature pyramid fusion (via bidirectional cross-scale interaction to integrate high-level semantics and low-level details), and boundary perception (three-stage architecture to extract edge heatmaps and reinforce weak boundaries). Experiments on WHU Satellite Datasets I/II and Qinqiang Dataset show superior performance over existing methods, effectively capturing boundary details in dense built-up and small-target regions while mitigating the "coarse shape" issue.

Keywords: Semantic Segmentation · Spiking Convolution · Boundary Awareness

1 Introduction

As one of the core technologies in the field of computer vision, image semantic segmentation aims to accurately identify and partition different semantic objects in digital images through pixel-level classification. This technology is crucial in scenarios like remote sensing for ground object information and image generation. However, with the advancement of imaging technology from the early low-resolution stage to today's ultra-high resolution, the demand for refined identification of microstructures in images has surged. Moreover, the multi-target interactions in complex scenes lead to the problem of blurred semantic boundaries, posing a dual challenge to traditional segmentation methods. The evolutionary process of image semantic segmentation technology can be divided

S. Yang et al. (Eds.): WISA 2025, LNCS 16299, pp. 386–398, 2026.
https://doi.org/10.1007/978-981-95-6426-2_35

into three landmark stages: semantic segmentation methods based on shallow features, deep learning, and spiking neural networks.

Semantic segmentation methods based on shallow features rely on hand - designed features. For example, one study achieved apple segmentation by integrating color - texture features [1], or improved the texture classification performance by combining Histogram of Oriented Gradients and Local Binary Pattern features [2], or extracted features through edge detection and texture analysis, then generated superpixels through clustering to enhance the segmentation accuracy [3]. These methods work well in structured scenes. However, their generalization in natural and complex scenes is limited: fixed thresholds struggle to adapt to illumination changes, preset morphological kernels fail to capture diverse shapes, and they rely heavily on manual feature engineering.

Semantic segmentation methods based on deep learning mainly use end-to-end architectures. For example, methods optimize local feature extraction via a mask attention mechanism; other approaches achieve high-precision segmentation in low-labeled scenarios using a diffusion model. However, these methods lack few-shot learning capability, face an efficiency-accuracy trade-off, and—more critically—rely on static feature fusion, which severely limits frequency-spatial collaboration. For instance, while SegNeXt [4]'s multi-scale attention captures spatial feature correlations, it fails to model frequency-domain information, leading to poor performance in ultra-high-resolution microstructure segmentation or complex-scene weak boundary distinction [5].

Semantic segmentation methods based on the Spiking Neural Network (SNN) [6] provide a new paradigm. Early studies, such as MNIST handwritten digit recognition [7], have verified the feasibility of SNN. Subsequent works, such as Spiking - YOLO [8], spiking neurons into the object detection framework, reducing energy consumption while maintaining accuracy; SEW - ResNet [9] addressed the problem of gradient attenuation in deep SNNs, and SNNFD [10] enhanced feature extraction through spike convolution in the frequency domain. However, most existing static image processing methods directly transplant the architecture, failing to fully utilize the characteristics of spike computation. In particular, there are deficiencies in the optimization of features in both the frequency domain and the spatial domain.

There are three key issues in the existing methods for frequency domain feature fusion: (1) The lack of modeling of the biological neural response mechanism leads to limited feature expression; (2) Insufficient collaboration between global and local features results in information imbalance; (3) Traditional edge detection has insufficient sensitivity to weak boundaries. In response to these issues, three innovations are proposed:

(1) Integrate the biological neural response mechanism with multi-scale frequency-spatial modeling, construct a dual-path feature extraction architecture, and achieve dynamic frequency domain feature expression;
(2) Design a multi-scale pyramid fusion module. It enables the complementarity of semantic and spatial features by channel alignment and bidirectional interaction.

(3) Propose a boundary-aware module, by gradient response modeling and dynamic attention modulation, adaptively enhance weak boundary features.

2 Method

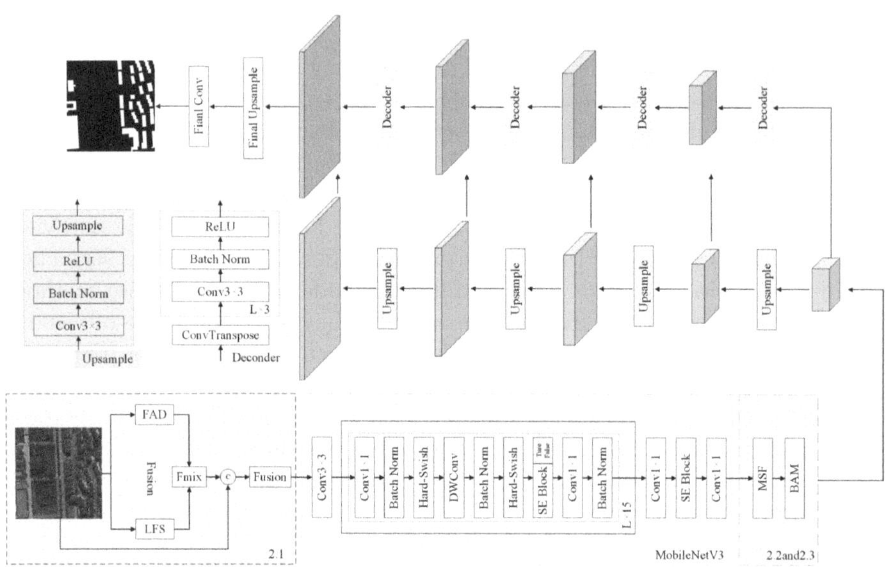

Fig. 1. Semantic Segmentation Framework Diagram of Multi-Scale Spatio-Temporal Boundary Perception Analysis.

By proposing a semantic segmentation framework with multi-scale spatio-tempor-al boundary perception analysis, a dual-path feature extraction is constructed to achieve accurate image parsing. As shown in Fig. 1, the input image passes through the Global Frequency Domain Enhancement Module (FAD) and the Local Frequency Domain Selection Module (LFS). After the features of the dual paths are fused in the frequency domain, they are reconstructed into spatial features through the inverse Discrete Cosine Transform and concatenated with the original RGB image. The fused features are formed through convolutional compression and input into the improved MobileNetV3-large backbone network. The first-layer convolution retains the detailed features, and the Multi-Scale Feature Pyramid Module (MSF) fuses multi-level semantic information through lateral connections and bidirectional paths. The Boundary Awareness Module (BAM) combines to strengthen the boundary representation. The decoder adopts a four-level enhancement structure to restore spatial details through transposed convolution and skip connections. Finally, the segmentation result of the same resolution as input is upsampled and output. Through adaptive learning of frequency domain kernels and joint spatio-temporal modeling, problem of blurred boundaries is effectively addressed, and semantic segmentation performance of dense building areas in complex scenes is significantly improved.

2.1 Frequency-Space Joint Representation Module

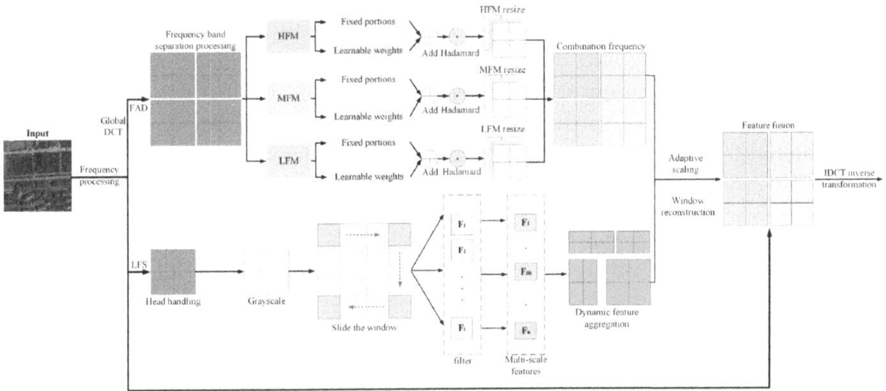

Fig. 2. Diagram of the Frequency-Space Joint Representation Module.

As shown in Fig. 2, after Image $X \in \mathbb{R}^{H \times W \times 3}$ is normalized, it is mapped from the spatial domain to the frequency domain via the Discrete Cosine Transform, generating the global frequency domain feature $F_{global} = DCT(X)$ which is then input into the FAD module. This module dynamically adjusts the multi-band feature information through three sets of learnable filters. Each set of filters is composed of a fixed band mask $M_b \in \{0, 1\}^{S \times S}$ (a binary matrix generated based on the band interval) and a learnable adjustment term $L_b \in \mathbb{R}^{S \times S}$. After summing in the channel dimension, the feature F_{fad} is obtained by scaling in the frequency domain using the learnable parameter k. The three types of band features, after being scaled, are concatenated along the channel, enabling the adaptive enhancement of key band information and the suppression of noise. As shown in Eq. (1):

$$F_{fad} = \frac{1}{k + \epsilon} \sum_{b=1}^{3} (M_b + (2\sigma(L_b) - 1)) \cdot F_{global} \tag{1}$$

At the same time, the image generates local frequency domain features $F_{local,p}$ through the sliding window DCT and inputs them into the LFS module. This module employs learnable filters to perform multi-scale filtering on the local frequency domain features. The structure of each group of filters is consistent with that of the FAD module. After the filtered local frequency domain features are aggregated by averaging, they are scaled in the frequency domain through the same learnable parameter k, which enhances the high-frequency details while preserving the low-frequency information. As shown in Eq. (2):

$$F_{lfs} = Fold(Unfold(\frac{1}{k+\epsilon} \cdot \frac{1}{M} \sum_{m=1}^{M} (M_m + (2\sigma(L_m) - 1))F_{local,p})) \tag{2}$$

After being processed by the FAD and LFS modules, F_{fad} and F_{lfs} are output respectively. Expand F_{lfs} to the corresponding dimension through channel replication, and directly add it to F_{fad} in the frequency domain to form a joint frequency domain representation. Then, reconstruct it into a spatial domain feature through the inverse discrete cosine transform. Finally, concatenate the reconstructed feature and the original RGB image along the channel dimension to form the cross-domain fusion tensor X_{fusion} (Eq. 3).

$$X_{fusion} = Concat(X, IDCT(F_{fad} + F_{fls})) \in \mathbb{R}^{H \times W \times 6} \tag{3}$$

2.2 Multi-Scale Feature Pyramid Information Fusion Module (MSF)

The cross-domain fusion feature tensor X_{fusion} is input into the improved Mobile-NetV3 backbone network for deep encoding after being compressed by convolution, Extract multi-scale features level by level through the inverted residual structure, Multi-scale features $\{F_l\}_l^L$ (L is the number of hierarchical levels, $F_l \in \mathbb{R}^{C_l \times H_l \times W_l}$) are extracted level by level through the inverted residual structure, and the MSF model is utilized for multi-scale information fusion.

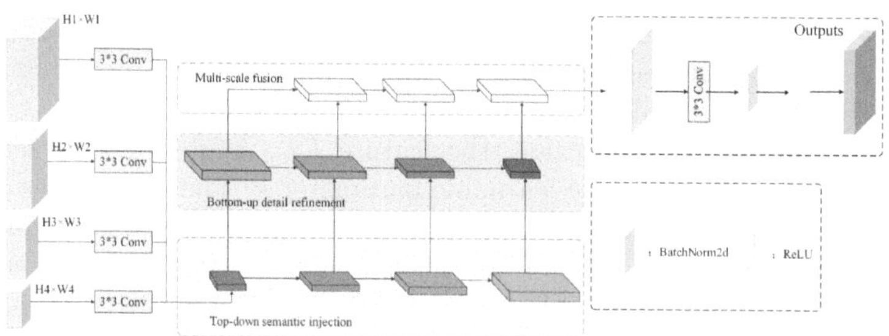

Fig. 3. Diagram of the Frequency-Space Joint Representation Module.

As shown in Fig. 3, the input features with any number of channels are uniformly mapped to a fixed-dimensional latent space through the horizontal connection layer, eliminating the dimensional differences among features at different levels, as shown in Eq. (4):

$$F_l' = Conv_{1\times1}(F_l) \in \mathbb{R}^{256 \times H_l \times W_l} \tag{4}$$

The top - down path starts from the highest semantic layer and upsampling through nearest neighbor interpolation. High - level semantic information is injected into the features of adjacent lower layers at the pixel level. **The bottom - up path** utilizes convolutional operations to process the low-level features, eliminating noise and extracting more representative features. The refined features are passed to the adjacent higher layer through element-wise addition. In the multi-scale fusion stage, a progressive resolution matching strategy is adopted. After uniformly scaling the features at each level to the input size, they are concatenated along the channel dimension. The composite convolutional layer cross-scale feature interaction through a learnable weight matrix, as shown in Eq. (5). This design achieves the cyclic enhancement of semantic and spatial information through bidirectional connections, breaking through the limitation of unidirectional propagation in the traditional feature pyramid.

$$F_{msf} = Conv_{final}(Concat(Upsample(F_l', scale = 2^{L-1}))_{l=1}^{L}) \tag{5}$$

2.3 Boundary Awareness Module (BAM)

The BAM module is employed to conduct multi-level feature compression and boundary attention processing on F_{msf}, precisely enhancing the edges of ground objects. The architecture of the BAM is shown in Fig. 4.

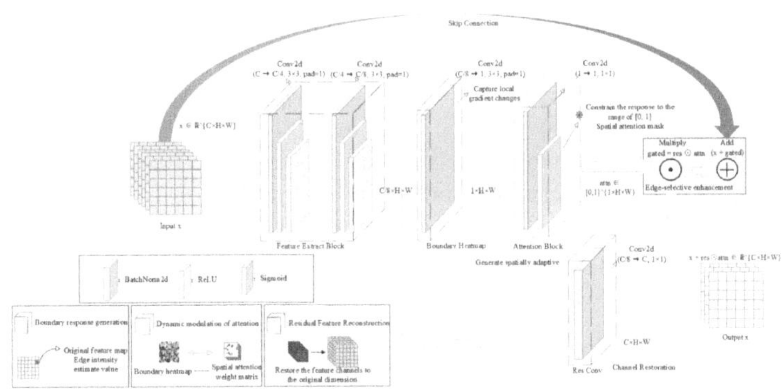

Fig. 4. Diagram of the Boundary Awareness Module.

Among them, the feature dimensionality reduction process involves implementing two levels of spatial information compression on the input features. By reducing the channel dimension layer by layer, a feature representation $F_{red} = Conv_{2\times2}(F_{msf}) \in \mathbb{R}^{\frac{C}{8}\times H\times W}$ with high information density is formed. In the boundary response generation stage, a single-channel convolutional kernel is used to construct a boundary heatmap. Local gradient changes are captured through convolutional operations, and each pixel point of the output tensor corresponds to the estimated edge intensity value $E = Conv_{3\times3,1}(F_{red}) \in \mathbb{R}^{1\times H\times W}$

of the original feature map.The attention dynamic modulation layer generates a spatially adaptive attention mask $\alpha = \sigma(Conv_{1\times1}(E)) \in \mathbb{R}^{1\times H\times W}$ by constraining the response values through non - linear transformation and activation functions.The high activation regions of the mask correspond to the edge positions of the feature map, guiding the network to focus on the learning of boundary semantics through the gradient mechanism. In the residual feature reconstruction stage, learnable convolution is used to restore the channel dimension, and element-wise multiplication is performed with the attention mask to achieve the spatial enhancement of edge features. Finally, through skip connections, the enhanced edge features are added to the original input, preserving both the deep semantics and the refined boundary information. as shown in Eq. (6):

$$F_{bam} = F_{msf} + Conv_{res}(F_{red}) \cdot \alpha \tag{6}$$

The decoding stage employs a four-level progressive structure.The final features are subjected to convolution to achieve pixel-level classification. The output resolution is aligned with the input size through bilinear upsampling, thereby completing the end-to-end image parsing.

3 Experiments and Analysis

3.1 Datasets

In order to evaluate the portability of this method in building extraction from remote sensing images, we used ISPRS Vaihingen dataset, LoveDA (Land-cOVEr Domain Adaptative) dataset, and WHU Building Data-Aerial Imagery dataset as the training datasets. The WHU Building Data-Satellite Dataset I (Global Cities) (with dense distribution) and Satellite Dataset II (East Asia) (with sparse distribution) were used as the test datasets. To further verify the semantic segmentation performance of the proposed method in complex scenarios with small targets, a Qinqiang Opera dataset was adopted for training and testing.

3.2 Experimental Setup and Evaluation Metrics

The experiments were conducted in a computing environment equipped with a single RTX 4090D graphics card (24GB video memory). The training set data was augmented by random rotation and brightness jitter, while the validation set remained unchanged. The AdamW optimizer was employed, and an early - stopping mechanism was used to control overfitting during training. The batch size was set to 4, and the maximum number of iterations was 100.

To evaluate the quantitative performance of different models, this paper adopts IoU, Precision, Recall, and F1-score as the evaluation metrics.

3.3 Ablation Experiments

Traditional unidirectional propagation networks have fragmentation, with high-level semantics rich but spatial positioning rough, and low-level spatial details abundant but semantic information ambiguous. However, MSF cyclically enhances semantic information and spatial details through bidirectional cross connections, thereby improving the accuracy of target positioning. Traditional models rely on the natural propagation of low-level features to model target boundaries, and lack an explicit boundary enhancement mechanism. BAM, however, expands the model's optimization objective from simple class classification to joint class classification and boundary positioning (via attention), forcing the network to enhance gradient responses in boundary regions during backpropagation and addressing boundary pixel misclassification—a common issue in segmentation tasks. Therefore, ablation experiments were conducted on Satellite Dataset I, Satellite Dataset II, and the Qinqiang Opera Dataset.

Table 1. Ablation Experiments on the Satellite Dataset I(Global Cities)

Method	IoU(%)	Precision(%)	Recall(%)	F1-score(%)
Mobilenetv3	40.83	52.98	62.18	57.21
Mobilenetv3 + MSF	41.12	54.83	63.37	58.79
Mobilenetv3 + MSF + BAM	42.51	55.77	64.12	59.66

Table 2. Ablation Experiments on the Satellite Dataset II(East Asia)

Method	IoU(%)	Precision(%)	Recall(%)	F1-score(%)
Mobilenetv3	16.72	28.41	27.96	28.18
Mobilenetv3 + MSF	17.88	29.10	30.98	30.01
Mobilenetv3 + MSF + BAM	18.26	30.13	31.68	30.88

Table 3. Ablation Experiments on the Qinqiang Opera Dataset

Method	IoU(%)	Precision(%)	Recall(%)	F1-score(%)
Mobilenetv3	61.37	70.58	78.92	74.52
Mobilenetv3 + MSF	61.86	71.75	79.99	75.65
Mobilenetv3 + MSF + BAM	62.02	72.91	80.59	76.56

The experimental data shows that, compared with the backbone network, the introduction of the MSF module increases the IoU by 0.29, 1.16, and 0.49

respectively, and the F1 by 1.58, 1.83, and 1.13 respectively. This confirms that the introduction of the MSF module enhances the target localization accuracy through bidirectional feature interaction. Further adding the BAM module increases the IoU by 1.39, 0.38, and 0.16 respectively, and the F1 by 0.87, 0.87, and 0.91 respectively, indicating that the BAM module effectively captures the geometric structure information of the target. Therefore, MSF and BAM solve the problems of scale fragmentation and boundary blurring in traditional models through the dual mechanism of "feature interaction - boundary enhancement".

3.4 Comparative Experiments

In order to fully verify the performance of the proposed method, PSPNet, DeepLabv3+, Unet, RAANet and SNNFD are employed for comparison.

(1) Semantic Segmentation Based on Satellite Dataset I(Global Cities)

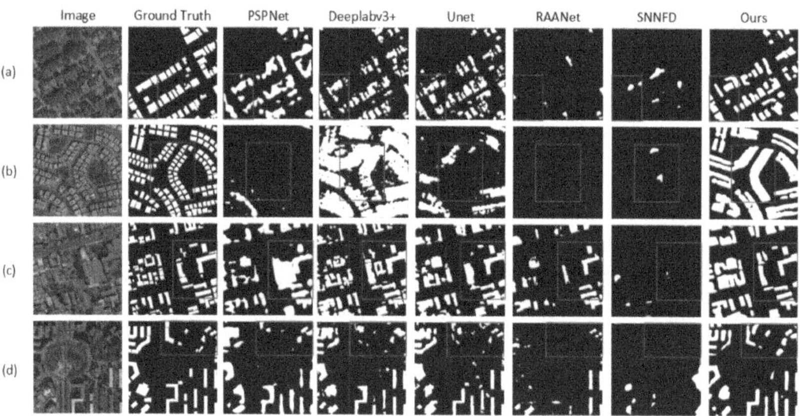

Fig. 5. Comparison Diagram of Segmentation Results of Global Cities Dataset.

As shown in Fig. 5, it can be observed that in subgraph (a), the buildings detected in the lower left corner by PSPNet, Deeplabv3+ and Unet are all blurry. RAANet and SNNFD fail to detect the buildings, while the proposed model can segment the buildings relatively clearly. In subgraph (b), only Deeplabv3+ and the proposed model can detect the curved part of the building, but the detection result of Deeplabv3+ is very blurry. In subgraphs (c,d), models other than SNNFD can roughly detect the positions of the buildings, but the proposed model provides the most accurate detection. Therefore, it can be seen that in the remote sensing semantic segmentation task of dense building areas, the proposed model can not only accurately locate the spatial coordinates of buildings, but also achieve the fine-grained segmentation of building contours.

Table 4. Comparison of Various Indicators of the Model on Global Cities Dataset

Method	IoU(%)	Precision(%)	Recall(%)	F1-score(%)
PSPNet	35.34	55.85	49.03	52.22
Deeplabv3+	40.00	52.71	62.40	57.14
Unet	40.01	**60.55**	54.11	57.15
RAANet	37.00	58.84	49.92	54.02
SNNFD	14.96	52.80	17.27	26.03
Ours	**42.51**	55.77	**64.12**	**59.66**

In Table 4, it can be concluded that the proposed model has achieved good results in IoU, recall, and F1-score. However, its precision is not satisfactory. This is because in the satellite dataset I, buildings are densely distributed and there are many interfering factors like a large amount of shadows. When the model tries to increase the overlap between predicted results and the ground truth and identify more real buildings, it will inevitably misclassify some shadow areas as buildings. When dealing with building boundaries, the model excessively includes surrounding building areas in the prediction scope, thus affecting precision.

(2) Semantic Segmentation Based on Satellite Dataset II(East Asia)

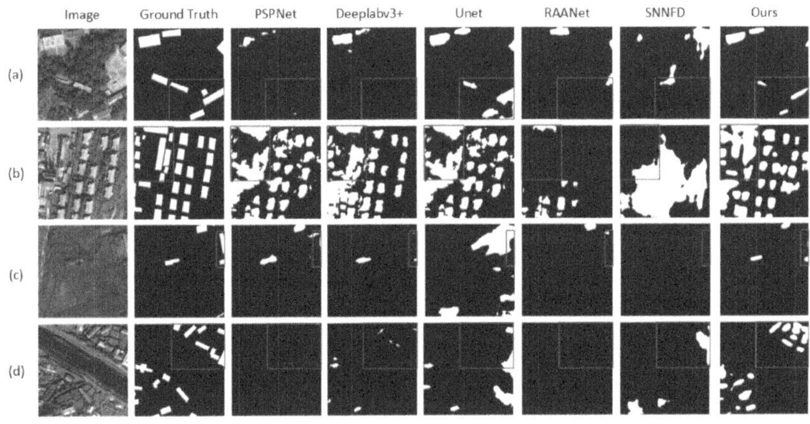

Fig. 6. Comparison Diagram of Segmentation Results of East Asia Dataset.

As shown in Fig. 6, it can be observed that in (a), PSPNet, Deeplabv3+, RAANet, and SNNFD all failed to detect the buildings except those in the upper left corner. In (b), only the proposed model can segment a large number of buildings on the left side relatively clearly. Both RAANet and SNNFD failed to detect the enclosed area. In (c), PSPNet, RAANet, and SNNFD all failed to detect the buildings in the upper right corner, and Unet had a large number of false detections. In (d), PSPNet and RAANet did not detect any buildings. Deeplabv3+, Unet, and SNNFD could detect a small number of buildings, while the proposed model can locate the approximate positions and perform relatively clear segmentation. Therefore, it can be seen that for small and sparse buildings under complex background objects, the proposed model is more sensitive and reliable in detecting buildings with detailed shape boundaries.

Table 5. Comparison of Various Indices of the Model on East Asia Dataset

Method	IoU(%)	Precision(%)	Recall(%)	F1-score(%)
PSPNet	6.04	42.64	6.57	11.39
Deeplabv3-plus	15.44	31.88	23.03	26.74
Unet	16.22	33.32	24.01	27.91
RAANet	3.67	**60.79**	3.76	7.09
SNNFD	5.67	12.91	9.18	10.73
Ours	**18.26**	30.13	**31.68**	**30.88**

Table 5 shows that our model excels in IoU, recall, and F1-score, but lags in precision. This is due to the complex backgrounds in satellite dataset II, where the spectral distinctions between buildings and elements like vegetation or roads in rural areas are minimal. To enhance the detection of small, scattered buildings, our model emphasizes edge details and low - contrast region information during feature extraction. Consequently, some background pixels with building - like spectral features are wrongly classified as targets, reducing precision.

As can be seen from Table 5, 6, in this experiment, since the test dataset used is of a different distribution from the training dataset, there are significant differences in imaging sensors, spatial resolutions, and the distribution areas of the targets. These differences make the evaluation scenarios more complex, resulting in relatively low overall performance indicators for all the methods.

(3) Semantic Segmentation Based on the Qinqiang Opera Dataset

As shown in Fig. 7, in (a), all the compared models can extract the approximate shape of the traditional Qinqiang opera official hat. However, the proposed model can better extract boundary details. In (b), only the last four models can extract the features of the three official hats, and the proposed model's prediction is closer to the label map. In (c), PSPNet, Deeplabv3+ and Unet all incorrectly detect the official hat, while RAANet and SNNFD also make misjudgments. In (d), only the proposed model can locate the wristband on the left hand. Therefore, it is concluded that for complex scenes with small targets, the proposed model can better preserve the shape details of the target categories.

As shown in Table 6, the proposed model has achieved good results in IoU, recall, and F1-score. However, precision is relatively low. This is because for small

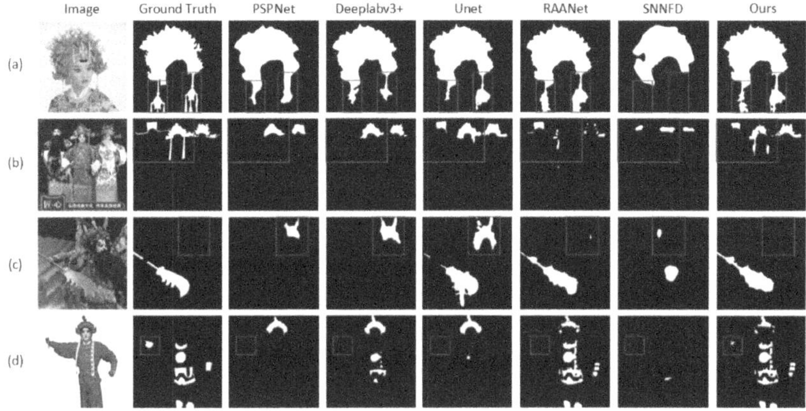

Fig. 7. Comparison Diagram of Segmentation Results of Qinqiang Opera Dataset.

targets, there is an imbalance in the proportion of background and target pixels. To capture the fragmented target areas as much as possible, the model misidentifies some of the background as the foreground. In the complex backgrounds of small targets like Qinqiang opera costumes and wristbands, by expanding the prediction range for target integrity, the recall has been significantly improved.

Table 6. Comparison of Various Indicators of the Model on Qinqiang Opera Dataset

Method	IoU(%)	Precision(%)	Recall(%)	F1-score(%)
PSPNet	6.04	42.64	6.57	11.39
Deeplabv3-plus	15.44	31.88	23.03	26.74
Unet	16.22	33.32	24.01	27.91
RAANet	3.67	**60.79**	3.76	7.09
SNNFD	5.67	12.91	9.18	10.73
Ours	**18.26**	30.13	**31.68**	**30.88**

4 Conclusion

This study proposes an improved deep learning network model and conducts an effective exploration for the task of image semantic segmentation. Based on the lightweight backbone network MobileNetV3, a multi-scale feature pyramid fusion module and a boundary perception module are introduced to enhance the multi-scale feature fusion ability and the perception ability of boundary details. The experimental results show that, in terms of the F1-score on three test datasets, our model is at least 2.51, 2.97, and 0.66 higher than the other five models, and the IOU is 2.50, 2.04, and 0.86 higher. It can be seen that our model has achieved good results in the semantic segmentation task and is able to understand and interpret the semantic information in images more accurately. Therefore, the research results have important practical significance for improving the accuracy of image semantic segmentation and can serve as an efficient and accurate solution for image semantic segmentation.

Acknowledgments. This work was supported by the National Social Science Foundation of China Art Project (No. 23EH232), the Key Research and Development Program of Shaanxi Province in 2023 (No. 2023-YBGY-404, No.2023-ZDLGY-48), Natural Science Basic Research Program of Shaanxi (No. 2023-JC-YB-558), Scientific Research Program Funded by Shaanxi Provincial Education Department (No. 23JS028), Research Project on Public Digital Cultural Services for the Year 2025 (GGSZWHFW2025-004).

References

1. Zhang, C., Zou, K., Pan, Y.: A method of apple image segmentation based on color-texture fusion feature and machine learning. Agronomy **10**(7), 972 (2020)
2. Hazgui, M., Ghazouani, H., Barhoumi, W.: Genetic programming-based fusion of hog and lbp features for fully automated texture classification. Vis. Comput. **38**(2), 457–476 (2022)
3. R Selection. Edtrs: A superpixel generation method for SARmages segmentation based on edge detection and texture region selection (2022)
4. Guo, M.-H., Cheng-Ze, L., Hou, Q., Liu, Z., Cheng, M.-M., Shi-Min, H.: Segnext: rethinking convolutional attention design for semantic segmentation. Adv. Neural. Inf. Process. Syst. **35**, 1140–1156 (2022)
5. Suryanto, N., Adiputra, A.A., Kadiptya, A.Y., Kim, Y., Kim, H.: Adversarial manhole: Challenging monocular depth estimation and semantic segmentation models with patch attack. In: Lee, JH., Emura, K., Lee, S. (eds.) International Conference on Information Security Applications, pp. 299–310. Springer (2024). https://doi.org/10.1007/978-981-96-1624-4_23
6. Bing, H., Gopalakrishnan, S., Kaushik, R.: RMP-SNN: Residual membrane potential neuron for enabling deeper high-accuracy and low-latency spiking neural network. In: Proceedings of the IEEE/CVF conference on computer vision and pattern recognition, pp. 13558–13567 (2020)
7. Shruti R.K., John, M.A., Bipin, R.: Learning and real-time classification of handwritten digits with spiking neural networks. In: 2017 24th IEEE International Conference on Electronics, Circuits and Systems (ICECS), pp. 128–131. IEEE (2017)
8. Kim, S., Park, S., Na, B., Yoon, S.: Spiking-yolo: spiking neural network for energy-efficient object detection. Proc. AAAI Conf. Artif. Intell. **34**, 11270–11277 (2020)
9. Fang, W., Zhaofei, Yu., Chen, Y., Huang, T., Masquelier, T., Tian, Y.: Deep residual learning in spiking neural networks. Adv. Neural. Inf. Process. Syst. **34**, 21056–21069 (2021)
10. Bo, Yu., Yang, A., Chen, F., Wang, N., Wang, L.: Snnfd, spiking neural segmentation network in frequency domain using high spatial resolution images for building extraction. Int. J. Appl. Earth Obs. Geoinf. **112**, 102930 (2022)

DiffSTRec: A Diffusion-Based Framework for Spatiotemporal Next POI Recommendation

Shun Wang[1], Xinyi Zhang[1], Di Wu[1], Junchao Zeng[1], Wei Liu[1(✉)], Jianxing Yu[1,2], Huaijie Zhu[1], and Jian Yin[1]

[1] School of Artificial Intelligence, Sun Yat-sen University, Zhuhai, China
{wangsh363,zhangxy758,wudi86,zengjch8}@mail2.sysu.edu.cn,
{liuw259,jxyu26,zhuhuaijie,issjyin}@mail.sysu.edu.cn
[2] Pazhou Lab, Guangzhou 510330, China

Abstract. Next point-of-interest (POI) recommendation has attracted increasing attention with the rapid development of location-based social networks. It aims to predict the next location a user is likely to visit based on their historical check-in behavior. Recent research indicates that diffusion models are gaining traction for their effectiveness in sequential recommendation tasks. However, these methods often overlook key spatiotemporal factors such as POI transition dynamics and target visit time. We propose DiffSTRec (Diffusion-based Spatiotemporal Recommendation), a diffusion-based framework that integrates a POI transition graph and multi-period time encoding to better capture POI semantics and users' temporal preference dynamics. By incorporating noisy target representations and visiting time into the preference extraction module, we dynamically estimate user preference signals to guide the denoising process at each step. This design allows the model to better capture latent and multifaceted user intent. Experiments show that DiffSTRec consistently outperforms competitive baselines on three real-world datasets.

Keywords: Next POI Recommendation · Diffusion Model · Spatial-Temporal Modeling · Preference Modeling

1 Introduction

With the growing popularity of location-based social networks, users can conveniently leave check-in records at the places they visit [22], providing a solid foundation for the development of point-of-interest (POI) recommendation systems. As a prominent variant, next POI recommendation task has become as a key component of location-based services, aiming to predict a user's next check-in location based on their historical mobility trajectories. This task places particular emphasis users' temporal behaviors and the dynamics of the transition between POIs, which distinguish it from typical recommendation scenarios.

S. Yang et al. (Eds.): WISA 2025, LNCS 16299, pp. 399–411, 2026.
https://doi.org/10.1007/978-981-95-6426-2_36

A wide range of methods have been proposed for the next POI recommendation task. Early approaches relied on techniques such as Matrix Factorization (MF) and Latent Dirichlet Allocation (LDA) to model the latent interactions between users and POIs [1]. The introduction of deep neural networks subsequently accelerated the development of this field. With the rise of deep learning, Recurrent Neural Networks (RNNs), particularly LSTM-based models [2], were introduced to capture temporal dependencies in user check-in sequences. More recently, Transformer-based architectures [4,5] have further improved performance by enabling more effective modeling of complex contextual and long-range sequential patterns.

User preferences are latent and multifaceted. For example, a user may simultaneously favor entertainment venues and sports facilities, with varying degrees of interest in each. Moreover, the decision to visit a particular place is inherently uncertain. Recent methods employ powerful sequence encoders such as RNNs or Transformers to model user preferences [8]; however, the resulting representations are typically treated as fixed outputs. Although efficient, this approach often limits the ability to capture the latent, multifaceted, and uncertain nature of user preferences. Diffusion models [3] have recently demonstrated strong capabilities in the field of computer vision. They have also gained attention in recommendation research [6,7], as their principled noise injection and denoising process effectively model item semantics and user preferences.

However, applying diffusion models to next POI recommendation task introduces new challenges beyond those in conventional recommendation tasks, due to the task's unique characteristics. The first challenge lies in capturing transition dynamics among POIs. Unlike general items (e.g., books, movies), POIs are tightly coupled with geographical mobility and transition behavior: Users move from one POI to another following latent spatial patterns. The second challenge is modeling multi-scale temporal dependencies. User check-in behavior exhibits clear periodicity and temporal sensitivity. For instance, users tend to visit restaurants during lunchtime or gyms after work. The third challenge is integrating future temporal intent into preference modeling. In real-world scenarios, the decision to visit a coffee shop depends not only on the user's past but also on what time they plan to go next. In addition, existing diffusion-based recommendation models inject noise into all item embeddings in the sequence [6], which distorts the original item information and hinders accurate preference modeling.

To address these challenges, we propose DiffSTRec. In summary, our contributions are as follows:

- We employ a diffusion-based approach to iteratively reconstruct user preferences from Gaussian noise. This transforms user intent from a fixed representation into a distribution, enabling the model to capture its multifaceted and uncertain nature.
- We enhance the diffusion framework with spatiotemporal contextual modeling. For spatial patterns, we construct a POI transition graph and apply a graph convolutional network (GCN) to capture both individual movement

dynamics and collaborative signals from other users. For temporal patterns, we introduce a multi-period time encoding mechanism to model daily and weekly periodicity in user behavior.
- Instead of injecting noise into all POI embeddings in the check-in sequence, we propose integrating the target visit time and current noise into a sequential encoder. This encoder jointly models users' historical behavior and future temporal intent, naturally bridging preference extraction and the denoising process.

Extensive experiments on three real-world check-in datasets demonstrate that DiffSTRec consistently outperforms competitive baselines.

2 Related Work

2.1 Next POI Recommendation

Next POI recommendation is widely studied in location-based social networks (LBSNs). Early approaches primarily relied on Markov chains or matrix factorization [9], while more recent advances have shifted towards deep learning-based sequential models [12]. Given that check-in trajectories can be naturally viewed as sequences, RNNs, especially LSTM and GRU variants, have been extensively applied to capture temporal dependencies in user trajectories. For instance, Graph-Flashback [11] leverages RNN-style modeling to learn user movement patterns. To incorporate contextual signals such as time, geography, and social influence, many models adopt context-aware mechanisms and attention modules. DeepMove [10], for example, applies RNN with attention to model complex mobility sequences. More recently, transformer-based architectures have demonstrated strong performance due to their ability to model long-range dependencies. GETNext [5] employs transformer-based sequence modeling with trajectory flow awareness, while ROTAN [8] introduces rotation-based temporal attention to capture user trajectories with fine-grained time sensitivity.

2.2 Diffusion Models for Recommendation

Generative modeling has been extensively studied in recommendation systems, with early work exploring variational autoencoders (VAEs) and generative adversarial networks (GANs). For instance, Liang et al. [13] applied a VAE for collaborative filtering, showing that a probabilistic latent factor model can improve top-N recommendation by better capturing uncertainty in userâĂŞitem interactions. In the GAN domain, IRGAN [14] was a seminal method that formulated recommendation as a minimax game between a generator and a discriminator.

In the domain of diffusion-based recommendation, CODIGEM [15] employs a denoising diffusion process to reconstruct userâĂŞitem interaction vectors for rating prediction. DiffRec [16] learns to denoise corrupted user interaction histories, modeling the generative process of implicit feedback. DiffuRec [6] introduces uncertainty into interaction sequences and uses a denoising mechanism to obtain target item representations. DreamRec [7] adopts guided diffusion to generate oracle items from user histories, reformulating sequential recommendation as a generation task without negative sampling.

3 Preliminaries

3.1 Problem Definition

Let $\mathcal{U} = \{u_1, u_2, \ldots, u_N\}$ denote the set of users, and $\mathcal{P} = \{p_1, p_2, \ldots, p_M\}$ denote the set of POIs. Each user $u_i \in \mathcal{U}$ has a historical check-in sequence represented as $\mathcal{S}_{u_i} = [(p_1, t_1), (p_2, t_2), \ldots, (p_l, t_l)]$, where $p_j \in \mathcal{P}$ denotes the POI visited at time t_j, and l is the length of the sequence. This sequence is also referred to as the check-in trajectory.

Given a user u_i and their historical check-in sequence $\mathcal{S}_{u_i}$, the goal of next POI recommendation is to predict the most probable POI the user will visit next. In this work, we focus on time-specific next POI recommendation [8], where the target visit time is explicitly incorporated into the prediction of the next location.

3.2 Diffusion Models

Diffusion models are a class of generative models that learn to map simple noise distributions to complex data distributions through a gradual denoising process. Formally, they define a Markov chain of diffusion steps that transform a data sample into Gaussian noise and then learn to reverse this process [3].

Let $\mathbf{x}_0$ denote a clean data sample, and let $q(\mathbf{x}_{1:T} \mid \mathbf{x}_0)$ represent the forward diffusion process, which gradually adds Gaussian noise over T steps. This process is defined as:

$$q(\mathbf{x}_t \mid \mathbf{x}_{t-1}) = \mathcal{N}(\mathbf{x}_t; \sqrt{1-\beta_t}\mathbf{x}_{t-1}, \beta_t \mathbf{I}), \quad (1)$$

where β_t is a small positive variance schedule that controls the amount of noise added at each step. By recursively applying the transition, we obtain a closed-form expression for sampling $\mathbf{x}_t$ conditioned on the original data $\mathbf{x}_0$:

$$q(\mathbf{x}_t \mid \mathbf{x}_0) = \mathcal{N}(\mathbf{x}_t; \sqrt{\bar{\alpha}_t}\mathbf{x}_0, (1-\bar{\alpha}_t)\mathbf{I}), \quad (2)$$

where $\alpha_t = 1 - \beta_t$, $\bar{\alpha}_t = \prod_{s=1}^{t}(1-\beta_s)$.

Using Bayes' theorem, the posterior distribution $q\left(\mathbf{x}_{t-1} \mid \mathbf{x}_t, \mathbf{x}_0\right)$ can be derived in closed form:

$$q(\mathbf{x}_{t-1} \mid \mathbf{x}_t, \mathbf{x}_0) = \mathcal{N}(\mathbf{x}_{t-1}; \tilde{\boldsymbol{\mu}}_t(\mathbf{x}_t, \mathbf{x}_0), \tilde{\beta}_t \mathbf{I}), \quad (3)$$

where the mean and variance are given by:

$$\tilde{\boldsymbol{\mu}}_t(\mathbf{x}_t, \mathbf{x}_0) = \frac{\sqrt{\bar{\alpha}_{t-1}}\beta_t}{1-\bar{\alpha}_t}\mathbf{x}_0 + \frac{\sqrt{\alpha_t}(1-\bar{\alpha}_{t-1})}{1-\bar{\alpha}_t}\mathbf{x}_t, \quad \tilde{\beta}_t = \frac{1-\bar{\alpha}_{t-1}}{1-\bar{\alpha}_t}\beta_t. \quad (4)$$

To approximate the reverse process during inference, a neural network $\epsilon_\theta(\cdot)$ is trained to predict the added noise. Using the reparameterization trick, the predicted mean can be expressed as:

$$\mu_\theta(\mathbf{x}_t, t) = \frac{1}{\sqrt{\alpha_t}} \left(\mathbf{x}_t - \frac{1-\alpha_t}{\sqrt{1-\bar{\alpha}_t}} \epsilon_\theta(\mathbf{x}_t, t) \right). \tag{5}$$

Training is typically performed by minimizing a variational bound or its simplified denoising objective. A widely used objective is to predict the noise $\boldsymbol{\epsilon}$ added in the forward process, where $\mathbf{x}_t = \sqrt{\bar{\alpha}_t}\mathbf{x}_0 + \sqrt{1-\bar{\alpha}_t}\boldsymbol{\epsilon}$, with $\boldsymbol{\epsilon} \sim \mathcal{N}(0, \mathbf{I})$. The simplified training loss is defined as:

$$\mathcal{L}_{\text{simple}} = \mathbb{E}_{\mathbf{x}_0, \boldsymbol{\epsilon}, t} \left[\|\boldsymbol{\epsilon} - \boldsymbol{\epsilon}_\theta(\mathbf{x}_t, t)\|_2^2 \right]. \tag{6}$$

4 Methodology

4.1 Framework Overview

Figure 1 illustrates the overall architecture of DiffSTRec, which consists of two main components: the *Preference Extraction Module* (PEM) and *Guided Denoising Module* (GDM). The PEM extracts user's raw preference for POI visits and transitions from the user's check-in sequence, and further incorporates future temporal intent to dynamically modulate the user's preference across different time slots. The GDM leverages the extracted raw preference to iteratively guide the denoising of Gaussian noise, producing a refined user preference representation that is both dynamic and expressive of the user's multifaceted intent.

4.2 Embedding Construction

Before extracting user preferences from check-in sequences, we first encode the user, POI, and time associated with each check-in. Each user is mapped to a learnable embedding in a lookup table, where a d-dimensional vector is used to preserve personalized information during training.

To model the transition patterns among POIs based on user behavior, we construct a POI transition graph $\mathcal{G} = (\mathcal{V}, \mathcal{E})$, where each node $v \in \mathcal{V}$ represents a POI, and each directed edge $(u, v) \in \mathcal{E}$ indicates a user transition from POI u to POI v. The edge weight reflects the transition frequency between two POIs.

Based on the constructed graph, we adopt GraphSAGE [17], which generates node embeddings by aggregating information from each node's local neighborhood in an inductive manner. Specifically, we use the SAGEConv layer with mean aggregation. At each layer k, the representation of node v is updated as:

$$h_v^{(k)} = \sigma \left(W^{(k)} \cdot \text{CONCAT} \left(h_v^{(k-1)}, \frac{1}{|\mathcal{N}(v)|} \sum_{u \in \mathcal{N}(v)} h_u^{(k-1)} \right) \right), \tag{7}$$

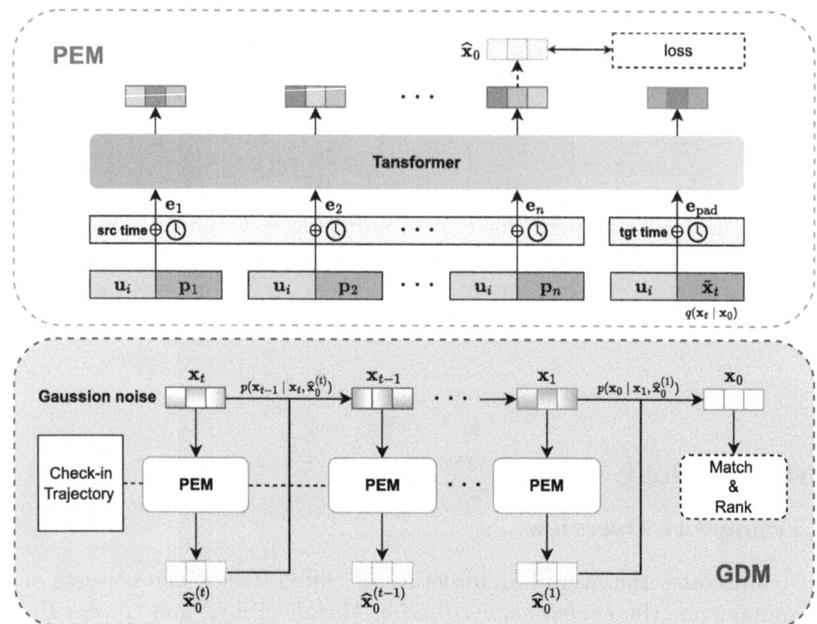

Fig. 1. An overview of the DiffSTRec architecture, which includes two main components: the Preference Extraction Module and the Guided Denoising Module.

where $h_v^{(k)} \in \mathbb{R}^d$ denotes the embedding of node v at layer k, $\mathcal{N}(v)$ is the set of its in-neighbors, $W^{(k)}$ is the trainable weight matrix, and σ is a non-linear activation function.

We encode temporal information using Time2Vec [18], capturing both daily and weekly periodic patterns in user check-in sequences. Each day is divided into 48 time slots, and each week into 7 days, with both temporal levels encoded as vector representations. The resulting embeddings are fused via a weighted sum to obtain a final time representation that captures multi-period temporal patterns in user behavior:

$$\mathbf{t}_i = \alpha \cdot \mathbf{t}_i^{(d)} + (1 - \alpha) \cdot \mathbf{t}_i^{(w)}, \tag{8}$$

where $\mathbf{t}_i^{(d)}$ and $\mathbf{t}_i^{(w)}$ denote the daily and weekly time embeddings of the i-th check-in, respectively, and $\alpha \in [0, 1]$ is the weight for daily periodicity.

4.3 Preference Extraction Module (PEM)

Following standard practice in prior work, we adopt the Transformer [19] as the backbone network for the PEM. For each check-in trajectory of length l, we first concatenate the user's embedding $\mathbf{u}$ with the POI embedding $\mathbf{p}_i$, and project the result through a linear layer. The resulting representation $\tilde{\mathbf{p}}_i$ is then enhanced by incorporating multi-period time embeddings, yielding unified check-in representations $\mathbf{e}_i$.

$$\tilde{\mathbf{p}}_i = \mathrm{Proj}([\mathbf{u}; \mathbf{p}_i]), \quad \mathbf{e}_i = \tilde{\mathbf{p}}_i + \mathbf{t}_i. \tag{9}$$

To explicitly incorporate the target visit time, we append a special placeholder token $\mathbf{e}_{\text{pad}}$ to the end of the trajectory. This token combines a noisy embedding $\tilde{\mathbf{x}}_t$ with the target visit time $\mathbf{t}^*$ to guide preference encoding with future intent. This process can be described as follows:

$$\mathbf{e}_{\text{pad}} = \mathrm{Proj}([\mathbf{u}, \tilde{\mathbf{x}}_t]) + \mathbf{t}^*, \quad \mathbf{E} = [\mathbf{e}_1, \mathbf{e}_2, \ldots, \mathbf{e}_l, \mathbf{e}_{\text{pad}}]. \tag{10}$$

Here, $\tilde{\mathbf{x}}_t = \lambda \odot (\mathbf{x}_t + r_t)$, where λ is a random parameter sampled from a Gaussian distribution, which controls the noise intensity, and $\odot$ denotes Hadamard multiplication. $\mathbf{x}_t$ denotes the target POI embedding after t steps of adding noise (as defined in Eq. 2) during training, or the denoised standard Gaussian vector at step t during inference. The noisy representation $\mathbf{x}_t$ is introduced to serve as a target signal throughout both training and inference. r_t is the embedding corresponding to the current diffusion (or reverse) step.

The embedding sequence $\mathbf{E}$ is fed into the Transformer, and the output corresponding to the final non-placeholder token is taken as the user's raw preference representation $\widehat{\mathbf{x}}_0$, which captures both historical behavior and target-aware intent. During training, the PEM is optimized using the following loss function to ensure that the extracted raw user preference better matches the target POI embedding $\mathbf{p}^*$.

$$\mathcal{L}_{\text{PEM}} = \mathbb{E}_{(\widehat{\mathbf{x}}_0, \mathbf{p}^\star)} \left[-\log \frac{\exp\left(\widehat{\mathbf{x}}_0^\top \cdot \mathbf{p}^\star\right)}{\sum_{p \in \mathcal{P}} \exp\left(\widehat{\mathbf{x}}_0^\top \cdot \mathbf{p}\right)} \right]. \tag{11}$$

During inference, the parameters of the PEM are fixed, and its output is used within the GDM.

4.4 Guided Denoising Module (GDM)

The denoising process is guided by the raw preference extracted from the PEM and is used to iteratively reconstruct user preferences from Gaussian noise. Specifically, we begin by sampling from a standard Gaussian distribution, i.e., $\mathbf{x}_T \sim \mathcal{N}(0, \mathbf{I})$, and then iteratively denoise it over a fixed number of steps T. Unlike most diffusion-based applications, and as shown in Eq. (3), we directly estimate $\mathbf{x}_0$ using the raw preference $\widehat{\mathbf{x}}_0$ obtained from the PEM, rather than predicting the noise at each step via a reparameterized model. Based on this design, the reverse process is formulated as:

$$p(\mathbf{x}_{t-1} \mid \mathbf{x}_t, \widehat{\mathbf{x}}_0) = \mathcal{N}(\mathbf{x}_{t-1}; \tilde{\boldsymbol{\mu}}_t(\mathbf{x}_t, \widehat{\mathbf{x}}_0), \tilde{\beta}_t \mathbf{I}). \tag{12}$$

At each denoising step $t \in \{T, T-1, \ldots, 1\}$, the intermediate latent vector $\mathbf{x}_t$ is fed into the PEM. This allows the PEM to recompute a revised user preference

$\widehat{\mathbf{x}}_0^{(t)}$ conditioned on the current denoising state. The updated preference then serves as guidance for the next step, formulated as:

$$\widehat{\mathbf{x}}_0^{(t)} = \mathrm{PEM}(\mathbf{x}_t, \cdot), \tag{13}$$

$$\mathbf{x}_{t-1} = \frac{\sqrt{\bar{\alpha}_{t-1}}\beta_t}{1-\bar{\alpha}_t}\widehat{\mathbf{x}}_0^{(t)} + \frac{\sqrt{\alpha_t}(1-\bar{\alpha}_{t-1})}{1-\bar{\alpha}_t}\mathbf{x}_t + \sqrt{\tilde{\beta}_t}\mathbf{z}, \quad \mathbf{z} \sim \mathcal{N}(0, \mathbf{I}), \tag{14}$$

here, the symbol $\cdot$ denotes static conditioning information, including the user's historical check-in sequence, and the target visit time.

This recurrent feedback loop between the diffusion steps and preference extraction enables dynamic adjustment of the denoising direction, effectively steering the process toward a more accurate and intent-aware preference representation. The final user preference obtained through guided denoising is more expressive, capturing both user history and future intent, and better reflecting the multifaceted nature of user interests.

Note that the GDM is only used during inference. After the final denoising step, we compute a relevance score s_i between the refined user preference representation $\mathbf{x}_0$ and each candidate POI embedding $\mathbf{p}_i$ via inner product. The final recommendation result $\hat{y}$ consists of the top-k POIs with the highest scores.

$$s_i = \mathbf{x}_0^\top \mathbf{p}_i, \quad \hat{y} = \text{Top-}k\left(\{s_i \mid p_i \in \mathcal{P}\}\right). \tag{15}$$

5 Experiments

5.1 Experiment Settings

Datasets. We evaluate our model on three real-world datasets widely used in next POI recommendation. Two are from the Foursquare [20] location-based social network, and one is from Gowalla [21]. Foursquare-NYC and Foursquare-TKY contain user check-in records collected from New York City and Tokyo, respectively. Gowalla-CA is a subset of the larger Gowalla dataset, containing check-ins within California. The statistics of them are summarized in Table 1. During preprocessing, we remove trajectories with fewer than 4 check-ins and truncate those exceeding 50 check-ins by retaining only the most recent 50.

Table 1. Statistics of the Foursquare and Gowalla datasets.

Datasets	NYC	TKY	CA
#users	1,048	2,282	3,957
#POIs	4,981	7,833	9,690
#check-ins	103,941	405,000	238,369

Baselines. We compare our model against four categories of baseline methods: (1) RNN-based methods, including **LSTM** [23], **DeepMove** [10] and **Graph-Flashback** [11], which leverage Recurrent Neural Networks to model user movement trajectories, with extensions to incorporate temporal and spatial information; (2) Transformer-based methods, such as **STAN** [4] and **GETNext** [5], which utilize self-attention mechanisms to capture long-range dependencies in check-in sequences enriched with contextual information; (3) Time-aware next POI recommendation methods, represented by **ROTAN** [8], which incorporate future temporal signals to enhance spatiotemporal prediction. To the best of our knowledge, ROTAN represents the state-of-the-art in this task; (4) Diffusion-based methods, such as **DiffuRec** [6], originally designed for item recommendation, which we slightly modify to suit the next POI recommendation task.

Training & Evaluation Protocol. Each dataset is split into training, validation, and test sets. For each user trajectory, the last check-in is held out as the test instance, and the second-to-last as the validation instance; the rest are used for training. This splitting strategy is widely adopted in sequential and POI recommendation tasks. We evaluate model performance using two standard metrics: Hit Rate (HR) and Normalized Discounted Cumulative Gain (NDCG).

Experimental Parameters. We set the number of diffusion steps to $T = 32$ with a linear variance schedule. The PEM consists of $L_{\text{PEM}} = 4$ Transformer blocks, each with 4 attention heads, while the POI transition graph is processed by 2 SAGEConv layers. All embeddings are set to a dimensionality of $d = 128$. The parameter λ in Eq. (10) is sampled from $\mathcal{N}(0.001, 0.001)$. A dropout rate of 0.1 is applied throughout the network, and the Adam optimizer is used with a learning rate of 0.001. Given these settings, the time complexity during inference can be expressed as $O\left(T \cdot L_{\text{PEM}} \cdot (l+1)^2 \cdot d\right)$, where l denotes the length of the historical trajectory.

5.2 Overall Performance

Table 2 summarizes the performance of all models on the three benchmark datasets. We observe that traditional RNN- and Transformer-based methods (e.g., LSTM, DeepMove, STAN) yield moderate performance, but generally fall short of the more recent approaches. (1) ROTAN ranks second on most metrics, which demonstrates the benefit of incorporating the target visit time into the prediction process. However, it still falls short compared to our method. (2) DiffuRec performs less competitively, likely due to its limited adaptation to the unique dynamics of POI recommendation. Its sequence-level noise injection may disrupt informative patterns, whereas our method introduces a dedicated placeholder token to carry noise and target time signals, guiding denoising while preserving historical behavior and maintaining sequence integrity. (3) Our model consistently achieves the best results across all datasets, significantly outperforming strong baselines such as ROTAN. This validates the effectiveness of our

Table 2. Comparison with baselines. Best results are in boldface, and second best results are underlined.

Model	NYC		TKY		CA	
	HR@5	NDCG@5	HR@5	NDCG@5	HR@5	NDCG@5
LSTM	42.85	30.08	37.52	24.16	22.47	16.55
DeepMove	55.72	41.11	46.36	34.75	27.03	19.39
Graph-flashback	56.87	40.31	<u>51.60</u>	37.84	28.07	20.56
STAN	53.68	35.27	44.76	34.63	27.85	20.23
GETNext	54.39	40.75	45.64	33.26	28.71	20.96
DiffuRec	55.51	43.03	47.22	35.42	33.27	24.76
ROTAN	<u>57.65</u>	<u>45.47</u>	51.58	<u>40.49</u>	<u>32.37</u>	<u>24.92</u>
Ours	**62.95**	**51.03**	**56.31**	**45.15**	**39.79**	**31.00**

guided diffusion design, which enables precise preference modeling conditioned on temporal intent and spatial transitions. The gain is particularly notable on CA, whose shorter user trajectories and sparser interactions further highlight the advantage of DiffSTRec in extracting rich preference signals from limited histories.

5.3 Ablation Study

To assess the contribution of each major component in our model, we perform an ablation study by selectively removing or altering individual modules. All variants are evaluated under identical settings, and the results are reported in Table 3.

- *w/o User*: We remove user embeddings from the input sequence, thus eliminating personalized information from the model.
- *w/o Temporal*: We remove all time-related embeddings, including both daily and weekly periodic components, rendering the model time-agnostic.
- *w/o Target*: We exclude the target visit time embedding, preventing the PEM from incorporating the user's future intent during preference extraction.
- *w/o Graph*: We remove the POI transition graph and instead use trainable POI embeddings, thereby omitting transition-aware spatial modeling.

As shown in Table 3, all ablated variants perform worse than the full model, confirming the effectiveness of each component. Removing user embeddings leads to the largest performance drop, highlighting the importance of personalized modeling. Excluding temporal features also results in consistent declines, indicating the value of capturing both periodic behavior and time-sensitive intent. Removing the target visit time causes a moderate drop, suggesting that explicitly modeling future intent brings additional benefits beyond general temporal patterns. Lastly, dropping the POI transition graph slightly reduces performance,

Table 3. The results of ablation studies.

Methods	TKY		CA	
	HR@5	NDCG@5	HR@5	NDCG@5
Original	**56.31**	**45.15**	**39.79**	**31.00**
w/o User	50.71	39.47	37.77	28.49
w/o Temporal	53.69	40.85	38.02	27.93
w/o Target	54.29	42.50	38.10	28.22
w/o Graph	55.18	44.31	37.82	29.27

showing that modeling spatial transitions contributes to better next-location prediction.

6 Conclusion

In this paper, we proposed DiffSTRec, a spatial-temporal diffusion-based framework for next POI recommendation. It employs a denoising diffusion process to iteratively refine the user's latent and multifaceted preferences. The model integrates a POI transition graph and multi-period temporal embeddings to capture users' spatial transition patterns and temporal periodic behaviors. Furthermore, by introducing noisy target signals and target visit time into the preference extraction module via a dedicated placeholder, our method enables more precise and target-aware preference modeling. Extensive experiments on three real-world datasets demonstrate that DiffSTRec consistently outperforms strong baselines, and ablation studies further confirm the effectiveness of each component.

Acknowledgements. This research was supported by Guangdong Basic and Applied Basic Research Foundation (2024A1515010241), Zhuhai City Industry-University-Research Project (2320004002797).

References

1. Yin, H., Cui, B., Sun, Y., Hu, Z., Chen, L.: LCARS: a spatial item recommender system. ACM Trans. Inf. Sys. (TOIS) **32**(3), 1–37 (2014)
2. Zhao, P., Zhu, H., Liu, Y., Li, Z., Xu, J., Sheng, V.S.: Where to go next: a spatio-temporal LSTM model for next POI recommendation. arXiv preprint arXiv:1806.06671 (2018)
3. Ho, J., Jain, A., Abbeel, P.: Denoising diffusion probabilistic models. Adv. Neural. Inf. Process. Syst. **33**, 6840–6851 (2020)
4. Luo, Y., Liu, Q., Liu, Z.: Stan: spatio-temporal attention network for next location recommendation. In: Proceedings of The Web Conference 2021, pp. 2177–2185 (2021)

5. Yang, S., Liu, J., Zhao, K.: GETNext: trajectory flow map enhanced transformer for next POI recommendation. In: Proceedings of the 45th International ACM SIGIR Conference on Research and Development in Information Retrieval, pp. 1144–1153 (2022)
6. Li, Z., Sun, A., Li, C.: Diffurec: a diffusion model for sequential recommendation. ACM Trans. Inf. Sys. **42**(3), 1–28 (2023)
7. Yang, Z., Wu, J., Wang, Z., Wang, X., Yuan, Y., He, X.: Generate what you prefer: reshaping sequential recommendation via guided diffusion. Adv. Neural. Inf. Process. Syst. **36**, 24247–24261 (2023)
8. Feng, S., Meng, F., Chen, L., Shang, S., Ong, Y.S.: Rotan: A rotation-based temporal attention network for time-specific next POI recommendation. In: Proceedings of the 30th ACM SIGKDD Conference on Knowledge Discovery and Data Mining, pp. 759–770 (2024)
9. Feng, S., Li, X., Zeng, Y., Cong, G., Chee, Y.M., Yuan, Q.: Personalized ranking metric embedding for next new POI recommendation. In: Proceedings of the 24th AAAI Conference on Artificial Intelligence, pp. 2069–2075. AAAI Press (2015)
10. Feng, J., et al.: DeepMove: Predicting human mobility with attentional recurrent networks. In: Proceedings of the 2018 World Wide Web Conference, pp. 1459–1468 (2018)
11. Rao, X., Chen, L., Liu, Y., Shang, S., Yao, B., Han, P.: Graph-flashback network for next location recommendation. In: Proceedings of the 28th ACM SIGKDD Conference on Knowledge Discovery and Data Mining, pp. 1463–1471 (2022)
12. Yang, Q., Hu, S., Zhang, W., Zhang, J.: Attention mechanism and adaptive convolution actuated fusion network for next POI recommendation. Int. J. Intell. Syst. **37**(10), 7888–7908 (2022)
13. Liang, D., Krishnan, R.G., Hoffman, M.D., Jebara, T.: Variational autoencoders for collaborative filtering. In: Proceedings of the 2018 World Wide Web Conference, pp. 689–698 (2018)
14. Wang, J., et al.: IRGAN: A minimax game for unifying generative and discriminative information retrieval models. In: Proceedings of the 40th International ACM SIGIR Conference on Research and Development in Information Retrieval, pp. 515–524 (2017)
15. Walker, J., Zhong, T., Zhang, F., Gao, Q., Zhou, F.: Recommendation via collaborative diffusion generative model. In: International Conference on Knowledge Science, Engineering and Management, pp. 593–605. Springer (2022)
16. Wang, W., Xu, Y., Feng, F., Lin, X., He, X., Chua, T.-S.: Diffusion recommender model. In: Proceedings of the 46th International ACM SIGIR Conference on Research and Development in Information Retrieval, pp. 832–841 (2023)
17. Hamilton, W., Ying, Z., Leskovec, J.: Inductive representation learning on large graphs. Adv. Neural. Inf. Process. Sys. **30** (2017)
18. Kazemi, S.M., et al.: Time2Vec: Learning a vector representation of time. arXiv preprint arXiv:1907.05321 (2019)
19. Vaswani, A., et al.: Attention is all you need. Adv. Neural. Inf. Process. Syst. **30**, 5998–6008 (2017)
20. Yang, D., Zhang, D., Zheng, V.W., Yu, Z.: Modeling user activity preference by leveraging user spatial temporal characteristics in LBSNs. IEEE Trans. Sys., Man, Cybern.: Sys. **45**(1), 129–142 (2014)
21. Cho, E., Myers, S.A., Leskovec, J.: Friendship and mobility: User movement in location-based social networks. In: Proceedings of the 17th ACM SIGKDD International Conference on Knowledge Discovery and Data Mining, pp. 1082–1090 (2011)

22. Meng, X., Zhang, Z., Cui, S., Huo, H.: Top-k collective spatial keyword approximate query. In: International Conference on Web Information Systems and Applications, pp. 227–238. Springer (2024)
23. Hochreiter, S., Schmidhuber, J.: Long short-term memory. Neural Comput. **9**(8), 1735–1780 (1997)

HGNNIM: A Hypergraph Neural Network-Based Approach to Maximize Influence in Social Networks

Runbin Yao[1,2], Wenli Fang[1,2], Chao Chang[2,3], Luyao Teng[2,3], Chengzhe Yuan[2,3], Hao Zhong[1,2(✉)], and Chengjie Mao[1,2]

[1] South China Normal University, Guangzhou, China
[2] Guangdong Polytechnic Normal University, Guangzhou, China
hzhong@m.scnu.edu.cn
[3] Guangzhou Panyu Polytechnic, Guangzhou, China

Abstract. Hypergraph which is used to model a social network can more accurately represent diverse relationships compared to general graph. Studying the influence maximization (IM for short) problem in a hypergraph helps to identify and optimize key nodes and strategies for information dissemination in social networks. In this paper, we proposed a hypergraph neural network-based approach for the IM problem (HGNNIM for short) in hypergraphs. First, HGNNIM integrates intrinsic attributes of node with the hyperedge attributes it belongs to, employing feature fusion techniques to generate high-dimensional embeddings. Next, these embeddings are subsequently fed into regression model in HGNNIM to predict node influence values, which are trained under both Independent Cascade (IC for short) and Susceptible-Infected-Recovered (SIR for short) diffusion models to simulate real-world information spread dynamics. Finally, the trained model is then deployed on target networks to identify high-impact seed nodes and predict their potential influence coverage. The experimental results indicate that the nodes identified by HGNNIM can achieve a wider range of influence in given hypergraph compared to the existing baselines.

Keywords: Hypergraph Neural Network · Complex Network · Influence maximization · Social network

1 Introduction

Influence Maximization (IM for short) seeks to identify optimal seed nodes that maximize information spread under specific diffusion models [1]. As social platforms proliferate, IM has emerged as a crucial tool for viral marketing and public opinion guidance [2]. Traditional graph-based approaches struggle to capture higher-order group behaviors (e.g., collaborative development, collective purchasing [3,4]), particularly missing semantic associations between node attributes and hyperedge contexts [5]. To address these limitations, we propose a Hypergraph Neural Network-based Approach to IM called HGNNIM that includes the following contributions:

S. Yang et al. (Eds.): WISA 2025, LNCS 16299, pp. 412–423, 2026.
https://doi.org/10.1007/978-981-95-6426-2_37

1. Propose a hypergraph neural network framework integrating node attributes with multi-scale structural features,
2. Design a pseudo-regression formulation of IM through hybrid IC/SIR label generation [6,7],
3. Propose a dual-channel fusion mechanism combining ordinary graph and hypergraph featuresd.

2 Related Work

Research on influence maximization in complex networks integrated three principal methodologies: topology-driven approaches, greedy algorithm frameworks, and emerging deep learning paradigms. Traditional topology-based methods leveraged structural characteristics to assess node significance, employing centrality metrics including degree, closeness, and betweenness centrality [8] for quantifying local influence potential. Hierarchical analysis through k-shell decomposition [9] systematically identified core-periphery structures by iteratively pruning low-degree peripheral nodes, while AOGC [10] introduced adaptive influence evaluation via dynamic gravity modeling that adjusted interaction ranges based on localized network topology. The greedy algorithm family, rooted in the seminal NP-hard formulation proposed by Kempe et al. [11] combining Independent Cascade and Linear Threshold models, evolved through multiple optimizations: CELF [12] exploited submodularity properties to achieve 700× efficiency gains by marginal gain prioritization, CELF++ [5] further reduced computational redundancy through cross-iteration marginal gain caching, and hybrid strategies proposed by Chen et al. [13] integrated subgraph construction with influence factor thresholds to balance accuracy and speed. Contemporary deep learning solutions addressed discrete optimization challenges through continuous representation learning - Zhang et al. [14] combined dynamic graph convolutional networks with pseudo-label distillation for influence prediction, Kumar et al. [6] synergized struc2vec graph embeddings with multi-layer GNNs to capture node relationships across structural scales, while reinforcement learning adaptations [15] reformulated topic-aware influence maximization as Markov decision processes. Advanced techniques like DGRL [1] employed learnable propagation matrices to model nonlinear diffusion patterns, and the multi-task framework proposed by Panagopoulos et al. [16] jointly optimized cascade representation learning and influence range prediction through neural architecture integration. Although the research on the issue of maximizing influence had been very rich, the analysis from the structure of hypergraphs was still relatively scarce. In addition, we introduced the hypergraph neural network to learn the eigenvectors of nodes, which could obtain higher quality node representations and also achieved better performance in the experiment.

3 Preliminaries

This section describes some background concepts that need to be understood for this work.

3.1 Problem Definition

The influence maximization problem on a hypergraph is defined as selecting the optimal seed node set $x \subseteq \mathcal{V}$ in the hypergraph $\mathcal{H} = (\mathcal{V}, \mathcal{E})$ in order to maximize the number of affected nodes $y \in \mathbb{R}^+$ evaluated by the information diffusion model M in the mathematical form: $y = M(x, H, \theta)$, where θ is the node infection probability parameter and x denotes the vectorized representation (node embedding/feature) of the seed node set. Thus, the mathematical definition of the influence maximization problem on hypergraphs is as

$$\tilde{x} = \arg\max_{|\mathrm{x}| \leq k} M(\mathrm{x}, H; \theta), \tag{1}$$

where $\tilde{x}$ is the best seed node set that can produce the maximum influence spread range in H.

3.2 Representation of Hypergraph

Hypergraphs generalize traditional graphs through hyperedges connecting arbitrary numbers of nodes, enabling natural representation of multi-body interactions in complex systems [17]. Typical applications include: modeling academic collaborations via paper-author hyperedges (Fig 1(a)) [18], analyzing traffic flow patterns in GIS [19], and capturing multi-gene/protein synergies in biological networks [20] (unachievable with pairwise graphs).

Formally, a hypergraph $\mathcal{H} = (\mathcal{V}, \mathcal{E})$ consists of node set $\mathcal{V}$ and hyperedge set $\mathcal{E}$ (non-empty node subsets), represented by incidence matrix $H \in \{0,1\}^{|\mathcal{V}| \times |\mathcal{E}|}$ where $H(v, e) = 1$ indicates membership of node v in hyperedge e (Fig 1(b)). This generalizes traditional graphs whose symmetric adjacency matrices $A(i, j) \in \{0, 1\}$ only encode pairwise connections. Hypergraph can directly represent a group (such as wechat group) through a super edge, so as to more accurately capture the collective effect of the group on the spread of influence, which can not be directly expressed by a pile of binary relationships in ordinary graphs.

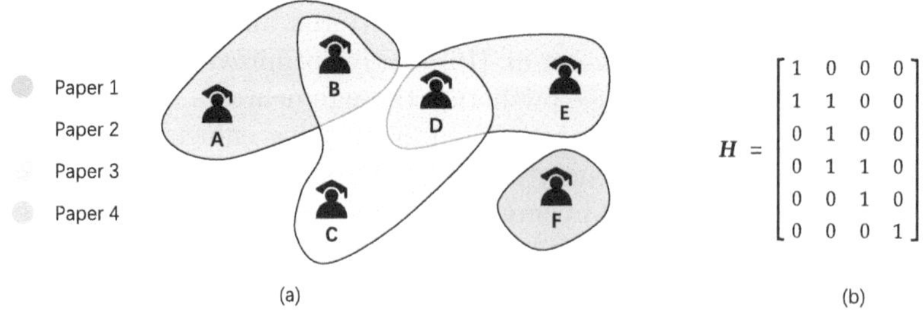

Fig. 1. (a) Academic collaboration hypergraph (A-E: scholars, hyperedges: co-authored papers); (b) Corresponding incidence matrix representation.

3.3 Hypergraph Neural Network

Hypergraph Neural Networks (HGNN for short) [7] extend GNNs by modeling multi-node interactions through hyperedges. Different from conventional GNNs limited to pairwise relations, HGNNs capture high-order associations via hyperedge feature aggregation and node feature updating. Let us first provide some symbols. (i) Vertex degree matrix: $D_v = \text{diag}(\sum_{e \in \mathcal{E}} w(e)h(v,e))$, (ii) Hyperedge degree matrix: $D_e = \text{diag}(\sum_{v \in \mathcal{V}} h(v,e))$, (iii) Hypergraph Laplacian: $\Delta = I - D_v^{-1/2} H W D_e^{-1} H^\top D_v^{-1/2}$, where W denotes hyperedge weights and H the incidence matrix. To preserve high-order correlations, HGNNs enforce feature smoothness through Laplacian regularization $\text{tr}(X^\top \Delta X)$. Spectral convolution is implemented via Chebyshev polynomial approximation [21] $g * x \approx \theta_0 x + \theta_1(\Delta - I)x$ with $\lambda_{\max} = 2$ simplification. The final convolution layer becomes $Y = \sigma\left(D_v^{-1/2} H W D_e^{-1} H^\top D_v^{-1/2} X \Theta\right)$, where Θ denotes learnable parameters and σ activation functions. This framework maintains $O(|\mathcal{E}|)$ complexity while capturing multi-modal interactions.

4 Methodology

HGNNIM framework tackles influence maximization through (i) a learning phase that extracts high-order connectivity through hypergraph convolutions, and (ii) an inference phase that selects optimal seed nodes using learned influence predictions. The model processes fused node features through an HGNN-based regressor trained on diffusion model-generated labels (Fig 2), establishing a pseudo-regression paradigm where continuous output values represent predicted influence spreads.

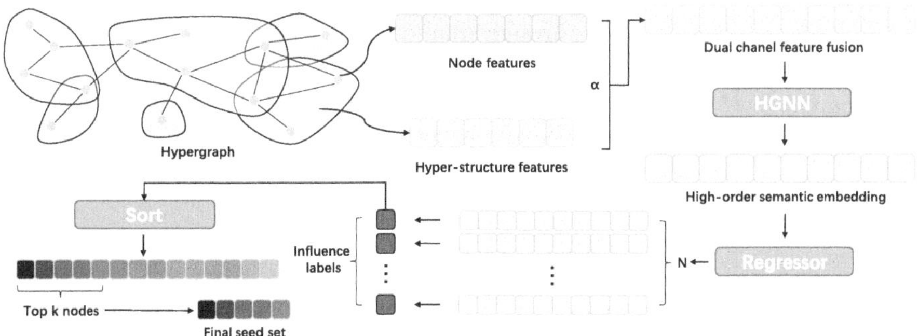

Fig. 2. HGNNIM architecture integrating hypergraph convolutions with dual-channel feature fusion.

4.1 Dual-Channel Feature Fusion

We construct enhanced node embeddings (Z) by synthesizing low-order (Zg) and hypergraph-based high-order (Zh) topological features:

$$Z_i = \alpha Zg_i + (1 - \alpha)Zh_i + b \tag{2}$$

where $\alpha \in [0, 1]$ balances structural preferences (ordinary graph vs. hypergraph), $b \in \mathbb{R}^k$ enables bias correction, and i is the index of the nodes. The fusion mechanism addresses (i) **First-order aggregation**: Node proximity in original graph space, (ii)**High-order collaboration**: Multi-node interactions via hyperedges, (iii)**Cross-domain complementarity**: Adaptive weighting of structural semantics. This joint representation overcomes single-modality limitations while preserving both local connectivity patterns and global collaborative relationships.

4.2 Influence Label Generation

To train our pseudo-regression model, we develop a hybrid labeling approach IDM($\cdot$) to generate labels. Generally, IDM adopts SIR or IC information diffusion model, 1) SIR describe disease spread by dividing the population into Susceptible, Infected, and Recovered groups. 2) IC is a model for information diffusion based on node influence and cascade probabilities to simulate propagation processes. Monte Carlo simulation generates multidimensional labels through standardized metrics (propagation range, efficiency, persistence), effectively combining the global attenuation mode of SIR and the local heterogeneity of IC for social network analysis. The labels used in this study are obtained by averaging 250 executions under SIR model and IC model respectively.

4.3 Feature Processing Under HGNN

HGNN process node features through dual-stage message passing: 1) Node-to-hyperedge aggregation $\mathbf{E} = \mathbf{H}^\top \mathbf{X}\mathbf{W}_e$; 2) Hyperedge-to-node diffusion $\mathbf{X}' = \mathbf{H}\mathbf{E}\mathbf{W}_v$, where $\mathbf{H}$ is the hyperedge incidence matrix and $\mathbf{W}_e/\mathbf{W}_v$ are learnable parameters. Two stacked HGNN layers followed by a fully-connected layer produce influence prediction scores, capturing multi-node collaboration patterns beyond traditional GNN capabilities. Its mathematical expression is as

$$\mathrm{SCORES} = \mathrm{HGNNIM}(X, H, d) \tag{3}$$

where d refers to the dimension of the hidden layer. Here, 128 is taken to improve the computing efficiency while preserving as much node information as possible.

4.4 Using a Regressor For Influence Prediction

The regressor establishes mapping $\hat{y}_i = f(X_i; \theta)$ between node features X_i and predicted influence $\hat{y}_i$, optimized through a composite loss

$$\mathcal{L} = (1 - \alpha)\mathcal{L}_{MSE} + \alpha\mathcal{L}_{Rank} \tag{4}$$

based on mean squared error $\mathcal{L}_{MSE} = \frac{1}{n}\sum_{i=1}^{n}(\hat{y}_i - y_i)^2$ and ranking loss $\mathcal{L}_{Rank} = \frac{1}{|S_{top}||S_{other}|}\sum_{i\in S_{top}}\sum_{j\in S_{other}}\max(0, 1-(\hat{y}_i - \hat{y}_j))$, where S_{top} indexes top-k nodes in ground truth y, S_{other} is the remainder. The hinge loss enforces margin which is not less than 1.0 between top nodes and others. Hyperparameter $\alpha \in [0,1]$ balances prediction accuracy and ranking discriminability, mitigating "ranking ambiguity" from pure MSE optimization.

4.5 Algorithms

The framework operates through two principal algorithms: **Algorithm 1 (Training)** trains the HGNNIM model on D_{train} networks to capture global propagation dynamics (e.g., temporal decay) and local activation patterns (e.g., node triggering), utilizing MIX-IDM labels, dual-channel feature fusion f_d, hypergraph convolution f, and output function o, ultimately yielding optimized parameters Opt_{model} and network configuration $Opt_{network}$.

Algorithm 1: Training the HGNN model

Input: Training Networks ($\mathcal{D}_{\text{train}}$), Test Networks ($\mathcal{D}_{\text{test}}$), Embedding dimension ($d$), convolution function (f), output function (o)
Output: Optimal Training Network ($Opt_{network}$), Optimal model (Opt_{model})

1 **for** $i \in range(1, |\mathcal{D}_{train}|)$ **do**
2 $X_{\text{train}} \leftarrow f_d(\mathcal{D}_{\text{train}})$;
3 $\text{label}_{\text{train}} \leftarrow \text{IDM}(\mathcal{D}_{\text{train}}[i])$;
4 $\text{train\_HGNN} \leftarrow \text{HGNN}(\mathcal{D}_{\text{train}}[i], X_{\text{train}}, H, \text{label}_{\text{train}}, f, o, d)$;
5 **for** $j \in range(1, |\mathcal{D}_{test}|)$ **do**
6 $X_{\text{test}} \leftarrow f_d(\mathcal{D}_{\text{test}})$;
7 $F_c[i][j] \leftarrow \text{trainHGNN}(\mathcal{D}_{\text{test}}[j], X_{\text{test}})$;
8 **end**
9 $Opt_{network} \leftarrow \text{argmax}\ \frac{1}{m}\sum F_c[i][j]$;
10 $X_{\text{opt}} \leftarrow f_d(H_{\text{opt}}, Opt_{network})$;
11 $\text{label}_{\text{opt}} \leftarrow \text{MIX-DIM}(Opt_{network})$;
12 $\text{Optmodel} \leftarrow \text{HGNN}(Opt_{network}, X_{\text{opt}}, H_{\text{opt}}, \text{label}_{\text{opt}}, f, o, d)$;
13 **end**
14 **return** Opt_{model},$Opt_{network}$;

Algorithm 2 (Prediction) deploys Opt_{model} on target network H_{target} with embedding dimension d and seed size k: extracts topological features, computes node influence scores via HGNNIM, selects top-k nodes as seed sets, and evaluates performance metrics against ground truth propagation data.

5 Experiment and Results Analysis

This section describes the experimental setup and performance evaluation results, and explains the performance advantages of this model over some contemporary influence maximization algorithms through side-by-side comparisons.

Algorithm 2: Influence Maximization using the trained HGNNIM model

Input: HGNNIM model, Target network (H_{target}), Embedding dimension (d), Seed set size (k)
Output: Seed set S

1 $X_{target} \leftarrow f_d(H_{target})$;
2 Pred_Inf $\leftarrow$ HGNNIM($X_{target}, H_{target}, d$);
3 Sorted_Inf $\leftarrow$ sort(Pred_Inf, desc);
4 $S \leftarrow$ top$_k$(Sorted_Inf);
5 **return** S;

5.1 Datasets and Evaluation Metrics

We divide the data set into three parts, the training data set, the test data set and the target data set. The training data set and the test data set are all generated by the Barabási-Albert (BA) model, and a total of six training networks and six test networks are generated. The training network is recorded as train_n_k, the test network is recorded as test_n_k (n represents the total number of nodes in the network, k represents the average degree), and the target network is a real world public data set, 1) Wiki-Vote [22]: Wikipedia admin election voting network; 2) Scholat [23]: Academic social network with 2,302 high-quality users; 3) Netscience [24]: Coauthorship network in physics; 4) CA-GrQc [25]: Collaboration network in general relativity. Hypergraphs are constructed via k-nearest neighbor hyperedge generation. Network statistics are detailed in Table 1.

Table 1. Details of datasets

Dataset	Nodes	Edges	Hyperedges
CA-GrQc [25]	5242	14496	2621
WikiVote [22]	889	2914	445
Netscience [24]	1461	2742	731
Scholat [23]	2302	11393	1151

This study uses the propagation scale $F(s) = \frac{Ns}{N}$ to measure the quality of the final propagation results of the model, where Ns denotes the number of nodes eventually infected and N denotes the total number of nodes in the network.

5.2 Parametric Analysis

This study evaluates the performance potential of the HGNN model and its adaptability to training network scale and structural complexity through parameter sensitivity analysis. Use the six models trained on the training network to verify the generalization ability of the model on six invisible test networks. Figure 3 compares propagation performance across training configurations, where the vertical axis represents influence spread $\sigma(S)$ and the horizontal axis denotes seed set size $|S|$. Key findings include: 1) Configuration Train_3000_4 ($n = 3000$, $k = 4$) demonstrates optimal robustness in cross-complexity tests, aligning with Eq. (5) theoretical predictions; 2) Influence spread increases monotonically with seed set size; 3) Model performance positively correlates with training network scale and connection density. Consequently, Train_3000_4 is selected as the optimal configuration ($\mathcal{T}_{opt}$). Comparative experiments on real-world network datasets validate the effectiveness of method through multidimensional evaluation metrics against existing influence maximization algorithms. The mathematical form for calculating $\mathcal{T}_{opt}$ is as

$$\mathcal{T}_{opt} = \arg \max_{r \in \mathcal{T}_{\text{train}}} \frac{1}{|\mathcal{T}_{test}|} \sum_{t \in \mathcal{T}_{test}} \left(\frac{1}{|S|} \sum_{s \in S} \sigma(r, t, s)\right), \tag{5}$$

where S is the set of seed set sizes and $\sigma(\cdot)$ is the propagation coverage calculated based on IC/SIR model.

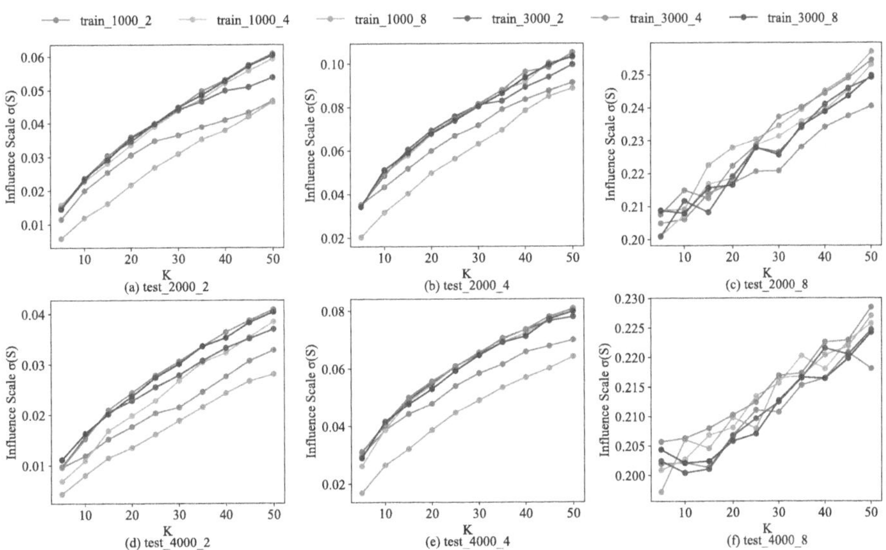

Fig. 3. (a)-(f) Final influence scale vs. Seed set size (K) for test networks. The infection probability α is set to 0.1 under the IC model.

5.3 Final Influenced Scale Vs Seed Set Size

We evaluate the performance of HGNNIM in comparison with several baseline methods on four real-world networks, employing both the SIR and IC propagation models. For all experiments, the infection or propagation probability is set to 0.1, and the recovery rate in the SIR model is set to 1. Seed set sizes vary from 5 to 50, with an increment of 5. To ensure robustness and reduce randomness, all propagation results are averaged over 100 independent Monte Carlo simulations. The baseline methods include

1)DPSO [26]—a discrete particle swarm optimization algorithm designed for combinatorial optimization problems;

2)ISF [27]—a sampling-based influence spread estimation framework;

3)KSN [27]—a key node selection method based on local-global influence propagation analysis.

Figure 4 illustrates the final influence diffusion sizes achieved by various algorithms across different datasets and propagation models, with seed set sizes ranging from 5 to 50 in steps of 5. Subfigures (a)-(d) show results under the IC model, while (e)-(h) show those under the SIR model. The diffusion size clearly increases with larger seed sets. In CA-GrQc networks (Fig. 4(a), (e)), HGNNIM performs comparably to DPSO for medium seed sets and outperforms others for small and large ones. In Wiki-Vote networks (Fig. 4(b), (f)), HGNNIM and DPSO are similar at small seed sets and outperform ISF and KSN, whose growth initially appears faster but plateaus, failing to surpass HGNNIM. In netscience networks (Fig. 4(c), (g)), all methods show similar trends, but HGNNIM consistently leads. In Scholat (Fig. 4(d), (h)), due to high average node degrees and wider propagation, all methods show close curves, yet HGNNIM still leads, demonstrating its superior ability to identify high-influence nodes. Overall, HGNNIM outperforms all baselines in final diffusion size and shows greater stability across seed set sizes, making it the best choice for selecting influential seed nodes.

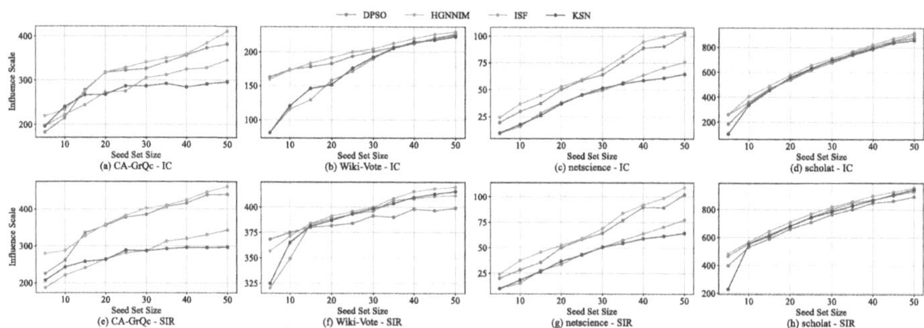

Fig. 4. Final influence scale in different network.

The performance of each algorithm is then compared quantitatively. The results are shown in Table 2. HGNNIM outperforms or approaches other algorithms on all datasets, especially in the CA-GrQc dataset, the score of HGNNIM is 0.076, which is significantly higher than that of DPSO (0.073), KSN (0.054), and ISF (0.055), reflecting its advantages in sparse networks; in the Wiki-Vote dataset, the score of HGNNIM scores 0.450, which is slightly higher than DPSO

(0.440) and ISF (0.445), indicating that it can also achieve more efficient influence dissemination in dense social networks; in addition, in the Scholat dataset, HGNNIM scores 0.356, which is better than DPSO (0.332) and KSN (0.341), further verifying that competitiveness of HGNNIM in academic social networks. Taken together, HGNNIM performs well in different types of networks, and has strong versatility and superiority.

Table 2. Final propagation size F(s) for different algorithms under different networks with seed set size k = 30 under the SIR model

Dataset	HGNNIM	DPSO	KSN	ISF
CA-GrQc [25]	**0.076**	0.073	0.054	0.055
Wiki-Vote [22]	**0.450**	0.440	0.448	0.445
Netscience [24]	**0.047**	0.044	0.035	0.035
Scholat [23]	**0.356**	0.332	0.341	0.348

6 Conclusion

In this paper, we propose a hypergraph neural network-based influence maximization solution (HGNNIM), which extends the traditional graph structure to the hypergraph space and achieves seed node selection through higher-order relationship modeling. The method reconstructs the influence maximization problem into a node influence prediction task: firstly, fusing the features of nodes in ordinary graph structure and hypergraph structure to construct enhanced feature vectors; then using hypergraph neural networks to capture the higher-order interactions among the nodes; and finally training the model to achieve influence prediction through a supervised learning paradigm. In the experimental part, we validate the model performance on four real network datasets, and the results show that HGNNIM significantly outperforms the benchmark method in terms of influence prediction accuracy. Future research proposes to improve the model in two directions: first, introducing an attention mechanism to enhance the expressiveness of key features; second, constructing a time-order-aware influence prediction model by combining the dynamic evolutionary properties of hypergraphs [28]; third, this research needs to construct pseudo labels for all nodes in the training phase, and the hypergraph structure is more complex than the ordinary graph, so the calculation cost is high, and there is room for improvement in model optimization or label generation.

Acknowledgments. This work was supported in part by the National Key Research and Development Program of China (Research and Demonstration Application of Key Technologies for Personalized Learning Driven by Educational Big Data) under Grant 2023YFC3341200, the Young Scientific and Technological Talent Support Project of

Guangzhou under grant QT202501, the Tertiary Education Scientific research project of Guangzhou Municipal Education Bureau under grant 2024312300, the scientific Research Platforms and Projects of Guangdong Provincial Education Department under grant 2024KQNCX232 and the Collaborative Innovation Center for Intelligent Educational Technology of Guangzhou under Grant 2025B04J0007.

References

1. Ling, C., Jiang, J., Wang, J., Thai, M.T., Xue, R., Song, J., Qiu, M., Zhao, L.: Deep graph representation learning and optimization for influence maximization. In Proceedings of International Conference on Machine Learning, Honolulu, Hawaii, USA, vol. 23–29, 21350–21361 (2023)
2. Nguyen, H., Zheng, R.: On budgeted influence maximization in social networks. IEEE J. Sel. Areas Commun. **31**(6), 1084–1094 (2013)
3. Xie, M., Zhan, X.-X., Liu, C., Zhang, Z.-K.: Influence maximization in hypergraphs. CoRR, abs/2206.01394 (2022)
4. Ye, F., Zhong, H., Lin, X., Yu, L., Mao, C.: Attribute multiplex network graph clustering: joint contrastive and high-order proximity. In: Web Information Systems and Applications - 21st International Conference, WISA 2024, Yinchuan, China, August 2-4, 2024, Proceedings, pages 290–297 (2024)
5. Goyal, A., Lu, W., Lakshmanan, L.V.S.: CELF++: optimizing the greedy algorithm for influence maximization in social networks. In: Proceedings of the 20th International Conference on World Wide Web, Hyderabad, India, March 28 - April 1, pages 47–48 (2011)
6. Kumar, S., Mallik, A., Khetarpal, A., Panda, B.S.: Influence maximization in social networks using graph embedding and graph neural network. Inf. Sci., **607**, 1617–1636 (2022)
7. Feng, Y., You, H., Zhang, Z., Ji, R., Gao, Y.: Hypergraph neural networks. In: Proceedings of the Thirty-Third AAAI Conference on Artificial Intelligence, the Thirty-First Innovative Applications of Artificial Intelligence Conference, the Ninth AAAI Symposium on Educational Advances in Artificial Intelligence, Honolulu, Hawaii, USA, January 27 - February 1, pp. 3558–3565 (2019)
8. Freeman, L.C., et al.: Centrality in social networks: conceptual clarification. Soc. Netw. Critical Concepts Sociol. Londres: Routledge, **1**(3), 238–263 (2002)
9. Kitsak, M., et al.: Identification of influential spreaders in complex networks. Nat. Phy., **6**(11), 888–893 (2010)
10. Yang, P., Meng, F., Zhao, L., Zhou, L.: Aogc: an improved gravity centrality based on an adaptive truncation radius and omni-channel paths for identifying key nodes in complex networks. Chaos, Solitons Fractals **166**, 112974 (2023)
11. Kempe, D., Kleinberg, J.M., Tardos, É.: Maximizing the spread of influence through a social network. In: Proceedings of the Ninth ACM SIGKDD International Conference on Knowledge Discovery and Data Mining, Washington, DC, USA, August, vol. 24–27, pp. 137–146 (2003)
12. Leskovec, J., Krause, A., Guestrin, C., Faloutsos, C., VanBriesen, J.M., Glance, N.S.: Cost-effective outbreak detection in networks. In: Proceedings of the 13th ACM SIGKDD International Conference on Knowledge Discovery and Data Mining, San Jose, California, USA, August 12-15, pp. 420–429 (2007)

13. Chen, W., Wang, Y., Yang, S.: Efficient influence maximization in social networks. In: Proceedings of the 15th ACM SIGKDD International Conference on Knowledge Discovery and Data Mining, Paris, France, June 28 - July 1, pp. 199–208 (2009)
14. Zhang, C., Li, W., Wei, D., Liu, Y., Li, Z.: Network dynamic GCN influence maximization algorithm with leader fake labeling mechanism. IEEE Trans. Comput. Soc. Syst. **10**(6), 3361–3369 (2023)
15. Tian, S., Mo, S., Wang, L., Peng, Z.: Deep reinforcement learning-based approach to tackle topic-aware influence maximization. Data Sci. Eng. **5**(1), 1–11 (2020)
16. Panagopoulos, G., Malliaros, F.D., Vazirgiannis, M.: Multi-task learning for influence estimation and maximization. IEEE Trans. Knowl. Data Eng. **34**(9), 4398–4409 (2022)
17. Battiston, F., et al.: Networks beyond pairwise interactions: structure and dynamics. CoRR, abs/2006.01764 (2020)
18. Gao, Y., Zhang, Z., Lin, H., Zhao, X., Shaoyi, D., Zou, C.: Hypergraph learning: methods and practices. IEEE Trans. Pattern Anal. Mach. Intell. **44**(5), 2548–2566 (2022)
19. Wang, K., Chen, J., Liao, S., Hou, J., Xiong, Q.: Geographic-semantic-temporal hypergraph convolutional network for traffic flow prediction. In: Proceedings of the 25th International Conference on Pattern Recognition, Virtual Event / Milan, Italy, January 10-15, pp. 5444–5450. IEEE (2020)
20. Gopalakrishnan, S., Sridharan, S., Nayak, S.R., Nayak, J., Venkatraman, S.: Central hubs prediction for bio networks by directed hypergraph - GA with validation to COVID-19 PPI. Pattern Recognit. Lett., **153**, 246–253 (2022)
21. Defferrard, M., Bresson, X., Vandergheynst, P.: Convolutional neural networks on graphs with fast localized spectral filtering. In: Proceedings of the Annual Conference on Neural Information Processing Systems, Barcelona, Spain, December 5-10, pp. 3837–3845 (2016)
22. Leskovec, J., Huttenlocher, D.P., Kleinberg, J.M.: Signed networks in social media. In: Proceedings of the 28th International Conference on Human Factors in Computing Systems, CHI 2010, Atlanta, Georgia, USA, April 10-15, pp. 1361–1370 (2010)
23. Lin, R., Tang, F., He, C., Zhengyang, W., Yuan, C., Tang, Y.: DIRS-KG: a kg-enhanced interactive recommender system based on deep reinforcement learning. World Wide Web **26**(5), 2471–2493 (2023)
24. Rossi, R.A., Ahmed, N.K.: The network data repository with interactive graph analytics and visualization. In: Proceedings of the Twenty-Ninth AAAI Conference on Artificial Intelligence, Austin, Texas, USA, January 25-30, pp. 4292–4293 (2015)
25. Leskovec, J., Kleinberg, J.M., Faloutsos, C.: Graph evolution: densification and shrinking diameters. ACM Trans. Knowl. Discov. Data **1**(1), 2 (2007)
26. Wang, Q., Gong, M., Song, C., Wang, S.: Discrete particle swarm optimization based influence maximization in complex networks. In: 2017 IEEE Congress on Evolutionary Computation, Donostia, San Sebastián, Spain, June 5-8, pp. 488–494 (2017)
27. Kuhnle, A., Alim, M.A., Li, X., Zhang, H., Thai, M.T.: Multiplex influence maximization in online social networks with heterogeneous diffusion models. IEEE Trans. Comput. Soc. Syst., **5**(2), 418–429, (2018)
28. Li, M., Xie, Z., Ding, L.: Reliable community search over dynamic bipartite graphs. In: Web Information Systems and Applications - 21st International Conference, WISA 2024, Yinchuan, China, August 2–4, 2024, Proceedings, pp. 298–307 (2024)

Representation Learning for Community Life Circle Structures with Quality-Aware POI Dependencies

Shangci Sun[1](✉) and Haotian Gao[2]

[1] Architectural Association School of Architecture (AA), London WC1B 3ES, UK
nicisun00@gmail.com
[2] School of Computer Science and Engineering, Shenyang Jianzhu University, Shenyang 110168, China

Abstract. Accurately identifying essential Points of Interest (POIs), modeling resident-POI dependencies, and quantifying spatial structures are fundamental to creating livable environments. We propose a representation learning framework for quality-aware modeling of community life circle structures. Our method integrates urban mobility patterns to resolve two key limitations of existing approaches: rigid spatial boundaries and ambiguous dependency relationships. First, we employ *k*-nearest neighbors (*k*NN) to adaptively identify candidate POIs, eliminating fixed-range constraints. Second, we design a POI quality index using multi-source urban data and compute community-POI dependency indices through synergistic integration of quality metrics and category weights. Third, temporal segmentation via KL-divergence constructs activity graphs reflecting time-variant resident mobility patterns. Finally, an autoencoder-based representation learning module embeds these graphs to generate structural feature vectors. For the convenience prediction application, experimental evaluations were conducted using real datasets to verify the effectiveness of the proposed representation method for community life circle structure.

Keywords: *k*NN · POI Quality · Community Life Circle · Life Circle Structure · Representation Learning

1 Introduction

The construction of service facilities in urban life circles plays a crucial role in enhancing residents' quality of life. A life circle is typically defined as the area accessible to residents within a short timeframe, such as 15 min. Points of Interest (POIs), as significant urban spatial facilities, possess strong capabilities for characterizing urban morphology and serve as fundamental data for exploring urban spatial structures. Research on urban structures based on POIs has yielded substantial achievements in urban infrastructure planning, commercial spatial layout, and residential distribution patterns.

With the rapid advancement of information technology, the influence of residents' spatio-temporal behaviors on urban spatial organization and structure has become

S. Yang et al. (Eds.): WISA 2025, LNCS 16299, pp. 424–436, 2026.
https://doi.org/10.1007/978-981-95-6426-2_38

increasingly pronounced. Therefore, in-depth analysis of urban spaces and residents' spatio-temporal behaviors holds significant importance [1]. By examining patterns in human mobility data, we can uncover hidden patterns that enhance our understanding of urban spatial structures. The spatial coverage of community life circles and the allocation of public service resources profoundly impact residents' spatial behaviors [2]. Current research on community life circles primarily focuses on two aspects: delineating life circle boundaries and discovering life circle structures [3]. Boundary delineation methods emphasize temporal and spatial scales, measuring life circle extents based on facility accessibility through fixed-time or fixed-distance approaches [4]. These methods, however, overlook the quality and scale of POIs—factors critically influencing a POI's contribution to communities. For instance, large-scale, high-quality POIs may influence distant communities and be shared across multiple neighborhoods, whereas nearby but small-scale, mediocre POIs may have limited impact.

Research on life circle structures falls into two categories: 1) exploring spatial configurations of urban facilities through static POI distributions, and 2) identifying urban functional areas and regional disparities by analyzing macro-level resident activity patterns [5–7]. Neither approach considers POI quality nor conducts fine-grained analysis of community structures. Residents may exhibit stronger dependence on distant, high-quality POIs than nearby, low-quality alternatives. Thus, delineating life circles according to actual living patterns and developing fine-grained quantitative representations of facility distributions within them enables better understanding of current conditions and optimization of facility layouts.

This paper proposes a representation method for community life circle structures with quality-aware POI dependencies, employing representation learning to model the precise relationships between life circles and POIs based on resident mobility patterns. Observations reveal complex life circle structures: activity ranges vary beyond 15-min boundaries, and spatial overlaps exist between circles. Constructing effective urban life circle structures faces three key challenges. First, shared POIs among numerous communities make simplistic delineation (e.g., Euclidean distance or walkable buffers) inadequate. Precisely identifying essential POIs remains challenging. Second, merely counting POIs or visit frequencies fails to capture structural characteristics. Quantifying POIs' importance to daily life is difficult. Third, periodic dynamics in resident activities complicate the extraction of temporal structural features.

To address these challenges, we first identify essential POIs using k-Nearest Neighbors (kNN) to establish community-POI connections, selecting shared POIs based on distance. Then we quantify POI quality using scale and review data, then compute category-specific dependency indices incorporating POI weights. Last, we segment daily periods via sliding windows using KL-divergence within topic-pattern frameworks, and learn embedded representations of temporal activity graphs to generate vector matrices encoding dynamic structural features.

Our contributions include:

(1) A novel method for identifying essential POIs and quantifying their community influence to delineate life circles and measure resident dependencies.
(2) Fine-grained representation learning capturing dynamic life circle structures via expressive vector matrices.

(3) Experimental validation on real-world datasets for applications like life circle convenience prediction, demonstrating method effectiveness.

2 Related Work

This section reviews related studies in urban functional zone identification, community life circle structures, and representation learning.

2.1 Urban Functional Zone Identification

With rapid urban development and maturation of sensing technologies and computational environments, urban computing based on big data has garnered significant attention. Within urban spatial computing, the proliferation of internet map services and location-based services (LBS) has positioned community structure and functional analysis using public service facilities as a key research direction. Functional zone identification primarily relies on POI data, where user behaviors determine regional functions to delineate static urban functional zones. Liu et al. [8] employed skip-gram models and t-SNE techniques with urban POI data to explore and visualize urban functional zones. Chao et al. [6] utilized taxi trajectory data to capture residents' visits to various POIs, thereby discovering regional functions. Fu et al. [7] proposed a probabilistic latent factor model to learn mixed urban functions.

2.2 Community Life Circle Structures

Spatial delineation of community life circles critically impacts structural analysis. Existing delineation methods include Euclidean distance-based approaches, walkable-range methods within fixed time thresholds [9], population-weighted delineation [10], regional area partitioning, and service facility accessibility-based methods [11]. With the rise of big data, Chai et al. [12] used GPS and mobile signaling data to statistically analyze resident activities for refined life circle delineation. However, the diversity and complexity of GPS behaviors necessitate large-scale sampling. Instead of fixed travel time/distance, we use k-Nearest Neighbors (kNN) to identify essential POIs for life circle delineation.

In terms of structural representation studies, Liu et al. [13] characterized community structures using POIs and street-view imagery. Liu [8] and Wang et al. [14] quantified mobility connectivity between communities and POIs via visitation frequency from travel data, constructing community structure graphs.

In this paper, we emphasize POI quality impact on communities by quantifying dependency relationships between residents and POIs activity graphs reflecting life circle structures. The dynamic structural features of life circles over time are captured by representation learning.

2.3 Representation Learning

Representation learning automatically discovers latent data features, eliminating manual feature engineering in machine learning. It comprises two domains including text representation learning and graph representation learning [15]. Graph representation learning aims to embed high-dimensional graph structures into low-dimensional vectors. Key approaches of graph representation learning include probabilistic graphical models, manifold learning and autoencoder-based reconstruction algorithms [16]. We used autoencoder-based reconstruction algorithms to project latent features into low-dimensional space via nonlinear mapping functions, minimizing loss between original and reconstructed features. Compared to alternatives, autoencoders do not require complex posterior distributions, and directly parameterize features/representation functions. We performed collective embedding representation learning on periodically varying life circle activity graphs to derive vector matrices that effectively characterize life circle structures.

3 Problem Definition

Definition 1. Candidate POIs for Community Life Circle. Given a community c and a POI set P_0, candidate POI set P for c's life circle consists of all POIs in P_0 whose distance to c does not exceed the distance to c's k-th nearest neighbor. Formally, $P = \{p \in P_0 | \text{dist}(c, p) \leq \text{dist}(c, p')\}$, where p' denotes the k-th nearest neighbor to c in P_0.

Definition 2. Community POI Dependency Relationship. Given a community c and its candidate POI set P, let E be the dependency relationship set between c and P. Each relationship $e \in E$ is a triplet denoted as $e(c, p_j, w)$, where w is the dependency weight between c and p_j, calculated as $w = \text{score}_j \times D_s$; score_j is the quality index of POI p_j; D_s is the category weight of p_j's POI type.

Residents' dependency on POIs is influenced by both spatial proximity and POI quality. This definition explicitly incorporates quality metrics.

Definition 3. Community Life Circle Activity Graph. For a community c, its single time slice activity graph is denoted $G^{(k)}(P, E)$, where P is the Candidate POI set for c's life circle, E is the dependency relationship set between c and P, and any edge $e(c, p, w) \in E$ indicates p is an essential POI with dependency weight w. The full activity graph is a time-series set of single-slice graphs $G^{(k)} = \{G_1{}^{(k)}, G_2{}^{(k)}, \ldots, G_n{}^{(k)}\}$, where $G_t{}^{(k)}$represents the activity graph during the i-th time period.

Definition 4. Representation Learning for Activity Graphs. Given a community life circle activity graph $G^{(k)}$, the representation learning aims to learn a spatial mapping function $f(c)$: $G^{(k)} \rightarrow R_k{}^d$ that encodes structural features of $G^{(k)}$ into a d-dimensional vector $R_k{}^d$.

4 Methodology

This section presents our method for representation learning of community life circle structures with quality-aware POI dependencies. First, we introduce the overall framework. Then, we detail three key steps: (1) single-time-slice activity graph construction, (2) multi-temporal activity graph integration, and (3) representation learning for quantitative structure embedding.

4.1 Framework

The core of this study is the quantitative representation learning of community life circle structures. The proposed framework emphasizes the impact of POI quality, avoids rigid boundary delineation for life circles, and enables fine-grained acquisition of dynamic structural characteristics.

Figure 1 illustrates the overall framework structure, which comprises four key steps. First, the k-nearest neighbors (kNN) method is employed to get candidate POIs within community life circles. Second, dependency indices between residents and POIs are calculated using POI review data, facility scale metrics, and category weights, thereby establishing community-POI relationships and constructing activity graphs. Next, temporal segmentation is performed using various behavioral datasets including bike-sharing records, taxi trajectories, and Baidu search queries to construct multi-temporal community life circle activity graphs. Finally, embedding representation learning is applied to these activity graphs to derive stable feature vectors that effectively represent the structural characteristics of community life circles.

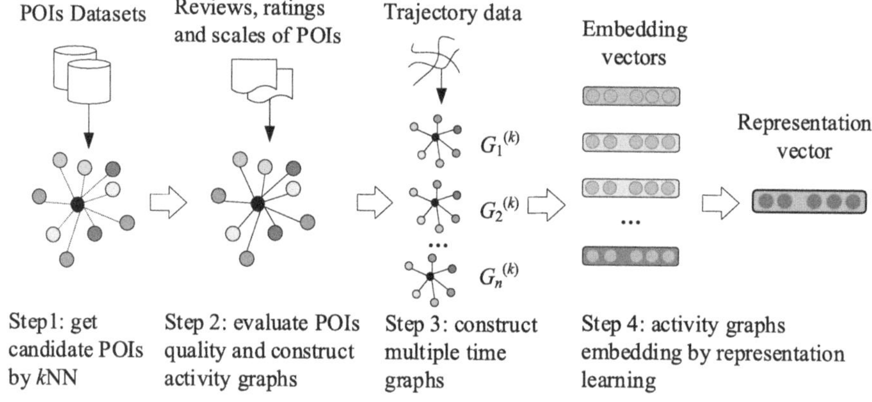

Fig. 1. The framework of representation learning for community life circle structure with quality-aware POI dependencies

4.2 Construction of Community Life Circle Activity Graphs

This paper adopts the following strategies when constructing community life circle activity graphs for individual time periods. First, POI quality indices are calculated using review data and scale data of POIs. Then, a community life circle activity graph construction algorithm is proposed, which utilizes the kNN algorithm to determine candidate POIs within community life circles. By combining quality indices with POI category weights, dependency indices between communities and POIs are synthesized, thereby establishing connections between them.

4.2.1 Generation of POI Quality Indices

This paper quantifies the influence of POIs by calculating their quality indices. The quality index is measured based on the following factors: the number of reviews (reflecting POI popularity), the rating score (reflecting satisfaction levels), and the floor area (reflecting facility scale). Review data is sourced from the Dianping website (www.dianping.com), while area data is the Area of Interest (AOI) dataset obtained from Baidu Maps API. Since AOI data and POI data do not always correspond one-to-one (e.g., a shopping mall contains numerous stores, each with POI markers but not necessarily AOI polygons), the quality index for such POIs is calculated solely based on review count and rating score.

Due to differing measurement units for review count, rating score, and floor area, normalization is first applied to these indicators. Given that their distributions are non-Gaussian and non-negative, the Min-Max normalization method is adopted to scale all three factors to a range of [0, 1]. The normalization formula is shown in Eq. (1):

$$X_{j,i}^{*} = \frac{X_{j,i} - \min\left(X_{*,i}\right)}{\max\left(X_{*,i}\right) - \min\left(X_{*,i}\right)} \tag{1}$$

Here, $X_{j,i}^{*}$ represents the normalized value of the i-th influencing factor for the j-th POI, while $\min\left(X_{*,i}\right)$ and $\max\left(X_{*,i}\right)$ denote the minimum and maximum values of the i-th factor across all POIs, respectively.

After normalization, a weighted sum of the three factors yields the quality index $score_j$ for each POI, as shown in Eq. (2), where w_i represents the weight assigned to each influencing factor.

$$score_j = \sum\nolimits_{i=1}^{3} w_i * X_{j,i}^{*} \tag{2}$$

The weights w_i for the POI quality index are determined using the Analytic Hierarchy Process (AHP). AHP employs a pairwise comparison matrix to evaluate the relative importance of factors, reducing comparison difficulties caused by differing factor properties and improving accuracy. This paper conducts pairwise comparisons for the three influencing factors, assigning qualitative ratings based on their importance.

The pairwise comparison results form a judgment matrix A, where element a_{ij} represents the relative importance of factor i compared to factor j. Equal importance is assigned a value of 1, while increasing importance is rated on a scale of 3, 5, 7, or 9, with intermediate values of 2, 4, 6, or 8.

After passing the consistency test, the eigenvector method is used to derive the weights for the three influencing factors. The consistency index (CI) is calculated as shown in Eq. (3):

$$CI = \frac{\lambda_{max} - n}{n - 1} \tag{3}$$

Here, $\lambda_{\max}$ is the largest eigenvalue of matrix A, and n is the order of the matrix. A CI value close to 0 indicates acceptable consistency, while larger values necessitate matrix adjustments.

To account for age-based differences in perceived importance of POI factors, this paper introduces age-group judgment matrices, dividing the population into four age groups: children, youth, middle-aged, and elderly. The final weight w_i is computed as shown in Eq. (4), where h_j is the weight for the j-th age group, and v_{ij} represents the judgment result of the j-th age group for the i-th factor.

$$w_i = \sum\nolimits_{j=1}^{4} h_j v_{ij} \tag{4}$$

4.2.2 Community Life Circle Activity Graph Construction Algorithm

This paper employs the *k*NN method to identify candidate POIs within community life circles and calculates dependency indices between communities and POIs to construct single-time-slice activity graphs. The algorithm is outlined in Algorithm 1.

The algorithm takes as input any community c and the urban POI set P. It proceeds through three main steps. Step 1 calculates the quality indices for all urban POIs (lines 1–2) using the method proposed in Sect. 4.2.1. Step 2 employs *k*NN to identify candidate POIs within the community life circle (line 3). Step 3 establishes connections between community c and its candidate POIs (lines 4–10). The calculation of POI dependency indices is differentiated based on whether a POI is shared by multiple communities or not (lines 6–9). Here, w_{cj} represents the dependency index of community c on candidate POI p_j, $score_j$ denotes the quality index of p_j, w_{Dj} indicates the category weight of the POI to which p_j belongs, and H_{kj} signifies the distance between community c and POI p_j. When the dependency index w_{kj} allocated to a community sharing a POI falls below the overall average, the corresponding edge is pruned (line 10). Finally, the algorithm outputs the activity graph $G_t^{(k)}$ within a single time period (line 11).

Algorithm 1. Community Life Circle Activity Graph Construction Algorithm

Input: Community c, urban POI set P.
Output: Single-time-slice activity graph $G_t^{(k)}$.

1. **for** (each POI p_j in P)
2. $\quad score_j = \sum_{i=1}^{3} w_i * X_{j,i}^*$;
3. Use *k*NN to determine the number of candidate POIs per category for c;
4. **for** (each category s in candidate POIs of c)
5. $\quad$ **for** (each candidate POI p_j in category s)
6. $\quad\quad$ **if** (p_j is not shared by multiple communities)
7. $\quad\quad\quad w_{cj} = score_j * w_{Dj}$;
8. $\quad\quad$ **else**
9. $\quad\quad\quad w_{cj} = \frac{1}{H_{kj}} score_j * w_{Dj}$;
10. $\quad\quad\quad$ **if** (w_{kj}<the average value) delete w_{kj};
11. **return** $G_t^{(k)}$;

The *k*NN query identifies all POIs within a distance not exceeding the k-th nearest neighbor. In practice, people tend to visit the nearest POIs, and experimental results validate this approach. The selection of k values follows prior research [17], as shown in Table 1.

Table 1. k values and weights of kNN for different POI categories

	Shopping	School	Medical	Dining	Leisure	Transportation	Daily Services	Sports & Culture
Quantity	36	5	20	70	20	5	40	10
Weight	0.3	0.1	0.06	0.2	0.04	0.2	0.06	0.04

The POI dependency index calculation adopts the dimensionless multiplication method proposed in [18], while category weights w_{Dj} are determined based on [19], focusing on eight POI categories related to resident convenience (Table 1). For POIs shared by multiple communities, dependency indices are allocated inversely proportional to distance. If a shared community's dependency index falls below the citywide average for that POI category, the edge is removed.

4.3 Construction of Multi-temporal Community Life Circle Activity Graphs

To capture dynamic structural features of resident activities over time, this paper divides time periods cyclically. While literature [8] generates seven dynamic graphs for Monday to Sunday, this study adopts a finer-grained approach inspired by [20], using topic modeling and sliding windows with KL-divergence for time segmentation shown in Eq. (5). KL-divergence is chosen over Wasserstein or JS-divergence due to its suitability for fine-grained, overlapping segmentation. Here, θ represents the dominant POI category in a time segment, and s_i denotes different POI categories.

$$S(\theta_2||\theta_1) = \sum_{i=1}^{|v|} p(s_i|\theta_2) log \frac{p(s_i|\theta_2)}{p(s_i|\theta_1)} \tag{5}$$

The process involves two steps: (1) Counting hourly visit frequencies to each POI category and ranking them by intensity, with the top category labeled as the topic θ; (2) Merging adjacent time segments with KL-divergence below a threshold. After segmentation, activity graphs are constructed for each time period, yielding a multi-temporal graph set G_k.

4.4 Representation Learning for Community Life Circle Structures

With multi-temporal activity graphs constructed, collective embedding representation learning is performed to derive structural feature matrices reflecting residents' daily routines. Dependency indices are used to vectorize activity graphs. To ensure each community has a distinct feature vector, individual community graphs $G_t^{(k)}$ are encoded separately. Each of the eight POI categories in $G_t^{(k)}$ is also vectorized for inter-category comparison.

Meanwhile, the vectorized community life circle activity graph $G_t^{(k)}$ incorporates eight distinct categories of POIs. To ensure each POI category possesses its own feature vector representation, facilitating inter-category comparisons, the quantified data of these eight POI types within $G_t^{(k)}$ are converted into vectors and separately fed as inputs to the autoencoder. Here, $\{y_{s,t}^{(k),1}, y_{s,t}^{(k),2}, ..., y_{s,t}^{(k),m}\}$ denotes the latent features in each hidden layer of the encoding step. The feature vectors $p_{s,t}^{(k)}$ of the eight POI categories after representation collectively constitute the intermediate layer, forming the representation vector $p_t^{(k)}$ for the community life circle structure. The hierarchical relationships between these vectors are defined in Eq. (6), where σ denotes the activation function, and w and b are weights and biases.

$$\begin{cases} y_{s,t}^{(k),1} = \sigma(w_{s,t}^{(k),1} g_{s,t}^{(k)} + b_{s,t}^{(k),1}), \forall t \in (1,2, ..., n), \\ y_{s,t}^{(k),r} = \sigma(w_{s,t}^{(k),r} y_{s,t}^{(k),r-1} + b_{s,t}^{(k),r}), \forall r \in (2,3, ..., m), \\ p_{s,t}^{(k)} = \sigma(w_{s,t}^{(k),m+1} y_{s,t}^{(k),m} + b_{s,t}^{(k),m+1}). \end{cases} \tag{6}$$

In the decoding step, the latent vector $p_{s,t}^{(k)}$ is reconstructed into $\widehat{g}_{s,t}^{(k)}$, with layer-wise relationships defined in Eq. (7).

$$\begin{cases} \widehat{y}_{s,t}^{(k),m} = \sigma(\widehat{w}_{s,t}^{(k),m+1} p_{s,t}^{(k)} + \widehat{b}_{s,t}^{(k),m+1}), \forall t \in (1,2, ..., n), \\ \widehat{y}_{s,t}^{(k),r} = \sigma(\widehat{w}_{s,t}^{(k),r} \widehat{y}_{s,t}^{(k),r-1} + \widehat{b}_{s,t}^{(k),r}), \forall r \in (2,3, ..., m), \\ \widehat{g}_{s,t}^{(k)} = \sigma(\widehat{w}_{s,t}^{(k),1} \widehat{y}_{s,t}^{(k),1} + \widehat{b}_{s,t}^{(k),1}). \end{cases} \tag{7}$$

This structured approach ensures comprehensive representation of community life circle dynamics while maintaining computational efficiency.

5 Experiment Evaluation

This section presents experimental evaluations of the proposed representation learning method for community life circle structures with quality-aware POI dependencies.

5.1 Datasets

All datasets used for experimental analysis are real-world datasets. The life circle dataset was obtained from the real estate website www.soufun.com, containing over 13,000 residential communities in Beijing with information including names, coordinates, addresses, and tags. The public service facility dataset was downloaded through the Baidu Maps API, comprising POI and AOI data totaling 1,871,019 POIs across 8 major categories and 22 subcategories. POI data includes names, coordinates, addresses, and tags; AOI data includes names, addresses, areas, and boundary point coordinates. Duplicate entries were removed using Baidu Maps' unique identifier code "uid". Review data was acquired from Dianping (www.dianping.com), containing names, addresses, prices, popularity metrics, and reviews for various POI categories. Trajectory data includes Baidu Maps route query trajectories, taxi trajectories, and bike-sharing trajectories. Specifically, Baidu Maps route query data originates from the Baidu Brain Transportation Dataset, while bike-sharing trajectory data (totaling 3,214,096 records) comes from the data competition community www.biendata.xyz.

5.2 Baselines and Evaluation Metrics

This paper evaluates the effectiveness of community life circle structure representation methods through the prediction of life circle convenience, defined as residents' satisfaction with surrounding public service facilities.

Zhang et al. [21] conducted a questionnaire-based survey on living convenience across Beijing districts, rating community convenience on a 1–5 scale. These survey results serve as ground-truth convenience labels. Existing representation methods and our proposed method generate feature vectors to quantify life circle structures, which are then used as predictive attributes for convenience level classification. Performance is evaluated based on prediction accuracy.

The specific representation methods are:

(1) **Explicit Feature of Fixed Range (ExpFix)**: Life circles are delineated using fixed distances, with POI counts per category as quantitative features.
(2) **Weekly Explicit Feature of Fixed Range (WExpFix)** [8]: Using trajectory data between residents and POIs, this method quantifies daily POI visitation counts per category within fixed-distance life circles across weekly cycles.
(3) **Daily Explicit Feature of Fixed Range (DExpFix)** [14, 20]: The method segments daily periods via KL-divergence-based sliding windows and quantifies time-sliced POI counts per category within fixed radii.
(4) **Daily Latent Feature of Fixed Range (DLatFix)** (Proposed): The method adopts the same time segmentation as DExpFix but quantifies POI quality metrics (Sect. 4.2.1) within fixed-distance life circles.
(5) **Daily Explicit and Latent Feature of Fixed Range (DExpLatFix)** (Proposed): The method combines DL-Fix and E-Fix methodologies.
(6) **Daily Explicit and Latent Feature of *k*NN (DExpLatKNN)** (Proposed): The method identifies essential POIs via *k*NN, segments days using sliding windows, and quantifies both quality and quantity of category-specific POIs per time slice.

Five existing ranking algorithms are selected to comparatively evaluate these six representation methods: (1) Multiple Additive Regression Trees (MART). Gradient-boosted regression trees performing gradient descent in function space. (2) RankBoost (RB). Boosting-based pairwise ranking that formalizes ranking as binary classification of instance pairs. (3) ListNet (LN). Listwise ranking using neural networks with gradient descent. (4) RankNet (RN). Models probabilistic cost functions via neural networks. (5) LambdaMART (LM). Enhanced boosted tree model combining LambdaRank and MART. Ranking results are measured by Normalized Discounted Cumulative Gain (NDCG@N). Higher NDCG@N values indicate superior ranking accuracy.

5.3 Experimental Results Comparison with Existing Algorithms

The comparison results are presented in Table 2. The results indicate that the ExpFix method performs the worst, primarily because it only considers POI counts while inadequately capturing other structural features of community life circles. In contrast, the DExpLatKNN method achieves the best performance

Table 2. Comparison of convenience of community life circle

NDCG	Rank	ExpFix	WExpFix	DExpFix	DLatFix	DExpLatFix	**DExpLatKNN**
@5	MART	0.468	0.405	0.438	0.462	0.525	**0.574**
	RN	0.405	0.3	0.435	0.441	0.474	**0.485**
	RB	0.354	0.216	0.345	0.363	0.504	**0.524**
	LM	0.438	0.462	0.375	0.465	0.573	**0.583**
	LN	0.414	0.3	0.414	0.441	0.470	**0.481**
@10	MART	0.471	0.411	0.459	0.48	0.594	**0.674**
	RN	0.456	0.3	0.486	0.492	0.516	**0.535**
	RB	0.375	0.384	0.381	0.39	0.546	**0.583**
	LM	0.483	0.477	0.42	0.528	0.564	**0.585**
	LN	0.477	0.327	0.498	0.492	0.512	**0.516**
@15	MART	0.489	0.42	0.462	0.507	0.621	**0.677**
	RN	0.495	0.336	0.537	0.546	0.558	**0.563**
	RB	0.39	0.462	0.375	0.411	0.612	**0.628**
	LM	0.471	0.522	0.522	0.543	0.636	**0.648**
	LN	0.516	0.336	0.537	0.546	0.554	**0.563**
@20	MART	0.495	0.489	0.471	0.525	0.642	**0.695**
	RN	0.516	0.366	0.588	0.594	0.600	**0.609**
	RB	0.441	0.471	0.474	0.474	0.627	**0.632**
	LM	0.564	0.582	0.555	0.624	0.666	**0.673**
	LN	0.576	0.366	0.579	0.594	0.596	**0.613**

This superiority stems from its use of kNN to identify essential POIs, incorporation of POI quality's impact on resident activities, extraction of deeper behavioral patterns, and elimination of redundant POIs weakly connected to residents that reflects the actual living conditions of residents in the community life circle.

Other methods exhibit performance levels between ExpFix and DExpLatKNN, showing progressively improving accuracy: (1) DExpFix outperforms WExpFix due to finer-grained temporal segmentation; (2) DLatFix surpasses DExpFix because it accounts for POI quality's influence on activities and differential impacts when POIs are shared across communities, enabling more rational allocation of POI influence; (3) DExpLatFix exceeds DLatFix by integrating ExpFix's static features with DLatFix's dynamic latent features, combining four dimensions: dynamic/static and latent/explicit characteristics.

The experiments confirm that the proposed learning framework effectively captures dynamic and latent features, significantly improving the accuracy of community life circle convenience evaluations.

6 Conclusion

This paper investigates a representation method for community life circle structures with quality-aware POI dependencies. The *k*NN method is employed to identify candidate POIs within community life circles, while POI review data, scale metrics, and category weights are utilized to compute dependency indices between residents and POIs, establishing community-POI relationships. By analyzing trajectory data reflecting resident activity patterns, temporal segmentation of activity graphs is performed. Collective embedding representation learning is then applied to these graphs to derive stable feature vectors characterizing life circle structures. For applications such as life circle convenience prediction, experimental evaluations on real-world datasets validate the effectiveness of the proposed method.

Acknowledgments. This study was funded by the Project of the Educational Department of Liaoning Province (JYTMS20231596).

References

1. Qiang, S.W., Chen, X.M., Jiang, K.D., et al.: Spatio-temporal behavior analysis in cities based on mobile network traffic logs. J. Comput. Res. Dev. **53**(4), 932–940 (2016). (In Chinese)
2. Pozder, N., Halilovi-Terzi, S., Pavlovi, A.: Understanding the principles of the 15-minute city concept and the repercussions on the city's health: the curious case of the city of Sarajevo. In: International Symposium on Annual Days of BHAAAS in Bosnia and Herzegovina. Springer, Cham (2024)
3. Aboulnaga, M., Ashour, F., Elsharkawy, M.,et al.: Urbanization and drivers for dual capital city: assessment of urban planning principles and indicators for a '15-minute city'.Land **14**(2) (2025)
4. Wei, W., Hong, M.Y., Xie, B.: Spatial delineation and optimization of 15-minute life circles in Wuhan based on supply-demand matching. Urban Plann. **35**(04), 11–17 (2019). (In Chinese)
5. Wang, P., Fu, Y., Zhang, J., et al.: Learning urban community structures: a collective embedding perspective with periodic spatial-temporal mobility graphs. ACM Trans. Intell. Syst. **9**(6), 63.1–63.28 (2018)
6. Chao, Z., Zhang, K., Quan, Y., et al.: Regions, periods, activities: uncovering urban dynamics via cross-modal representation learning. In: The 26th International Conference. International World Wide Web Conferences Steering Committee, pp. 361–370 (2017)
7. Fu, Y., Liu, G., Papadimitriou, S., et al.: Real estate ranking via mixed land-use latent models. In: ACM International Conference on Knowledge Discovery and Data Mining, pp. 299–308 (2015)
8. Liu, K., Yin, L., Lu, F., et al.: Visualizing and exploring POI configurations of urban regions on POI-type semantic space. Cities **99**, 102610 (2020)
9. Chen, L.F., Li, C.M., Dai, Z.X., et al.: Research on public service facilities allocation in 15-minute life circles of residential areas. Geom. Sci. **47**(01), 236–244 (2022). (In Chinese)
10. Tan, P.Y., Samsudin, R.: Effects of spatial scale on assessment of spatial equity of urban park provision. Landsc. Urban Plan. **158**, 139–154 (2017)
11. Yu, Z., Shangguan, Z.S., Zhou, D.D., et al.: Optimal allocation of public service facilities in future community life circles based on spatio-temporal impact factors. Bull. Surv. Mapping (08), 143–148+154 (2022). (In Chinese)

12. Chai, Y., Chunjiang, L.I., Xia, W., et al.: Study on the delineation model of urban community life circle: based on Qinghe district in Haidian district, Beijing. Urban Dev. Stud. (2019)
13. Liu, X., Long, Y., et al.: Automated identification and characterization of parcels with OpenStreetMap and points of interest. Environ. Plann. B. Plann. Des. **43**(2), 341–360 (2015)
14. Wang, H., Liu, J., Peng, C., Sun, H. Representation learning of multi-layer living circle structure. In: Proceeding of WISA 2023. Lecture Notes in Computer Science, vol. 14094, pp. 125–136 (2023)
15. Zhang, Y., Licia, C., et al.: Urban computing: concepts, methodologies, and applications. ACM Trans. Intell. Syst. Technol. 5(3), 38.1–38.55 (2014)
16. Ou, M., Cui, P., Pei, J., et al.: Asymmetric transitivity preserving graph embedding. In: ACM International Conference on Knowledge Discovery and Data Mining, pp. 1105–1114 (2016)
17. Luo, X.Y., Zhang, W.J., Chai, Y.W.: Threshold effect of built environment in 15-minute life circles. Geogr. Res. **41**(8), 2155–2170 (2022). (In Chinese)
18. Erath, A., Maheshwari, T., Joos, M., et al.: Visualizing transport futures: the potential of integrating procedural 3D modelling and traffic micro-simulation in virtual reality applications. In: Transportation Research Board Annual Meeting (2017)
19. Fu, Y., Hui, X., Yong, G., et al.: Exploiting geographic dependencies for real estate appraisal: a mutual perspective of ranking and clustering. In: ACM International Conference on Knowledge Discovery and Data Mining, pp. 1047–1056 (2014)
20. Sun, H.L., Peng, C., Liu, J.L., et al.: Representation learning for community structures oriented to 15-minute life circles. J. Comput. Appl. **42**(06), 1782–1788 (2022). (In Chinese)
21. Zhang, X., Du, S., Zhang, J.: How do people understand convenience-of-living in cities? A multiscale geographic investigation in Beijing. ISPRS J. Photogram. Remote Sens. **148**(FEB.), pp. 87–102 (2019)

GA-CLIP: A Multimodal POI Classification System Based on CLIP with Gated Attention Mechanism

Junling Liu[1,2], Tianyu Sun[1,2](✉), Huanliang Sun[1,2], and Jingke Xu[1,2,3]

[1] School of Computer Science and Engineering, Shenyang Jianzhu University, Shenyang 110168, China
sunhl@sjzu.edu.cn

[2] Liaoning Province Big Data Management and Analysis Laboratory of Urban Construction, Shenyang 110168, China

[3] Shenyang Branch of National Special Computer Engineering Technology Research Center, Shenyang 110168, China

Abstract. Existing multimodal Point of Interest (POI) classification methods mainly used convolutional neural networks to extract visual features and employed simple strategies for feature fusion, neglecting the impact of contextual semantics. This paper proposes a POI classification system based on the large multimodal model CLIP, named GA-CLIP, which includes two stages: knowledge-enhanced feature learning and POI classification decision. Among them, in the multimodal feature learning stage, the system employs a gated attention mechanism to integrate textual, visual and emotional features for fine-grained cross-modal interaction, while incorporating a prompt engineering component that leverages a local knowledge base with dictionary definitions as prior knowledge to enrich sparse textual representations and enhance classification robustness. In the classification decision stage, GA-CLIP leverages the large multimodal model to analyze emotional features to enhance the interpretability of classification decisions. Experimental results validate the GA-CLIP system's robust classification performance (F1 score: 65.82) and its applicability in POI classification tasks.

Keywords: POI Classification · Prompt Engineering · Gated Attention

1 Introduction

With the rapid development of Internet technology, social media platforms have become an important channel for users to share their experiences and feelings. The rich POI information contained in these social media posts can be leveraged to enhance information retrieval and personalized recommendations, as well as to support applications such as commercial site selection analysis, urban functional zoning research and intelligent transportation planning [1, 2].

Traditional POI classification methods rely on social media texts, but short texts often suffer from feature sparsity and ambiguity, limiting classification accuracy. Multimodal

S. Yang et al. (Eds.): WISA 2025, LNCS 16299, pp. 437–441, 2026.
https://doi.org/10.1007/978-981-95-6426-2_39

data can alleviate this issue, yet existing models still depend on conventional visual feature extractors and simple fusion strategies, resulting in weak semantic comprehension [3]. Therefore, we propose GA-CLIP, a POI classification system that integrates prompt engineering for knowledge enhancement and incorporates emotional features into the decision-making process to achieve POI classification under complex semantic conditions.

The main contributions of this paper include:(1) We propose a POI classification system called GA-CLIP, which integrates a gated attention mechanism and the CLIP model to perform POI classification effectively; (2) Prompt engineering templates based on the local knowledge base are designed, where definitions from a dictionary are used as prior knowledge to guide the model's learning; (3) In order to demonstrate the functionality of the proposed system, we implemented a Web demo. Code are available on https://github.com/sty222/sty/POI.

2 System Overview

The GA-CLIP system proposed in this paper consists of three modules: an input processing module, a knowledge-enhanced feature learning module and a classification decision module (see Fig. 1).

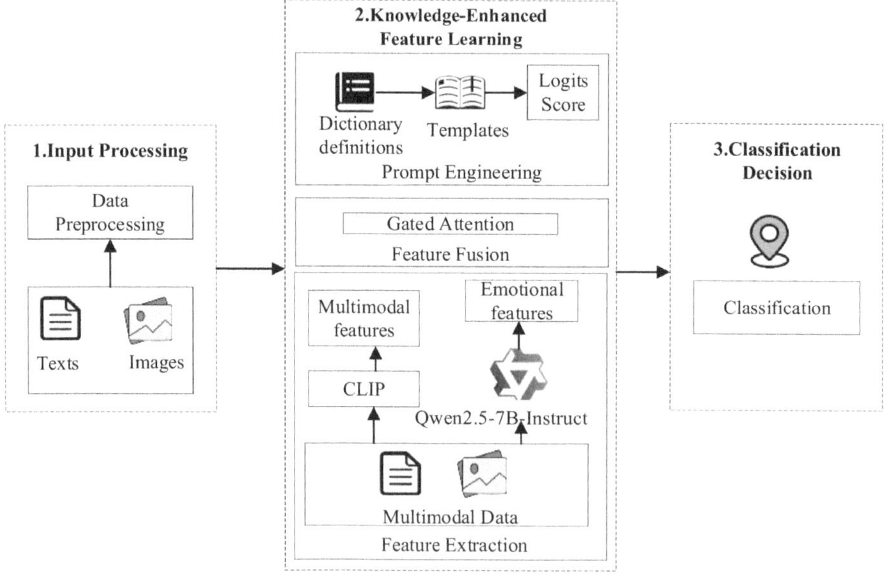

Fig. 1. GA-CLIP system structure

Input Processing Module

The input processing module acquires raw texts and images from social media posts and performs necessary preprocessing steps, including tokenization of the texts and resizing and normalization of the images. This ensures that both modalities are properly formatted and ready for subsequent multimodal feature learning.

Knowledge-Enhanced Feature Learning Module. The module contains three submodules, the specific realization is as follows:

(1) Feature Extraction Submodule. In the feature extraction submodule, the Transformer model and ViT-L/14 model are employed to extract textual semantic features and visual features respectively. For visual feature extraction, the ViT-L/14 processes 224×224 RGB images by segmenting them into a sequence of 14×14 patches, which are embedded through linear projection and positional encoding. In addition, the multimodal large model Qwen2.5 is introduced to analyze the semantic analysis of sentiment of graphic data to capture the implicit features of sentiment in social networks.
(2) Feature Fusion Submodule. In the feature fusion submodule, a gated fusion strategy is proposed. The textual, visual and emotional features are mapped to the shared space by linear transformation, combined with the GELU activation function to enhance the feature expression ability, and the Sigmoid Gating is introduced to dynamically adjust the fusion weights of each modal feature. Emotional features extracted by the Qwen2.5 model are projected and integrated to enhance emotion-relevant dimensions while suppressing irrelevant noise. This fine-grained fusion strategy strengthens multimodal feature complementarity and improves both the accuracy and interpretability of POI classification.
(3) Prompt Engineering Submodule. The prompt engineering submodule is built upon a local knowledge base created by transforming textual definitions from the Oxford Advanced Learner's English-Chinese Dictionary and expanded label vocabulary from the Collins English Dictionary into structured prompt templates [4].

Classification Decision Module
The classification decision module utilizes multimodal features, including sentiment analysis [5], to improve the transparency and interpretability of classification results. This paper divides POI types into eight major categories, as follows: "Food", "Arts & Entertainment", "Nightlife Spot", "College & University", "Professional & Other Places", "Shop & Service", "Great Outdoors" and "Travel & Transport".

3 Demonstration

In this section, we construct a demonstration environment to evaluate the POI classification capability of our system, which is implemented as the training process (see Fig. 2a) and a web demonstration (see Fig. 2b), and the overall capability of the system is demonstrated through the mode of image-text input classification.

In the image-text input classification mode, the user uploads a photo and an associated textual description, and the system automatically identifies the corresponding POI category, such as "Food", "Arts & Entertainment", etc. This demonstrates the system's ability to understand and fuse multimodal information for POI classification. To evaluate the performance of classification, we measured Precision, Recall and F1-score. The proposed system achieves improvements of 20.64%, 13.79% and 18.61% respectively over the best existing multimodal POI classification system.

(a) Training process (b) Web demonstration

Fig. 2. Demonstration of POI classification system

4 Conclusion

This paper introduces GA-CLIP, a POI classification system based on the CLIP framework. It leverages a gated attention mechanism and integrates emotional features to enhance the representation of multimodal data, significantly improving classification accuracy. Overall, the proposed multimodal POI classification system demonstrates strong effectiveness, robustness, and high accuracy in practical scenarios.

5 Disclosure of Interests.

The authors have no competing interests to declare that are relevant to the content of this article.

Acknowledgments. This study was supported by the National Natural Science Foundation of China under Grant No.62073227, and the Liaoning Provincial Department of Education Research Program under Grant No. LJ212510153014.

References

1. Kim, H., Lee, S.: POI GPT: extracting POI information from social media text data. In: The International Archives of the Photogrammetry, Remote Sensing and Spatial Information Sciences, vol. 48, pp. 113–118 (2024)
2. Li, X.L., Wei, F.: Research on urban functional area and its mixing degree identification based on POI: A case study of the core area of Chongqing. J. Southwest Univ. (Nat. Sci. Edn.) **43**(1), 164–173 (2021)
3. Villegas, D.S., Aletras, N.: Point-of-interest type prediction using text and images. In: Proceedings of the 2021 Conference on Empirical Methods in Natural Language Processing, pp. 7785–7797 (2021)

4. Huang, W., Wang, J., Cong, G.: Zero-shot urban function inference with street view images through prompting a pretrained vision-language model. Int. J. Geograph. Inf. Sci. 1–29 (2024)
5. Liu, J., Chang, T., Shi, X., Sun, H., Xu, J.: Dataset construction for fine-grained emotion analysis in catering review data. In: WISA 2024. Lecture Notes in Computer Science, vol.14883, pp. 264-276, Springer, Singapore (2024)

Knowledge Construction

A GuwenBERT-BiGRU-Focal-CRF Model for Ancient Chinese Named Entity Recognition with PGD-Based Adversarial Training

Zhengyu Chen, Fangjiao Jiang(✉), and Yong Ma

School of Physics and Electronic Engineering, Jiangsu Normal University, Xuzhou 221000, China
jiangfj@jsnu.edu.cn

Abstract. Named Entity Recognition (NER) in classical Chinese is challenging due to ambiguous word boundaries, variant-standard character coexistence, and contextual polysemy. To address these issues, we propose a NER method based on the GuwenBERT-BiGRU-FOCAL-CRF framework enhanced with PGD adversarial training. GuwenBERT, pre-trained on classical Chinese corpora, provides character-level contextual representations, improving the capture of linguistic features. A Bidirectional Gated Recurrent Unit (BiGRU) models long-range dependencies, while a Conditional Random Field (CRF) captures label relations. Focal Loss optimizes training by emphasizing hard-to-classify samples. Robustness is further improved through adversarial training, where learnable perturbations are added to embeddings or intermediate features, enabling stable predictions under minor perturbations. Experiments show that our model outperforms the BERT-BiLSTM-CRF baseline on classical Chinese NER, with F1-score gains of 12.40% and 15.11%. These results confirm the effectiveness and superiority of our approach for named entity recognition in ancient Chinese texts.

Keywords: Named Entity Recognition · Adversarial Training · Focal Loss

1 Introduction

Ancient book Named Entity Recognition (NER) aims to automatically extract entities such as place names, personal names, official titles, institutional names, and book titles from rare ancient texts. This technology not only improves the readability of these texts but also provides efficient data support for academic research [1]. As a foundation for knowledge mining and organization in digital humanities, ancient book NER holds significant academic value and practical importance for constructing historical and cultural databases.

However, ancient texts commonly exhibit complex linguistic phenomena such as blurred boundaries between multiple characters or words, mixed usage of variant and simplified characters, and contextual ambiguities, which pose substantial challenges to entity recognition.

S. Yang et al. (Eds.): WISA 2025, LNCS 16299, pp. 445–456, 2026.
https://doi.org/10.1007/978-981-95-6426-2_40

To address these challenges, this paper employs the CCLUE and CHisIEC datasets as experimental corpora and proposes a deep learning model tailored for named entity recognition in classical Chinese texts. The model consists of four key components: first, semantic features are extracted using GuwenBERT, a pretrained model specialized in classical Chinese; second, Bidirectional Gated Recurrent Units (BiGRU) capture long-range dependencies in sequences; third, a Conditional Random Field (CRF) layer models label dependencies; and fourth, the Focal Loss function is integrated to emphasize learning from hard-to-classify samples. To further enhance robustness, an adversarial training strategy introduces learnable perturbations into word embeddings or intermediate features, guiding the model to produce stable predictions under slight input disturbances.Extensive experiments demonstrate that the proposed method significantly improves recall and F1 scores in ancient named entity recognition. The main contributions are as follows:

(1) A Named Entity Recognition in Ancient Chinese Texts using adversarial training combined with GuwenBERT-BiGRU-FOCAL-CRF is proposed.Adversarial training with GuwenBERT-BiGRU-FOCAL-CRF is used to better handle the task of recognizing ancient Chinese named entities.
(2) By incorporating four adversarial training methods—FGSM, FGM, PGD, and FreeLB—and conducting experiments on two publicly available datasets with significant category differences, the results demonstrate that these strategies effectively enhance the robustness and generalization ability of the model.
(3) We conducted experiments on two publicly released datasets to evaluate the effectiveness and accuracy of our proposed model. The results demonstrate that our model achieves significant improvements over baseline models.

2 Related Work

With the advancement of deep learning technologies, Named Entity Recognition (NER) has been widely applied in fields such as information extraction, question answering systems, and machine translation. As a fundamental task in digital humanities research, ancient text NER aims to automatically identify proper nouns like person names, place names, and organization names within historical texts, where current mainstream methods can be broadly categorized into four types: rule and dictionary-based NER, statistical machine learning-based NER, deep learning-based NER, and Transformer-based NER [2].

In recent years, numerous studies have focused on NER for various categories of ancient texts including historical records, local gazetteers, poetry, and traditional Chinese medical texts. Huangfu Jing et al. [3] compiled a *Dictionary of Chinese Surnames* based on surname statistics from *the Index of Ming-Qing Imperial Examination Stone Inscription Records*, developing boundary rules for personal names to identify names in the *Records of the Three Kingdoms: Shu Annals.* Huang Shuiqing et al. [4] analyzed features of place names in *Zuo Commentary*, combining a CRF model to recognize place name entities in *Zuo Commentary* and *Discourses of the States* corpora. Xie Tao [5] integrated an LSTM [6] model within a CRF framework, achieving good results on annotated *Zuo Commentary* corpus though requiring substantial high-quality annotated data to capture accurate contextual semantics. Li Zhangchao et al. [7] employed a RoBERTa-CRF model to extract arguments from *Zuo Commentary* events from a knowledge organization perspective, finding that argument data scale significantly impacted recognition performance. In order to enhance the textual representation of classical Chinese, researchers have proposed a series of pre-trained language models based on ancient book corpora, such as SikuBERT [8], SikuRoBERTa, GuwenBERT, and BERT-Ancient-Chinese.

3 The Proposed Model

Our proposed named entity recognition architecture first employs GuwenBERT to encode classical Chinese input texts, generating contextually enriched embeddings. Subsequently, a bidirectional GRU (BiGRU) network captures deeper contextual dependencies to extract hierarchical semantic features. During decoding, a CRF layer models label transition constraints for optimal sequence prediction. Concurrently, Focal Loss mitigates class imbalance by emphasizing challenging samples during training. Finally, PGD adversarial training with learnable perturbations improves robustness to input variations and significantly enhances model generalization. Below, we briefly introduce each component of the model (Fig. 1).

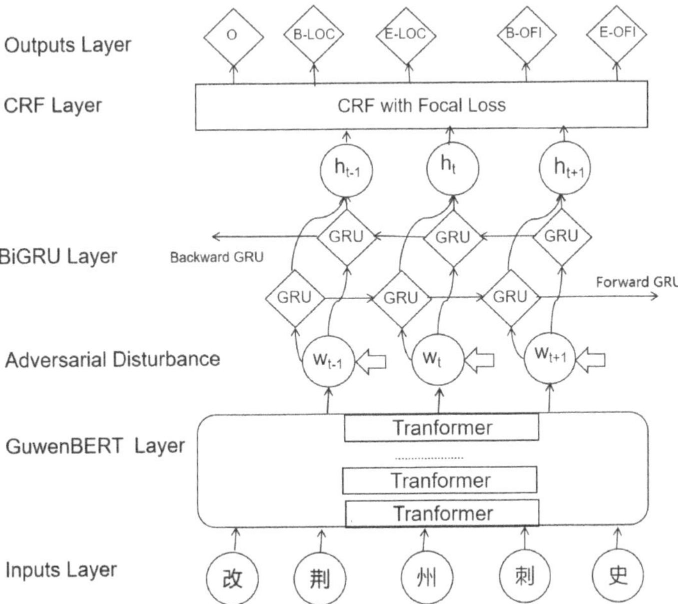

Fig. 1. The model structure diagram

3.1 GuwenBERT Layer

GuwenBER is a model proposed by Yan Qin in 2020, which is the first time this work has been transferred to the field of ancient Chinese. The GuwenBERT model is based on the training of the ancient Chinese literature corpus of Daizhi Pavilion, combined with the RoBERTa [9] weights of modern Chinese and a large amount of ancient Chinese corpus, transferring some language features of modern Chinese to ancient Chinese to improve performance. And focus on sentence segmentation, punctuation and named entity recognition [10].

RoBERTa pre-trained model RoBERTa (Robustly Optimized BERT Approach) is an improved model of BERT, proposed by FacebookAI in 2019 [9]. RoBERTa improves upon BERT through adjustments in training strategies and model structure, achieving superior performance across NLP tasks. In entity recognition, these optimizations enhance its ability to capture contextual information and understand complex language patterns, which is crucial for identifying named entities that rely on surrounding context. With its stronger and more refined language representations, RoBERTa can more accurately identify and classify entities, thereby improving overall NER performance.

3.2 BiGRU Layer

The encoding layer of this article adopts a bidirectional GRU recurrent neural network to extract global features from vectors output by the word embedding layer. GRU (Gated Recurrent Unit), a variant of recurrent neural networks (RNNs), employs gating mechanisms similar to LSTM to regulate information flow. This architecture effectively mitigates vanishing and exploding gradient problems inherent in traditional RNNs through controlled information retention and forgetting. While LSTM uses three gates (forget, input, output), GRU simplifies this to two fundamental gates: Update gate: merging LSTM's input and forget gates, Reset gate: governing historical information relevance [11]. The GRU network structure is simpler than LSTM's, thus it has fewer parameters and is easier to train, greatly improving training efficiency. Moreover, entity recognition models based on GRU achieve similar performance to LSTM.

The parameter update calculation method for the GRU network is shown in Eq. (1) below:

$$\begin{cases} z_t &= \sigma(W_z \cdot [h_{t-1}, x_t]) \\ r_t &= \sigma(W_r \cdot [h_{t-1}, x_t]) \\ w_t &= \tanh(W_w \cdot [r_t \odot h_{t-1}, x_t]) \\ h_t &= (1 - z_t) \odot h_{t-1} + z_t \odot w_t \end{cases} \tag{1}$$

In the equations, $\boldsymbol{W}$ denotes the weight matrix; $\mathrm{x_t}$ represents the input information at the current time step; h_{t-1} is the hidden state from the previous time step; $\mathrm{r_t}$ represents the reset gate; $\mathrm{z_t}$ denotes the update gate; $\hat{\mathrm{w}}_\mathrm{t}$ is the candidate hidden state; h_t is the hidden state passed to the next time step; t indicates the time step t; σ is the sigmoid activation function; tanh is the hyperbolic tangent activation function; and $\odot$ denotes the Hadamard product (element-wise multiplication).

3.3 CRF Layer

Conditional Random Fields (CRF) constitute a fundamental model in Natural Language Processing (NLP), demonstrating particular efficacy for sequence labeling tasks. These include text segmentation, part-of-speech tagging, and named entity recognition. By modeling sentence-level contextual dependencies, CRF predicts globally optimal label sequences [12].

A key feature of the CRF model is its ability to capture the dependencies between tags, that is, to model the relationships between adjacent tags in a sequence by learning transition probabilities. In NER, an entity tag usually does not follow another unrelated entity tag. CRF can improve the accuracy of tagging by learning these constraints.

In the model, CRF utilizes the feature vectors from the previous layer by calculating the score of the entire sequence to integrate context information and determine the most appropriate label sequence. In this paper, the output sequence of BiGRU provides the necessary context information for the CRF layer to determine the most likely label sequence. The calculation formula of the CRF layer is shown in (2) [13].

$$S(X, y) = \sum_{i-1}^{n} \boldsymbol{W}_{y_{i-1}, y_i} + \sum_{i-1}^{n} \boldsymbol{P}_{i, y_i} \quad (2)$$

In the formula, X represents the input sequence; y represents the label sequence; $\boldsymbol{W}_{yi\text{-}1,yi}$ is an element in the transition feature score matrix, indicating the score of transitioning from the previous label $y_{i\text{-}1}$ to the current label y_i. The transition feature is used to model the dependency relationship between adjacent labels in the sequence. $\boldsymbol{P}_{i,yi}$ is the emission feature score, representing the score of the feature at position i in the BiGRU output being labeled as y_i. The emission feature is used to model the relationship between the input feature and the label.

3.4 Loss Function

The predominance of non-entity tokens (O-tags) in classical Chinese texts causes conventional binary cross-entropy loss to exhibit slow convergence and optimization bias, failing to achieve optimal performance. Furthermore, severe class imbalance persists in classical Chinese NER tasks. To minimize loss functions effectively while enhancing training data fitting and test generalization, we introduce Focal Loss [14] for model optimization. This improved loss function addresses class imbalance by modulating relative frequencies of positive/negative samples and down-weighting easy samples' contributions. Its formula is:

$$FL(p_t) = -\alpha_t (1 - p_t)^r \log(p_t) \quad (3)$$

Among them, P_t represents the predicted probability of the model for positive samples; α_t is the balance coefficient of positive and negative samples, used to balance the weights of positive and negative samples; while γ is a regulating factor, used to control the weights of difficult and easy-to-classify samples. By configuring an elevated γ value, the model prioritizes hard-to-classify samples while down-weighting the loss contribution from easily classifiable instances.

3.5 Adversarial Training

Adversarial training is a regularization approach that enhances model robustness by introducing controlled perturbations during training, particularly effective in low-resource or noisy environments. It perturbs input embeddings to generate adversarial examples, which are processed alongside original inputs, encouraging the model to remain accurate under adversarial conditions.

Initially developed in computer vision, adversarial examples are minimally altered inputs that appear unchanged to humans but mislead models. Miyato et al. [15] first applied this idea to text classification, adding embedding-level perturbations to improve generalization. Li et al. [16] proposed the Ad-Attention-BiLSTM model, which combines adversarial learning, attention mechanisms, and BiLSTM to enhance word-level feature discrimination. Wang et al. [17] integrated adversarial training with label attention to mitigate misclassification in recommendation systems.

This study compares four adversarial methods—FGSM, FGM, PGD, and FreeLB—and ultimately selects PGD (Projected Gradient Descent) for its superior performance. FGSM (Fast Gradient Sign Method) uses a single-step perturbation guided by the gradient sign. FGM (Fast Gradient Method) improves this by computing and normalizing the embedding gradients to form a perturbation Δx, then updating model parameters using gradients from x + Δx. PGD is an improved version of FGM (Fast Gradient Method). Unlike FGM's one-step perturbation approach, PGD adopts a "small-step, multi-iteration" strategy, where gradients are updated by accumulating perturbations over t steps [18]. FreeLB (Free Large Batch) averages gradients from multiple adversarial steps, simulating a virtual batch of size K to stabilize training.The computation formula for PGD is defined as:

$$\boldsymbol{x}_{i+1} = \prod_{x+\mathcal{S}} \left(\boldsymbol{x}_i + \alpha \frac{\boldsymbol{g}(\boldsymbol{x}_i)}{\|\boldsymbol{g}(\boldsymbol{x}_i)\|_2} \right) \tag{4}$$

$$\boldsymbol{g}(\boldsymbol{x}_i) = \nabla_{\boldsymbol{x}} \mathcal{L}(\theta, \boldsymbol{x}_i, y) \tag{5}$$

In the formula, $\boldsymbol{x}_{i+1}$ represents the adversarial example at the $(i + 1)$-th iteration; S denotes the constraint space of perturbations; α is the step size, controlling the magnitude of perturbation changes at each iteration; $g(x_i)$ denotes the gradient of the loss function L with respect to the input x at step i; and $\|\boldsymbol{g}(\boldsymbol{x}_i)\|_2$ is the L_2 norm of the gradient, which ensures that the update direction at each iteration follows the normalized gradient direction.

4 Experiment and Result Analysis

4.1 Experimental Dataset

In this study, two datasets were employed to evaluate the performance of the proposed model on ancient Chinese named entity recognition tasks. The first is a subtask dataset from CCLUE, a benchmark suite for Classical Chinese Language Understanding developed by Ethan-yt and collaborators. CCLUE encompasses multiple tasks such as sentence segmentation and punctuation recovery, NER, text classification, poetry sentiment analysis, and wenyan-baihua retrieval, each with designated training, development, and test sets. It serves as a comprehensive evaluation platform that significantly advances research in classical Chinese NLP. The second dataset is CHisIEC (Chinese Historical

Information Extraction Corpus), curated by the Department of Information Management and the Center for Digital Humanities at Peking University. It focuses on NER and relation extraction (RE) in historical Chinese texts. The dataset defines four entity types and twelve relation types, annotated under a detailed schema. CHisIEC provides valuable resources for structured analysis of ancient historical documents in digital humanities and historical research.

The named entity recognition task typically adopts one of three annotation schemes: BIO, BMES, or BIOES. The CHisIEC dataset uses the BIOES format, where "B" marks the beginning of an entity, "I" the inside, "O" non-entity tokens, "E" the end, and "S" a single-token entity. CHisIEC defines four entity types in its training set: person names (PER), locations (LOC), book titles (BOOK), and organizations (OFI).In contrast, the CCLUE dataset adopts the BIO scheme, where tokens are tagged as the beginning (B), inside (I), or outside (O) of an entity. It defines two entity types: book names (BOOKNAME) and others (OTHER) (Table 1).

Table 1. Statistics of CHisIEC and CCLUE

CCLUE		CHisIEC	
Entity	Numbers	Entity	Numbers
Bookname	16721	Person	5286
Other	11694	Location	3101
		Book Title	2721
		Organization	176

4.2 Experimental Environment and Parameter Settings

The experiments in this section were implemented using the Python programming language and the PyTorch deep learning framework. The software and hardware environment includes: Windows 10 operating system, NVIDIA RTX 4090 GPU with 24 GB memory, PyTorch 2.4.1, Python 3.10.To ensure the fairness of the experiments, all hyperparameters were kept consistent across multiple training runs, except for certain model-specific settings such as the number of Transformer layers, hidden units, and vocabulary size of different pre-trained models.The embedding dimensions of the six pre-trained models are all 768, except for BERT-CCPoem, which has an embedding size of 512. This difference is determined by the architectural design of the models. The models with 768-dimensional embeddings typically consist of 12 Transformer layers, each with 12 attention heads, whereas BERT-CCPoem is configured with 8 Transformer layers, each containing 8 attention heads, resulting in a smaller embedding size. Detailed configurations are presented in the following sections (Table 2).

Table 2. Model parameters

Hyperparameter	value
Learing_rate	3e−5
Batch_size	16
rnn_size	256
Dropout_rate	0.5
Optimizer	Adam

4.3 Evaluation Indicators

In this paper's experiment, three conventional evaluation indicators are adopted to assess the recognition effect of text naming recognition, namely Precision (P), Recall (R), and F1 score (F1 score) as the evaluation criteria. The calculation method is shown as below.

$$P = \frac{TP}{TP + FP} \times 100\% \tag{6}$$

$$R = \frac{\mathrm{TP}}{\mathrm{TP+FN}} \times 100\% \tag{7}$$

$$F_1 = \frac{2 \times P \times R}{P + R} \times 100\% \tag{8}$$

Among them, TP (True Positive) represents the correct value identified by the positive sample; FP (False Positive) represents the value of the positive sample recognition error; TN (True Negative) represents the correct value identified by negative samples; FN (False Negative) represents the value of negative sample identification errors [19].

4.4 Comparative Experiments on Four Adversarial Training Methods

To evaluate the effectiveness of four adversarial training algorithms in the task of named entity recognition for classical Chinese texts, we conducted comprehensive experiments and analysis on two datasets: CCLUE and CHisIEC. The experimental results demonstrate that all four adversarial training methods significantly enhance the performance of classical Chinese NER models, thereby validating the effectiveness of adversarial training in improving both the robustness and accuracy of the models.

As shown in Table 3 and Table 4, the introduction of PGD (Projected Gradient Descent) yields the highest F1-scores among the four methods, achieving 85.50% on the CCLUE dataset and 92.28% on the CHisIEC dataset. These results confirm that PGD outperforms other adversarial strategies and is the most suitable and effective adversarial training method for classical Chinese named entity recognition tasks.

Table 3. Experimental comparison results on CCLUE

Model	Precision	Recall	F1-Score
GuwenBERT-BiGRU-FOCAL-CRF	80.60%	81.70%	80.65%
GuwenBERT-BiGRU-FOCAL-CRF + FGSM	81.66%	84.55%	81.78%
GuwenBERT-BiGRU-FOCAL-CRF + FGM	83.32%	81.24%	84.89%
GuwenBERT-BiGRU-FOCAL-CRF + FreeLB	82.67%	83.94%	82.28%
GuwenBERT-BiGRU-FOCAL-CRF + PGD	84.79%	86.28%	85.50%

Table 4. Experimental comparison results on CHisIEC

Model	Precision	Recall	F1-Score
GuwenBERT-BiGRU-FOCAL-CRF	84.57%	83.44%	84.28%
GuwenBERT-BiGRU-FOCAL-CRF + FGSM	82.15%	85.96%	85.70%
GuwenBERT-BiGRU-FOCAL-CRF + FGM	85.87%	88.65%	86.75%
GuwenBERT-BiGRU-FOCAL-CRF + FreeLB	86.44%	84.46%	88.57%
GuwenBERT-BiGRU-FOCAL-CRF + PGD	90.67%	91.94%	92.28%

4.5 Experimental Results and Analysis

To validate the effectiveness of the proposed model, we selected four deep neural network-based baselines: BiGRU-CRF, BiLSTM-CRF, BERT-BiGRU-CRF, and BERT-BiLSTM-CRF. Additionally, four classical Chinese NER models were included for comparison: SikuBERT-BiGRU-CRF, SikuRoBERTa-BiGRU-CRF, BERT-CCPoem-BiGRU-CRF, and bert-ancient-chinese-BiGRU-CRF. These models serve as baselines against our proposed GuwenBERT-BiGRU-FOCAL-CRF + PGD architecture. Table 5 presents the performance of the eight models on the CCLUE dataset, while Table 6 reports their results on the CHisIEC dataset.

Table 5. Experimental comparison results on CCLUE

Model	Precision	Recall	F1-Score
BiGRU-CRF	43.33%	27.20%	33.42%
BiLSTM-CRF	35.68%	19.41%	25.14%
BERT-BiGRU-CRF	59.69%	64.71%	62.10%
BERT-BiLSTM-CRF	61.64%	65.29%	63.10%
SikuBERT-BiGRU-CRF	50.57%	63.56%	56.33%
SikuRoBERTa-BiGRU-CRF	55.16%	54.33%	54.74%

(continued)

Table 5. (*continued*)

Model	Precision	Recall	F1-Score
BERT-CCPoem-BiGRU-CRF	49.96%	44.81%	47.24%
bert-ancient-chinese -BiGRU-CRF	64.93%	62.91%	63.91%
GuwenBERT-BiGRU-FOCAL-CRF + PGD	84.79%	86.28%	85.50%

Table 6. Experimental comparison results on CHisIEC

Model	Precision	Recall	F1-Score
BiGRU-CRF	46.31%	34.17%	39.33%
BiLSTM-CRF	32.27%	21.44%	25.77%
BERT-BiGRU-CRF	78.95%	78.63%	79.22%
BERT-BiLSTM-CRF	75.79%	78.59%	77.17%
SikuBERT-BiGRU-CRF	65.21%	71.20%	68.08%
SikuRoBERTa-BiGRU-CRF	66.03%	68.14%	67.07%
BERT-CCPoem-BiGRU-CRF	50.30%	56.64%	53.28%
bert-ancient-chinese-BiGRU-CRF	75.21%	72.91%	74.04%
GuwenBERT-BiGRU-FOCAL-CRF + PGD	90.67%	91.94%	92.28%

As shown in Tables 5 and 6, the proposed GuwenBERT-BiGRU-FOCAL-CRF + PGD model achieves 84.79% precision, 86.28% recall, and an F1-score of 85.50% on the CCLUE dataset. On the CHisIEC dataset, it achieves 90.67% precision, 91.94% recall, and a 92.28% F1-score. Compared with the selected baseline models, our model demonstrates consistent performance improvements. Moreover, using general-purpose BERT or domain-specific pre-trained models—such as GuwenBERT or SikuBERT—leads to a significant improvement in F1 score, demonstrating that pre-trained semantic representations are crucial for NER tasks.Specifically, the proposed model surpasses the BERT-BiLSTM-CRF model by 12.40% in F1-score on the CCLUE dataset and by 15.11% on the CHisIEC dataset. These comparative results across the eight models on both datasets indicate that, among general-purpose architectures, GuwenBERT performs better in recognizing named entities in classical Chinese texts than standard BERT and several domain-specific pretrained language models. The experimental results validate that our proposed model achieves superior performance in classical Chinese named entity recognition tasks compared to both general-domain and domain-adapted models.

5 Conclusion

To address the challenges of NER in classical Chinese texts and mitigate the issue of imbalanced entity distribution, this study introduces a GuwenBERT-BiGRU-FOCAL-CRF model enhanced with PGD-based adversarial training. The architecture integrates

GuwenBERT, bidirectional gated recurrent units (BiGRU), and a CRF layer, optimized jointly with Focal Loss. The integration of PGD adversarial training effectively strengthens the model's resilience to input perturbations and improves its generalization across datasets. Extensive experiments on two classical Chinese corpora and multiple deep neural network baselines demonstrate that domain-specific models better suit domain-specific NLP tasks, and self-attention mechanisms selectively capture key sequence information. Future work will explore advanced techniques to further improve NER in classical Chinese and extend solutions to other domains with diverse entity recognition challenges.

References

1. Su, W., Zhao, D., Meng, J., et al.: Named entity recognition for ancient Chinese based on knowledge embedding. In: 2023 3rd International Conference on Digital Society and Intelligent Systems (DSInS), pp. 429–432. IEEE (2023)
2. Bao, Z., Song, B., Zhang, W., et al.: Named entity recognition in traditional Chinese medicine books combining semi-supervised learning and rule-based approach. J. Chin. Inf. Process. **36**(6), 90–100 (2022)
3. Jing, H., Wang, L.: Rule-based Chinese person names identification in ancient Chinese literature of annals-biography (Jizhuan) style. Libr. Inf. Serv. **37**(3), 118–124 (2013)
4. Huang, S., Wang, D., He, L.: Research on constructing automatic recognition model for ancient Chinese place names based on pre-qin corpus. Libr. Inf. Serv. **59**(12), 135–140 (2015)
5. Xie, T.: Reseach and implementation of named entity recognition based on ancient literature. Beijing University of Posts and Telecommunications, Beijing (2018)
6. Siami-Namini, S., Tavakoli, N., Namin, A.S.: The performance of LSTM and BiLSTM in forecasting time series. In: 2019 IEEE International Conference on Big Data, pp. 3285–3292. IEEE (2019)
7. Li, Z., He, L., Yu, X.: Contextual knowledge organization and application of classics based on event knowledge graph—taking ZuoZhuan as an example. Libr. Tribune **44**(4), 125–137 (2024)
8. Wang, D., et al.: SikuBERT and SikuRoBERTa: construction and application research of pre training models for digital humanities in the complete library of four branches (in chinese). Lib. Forum **42**, 31–43 (2022)
9. Liu,Y., et al.: RoBERTa: a robustly optimized BERT pretraining approach. arXiv preprint arXiv:1907.11692 (2019)
10. Xinqi, P., Wei, W., Yefeng, C.: Theme classification of Chinese classical poetry based on the GB-MSA-BGTCN hybrid model. In: CISAI 2024 Proceedings of the 2024 7th International Conference on Computer Information Science and Artificial Intelligence, pp. 276–280 (2024)
11. Xu, C., Gao, Y.: Multi-modal transformer with multi-head attention for emotion recognition. In: 2023 IEEE International Conference on Sensors, Electronics and Computer Engineering (ICSECE), Jinzhou, China, pp. 826–831. IEEE (2023)
12. Wang, B., Jiang, F.: Named entity recognition using EHealth-BiLSTM-CRF combine with multi-head self-attention for Chinese medical information. In: Jin, C., Yang, S., Shang, X., Wang, H., Zhang, Y. (eds.) WISA 2024. LNCS, vol. 14883, pp. 451–462. Springer, Singapore (2024). https://doi.org/10.1007/978-981-97-7707-5_37
13. Yang, P., Dong, W.: Chinese named entity recognition method based on BERT embedding. Comput. Eng. **46**(4), 40–45, 52 (2021)
14. Lin, T.Y., Goyal, P., Girshick, R., et al.: Focal loss for dense object detection. IEEE Trans. Pattern Anal. Mach. Intell. **42**(2), 318–327 (2020)

15. Lhamo-kyap, Pema-Tseden, Yongtso, Gao, X., et al.: Study on the tibetan medicine named entity recognition method integrating adversarial training and iterative dilated convolution. Plateau Sci. Res. **9**(01), 105–118 (2025). https://doi.org/10.16249/j.cnki.2096-4617.2025.01.011
16. Liwh, Z., Pan, L.H.: Short texts classification method based on biLSTM. Comput. Eng. Design **41**(3), 880–886 (2020)
17. Wang, Q., Wu, W., Zhao, Y., et al.: Combining label-wise attention and adversarial training for tag prediction of web services. In: 2021 IEEE International Conference on Web Services (ICWS), pp. 358–363 (2021)
18. Yang, J.H., Liu, B.B.: Chinese named entity recognition based on lexical enhancement and adversarial training. Comput. Eng. Design **45**(12), 3712–3718 (2024). https://doi.org/10.16208/j.issn1000-7024.2024.12.025
19. Ying, S., Li, P., Chen, J., et al.: An EEG-based single-channel dual stream automatic sleep staging network with transfer learning. Appl. Soft Comput. **170**, 112722 (2025)

AFR-Agent: Artifact Fragment Reassembly Framework Based on Multi-modal Fusion

Zhongqi Wang[1], Weifan Wang[2(✉)], Lidong Zhang[1], Xinghao Huang[3], Peng Reng[4(✉)], and Chunxiao Xing[4(✉)]

[1] School of History and Culture, Henan University, Kaifeng 475004, China
[2] School of Artificial Intelligence, Henan University, Zhengzhou 450046, China
henuwwf@henu.edu.cn
[3] Staff Development Institute, China National Tobacco Corporation, Zhengzhou 450008, China
[4] BNRist, DCST, RIIT, Tsinghua University, Beijing 100084, China
{renpeng,xingcx}@tsinghua.edu.cn

Abstract. Artifact fragment reassembly is a crucial component in cultural heritage preservation, but traditional manual methods highly rely on specialized domain knowledge and consume considerable time. While recent advances in computer-aided methods have shown promise, limitations such as insufficient utilization of multi-modal prior knowledge and weak scalability caused by strong rule constraints still cannot be ignored. For these challenges, this paper proposes AFR-Agent, an artifact fragment auxiliary reassembly framework based on multi-modal feature fusion representation. During training, AFR-Agent initially derives vector representation for multi-level visual feature and structured prior knowledge, followed by realizing multi-modal semantic fusion through encoder-based contrastive learning. In inference process, AFR-Agent first performs similarity search in metric space to preliminarily filter high-confidence candidate set. Then, a text-image LLM agent conducts secondary inferences, thus obtaining the final auxiliary reassembly result. Experimental validation on real-world dataset of 486 pottery fragments demonstrates that AFR-Agent achieves non-iterative precision of 71.8%. The code of AFR-Agent is available on GitHub.

Keywords: Artifact Fragment Reassembly · Multi-modal Feature Fusion · Contrastive Learning · Text-image Agent

1 Introduction

Artifacts, as significant material bearers of historical information, constitute one of the key objects in cultural heritage preservation. Due to the impact of burial environments and anthropogenic activities, artifacts frequently survive in fragmented states [1,2]. The reassembly of fragments is required to facilitate research,

S. Yang et al. (Eds.): WISA 2025, LNCS 16299, pp. 457–467, 2026.
https://doi.org/10.1007/978-981-95-6426-2_41

conservation, and exhibition. Unfortunately, traditional manual reassembly techniques are labor-intensive and heavily reliant on expert experience, rendering them inefficient for processing large-scale fragments. Therefore, the development of intelligent, computer-aided reassembly method is essential to overcome these efficiency bottlenecks and advance the preservation of cultural heritage.

Recent advancements in information technology have significantly enhanced the potential for computer-aided reassembly. Early methods primarily used 2D images, pioneering the automation of edge feature extraction and iterative screening for the reconstruction of planar geometries [3]. However, 2D-based methods have persistent limitations when it comes to handling complex fragment morphologies and real-world data variability. The adoption of 3D scanning technology paved the way for the development of point cloud-based methods [4,5], allowing for enhanced geometric and textural representation. But such techniques are subject to critical operational constraints, such as stringent fidelity requirements and substantial computational overhead [6,7]. Minor surface imperfections or algorithmic parameters perturbation induce significant accuracy instabilities, while computational demands preclude efficient large-scale processing [8]. Furthermore, this reliance on idealized conditions inherently limits scalability, confining robust performance to controlled environments and impeding generalization across fragments with varying degradation.

Despite advancements in artifact fragment reassembly, state-of-the-art methods face two fundamental limitations: (1) uni-modal dependency, wherein reliance on isolated 2D geometric contours or 3D point clouds neglects complementary multi-modal information (e.g., surface texture, material properties, decorative patterns and contextual archaeological metadata), leading to suboptimal feature representation; and (2) scalability deficits, as real-world scenarios involve complex fragment variability—including surface degradation, missing components, and diverse artifact typologies—that overwhelms existing models' generalization capacity. To address these challenges, this paper introduces AFR-Agent, a novel auxiliary reassembly framework based on multi-modal semantic fusion. AFR-Agent uses contrastive learning in a shared metric space to integrate multi-modal information, such as edge geometry and surface texture, with structured archaeological priors (e.g. vessel typology and excavation background). After that, it adopts a two-stage inference paradigm that first leverages efficient metric-space similarity search to generate high-confidence candidate matches, then employs a text-image LLM agent to resolve ambiguities through semantic reasoning. Experimental evaluations demonstrate that AFR-Agent achieves significant improvements in reassembly accuracy and efficiency across real-world scenarios. By robustly handling complex fragment variability, the framework unlocks the full potential of multi-modal data integration.

Roadmap: The rest of this paper is organized as follows: Sect. 2 introduces related work. Section 3 provides a formal description of the research problem. Section 4 presents the complete AFR-Agent framework. Section 5 shows and analyzes the experimental results. Section 6 concludes whole paper and discusses future work.

2 Related Work

The process of automatically reassembling artifact fragments has advanced through successive technical iterations, with existing approaches predominantly falling into two categories: 2D geometric modeling and 3D point cloud. Early 2D-based methods, such as multiscale dynamic programming framework [3], automated contour matching via curvature encoding but struggled with edge degradation. Eslami et al. [9] introduced wavelet-transformed edge analysis for fragment clustering. However, the over-reliance on manual segmentation limits its scalability. Shang et al. [10] developed a circular contour-specific algorithm, achieving efficient rigid-body matching for symmetric fragments but lacking generality for complex shapes. Generally, these techniques laid groundwork for feature-based assembly but remained constrained by 2D image subjectivity and contour integrity assumptions.

The adoption of 3D scanning enabled more detailed geometric modeling. Liang et al. [11] proposed a Siamese-based network combined with region-growing segmentation and iterative closest point(ICP) registration, achieving high-precision alignment for intact fracture surfaces but failing to address multi-fragment global optimization or texture-deficient samples. Pinho et al. [12] trained the PotNet architecture on synthetic 3D fragments to predict transformation parameters, though its performance relied heavily on pre-trained artifact types. While 3D methods improved geometric fidelity, they faced computational overhead and strict surface integrity requirements, limiting applicability to large-scale or degraded fragment collections.

Collectively, current approaches exhibit two critical limitations: (1) Over-reliance on single modality, neglecting crucial aspects such as surface texture, decorative patterns and archaeological metadata. (2) Inadequate robustness to real-world dataset complexity, such as surface degradation and mixed typologies.

3 Problem Formulation

To facilitate formal description, the artifact fragment reassembly problem addressed by AFR-Agent is formulated as follow. Given a fragment dataset $\mathcal{D} = \{F_1, F_2, \ldots, F_n\}$, Each fragment F_i is represented by multi-modal data, comprising visual attributes V_i (encoding shape, texture, and color derived from visual data) and structured prior knowledge S_i(encompassing discrete attributes such as artifact typology, decoration motifs, and vessel parts, as well as continuous attributes like thickness). The fundamental objective of AFR-Agent is to learn a latent embedding function $f(V_i, S_i) \rightarrow h_i \in \mathcal{H}$, mapping each fragment into a metric embedding space $\mathcal{H}$. In particular, $f(V_i, S_i)$ is optimized such that the metric distance(such as the Euclidean) $d(h_i, h_j)$ in $\mathcal{H}$ is minimized for fragment pairs (F_i, F_j) known to be compatible, while maximized for incompatible pairs.

For any target fragment F_j, the reassembly algorithm first computes similarity metric $s(h_j, h_i)$ between its embedding h_j and other $h_i (i \neq j)$ in $\mathcal{D}$. The algorithm subsequently selects the top-K fragments $\mathcal{P}_j^k = \{F_{i1}, F_{i2}, \ldots, F_{ik}\}$ exhibiting the highest similarity to F_j as the predicted set of compatible fragments. The effectiveness of reassembly framework is rigorously evaluated via the Precision@k metric. Given $\mathcal{T}_j$ as the ground-truth set of fragments archaeologically verified to be compatible with F_j, Precision@k is formally defined as the ratio of correctly retrieved compatible fragments to the total number of true compatible fragments:

$$\text{Precision@}k(F_j) = \frac{\mathcal{P}_j^k \cap \mathcal{T}_j}{\mathcal{T}_j} \quad (1)$$

The overarching optimization goal of AFR-Agent is therefore to maximize Precision@k through effective multi-modal representation learning and fusion strategies within this formalism.

4 AFR-Agent Framework

This section introduces the components of AFR-Agent. As shown in Fig 1, AFR-Agent first extracts multi-modal semantic features of fragments in the Multi-modal Embedding module. Subsequently, the Feature Fusion and Contrastive Learning module employs an encoder to project multi-modal features into a unified metric space, where contrastive learning over positive and negative fragment pairs optimizes semantic alignment. In the Reassembly Process module, AFR-Agent conducts iterative inference via similarity search and a Text-image LLM Agent, enabling efficient auxiliary reassembly of fragments.

Multi-modal Embedding. Existing methods predominantly focus on extracting and analyzing visual features, particularly boundary contours, and subsequently design complex matching rules based on these features. The advantages demonstrated by these methods in previous specific work are usually based on idealized assumptions about the debris situation. For example, they assume that fragments with single breaks and regular edges can only match one specific counterpart with fixed edge widths. This overlooks real-world scenarios where multiple compatible fragments may exist for the same edge, resulting in suboptimal performance.

To address this issue, AFR-Agent employs a multi-modal strategy to reveal the latent relationships of fragments. It first uses a pre-trained model to accurately segment the boundary mask and extract geometric features, such as contour perimeter, area, and bounding box ratio, as well as directional descriptors from consecutive contour points. Global shape invariance is encoded using

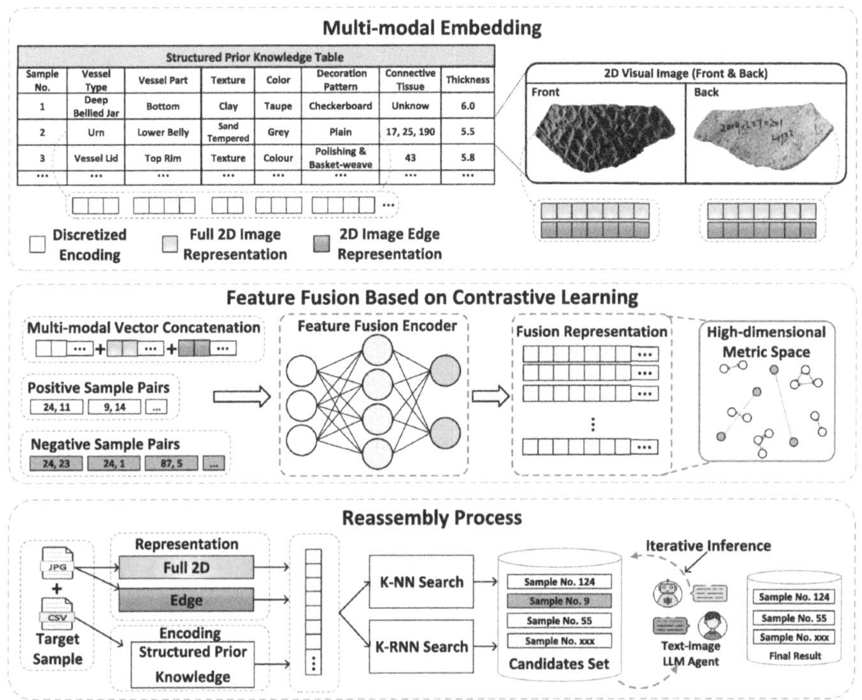

Fig. 1. AFR-Agent Framework

Fourier descriptors, which are derived from the fast Fourier transform(FFT) of contour coordinates with normalize low-frequency components [13]. At the same time, full 2D image representations are embedded using visual models to capture holistic texture and structural contexts. The boundary-specific geometric features and global image embeddings are concatenated to form unified feature vectors. These vectors integrate local edge metrics with global semantic representations, enabling robust characterization of fragments. The AFR-Agent also encodes structured prior knowledge of fragments. Discrete features are encoded using a one-hot-like approach, while continuous features such as average thickness are normalized and added as separate dimensions. Due to the uncertainties in archaeology, such as unknown artifact types or weathered features, a fault-tolerance mechanism has been incorporated into the encoding process. For example, if the position of a fragment in a complete artifact is unknown (out of five known positions), its position code is (0, 0, 0, 0, 0). If it has plain and checkered patterns (out of three pattern types), its pattern code is (1, 0, 1).

Feature Fusion Based on Contrastive Learning. AFR-Agent employs a contrastive learning [14] framework to project multi-modal fragment features into a unified semantic metric space. After geometric descriptors, visual embeddings and structured coding vectors are concatenated, the fused features are fed into an encoder network that optimizes contrastive loss:

$$L = \frac{1}{2N} \sum_{i=1}^{N} \sum_{j=1}^{N} \Big[y_{ij}\, d(f(x_i), f(x_j))^2 + (1 - y_{ij}) \max\big(m - d(f(x_i), f(x_j)), 0\big)^2 \Big] \tag{2}$$

Here,$f(\cdot)$ denotes the encoder mapping, y_{ij} labels positive(or negative) fragment pairs, $d(\cdot)$is the Euclidean distance, and m is the margin parameter. Positive pairs represent fragments that are compatible with the ground truth, while negative pairs are randomly sampled from combinations that are not compatible. This strategy brings semantically similar fragments close together in the metric space while separating dissimilar ones, thus enabling a robust representation of fragment relationships. The contrastive framework integrates structured prior knowledge, with visual features in order to learn context-aware similarity metrics. AFR-Agent uses cross-modal collaborative learning to optimize feature representations, aiming to encode both local geometric compatibility and global cultural context. The learned embeddings facilitate the efficient, large-scale reassembly of fragments by capturing micro-level edge fits and macro-level archaeological semantics.

Fragment Reassembly Process. AFR-Agent overcomes the limitations of traditional rule-based reassembly, such as single-modal dependency and weak robustness to fragmented edges, by using a two-stage inference paradigm that integrates metric space similarity search with multi-modal LLM reasoning. The process begins by projecting fragments into $\mathcal{H}$, which encodes geometric features and archaeological metadata. For a target fragment F_j, an initial candidate set $\mathcal{P}_j$ is constructed using a hybrid strategy: K-Nearest Neighbors(KNN) retrieves the K most similar fragments in$\mathcal{H}$, while Reverse K-Nearest Neighbors(R-KNN) captures reciprocal similarity relationships (i.e., fragments for which F_j is among their top-K neighbors). Based on above, the AFR-Agent obtains the initial collocation candidate set $\mathcal{P}_j = \mathcal{P}_j^{\text{KNN}} \cup \mathcal{P}_j^{\text{R-KNN}}$ of the target debris samples.

Subsequently, AFR-Agent processes the candidate set $\mathcal{P}_j$ by pairwise inference. In each iteration, the input comprises the target fragment and one candidate from $\mathcal{P}_j^{\text{ranked}}$, encoded as a unified multi-modal representation that couples the fragment image with its structured metadata (see Fig. 2). Under a fixed prompt template, the Agent assesses compatibility along three axes: local geometric fit, covering edge continuity and thickness consistency; global semantic coherence, covering vessel typology and decorative motifs; and contextual validity, including stratigraphic association and excavation provenance.

This workflow yields returns a structured record consisting of three elements: a binary compatibility label, a concise justification, and a termination indicator. Short-term context is preserved through a conversational memory buffer implemented with LangChain's ConversationBufferMemory [15], which ensures con-

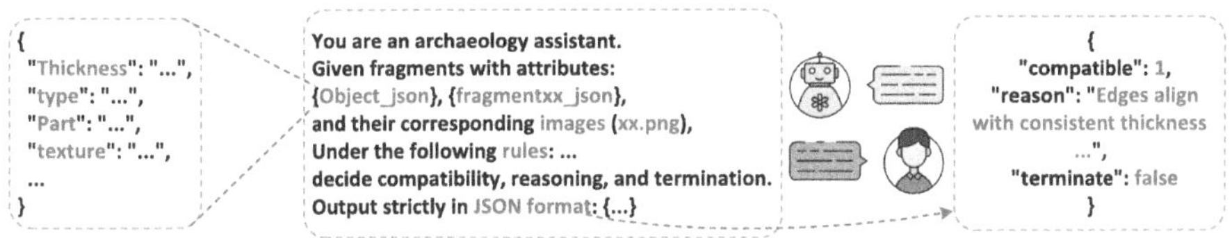

Fig. 2. Example of prompt engineering in AFR-Agent workflow

sistency across iterations and reduces the likelihood of generating spurious metadata. Output validity is further guaranteed by schema-validation mechanisms, including JSON schema checks, constrained function calls, and guardrails, with bounded retries applied in the case of malformed responses. The iterative process terminates once the termination indicator is set, the candidate list is exhausted, or a predefined limit of consecutive incompatibility judgments is reached. At that point, the accumulated set of compatible fragments is returned as $\mathcal{C}_j$. For large-scale settings, batch evaluation of the top-k candidates may improve efficiency, although pairwise serial inference remains the default design to preserve contextual isolation and reproducibility.

5 Experiment Evaluation

Dataset Description. The field of archaeology currently lacks an established convention for sharing data on fragmented pottery, which poses significant challenges to obtaining pottery fragment data from other archaeological sites. Nevertheless, our proposed workflow can be applied to various cultural relic fragment datasets, provided that a data preprocessing procedure similar to the one aforementioned in this paper is adopted and a certain proportion of positive and negative sample pairs is determined. In our experiments, we used a real-world multimodal dataset of pottery fragments. The materials originate from the Longshan Culture period, excavated in 2019 by relevant institutions in collaboration at the Xiuwu Ligu Site, comprising 486 identifiable pottery fragments [16]. This dataset contains rich information on the physical attributes and categorizations of the fragments, which is critical for our research on fragment reassembly and cultural analysis. Detailed dataset information is presented in the following Table 1. Due to policy restrictions, the full dataset cannot be directly open sourced. But, a demo version of the original dataset and the data pre-processing code are publicly available on the GitHub repository of AFR-Agent [17].

Experimental Setup. The experimental design rigorously evaluates AFR-Agent against two state-of-the-art baselines: a 2D geometric matching algorithm (Shang et al. [10]) leveraging circular contour constraints and a wavelet transformation method (Eslami et al. [9]). Performance is measured using two key metrics: non-iterative precision, measuring the accuracy of top-K matches without refinement, and final iterative precision, evaluating the global optimization

Table 1. Dataset Description

Dataset Variable	Description
Fragment Count	486
Vessel Part	13
Pottery Texture	2
Pottery Color	7
Decoration Pattern	12
Individual ID	4
Vessel Type	12
Average Thickness	(mm)
Visual Features	972 (Front & back)
Positive Pairs for Training	127 (13.1%)

capability after meeting termination criteria (e.g., confidence score decay). All experiments were conducted on a 64-bit Ubuntu 22.04 server equipped with Intel Xeon Gold 6226R CPU, and NVIDIA A5000 GPU. All experiments are implemented in Python.

The experimental workflow encompasses three phases. First, we perform hyperparameter optimization for AFR-Agent, encompassing contrastive learning margins, LLM confidence thresholds, and similarity search parameters. Second, we conduct a comparative performance evaluation against the baselines using identical training-testing splits to ensure a fair assessment. To systematically quantify the contribution of each architectural component, we also execute an ablation study. Specifically, this ablation isolates the impact of multi-modal integration by modifying the feature fusion process: structural prior knowledge is removed during training, retaining only 2D image and edge features.

About implementation, AFR-Agent leverages a Qwen2-VL-72B [18]-based multi-modal LLM for reasoning agent, with YOLOv8 fine-tuned [19] on 60 manually annotated fragment edges to enhance edge feature extraction. CLIP(ViT-B/32) [20] is employed for 2D image embedding. The contrastive learning framework uses a four-layer autoencoder to project multi-modal inputs into a shared semantic space(512-dimensional), while the two-stage inference combines FAISS [21] vector search with LLM validation to resolve ambiguous matches. This architecture ensures robust integration of geometric, visual, and archaeological knowledge for accurate fragment reassembly.

Results and Analysis. Experimental evaluation substantiates the efficacy of the AFR-Agent framework through quantitative metrics and qualitative visualizations. Figure 2 presents the t-SNE projection of our 512-dimensional semantic metric space, where fragmentation groups (left and right) connect via dashed vectors to corresponding clusters in the latent space. Crucially, correctly reassembled cases (star and box annotations) demonstrate joint learning of local geometric compatibility and global contextual patterns. This spatial organization emerges from the contrastive learning framework's ability to fuse heterogeneous features, though it should be noted that while target fragments appear proximally closer to correct matches in 2D visualization, actual distance metrics are computed in the original 512-dimensional space, this projection serves strictly as an interpret ability aid (Fig. 3).

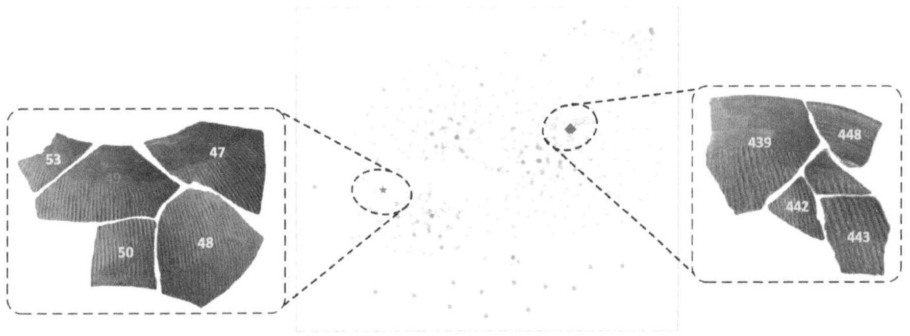

Fig. 3. Reassembly Reasoning in Semantic Fusion Metric Space

Quantitative superiority of AFR-Agent is decisively established in Table 2, where AFR-Agent achieves 71.8% non-iterative precision and 79.6% iterative precision. This represents an absolute improvement of 28.9% over Shang et al.'s 2D method (42.9%) and 12.1% points over Eslami et al.'s wavelet approach (59.7%) in non-iterative precision. For iterative precision, AFR-Agent outperforms these baselines by 23.3% and 9.2% respectively. Statistical significance confirms these performance differentials originate from inherent baseline limitations: Shang's uni-modal approach cannot integrate structured archaeological priors (e.g., decorative motif compatibility), while Eslami's method lacks semantic reasoning to resolve texture ambiguities. Ablation analysis also reveals critical dependencies within our method: Exclusion of structured prior knowledge causes a 24.9% reduction in non-iterative precision, demonstrating the necessity of multi-modal fusion. The two-stage inference architecture offers complementary advantages: 1) Metric space similarity search effectively prunes geometrically implausible correspondences; 2) The text-image LLM agent resolves ambiguities via cross-modal constraint enforcement (e.g., verifying thickness consistency among ceramic body shreds). Collectively, these mechanisms show that it is necessary to use hierarchical reasoning over integrated visual-archaeological features to address real-world fragment variability. This is a capability lacking in existing methodologies.

It should be noted that although AFR-Agent consistently outperforms existing baselines, its overall precision still leaves room for improvement. Two main factors contribute to this limitation: the current feature fusion strategy directly concatenates visual embeddings with structured encoding, which provides coarse alignment across modalities; and the framework shows reduced efficacy for small fragments, highly curved pieces, or artifacts with complex morphology.

Table 2. Comparative Experimental Results of AFR-Agent.

Method	P@k (Non-iterative)	P@k (Iterable)
Shang et al. [10]	42.9%	56.3%
Eslami et al. [9]	59.7%	70.4%
AFR-Agent(Only 2D)	46.9%	62.1%
AFR-Agent(Ours)	71.8%	79.6%

6 Conclusions

AFR-Agent presents a multi-modal fusion framework for artifact fragment reassembly, addressing the limitations of uni-modal feature reliance and iterative error accumulation in traditional methods. By integrating visual geometric features with archaeological metadata via contrastive learning, the framework enables robust representation of complex or degraded fragments. Its two-stage inference paradigm enables non-iterative global optimization, circumventing reliance on intact contours and contextualizing geometric compatibility with archaeological knowledge systems. Notably, experiments demonstrate that AFR-Agent achieves comparable performance to 3D-based methods using only 2D visual data, making it particularly valuable for scenarios where 3D point cloud acquisition is impractical or resource-intensive. This capability extends its applicability to a broader range of heritage preservation contexts. Experimental validation on real-world datasets confirms its effectiveness, though current limitations include domain-specific fine-tuning requirements and computational overhead in LLM reasoning.

Future work will focus on the development of lightweight cross-modal architectures that enable unsupervised adaptation, thereby improving adaptability in unlabeled scenarios while reducing computational overhead. In addition, we aim to enhance robustness under extreme fragmentation to support efficient processing of highly deteriorated artifacts, and to integrate the framework into automated restoration workflows. These extensions will broaden the applicability of AFR-Agent from pottery ware to a wider range of cultural relics, including bronze artifacts and murals, thereby advancing the ongoing evolution of digital heritage preservation technologies.

References

1. Kleber, F., Sablatnig, R.: A survey of techniques for document and archaeology artefact reconstruction. In: 2009 10th International Conference on Document Analysis and Recognition, pp. 1061–1065. IEEE (2009)

2. Di Angelo, L., Di Stefano, P., Guardiani, E.: A review of computer-based methods for classification and reconstruction of 3D high-density scanned archaeological pottery. J. Cult. Herit. **56**, 10–24 (2022)
3. da Gama Leitao, H.C., Stolfi, J.: A multiscale method for the reassembly of two-dimensional fragmented objects. IEEE Trans. Pattern Anal. Mach. Intell. **24**(9), 1239–1251 (2002)
4. Zhao, J., Yin, L., Yang, J., Hua, X., Liu, Z., Wang, X.: Research progress in the splicing and restoration of artifact fragments based on point cloud. ISPRS Ann. Photogram. Remote Sens. Spat. In. Sci. **10**, 255–262 (2023)
5. Qi, C.R., Su, H., Mo, K., Guibas, L.J.: PointNet: deep learning on point sets for 3d classification and segmentation. In: Proceedings of the IEEE Conference on Computer Vision and Pattern Recognition, pp. 652–660 (2017)
6. Geng, G., Zhang, P., Liu, Y., Zhou, M., Yao, W.M., Li, K.: Reassembly method of cultural relic fragments based on the neighborhood characteristics of fracture surface. Opt. Precis. Eng. **29**, 1169–1179 (2021)
7. Zhang, Y., Li, K., Chen, X., Zhang, S., Geng, G.: A multi feature fusion method for reassembly of 3D cultural heritage artifacts. J. Cult. Herit. **33**, 191–200 (2018)
8. Wang, P., Geng, G., Wang, Y., Hu, J., Zhang, Y.: Fragments splicing method based on surface texture and fracture boundary. In: 2018 International Computers, Signals and Systems Conference (ICOMSSC), pp. 644–648. IEEE (2018)
9. Eslami, D., Di Angelo, L., Di Stefano, P., Guardiani, E.: A semi-automatic reconstruction of archaeological pottery fragments from 2D images using wavelet transformation. Heritage **4**(1), 76–90 (2021)
10. Chong, S., Cao Ligang, F.H.: Two-dimensional stitching algorithm of potsherd based on shape constraint. J. Ceram. **42**(2), 320–324 (2021)
11. Liang, Q., Yang, L., Luo, Z., Jiang, W., Hong, C.: A Siamese network-based method for automatic stitching of artifact fragments. IEEE Trans. Instrum. Meas. **72**, 1–13 (2023)
12. Pinho, M.F.C., Mota, G.L.A., da Costa, G.A.O.P.: A deep learning approach to assist in pottery reconstruction from its sherds. Heritage **8**(5), 167 (2025)
13. Reddy, B.S., Chatterji, B.N.: An FFT-based technique for translation, rotation, and scale-invariant image registration. IEEE Trans. Image Process. **5**(8), 1266–1271 (1996)
14. Sun, S., Cai, T., Yan, F., Ju, S.: Two-stage enhancement for recommendation systems based on contrastive learning. In: Jin, C., Yang, S., Shang, X., Wang, H., Zhang, Y. (eds.) WISA 2024. LNCS, vol. 14883, pp. 150–162. Springer, Cham (2024). https://doi.org/10.1007/978-981-97-7707-5_13
15. Chase, H.: LangChain (2022). https://github.com/langchain-ai/langchain
16. Zhang, L., Li, J.: Site of Li Gu from Longshan culture to Erligang culture period in Xiuwu county. In: Society, C.A. (ed.) Yearbook of Archaeological in China 2020, p. 348. China Social Sciences Press, Beijing (2021)
17. Wang, W.: AFR-agent (2025). https://github.com/Shadowwf/AFR-Agent
18. Wang, P., et al.: Qwen2-VL: enhancing vision-language model's perception of the world at any resolution. arXiv preprint arXiv:2409.12191 (2024)
19. Jocher, G., Qiu, J., Chaurasia, A.: Ultralytics YOLO (2023). https://github.com/ultralytics/ultralytics
20. Radford, A., et al.: Learning transferable visual models from natural language supervision. In: International Conference on Machine Learning, pp. 8748–8763. PMLR (2021)
21. Douze, M., et al.: The FAISS library (2024)

Chinese Relation Extraction Based on Global–Local Perception and Contrastive Learning

Chengyuan Cao[1], Mei Song[1,2(✉)], Yi Zhu[1,2], and Li Zhao[1,2]

[1] School of Artificial Intelligence and Computer Science, Jiangsu Normal University, Xuzhou 221116, Jiangsu, China
msong@jsnu.edu.cn

[2] Jiangsu Provincial Key Laboratory of Educational Intelligent Technology, Xuzhou 221116, Jiangsu, China

Abstract. Relation extraction (RE) plays a crucial role in extracting information from Chinese texts but faces notable challenges, including difficulty in accurately recognizing overlapping triples and balancing the modeling of global and local semantic features. To address these issues, we propose a novel RE method that integrates a global–local perception mechanism and contrastive learning. The proposed mechanism comprises a local feature extraction module designed to capture fine-grained semantic distinctions between entity pairs and a global semantic association module that encodes sentence-level contextual information, thereby improving the overall feature representation. Moreover, the method introduces a contrastive optimization strategy based on contrastive learning, which enhances the model's ability to cluster semantically similar relations while effectively distinguishing dissimilar ones. This strategy further improves the model's discriminative ability to relation. Experimental results on two public Chinese datasets indicate that our model outperforms previous methods, confirming the effectiveness of our proposed approach in RE.

Keywords: Relation extraction · Global–local perception · Contrastive learning · Information extraction

1 Instruction

Relation extraction (RE) is a fundamental task in natural language processing (NLP), aimed at extracting structured relational triples from unstructured text [1]. It has been widely applied in various domains such as knowledge graph construction and machine translation. However, RE in Chinese presents unique challenges.

Compared to English, Chinese sentences are generally longer, exhibit more complex entity and relation types, and often contain multiple entity pairs [2],

S. Yang et al. (Eds.): WISA 2025, LNCS 16299, pp. 468–480, 2026.
https://doi.org/10.1007/978-981-95-6426-2_42

placing higher demands on models to capture fine-grained features. The "overlapping triples" phenomenon is particularly prevalent in Chinese, wherein multiple entities may share the same text span or co-occur within a single sentence. Common cases include single entity overlap (SEO) and entity pair overlap (EPO), as shown in Fig. 1. The unique syntactic and semantic characteristics of Chinese further aggravate these challenges [3]. Consequently, models need to accurately identify entity boundaries and relations while distinguishing subtle semantic differences to prevent relation confusion. Enhancing the ability of models to handle overlapping triples remains a central challenge in Chinese RE.

Normal	Text	乔布斯创立了苹果公司，沃兹尼亚克是联合创始人。 Steve Jobs founded Apple, and Steve Wozniak was the co-founder.
	Triples	(乔布斯, 创立, 苹果公司) (沃兹尼亚克, 担任, 联合创始人) (Steve Jobs, founded, Apple) (Steve Wozniak, held position, Co-founder)
SEO	Text	乔布斯创立了苹果公司，同时获得了多个技术专利。 Steve Jobs founded Apple and obtained multiple technology patents at the same time.
	Triples	(乔布斯, 创立, 苹果公司) (乔布斯, 获得, 技术专利) (Steve Jobs, founded, Apple) (Steve Jobs, obtained, technology patents)
EPO	Text	乔布斯是苹果公司的创始人，并担任过苹果公司的CEO。 Steve Jobs is the founder of Apple and has served as the CEO of Apple.
	Triples	(乔布斯, 创立, 苹果公司) (乔布斯, 领导, 苹果公司) (Steve Jobs, founded, Apple) (Steve Jobs, led, Apple)

Fig. 1. Examples of normal, single entity overlap (SEO), and entity pair overlap (EPO) scenarios.

Early RE methods typically followed a pipeline architecture, performing named entity recognition (NER) before RE. However, this separation weakened the connection between subtasks and was prone to error propagation [4], a problem exacerbated in Chinese because of the ambiguous entity boundaries and semantic diversity [5]. Joint extraction methods addressed this by unifying NER and RE, enhancing information sharing and reducing error propagation, thus improving performance [6]. However, establishing a balance between the modeling of global and local features remained a challenge. Global features offer contextual understanding from sentence-level semantics, while local ones focus on fine-grained semantics around entity pairs to detect relation triggers. Overemphasis on global features may obscure crucial local cues, hindering relation identification under vague boundaries [7], while excessive focus on local details may impair contextual reasoning [8]. Therefore, balancing between global and local feature modeling is essential to advancing Chinese RE, particularly in the extraction of overlapping triples.

Pre-trained language models have demonstrated strong contextual modeling capabilities in RE; however, they still struggle to capture both global and local features in complex Chinese texts. Contrastive learning can enhance semantic representation by distinguishing positive and negative samples [9], improving both global understanding and fine-grained relational discrimination. Therefore, we propose an RE framework that integrates contrastive learning and a global–local perception mechanism.

The following are the main contributions of this study:

1. To balance global and local feature modeling, we propose a global–local perception mechanism that enhances multi-level semantic representation and improves the recognition of overlapping triples.
2. Based on contrastive learning, we propose a contrastive optimization strategy that improves relational discrimination and semantic representation, substantially enhancing the extraction performance for overlapping triples.
3. Experimental results on two public Chinese datasets show that the proposed method outperforms existing ones, demonstrating its superiority in the extraction of overlapping relations.

2 Related Work

RE aims to identify semantic relationships between entities from unstructured text [10]. Traditional pipeline methods separate the NER and RE tasks [11], allowing for modular design but increasing susceptibility to error propagation, particularly in overlapping-relation scenarios. Some studies introduced syntactic structures [12] or BERT-based two-stage frameworks [13]; however, task separation often led to suboptimal performance.

Joint extraction methods have emerged to address these limitations, considering entity and relation extraction as a unified task and performing end-to-end joint prediction [14]. Zeng et al. [15] introduced a copy mechanism to alleviate BIO tag conflicts and further incorporated a bidirectional long short-term memory network (BiLSTM) and a graph convolutional network (GCN) for token-level feature modeling [16]. Nevertheless, their model suffered limitations in capturing global information in long texts. Wei et al. [17] designed a cascaded binary tagging framework, which performed well in extracting overlapping triples; however, the decoupled modeling of entities and relations hindered the full capture of their deep interactions. Wang et al. [18] proposed a one-stage extraction framework with a novel handshaking tagging scheme to improve entity pair matching efficiency. However, its capability to represent local features remains insufficient in complex contexts. Xu et al. [19] enhanced semantic representation by explicitly modeling the interaction between entities and relations; however, they lacked a hierarchical strategy to encode contextual information. Li et al. [20] combined pre-trained language models and multiturn conversational context to improve RE in dialogue scenarios; however, the performance tended to be unstable on non-conversational texts.

In summary, although existing methods have alleviated label conflicts and enhanced the modeling of overlapping triples, they often emphasize either global context or local relational detail in isolation.

In recent years, contrastive learning has been increasingly applied to RE. For instance, Gao et al. [21] optimized representation learning through contrastive

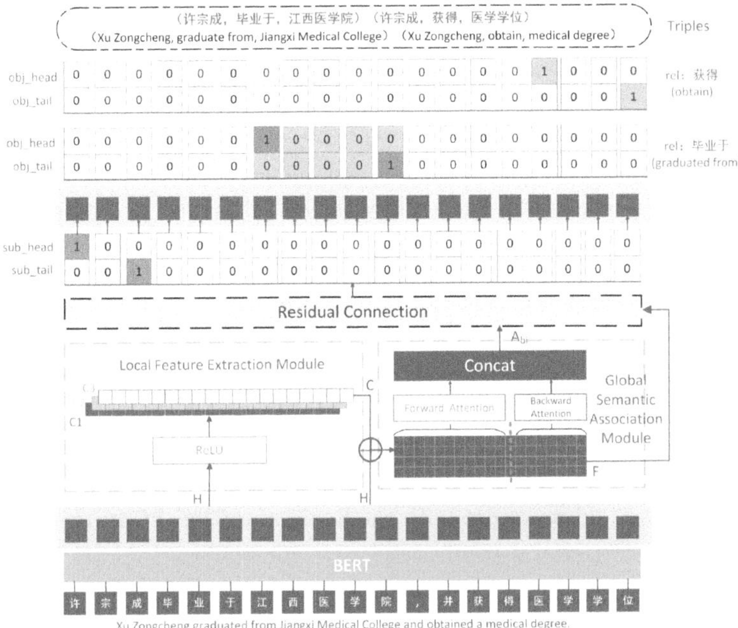

Fig. 2. Framework of our model.

objectives, while Wu et al. [22] introduced data augmentation to generate diverse sample pairs, improving semantic discrimination. However, most of these methods focused on English datasets. In the context of Chinese RE, particularly when dealing with overlapping triples, the application and effectiveness of contrastive learning have remained relatively underexplored.

3 Methodology

This study proposes a Chinese RE model that combines a global–local perception mechanism and contrastive learning, aiming to strike a balance between the capturing of global information and local features while effectively identifying overlapping triples. The framework of the model is shown in Fig. 2.

3.1 Task Definition

The objective of RE is to identify entities and their relationships in a sentence, representing them as triples (s, r, o), where s and o denote the subject and object entities, respectively, and r the relationship. Given an input sentence $X = [x_1, x_2, \ldots, x_n]$ containing n words and a set of m predefined relations $R = [r_1, r_2, \ldots, r_m]$, the RE task aims to predict the set of triples $T = \{(s, r, o) \mid s, o \in E, r \in R\}$, where E denotes the set of entities in sentence X.

3.2 Encoder

We used BERT as the text encoder to obtain contextual representations of the input sentences:

$$H = \mathrm{BERT}(x_1, x_2, \ldots, x_n) \tag{1}$$

Given an input sentence $X = [x_1, x_2, \ldots, x_n]$, where n denotes the length of the sentence, the sentence is first encoded by the BERT model to obtain the global context representation $H = [h_1, h_2, \ldots, h_n]$. Here, $H \in \mathbb{R}^{n \times d}$ denotes the hidden state matrix output by BERT, where d denotes the hidden layer dimension of BERT and h_i the encoding vector of the i-th word.

3.3 Global–Local Perception Mechanism

To better capture and integrate information from different semantic levels, we designed a global–local perception mechanism aimed at balancing between the modeling of global and local features. This mechanism enhanced the model's ability to perceive and represent both the fine-grained local features and overall contextual semantics, and it comprised two components:

Local Feature Extraction Module. Because of the prevalent ambiguity of entity boundaries in Chinese text, accurately identifying entity spans remains challenging. To address this issue, we introduced a local feature extraction module that applied convolution operations to BERT outputs, capturing phrase-level contextual cues and enhancing fine-grained representations while reducing interference from global semantics. Convolution was effective at modeling local semantic patterns, making it well-suited for distinguishing similar entity pairs with different relation types, thereby improving the model's ability to handle overlapping relations:

$$C_j = ReLU(W_c * H_{j+k-1} + b_c) \tag{2}$$

where $W_c \in \mathbb{R}^{k \times d \times m}$ denotes the convolution kernel, where k represents the kernel size and m the number of output channels. H_{j+k-1} denotes the local feature segment from position j to $j+k-1$, where $j \in [1, n-k+1]$ denotes the starting token position of the convolution window in the input sequence. b_c denotes the bias term, $*$ the convolution operation, and ReLU the activation function.

The resulting local feature matrix $C = [C_1, C_2, \ldots, C_n]$ enhanced local semantic representations, aiding both in complex relation recognition and more precise entity boundary identification, thus improving the overall extraction performance.

Global Semantic Association Module. For integration of global features H and local features C, we concatenated them along the feature dimension to obtain a fused feature matrix F, which contained both global and local semantics:

$$F = \mathrm{concat}(H, C) \tag{3}$$

The matrix F was then fed into the global semantic association module for further processing via a bidirectional attention mechanism, which captured global contextual information from both directions: forward attention captured dependencies from the beginning of the sentence to the current position, while backward attention captured context from the current position to the end. This resulted in a comprehensive sentence representation, enhancing the ability to recognize global semantic relations. The attention was computed as follows:

$$\text{Attention}(F) = \text{softmax}\left(\frac{(W^Q F)(W^K F)^T}{\sqrt{d_k}}\right)(W^V F) \tag{4}$$

Here, W_Q, W_K, and W_V denote projection matrices and d_k the dimension of key vectors. Attention weights were computed via the dot product of Q and K, followed by softmax normalization. This allowed for the fused feature matrix F to capture contextual dependencies through the forward (fwd) and backward (bwd) attentions, producing respective outputs:

$$A_{fwd/bwd} = \text{Attention}(Q_{fwd/bwd}, K_{fwd/bwd}, V_{fwd/bwd}) \tag{5}$$

The outputs of the forward (A_{fwd}) and backward (A_{bwd}) attention were concatenated to form a comprehensive feature representation, denoted by A_{bi}, which captured the complete semantic information from both the preceding and succeeding contexts:

$$A_{bi} = \text{concat}(A_{fwd}, A_{bwd}) \tag{6}$$

3.4 Linear Relation Extraction Layer

To enhance training stability and preserve original semantic features, we introduced a residual connection mechanism after the global–local perception module. In particular, a bidirectional attention output A_{bi} was added to the original input representation F to obtain the final fused representation:

$$A = A_{bi} + F \tag{7}$$

This design preserved original global and local features. The model then predicted the subject position by feeding the feature matrix A into two linear layers so as to estimate the head and tail positions:

$$\hat{y}_{sub\_head/tail} = \sigma(W_{sub\_head/tail} A + b_{sub\_head/tail}) \tag{8}$$

where $W_{sub\_head/tail}$ and $b_{sub\_head/tail}$ denote the weight matrices and biases for predicting the subject's head and tail positions, respectively. The sigmoid function $\sigma(\cdot)$ mapped the outputs to probabilities, from which the subject position was selected based on the highest score.

Thereafter, the model predicted the object and relation, starting by deriving the subject representation via a linear transformation; W_{sub} denotes a learnable projection matrix that transformed the subject representation to interact with the original feature matrix A:

$$F_{rel} = (W_{sub}A) \oplus A \tag{9}$$

The fused representation F_{rel} was then used to predict the object's head and tail positions:

$$\hat{y}_{obj\_head/tail} = \sigma(W_{obj\_head/tail}F_{rel} + b_{obj\_head/tail}) \tag{10}$$

Finally, the subject representation and contextual information were used to predict the relation type:

$$\hat{R} = softmax(W_{rel}F_{rel} + b_{rel}) \tag{11}$$

W_{rel} and b_{rel} denote the weight matrix and bias vector for relation classification, respectively. $\hat{R}$ denotes the probability distribution over relation types. This multilevel prediction framework enabled accurate extraction of entities and relations, particularly for overlapping triples in Chinese texts.

3.5 Contrastive Optimization Strategy

We proposed a contrastive optimization strategy to enhance the model's ability to distinguish relation types in Chinese RE tasks. For each input sentence, an anchor a (the feature representation of the current sentence), a positive sample p (with the same relation type as the anchor), and multiple negative samples n (from other samples in the same training batch) were constructed. The contrastive loss was defined as follows:

$$L_{contrastive} = -\log \frac{\exp(\text{sim}(a,p)/\tau)}{\exp(\text{sim}(a,p)/\tau) + \sum_{n=1}^{N} \exp(\text{sim}(a,n)/\tau)} \tag{12}$$

Here, $sim(x,y)$ denotes the cosine similarity between feature vectors and τ the temperature. This loss function pulled semantically similar samples closer while separating dissimilar ones. We introduced dynamic temperature adjustment, wherein τ started with a high value to focus on global features and decreased with time to emphasize fine-grained features:

$$\tau = \tau_0 \cdot e^{-\lambda t} \tag{13}$$

where τ_0 denotes the initial temperature, λ the decay factor, and t the training iteration. This strategy enabled the model to adapt at different training stages, optimizing RE performance.

To address class imbalance in RE, we adopted a focal loss for predicting the head and tail positions of entities and relations. By increasing the weight of hard examples and reducing the impact of easy ones, the focal loss improved performance on minority classes. The focal loss was defined as follows:

$$\mathcal{L}_{focal} = -\alpha(1-y)^{\gamma}\log(y) \tag{14}$$

where α balanced positive and negative samples, γ emphasized hard examples, and y denotes the predicted probability. This loss enhanced robustness in entity and relation boundary prediction. The total loss combined the head and tail prediction loss and the contrastive loss:

$$\mathcal{L} = \mathcal{L}_{sub\_head/tail} + \mathcal{L}_{obj\_head/tail} + \mathcal{L}_{contrastive} \tag{15}$$

4 Experiments

4.1 Datasets

Experiments were conducted on two widely used, public Chinese datasets, DuIE [23] and CMeIE [24]. DuIE, from the 2019 Baidu Information Extraction competition, covered diverse real-world domains such as news and entertainment. CMeIE comprises medical texts from clinical records and textbooks such as those on pediatrics and clinical pediatrics. Dataset statistics are shown in Table 1.

Table 1. Statistics of datasets.

Dataset	Training	Validation	Test	Entities	Relations
DuIE	153429	19178	19180	26	48
CMeIE	14339	3585	4482	11	44

We analyzed the overlap types in the datasets so as to assess the model's ability to handle overlapping triplets, as shown in Table 2:

Table 2. Details of overlapping pattens in datasets.

Category	DuIE			CMeIE		
	Training	Validation	Test	Training	Validation	Test
Normal	107058	13520	13366	5808	1520	1832
EPO	5763	671	699	214	43	64
SEO	40608	4987	5115	8317	2022	2586

4.2 Experimental Environment

The experimental environment was based on Python 3.8, with PyTorch 1.10.1 as the deep learning framework. The pretrained model BERT-Base-Chinese was employed, using the Adam optimizer with a learning rate of 1e-5. The maximum sentence lengths were set to 256 and 300 for the DuIE and CMeIE datasets, respectively, with a batch size of 8. The model was trained for 100 epochs.

4.3 Evaluation Metrics

We evaluated the model's performance across three standard metrics: precision, recall, and F1-score. Precision represents the proportion of correctly predicted positive instances among all instances predicted as positives, while recall measures the proportion of true positives among all actual positives. F1-score, which is the harmonic mean of precision and recall, provided a balanced assessment of the model's effectiveness.

4.4 Experimental Results and Analysis

We compared our model with the following five representative methods: (1) CopyRE [15]: a Seq2Seq model that used a copy mechanism to generate triplets. (2) GraphRel [16]: a GCN-based method that constructs text graphs for relation reasoning. (3) DERP [25]: a two-stage framework with head–tail entity recognition and a triple prediction module for more accurate relational triple extraction.(4) PURE [13]: a pipeline framework with dual encoders for separate entity recognition and relation classification. (5) ChatIE [20]: a dialogue-based approach leveraging pretrained models and multi-turn context to enhance extraction in conversational scenarios.(6) CasRel [17]: A cascade framework that first extracted subjects and then identified objects by relation type. It effectively handled overlapping relations and complex entity dependencies.

Table 3. Results of baseline models on the DuIE and CMeIE datasets. Bold marks the best result, while underline marks the second best result.

Model	DuIE			CMeIE		
	Precision	Recall	F1	Precision	Recall	F1
CopyRE	60.0	54.4	57.1	45.8	36.8	40.8
GraphRel	64.9	66.5	61.2	50.3	36.4	42.2
DEPR	71.1	65.4	68.1	47.5	46.1	46.8
PURE	72.8	64.3	68.3	51.9	41.3	46.0
ChatIE	<u>74.6</u>	67.5	70.8	–	–	–
CasRel	72.4	**82.1**	<u>76.9</u>	<u>52.9</u>	<u>49.5</u>	<u>51.1</u>
Ours	**81.9**	<u>78.0</u>	**79.9**	**65.5**	**51.2**	**57.5**

Table 3 compares the performances of different methods on the DuIE and CMeIE datasets. Since ChatIE was designed for conversational datasets, its results on CMeIE are not available. Our model achieved an F1 score of 79.9% on the DuIE dataset, improving by 3% over the best-performing baseline, CasRel. On the specialized domain dataset CMeIE, which contained a high proportion of semantically similar relation types, our model improved the F1 score by 6.4% over CasRel. This improvement was primarily attributed to the application of the global–local perception mechanism, which captured fine-grained features through the local feature perception module and modeled contextual information more effectively through the bidirectional attention mechanism. However, CasRel solely relied on local features and lacked access to the global context, which led to more confusion in entity boundary recognition. Meanwhile, the contrastive optimization strategy enhances the model's ability to distinguish semantically similar relations and complements the global–local perception mechanism, thereby further improving performance in complex domain-specific RE tasks.

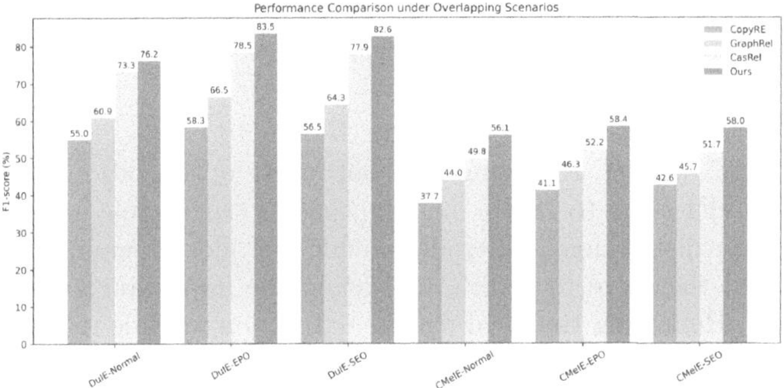

Fig. 3. F1 scores of different overlapping pattern.

Figure 3 shows the performances of different models on overlapping triplets in the DuIE and CMeIE datasets. Our method outperformed other comparison models in all the Normal, SEO, and EPO scenarios. In the Normal scenario, our method was able to accurately capture relationship features even when there were no overlaps. In the EPO and SEO scenarios, although CasRel performed well in local feature extraction, its ability to model global context and ability to distinguish overlapping relations were relatively poor, leading to suboptimal performance when handling multiple triplets. Furthermore, GraphRel and CopyRE, because of their reliance on graph structures or copy mechanisms, failed to effectively handle overlapping triplets and complex semantic structures, often resulting in misclassification. However, our method more effectively managed these complex situations through the global–local perception mechanism and the contrastive optimization strategy.

4.5 Ablation Study

Table 4 presents the results of ablation experiments, highlighting the importance of each module in the model's performance.

Table 4. Results of ablation study

Dataset	Strategy	Precision	Recall	F1
DuIE	Origin	**81.9**	78.0	**79.9**
	w/o Contrastive optimization Strategy	73.8	**84.3**	78.6
	w/o Global Semantic Relation Module	75.1	80.8	77.8
	w/o Local Feature Extraction Module	72.2	83.2	77.3
CMeIE	Origin	**65.5**	**51.2**	**57.5**
	w/o Contrastive optimization Strategy	64.7	50.3	56.6
	w/o Global Semantic Relation Module	60.3	49.3	54.2
	w/o Local Feature Extraction Module	57.2	49.3	53.0

On the DuIE dataset, removing the contrastive optimization strategy module reduced the F1 score to 78.6%, highlighting its crucial role in enhancing the model's relation distinction ability. Removing the global semantic association module lowered the score to 77.8%, indicating the importance of attending to global context. Removing the local feature extraction module reduced the score to 77.3%, demonstrating its contribution to fine-grained feature representation. On the CMeIE dataset, removing the contrastive optimization strategy module reduced the F1 score to 56.6%. Further removing the bidirectional global attention and local feature extraction modules reduced the score to 54.2% and 53.0%, respectively, confirming the key role of the proposed method in specialized domain RE.

5 Conclusion

This study presents a novel Chinese RE method that integrates a global–local perception mechanism and contrastive learning, significantly enhancing the extraction of overlapping triplets. The global–local perception mechanism extracts fine-grained features through local convolution and captures global context via a semantic association mechanism, ensuring a balance between the capturing of local and global features. The contrastive optimization strategy enables better differentiation of relation types, thereby effectively improving the model's discriminative ability and its ability to recognize overlapping triplets. Experimental results demonstrated superior performance of the model on multiple datasets. However, challenges remain in handling domain-specific terminology and complex relationships, particularly in low-resource settings. Accordingly, future research is expected focus on leveraging large language models, combined with few-shot learning and data augmentation techniques, to enhance performance in low-resource scenarios.

Acknowledgments. This work was supported by the Jiangsu Educational Science Planning Project (B-b/2024/01/47), and the Jiangsu Graduate Scientific Research and Practice Innovation Program (SJCX2_1522).

References

1. Zhao, X., Deng, Y., Yang, M., et al.: A comprehensive survey on relation extraction: recent advances and new frontiers. ACM Comput. Surv. **56**(11), 1–39 (2024)
2. Wang, P., Xie, J., Chen, X., et al.: PasCore: a Chinese overlapping relation extraction model based on global pointer annotation strategy. In: Proceedings of the 32th International Joint Conference on Artificial Intelligence Main Track, pp. 5215–5223 (2023)
3. He, M., Bai, Y., Wei, D.: CJE-PCHF: Chinese joint entity and relation extraction model based on progressive contrastive learning and heterogeneous feature fusion. Appl. Sci. **14**(20), 9256 (2024)
4. Tuo, M., Yang, W., Wei, F., et al.: A novel Chinese overlapping entity relation extraction model using word-label based on cascade binary tagging. Electronics **12**(4), 1013 (2023)
5. Cheng, J., Zhang, T., Zhang, S., et al.: A cascade dual-decoder model for joint entity and relation extraction. IEEE Trans. Emerg. Top. Comput. Intell. **9**(2), 1130–1142 (2024)
6. Zheng, S., Wang, F., Bao, H., et al.: Joint extraction of entities and relations based on a novel tagging scheme. arXiv preprint arXiv:1706.05075 (2017)
7. Li, Z., Song, M., Zhu, Y., Zhang, L. Chinese nested named entity recognition based on boundary prompt. In: Yuan, L., Yang, S., Li, R., Kanoulas, E., Zhao, X. (eds.) WISA 2023. LNCS, vol. 14094, pp. 331–343. Springer, Singapore (2023). https://doi.org/10.1007/978-981-99-6222-8_28
8. Dai, D., Xiao, X., Lyu, Y., et al.: Joint extraction of entities and overlapping relations using position-attentive sequence labeling. In: Proceedings of the AAAI Conference on Artificial Intelligence, vol. 33(1), pp. 6300–6308 (2019)
9. Ge, S., Mishra, S., Li, C.L., et al.: Robust contrastive learning using negative samples with diminished semantics. Adv. Neural. Inf. Process. Syst. **34**, 27356–27368 (2021)
10. Pawar, S., Palshikar, G.K., Bhattacharyya, P.: Relation extraction: a survey. arXiv preprint arXiv:1712.05191 (2017)
11. Chan, Y.S., Roth, D.: Exploiting syntactico-semantic structures for relation extraction. In: Proceedings of the 49th Annual Meeting of the Association for Computational Linguistics: Human Language Technologies, pp. 551–560 (2011)
12. Xu, K., Feng, Y., Huang, S., Zhao, D.: Semantic relation classification via convolutional neural networks with simple negative sampling. In: Proceedings of the 2015 Conference on Empirical Methods in Natural Language Processing, pp. 536–540 (2015)
13. Zhong, Z., Chen, D.: A frustratingly easy approach for entity and relation extraction. arXiv preprint arXiv:2010.12812 (2020)
14. Sui, D., Zeng, X., Chen, Y., et al.: Joint entity and relation extraction with set prediction networks. IEEE Trans. Neural Netw. Learn. Syst. (2023)
15. Zeng, X., Zeng, D., He, S., et al.: Extracting relational facts by an end-to-end neural model with copy mechanism. In: Proceedings of the 56th Annual Meeting of the Association for Computational Linguistics, pp. 506–514 (2018)
16. Zeng, X., He, S., Zeng, D., et al.: Learning the extraction order of multiple relational facts in a sentence with reinforcement learning. In: Proceedings of the 57th Conference on Empirical Methods in Natural Language Processing and the 9th International Joint Conference on Natural Language Processing, pp. 367–377 (2019)

17. Wei, Z., Su, J., Wang, Y., Tian, Y., Chang, Y.: A novel cascade binary tagging framework for relational triple extraction. In: Proceedings of the 58th Annual Meeting of the Association for Computational Linguistics, pp. 1476–1488 (2020)
18. Wang, Y., Yu, B., Zhang, Y., Liu, T., Zhu, H., Sun, L.: TPLinker: single-stage joint extraction of entities and relations through token pair linking. In: Proceedings of the 28th International Conference on Computational Linguistics, pp. 1572–1582 (2020)
19. Xu, B., et al.: EMREL: joint representation of entities and embedded relations for multi-triple extraction. In: Proceedings of the 2022 Conference of the North American Chapter of the Association for Computational Linguistics: Human Language Technologies, pp. 659–665 (2022)
20. Li, G., Wang, P., Ke, W.: Revisiting large language models as zero-shot relation extractors. arXiv preprint arXiv:2310.05028 (2023)
21. Gao, T., Yao, X., Chen, D.: SimCSE: simple contrastive learning of sentence embeddings. arXiv preprint arXiv:2104.08821 (2021)
22. Wu, Z., Wang, S., Gu, J., et al.: CLEAR: contrastive learning for sentence representation. arXiv preprint arXiv:2012.15466 (2020)
23. Li, S., He, W., Shi, Y., Jiang, W., Liang, H., Jiang, Y., Zhang, Y., Lyu, Y., Zhu, Y.: DuIE: a large-scale Chinese dataset for information extraction. In: Tang, J., Kan, M.-Y., Zhao, D., Li, S., Zan, H. (eds.) NLPCC 2019. LNCS (LNAI), vol. 11839, pp. 791–800. Springer, Cham (2019). https://doi.org/10.1007/978-3-030-32236-6_72
24. Guan, T., Zan, H., Zhou, X., Xu, H., Zhang, K.: CMeIE: construction and evaluation of Chinese medical information extraction dataset. In: Zhu, X., Zhang, M., Hong, Yu., He, R. (eds.) NLPCC 2020. LNCS (LNAI), vol. 12430, pp. 270–282. Springer, Cham (2020). https://doi.org/10.1007/978-3-030-60450-9_22
25. Xiao, Y., Chen, G., Du, C., et al.: A study on double-headed entities and relations prediction framework for joint triple extraction. Mathematics **11**(22), 4583 (2023)

A Survey on Zeroth-Order Optimization for Machine Learning

Liting Lin, Hansong Ma, Junxiao Wang, and Shiyu Yang(✉)

Guangzhou University, Guangzhou, China
{litinglin,hansongma}@e.gzhu.edu.cn, {junxiao.wang,syyang}@gzhu.edu.cn

Abstract. Zeroth-order (ZO) optimization has become a popular technique for solving machine learning problems, especially when first-order (FO) information is difficult or impossible to obtain. However, there remains a lack of extensive surveys covering recent advances and thorough analysis of ZO optimization for machine learning. In this paper, we present a comprehensive survey on ZO optimization, aiming to summarize the cutting-edge research and broaden the horizons for different domains. Firstly, we introduce the definition of ZO optimization by characterizing its inherent properties with respect to machine learning problems. In particular, we dig out some critical ingredients from the properties including no gradient dependence, dimension sensitivity, noise robustness, and implicit smoothing. Second, we summarize emerging strategies by combining the properties of ZO optimization and specific machine learning problems like adversarial examples, fine-tuning of LLMs, federated learning and reinforcement learning. Finally, we discuss some promising directions and open problems for further research.

Keywords: Zeroth-order optimization · Machine Learning

1 Introduction

Zeroth-order (ZO) optimization, also known as gradient-free optimization, has emerged as a powerful tool in modern machine learning, especially in scenarios where gradient information is inaccessible, unreliable, or computationally expensive [10,38]. Unlike traditional first-order methods that require analytical or backpropagated gradients, ZO algorithms rely solely on function value queries, making them well suited for black-box, non-differentiable, and simulation-based tasks. The appeal of ZO methods stems from four intrinsic properties. They are gradient-independent, allowing application to models without explicit derivatives; robust to noise and stochasticity in function evaluations; capable of implicitly smoothing rugged loss landscapes via random perturbations; and naturally extensible to high-dimensional or resource-constrained settings. These advantages make ZO optimization particularly useful for complex, privacy-sensitive, or constrained environments where gradient-based methods struggle.

S. Yang et al. (Eds.): WISA 2025, LNCS 16299, pp. 481–497, 2026.
https://doi.org/10.1007/978-981-95-6426-2_43

Recent progress highlights ZO optimization's versatility across key domains. In adversarial learning, ZO methods generate black-box perturbations without gradient access [25]. In fine-tuning large language models (LLMs), they enable forward-only, memory-efficient adaptation compatible with edge deployment [34]. In federated learning (FL), ZO techniques reduce communication overhead and increase robustness to client heterogeneity [19]. In reinforcement learning (RL), they support policy optimization via simulation feedback, without requiring differentiable policies [60]. Despite promising advances, challenges remain. Theoretical guarantees often depend on restrictive assumptions (e.g., convexity, smoothness) that rarely hold in practice. In practice, objectives can be non-convex, non-smooth, or subject to biased and heavy-tailed noise. Smoothing parameter selection and adaptation are still poorly understood, affecting convergence and stability. Algorithmically, ZO methods struggle to scale in high dimensions, and existing task-specific heuristics (e.g., sparsity, quantization) lack general applicability. Achieving dimension-agnostic, sublinear query complexity under noisy conditions remains an open problem.

To provide a structured understanding of this growing field, this survey presents a comprehensive overview of ZO optimization across four representative areas: adversarial examples, LLM fine-tuning, federated learning, and reinforcement learning. As illustrated in Fig. 1, we organize the literature by mapping domain-specific challenges to methodological strategies such as transfer-based estimation and coordinate-wise perturbations. We hope this survey serves as both a conceptual reference and a practical guide, offering a unified perspective that supports future research into more robust, scalable, and theoretically sound gradient-free optimization methods.

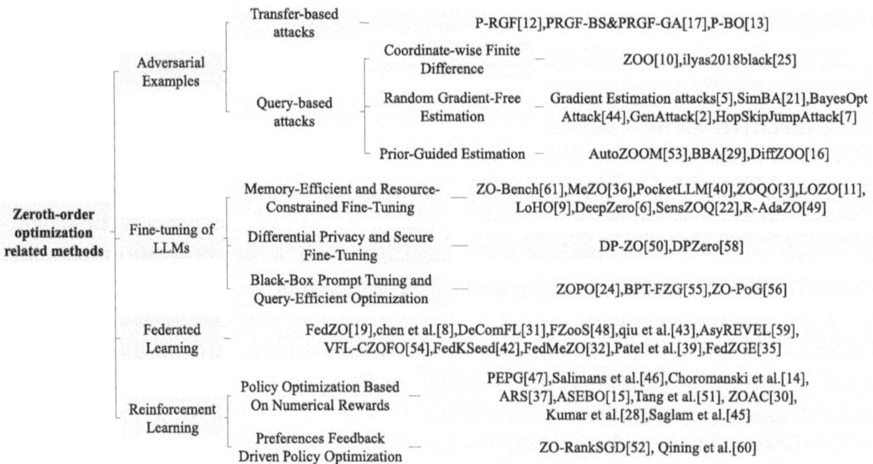

Fig. 1. The proposed taxonomy and structured overview of existing research on ZO optimization in machine learning.

2 Definition of Zeroth-Order Optimization

ZO optimization offers a gradient-free alternative to first-order (FO) methods by estimating descent directions using only function value queries. This property makes ZO optimization particularly useful for black-box optimization, privacy-preserving learning, and scenarios where gradient computation is impractical.

The most commonly used estimation technique is the Randomized Gradient Estimator (RGE), originally proposed by Nesterov and Spokoiny [38]. This method approximates the FO gradient using finite differences of function values along randomly chosen direction vectors and has been used widely in the ZO optimization literature. Given a loss function $L(\cdot)$ and a model with parameter $\theta \in \mathbb{R}^d$, the RGE employed by studies like [36], referred to as $\hat{\nabla} L(\theta)$, is to approximate $\nabla L(\theta)$ and is expressed using central difference:

$$\hat{\nabla} L(\theta) = gz \in \mathbb{R}^d, \tag{1}$$

$$g = \frac{L(\theta + \epsilon z) - L(\theta - \epsilon z)}{2\epsilon} \in \mathbb{R}^1, \tag{2}$$

where z is a random direction vector drawn from the standard Gaussian distribution $\mathcal{N}(0, \mathrm{I})$, and $\epsilon > 0$ is a small perturbation step size, also known as the smoothing parameter. g represents the **projected gradient** computed using the model's dual-forward passes. Notably, $g \in \mathbb{R}^1$ is just a scalar value and requires minimal memory space. $\hat{\nabla} L(\theta)$ is a d-dimensional vector. This method is grounded in the theory of directional derivatives [18], and provides an unbiased approximation of the true gradient $\nabla f(x)$ as $\epsilon \to 0$ [61].

RGE has been widely applied in ZO Optimization algorithms, such as ZO Stochastic Gradient Descent (ZO-SGD), where the parameter update rule is given by:

$$\theta = \theta - \eta \cdot g \cdot z, \tag{3}$$

where η is the learning rate. It is important to note that the variable z in Eq. 3 should be identical to the z in Eq. 2.

2.1 No Gradient Dependence

In the standard ZO optimization setting, the optimizer can only access a ZO oracle, defined as $\mathcal{O}_0(x) = f(x)$, and cannot access a FO oracle $\mathcal{O}_1(x) = \nabla f(x)$. To estimate a descent direction, two common approaches are: Finite difference approximation along coordinate directions:

$$\frac{f(x + he_i) - f(x)}{h}, \tag{4}$$

where $h > 0$ is a small finite difference step size and e_i is the i-th standard basis vector. Random directional gradient estimation:

$$\frac{f(x + \mu u) - f(x)}{\mu} \cdot u, \tag{5}$$

where $\mu > 0$ is a smoothing parameter and u is a random unit vector sampled uniformly from the unit sphere. These methods construct approximate descent directions without directly accessing $\nabla f(x)$.

Under mild assumptions such as Lipschitz continuity, ZO optimization can achieve provable convergence given sufficient function evaluations [20]. This makes it well-suited for black-box scenarios—including adversarial attacks, RL, hyperparameter optimization, and differentially private training—where reliance on function-only queries not only reduces memory overhead but also aligns with privacy-preserving requirements [23].

2.2 Dimension Sensitivity

The effectiveness of ZO optimization critically depends on the dimensionality of the problem. For instance, the commonly used two-point random estimator:

$$\hat{\nabla} f(x) = \frac{f(x + \delta u) - f(x - \delta u)}{2\delta} \cdot u, \tag{6}$$

where $u \in \mathbb{R}^d$ is a random vector uniformly sampled from the unit sphere, and $\delta > 0$ is a smoothing parameter. Berahas et al. [4]showed that to ensure relative gradient estimation error:

$$\|\hat{\nabla} f(x) - \nabla f(x)\| \leq \theta \|\nabla f(x)\|, \tag{7}$$

the required number of samples is

$$N = O\left(\frac{d}{\theta^2}\right), \tag{8}$$

illustrating ZO optimization's sensitivity to high-dimensional settings. This motivates researchers to design structure-aware algorithms (e.g., sparsity or low-rank modeling) or apply dimensionality reduction techniques to alleviate the computational burden caused by high dimensionality. While random estimators reduce dimensional dependence, they may introduce high variance. Thus, there exists a trade-off between scalability and estimation accuracy.

2.3 Noise Robustness

In practical scenarios, function evaluations often contain noise, and ZO gradient estimations must remain effective in the presence of such noise. Assume the function evaluations are corrupted by bounded noise:

$$\tilde{f}(x) = f(x) + \xi, \quad |\xi| \leq \varepsilon_f, \tag{9}$$

where ξ denotes the noise in the function value and satisfies the bounded noise assumption with noise level ε_f. Berahas et al. [4] derived an upper bound on the gradient estimation error that depends on the noise level ε_f and the smoothing parameter δ:

$$\|\hat{\nabla} f(x) - \nabla f(x)\| \leq \mathcal{O}\left(\frac{\varepsilon_f}{\delta} + \delta L\right), \tag{10}$$

where L is the Lipschitz constant of the gradient. By appropriately choosing δ, a balance between noise amplification and smoothing bias can be achieved. Work by Akhavan and Tsybakov [1] further establishes convergence guarantees under sub-Gaussian or bounded noise settings.

2.4 Implicit Smoothing

ZO optimization algorithms do not directly optimize the original objective function f, but instead optimize a smoothed version f_δ obtained by convolving f with a smoothing kernel. This smoothing process not only enhances the stability of gradient estimation but also effectively reduces sensitivity to local irregularities and noise in the function, thereby improving the robustness of the algorithm. Formally, let u be a random vector drawn from the standard multivariate normal distribution $\mathcal{N}(0, I)$. The smoothed function is defined as

$$f_\delta(x) = \mathbb{E}_{u \sim \mathcal{N}(0,I)}\left[f(x + \delta u)\right], \tag{11}$$

where $\delta > 0$ is the smoothing parameter controlling the scale of smoothing. This expression essentially corresponds to the Gaussian convolution of the function f over a ball of radius δ. The smoothed function f_δ possesses better smoothness properties, and its gradient can be expressed as

$$\nabla f_\delta(x) = \frac{1}{\delta}\mathbb{E}_u\left[f(x + \delta u) \cdot u\right], \tag{12}$$

As noted in Liu et al. [33], implicit smoothing improves robustness to local noise and non-smoothness but introduces bias that decreases as $\delta \to 0$. Thus, selecting δ is crucial to balancing estimation variance and bias. Implicit smoothing improves convergence stability but introduces bias. Careful parameter tuning is essential to balance variance reduction with estimation accuracy.

3 Zeroth-Order Optimization for Adversarial Attacks

In black-box adversarial attacks, where gradients of the target model are inaccessible and only output queries are allowed, traditional gradient-based methods become inapplicable. ZO optimization overcomes this limitation by estimating gradient directions from function value differences, without requiring access to model internals. Beyond its gradient-free nature, ZO optimization is robust to query noise, employing random sampling and smoothing to stabilize gradient estimates and enhance attack success rates. Additionally, by optimizing a smoothed surrogate of the objective, ZO methods can more effectively navigate complex, non-smooth loss landscapes. Table 1 lists the relevant approaches regarding ZO optimization in adversarial samples.

Table 1. Overview of ZO optimization methods for adversarial attacks.

Attack Category		Method	Key Features
Transfer-based Attacks		P-RGF [12]	Guided search, improved attack efficiency
		PRGF-BS&PRGF-GA [17]	Adaptive fusion mechanism, enhanced adaptability
		P-BO [13]	Adaptive confidence adjustment, enhanced query efficiency
Query-based Attacks	Coordinate-wise Finite Difference	ZOO [10]	ZO gradient estimation using function value differences, robust to noise
		ilyas2018black [25]	Gradient estimation in limited-query and label-only settings
	Random Gradient-Free Estimation	Gradient Estimation attacks [5]	Uses random perturbations for gradient-free optimization, improves query efficiency
		SimBA [21]	High query efficiency, output-based update directions
		BayesOpt Attack [44]	Bayesian optimization, adaptive sampling
		GenAttack [2]	Uses genetic algorithms, avoids gradient noise
		HopSkipJumpAttack [7]	Binary feedback, geometric steps, binary search
	Prior-Guided Estimation	AutoZOOM [53]	Dimensionality reduction, query optimization
		BBA [29]	Block-based Bayesian optimization, improves query efficiency
		DiffZOO [16]	Pure query-based attack for generative diffusion models, effective prompt optimization

3.1 Transfer-Based Attacks And Hybrid Attacks

Early transfer-based attacks relied on training surrogate models locally to generate adversarial examples that directly attack the target model. Although these methods can partially succeed, they are limited by the mismatch between the decision boundaries of the surrogate and the target models, leading to unstable attack success rates and query efficiency. To address this issue, researchers have progressively developed hybrid methods that integrate surrogate information with ZO optimization. For instance, the P-RGF method [12] incorporates the gradient of the surrogate model as a prior for random gradient estimation, enabling guided search and improving attack efficiency while retaining exploration capability. Further, PRGF-BS and PRGF-GA [17] introduced an optimal fusion mechanism based on theoretical derivations, allowing for a more adaptive balance between surrogate priors and query feedback. Subsequently, the P-BO method [13] embedded the global loss information of the surrogate model into the mean function of the Gaussian process in Bayesian optimization, proposing adaptive confidence weighting to automatically assess the reliability of the surrogate prior, thereby enhancing the robustness and stability of the attacks. These transfer-based approaches improve efficiency but are fragile when the surrogate poorly matches the target model.

3.2 Query-Based Attacks

When surrogate models are unavailable or inaccurate, attackers rely on the target model's output. Query-based attacks address this by efficiently approaching the decision boundary in black-box settings, especially in high-dimensional spaces where minimizing query numbers is crucial.

Coordinate-wise Finite Difference. ZO optimization initially estimates gradients along each input dimension using coordinate-wise finite difference estimation. Chen et al. [10] introduced this approach in black-box settings, enabling gradient-based attacks like Carlini & Wagner (C&W) using central difference formulas. However, this method incurs linear query costs with input dimensionality, making it impractical for high-dimensional inputs. To address this, Ilyas et al. [25] refined the gradient estimation procedure, reducing the query dimensions and broadening ZO attack applicability under limited feedback.

Random Gradient-Free Estimation. To mitigate the high query cost in coordinate-wise methods, researchers adopted random direction perturbations. Bhagoji et al. [5] used Gaussian random vectors for central difference estimation, reducing query dependence on input dimensionality. Guo et al. [21] further structured this with sequential perturbations along fixed orthogonal bases, improving query efficiency. Ru et al. [44] enhanced this by applying Bayesian optimization to sample high-loss regions, boosting efficiency. Alzantot et al. [2] bypassed gradient estimation entirely, using genetic algorithms to evolve adversarial examples, balancing low query cost with high success rates. Chen et al. [7] also demonstrated random gradient-free estimation by using binary feedback alone, achieving efficient convergence with decision-based attacks.

Prior-Guided Estimation. Researchers have also incorporated structural priors, such as autoencoder-based dimension reduction, to further enhance efficiency. Tu et al. [53] used autoencoders to reduce data dimensionality, significantly lowering query numbers. Lee et al. [29] extended ZO attacks to discrete sequential data, using block decomposition and Bayesian optimization for adaptive query allocation. Dang et al. [16] adapted ZO attacks for the prompt space in generative models, improving success rates in multimodal security testing, such as text-to-image diffusion models.

4 Zeroth-Order Optimization for LLM Adaptation

As the parameter scale of large language models (LLMs) continues to increase, traditional first-order optimization methods (such as Adam or SGD) face significant memory and scalability challenges in fine-tuning or personalized adaptation. This is particularly true in scenarios such as edge devices and personalized tuning of large-scale models [26], where explicit gradient backpropagation (BP) requires substantial memory, leading to high memory overhead. In contrast, ZO optimization, which relies solely on forward function evaluations, is inherently suited for memory-constrained and black-box settings, offering key advantages such as no gradient dependence, scalability to high-dimensional parameter spaces, noise robustness, and implicit smoothing, making it particularly well-suited for efficient adaptation and optimization of LLMs.

4.1 Memory-Efficient and Resource-Constrained Fine-Tuning

To systematically evaluate the performance of various ZO Optimization methods in fine-tuning LLMs, Zhang et al. [61] introduced the first standardized benchmark framework for evaluating ZO optimization methods in LLM fine-tuning, covering a wide range of models and tasks. Their results highlighted the advantages of ZO methods in memory-constrained environments. To further reduce memory usage, Malladi et al. [36] proposed MeZO, a memory-efficient ZO optimizer based on SPSA, which uses only two forward passes for parameter updates, eliminating the need for backpropagation and making it suitable for models with up to 30 billion parameters. Building on this, Peng et al. [40] introduced PocketLLM, a ZO-based on-device fine-tuning method designed for personalized adaptation on mobile devices. It offers a favorable balance between privacy and memory efficiency, making it ideal for edge deployment. Recent work has also integrated ZO methods with model compression techniques. As shown in Fig. 2, ZO has been combined with methods like low-rank decomposition and model pruning to further reduce memory consumption. For example, Bar et al. [3] proposed a quantization-aware perturbation strategy, enhancing optimization stability and efficiency under low-precision settings (e.g., INT8, INT4). Chen et al. [11] incorporated low-rank structures into the ZO framework to reduce parameter updates and computational complexity, making it suitable for large-scale LLM fine-tuning. Chen et al. [9] also introduced LoHO, a hybrid approach that accelerates convergence by integrating FO and ZO optimization, while maintaining memory efficiency. Similarly, DeepZero [6], by Chen et al., improves ZO scalability through model pruning and feature reuse, overcoming traditional ZO's performance bottleneck in high-dimensional spaces. Guo et al. [22] introduced SensZOQ, combining quantization with static sparsity. By updating only the top 0.1% most sensitive parameters, SensZOQ significantly reduces computational and memory costs in LLM fine-tuning. To further improve convergence speed and stability, Shu et al. [49] proposed R-AdaZO, which refines momentum and variance estimation in adaptive ZO optimization, outperforming traditional ZO-AdaMM. As LLM parameter scales grow, traditional FO methods like Adam and SGD face significant challenges in memory and scalability, particularly in scenarios like edge devices or personalized tuning [26]. ZO optimization, relying solely on forward evaluations, is well-suited for memory-constrained and black-box settings, offering advantages like no gradient dependence, scalability to high-dimensional spaces, noise robustness, and implicit smoothing, making it ideal for efficient LLM adaptation and optimization.

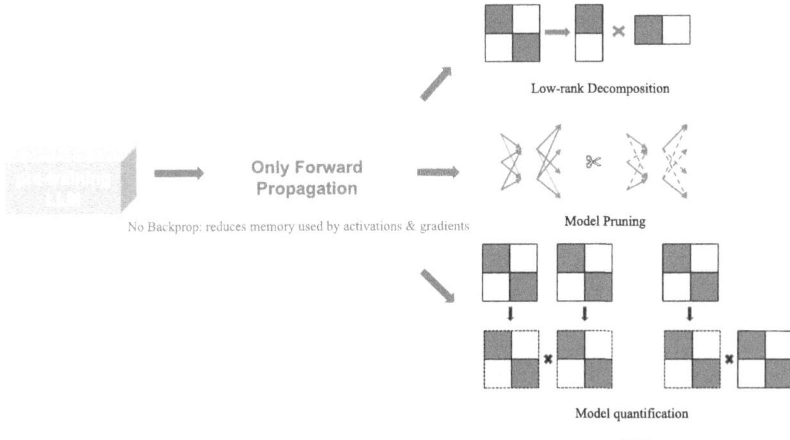

Fig. 2. ZO optimization fusion for memory-efficient fine-tuning of LLMs.

4.2 Differential Private and Secure Fine-Tuning

As privacy protection and security issues become increasingly important, the combination of differential privacy (DP) techniques with ZO optimization has emerged as an effective approach to protect data privacy. By incorporating differential privacy mechanisms into ZO optimization, researchers can fine-tune LLMs while ensuring privacy. Tang et al. [50] proposed a differentially private ZO optimization method, DP-ZO, which introduces noise solely to scalar updates, thereby avoiding the high memory overhead associated with traditional gradient clipping techniques. DP-ZO enables privacy-preserving fine-tuning of LLMs under memory-constrained settings, offering privacy guarantees comparable to those of conventional DP-SGD while achieving higher computational efficiency. Building on this approach, Zhang et al. [58] introduced DP-Zero, a refinement that eliminates the need for backpropagation. By leveraging ZO optimization to enforce differential privacy, DP-Zero effectively mitigates the risk of gradient leakage and further improves both memory and computational efficiency compared to traditional differentially private training methods.

4.3 Black-Box Prompt Tuning and Query-Efficient Optimization

In black-box prompt optimization for language models, the inability to access gradient information poses a significant challenge, particularly under limited query budgets. ZO optimization, which does not rely on gradient information, has emerged as an effective solution to this problem. Hu et al. [24] proposed the ZOPO method, which integrates Neural Tangent Kernels (NTKs) with Gaussian Processes to perform efficient localized search, significantly reducing query costs in black-box prompt tuning. Zhang et al. [55] combined FO and ZO gradients to propose a new minimax optimization method, which improves performance on imbalanced data. This method is particularly suited for black-box models where gradient information cannot be directly obtained. By utilizing effective gradient estimation, the method significantly enhances the model's performance. Additionally, Zhang et al. [56] proposed a collaborative discrete-continuous prompt optimization approach for black-box settings, which leverages ZO techniques to jointly optimize discrete tokens and continuous embeddings. This hybrid strategy further enhances the query efficiency and optimization performance in black-box prompt learning.

5 Zeroth-Order Optimization for Federated Learning

Federated Learning (FL), a distributed machine learning paradigm, faces key challenges such as inaccessible gradients, high communication overhead, and client heterogeneity due to the decentralized nature of private data. To mitigate these issues, ZO Optimization has been integrated into FL, providing a gradient-free, communication-efficient, and privacy-preserving optimization framework that has significantly advanced the field.

The earliest introduction of ZO optimization in FL was by Fang et al. [19], who developed the FedZO algorithm. By combining local updates with AirComp aggregation, FedZO enables communication-efficient federated training over wireless channels without the need for gradient access. To address theoretical limitations, Chen et al. [8] analyzed generalization and optimization error bounds, strengthening FedZO's foundation for asynchronous and heterogeneous environments. As high-dimensional models exacerbate communication overhead, Li et al. [31] proposed DeComFL, which, based on a low-rank assumption, achieved dimension-free communication protocols. To tackle client heterogeneity, Shu et al. [48] introduced the FZooS framework, which constructs surrogate gradients using historical trajectory information, leveraging random Fourier feature approximations and an adaptive gradient correction mechanism to enhance convergence and communication efficiency in heterogeneous settings.

For non-smooth and hierarchical problems, Qiu et al. [43] proposed a randomized smoothing framework using Clarke subdifferentials, addressing the challenges of non-differentiable settings. In Vertical Federated Learning (VFL), Zhang et al. [59] introduced AsyREVEL, combining asynchronous communication with ZO optimization to enhance privacy and scalability, while Wang et al. [54] developed a hybrid framework (VFL-CZOFO) combining ZO in key layers with FO optimization elsewhere, ensuring differential privacy and efficient convergence. For large-scale black-box model tuning, Qin et al. [42] and Lin et al. [32] provided convergence and communication complexity analyses for full-parameter ZO federated tuning, offering theoretical support for privacy-preserving fine-tuning. Patel et al. [39] extended ZO optimization to high-dimensional bandit settings, demonstrating tighter regret bounds through federated collaboration. Finally, Ma et al. [35] highlighted ZO optimization's potential for privacy preservation in data-free, fully black-box environments, extending its applicability in extreme scenarios. While ZO methods reduce communication and preserve privacy, they often converge more slowly than FO-based federated approaches.

6 Zeroth-Order Optimization in Reinforcement Learning

The goal of reinforcement learning (RL) is to allow an agent to learn an optimal strategy through interaction with the environment to maximize long-term cumulative rewards. Mainstream RL algorithms often explore in the action space and use the information of the FO gradient of the policy function to update parameters to maximize rewards. However, FO methods often show high variance, sensitivity to local optimality, or require differentiable policies in long-term tasks. In order to alleviate these problems, policy search methods based on ZO optimization have emerged. Such methods often perform random searches in parameter space and estimate the gradient of policy parameters based on the idea of finite differences.

6.1 Policy Optimization Based On Numerical Rewards

Early applications of ZO optimization in reinforcement learning (RL) emerged as alternatives to policy gradient methods, where agents directly receive numerical rewards from the environment and update parameters via policy search. Sehnke et al. [47] proposed the Parameter-Exploring Policy Gradient (PGPE) method, which reduces gradient variance through symmetric sampling in the parameter space, leading to more efficient and stable learning. Since ZO methods rely solely on value queries, they naturally align with evolutionary strategies. Salimans et al. [46] introduced evolutionary strategies into ZO-based RL, demonstrating strong scalability. Choromanski et al. [14] further improved efficiency via structured exploration and compact architectures, while Mania et al. [37] proposed Augmented Random Search (ARS), applying ZO optimization to simple linear policies and challenging the trend of increasing complexity in RL algorithms. However, methods like ARS conduct isotropic random search in high-dimensional spaces, resulting in sample inefficiency. To address this, Choromanski et al. [15] proposed ASEBO, an adaptive ZO algorithm that identifies effective subspaces to improve sample efficiency in high-dimensional black-box optimization tasks, particularly in RL.

Due to the global nature of parameter exploration and the high sample efficiency of FO policy gradient methods, researchers try to combine them to obtain better performance [27,41]. Tang et al. [51] proposed using FO gradients as control variates for ZO gradients, significantly reducing variance while preserving the ZO nature of the method. However, most hybrid methods still rely on FO gradient-based policy updates and require differentiable policies. To overcome this, Lei et al. [30] proposed Zeroth-Order Actor-Critic (ZOAC), which integrates ZO-based policy updates with FO-based value estimation within the Actor-Critic framework, achieving superior performance over FO baselines. While ZO optimization is traditionally applied in parameter space, some studies have extended it to the action space. Kumar et al. [28] replaced the critic network with ZO optimization over actions to implement a deterministic policy gradient method with an actor-only structure. Similarly, Saglam et al. [45] used ZO optimization in the action space to approximate action-value gradients, addressing compatibility issues in deterministic policy gradient algorithms for continuous control tasks. Table 2 summarizes typical ZO approaches in reinforcement learning, highlighting their reward signals, exploration space, and key features.

6.2 Preferences Feedback Driven Policy Optimization

The success of LLM has brought RL from human feedback (RLHF) into the spotlight. However, traditional RLHF relies on "reward modeling", which is a complex and bottleneck process. Researchers began to look for more direct and end-to-end solutions. Tang et al. [52] proposed ZO-RankSGD, the first theoretically guaranteed ZO algorithm for optimization with only ranking feedback. This method introduces a new paradigm for RL (especially RLHF) to directly use online human ranking to align AI policies without training reward models. However, the theoretical framework of ZO-RankSGD implies a relatively idealized assumption that ranking feedback is error-free and deterministic. In real RLHF scenarios, human preferences are often random and noisy. Qining et al. [60] designed a theoretically guaranteed method to optimize AI strategies directly from human preferences, completely skipping reward inference, and this method can be applied to general random RL environments. Two new algorithms based on ZO optimization ideas, Zeroth-Order Policy Gradient (ZPG) and Zeroth-Order Block-Coordinate Policy Gradient (ZBCPG), were proposed, and proved to have polynomial convergence speed, theoretically proving the effectiveness and sample efficiency of this method.

Table 2. ZO optimization in reinforcement learning.

Nature of Reward Singnal	Method	Exploration Space of ZO Optimization	Gradient Source of policy updates	Key Feature
Policy optimization based on numerical rewards	PEPG [47]	Parameter space	ZO	Early parameter space gradients
	Salimans et al. [46]	Parameter space	ZO	introducing Evolutionary Strategies into gradient estimation
	Choromanski et al. [14]	Parameter space	ZO	Using structured random orthogonal matrices, providing more accurate estimators
	ARS [37]	Parameter space	ZO	Random Search Algorithm for Static Linear Strategies
	ASEBO [15]	Parameter space	ZO	Adaptive subspace search for guided exploration
	Tang et al. [51]	Parameter space	ZO+FO	Variance reduction based on policy gradient control variables
	ZO-AC [30]	Parameter space	ZO	Critic-guided ZO gradient for low variance
	Kumar et al. [28]	Action space	FO+ZO	Critic-free Deterministic Policy Gradients via ZO Optimization in Action Space
	Saglam et al. [45]	Action space	FO+ZO	To solve the gradient incompatibility problem of deterministic policy gradient
Preferences Feedback Driven Policy Optimization	ZO-RankSGD [52]	Parameter space	ZO	Estimating gradients from non-numerical ranking feedback
	Qining et al. [60]	Parameter space	ZO	Direct human feedback learning without reward model

7 Open Problems

Although ZO optimization has demonstrated remarkable progress in various machine learning applications, such as adversarial attacks, LLM fine-tuning, FL, and reinforcement learning-several fundamental theoretical and practical challenges remain unsolved. From a theoretical standpoint, most existing convergence guarantees are established under relatively strong assumptions, including smoothness or Lipschitz continuity of the objective function (e.g., the R-AdaZO algorithm by Shu et al. [49]). However, real-world optimization problems are often non-convex, non-smooth, and subject to stochastic or biased noise. This gap suggests two directions: (i) developing assumption-robust techniques that remain valid under weak convexity, non-smoothness, or heavy-tailed oracles; and (ii) designing dimension-adaptive strategies that generalize beyond task-specific heuristics such as sparsity or quantization. While recent studies, such as the work by Liu et al. [33], have investigated the implicit smoothing effect to improve stability, there is still a lack of systematic understanding regarding the selection of the smoothing parameter, its dynamic adjustment, and its overall impact on convergence. These gaps highlight the need for new theoretical frameworks that can better reflect the behavior of ZO optimization algorithms under more general and realistic conditions.

From an algorithmic perspective, the scalability of ZO optimization in high-dimensional settings remains a critical bottleneck. Although several recent approaches, such as SensZOQ [22] and DPZero [57] have proposed techniques like static sparsity, quantization, and low-rank approximations to reduce query complexity, these methods often rely on structural priors or specific assumptions tied to particular tasks. As a result, they may not generalize well to unstructured or task-agnostic high-dimensional problems. Therefore, a key open direction is to develop more general-purpose, dimension-adaptive strategies that can dynamically identify and leverage the intrinsic structure of the data or task. Achieving sublinear, or ideally dimension-independent, query complexity under realistic settings remains one of the most important and challenging goals for future research in ZO optimization. In addition, systematic comparisons between ZO and FO in terms of computational cost, query complexity, and scalability remain scarce, especially in high-dimensional settings.

Overall, a central challenge is to systematically understand and quantify the trade-offs across different ZO methods, particularly between query efficiency, robustness, and convergence speed, to guide practical algorithm selection.

8 Conclusion

This survey provides a comprehensive review of the recent advances in ZO optimization across several representative machine learning tasks, covering its fundamental principles, core methodologies, and key applications in adversarial attacks, LLM fine-tuning, FL, and RL. Based on this foundation, we also briefly outline some prevailing open problems and potential future research directions.

We hope this survey may offer useful insights for subsequent studies in the field and serve as a preliminary reference for further exploration and broader deployment of ZO optimization in practical scenarios.

References

1. Akhavan, A., Tsybakov, A.B.: Gradient-free stochastic optimization for additive models. arXiv preprint arXiv:2503.02131 (2025)
2. Alzantot, M., Sharma, Y., Chakraborty, S., Zhang, H., Hsieh, C.J., Srivastava, M.B.: Genattack: practical black-box attacks with gradient-free optimization. In: Proceedings of the Genetic and Evolutionary Computation Conference, pp. 1111–1119 (2019)
3. Bar, N., Giryes, R.: ZOQO: zero-order quantized optimization. arXiv preprint arXiv:2501.06736 (2025)
4. Berahas, A.S., Cao, L., Choromanski, K., Scheinberg, K.: A theoretical and empirical comparison of gradient approximations in derivative-free optimization. Found. Comput. Math. **22**(2), 507–560 (2022)
5. Bhagoji, A.N., He, W., Li, B., Song, D.: Practical black-box attacks on deep neural networks using efficient query mechanisms. In: Proceedings of the European Conference on Computer Vision (ECCV), pp. 154–169 (2018)
6. Chen, A., et al.: Deepzero: scaling up zeroth-order optimization for deep model training. In: The Twelfth International Conference on Learning Representations
7. Chen, J., Jordan, M.I., Wainwright, M.J.: Hopskipjumpattack: a query-efficient decision-based attack. In: 2020 IEEE Symposium on Security and Privacy (SP), pp. 1277–1294. IEEE (2020)
8. Chen, J., Chen, H., Gu, B., Deng, H.: Fine-grained theoretical analysis of federated zeroth-order optimization. Adv. Neural. Inf. Process. Syst. **36**, 54496–54508 (2023)
9. Chen, M., Huang, Y.L., Wen, Z.: Towards efficient low-order hybrid optimizer for language model fine-tuning (2025)
10. Chen, P.Y., Zhang, H., Sharma, Y., Yi, J., Hsieh, C.J.: Zoo: Zeroth order optimization based black-box attacks to deep neural networks without training substitute models. In: Proceedings of the 10th ACM Workshop on Artificial Intelligence and Security, pp. 15–26 (2017)
11. Chen, Y., Zhang, Y., Cao, L., Yuan, K., Wen, Z.: Enhancing zeroth-order fine-tuning for language models with low-rank structures. In: International Conference on Learning Representations (2025)
12. Cheng, S., Dong, Y., Pang, T., Su, H., Zhu, J.: Improving black-box adversarial attacks with a transfer-based prior. In: Advances in Neural Information Processing Systems, vol. 32 (2019)
13. Cheng, S., Miao, Y., Dong, Y., Yang, X., Gao, X.S., Zhu, J.: Efficient black-box adversarial attacks via Bayesian optimization guided by a function prior. In: Forty-first International Conference on Machine Learning
14. Choromanski, K., Rowland, M., Sindhwani, V., Turner, R., Weller, A.: Structured evolution with compact architectures for scalable policy optimization. In: International Conference on Machine Learning, pp. 970–978. PMLR (2018)
15. Choromanski, K.M., Pacchiano, A., Parker-Holder, J., Tang, Y., Sindhwani, V.: From complexity to simplicity: adaptive es-active subspaces for blackbox optimization. In: Advances in Neural Information Processing Systems, vol. 32 (2019)

16. Dang, P., Hu, X., Li, D., Zhang, R., Guo, Q., Xu, K.: DiffZOO: a purely query-based black-box attack for red-teaming text-to-image generative model via zeroth order optimization. In: Chiruzzo, L., Ritter, A., Wang, L. (eds.) Findings of the Association for Computational Linguistics: NAACL 2025, pp. 17–31. Association for Computational Linguistics, Albuquerque, New Mexico (2025). https://doi.org/10.18653/v1/2025.findings-naacl.2, https://aclanthology.org/2025.findings-naacl.2/
17. Dong, Y., Cheng, S., Pang, T., Su, H., Zhu, J.: Query-efficient black-box adversarial attacks guided by a transfer-based prior. IEEE Trans. Pattern Anal. Mach. Intell. **44**(12), 9536–9548 (2021)
18. Duchi, J.C., Jordan, M.I., Wainwright, M.J., Wibisono, A.: Optimal rates for zero-order convex optimization: the power of two function evaluations. IEEE Trans. Inf. Theory **61**(5), 2788–2806 (2015)
19. Fang, W., Yu, Z., Jiang, Y., Shi, Y., Jones, C.N., Zhou, Y.: Communication-efficient stochastic zeroth-order optimization for federated learning. IEEE Trans. Signal Process. **70**, 5058–5073 (2022)
20. Golovin, D., Karro, J., Kochanski, G., Lee, C., Song, X., Zhang, Q.: Gradientless descent: high-dimensional zeroth-order optimization. In: International Conference on Learning Representations
21. Guo, C., Gardner, J., You, Y., Wilson, A.G., Weinberger, K.: Simple black-box adversarial attacks. In: International Conference on Machine Learning, pp. 2484–2493. PMLR (2019)
22. Guo, W., et al.: Zeroth-order fine-tuning of LLMs with transferable static sparsity. In: The Thirteenth International Conference on Learning Representations
23. Gupta, D., Razaviyayn, M., Sharan, V.: On the inherent privacy of two point zeroth order projected gradient descent. In: OPT 2024: Optimization for Machine Learning (2024)
24. Hu, W., et al.: Localized zeroth-order prompt optimization. Adv. Neural. Inf. Process. Syst. **37**, 86309–86345 (2024)
25. Ilyas, A., Engstrom, L., Athalye, A., Lin, J.: Black-box adversarial attacks with limited queries and information. In: International Conference on Machine Learning, pp. 2137–2146. PMLR (2018)
26. Jiang, J., Jiang, H., Ma, Y., Liu, X., Fan, C.: Low-parameter federated learning with large language models. In: International Conference on Web Information Systems and Applications, pp. 319–330. Springer, Cham (2024)
27. Khadka, S., Tumer, K.: Evolution-guided policy gradient in reinforcement learning. In: Advances in Neural Information Processing Systems, vol. 31 (2018)
28. Kumar, H., Kalogerias, D.S., Pappas, G.J., Ribeiro, A.: Actor-only deterministic policy gradient via zeroth-order gradient oracles in action space. In: 2021 IEEE International Symposium on Information Theory (ISIT), pp. 1676–1681. IEEE (2021)
29. Lee, D., Moon, S., Lee, J., Song, H.O.: Query-efficient and scalable black-box adversarial attacks on discrete sequential data via Bayesian optimization. In: International Conference on Machine Learning, pp. 12478–12497. PMLR (2022)
30. Lei, Y., et al.: Zeroth-order actor-critic: an evolutionary framework for sequential decision problems. IEEE Trans. Evol. Comput. (2025)
31. Li, Z., Ying, B., Liu, Z., Dong, C., Yang, H.: Achieving dimension-free communication in federated learning via zeroth-order optimization. arXiv preprint arXiv:2405.15861 (2024)
32. Ling, Z., Chen, D., Yao, L., Li, Y., Shen, Y.: On the convergence of zeroth-order federated tuning for large language models. In: Proceedings of the 30th ACM

SIGKDD Conference on Knowledge Discovery and Data Mining, pp. 1827–1838 (2024)
33. Liu, S., Chen, P.Y., Kailkhura, B., Zhang, G., Hero, A.O., III., Varshney, P.K.: A primer on zeroth-order optimization in signal processing and machine learning: principals, recent advances, and applications. IEEE Signal Process. Mag. **37**(5), 43–54 (2020)
34. Liu, Y., Zhu, Z., Gong, C., Cheng, M., Hsieh, C.J., You, Y.: Sparse mezo: less parameters for better performance in zeroth-order LLM fine-tuning. arXiv preprint arXiv:2402.15751 (2024)
35. Ma, X., Wang, J., Zhang, X.: Data-free black-box federated learning via zeroth-order gradient estimation. In: Proceedings of the AAAI Conference on Artificial Intelligence, vol. 39, pp. 19314–19322 (2025)
36. Malladi, S., et al.: Fine-tuning language models with just forward passes. Adv. Neural. Inf. Process. Syst. **36**, 53038–53075 (2023)
37. Mania, H., Guy, A., Recht, B.: Simple random search of static linear policies is competitive for reinforcement learning. In: Advances in Neural Information Processing Systems, vol. 31 (2018)
38. Nesterov, Y., Spokoiny, V.: Random gradient-free minimization of convex functions. Found. Comput. Math. **17**(2), 527–566 (2017)
39. Patel, K.K., Wang, L., Saha, A., Srebro, N.: Federated online and bandit convex optimization. In: International Conference on Machine Learning, pp. 27439–27460. PMLR (2023)
40. Peng, D., Fu, Z., Wang, J.: Pocketllm: enabling on-device fine-tuning for personalized LLMs. In: Proceedings of the Fifth Workshop on Privacy in Natural Language Processing, pp. 91–96 (2024)
41. Pourchot, A., Sigaud, O.: CEM-RL: combining evolutionary and gradient-based methods for policy search. In: 7th International Conference on Learning Representations, ICLR 2019 (2019)
42. Qin, Z., Chen, D., Qian, B., Ding, B., Li, Y., Deng, S.: Federated full-parameter tuning of billion-sized language models with communication cost under 18 kilobytes. In: Proceedings of the 41st International Conference on Machine Learning, pp. 41473–41497 (2024)
43. Qiu, Y., Shanbhag, U., Yousefian, F.: Zeroth-order methods for nondifferentiable, nonconvex, and hierarchical federated optimization. In: Advances in Neural Information Processing Systems, vol. 36 (2023)
44. Ru, B., Cobb, A., Blaas, A., Gal, Y.: Bayesopt adversarial attack. In: International Conference on Learning Representations (2019)
45. Saglam, B., Kalogerias, D.: Compatible gradient approximations for actor-critic algorithms. In: ICML 2024 Workshop: Foundations of Reinforcement Learning and Control–Connections and Perspectives
46. Salimans, T., Ho, J., Chen, X., Sidor, S., Sutskever, I.: Evolution strategies as a scalable alternative to reinforcement learning. arXiv preprint arXiv:1703.03864 (2017)
47. Sehnke, F., Osendorfer, C., Rückstieß, T., Graves, A., Peters, J., Schmidhuber, J.: Parameter-exploring policy gradients. Neural Netw. **23**, 551–559 (2010)
48. Shu, Y., Lin, X., Dai, Z., Low, B.K.H.: Heterogeneous federated zeroth-order optimization using gradient surrogates. In: ICML 2024 Workshop on Differentiable Almost Everything: Differentiable Relaxations, Algorithms, Operators, and Simulators (2024)
49. Shu, Y., Zhang, Q., He, K., Dai, Z.: Refining adaptive zeroth-order optimization at ease. arXiv preprint arXiv:2502.01014 (2025)

50. Tang, X., Panda, A., Nasr, M., Mahloujifar, S., Mittal, P.: Private fine-tuning of large language models with zeroth-order optimization. In: ICML 2024 Workshop on Foundation Models in the Wild
51. Tang, Y., Choromanski, K., Kucukelbir, A.: Variance reduction for evolution strategies via structured control variates. In: International Conference on Artificial Intelligence and Statistics, pp. 646–656. PMLR (2020)
52. Tang, Z., Rybin, D., Chang, T.H.: Zeroth-order optimization meets human feedback: Provable learning via ranking oracles. In: ICML 2023 Workshop The Many Facets of Preference-Based Learning
53. Tu, C.C., et al.: Autozoom: autoencoder-based zeroth order optimization method for attacking black-box neural networks. In: Proceedings of the AAAI Conference on Artificial Intelligence, vol. 33, pp. 742–749 (2019)
54. Wang, G., Gu, B., Zhang, Q., Li, X., Wang, B., Ling, C.X.: A unified solution for privacy and communication efficiency in vertical federated learning. Adv. Neural. Inf. Process. Syst. **36**, 13480–13491 (2023)
55. Zhang, H., Liu, Z., Gu, B., Chang, Y.: Leveraging first and zeroth-order gradient to address imbalanced black-box prompt tuning via minimax optimization. In: Proceedings of the AAAI Conference on Artificial Intelligence, vol. 39, pp. 22407–22415 (2025)
56. Zhang, H., Zhang, H., Liu, Z., Gu, B., Chang, Y.: Collaborative discrete-continuous black-box prompt learning for language models. In: The Thirteenth International Conference on Learning Representations (2025)
57. Zhang, L., Li, B., Thekumparampil, K.K., Oh, S., He, N.: Dpzero: private fine-tuning of language models without backpropagation. arXiv preprint arXiv:2310.09639 (2023)
58. Zhang, L., Li, B., Thekumparampil, K.K., Oh, S., He, N.: Dpzero: private fine-tuning of language models without backpropagation. In: International Conference on Machine Learning, pp. 59210–59246. PMLR (2024)
59. Zhang, Q., Gu, B., Dang, Z., Deng, C., Huang, H.: Desirable companion for vertical federated learning: New zeroth-order gradient based algorithm. In: Proceedings of the 30th ACM International Conference on Information & Knowledge Management, pp. 2598–2607 (2021)
60. Zhang, Q., Ying, L.: Zeroth-order policy gradient for reinforcement learning from human feedback without reward inference. arXiv preprint arXiv:2409.17401 (2024)
61. Zhang, Y., et al.: Revisiting zeroth-order optimization for memory-efficient LLM fine-tuning: a benchmark. In: Proceedings of the 41st International Conference on Machine Learning, pp. 59173–59190 (2024)

A Temporal Knowledge Completion Model Driven by Dual-Module for Evolutionary Feature Perception

Tianyi Xu[1], Jianhang Song[1], Jiujiang Guo[1], Mei Yu[1], Abdelbari Redouane[2], Mustapha Kchikach[2], and Mankun Zhao[1(✉)]

[1] College of Intelligence and Computing, Tianjin University, Tianjin 300354, China
{tianyi.xu,2022229046,1335130740,yumei,zmk}@tju.edu.cn

[2] National Higher School of Mines of Rabat, 10000 Rabat, Morocco
{redouane,kchikach}@enim.ac.ma

Abstract. Knowledge, as the core driving force of intelligent applications, directly impacts the reliability and effectiveness of downstream tasks such as recommendation system and wind energy forecast through its completeness. Real-world knowledge evolves over time, and Temporal Knowledge Graph Completion (TKGC) aims to complete missing entities or relations under temporal constraints to enhance the completeness of knowledge. However, Existing research mostly focuses on the local temporal dependencies of entities, neglecting the combined influence of different time domains during their long-term evolution processes. To address this challenge, We propose a dual-module-driven completion model. The model employs a dual-module joint mechanism to capture the time-dependent features of relations and entities during evolution. Specifically, the evolutionary feature decomposition module utilizes the temporal decomposition capability of time series decomposition methods to decompose the temporal evolution of relations into periodic and trend components, enabling explicit modeling for precise perception of relational evolutionary features. The global correlation optimization module compresses global temporal information using discrete cosine transform to extract cross-temporal association features. It further integrates entity information to dynamically adjust and optimize the current temporal context, thereby capturing cross-temporal entity correlations. Extensive experiments conducted on four public temporal knowledge graph datasets demonstrate the effectiveness of the proposed models in enhancing temporal knowledge completion performance.

Keywords: Temporal knowledge graph completion · Time series decomposition · Discrete cosine transform · Evolving knowledge

1 Introduction

Knowledge graphs (KGs) can be broadly classified into two categories based on data format: static knowledge graphs (SKGs) and temporal knowledge graphs

S. Yang et al. (Eds.): WISA 2025, LNCS 16299, pp. 498–509, 2026.
https://doi.org/10.1007/978-981-95-6426-2_44

(TKGs). SKGs and TKGs represent facts as triples and quadruples, respectively. Existing complex knowledge graphs (such as BMKG [1], AceKG [2] and XLORE3 [3]) have acquired a vast amount of rich information through intelligent means and manual collection. They have been successfully applied in various domains including information retrieval [4], intelligent healthcare, and natural language processing [5].

As the volume of information in knowledge graphs (KGs) continues to rise, various challenges emerge, including sparsity, quality control, and imbalance. Past research has developed various powerful models, but most are designed for static knowledge graphs (SKGs), with relatively less attention given to temporal knowledge graphs (TKGs). In TKGs, timestamps are attached to each fact, indicating its validity period. Therefore, compared to SKGC models, TKGC models need to additionally learn temporal properties, which poses a key challenge in effectively leveraging these properties. Existing research maps facts in TKGs to high-dimensional vectors, with each fact mapping considering the graph structure as a constraint. These models have gained popularity due to their simplicity, scalability, and expressiveness. However, these methods fail to explicitly model the unique time related characteristics of relations and entities in the evolution process. Based on the in-depth observation of the complex temporal knowledge network structure, the analysis shows that the relation and entity show the following unique time related characteristics in the evolution process: Entities are jointly affected by different time domains in the evolution process of long sequence period or long time span. As shown in Fig. 1, take the career of famous football player Ronaldo as an example. He moved to Juventus from *2009* to *2018* in Real Madrid, then from *2018* to *2021*, then returned to Manchester United from *2021* to *2022*, and finally joined Riyadh to win in *2023*. In the 16 year long evolution process from *2009* to *2025*, Ronaldo played for different teams in different time domains, and these experiences together shaped Ronaldo's career path.

Fig. 1. Evolving knowledge graph.

We evaluated our work based on various established TKGC benchmarks. Through experiments and a series of ablation tests, we found that DM-EFP achieves outstanding results on all four widely recognized datasets.

2 Related Work

2.1 SKGC Models

Research on SKGC has become very rich. Generally, mainstream research can be divided into the following series: **1) Translation Series**. TransE [6] is the most classic translation model that embodies this assumption, grounded in the principle $s + r \approx o$. A series of extended works have been proposed in this category, including TransH [7], TransR [8], and TransD [9]. **2) Complex Series**. Learning KGs in complex spaces has garnered attention from scholars, such as ComplEx [10] measures plausibility by matching the latent semantics of facts. RotatE [11] embeds relations into complex rotations, thus leveraging rotational capabilities to model different relation patterns. Additionally, QuatE [12] extends the complex space to quaternion space, integrating two rotational surfaces to facilitate rich semantic matching. **3) Tensor Series** utilizes tensor decomposition properties to capture latent interactions within fact vectors. RESCAL [13] performs bilinear product between the embeddings of head and tail entities and the full-rank matrix of each relation. TuckER [14] leverages the structural advantages of Tucker decomposition to integrate facts into three member matrices and uses core tensors to facilitate interaction between entities and relations. Recently, BDRI [15] enhances the interaction between forward and inverse relations leveraging the advantages of BTD structure to capture latent semantics. **4) Deep learning Series**. Some neural network models have achieved significant results, yet the main issues revolve around instability and lack of interpretability. For instance, ConvKB [16], InteractE [17].

2.2 TKGC Models

In recent years, research on TKGC has gradually gained attention. Similar to SKGC models, it is generally divided into the following series: **1) Translation Series**. TTransE [18] expands upon TransE by incorporating temporal information, where temporal information is represented as translation operations. Following the translational approach, scholars continue to develop a series of works, such as HYTE [19]. **2) Complex Series**. TComplEx [20] model temporal facts as fourth-order tensors, representing them as such through matrix factorization models. Unlike TComplEx, TNTComplEx divides the decomposition into temporal and non-temporal components. ChronoR [21], derived from RotatE, acquires a rotation transformation parametrized by relation and timestamp. This ensures that, following the application of the rotation to the head entity, it aligns closely with its associated tail entity. Next, RotateQVS [22] and EHPR [23] harnesses quaternions to represent temporal information as quaternion rotations. Besides, TeRo transforms the temporal evolution of embedded entities into rotations from beginning to ending. **3) Tensor Series**. ASALSAN [24] analyzes potential information within TKGs by computing DEDICOM. It utilizes these components to assist in modeling various relation patterns. TeLM [25] uses multi-embeddings to perform 4th-order tensor decomposition. Recently,

TBDRI [26], based on BTD, utilize different methods to represent temporal facts, respectively. **4) Deep learning Family**. Some neural network models tailored for TKGs have been gradually developed, yet the outcomes often prove challenging to interpret. For instance, CyGNet [27], RE-GCN [28], Neo-TKGC [29].

3 Preliminaries

3.1 Problem Definition

To formally define the problem and describe the solution, we use consistent notations in the rest of the paper.

Temporal knowledge graph $\mathcal{G}$ is composed of quadruples consisting of entities, relations, and timestamps. We formulate denote $\mathcal{E}$ as the finite entity set, $\mathcal{R}$ as the finite relation set, and $\mathcal{T}$ as the finite timestamp set. A quadruple is represented as $(s, r, o, t) \subset \mathcal{E} \times \mathcal{R} \times \mathcal{E} \times \mathcal{T}$. We use Π and $\Pi^- = \mathcal{E} \times \mathcal{R} \times \mathcal{E} \times \mathcal{T} - \Pi$ to represent the set of observed quadruples and the set of unobserved quadruples, respectively.

3.2 Discrete Cosine Transform (DCT)

DCT expresses a finite sequence of data points in terms of a sum of cosine functions oscillating at different frequencies. The basis function of one-dimensional (1D) DCT is:

$$B_u^i = \cos\left[\frac{\pi(2i+1)u}{2N}\right], \tag{1}$$

Then 1D DCT can be written as:

$$\begin{aligned} f_u &= \sum_{i=0}^{N-1} x(i) B_u^i \\ s.t.\ \ u &\in \{0, 1, \cdots, N-1\}, \end{aligned} \tag{2}$$

in which $f \in \mathbb{R}^N$ is 1D DCT frequency spectrum, x is a 1-D sequence of length N.

4 Model

Symbol description. Suppose that $\mathcal{G}$ consists of N_e entities, N_r relations, and $N_\mathcal{T}$ timestamps. We use the matrix $\mathbf{E} \in \mathbb{R}^{N_e \times 2k}$ to denote all entity embeddings, $\mathbf{R} \in \mathbb{R}^{N_r \times 2 \times 2k}$ to denote all relation embeddings, $\mathbf{T} \in \mathbb{R}^{N_\mathcal{T} \times 2 \times 2k}$ to denote all timestamp embeddings. Taking the relation matrix R as an example, each row is a d-dimensional embedding vector for a specific relation. Given a quadruple (s, r, o, τ), the head entity s and the tail entity o correspond to $\mathbf{E}_s \in \mathbf{E}$ and $\mathbf{E}_o \in \mathbf{E}$, respectively, while the relation r and the timestamp τ correspond to $\mathbf{R}_r \in \mathbf{R}$ and $\mathbf{T}_\tau \in \mathbf{T}$ (Fig. 2).

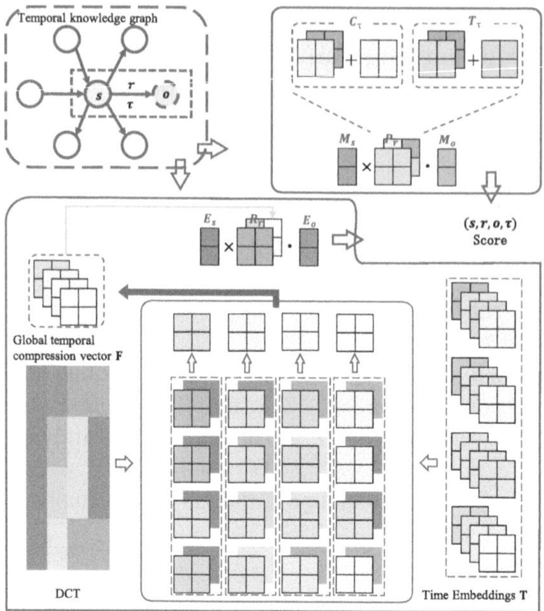

Fig. 2. DM-EFP architecture.

4.1 DM-EFP

In order to fully capture the information contained in the timestamps, an efficient compression method is required. The multi-frequency component of DCT can compress information in different ways. Because The multi-spectrum of DCT can be viewed as different weighted sum of inputs, and its different weightings are independent of each other.

First, all timestamp embeddings $\mathbf{T}$ are split into n parts along the embedding dimension. Denote $[\mathbf{T}^0, \mathbf{T}^1, \cdots, \mathbf{T}^{n-1}]$ as the parts, in which $\mathbf{T}^i \in \mathbb{R}^{N_\tau \times 2 \times 2k'}$, $i \in \{0, 1, \cdots, n-1\}$, $k' = \frac{k}{n}$, and k should be divisible by n. For each part of $\mathbf{T}$, a corresponding 1D DCT frequency component is allocated, which means that the global time is compressed in different ways. The compression vector is defined as follows:

$$\begin{aligned} \mathbf{F}^i &= DCT_i(\mathbf{T}^i) \\ &= \sum_{j=0}^{N_\tau - 1} B_i^j \mathbf{T}_j^i, \\ s.t.\ \ & i \in \{0, 1, \cdots, n-1\}, \end{aligned} \tag{3}$$

whole compression vector can be obtained by concatenation:

$$\begin{aligned} \mathbf{F} &= compress(\mathbf{T}) \\ &= \text{concat}([\mathbf{F}^0, \mathbf{F}^1, \cdots, \mathbf{F}^{n-1}]), \end{aligned} \tag{4}$$

in which $compress : \mathbb{R}^{N_\tau \times 2 \times 2k} \mapsto \mathbb{R}^{2 \times 2k}$ represents the purpose for which we use DCT. $\mathbf{F} \in \mathbb{R}^{2 \times 2k}$ is the obtained vector. The whole attention framework can be written as:

$$\mathbf{att} = sigmoid(fc(\mathbf{E}_s \times \mathbf{F})), \tag{5}$$

where $\mathbf{att} \in \mathbb{R}^k$ is the attention vector, *sigmoid* is the Sigmoid function, *fc* represents the fully connected layer. After obtaining the attention vector of all k dimensions, each dimension of input timestamp embedding $\mathbf{T}_\tau$ is scaled by the corresponding attention value:

$$\mathbf{T}'_\tau = \mathbf{att} \odot \mathbf{T}_\tau. \tag{6}$$

4.2 Score Function

After obtaining the scaled time $\mathbf{T}'_\tau$, DM-EFP considers the completion of temporal knowledge graph as a fourth-order tensor decomposition, and fully interacts with each embedded representation when scoring facts (s, r, o, τ).

Specifically, first, the subject embedding interacts with the relation embedding, then interacts with the time embedding, and finally measures the cosine similarity between the result vector and the object vector. The scoring function for the DM-EFP is shown in the formula:

$$\phi(s, r, o, \tau) = ((\mathbf{E}_s + \mathbf{S}_\tau) \times \mathbf{R}_r \times \mathbf{T}'_\tau) \cdot (\mathbf{E}_o + \mathbf{O}_\tau), \tag{7}$$

where $\times$ represents multiplication between vectors and matrices, as well as between matrices. $\cdot$ represents the inner product of vectors.

4.3 Learning Process

We employ the Adagrad optimizer [30] to train our proposed DM-EFP by minimizing the following loss function [10] with the regularization on model parameters θ as:

$$\begin{gathered}\mathcal{L} = \sum_{(s,r,o,\tau) \in \{\Pi \cup \Pi^-\}} \log \left(1 + \exp\left(-l_{(s,r,o,t)} \cdot \phi(s, r, o, \tau)\right)\right) + \lambda \|\boldsymbol{\theta}\|_2^2 \\ \text{in which, } l_{(s,r,o,\tau)} = \begin{cases} 1 & \text{for } (s, r, o, \tau) \in \Pi \\ -1 & \text{for } (s, r, o, t) \in \Pi^- \end{cases}\end{gathered} \tag{8}$$

where we use l_2-norm with the regularization rate λ.

5 Experiments

5.1 Datasets and Evaluation Metrics

To assess our works, we select four common TKGC benchmarks.

ICEWS14 and ICEWS05-15 are subsets of the Integrated Crisis Early Warning System (ICEWS) [31]. ICEWS acquires and processes millions of data from various sources such as digital news, social media, etc., to aid in monitoring and responding to global events.

YAGO15K is an upgraded version of FB15k [6], with the addition of timestamps for each fact. It is noteworthy that the absence of temporal information for some facts in YAGO results in an increased complexity. Thus, YAGO15K presents a greater challenge for us.

GDELT [32] is a global event database that offers a substantial collection of worldwide facts from *2015-04-01* to *2016-03-31*. It encompasses various events, primarily focusing on political, social, economic, and environmental occurrences. The database aims to monitor and comprehend the development of global events during this specified timeframe.

To assess the performance of DM-EFP accurately and comprehensively, we make some preparations as follows: Using commonly used metrics to evaluate the model more precisely. **MRR (Mean Reciprocal Rank)** is used to calculate the average of the reciprocals of ranks. **Hit@n** represents the percentage of correct items within the top n ranks. A higher value indicates better performance. This paper selects Hit@10, Hit@3, and Hit@1 to ensure the comprehensiveness of the evaluation.

5.2 Results

Compared to existing models, DM-EFP achieves excellent results on four publicly available datasets. Table 1 and Table 2 show the completion results. The results of other models come from the original paper (Tables 1 and 2).

Specifically, For ICEWS14, compared with TBDRI of tensor category, DM-EFP increases 1.4%, 0.2%, 0.4% on MRR, Hit@1, Hit@10 apart.

For ICEWS05-15, compared with TBDRI, DM-EFP increases 0.7%, 0.6%, 1.7% on MRR, Hit@1, Hit@10 apart.

For YAGO15k, compared with existing SOTA models, DM-EFP achieves impressive results. Specifically, compared with TBDRI, DM-EFP increases 1.1%, 6.0%, 12.1%, 2.2% on MRR, Hit@1, Hit@3, Hit@10 apart. Compared with TNT-ComplEx, DM-EFP increases 0.5%, 10.0%, 6.9%, 4.8% on MRR, Hit@1, Hit@3, Hit@10 apart.

For GDELT, DM-EFP also achieves impressive results on all metrics. Compared with TeAST, DM-EFP obtains an improvement of 4.9%, 5.3%, 2.7%, and 5.5% on MRR, Hit@1, Hit@3, Hit@10 apart.

Intuitively, DM-EFP exhibits the most significant impact on the YAGO15K and GDELT. Our preliminary analysis suggests that the largest contribution stems from the Attention Module's effective capture of global relation information and global time information.

Table 1. Experimental results on ICEWS14 and ICEWS05-15.

	ICEWS14				ICEWS05-15			
	MRR	H@1	H@3	H@10	MRR	H@1	H@3	H@10
TransE(2013)	28.0	9.4	–	63.7	29.4	8.4	–	66.30
ComplEx(2016)	47.0	35.0	54.0	71.0	49.0	37.0	55.0	73.0
RotatE(2019)	41.8	29.1	47.8	69.0	30.4	16.4	35.5	59.5
QuatE(2019)	47.1	35.3	53.0	71.2	48.2	37.0	52.9	72.7
TTransE(2018)	25.5	7.4	–	60.1	27.1	8.4	–	61.6
HyTE(2018)	29.7	10.8	41.6	65.5	31.6	11.6	44.5	68.1
TA-DistMult(2018)	47.7	–	36.3	68.6	47.4	34.6	–	72.8
DE-SimplE(2020)	52.6	41.8	59.2	72.5	51.3	39.2	57.8	74.8
TNTComplEx(2020)	60.7	51.9	65.9	77.2	66.6	58.3	71.8	81.7
TeRo(2020)	56.2	46.8	62.1	73.2	58.6	46.9	66.8	79.5
TeMP(2020)	60.7	48.4	68.4	**84.0**	68.0	55.3	76.9	**91.3**
TeLM(2021)	62.5	54.5	67.3	77.4	67.8	59.9	72.8	82.3
ChronoR(2021)	62.5	54.7	66.9	77.3	67.5	59.6	72.3	82.0
RotateQVS(2022)	59.1	50.7	64.2	75.4	63.3	52.9	70.9	81.3
BTDG(2022)	60.1	51.6	65.6	75.3	62.7	53.4	68.7	79.8
TBDRI(2023)	65.2	55.2	69.7	78.5	70.9	64.6	75.7	82.1
TeAST(2023)	63.7	56.0	68.2	78.2	68.3	60.4	73.2	82.9
HGE(2024)	63.4	55.0	68.5	78.8	68.8	60.8	74.0	83.5
MvTuckER(2024)	65.4	57.7	69.9	79.7	69.8	61.8	74.7	84.1
Neo-TKGC(2025)	64.2	53.7	70.9	83.6	72.2	61.7	**79.4**	90.8
DM-EFP (ours)	**73.9**	**68.5**	**77.3**	83.8	**72.3**	**65.9**	76.1	84.2

5.3 Case Study

To specifically evaluate the performance of DM-EFP in terms of perceptual relations and entity evolution characteristics, we extracted quadruple samples with fixed subjects and relationships from the dataset for experimentation. Specifically, taking ICEWS14 as an example, events in the form of (Barack Obama, Make a visit, *, *) have different objects and span different time periods, thus exhibiting different evolutionary characteristics. The experimental results are shown in the Fig. 3.

By comparing the classical models TeLM and TComplEx, DM-EFP performs the best on different constrained data and significantly outperforms other models. The analysis results indicate that DM-EFP, based on fully constructing the interaction between entities, relationships, and timestamps, focuses on reconstructing the fact that the model evolves over time to perceive the evolutionary characteristics that are commonly present in the evolving facts. However, the classic temporal knowledge graph completion models TeLM and TComplEx lack

Table 2. Experimental results on YAGO15K and GDELT.

	YAGO15K				GDELT			
	MRR	H@1	H@3	H@10	MRR	H@1	H@3	H@10
TransE(2013)	29.6	22.8	–	46.8	11.3	0	15.8	31.2
ComplEx(2016)	35.0	28.0	35.0	52.0	23.1	14.4	24.8	40.3
TTransE(2018)	32.1	23.0	–	51.0	11.5	0.0	16.0	31.8
HyTE(2018)	–	–	–	–	11.8	0.0	16.5	32.6
TA-DistMult(2018)	29.1	21.6	–	47.6	20.6	12.4	21.9	36.5
DE-SimpIE(2020)	–	–	–	–	23.0	14.1	24.8	40.3
TNTComplEx(2020)	37.0	29.0	39.0	54.0	30.0	21.4	32.3	46.7
TeRo(2020)	–	–	–	–	24.5	15.4	26.4	42.0
ChronoR(2021)	36.6	29.2	37.9	53.8	–	–	–	–
RotateQVS(2022)	–	–	–	–	27.0	17.5	29.3	45.8
TBDRI(2023)	36.8	30.1	37.2	55.4	26.9	16.4	28.0	44.1
TeAST(2023)	–	–	–	–	37.1	28.3	40.1	54.4
HGE(2024)	–	–	–	–	37.1	28.3	40.2	55.6
DM-EFP (ours)	**37.2**	**31.9**	**41.7**	**56.6**	**39.0**	**29.8**	**41.2**	**57.4**

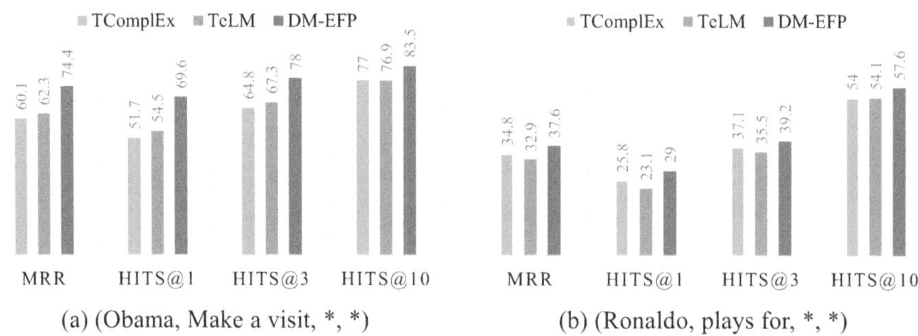

(a) (Obama, Make a visit, *, *)

(b) (Ronaldo, plays for, *, *)

Fig. 3. Case Study Results.

deeper modeling of the temporal dimension, resulting in a slightly insufficient understanding of the evolutionary characteristics of strongly constrained data.

6 Conclusion

In this paper, we propose a novel TKGC model DM-EFP. The model is based on two time series analysis methods, time series decomposition and discrete cosine transform, and constructs an evolutionary property decomposition module and a global correlation optimization module, respectively. DM-EFP focuses on explicitly modeling the evolutionary characteristics of relationships and entities in temporal knowledge. Among them, the evolutionary characteristic decomposition module focuses on the evolutionary characteristics of relationships, such as periodicity and trendiness; The global association optimization module focuses on the potential associations between entities in different time periods. Finally, the model combines the outputs of the two modules to jointly score the authenticity of the facts. Empirically, DM-EFP can boost the performance over four TKGC benchmarks.

Acknowledgements. This work is supported by the National Key Research and Development Program of China (No. 2024YFE0115700) and National Natural Science Foundation of China (No.62402335)

References

1. Cong, Q., Feng, Z., Li, F., Zhang, L., Rao, G., Tao, C.: Constructing biomedical knowledge graph based on SemMedDB and linked open data. In: IEEE International Conference on Bioinformatics and Biomedicine, pp. 1628–1631. IEEE, Piscataway (2018)
2. Wang, R., et al.: AceKG: a large-scale knowledge graph for academic data mining. In: Proceedings of the 27th ACM International Conference on Information and Knowledge Management, pp. 1487–1490. ACM, New York (2018)
3. Zeng, K., et al.: XLORE 3: a large-scale multilingual knowledge graph from heterogeneous wiki knowledge resources. ACM Trans. Inf. Syst. **42**(6), 1–47 (2024)
4. Li, Z., Liu, H., Zhang, Z., Liu, T., Xiong, N.N.: Learning knowledge graph embedding with heterogeneous relation attention networks. IEEE Trans. Neural Networks Learn. Syst. **33**(8), 3961–3973 (2022)
5. Yang, B., Mitchell, T.: Leveraging knowledge bases in LSTMs for improving machine reading. In: Barzilay, R., Kan, M.-Y. (eds.) Proceedings of the 55th Annual Meeting of the Association for Computational Linguistics, ACL 2017, Vancouver, Canada, July 30 - August 4, Volume 1: Long Papers, pp. 1436–1446. Association for Computational Linguistics (2017)
6. Bordes, A., Usunier, N., García-Durán, A., Weston, J., Yakhnenko, O.: Translating embeddings for modeling multi-relational data. In: Burges, C.J.C., Bottou, L., Ghahramani, Z., Weinberger, K.Q. (eds.) Advances in Neural Information Processing Systems 26: 27th Annual Conference on Neural Information Processing Systems 2013. Proceedings of a meeting held December 5-8, 2013, Lake Tahoe, Nevada, United States, pp. 2787–2795 (2013)
7. Wang, Z., Zhang, J., Feng, J., Chen, Z.: Knowledge graph embedding by translating on hyperplanes. In: Proceedings of the AAAI Conference on Artificial Intelligence, vol. 28 (2014)
8. Lin, Y., Liu, Z., Sun, M., Liu, Y., Zhu, X.: Learning entity and relation embeddings for knowledge graph completion. In: Twenty-ninth AAAI Conference on Artificial Intelligence (2015)
9. Ji, G., He, S., Xu, L., Liu, K., Zhao, J.: Knowledge graph embedding via dynamic mapping matrix. In: Proceedings of the 53rd Annual Meeting of the Association for Computational Linguistics and the 7th International Joint Conference on Natural Language Processing (volume 1: Long papers), pp. 687–696 (2015)

10. Trouillon, T., Welbl, J., Riedel, S., Gaussier, É., Bouchard, G.: Complex embeddings for simple link prediction. In: Balcan, M.-F., Weinberger, K.Q. (eds.) Proceedings of the 33nd International Conference on Machine Learning, ICML 2016, New York City, NY, USA, 19–24 June 2016, volume 48 of JMLR Workshop and Conference Proceedings, pp. 2071–2080. JMLR.org (2016)
11. Sun, Z., Deng, Z.-H., Nie, J.-Y., Tang, J.: RotatE: knowledge graph embedding by relational rotation in complex space. In: 7th International Conference on Learning Representations, ICLR 2019, New Orleans, LA, USA, 6–9 May 2019. OpenReview.net (2019)
12. Zhang, S., Tay, Y., Yao, L., Liu, Q.: Quaternion knowledge graph embeddings. In: Wallach, H.M., Larochelle, H., Beygelzimer, A., d'Alché-Buc, F., Fox, E.B., Garnett, R. (eds.) Advances in Neural Information Processing Systems 32: Annual Conference on Neural Information Processing Systems 2019, NeurIPS 2019, 8–14 December 2019, Vancouver, BC, Canada, pp. 2731–2741 (2019)
13. Nickel, M., Tresp, V., Kriegel, H.-P.: A three-way model for collective learning on multi-relational data. In: ICML, vol. 11, pp. 809–816 (2011)
14. Balazevic, I., Allen, C., Hospedales, T.M.: TuckER: tensor factorization for knowledge graph completion. In: Inui, K., Jiang, J., Ng, V., Wan, X. (eds.) Proceedings of the 2019 Conference on Empirical Methods in Natural Language Processing and the 9th International Joint Conference on Natural Language Processing, EMNLP-IJCNLP 2019, Hong Kong, China, 3–7 November 2019, pp. 5184–5193. Association for Computational Linguistics (2019)
15. Yu, M., et al.: BDRI: block decomposition based on relational interaction for knowledge graph completion. In: He, J. (ed.) Data Mining and Knowledge Discovery, pp. 1–21. Springer (2023)
16. Nguyen, D.Q., Nguyen, T.D., Nguyen, D.Q., Phung, D.Q.: A novel embedding model for knowledge base completion based on convolutional neural network. In: Walker, M.A., Ji, H., Stent, A. (eds.) Proceedings of the 2018 Conference of the North American Chapter of the Association for Computational Linguistics: Human Language Technologies, NAACL-HLT, New Orleans, Louisiana, USA, 1–6 June 2018, Volume 2 (Short Papers), pp. 327–333. Association for Computational Linguistics (2018)
17. Vashishth, S., Sanyal, S., Nitin, V., Agrawal, N., Talukdar, P.P.: InteractE: improving convolution-based knowledge graph embeddings by increasing feature interactions. In: AAAI, pp. 3009–3016 (2020)
18. Jiang, T., et al.: Towards time-aware knowledge graph completion. In: Proceedings of COLING 2016, the 26th International Conference on Computational Linguistics: Technical Papers, pp. 1715–1724 (2016)
19. Dasgupta, S.S., Ray, S.N., Talukdar, P.P.: HyTE: hyperplane-based temporally aware knowledge graph embedding. In: Riloff, E., Chiang, D., Hockenmaier, J., Tsujii, J. (eds.) Proceedings of the 2018 Conference on Empirical Methods in Natural Language Processing, Brussels, Belgium, October 31 - November 4, 2018, pp. 2001–2011. Association for Computational Linguistics (2018)
20. Lacroix, T., Obozinski, G., Usunier, N.: Tensor decompositions for temporal knowledge base completion. In: 8th International Conference on Learning Representations, ICLR 2020, Addis Ababa, Ethiopia, 26–30 April 2020. OpenReview.net (2020)

21. Sadeghian, A., Armandpour, M., Colas, A., Wang, D.Z.: ChronoR: rotation based temporal knowledge graph embedding. In: Thirty-Fifth AAAI Conference on Artificial Intelligence, AAAI 2021, Thirty-Third Conference on Innovative Applications of Artificial Intelligence, IAAI 2021, The Eleventh Symposium on Educational Advances in Artificial Intelligence, EAAI 2021, Virtual Event, 2–9 February 2021, pp. 6471–6479. AAAI Press (2021)
22. Chen, K., Wang, Y., Li, Y., Li, A.: RotateQVS: representing temporal information as rotations in quaternion vector space for temporal knowledge graph completion. In: Muresan, S., Nakov, P., Villavicencio, A. (eds.) Proceedings of the 60th Annual Meeting of the Association for Computational Linguistics (Volume 1: Long Papers), ACL 2022, Dublin, Ireland, 22–27 May 2022, pp. 5843–5857. Association for Computational Linguistics (2022)
23. Guo, J., et al.: EHPR: learning evolutionary hierarchy perception representation based on quaternion for temporal knowledge graph completion. Inf. Sci. **688**, 121409 (2025)
24. Bader, B.W., Harshman, R.A., Kolda, T.G.: Temporal analysis of semantic graphs using ASALSAN. In: Proceedings of the 7th IEEE International Conference on Data Mining (ICDM 2007), 28–31 October 2007, Omaha, Nebraska, USA, pp. 33–42. IEEE Computer Society (2007)
25. Xu, C., Chen, Y.-Y., Nayyeri, M., Lehmann, J.: Temporal knowledge graph completion using a linear temporal regularizer and multivector embeddings. In: Proceedings of the 2021 Conference of the North American Chapter of the Association for Computational Linguistics: Human Language Technologies, pp. 2569–2578, Online. Association for Computational Linguistics (2021)
26. Mei, Y., et al.: TBDRI: block decomposition based on relational interaction for temporal knowledge graph completion. Appl. Intell. **53**(5), 5072–5084 (2023)
27. Zhu, C., Chen, M., Fan, C., Cheng, G., Zhang, Y.: Learning from history: modeling temporal knowledge graphs with sequential copy-generation networks. In: Thirty-Fifth AAAI Conference on Artificial Intelligence, AAAI 2021, Thirty-Third Conference on Innovative Applications of Artificial Intelligence, IAAI 2021, The Eleventh Symposium on Educational Advances in Artificial Intelligence, EAAI 2021, Virtual Event, 2–9 February 2021, pp. 4732–4740. AAAI Press (2021)
28. Xu, C., Nayyeri, M., Alkhoury, F., Lehmann, J., Yazdi, H.S.: Temporal knowledge graph embedding model based on additive time series decomposition. CoRR, abs/1911.07893 (2019)
29. Qiu, Z., Zhou, X., An, C., Yang, Q., Li, Z.: Neo-TKGC: enhancing temporal knowledge graph completion with integrated node weights and future information. In: Proceedings of the 18th ACM International Conference on Web Search and Data Mining, pp. 373–381. ACM (2025)
30. Duchi, J., Hazan, E., Singer, Y.: Adaptive subgradient methods for online learning and stochastic optimization. J. Mach. Learn. Res. **12**(7) (2011)
31. Boschee, E., Lautenschlager, J., O'Brien, S., Shellman, S., Starz, J., Ward, M.: ICEWS coded event data. Harvard Dataverse **12** (2015)
32. Trivedi, R., Dai, H., Wang, Y., Song, L.: Know-Evolve: deep temporal reasoning for dynamic knowledge graphs. In: Precup, D., Teh, Y.W. (eds.) Proceedings of the 34th International Conference on Machine Learning, volume 70 of Proceedings of Machine Learning Research, pp. 3462–3471. PMLR (2017)

Chain of Thought and Reinforcement Learning Based Low-Resource Named Entity Recognition Model

Baoxing Jiang[1], Chunyan Han[2], Jingjing Zhu[1], Xin Xia[1], and Shenggen Ju[1(✉)]

[1] College of Computer Science, Sichuan University, Chengdu 610065, China
jsg@scu.edu.cn

[2] School of Intelligent Science and Technology, Sichuan Minzu College, Kangding 626001, China

Abstract. Deep learning has significantly advanced named entity recognition (NER), but its performance gains largely depend on large-scale annotated data. In most domains, however, annotated resources remain scarce, and building integrated datasets is costly. The emergence of large language models (LLMs) and prompt learning offers new approaches to mitigate this dependence and address NER tasks in low-resource settings. Yet, existing prompt-based methods often lack sufficient contextual semantic information for entities and fail to fully exploit the relations between entities and their types, making it difficult to capture and optimize entity boundaries and quantity information explicitly. To address these challenges, we design an entity knowledgeâĂŞenhanced prompt framework that integrates the chain-of-thought paradigm with LoRA fine-tuning. This approach enriches contextual information and guides the model to acquire entity knowledge before prediction. Furthermore, by constructing entity-level preference data and applying the DPO algorithm, we incorporate boundary and quantity information to alleviate category confusion and boundary offset issues. Extensive experiments on four benchmark datasets demonstrate the effectiveness of the proposed method.

Keywords: Low-resource named entity recognition · Prompt learning · Optimization based on human preference

1 Introduction

Named entity recognition(NER) is a fundamental task in the field of natural language processing(NLP), aimed at identifying text ranges corresponding to named entities and assigning them to predefined entity categories such as person, location, organization, etc. [13]. In most real-world fields, such as medicine and law, large-scale annotated data is scarce. The lack of training data can lead to a decrease in model performance. Recently, prompt learning methods have

S. Yang et al. (Eds.): WISA 2025, LNCS 16299, pp. 510–522, 2026.
https://doi.org/10.1007/978-981-95-6426-2_45

become an important new paradigm in the NLP community. By designing specific forms of input prompts, downstream tasks are transformed into language modeling tasks to bridge the gap between pre-trained and downstream tasks, effectively utilizing the capabilities of language models. For example, Cui et al. [2] proposed a template based approach that constructs prompts for each potential entity span and then predicts its entity type separately. Kondragunta et al. [7] used templates to generate synthetic sentences, transforming NER tasks into sequence-to-sequence tasks and providing a faster inference transformation framework. However, the prompt templates designed by existing methods lack contextual information related to entities, and it is difficult to guide the model to think and learn entity related information better. At the same time, existing methods are difficult to effectively capture and optimize entity boundaries and quantity information, ignoring the potential relation between entity types and entities.

To address the challenges of low-resource NER, we propose CRL-NER, a Chain-of-Thought and Reinforcement Learning based model. The main contributions are as follows: (1) We design an entity knowledge enhancement prompt grounded in the chain-of-thought paradigm, consisting of three parts: entity knowledge, knowledge enhancement prompt, and task instructions. Entity knowledge incorporates background knowledge, general examples, and recognition rules, providing rich contextual semantics. By linking this knowledge in a structured chain, the prompt guides the model to "learn before thinking" and reason more effectively in entity recognition. (2) To better capture relations between entities and entity types, we explicitly optimize entity boundaries and quantity information. We also construct reinforcement learning data based on human preference—entity-level preference data—yielding positive and negative examples that improve entity diversity and robustness. Through two-stage fine-tuning, the model alleviates category confusion and boundary shift problems. (3) Experiments on four benchmark datasets demonstrate that CRL-NER consistently outperforms baseline models, validating its effectiveness.

2 Related Work

2.1 Prompt Learning

The core idea of prompt learning is to design prompt templates for specific tasks, convert input into natural language form, and guide pre-trained models to generate expected prediction results. The method based on prompt learning is an innovative solution that has gained attention in the field of low-resource NER due to the success of large language models in various NLP downstream tasks [9,20]. Cui et al. [2] constructed a template based NER framework that successfully transformed NER tasks into language model sorting tasks within a sequence to sequence framework. Li et al. [8] rephrased NER as a text implication task to better utilize knowledge in PLM. However, existing prompt learning methods cannot better utilize contextual information of entities and guide models to learn entity knowledge according to human thinking to complete tasks. It is also difficult to explicitly capture and optimize entity boundary and quantity information, which limits model performance in low-resource environments.

2.2 Direct Preference Optimization

Direct Preference Optimization (DPO) [12] is a recent method that directly aligns model behavior with human or expert preferences. Unlike traditional reinforcement learning based on cumulative rewards, DPO adjusts model parameters using pairwise human feedback, ensuring consistency between the optimized model and human expectations. The algorithm involves two models: the reference model (π_{ref}), typically a supervised fine-tuned model with frozen parameters, and the policy model (π_θ), which is continuously updated. Training begins by constructing a preference dataset $\mathcal{D} = x^i, y_w^i, y_l^i{}_{i=1}^N$, where each triplet encodes a preferred (y_w) and less-preferred (y_l) response. The optimization objective increases the likelihood of preferred responses while constraining divergence from the reference distribution:

$$\max \mathbb{E}_{x \sim D, y \sim \pi_\theta(y|x)} \left[r_\phi(x, y) \right] - \beta \mathbb{D}_{\mathrm{KL}} \left(\pi_\theta(y|x) \,||\, \pi_{\mathrm{ref}}(y|x) \right). \tag{1}$$

Here, $r_\phi(x, y)$ is the implicit reward function, and the KL divergence term regularizes π_θ to remain close to π_{ref}, with β controlling the deviation strength.

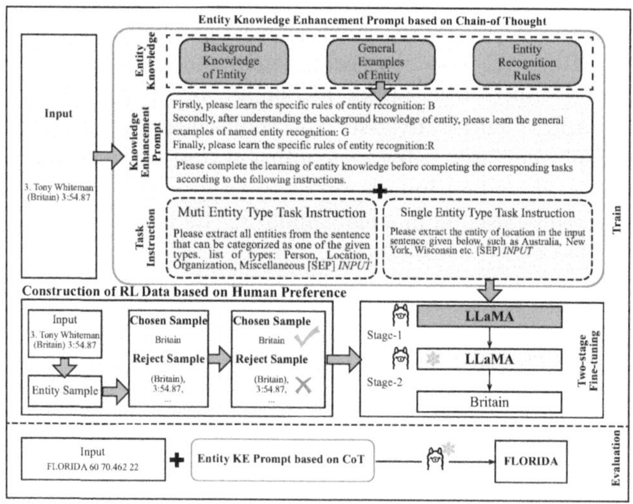

Fig. 1. The overview of our proposed model.

3 Method

3.1 Model Architecture

The overall structure is shown in Fig. 1. The model consists of three parts, including entity knowledge enhancement prompt based on chain of thought, reinforcement learning data construction based on human preference, and two-stage fine-tuning.

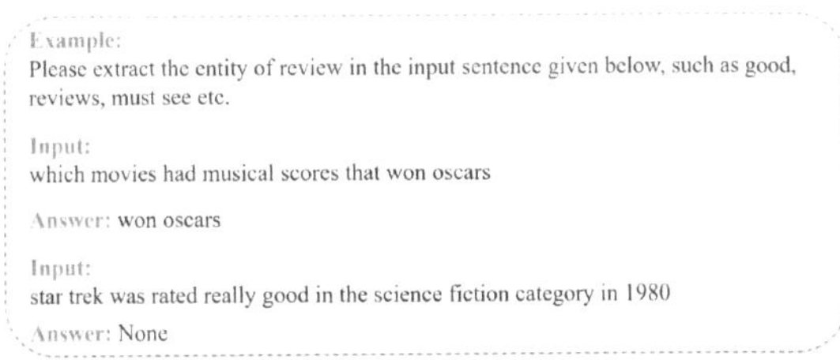

Fig. 2. Example of positive and negative answers for entity categories.

This article designs the displayed prompt template to guide the model to follow the human problem-solving train of thought, that is, the thinking logic of learning first and then solving the problem, and finally complete the NER task according to the task instructions. The entity knowledge enhancement prompt (T) based on chain of thought includes three parts: entity knowledge (E), knowledge enhancement prompt (K), and task instruction (I), that is, $T = (E, K, I)$. Entity knowledge consists of three parts: Background Knowledge of Entity (B), General Examples of Entity (G), and Entity Recognition Rules (R).

B mainly explains the basic concepts of NER tasks, textual introductions of common entity types and their related entities. For example, providing an explanation of named entity types, "Location: Indicating geographical places, e.g., New York City, Mount Everest.". G provides specific and universal entity recognition cases, enabling the model to clarify NER examples and entity classification methods. It should be noted that the entity classification and recognition method is general and has not been constructed or constrained by specific entity recognition rules.

For example, the input statement "The Lord of the Rings trilogy was filmed in New Zealand between 1999 and 2003." has an output format of "New Zealand $\rightarrow$ Location". R provide a detailed explanation of the rules and constraints for NER, such as the principles for determining entity boundaries and classification standards. In addition, to enhance the model's rejection ability, when there is no entity corresponding to the entity type in the input sentence, the model is required to output None. The positive and negative answer examples are shown in Fig. 2, and the R specified in this article are consistent with the positive and negative answer examples.

In order to more clearly guide the model to learn and utilize entity knowledge in the way of human thinking, K based on chain of thought is designed to guide the model to think based on the learned entity knowledge before giving a response. The K designed in this article is as follows:

> K=(Firstly, please learn the specific rules of entity recognition: B.
> Secondly, after understanding the background knowledge of entity, please learn the general examples of named entity recognition: G.
> Finally, please learn the specific rules of entity recognition: R.).

The task instruction I is divided into multi entity type tasks and single entity type tasks. The multi entity type task consists of multi entity examples and multi entity type task instructions. By providing all entity types, the model is required to recognize all entity types and their corresponding entities in the input sentence. The single entity type task consists of a single entity example and a single entity type task instruction, with the main purpose of outputting the corresponding entity based on the entity type specified in the instruction to complete the NER task.

3.2 Construction of Reinforcement Learning Data Based on Human Preference

Firstly, an entity level preference dataset is constructed for low-resource NER tasks. The data format consists of three parts, defined as $D = (P, y_w, y_l)$, where P represents contextual input, y_w represents chosen samples, and y_l represents rejected samples. For NER tasks, the prompt is consistent with the instructions for single entity type tasks. Input text $S = (s_1, s_2, \ldots, s_n)$, where n represents the length of the text and s_i represents the i-th character in the text. Words composed of characters can be represented as $w_{i:j} = \{s_i, s_{i+1}, \ldots, s_j\}$, where i represents the starting position of the word and j represents the ending position of the word. Divide the chosen samples into real entity chosen samples, if there is an entity of the specified entity type in S, the chosen sample is a real entity, and if there is no entity of the specified entity type in S, the chosen sample is None entity chosen samples. The corresponding rejected samples are divided into real entity rejected samples and None entity rejected samples. If the chosen sample is a real entity, it is divided into multiple and single real entity rejected samples based on their quantity. The selection of rejected samples for multiple real entities is either a single word randomly selected from S or None. The single real entity rejected samples may take any of the following forms:(None,$\{w_{i_1:j_1}, \ldots, w_{i_m:j_m}\}$,$w_i$). If the chosen sample is None, the rejected samples is a randomly selected single word (Table 1).

3.3 Two-Stage Fine-Tuning

In the first stage of fine-tuning, namely the supervised fine-tuning stage. Firstly, the input of the model is $X = (T, S)$, which is fed into an Embedding module to obtain the corresponding embedding vector. Subsequently, the obtained

Table 1. Chosen sample and rejected sample classification final results

Is there a real entity	Number of entity	Chosen sample	rejected sample
Yes	single	w_i	w_i, $\{w_{i_1:j_1}, \ldots, w_{i_m:j_m}\}$, None
Yes	multiple	$\{w_{i_1:j_1}, \ldots, w_{i_m:j_m}\}$	w_i
No	single	None	w_i

embedding vector h_i is fed into a Transformer Decoder. For each time step t, the decoded input is the output marker information $x_{1:t-1}$ from the previous step. The process is as follows:

$$h_i = \text{Embedding}(x_i), \quad i \in n, \qquad h_i^{\text{dec}} = \text{DECODER}(x_{1:t-1}). \tag{2}$$

Among them, $X = (x_1, x_2, \ldots, x_n)$ is input sentence, n is the total number of characters in X, and x_i represents the i-th character in X. h_i represented as the embedding vector corresponding to x_i.

Next, based on h_i^{dec}, the word with the highest probability is selected as the output of the current time step. Finally,cross entropy loss is used as the objective function to measure the difference between the entity predicted by the model and the target entity. The process can be formalized as:

$$p(x_t|x_{1:t-1}, X) = \sigma(h_t^{dec} W_{lm} + b_{lm}), \quad \mathcal{L} = -\sum_{t=1}^{n} \log p(x_t|x_{t-1}, X). \tag{3}$$

Among them, w_{lm} and b_{lm} are represented as trainable matrix parameters and bias parameters, and $softmax$ is selected as the activation function σ.

In the second stage of fine-tuning, namely the human preference fine-tuning stage, reinforcement learning data based on human preference is used, and the DPO algorithm is employed for supervised preference optimization. During the fine-tuning process, it is required that the model's predicted results after receiving input are closer to the chosen sample and farther away from the rejected sample. At the same time, it is necessary to ensure that the optimized model distribution does not deviate too much from the original model. The application form of DPO loss function in the model is as follows:

$$\begin{aligned} \mathcal{L}_{DPO_i}(\pi_\theta, \pi_{ref}) = -\mathbb{E}_{(x, y_{w_i}, y_{l_i}) \sim \mathcal{D}} \Big[& \log \sigma \big(\beta (\log p(y_{w_i}) - \log p_{\text{ref}}(y_{w_i})) \\ & - \beta (\log p(y_{l_i}) - \log p_{\text{ref}}(y_{l_i})) \big) \Big], \end{aligned} \tag{4}$$

$$\log p(y_{w_i}) = \pi_\theta(y_{w_i}|X), \qquad \log p_{\text{ref}}(y_{w_i}) = \pi_{\text{ref}}(y_{w_i}|X), \tag{5}$$

$$\log p(y_{l_i}) = \pi_\theta(y_{l_i}|X), \qquad \log p_{\text{ref}}(y_{l_i}) = \pi_{\text{ref}}(y_{l_i}|X), \tag{6}$$

Among them, i represents the chosen and rejected samples of the two styles shown above. σ is the activation function for *sigmoid*, and β is the hyperparameter.

The final model uses cross entropy loss for the first stage fine-tuning during parameter updates, and DPO loss for the human preference fine-tuning stage. The specific formula for the joint loss of the two stages is shown as follows:

$$\mathcal{L}_{LOSS} = \mathcal{L}_{CEL} + \mathcal{L}_{DPO}. \tag{7}$$

4 Experiments

4.1 Dataset and Evaluation

In order to conduct a fair comparison of the proposed models, experiments were conducted on four NER benchmark datasets: CoNLL2003 [13], MIT Movie [3], MIT Restaurant [3], and ATIS [5]. Continuing with previous work methods, this article adopts recall (R), precision (P), and F1 score (F1) as evaluation metrics.

4.2 Baselines

Representative low-resource NER methods can be grouped as follows. Nearest-neighbor approaches include NeighborTagger [17], which transfers tags from retrieved sentences, and NNShot/StructShot [18], the latter extending NNShot with a Viterbi decoder. Span- and example-based methods (e.g., Example-based [21]) detect spans first and then classify entities by comparison with labeled instances. Noisy-data utilization is explored by MP-NSP [5], which leverages web-scale noisy NER data for pre-training and self-training. Template- and prompt-based approaches are widely studied: TemplateNER [2], PTE [8], Light-NER [1], EntLM [11], InstructionNER [16], PromptNER [14], TKDP [10], and EKEP [4] each reformulate NER as a template or prompt-learning task, with innovations such as dual-slot prompts, soft prompt embeddings, and knowledge-enriched prompts. Seq2seq and generation frameworks (e.g., BART-based [7]) simplify NER as sequence-to-sequence learning, while OADA [15] focuses on data augmentation. Finally, MRC-based methods (PMRC [6], BERT-MRC [19]) reformulate NER as machine reading comprehension tasks.

4.3 Low-Resource Experiment

A portion of training instances were randomly sampled from the MIT Movie, MIT Restaurant, CoNLL2003, and ATIS datasets as training data for the target domain. k instances were randomly sampled for each entity type and non-existent entity type, $k \in \{10, 20, 50\}$. If the number of instances of a specific type is less than the number extracted, all available instances are used for training. The final F1 values for the MIT, ATIS, CoNLL2003 datasets with 10, 20, and 50 shot settings are shown in Table 2, Table 3 and Table 4 respectively.

Table 2. Experiment results of MIT Movie and MIT Restaurant datasets

Models	MIT Movie			MIT Restaurant		
	10	20	50	10	20	50
NeighborTagger	3.10	4.50	4.10	4.10	3.60	4.00
Example-based	40.10	39.50	40.20	27.60	29.50	31.20
MP-NSP	36.40	36.80	38.00	46.10	48.20	49.60
TemplateNER	42.40	54.20	59.60	53.10	60.30	64.10
LightNER	41.70	57.80	73.10	48.50	58.00	62.00
InstructionNER	64.40	70.00	74.10	58.70	65.50	71.20
PTE (soft)	47.80	60.80	68.10	-	-	-
NNShot	50.47	58.94	71.17	-	-	-
StructShot	53.19	61.42	72.07	-	-	-
EntLM	57.31	62.36	71.93	-	-	-
BART +OADA	57.42	64.04	72.19	-	-	-
Improving and Simplifying Template-based BART	53.29	61.89	75.75	47.09	60.25	67.02
PromptNER	55.60	68.20	76.50	56.10	62.60	69.30
TKDP	59.56	62.87	-	56.80	61.76	-
BERT-MRC	18.70	48.30	55.50	12.30	37.10	53.50
PMRC	58.20	**75.20**	**78.70**	49.50	58.20	65.40
EKEP+random	64.19	71.15	76.12	56.25	61.03	66.69
Ours	**67.92**	73.21	78.36	**62.26**	**66.68**	**71.80**

Compared to the NeighborTagger method, the CRL-NER model significantly boosted the F1 score by 74.26%, especially under the 50-shot setting. In the 10-shot setting on the ATIS dataset, CRL-NER outperformed NeighborTagger and MP-NSP by 86.7% and 22.2%, respectively. For the CoNLL2003 dataset, CRL-NER improved F1 scores by 4.12% and 7.25% over StructShot and NNShot methods, respectively, on the 10-shot setting. On the MIT Restaurant dataset, CRL-NER showed F1 score improvements of 12.76%, 8.48%, and 6.4% over PMRC in the 10, 20, and 50-shot settings, respectively. For the MIT Movie dataset, CRL-NER outperformed EKEP by 3.73%, 2.06%, and 2.24% in the 10, 20, and 50-shot settings. Lastly, CRL-NER achieved the highest improvement of 8.43% over OADA on the 10-shot setting of the CoNLL2003 dataset.

4.4 Imbalanced Domain Instance Experiment

Random sampling was performed on the CoNLL2003 training set to construct a learning scenario with imbalanced instances, resulting in 632 training instances, including 456 ORG type entities, 1010 PER type entities, 20 LOC type entities, and 20 MISC type entities. The test set was kept unchanged for testing. The experiment results as shown in Table 5, our model has an improvement of 1.82% and 23.12% compared to the fine-tuned LlaMA-3 model on entity types PER

Table 3. Experiment results of ATIS dataset

Methods	ATIS		
	10	20	50
NeighborTagger	6.70	8.80	11.10
MP-NSP	71.20	74.80	76.00
TemplateNER	77.30	88.90	93.50
LightNER	76.30	85.30	92.80
InstructionNER	90.70	93.00	95.10
Improving and Simplifying Template-based BART	91.90	93.80	94.58
PromptNER	91.50	94.30	**95.50**
BERT-MRC	35.30	63.20	90.20
PMRC	92.00	94.40	95.30
Ours	**93.40**	**94.57**	95.01

Table 4. Experiment results of CoNLL2003 dataset

Methods	CoNLL2003		
	10	20	50
EKEP+random	63.94	71.83	75.18
NNShot	59.24	66.89	72.63
StructShot	62.37	69.51	74.73
TemplateNER	57.86	66.38	72.71
InstructionNER	61.30	67.20	76.30
EntLM	64.79	69.52	73.66
BART +OADA	58.06	65.20	73.04
Ours	**66.49**	**72.36**	**77.28**

and ORG with sufficient resources, respectively. On the resource scarce entity type LOC, there was a 9.54% increase, especially on the MISC entity type, where the improvement was particularly significant, reaching a 29.37% increase.

4.5 Abundant Resources Experiment

Using the entire training set of CoNLL2003 to train the model and evaluate its performance on the test set, the results are shown in Table 6. From the experimental results, it was found that the model had the highest F1 value in a rich resource environment.

Table 5. Experiment results of imbalanced instances within the domain

Models	PER	ORG	LOC*	MISC*	Overall
LlaMA-3	94.85	56.64	66.49	11.15	68.64
Ours	96.67	79.76	76.03	40.52	77.82

Table 6. Model results in rich resource experiments

Few-shot Friendly Models	P	R	F1
NeighborTagger	-	-	89.94
TemplateNER	90.51	93.34	91.90
LightNER	92.39	93.48	92.93
PTE (soft)	92.01	92.01	92.23
Improving and Simplifying Template-based BART (T5-base)	91.00	89.00	90.00
PMRC	92.42	91.31	91.86
Ours	95.03	94.29	94.71

4.6 Ablation Study

In order to further explore the impact of entity knowledge enhancement prompt based on chain of thought and reinforcement learning fine-tuning based on human preference on the performance of the model, ablation experiments were designed for verification. The ablation experiment results are shown in Table 7.

Table 7. Experiment results of low-resource ablation

Models	MIT Movie			MIT Restaurant			CoNLL2003			ATIS		
	10	20	50	10	20	50	10	20	50	10	20	50
w/o CoT-EKP+ DPO	65.04	66.04	70.45	59.41	62.41	62.67	61.63	65.71	67.54	88.23	90.08	91.45
w/o DPO	66.28	71.88	75.29	61.20	64.42	68.56	64.00	67.77	75.24	90.02	92.21	93.26
Ours	67.92	73.21	78.36	62.26	66.68	71.80	66.49	72.36	77.28	93.40	94.57	95.01

"W/o DPO" refers to the model fine-tuned based on human preferences without reinforcement learning. On the MIT Movie dataset, the F1 score of the original model decreased by up to 3.07% under the 50-shot setting. For the MIT Restaurant dataset, F1 decreased by at least 1.06% on the 10-shot setting, and the ATIS dataset showed a decrease of 3.38% on the same setting. The CoNLL2003 dataset experienced the largest impact on the 10 and 20-shot settings, with decreases of 2.49% and 4.59%, respectively. "W/o CoT-EKP+DPO" represents the simultaneous removal of entity knowledge enhancement prompts based on the chain of thought and human preference-based reinforcement learning fine-tuning. Compared to the original model, F1 values dropped significantly by 2.88% and 2.85% on the 10-shot settings of the MIT Movie and MIT Restaurant datasets, respectively. For the CoNLL2003 dataset, F1 dropped by 9.74% on the 50-shot setting, and for the ATIS dataset, there was a 5.17% decrease on the 10-shot setting.

Table 8. Experiment results of imbalanced ablation of instances within the domain

Models	PER	ORG	LOC*	MISC*
LlaMA-3	94.85	56.64	66.49	11.15
with CoT-EKP	94.88	66.96	75.11	29.76
Ours	96.67	79.76	76.03	40.52

The experiment results under imbalanced instance settings are shown in Table 8. "CoT-EKP" refers to a model that enhances prompts using entity knowledge based on the chain of thought. Compared to the baseline LLaMA-3 model, the entity knowledge-enhanced model showed improvements in resource-rich entity types (PER and ORG), with increases of 0.03% and 10.32%, respectively. For resource-deficient entity types (LOC and MISC), the increases were 8.62% and 18.61%, respectively. After incorporating reinforcement learning fine-tuning based on human preferences, the model's performance improved further: PER and ORG for resource-rich entity types increased by 1.79% and 12.8%, while LOC and MISC for resource-deficient entity types grew by 0.92% and 10.76%, respectively.

case	**Input**: Although there are no known precedents in the country , the action is envisaged in Spanish legislation governing sports events . **Golden Entity**: (Spanish) MISC
w/o CoT-EKP+DPO	**Input**: Although there are no known precedents in the country , the action is envisaged in Spanish legislation governing sports events . **Predict Entity**: None
w/o DPO	**Input**: Although there are no known precedents in the country , the action is envisaged in Spanish legislation governing sports events . **Predict Entity**: (Spanish legislation) MISC
Ours	**Input**: Although there are no known precedents in the country , the action is envisaged in Spanish legislation governing sports events . **Predict Entity**: (Spanish) MISC

Fig. 3. Case analysis diagram.

4.7 Case Study

Consider the input sentence "Although there are no known precedents in the country, the action is envisaged in Spanish legislation governing sports events." As shown in Fig. 3, the "w/o CoT-EKP+DPO" model lacks sufficient contextual semantic understanding, resulting in its failure to identify entities of the miscellaneous (MISC) type in the input sentence. On the other hand, the "w/o DPO" model can capture some contextual semantic information related to entities, but it struggles with accurately defining entity boundaries. This leads to fuzzy entity detection and incorrect predictions.

In contrast, the CRL-NER model benefits from a structured prompt template that helps it learn more effectively. Additionally, by fine-tuning based on human preferences, the model is able to better capture and display the boundaries of entities. This reduces ambiguity in entity recognition, ultimately allowing for more accurate entity identification in the sentence.

5 Conclusion

This article introduces a new low-resource NER method based on chain of thought and reinforcement learning. The method aims to more fully explore the relation between entity types and entities, more explicitly capture and optimize entity boundary and quantity informations, and solve the problem of fuzzy boundaries when extracting entities. At the same time, it can effectively stimulate and utilize the contextual learning ability of the model itself, guiding the model to think effectively about the contextual information of learning entities through prompt template. The experimental results on four NER benchmark datasets, MIT Movie, MIT Restaurant, ATIS, and CoNLL2003, further demonstrate the efficiency of the method.

Acknowledgments. This work was supported by National Natural Science Foundation of China major projects under grant numbers (No.62137001).

References

1. Chen, X., et al.: Lightner: a lightweight tuning paradigm for low-resource NER via pluggable prompting. In: COLING. International Committee on Computational Linguistics, pp. 2374–2387 (2022)
2. Cui, L., Wu, Y., Liu, J., Yang, S., Zhang, Y.: Template-based named entity recognition using BART. In: ACL/IJCNLP (Findings). Findings of ACL, vol. ACL/IJCNLP 2021, pp. 1835–1845. Association for Computational Linguistics (2021)
3. Hakkani-Tür, D., et al.: Multi-domain joint semantic frame parsing using bi-directional RNN-LSTM. In: INTERSPEECH, pp. 715–719. ISCA (2016)
4. Hou, W., Zhao, W., Liu, X., Guo, W.: Knowledge-enriched prompt for low-resource named entity recognition. ACM Trans. Asian Low Resour. Lang. Inf. Process. **23**(5), 72 (2024)

5. Huang, J., et al.: Few-shot named entity recognition: a comprehensive study. CoRR **abs/2012.14978** (2020)
6. Huang, J., Yan, D., Cai, Y.: PMRC: prompt-based machine reading comprehension for few-shot named entity recognition. In: AAAI, pp. 18316–18326. AAAI Press (2024)
7. Kondragunta, M., Perez-de-Viñaspre, O., Oronoz, M.: Improving and simplifying template-based named entity recognition. In: EACL (Student Research Workshop), pp. 79–86. Association for Computational Linguistics (2023)
8. Li, D., Hu, B., Chen, Q.: Prompt-based text entailment for low-resource named entity recognition. In: COLING, pp. 1896–1903. International Committee on Computational Linguistics (2022)
9. Li, Z., Pang, N., Zhao, X.: Instruction tuning large language models for multimodal relation extraction using lora. In: Web Information Systems and Applications - 21st International Conference, WISA 2024, Yinchuan, China, August 2-4, 2024, Proceedings, pp. 364–376. Springer (2024)
10. Liu, J., et al.: TKDP: threefold knowledge-enriched deep prompt tuning for few-shot named entity recognition. IEEE Trans. Knowl. Data Eng. **36**(11), 6397–6409 (2024)
11. Ma, R., et al.: Template-free prompt tuning for few-shot NER. In: NAACL-HLT, pp. 5721–5732. Association for Computational Linguistics (2022)
12. Rafailov, R., Sharma, A., Mitchell, E., Manning, C.D., Ermon, S., Finn, C.: Direct preference optimization: your language model is secretly a reward model. In: NeurIPS (2023)
13. Sang, E.F.T.K., Meulder, F.D.: Introduction to the conll-2003 shared task: language-independent named entity recognition. In: CoNLL, pp. 142–147. ACL (2003)
14. Shen, Y., et al.: Promptner: prompt locating and typing for named entity recognition. In: ACL, vol. 1, pp. 12492–12507. Association for Computational Linguistics (2023)
15. Wang, H., Cheng, L., Zhang, W., Soh, D.W., Bing, L.: Order-agnostic data augmentation for few-shot named entity recognition. In: ACL, vol. 1, pp. 7792–7807. Association for Computational Linguistics (2024)
16. Wang, L., et al.: Instructionner: a multi-task instruction-based generative framework for few-shot NER. CoRR **abs/2203.03903** (2022)
17. Wiseman, S., Stratos, K.: Label-agnostic sequence labeling by copying nearest neighbors. In: ACL, vol. 1, pp. 5363–5369. Association for Computational Linguistics (2019)
18. Yang, Y., Katiyar, A.: Simple and effective few-shot named entity recognition with structured nearest neighbor learning. In: EMNLP, vol. 1, pp. 6365–6375. Association for Computational Linguistics (2020)
19. Yu, J., Bohnet, B., Poesio, M.: Named entity recognition as dependency parsing. In: ACL, pp. 6470–6476. Association for Computational Linguistics (2020)
20. Zeng, G., Xue, Q., Lu, X.: Instruction tuning with LLMs for programming exercise generation. In: Web Information Systems and Applications - 21st International Conference, WISA 2024, Yinchuan, China, August 2–4, 2024, Proceedings, pp. 377–385. Springer, Cham (2024)
21. Ziyadi, M., Sun, Y., Goswami, A., Huang, J., Chen, W.: Example-based named entity recognition. CoRR **abs/2008.10570** (2020)

DSRAG: A Domain-Specific Retrieval Framework Based on Document-Derived Multimodal Knowledge Graph

Mengzheng Yang[1], Yanfei Ren[1](✉), David Osei Opoku[2], Ruochang Li[1], Peng Ren[2], and Chunxiao Xing[2]

[1] School of Software, Henan University, Kaifeng 475004, China
renyanfei@henu.edu.cn
[2] BNRist, DCST, RIIT, Tsinghua University, Beijing 100084, China

Abstract. Current general-purpose large language models (LLMs) commonly exhibit knowledge hallucination and insufficient domain-specific adaptability in domain-specific tasks, limiting their effectiveness in specialized question answering scenarios. Retrieval-augmented generation (RAG) effectively tackles these challenges by integrating external knowledge to enhance accuracy and relevance. However, traditional RAG still faces limitations in domain knowledge accuracy and context modeling. To enhance domain-specific question answering performance, this work focuses on a graph-based RAG framework, emphasizing the critical role of knowledge graph quality during the generation process. We propose DSRAG (Domain-Specific RAG), a multimodal knowledge graph-driven retrieval-augmented generation framework designed for domain-specific applications. Our approach leverages domain-specific documents as the primary knowledge source, integrating heterogeneous information such as text, images, and tables to construct a multimodal knowledge graph covering both conceptual and instance layers. Building on this foundation, we introduce semantic pruning and structured subgraph retrieval mechanisms, combining knowledge graph context and vector retrieval results to guide the language model towards producing more reliable responses. Evaluations using the Langfuse multidimensional scoring mechanism show that our method excels in domain-specific question answering, validating the efficacy of integrating multimodal knowledge graphs with retrieval-augmented generation.

Keywords: Retrieval-augmented generation · Knowledge graph · Large language model

1 Introduction

Current general-purpose large language models (LLMs) face two main challenges in domain-specific tasks: insufficient coverage of specialized knowledge in pretraining data [1], and frequent factual hallucinations or logical inconsistencies

S. Yang et al. (Eds.): WISA 2025, LNCS 16299, pp. 523–530, 2026.
https://doi.org/10.1007/978-981-95-6426-2_46

in outputs [2]. Although models like ChatGPT, DeepSeek, GLM, and Qwen demonstrate strong semantic understanding in general domains, they struggle in specialized applications. This gap highlights the limitations of purely data-driven statistical learning and the need to integrate external knowledge to enhance reliability in domain-specific contexts [3].

To mitigate this gap, two major approaches have been studied: unsupervised incremental pre-training [4] and retrieval-augmented generation (RAG) [5]. The former injects domain corpora but suffers from high training costs and slow updates; the latter dynamically retrieves from external knowledge bases but still faces issues in semantic precision and retrieval quality [6].

GraphRAG [7] has recently attracted attention by organizing semantic units into knowledge graphs and leveraging entity relationships for retrieval and generation. However, its performance depends on the accuracy and completeness of the underlying knowledge graph, which remains challenging to construct in domain-specific settings [8]. Since multimodal documents (text, images, tables) are the main carriers of such knowledge, building high-quality multimodal knowledge graphs is critical for enabling reliable Graph-based RAG.

In this paper, we propose DSRAG, a multimodal knowledge graph-driven RAG framework for domain-specific tasks. We utilize document tables of contents, LLMs, and expert annotations to generate summaries, extract domain concepts, and construct a concept knowledge graph. On this basis, entities, relations, and multimodal data are extracted to build an instance knowledge graph. Finally, a graph-enhanced retrieval method that combines semantic subgraph retrieval with vector matching is designed, improving both accuracy and efficiency in domain-specific question answering.

Contributions. The main contributions are summarized as follows:

1. We propose DSRAG, a framework integrating multimodal knowledge graphs with RAG for domain-specific tasks.
2. We construct DSKG, a multimodal knowledge graph combining text, images, and tables, with conceptual and instance layers for systematic modeling.
3. We design DSKG-Enhanced Retrieval, combining semantic subgraph retrieval with vector matching to improve accuracy and robustness in domain-specific QA.

2 Related Work

Traditional retrieval-augmented generation (RAG) methods rely on external knowledge bases to enhance generation models. The RAG framework by Lewis et al. achieved notable success in open-domain tasks [5], but struggles with structured semantic representations. Graph-based RAG (GraphRAG) [7] addresses this by leveraging knowledge graphs to organize semantic units, enabling retrieval to exploit entity relationships and incorporating graph constraints during generation to mitigate hallucinations and logical conflicts. Compared with traditional

RAG, GraphRAG emphasizes structured entities and relations to improve accuracy and interpretability.

Knowledge graphs, central to GraphRAG, store domain knowledge via graph structures. Traditional methods rely on structured data for entity and relation extraction [9] but face limitations with multimodal information. In technical documents, images and tables are as important as text. Research has explored integrating multiple modalities, e.g., Mukea [10] integrates image and text for visual QA, and MORLD [11] enhances multimodal retrieval, yet challenges remain in efficiently fusing heterogeneous sources in domain-specific contexts.

GraphRAG applications are expanding: HybridRAG [12] combines knowledge graphs with vector retrieval for specialized terminology and complex formats; DO-RAG [13] improves domain-specific QA by enhancing retrieval with knowledge graphs. Industrial systems like AutoFlow [14] and RAGFlow [15] further exemplify this trend. Nevertheless, graph-based RAG still faces critical issues: (1) lack of systematic conceptual and instance layer modeling; (2) limited integration of text, images, and tables; and (3) incomplete graphs introducing noise and degrading generation quality.

To address these challenges, this paper proposes DSRAG, which constructs a multimodal knowledge graph with conceptual and instance layers from domain-specific documents. By systematically integrating text, images, and tables, DSRAG enhances retrieval and generation reliability in domain-specific question answering.

3 Methodology

3.1 Framework of DSRAG

To address semantic heterogeneity and factual hallucinations in domain-specific QA, we propose DSRAG, a retrieval-augmented generation framework driven by a domain-specific multimodal knowledge graph (DSKG) (Fig. 1). DSRAG constructs a hierarchical knowledge graph comprising a Concept KG, which captures core domain concepts and workflows via summaries and keywords extracted from document structures using LLMs and expert annotations, and an Instance KG, representing fine-grained triples from text and multimodal data. The DSKG-Enhanced Retrieval employs a two-stage strategy, combining graph traversal for structured semantic retrieval with vector similarity search for contextual relevance. Structured graph, vector-based, and historical dialogue contexts are integrated into a unified LLM prompt, producing answers with enhanced factual accuracy and semantic consistency.

3.2 DSKG Construction

As shown in Fig. 2, the proposed DSKG construction consists of three stages: data preprocessing, Concept KG (CKG) construction, and Instance KG (IKG) construction, forming a pipelme from raw domain specific documents to a multimodal knowledge graph. In the data preprocessing stage, Mineru and a hybrid

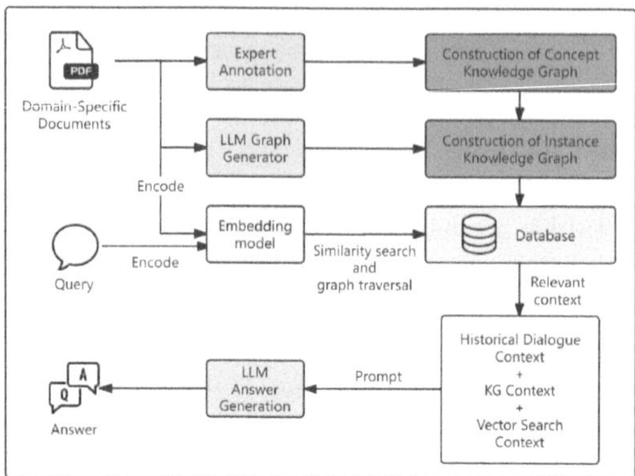

Fig. 1. The overall framework of DSRAG.

OCR engine parse PDFs at the pixel level and convert them into standardized JSON-LD. Text is denoised, tables are structurally detected and converted to Markdown, and vision models extract image features and generate descriptions. All outputs are unified into a standard text format for knowledge extraction, with paragraph-based segmentation preserving semantic continuity and tables/images treated as independent chunks for multimodal integration.

In the CKG and IKG construction stage, the document hierarchy guides concept KG creation, with Part, Chapter, and Section titles as concept nodes and section summaries as semantic cores. Expert annotations and large language models generate subgraphs and extract entities and relations, while hierarchical edges (e.g., subTopic, hasKeyword) form a DAG preserving organizational logic and providing pruning paths for QA. Building on this, the instance KG handles multimodal heterogeneity via structured chunking of text and independent integration of figures and tables. A three-layer extractor captures macro semantic relations, core entities and operations, and fine-grained technical entities, enriched by attribute completion. Extracted knowledge is aligned with CKG subgraphs to yield a multimodal, hierarchical KG that supports reliable domain-specific QA.

3.3 DSKG-Enhanced Retrieval

As shown in Fig. 3, the proposed DSKG-enhanced retrieval integrates structured graph semantics with unstructured corpus search, forming a multi-source framework that links conceptual abstraction with instance-level detail. In the graph-guided focusing stage, the knowledge graph decomposes the query into sub-queries and applies layer-wise pruning: at the concept KG, semantic matching filters irrelevant parts and locates relevant chapters and sections; at the instance KG, the focused subgraph constrains entity, attribute, and relation matching. The merged results yield a structured semantic context closely aligned with the query.

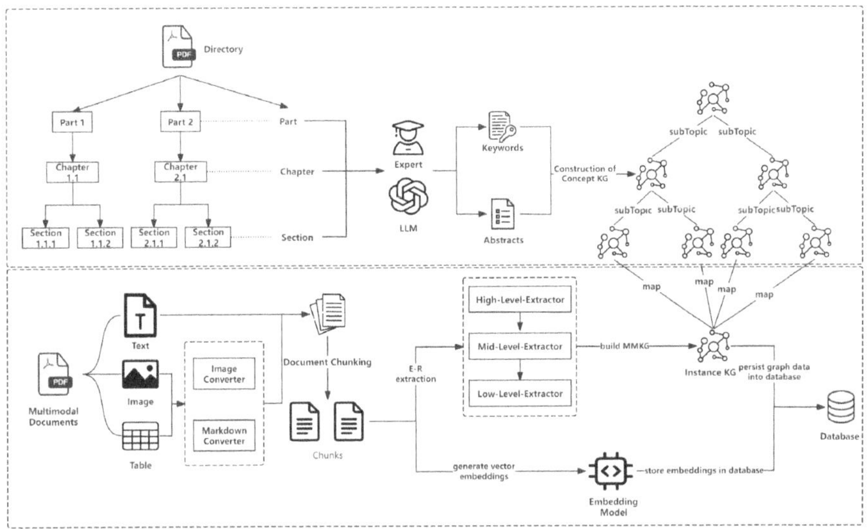

Fig. 2. Flow Diagram of DSKG Construction from Original Documents to MMKG.

In the vector retrieval stage, embedding-based search complements structured focusing to enhance coverage and granularity. Candidate chunks from pruned chapter nodes are aligned with the embedding space, while refined queries from the graph context are encoded into high-dimensional vectors for similarity search. Finally, structured semantics, retrieved text, and dialogue history are consolidated into a Unified Search Context for the language model, ensuring factual accuracy, semantic consistency, and improved precision in domain-specific QA.

4 Experiment and Results

4.1 Experiment Setup

Dataset. The experiments use domain-specific multimodal database documentation comprising over 4,500 pages of text, tables, images, and code snippets. Organized into seven parts, 54 chapters, and roughly 900 knowledge units, it provides a hierarchical structure for knowledge graph construction. To evaluate DSRAG, domain experts curated 100 high-quality questions with ground truths, covering database operations, configuration management, and performance optimization.

Environments. Experiments were conducted on a high-performance Ubuntu 22.04.5 LTS workstation equipped with an Intel Xeon Gold 6226R CPU (32 cores), 256 GB RAM, and an NVIDIA RTX A5000 GPU. The software environment included CUDA 12.8 and Python 3.11. The experimental pipeline utilized GPT-4o-mini for generation and graph construction, GPT-4o for evaluation,

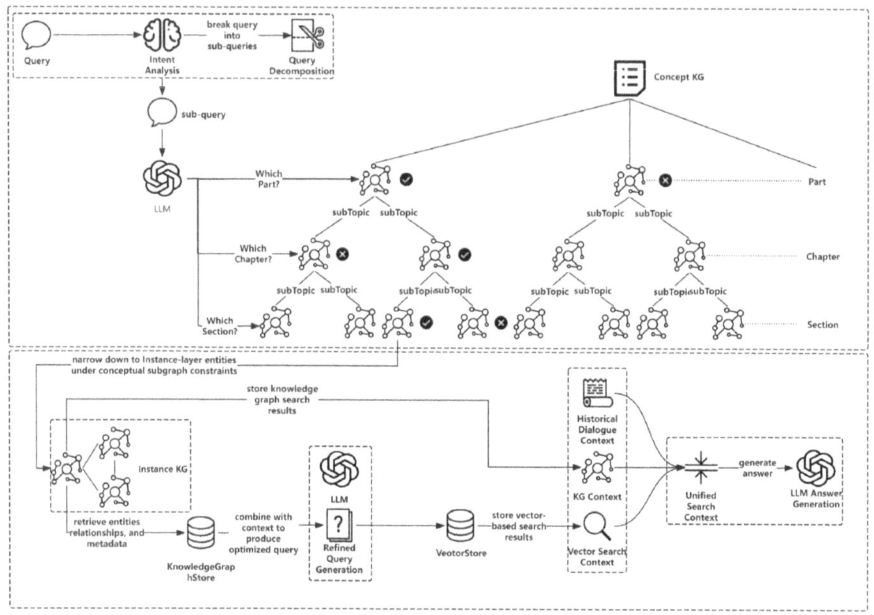

Fig. 3. DSKG-enhanced Retrieval Process.

text-embedding-3-small for embeddings, and jina-reranker-v2-base for reranking, ensuring efficient processing and accurate domain-specific knowledge graph construction and retrieval-augmented generation.

Baselines. Three baselines were established to evaluate the impact of domain knowledge on factuality, contextual relevance, and cross-modal reasoning. NativeRAG [5] is a standard RAG framework without knowledge graph support. TiDB AutoFlow [14] integrates structured domain knowledge to enhance factual consistency and semantic alignment. RAGFlow [15] is a multimodal RAG engine supporting heterogeneous data and hybrid retrieval strategies for broader context coverage.

Metrics. Performance was evaluated using Faithfulness, Answer Relevancy, and Contextual Precision. Faithfulness measures the proportion of context-supported statements in the generated answer, assessing factual consistency. Answer Relevancy evaluates alignment with the query, typically via cosine similarity, while Contextual Precision checks whether relevant facts are ranked higher within the retrieved context. Together, these metrics provide a comprehensive assessment of answer quality, factual grounding, and retrieval effectiveness.

4.2 Performance Comparison

Overall Performance. As shown in Table 1, DSRAG outperforms all baselines across Faithfulness, Answer Relevancy, and Contextual Precision. It generates factually consistent answers, provides highly relevant responses, and accurately ranks contextual knowledge, demonstrating its effectiveness in domain-specific QA. These results highlight the value of integrating domain-specific multimodal knowledge graphs with retrieval-augmented generation to enhance intelligent question-answering performance.

Table 1. Performance Comparison between DSRAG and Baseline Methods

Method	Faithfulness	Answer Relevancy	Context Precision
Naive RAG	0.63	0.58	0.55
TiDB AutoFlow	0.72	0.65	0.62
RAGFlow	0.69	0.67	0.70
DSRAG	**0.83**	**0.87**	**0.85**

Ablation Study. We conducted an ablation study to assess the contribution of different components: (1) baseline RAG without knowledge graphs, (2) RAG with only the IKG, and (3) the full DSRAG framework with both CKG and IKG (DSKG). As shown in Table 2, the complete DSRAG achieved the highest scores across all metrics—faithfulness 0.83, answer relevance 0.87, and contextual precision 0.85—demonstrating the critical role of integrating structured knowledge in enhancing factual consistency, semantic relevance, and retrieval accuracy.

Table 2. Ablation Study on RAG Configurations with Different Levels of Knowledge Graph Incorporation

Method	Faithfulness	Answer Relevancy	Context Precision
RAG	0.68	0.70	0.58
IKG-RAG	0.72	0.80	0.69
DSKG-RAG	**0.83**	**0.87**	**0.85**

5 Conclusion

This paper presents DSRAG, a multimodal knowledge graph-driven retrieval-augmented generation framework for domain-specific tasks, enhancing response accuracy and adaptability. By integrating concept and instance KGs, DSRAG optimizes retrieval and generation through semantic pruning and structured subgraph search. Experiments show it surpasses traditional baselines in faithfulness, relevance, and contextual accuracy, with ablation studies confirming the complementary value of combining concept and instance KGs. This work provides a novel approach to domain-specific QA, and future research may explore advanced graph structures and cross-modal fusion for more complex tasks.

References

1. Gu, Y., Tinn, R., Cheng, H., et al.: Domain-specific language model pretraining for biomedical natural language processing. ACM Trans. Comput. Healthc. (HEALTH) **3**(1), 1–23 (2021)
2. Ji, Z., Lee, N., Frieske, R., et al.: Survey of hallucination in natural language generation. ACM Comput. Surv. **55**(12), 1–38 (2023)
3. Hu, L., Liu, Z., Zhao, Z., et al.: A survey of knowledge enhanced pre-trained language models. IEEE Trans. Knowl. Data Eng. **36**(4), 1413–1430 (2023)
4. Singhal, K., Azizi, S., Tu, T., et al.: Large language models encode clinical knowledge. Nature **620**(7972), 172–180 (2023)
5. Lewis, P., Perez, E., Piktus, A., et al.: Retrieval-augmented generation for knowledge-intensive NLP tasks. Adv. Neural. Inf. Process. Syst. **33**, 9459–9474 (2020)
6. Barnett, S., Kurniawan, S., Thudumu, S., et al.: Seven failure points when engineering a retrieval augmented generation system. In: Proceedings of the IEEE/ACM 3rd International Conference on AI Engineering-Software Engineering for AI, pp. 194–199 (2024)
7. Peng, B., Zhu, Y., Liu, Y., et al.: Graph retrieval-augmented generation: a survey. arXiv preprint arXiv:2408.08921 (2024)
8. Abu-Salih, B.: Domain-specific knowledge graphs: a survey. J. Netw. Comput. Appl. **185**, 103076 (2021)
9. Yin, H., Huang, S., Wang, Z., Ye, Y., Zhu, W.: An generative entity relation extraction model based on UIE for legal text. In: International Conference on Web Information Systems and Applications, pp. 91–99. Springer, Cham (2024)
10. Ding, Y., Yu, J., Liu, B., et al.: Mukea: multimodal knowledge extraction and accumulation for knowledge-based visual question answering. In: Proceedings of the IEEE/CVF Conference on Computer Vision and Pattern Recognition, pp. 5089–5098 (2022)
11. Sheng, M., Wang, S., Zhang, Y., et al.: MQRLD: a multimodal data retrieval platform with query-aware feature representation and learned index based on data lake. Inf. Process. Manage. **62**(4), 104101 (2025)
12. Sarmah, B., Mehta, D., Hall, B., et al.: Hybridrag: integrating knowledge graphs and vector retrieval augmented generation for efficient information extraction. In: Proceedings of the 5th ACM International Conference on AI in Finance, pp. 608–616 (2024)
13. Opoku, D.O., Sheng, M., Zhang, Y.: Do-rag: a domain-specific qa framework using knowledge graph-enhanced retrieval-augmented generation. arXiv preprint arXiv:2505.17058 (2025)
14. PingCAP: Tidb autoflow (2023). https://github.com/pingcap/autoflow. Accessed 28 Apr 2025
15. InfiniFlow: Ragflow (2024). https://www.ragflow.io/. Accessed 28 Apr 2025

Phase Transition Behavior in Random Constraint Satisfaction Problems with Heterogeneous Constraints

Chunyan Zhao[1(✉)], Di Jin[2], and Yutao Li[2]

[1] School of Data Science and Engineering, East China Normal University, Shanghai 200062, China
cyzhao@dase.ecnu.edu.cn

[2] College of Science, University of Shanghai for Science and Technology, Shanghai 200093, China

Abstract. The phase transition phenomenon in random constraint satisfaction problems (CSPs) is a pivotal topic at the intersection of computational complexity theory and theoretical computer science. We propose an extended random CSP model ($\mathrm{RB}^{\mathrm{mix}}$ model) by introducing a hierarchical hybrid structure combining multi-length local constraints and all-different global constraints, systematically investigating the impact of multiscale heterogeneous constraints on computational hardness phase transitions in NP-complete problems under the theoretical framework of exact phase transitions in RB model. Theoretically, we rigorously prove the existence of a sharp satisfiability phase transition using first- and second-moment methods, deriving analytical expressions for critical threshold parameters. Numerical experiments demonstrate that the hybrid constraint mechanism preserves the sharp phase transition of RB model while significantly enhancing theoretical inclusiveness. Moreover, all-different global constraint integration maintains critical phase transition characteristics. These results establish a universal framework for phase transition analysis under heterogeneous constraints, providing new theoretical tools for understanding computational hardness boundaries of NP-complete problems, optimizing constraint design, and generating algorithm testing benchmarks.

Keywords: Constraint satisfaction problem · model $\mathrm{RB}^{\mathrm{mix}}$ · Phase transition · Mixed constraints

This work was supported by the National Key Research and Development Program of China (No. 2023YFC3304700), the Shanghai 2024 "Science and Technology Innovation Action Plan" Project (No. 24BC3201100), the Program of Shanghai Academic/Technology Research Leader (No. 23XD1401100), the National Natural Science Foundation of China (Nos. 11301339, and 11491240108).

S. Yang et al. (Eds.): WISA 2025, LNCS 16299, pp. 531–539, 2026.
https://doi.org/10.1007/978-981-95-6426-2_47

1 Introduction

Phase transitions in NP-complete problems has attracted much interest in statistical physics, discrete mathematics and theoretical computer science [1]. It is generally thought that most of random CSPs belong to the class of NP-complete problems. An instance of a random CSP consists of a set of m constraints over a set of n variables, and each variable takes values from a nonempty domain D with size $|D| = d$. The instance is satisfiable if there exists a variable assignment that simultaneously satisfies all constraints and we call such an assignment a solution to the problem. Accumulative empirical evidences suggest that CSPs undergo phase transitions at some critical values.

Random k-satisfiability problem (k-SAT) is a canonical version of CSPs, and a lot of efforts for the location of threshold points have been devoted to it. The variables of k-SAT are Boolean with $d = 2$. For $k = 2$, the problem is in P and it has been analytically shown that the threshold locates at $\alpha = 1$ ($\alpha \equiv m/n$) [2]. However, for $k \geq 3$, which is NP-complete, rigorous analysis appears to be difficult to locate the exact values of critical points [3]. Monasson et al. proposed the random $(2+p)$-SAT ($p \in [0,1]$) model with mixtures of $(1-p)m$ 2-clauses and pm 3-clauses, which can smoothly interpolate between the polynomial 2-SAT ($p = 0$) problem and the NP-complete 3-SAT ($p = 1$) problem [4]. Model RB is another typical CSP model with growing domains proposed by Xu and Li [5] to overcome the trivial unsatisfiability of standard CSP models. The satisfiability phase transition of model RB was exactly established. Moreover, it has been shown both theoretically and experimentally that almost all instances of model RB are hard to solve at the thresholds [6,7]. In recent years, model RB has received increasing attention and research in algorithm design [8–11] and extended model construction [12].

We propose an extension of model RB (named model RB^{mix}) in terms of constraint structure to explore the behavioral characteristics of phase transitions under mixed constraint lengths. Through rigorous theoretical analysis and numerical experiments, we demonstrate that RB^{mix} model exhibits a sharp satisfiability phase transition under mixed constraint lengths with an exact expression for the critical threshold, nd the solving hardness reaches the peak in the phase transition region. This theoretical framework establishes a new paradigm for studying phase transition mechanisms in complex constraint combinations, offering significant implications for algorithmic complexity analysis and phase transition-based algorithm design.

2 Model RB^{mix} and Main Results

2.1 Model RB^{mix}

- A set of variables $\{x_i\}_{i=1,\ldots,n}$, each variable x_i is associated with a nonempty domain D of $d = n^{\alpha}$ (where $\alpha > 0$ is a constraint) possible values.
- A set of constraints $\{C_i\}_{i=1,\ldots,rn\ln n}$, among which $r_i n \ln n$ ($i = 1, \ldots, l$) constraints are formed by randomly selecting k_i distinct variables from the n

variables, where $\sum_{i=1}^{l} r_i = r$, l is a finite positive integer, $k_{i_1} \neq k_{i_2}$ if $i_1 \neq i_2$, each $k_i \geq 2$ is an integer.
- A set of incompatible assignment set Q_i, which restricts the values of each constraint C_i, $Q_i \subseteq D^{k_i}$ with $|Q_i| = pd^{k_i}$, where $0 < p < 1$ determines the proportion of incompatible tuples to possible d^{k_i} assignments.

We take the logical conjunction of constraints as an instance of model RB^{mix}. An assignment of the n variables satisfying the instance is a solution.

2.2 Main Results

Given a random instance I of model RB^{mix}, let $S(I) = \{\sigma : \sigma \text{ satisfies } I\}$ denote the solution space of I, where σ is an assignment to the n variables from their domains. Let $X = |S(I)|$ be the number of solutions of I. We can obtain the following results.

Theorem 1. *Let $r_s = -\frac{\alpha}{\ln(1-p)}$. If $\alpha > \frac{1}{k}$, where $k = \min\{k_1, k_2, \ldots, k_l\}$, $0 < p < 1$ are two constants and k, p satisfy the inequality $k \geq \frac{1}{1-p}$, then*

$$\lim_{n\to\infty} \Pr(X > 0) = \begin{cases} 1 & \textit{for } r < r_s, \\ 0 & \textit{for } r > r_s. \end{cases}$$

Theorem 2. *Let $p_s = 1 - e^{-\frac{\alpha}{r}}$. If $k = \min\{k_1, k_2, \ldots, k_l\}$, $r > 0$ are two constants and k, α, and r satisfy the inequality $ke^{-\frac{\alpha}{r}} \geq 1$, then*

$$\lim_{n\to\infty} \Pr(X > 0) = \begin{cases} 1 & \textit{for } p < p_s, \\ 0 & \textit{for } p > p_s. \end{cases}$$

3 Proof of Theorem 1 and Theorem 2

Proof. We will prove Theorem 1 using first- and second-moment method.

$$E(X) = E\left(\sum_{\sigma} 1_{\sigma \in S(I)}\right) = d^n \prod_{i=1}^{l} (1-p)^{r_i n \ln n} = e^{[\alpha + r\ln(1-p)]n\ln n}. \quad (1)$$

If $r > r_s$, it is not difficult to obtain that $E(X) \to 0$ $(n \to \infty)$. By the Markov inequality $\Pr(X > 0) \leq E(X)$, we have $\Pr(X > 0) \to 0$ $(n \to \infty)$.

Let both $\sigma = (\sigma_1, \sigma_2, \ldots, \sigma_n)$ and $\tau = (\tau_1, \tau_2, \ldots, \tau_n)$ be a set of assignment, let $S = \sum_{i=1}^{n} 1_{\sigma_i, \tau_i}$ denote the similarity number of σ and τ, where $1_{\sigma_i, \tau_i} = 1$ if $\sigma_i = \tau_i$, otherwise $1_{\sigma_i, \tau_i} = 0$. Let $\beta = \frac{S}{n}$, by the condition $\alpha > \frac{1}{k}$, we can get

$$E(X^2) = E\left(\sum_{\sigma,\tau} 1_{\sigma,\tau\in S(I)}\right) = \sum_{\sigma,\tau}\prod_{C_i} P(1_{\sigma,\tau\in S(C_i)})$$
$$= d^n \sum_{S=0}^{n} \binom{n}{S}(d-1)^{n-S}\prod_{i=1}^{l}\left[(1-p)\frac{\binom{S}{k_i}}{\binom{n}{k_i}} + \frac{\binom{d^{k_i}-2}{pd^{k_i}}}{\binom{d^{k_i}}{pd^{k_i}}}\left(1-\frac{\binom{S}{k_i}}{\binom{n}{k_i}}\right)\right]^{r_i n\ln n}$$
$$= E^2(X)\sum_{S=0}^{n} f(\beta)\left(1+O\left(\frac{1}{n}\right)\right), \tag{2}$$

where

$$f(\beta) = \binom{n}{n\beta}\frac{(n^\alpha-1)^{n-n\beta}}{n^{\alpha n}}\prod_{i=1}^{l}\left[1+\frac{p}{1-p}\left(\beta^{k_i}+\frac{g_i(\beta)}{n}\right)\right]^{r_i n\ln n}, \tag{3}$$

$$g_i(\beta) = k_i(k_i-1)(\beta^{k_i}-\beta^{k_i-1})/2. \tag{4}$$

When n is sufficiently large, $f(\beta)$ is approximately evaluated as follows

$$f(\beta) = \prod_{i=1}^{l}\left[1+\frac{p}{1-p}\beta^{k_i}\right]^{r_i n\ln n}\binom{n}{n\beta}\left(1-\frac{1}{n^\alpha}\right)^{n-n\beta}\left(\frac{1}{n^\alpha}\right)^{n\beta}. \tag{5}$$

We divide β into $[0,\frac{1}{n^\varepsilon}]$, $[\frac{1}{n^\varepsilon},1-\theta]$, and $[1-\theta,1]$ to calculate $f(\beta)$ respectively. In the first interval $[0,\beta_1]$, $\beta_1=\frac{1}{n^\varepsilon}$ $(\varepsilon>\frac{1}{k})$. Let $f(\beta)=G(\beta)H(\beta)$,where

$$G(\beta) = \prod_{i=1}^{l}\left[1+\frac{p}{1-p}\beta^{k_i}\right]^{r_i n\ln n}, H(\beta) = \binom{n}{n\beta}\left(1-\frac{1}{n^\alpha}\right)^{n-n\beta}\left(\frac{1}{n^\alpha}\right)^{n\beta}. \tag{6}$$

Since $\lim_{n\to\infty} G(\beta_1)=1$ and $G(\beta)$ is an increasing function, then

$$\sum_{[0,\beta_1]} f(\beta) = \sum_{[0,\beta_1]} G(\beta)H(\beta) \le \sum_{[0,1]} H(\beta) = 1. \tag{7}$$

In the second interval $[\beta_1,\beta_2]$, $\beta_2=1-\theta$, $\theta>0$ is a sufficiently small constant. When n is sufficiently large, we have

$$f(\beta) \le e^{\sum_{i=1}^{l} r_i n\ln n\ln\left(1+\frac{p}{1-p}\beta^{k_i}\right)+n\beta\ln\frac{1}{n^\alpha}+n[-\beta\ln\beta-(1-\beta)\ln(1-\beta)]}$$
$$\le e^{n\ln n\varphi(\beta)+n\psi(\beta)}, \tag{8}$$

where $\varphi(\beta)=\sum_{i=1}^{l} r_i\ln(1+\frac{p}{1-p}\beta^{k_i})-\alpha\beta$, $\psi(\beta)=-\beta\ln\beta-(1-\beta)\ln(1-\beta)$.

By the condition $k\ge\frac{1}{1-p}$, we have $\varphi''(\beta)\ge 0$. It is easy to see that $\varphi(0)=0$, and $\varphi(1)=-r\ln(1-p)-\alpha<0$ for $r<r_s$. So $\varphi(\beta)<0$ for $0<\beta\le 1$. Therefore, there exists a $\beta_0\in[\beta_1,\beta_2]$ and a constant $M_1>0$ such that $\varphi(\beta)\le -M_1$.

We can obtain that the maximum value of $\psi(\beta)$ is $\ln 2$ at $\beta = \frac{1}{2}$. Consequently,

$$\sum_{[\beta_1,\beta_2]} f(\beta) \le e^{n \ln n\varphi(\beta)+n\psi(\beta)} \le e^{-M_1 n \ln n+n\ln 2} = \left(\frac{2}{n^{M_1}}\right)^n. \quad (9)$$

When $n \to \infty$, $\sum_{[\beta_1,\beta_2]} f(\beta) = o(1)$.

In the last interval $[\beta_2, 1]$, we need to prove $\lim_{n\to\infty} \sum_{[\beta_2,1]} f(\beta) = 0$, i.e., for any $\delta > 0$, there exists a positive integer N such that $\sum_{[\beta_2,1]} f(\beta) < \delta, \quad \forall n > N$.

Since $\varphi(\beta) < 0$ for $\beta \in [\beta_2, 1]$, so there is a positive number τ such that $\varphi(\beta) < -\tau, \quad \forall \beta \in [\beta_2, 1]$.

Since $\lim_{n\to\infty} \frac{\psi(\beta)}{\ln n} = 0$, so there is a positive integer N_1 such that $\frac{\psi(\beta)}{\ln n} < \frac{\tau}{3}, \quad \forall \beta \in [\beta_2, 1]$ when $n > N_1$.

Since $\lim_{n\to\infty} \frac{(1-\beta)\ln(1-\frac{1}{n^\alpha})}{\ln n} = 0$, so there is a positive integer N_2 such that $\frac{(1-\beta)\ln(1-\frac{1}{n^\alpha})}{\ln n} < \frac{\tau}{3}$ when $n > N_2$.

Now, let $N_3 = \max\{N_1, N_2\}$, then $\frac{\psi(\beta)}{\ln n} + \frac{(1-\beta)\ln(1-\frac{1}{n^\alpha})}{\ln n} + \varphi(\beta) < -\frac{\tau}{3}$ when $n > N_3$.

Then we take $N_4 = \exp\left\{\sqrt{-\frac{3\ln\delta}{\tau}}\right\}$ such that $\exp\left\{-\frac{n\ln n\cdot\tau}{3}\right\} < \exp\left\{-\frac{(\ln n)^2\cdot\tau}{3}\right\} < \exp\{\ln\delta\} = \delta$ when $n > N_4$.

At last, let $N = \max\{N_3, N_4\}$, then we can get when $n > N$,

$$\sum_{[\beta_2,1]} f(\beta)f(\beta) \le \exp\left\{n\ln n\left[\frac{\psi(\beta)}{\ln n} + \frac{(1-\beta)\ln(1-\frac{1}{n^\alpha})}{\ln n} + \varphi(\beta)\right]\right\} < \delta. \quad (10)$$

So when $n \to \infty$, we can have $\sum_{[\beta_2,1]} f(\beta) = o(1)$,

Therefore, $E(X^2) = E^2(X)\sum_{[0,1]} f(\beta) = E^2(X)(1 + o(1))$.

So we can have

$$\lim_{n\to\infty} \frac{E^2(X)}{E(X^2)} = 1 \quad \text{if} r < r_s. \quad (11)$$

Using the Cauchy inequality $\Pr(X > 0) \ge \frac{E^2(X)}{E(X^2)}$, it is not hard to prove that $\lim_{n\to\infty} \Pr(X > 0) = 1$, if $r < r_s$.

Hence, Theorem 3 are proved. Theorem 4 can be proved similarly.

4 Experimental Result

We have proved that model RB^{mix} exhibits a distinct phase transition phenomenon, precisely locating critical values which become sharper with larger instance sizes. As control parameter r increases, the satisfiability probability of random instances shifts sharply from almost 1 to almost 0. Experimentally, the Abscon platform solves 50 random RB^{mix} instances per configuration—defined by tuples (n, r, α, p, K, L) where K specifies constraint group sizes and L their proportions—to generate phase transition and complexity curves. Four numerical experiments explore different K values for $n = 20, 30, 40$. Results show paired graphs: satisfiability probability (left, axes: r vs probability) and solution difficulty (right, axes: r vs log(time)). The first experiment ($\alpha = 0.8$, $p = 0.4$, $K = (2, 3)$, $L = (0.6, 0.4)$) confirms the theoretical critical point $r_s = 1.56$ (Figs. 1, 2), with high satisfiability below r_s and high unsatisfiability above it (Figs. 4, 5, 6 and 8).

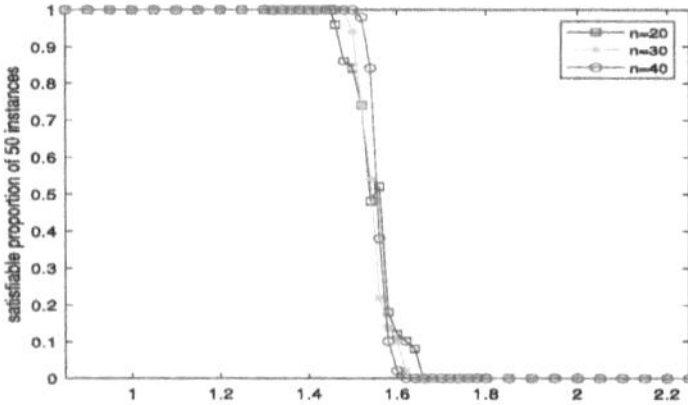

Fig. 1. Satisfiability phase transition diagram of $\mathrm{RB}^{\mathrm{mix}}(n, r, 0.8, 0.4, K1, L)$

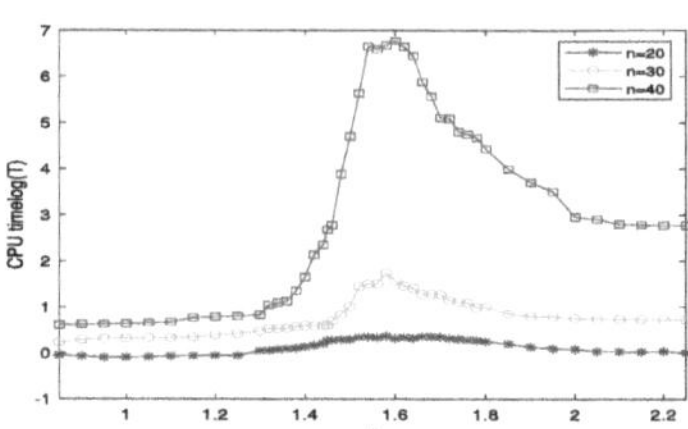

Fig. 2. Intractability curves of $\mathrm{RB}^{\mathrm{mix}}$ $(n, r, 0.8, 0.4, K1, L)$

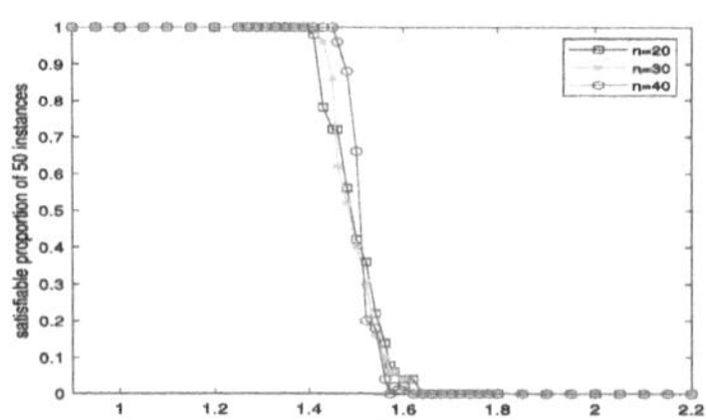

Fig. 3. Satisfiability phase transition diagram of $\mathrm{RB}^{\mathrm{mix}}(n, r, 0.8, 0.4, K2, L)$

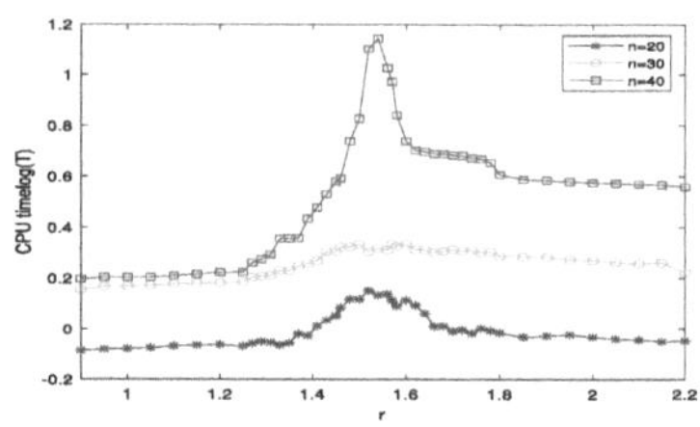

Fig. 4. Intractability curves of $\mathrm{RB}^{\mathrm{mix}}$ $(n, r, 0.8, 0.4, K2, L)$

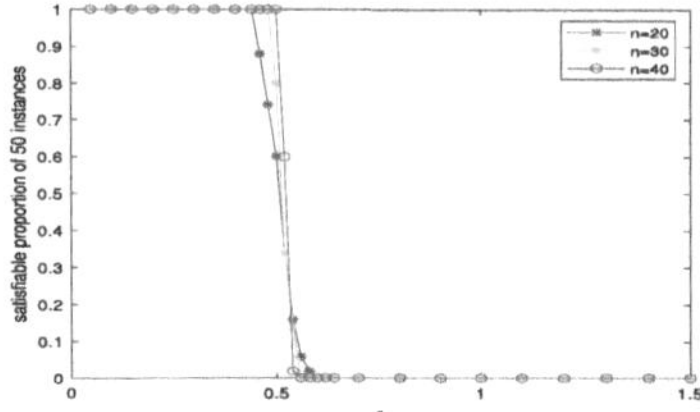

Fig. 5. Satisfiability phase transition diagram of $\mathrm{RB}^{\mathrm{mix}}(n, r, 0.5, 0.6, K3, L)$

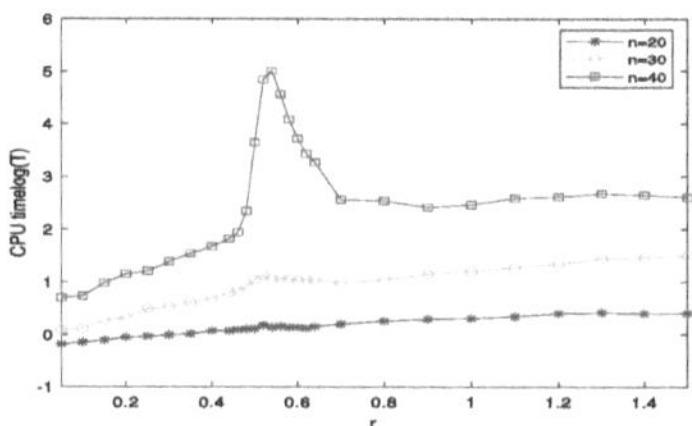

Fig. 6. Intractability curves of $\mathrm{RB}^{\mathrm{mix}}$ $(n, r, 0.5, 0.6, K3, L)$

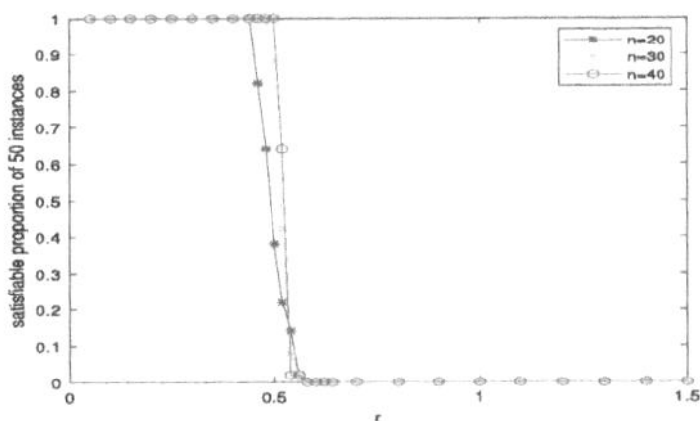

Fig. 7. Satisfiability phase transition diagram of $\mathrm{RB}^{\mathrm{mix}}(n, r, 0.5, 0.6, K4, L)$

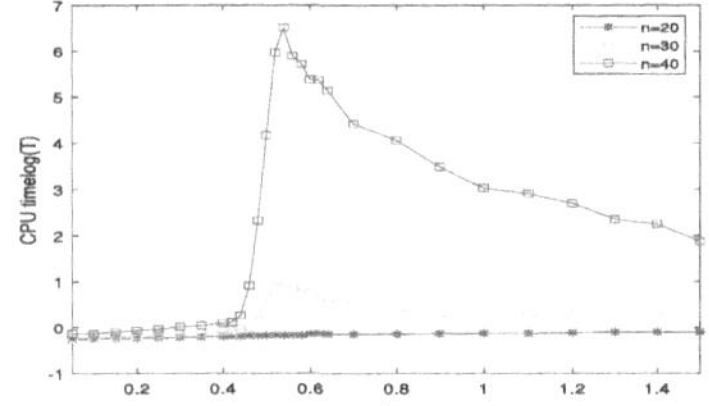

Fig. 8. Intractability curves of $\mathrm{RB}^{\mathrm{mix}}$ $(n, r, 0.5, 0.6, K4, L)$

When all constraints share the same arity (uniform K components), model $\mathrm{RB}^{\mathrm{mix}}$ degenerates to the standard RB model, demonstrated numerically in Figs. 3 and 7. All experiments show satisfiability probability sharply drops from 1 to 0 near the phase transition point, with threshold behavior sharpening as $n \to \infty$. Solving difficulty peaks in this transition region. The four experimental configurations confirm model $\mathrm{RB}^{\mathrm{mix}}$ exhibits robust threshold behavior, with results across different α values matching Theorem 3's predicted critical points. This verifies model RB is a special case of the more general model $\mathrm{RB}^{\mathrm{mix}}$ framework.

5 Discussion

By adding an all-different global constraint, which requires all variables to take distinct values, alongside the m existing constraints (where $\alpha > 1$). We find that phase transitions still exist and the critical values remain unchanged. We prove the conclusion as follows:

$$E(X) = d(d-1)\cdots(d-n+1)(1-p)^{rn\ln n} \le \mathrm{e}^{nlnn(\alpha + r\ln(1-p))}. \tag{12}$$

Using the Markov inequality $\Pr(X > 0) \le E(X)$, we have $\lim_{n\to\infty} \Pr(X > 0) = 0$ when $r > \frac{-\alpha}{\ln(1-p)}$, and $E(X)$ grows exponentially with n when $r < \frac{-\alpha}{\ln(1-p)}$, then we use the second moment to find the relations of $E(X)$ to $E(X^2)$.

$$\left|1_{\sigma,\tau\in S(I)}\right| = d(d-1)\cdots(d-n+1)\tbinom{n}{S}(d-S-1)(d-S-2)\cdots(d-n). \tag{13}$$

$$\sum_{S=0}^{n} \left|1_{\sigma,\tau\in S(I)}\right| \mathrm{P}(\langle t_i, t_j\rangle) = E^2(X) \sum_{\beta\in[0,1]} \prod_{i=1}^{l} [(1 + \tfrac{p}{1-p}(\beta^{k_i} + \tfrac{g_i(\beta)}{n})]^{r_i n\ln n} \cdot \tbinom{n}{n\beta} \cdot (\tfrac{1}{n^\alpha})^{n\beta}(1-\tfrac{1}{n^\alpha})^{n-n\beta}(1+O(\tfrac{1}{n})). \tag{14}$$

So

$$f(\beta) = \prod_{i=1}^{l} [(1 + \frac{p}{1-p}(\beta^{k_i} + \frac{g_i(\beta)}{n})]^{r_i n\ln n} \cdot \binom{n}{n\beta} \cdot (\frac{1}{n^\alpha})^{n\beta}(1-\frac{1}{n^\alpha})^{n-n\beta}. \tag{15}$$

We can see that $f(\beta)$ is the same as the one in the Sect. 3. By using the Markov inequality $\Pr(X>0) \geq \frac{E^2(X)}{E(X^2)}$, we have $\lim_{n\to\infty} Pr(X>0)=1$ when $r<\frac{-\alpha}{\ln(1-p)}$, so model $\mathrm{RB}^{\mathrm{mix}}$ exhibits the sharp phase transitions from satisfiability to unsatisfiability after adding the global constraint, and the critical values is also located at $r_s=\frac{-\alpha}{\ln(1-p)}$.

6 Conclusion

This paper introduces $\mathrm{RB}^{\mathrm{mix}}$, a new random CSP model extending RB with mixed-constraint lengths. Theoretically, we use the second moment method to prove $\mathrm{RB}^{\mathrm{mix}}$ exhibits a sharp phase transition from satisfiability to unsatisfiability, precisely locating its critical point. Experimentally, the Abscon platform (see http://www.cril.univ-artois.fr/~lecoutre) validates this threshold behavior and collects precise data. The phase transition persists even when adding global constraints with varied variable assignments. Future work will develop efficient solvers—combining belief propagation, simulated annealing, and backtracking algorithms—to reduce solving time and uncover new phenomena.

References

1. Cheeseman, P., Kanefsky, B., Taylor, W.: Where the really hard problems are. In: Proceedings of 12th International Joint Conference on Artificial Intelligence, pp. 331–337 (1991)
2. Cook, S.: The complexity of theorem-proving procedures. In: Proceedings of 3rd ACM Symposium on Theorey of Computing, pp. 151–158 (1971)
3. Mertens, S., Mézard, M., Zecchina, R.: Threshold values of random k-SAT from the cavity method. Random Struct. Algorithms **28**, 304–373 (2006)
4. Monasson, R., Zecchina, R., Kirkpatrick, S., Selman, B., Troyansky, L.: Phase transition and search cost in the $(2+p)$-SAT problem. In: 4th Workshop on Physics and Computation, Boston (1996)
5. Xu, K., Li, W.: Exact phase transitions in random constraint satisfaction problems. J. Artif. Intell. Res. **12**, 93–103 (2000)
6. Xu, K., Li, W.: Many hard examples in exact phase transitions. Theoret. Comput. Sci. **355**, 291–302 (2006)
7. Xu, K., Boussemart, F., Hemery, F., Lecoutre, C.: Random constraint satisfaction: easy generation of hard (satisfiable) instances. Artif. Intell. **171**, 514–534 (2007). Earlier version appeared in Proc. of 19th IJCAI, Scotland, 2005, 337-342
8. Zhao, C., Zheng, Z.: Threshold behaviors of a random constraint satisfaction problem with exact phase transitions. Inform. Process. Lett. **111**, 985–988 (2011)
9. Zhao, C., Zhou, H., Zheng, Z.: A message-passing approach approach to random constraint satisfaction problems with growing domains. J. Stat. Mech. Theory Exp. P02019 (2011)
10. Zhao, C., Zhang, P., Zheng, Z., Xu, K.: Analytical and belief-propagation studies of random constraint satisfaction problems with growing domains. Phys. Rev. E **85**, 016106 (2012)

11. Zhao, C., Fu, Y.: Belief propagation guided decimation algorithms for random constraint satisfaction problems with growing domains. Interdisc. Stat. Mech. 033408 (2021)
12. Fan, Y., Shen, J., Xu, K.: A general model and thresholds for random constraint satisfaction problems. Artif. Intell. **193**, 1–17 (2012)

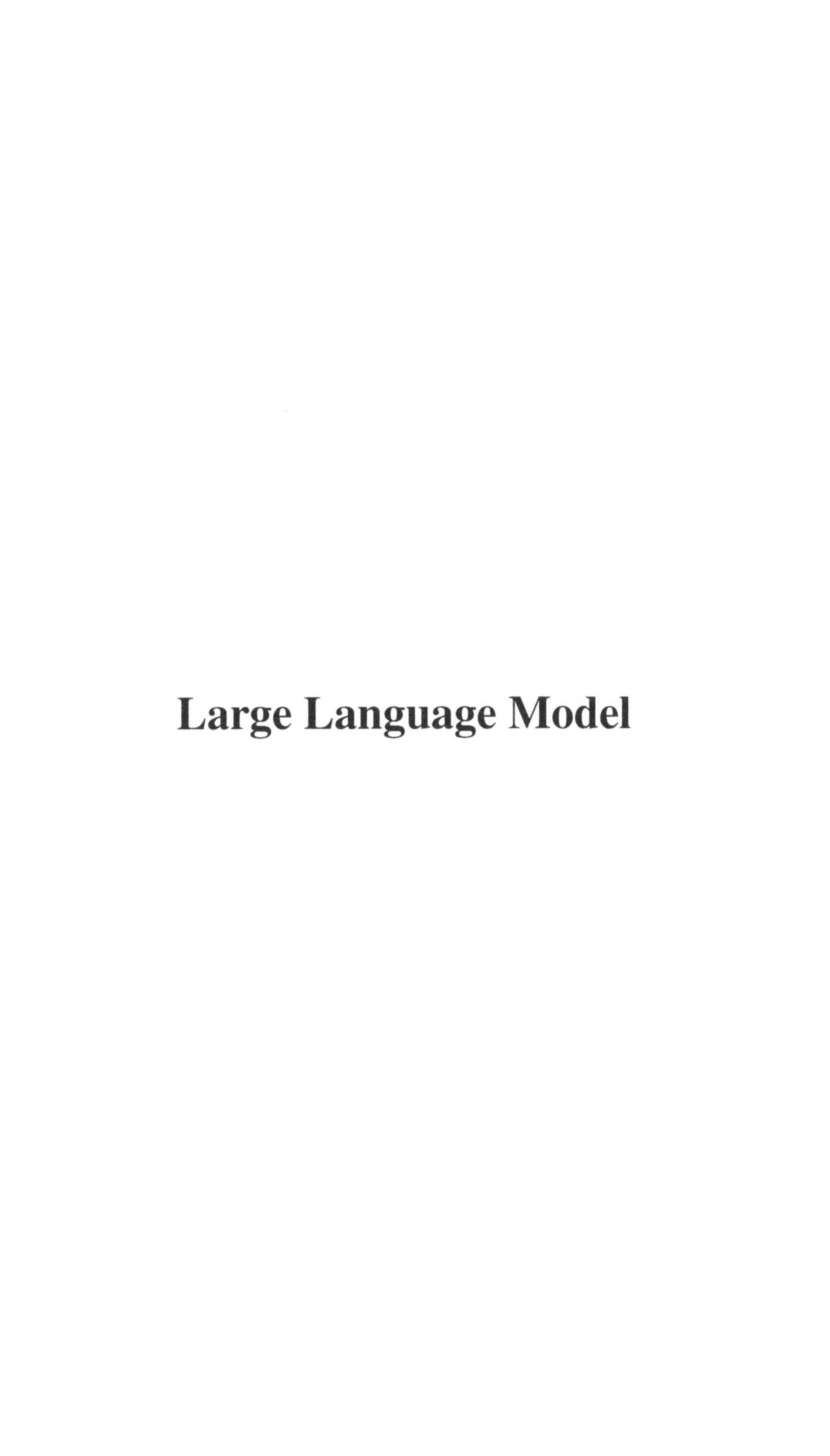

Large Language Model

MARAG: Multi-agent Retrieval-Augmented Generation for Mitigating Knowledge Conflicts in Large Language Models

Jiaming Tian(✉), Weixin Zeng, Jibing Wu, Lihua Liu(✉), and Xiang Zhao

Laboratory for Big Data and Decision, National University of Defense Technology, Changsha, China
tianjiaming19@nudt.edu.cn

Abstract. Large language models (LLMs) are pretrained on vast amounts of data, enabling them to amass extensive knowledge within their internal parameters. As knowledge is continually evolving and updating, retrieval-augmented generation (RAG) becomes a popular technique to inject the latest external information into LLMs to generate more accurate responses. However, a notable issue with RAG is the conflict between the external information and LLM's internal parametric knowledge. Existing solutions to mitigate this issue primarily focus on relying solely on external context or internal knowledge, mitigating conflicts by decoding and amplifying the influence of external context or enhancing the weight of one side of the knowledge, which may lead to ambiguous erroneous answers in scenarios with multiple mutually exclusive sources. To address these limitations, we propose MARAG, which adopts an end-to-end reasoning process that involves 'knowledge refinement, conflict detection, answer generation' to effectively resolute the conflicts and generate more accurate responses. MARAG consists of two major collaborative agents: the **Focuser Agent** and the **Option Helper**. The Focuser Agent executes relevance selection and refinement of external context and the model's internal implicit knowledge, while assessing consistency among different information sources to explicitly flag potential conflicting segments as it removes noise from the text. The Option Helper fully utilizes the option structures in the dataset to transform the open-domain question answering (QA) task into semantic consistency matching, while assessing conflict types and generating the final answer based on evidence refined by the Focuser Agent. Through extensive experiments on the mainstream benchmarks, MARAG demonstrates an average accuracy improvement of 2% compared to existing methods across various LLM backbones, significantly reducing the rate of erroneous responses without the need of fine-tuning the LLM parameters.

Keywords: Knowledge conflicts · retrieval-augmented generation · multi-agent cooperation

S. Yang et al. (Eds.): WISA 2025, LNCS 16299, pp. 543–555, 2026.
https://doi.org/10.1007/978-981-95-6426-2_48

1 Introduction

In recent years, Large Language Models (LLMs) have made significant breakthroughs in tasks such as machine translation, code generation, and open-domain question answering (QA), thanks to their vast amounts of pre-training data and powerful language generation capabilities [5]. However, relying solely on parameterized memory makes it difficult for models to keep up with the rapidly evolving world knowledge in real-time. Retrieval-augmented generation (RAG) is considered a mainstream paradigm to alleviate the problem of "stale knowledge" by explicitly retrieving external documents during the reasoning phase and inputting them together with the original query into the model, thereby injecting the latest information without the need for retraining [1, 2].

Although RAG effectively broadens the boundaries of new knowledge, in real applications, external knowledge often faces multifaceted conflicts with the model's internal old knowledge [6]. When conflicts are not properly resolved, the model struggles to focus on key information and may produce contradictory or ambiguous responses [3]. For instance, as shown in Fig. 1, when the retrieved context incorrectly states that "*Donald Trump sworn in as the 47th president of the United States.*" it clashes with the model's internal memory that "*Joe Biden is the 46th president.*" Faced with these mutually exclusive cues, the model may either adopt the erroneous external claim or vacillate between the two, producing an uncertain answer such as "*Donald Trump or Joe Biden*". Therefore, how to automatically identify and handle conflicts in retrieval evidence, fully leverage the complementary nature of new and old knowledge rather than their interference, has become a key issue in improving the reliability and interpretability of RAG systems.

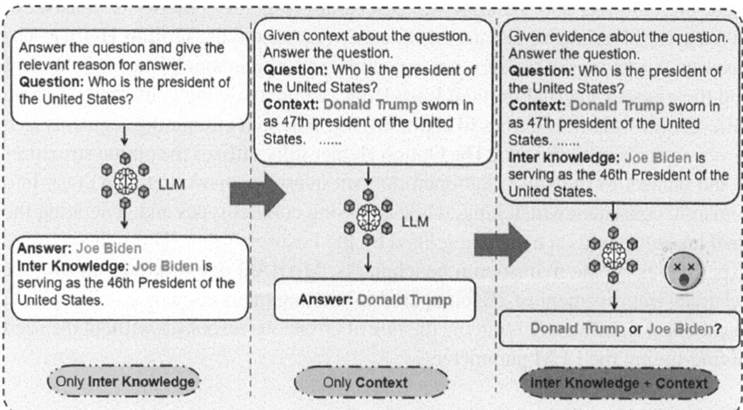

Fig. 1. An example illustrating knowledge conflicts arising from a LLM when answering questions using internal knowledge or context.

Existing work primarily addresses the resolution of conflicts between internal knowledge of large models and external context through two pathways. One is detecting factors that may influence the output during the decoding process of the large model's input, rearranging or suppressing tokens that are inconsistent with the evidence during answer generation [4, 5]. The other method is the modulation of the LLM at the neuron level, which weakens the influence of old parameter knowledge by locally reweighting the weights of the model's internal nodes [6]. However, these methods generally assume that the retrieved content is correct and lack the utilization of correct knowledge within the large model, which can lead to incorrect answers when faced with erroneous external information or multiple conflicting information.

To address the aforementioned issues, we propose a multi-agent retrieval-augmented generation framework called MARAG. MARAG adheres to a "division-of-labor, collaboration, arbitration" workflow and realizes an end-to-end "retrieve-refine-arbitrate-generate" reasoning chain without fine-tuning the backbone LLM. Specifically, it deploys two auxiliary agents: (1) Focuser Agent, which screens and refines external evidence as well as the model's implicit knowledge, labels potential conflicts between them, and forwards a conflict-marked context. (2) Option Helper, which leverages the option structure of multiple-choice questions to cast answering as semantic-consistency matching, performs conflict-aware arbitration across candidate options, and produces the final answer. By strictly separating roles and constraining information flow, MARAG markedly reduces the disturbance of conflicting noise, thereby improving both reasoning accuracy and interpretability in knowledge-conflict scenarios while keeping all LLM parameters frozen.

We empirically evaluate the effectiveness of our proposal on the mainstream dataset against four representative baselines. The experimental results show an average accuracy improvement of 2% compared to existing advanced methods across various LLM backbones in alleviating knowledge conflicts and improving the reliability of LLM reasoning.

The main contributions of this paper can be summarized as follows:

1. We propose a multi-agent framework MARAG that does not require fine-tuning to achieve an end-to-end reasoning process to automatically identify and reconcile knowledge conflicts between retrieved evidence and the LLM's parametric knowledge in open-domain multiple-choice question answering, thereby boosting factual accuracy and interpretability.
2. We design two types of agents, i.e., Focuser Agent and Option Helper, to work together on conflict extraction and semantic consistency matching respectively, which can collaboratively filter out noisy or mutually inconsistent evidence, perform conflict-aware arbitration, and thus improve answer accuracy and interpretability.
3. We conduct comprehensive experiments on the mainstream dataset, demonstrating the effectiveness of MARAG.

2 Related Work

Retrieval-Augmented Generation(RAG). RAG aims to improve the factual consistency of LLMs by explicitly retrieving external documents during the reasoning phase

[1, 2], while reducing hallucinations [5] and the risk of stale knowledge [6]. Subsequent works REALM [2], FiD [4], and RETRO [3] have extended the retriever structure and document encoding strategy. In order to suppress inconsistent information, SELF-RAG [7] introduced a model self-criticism mechanism, and Astute-RAG [8] used the parameterized knowledge of the model for consistency alignment and cluster filtering [9].

Multi-agent Collaboration and Debate Framework. The multi-agent framework provides a new approach to modeling conflicting information. Main-RAG [10] and MADAM-RAG [11] assign different documents to independent agents, and then aggregate the answers through multiple rounds of debate, significantly improving the robustness to ambiguous queries and false documents. Similarly, work such as Debate-QA [12] shows that the clash of opinions between agents can prompt the model to expose potential errors and self-correct.

Research on Knowledge Conflict and Consistency. There are three types of conflicts between the parameter knowledge of LLM and the external retrieval knowledge: *parameter conflict, single document conflict, and cross-document conflict* [13]. Existing studies have focused on parameter conflicts or single document contradictions, while the analysis of multi-document and multi-source conflicts is still insufficient. These approaches have only been validated on single-source evidence, often resulting in compromise answers or conservative expressions when faced with mutually exclusive information from multiple sources. A few studies have attempted fact-level voting on the retrieval aspect [14], but they do not define conflict categories and lack fine-grained annotated data, making it difficult to assess robustness to different forms of conflict.

The RAG Benchmarks. AmbigNQ [17] and AmbigDocs [18] focus on query ambiguity. FaithEval [19] measures the model's immunity to misleading evidence. RGB [14] and CRAG [15] introduce domain diversity and noisy documents. These datasets usually only cover a single conflict factor. RAMDocs [11] combines ambiguity, noise, and misinformation, reflecting the complexity of real-world retrieval scenarios, but has not yet been subdivided into conflict dimensions such as time and semantics. ConflictBank [16] juxtaposes correct information, incorrect information, temporal conflicts, and semantic conflicts at the data level for the first time, providing a benchmark for analyzing the impact of conflict types on LLM. RAMDocs is not as refined as ConflictBank in terms of conflict type segmentation, explicit confrontation labeling, and interpretive support, making it more suitable as a benchmark for robust retrieval and filtering for conflict analysis.

3 Methodology

Given a natural-language multiple-choice question q with option set $A = \{A_1...A_m\}$, we first retrieve a candidate evidence set $C = \{C_1...C_m\}$ from an external corpus via a retriever $R(\cdot)$ and pair it with the frozen backbone LLM's internal knowledge K_θ. Because facts in K_θ and C may conflict, MARAG detect and label contradictory fragments, discard or down-weight them, and then select the option most semantically consistent with the remaining, non-conflicting context.

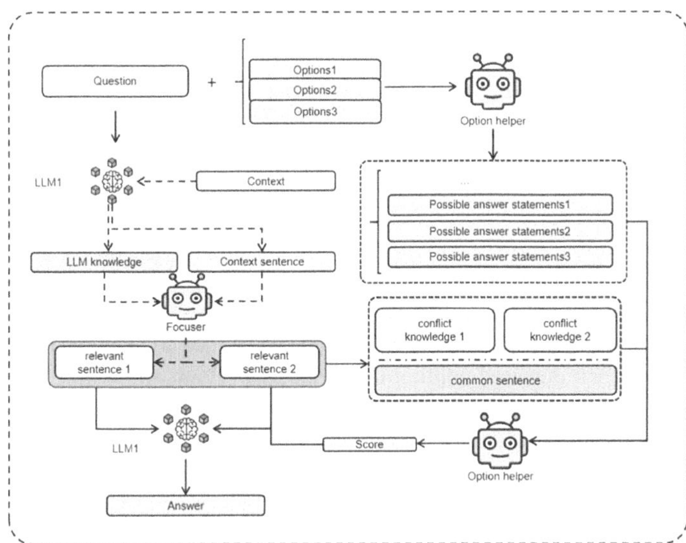

Fig. 2. Overall framework of MARAG. Mainly includes **Focuser Agent** and **Option Helper.**

The core innovation of the MARAG framework lies in resolving knowledge conflicts through multi-agent collaboration. A clear information flow and division of labor is established between the two agents: the **Focuser Agent** concentrates on evidence quality control and conflict detection, whereas the **Option Helper** is responsible for decision making based on the refined evidence. The workflow proceeds as Fig. 2 follows: once the system receives the question and its raw options, the Option Helper first rewrites each option into a complete natural-language statement. The question is then passed to a frozen LLM to trigger retrieval, after which the Focuser selects the most relevant sentences from the returned context and tags mutually exclusive information as either "conflict" or "common," yielding a distilled evidence set. The Option Helper next combines each rewritten statement with this evidence and feeds the prompt back into the same LLM to obtain a three-way probability vector "Supported, Contradicted, or Irrelevant". Any statement that clashes with a conflict sentence has its support probability dynamically down-weighted according to conflict severity. Finally, the helper computes a "support minus contradiction" score, selects the statement with the highest score as the answer, and returns the key supporting sentences, the conflict sentences, and their probability weights, providing a reasoning chain that is both robust and transparent, all without fine-tuning the LLM.

3.1 Focuser Agent: Evidence Refinement and Conflict Annotation.

The Focuser Agent examines C and the internal knowledge fragments of LLM in parallel, filters noise, annotates conflicts, and obtains refined evidence C'. Focuser Agent uses "Question + Context sentence" as the query, then call LLM twice to calculate sentence-level similarity and select the highest-scoring sentences. It performs adversarial reasoning on each pair of candidate sentences S_i, S_j and LLMs own knowledge paragraphs. If the models give mutually exclusive conclusions, they are recorded as potential conflicts and labeled by category. Finally, three lists are returned: relevant sentences, conflict knowledge, and common sentences, for explicit reference by subsequent modules. Focuser Agent acts as a dedicated "evidence gatekeeper". It receives the *top_K* passages $C = \{C_1...C_K\}$ given and the LLM's own parametric recall K_{int} ($\approx$150 tokens elicited with the prompt "Please recall before answering"). Focuser first performs fine-grained relevance filtering on the bulk of sentences returned by retrieval leveraging question keywords, entity matching, and semantic-similarity scores. Then it retains only the core sentences most capable of explaining or answering the question. After that it enters a conflict-mining stage, comparing these core sentences pair-wise with one another and with the candidate answer statements to surface explicit contradictions. Finally, each detected conflict is labeled, supplying an interpretable basis for the Option Helper's subsequent conflict-penalty allocation. The entire work is depicted in Fig. 3 and formally defined below.

During the entire process, Focuser lets $K_{int} = \{k^{(1)}, ..., k^{(m)}\}$ be the set of sentences in the model-recall draft. Then for every pair (s, k) with $s \in S$, $k \in K_{int}$, Focuser calculate the conflict score of each pair. After that, it returns a triplet $(S_{com}, S_{con}, \{(s, k, T(s, k)\})$ where S_{con} is the subset of S that participates in at least one conflict, S_{com} is the common sentence and $T(s, k)$ is the type of conflicts between s and k. These three lists respectively serve the Option Helper as clean evidence, conflict evidence, and typed conflict descriptors.

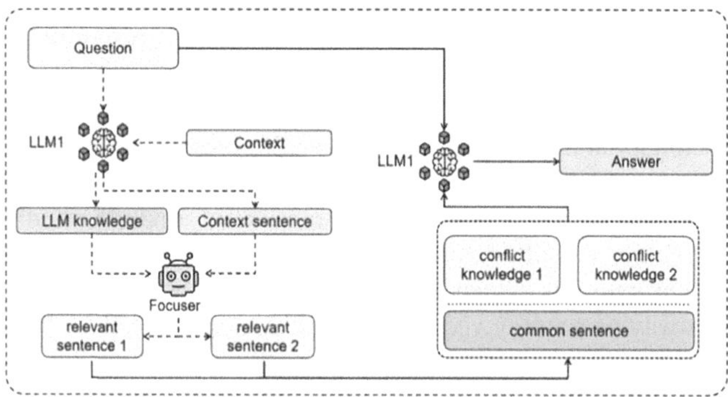

Fig. 3. The framework of Focuser Agent. Using LLMs for explicit evidence revision and conflict annotation.

3.2 Option Helper: Consistency Arbitration and Answer Generation

As shown in Fig. 4, the Option Helper transcribes the multiple options $A = \{A_1 \dots A_4\}$ into declarative sentences $E = \{E_1, \dots E_4\}$, and then scores each E_i using semantic consistency scores and conflict markers. The highest-scoring $E*$ is backfilled as the final answer, and is accompanied by a basis sentence and conflict type description to enhance interpretability. The Option Helper executes a fine-grained, conflict-aware ranking routine that converts the multiple-choice QA task into sentence-level entailment estimation while leaving the backbone LLM untouched. Every candidate option A is normalized into a full declarative statement E_i that follows the dataset's template. This step eliminates ambiguity created by terse answer labels and allows the downstream model to judge factual consistency in a natural-language setting. Then the original question Q, the rewritten statement E_i, and the refined, conflict-annotated evidence bundle supplied by Focuser Agent are concatenated into a single prompt block with an instruction such as: "*Classify the statement as Supported, Contradicted, or Irrelevant with respect to the question and the evidence.*" The LLM returns a sentence-level probability vector $p_i = (p_i^{sup}, p_i^{con}, p_i^{irr})$ representing the likelihood of the three semantic relations. After that, Option Helper computes a final consistency score $F = p_i^{sup} - p_i^{con}$. The statement achieving the highest score $F*$ is selected, its originating option is returned as the predicted answer.

By chaining relevance filtering(performed by Focuser Agent) with explicit conflict penalisation, Option Helper substantially attenuates retrieval noise. Simultaneously, reformulating open-domain QA as discriminative semantic-consistency matching rather than direct answer generation yields stable, interpretable reasoning traces even under dense, high-conflict evidence streams, all while keeping LLM parameter frozen.

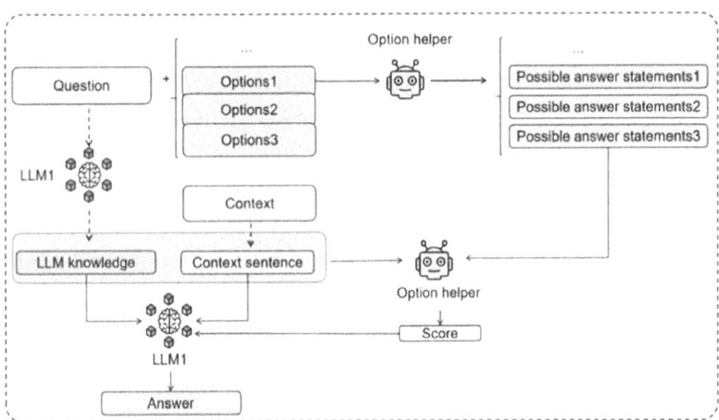

Fig. 4. The framework of Option Helper. Consistency arbitration and answer generation for contextual knowledge and internal parameter knowledge.

Notably, the aforementioned modular design allows each agent to specialize in its primary task, alleviating the burden that a single model would face when handling multiple complex subtasks simultaneously. The Focuser Agent can fully exploit its strengths in textual relevance assessment and semantic analysis, while the Option Helper can focus on accurate answer selection by leveraging the option structure. The explicit conflict-labeling mechanism enables the system to proactively identify and address knowledge discrepancies instead of passively expecting the model to resolve them on its own. By treating conflict detection as an independent processing stage, MARAG can apply targeted strategies to different types of conflicts. Meanwhile, the principle of keeping the base model's parameters frozen guarantees the framework's generality and scalability: MARAG can be seamlessly applied to LLMs of varying sizes and architectures while consistently delivering performance gains.

4 Experiments and Results

4.1 Experimental Setup

Datasets. We use ConflictBank,which is the first benchmark specifically designed to systematically evaluate knowledge conflicts in LLMs [16]. It performs fine-grained annotations on the three conflict dimensions of internal retrieval knowledge conflicts, model-encoded knowledge conflicts, and the interaction between the two. Each sample in the ConflictBank contains: question Q, context C returned by the retrieval system, four options $A = \{A_1 - A_4\}$, , and corresponding golden evidence and conflict type annotations. The retrieval context is subdivided into four categories: (1) Correct information, (2) Wrong information, (3) Temporal conflict, (4) Semantic conflict. To simulate the interference of noise intensity in context content during actual use, we randomly sampled 1,000 data sets from the official test set to form a data subset, the context content of the subset is composed of 70% correct information, 10% incorrect information, 10% temporal conflict, and 10% semantic conflict.

Backbone LLM Models. We evaluate MARAG on four publicly reproducible large-scale decoding language models to cover the parameter size range from 0.5B to 32B, verifying the adaptability of the framework to different capacities and architectures. To more comprehensively study how large models with different parameter scales perform, we selected four representative models **Qwen-1.5–0.5B, Gemma2-2B, Ministral-8B, and QWQ-32B.** This study directly adopts its official weight and prompt format without any fine-tuning or incremental training; the decoding hyperparameters are uniformly set to temperature = 0.2, top-p = 0.9, and max tokens = 256 to ensure that all performance differences are derived from MARAG's evidence screening and conflict modeling mechanism rather than additional optimization factors.

Evaluation Metrics. In all experiments, we uniformly use accuracy as the core evaluation indicator [17]. Accuracy is defined as the proportion of samples whose model output is completely consistent with the reference answer to the total number of samples. The reason why accuracy is chosen instead of indicators such as F1 and BLEU is that the task requires the model to give a unique and unambiguous factual judgment when facing potential conflicts. In addition, any deviation should be regarded as an error, accuracy is comparable for different solve strategies, which facilitates horizontal analysis of the performance gains brought by each component.

Baselines. No-RAG, which directly input the original question into the frozen LLM without any retrieval text attached. The model relies entirely on parameterized memory to answer, providing a "zero noise" upper limit to measure the effect of pure internal knowledge and the real gain brought by the subsequent introduction of retrieval or conflict handling. **Inter-Knowledge**[6], which explicitly compare "model recall" and "external evidence", first prompt LLM "Please recall before answering", generate an internal knowledge draft of about 150 tokens. Combine the context, splice the retrieved context and the draft into a complete context, and then return it to the same LLM to generate the answer. By explicitly writing the model's prior knowledge into the prompt, Inter-Knowledge can find the inconsistency between "model memory" and "external evidence", but still does not handle conflicts between documents. **Standard RAG** [1] is the classic "one piece of evidence splicing" retrieval enhancement. The external context and the questions are directly concatenated and sent to the LLM. This method completely ignores the conflicts. **IRCAN [22]** does not filter conflicts on the retrieval side, but starts from the inside of the model, suppressing parameterized old knowledge and emphasizing new knowledge by identifying and reweighting context-aware neurons. This strategy does not require additional parameter training and is a one-time, pluggable local weight modulation method that can be directly applied to existing LLMs.

Notably, the four baselines are completely consistent with MARAG in external conditions such as the retriever, input length, and decoding hyperparameters. The performance differences are only caused by their respective conflict handling and reasoning mechanisms, not the context of retrievals or other configuration differences.

4.2 Main Results

On four LLMs of different capacities, MARAG's accuracy is always high: compared with the classic RAG framework, the lightweight model **Qwen1.5–0.5B increased by 6.6%, Gemma2-2B and Ministral-8B increased by 6.8% and 7%** respectively, and the **QWQ-32B** achieved a substantial gain of **21.1%**. It shows that even if the base model is larger and the parameter knowledge is richer, conflict is still the main bottleneck, and MARAG's explicit conflict annotation and consistency arbitration mechanism can effectively remove this bottleneck. On the 0.5B and 2B models, the traditional RAG increased the accuracy from about 25% to about 30%, proving that external evidence can make up for the knowledge gap of small models to a certain extent. After explicitly splicing the knowledge within the model, the LLM still lacks the ability to distinguish between true and false. Regarding the inter knowledge method, the performance of four LLMs is all lower than RAG, verifying that simply exposing internal knowledge cannot solve conflicts. The IRCAN method suppresses old knowledge by reweighting local neurons, which is better than RAG by 3% at least overall, but slightly worse than MARAG about 2% on the 0.5B and 8B model. In addition, since the method consumes too much memory resources, the single A800 80G cannot complete the experiment in 24 h for training the 32B LLM (Table 1).

Table 1. The accuracy performance (%) of the four LLMs under the MARAG framework.

Heading level	none	Inter knowledge	RAG	IRCAN	MARAG
Qwen1.5–0.5B	25.2	23.8	30.6	34.1	**37.2**
Gemma2-2B	25.9	24.1	30.1	**38.5**	36.9
Ministral-8B	28.3	20.8	38.6	44.2	**44.6**
QwQ-32B	30.2	28.4	48.1	-	**69.2**

From the perspective of absolute value improvement, the larger the model size, the more obvious the gain of MARAG relative to RAG. The reason is that LLMs themselves carry a lot of old or wrong parameterized knowledge, which is more likely to cause conflicts when encountering new evidence. LLMs with large parameters have stronger conflict identification capabilities under the MARAG framework, thus releasing greater potential.

4.3 Ablations and Analysis

To quantitatively measure the contribution of each submodule, we conducted a two-dimensional ablation experiment: ①Remove the Focuser Agent, that is, the retrieved documents are directly input into the big model without evidence refinement and conflict annotation. ②Remove the Option Helper, retain the refined evidence provided by the Focuser Agent but delete the semantic consistency scoring mechanism and explicit conflict determination steps, and the answer is generated by the big model at one time. The rest of the process, prompt templates, and decoding hyperparameters remain unchanged.

The results show that the accuracy of the complete framework based on QWQ-32B on the ConflictBank test set is **69.2%**, while case① drops to 53.8% and case② drops to **63.5%**, which is only slightly better than the traditional RAG. It can be seen that the text refinement of Focuser Agent lays a reliable input for subsequent reasoning, while the consistency judgment of Option Helper effectively suppresses the wrong answers caused by document contradictions. The two modules together constitute the key factors for improving the performance of MARAG in conflict scenarios.

4.4 Case Study

To illustrate how MARAG outperforms conventional RAG pipelines in high-conflict settings, we analyze an example taken from the ConflictBank test split based on Qwen1.5–0.5B. As a comparison, a standard RAG system concatenates Question + Evidence and directly feeds the block to the LLM to answer the question (Table 2).

Table 2. Performance of accuracy of components removed from the MARAG framework.

LLM	w/o Focuser	w/o Option Helper	MARAG
Qwen1.5–0.5B	30.1	35.9	**37.2**
Gemma2-2B	29.6	34.7	**36.9**
Ministral-8B	35.0	41.2	**44.6**
QwQ-32B	53.8	63.5	**69.2**

Table 3. An example of processed data from the ConflictBank dataset.

Question	Which award did Arnulf Herrmann receive?
Evidence	Berlin, Germany - In a ceremony held at the Prinzregententheater in......
Inter-knowledge	Here is some information about Arnulf Herrmann......
Options	A. Ernst von Siemens Composers' Prize B. Pulitzer Prize for Music C. Royal Philharmonic Society Award D. Uncertain

Table 3 shows an example data in experiments. After retrieval relevant knowledge in evidence, the Focuser keeps three high-scoring snippets and attaches conflict labels as shown in Table 4.

Table 4. Some relevant sentences in evidence.

Source	Evidence sentence	**Relevance**	**Conflict tag**
Evidence	In 2010, Arnulf Herrmann received the Ernst von Siemens Composers' Prize for his work Terzenseele	√	None
Inter-knowledge	Herrmann was honoured with the German Music Authors' Prize in 2019	√	Conflict
Evidence	The Pulitzer Prize for Music went to Kendrick Lamar in 2018, not Herrmann	√	Conflict

The Focuser outputs the packed context $C\prime = \{C_1(conflict), C_2(none)\}$ by the score. Then Option Helper makes the option to be declarative sentence (option A as "*Arnulf Herrmann received the Ernst von Siemens Composers' Prize*"). After that, Option Helper calculate the conflict score between relevant sentences and declarative option sentences based on knowledge source or other reasons in Table 5. Option Helper choose the final answer based on the last conflict score.

Table 5. The conflict score between relevant sentences and options from LLM.

Option	p_i^{sup}	p_i^{con}	p_i^{irr}	Conflict score
A	0.81	0.05	0.14	**0.76**
B	0.10	0.70	0.20	−0.60
C	0.33	0.50	0.17	−0.17
D	0.18	0.42	0.40	−0.24

5 Conclusion and Future Work

We propose a multi-agent retrieval enhanced generation framework MARAG for scenarios of external and inter knowledge conflicts in LLMs, and verifies it with zero fine-tuning on four open source LLMs. Experimental results show that MARAG improves the average accuracy of traditional RAG by 6% and the average accuracy of IRCAN by 2% on ConflictBank, while significantly improving the interpretability to noisy evidence, demonstrating the effectiveness of multi-agent collaboration and explicit conflict modeling. In the future, we plan to introduce adaptive agent orchestration, dynamically adjust the collaboration level, and incorporate external knowledge graphs and graph-structure fact verifiers into evidence screening to obtain traceable conflict locations.

Acknowledgement. This work was partially supported by National Natural Science Foundation of China under Grant No. 62302513.

References

1. Lewis, P., Perez, E., Piktus, A., et al.: Retrieval-augmented generation for knowledge-intensive NLP tasks. In: Advances in Neural Information Processing Systems, vol. 33, pp. 9459–9474 (2020)
2. Guu, K., Lee, K., Tung, Z., Pasupat, P., Chang, M.: Retrieval-augmented language model pre-training. Proc. ICML **37**, 3929–3938 (2020)
3. Borgeaud, S., Mensch, A., Hoffmann, J., et al.: Improving language models by retrieving from trillions of tokens. Proc. ICML **39**, 2206–2240 (2022)
4. Izacard, G., Grave, E.: Leveraging passage retrieval with generative models for open-domain question answering. Proc. EACL **16**, 874–880 (2021)
5. Liu, Y., et al.: The journey of language models in understanding natural language. Web Inf. Syst. Appl. WISA (2024)
6. Kasai, J., Sakaguchi, K., Takahashi,Y., et al.: Realtime QA: what's the answer right now? In: NeurIPS Datasets & Benchmarks Track (2023)
7. Asai, A., Wu, Z., Wang, Y., Sil, A., Hajishirzi, H.: SELF-RAG: learning to retrieve, generate, and critique through self-reflection. In: ICLR 2024 (2024)
8. Wang, F., Wan, X., Sun, R., Chen, J., Arık, S.Ö.: Astute-RAG: overcoming imperfect retrieval augmentation and knowledge conflicts for large language models (2024a). arXiv:2410.07176
9. Liu, N.F., Lin, K., Hewitt, J., et al.: Lost in the middle: how language models use long contexts. Trans. ACL **12**, 157–173 (2024)
10. Chang, C.-Y., Jiang, Z., Rakesh, V., et al.: MAIN-RAG: multi-agent filtering retrieval-augmented generation (2024). arXiv:2501.00332
11. Wang, H., Prasad, A., Stengel-Eskin, E., Bansal, M.: MADAM-RAG: multi-agent debate for ambiguity and misinformation in RAG (2025). arXiv:2504.13079
12. Stengel-Eskin, E., Rawlins, K., Van Durme, B.: Zero- and few-shot semantic parsing with ambiguous inputs. In: ICLR 2024 (2024)
13. Xu, R., Qi, Z., Guo, Z., et al.: Knowledge conflicts for LLMs: a survey, pp. 8541–8565. In: EMNLP 2024 (2024b)
14. Chen, J., Lin, H., Han, X., Sun, L.: RGB: benchmarking large language models in retrieval-augmented generation. In: AAAI, vol. 38, pp. 17754–17762 (2024)
15. Yang, X., Sun, K., Xin, H., et al.: CRAG: A Comprehensive RAG Benchmark. In: NeurIPS, vol. 37, pp. 10470–10490 (2024)

16. Su, Z., Zhang, J., Qu, X., et al.: ConflictBank: a benchmark for evaluating the influence of knowledge conflicts in LLMs. In: NeurIPS Datasets & Benchmarks Track, vol. 38 (2024)
17. Min, S., Michael, J., Hajishirzi, H., Zettlemoyer, L.: AmbigQA: answering ambiguous open-domain questions. In: EMNLP 2020, pp. 5783–5797 (2020)
18. Lee, Y., Ye, X., Choi, E.: AmbigDocs: reasoning across documents on different entities under the same name. In: LM 2024 (2024)
19. Ming, Y., Purushwalkam, S., Pandit, S., et al.: FaithEval: can your language model stay faithful to context? (2024). arXiv:2410.03727

A Self-enhancement Framework to Improve Zero-Shot Information Extraction Ability for Large Language Models

Yuli Feng[1,5], Feng Jin[1], Hao Wang[2], Jiadi Xu[3], Yuanyuan Zhou[4], and Guanghao Ren[5](✉)

[1] School of Automation, Beijing Institute of Technology, Beijing, China
[2] Traffic Management Bureau, Beijing Municipal Public Security Bureau, Beijing, China
[3] COMAC Shanghai Aircraft Customer Service Co., Ltd., Shanghai, China
xujiadi@comac.cc
[4] BNRist, DCST, RIIT, Tsinghua University, Beijing, China
west@tsinghua.edu.cn
[5] Institute of Automation, Chinese Academy of Sciences, Beijing, China
guanghao.ren@ia.ac.cn

Abstract. In the task of fault diagnosis, information extraction (IE) serves as a key technology in the data preparation process. However, the operating environment of aviation equipment (e.g., aircraft) is complex and highly variable, and new types of fault patterns keep emerging, making it extremely difficult to obtain comprehensive and accurate labeled data for traditional supervised methods. Large language models (LLMs) pre-trained on massive corpora have acquired extensive prior knowledge and powerful zero-shot learning abilities, presenting significant opportunities for IE tasks in low-resource scenarios. Zero-shot IE, especially zero-shot named entity recognition (NER), still faces some challenges, including *the failure to fully capture the unique features of NER tasks*, *oversight of correlations between contexts surrounding entities*, and *indiscriminate utilization of irrelevant task demonstrations*. This paper presents a self-enhancement framework to better address these issues. Specifically, we propose a two-stage self-annotation strategy for unlabeled corpus to obtain pseudo in-context learning (ICL) demonstrations. Additionally, use mutual information criteria to extract class-level type-related features (TRFs) and take them to retrieve relevant ICL demonstrations. The results of ablation experiments show that the proposed method is effective for zero-shot NER tasks based on large language models.

Keywords: Information Extraction · Named Entity Recognition · Large Language Model · Zero-shot Learning

1 Introduction

In the aviation field, the operation of aircraft generates a vast amount of data, covering flight parameters, maintenance records, etc. The NER task can accu-

S. Yang et al. (Eds.): WISA 2025, LNCS 16299, pp. 556–566, 2026.
https://doi.org/10.1007/978-981-95-6426-2_49

rately extract fault patterns from such complex unstructured data, playing a crucial role in aircraft health management. But supervised methods have poor flexibility and scalability and rely heavily on a large amount of labeled data.

LLMs, including LLaMA, Qwen, and DeepSeek, etc., possess superior capabilities in text understanding and generation, which enable them to provide high-quality explanations for the output results and achieve promising performance on a wide range of downstream NLP applications. LLMs have the potential to overcome the challenges of data sparsity faced by traditional supervised models.

However, fine-tuning them to address IE subtasks like named entity recognition (NER) faces the issues of unlabeled corpus and high training cost. This situation has motivated research into designing prompting methods for low-resource IE tasks with LLMs. Chain-of-Thoughts (CoT) [1] and in-context learning (ICL) [2] are advanced prompt engineering techniques that utilize the inherent semantic understanding and reasoning capabilities of LLMs, which do not require updating model parameters and can quickly adapt to downstream tasks, and present significant opportunities for IE tasks in low-resource scenarios.

NER is a sequence labeling task in nature, where the model needs to assign an entity-type label, such as person names, organizations, and locations, to each token within a sentence. However, the semantics of human instructions comprehended by LLMs might deviate from the target NER concepts, making it difficult to obtain correct results. Due to the gap between generation models and sequence labeling tasks, LLMs without task-specific fine-tuning still lag behind supervised SOTA models. However, supervised NER [3] approaches that require numerous annotated data are not available in low-resource scenarios. Zero-shot NER reduces the time and effort needed for data labeling, which is a more universal and feasible approach. LLMs still encounter some challenges in zero-shot NER: (i)overlooking correlations between contexts surrounding entities; (ii)difficult to effectively define the scope of each entity type; (iii)irrelevant demonstrations retrieved affect the LLM's inference performance.

To address the aforementioned problems, we propose a self-enhancement framework to push the performance boundary of zero-shot NER with LLMs. To sum up, the contributions of this paper can be summarized as follows:

- We use the cluster algorithm to obtain point entities, which provide more detailed and clear entity definitions for LLMs.
- We propose a two-stage self-annotation strategy to obtain pseudo ICL demonstrations.
- We use mutual information criteria to extract class-level TRFs and take them to serve as the guideline to retrieve more relevant ICL demonstrations from self-annotated data.

2 Related Work

2.1 In-Context Learning

In-context learning (ICL) refers to a scenario where the model is conditioned on a few task demonstrations and then completes further task instances by simply

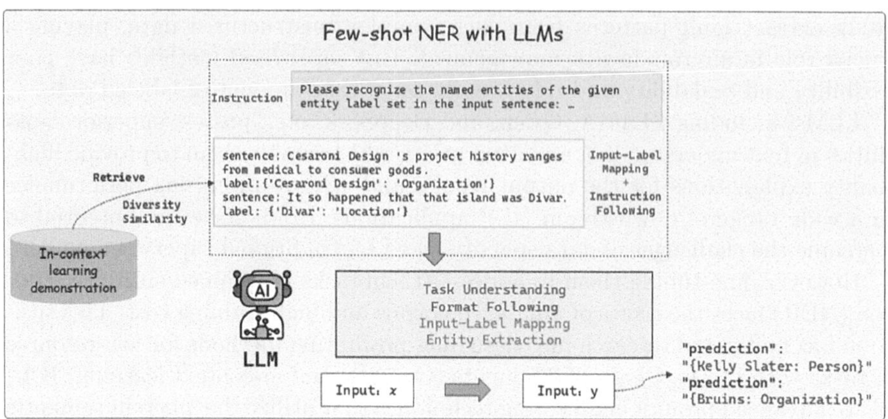

Fig. 1. The overview framework of in-context learning for NER task.

predicting subsequent content. And according to the number of examples, ICL can be classified into three types: zero-shot, one-shot and few-shot paradigms. The selection of demonstration examples affects the performance of ICL, and the two main criteria for selecting demonstration examples are similarity and diversity. Liu et al. [4] propose KATE which sort the examples according to the distances in the sentence encoder's embedding space, and then select the top-K examples. In order to enhance diversity of selected examples, Li et al. [5] divide all examples into K-clusters and group similar examples together, and then select the most similar example from each cluster. RetICL [6] proposes an iterative retrieval process, where the selection of the next example depends on the current problem and the selected examples. However, incorrect input-label mappings of examples can degrade the performance of ICL, and LLMs with larger parameter quantity generally exhibit better instruction-following capabilities and also have higher sensitivity to label errors of ICL examples [7]. Self-annotated ICL examples are more suitable for improving the zero-shot NER performance of weaker LLMs (Fig. 1).

2.2 LLMs for Few-Shot NER

Few-shot demonstrations provide the LLM with direct evidence about the NER tasks and references to make predictions. GPT-NER [8] marks entities with special tokens as few-shot examples for reference, so as to format the LLM's output to predict the target entities. This approach works well for sentences containing only a single entity, transforming the sequence labeling task into text generation. LTNER [9] not only marks entities with special tokens but also labels the entity category, thus achieving the prediction of multiple categories of entities in the texts. Besides a set of annotated entity definitions as demonstrations, PromptNER [10] prompts an LLM to produce a list of potential entities along with explanations justifying that they are matched with the provided entity type definitions. To achieve domain knowledge transfer, Wang et al. [11] propose a framework named PLTR to extract relevant entity type-related features (TRFs) based on mutual information criteria. This framework bridges the gap between training data and out-of-domain examples, enabling generalization to an unseen target domain with only a few labeled examples. Standard ICL demonstrations often neglect the particularity of the NER task itself. Jiang et al. [12] propose a point entity selection method based on K-Means clustering and use point entities as the ICL demonstrations to recognize each entity type.

2.3 LLMs for Zero-Shot NER

Without human-annotated data as ICL demonstrations, the zero-shot NER approach is typically divided into two steps: self-annotation and relevant ICL examples retrieval. Xie et al. [13] propose a two-stage majority voting (TSMV) strategy to adapt self-consistency approach [14] to NER, which first votes for the most consistent mentions, then the most consistent types. Furthermore, to select reliable samples from self-annotated data as ICL demonstrations for zero-shot NER, Xie et al. [15] explore diverse strategies, such as entity-level threshold filtering, sample-level threshold filtering. Wang et al. [16] utilize LLMs to generate entities from their definitions and subsequently expand them into full entity-labeled sentences as references to facilitate the LLM's entity recognition. Tong et al. [17] leverage LLMs for entity definition and type distinction summarization to serve as a continuous refinement guideline, addressing the misalignment between model and human definitions of entity types and the confusion of similar types. ChatIE [18] transform the zero-shot NER task into a multi-turn question-answering problem with a two-stage framework to extract entities. In addition to using a single model for the zero-shot NER task, Wang et al. [19] design a cooperative multi-agent system for zero-shot NER that harnesses the collaborations and unique roles of multiple agents to integrate contextual correlations and the self-reflection mechanism.

In this paper, we focus on the zero-shot NER task with LLMs and the works most closely related to ours are [11,12,15]. These previous works provide the guidance for further improving the accuracy of self-annotated data and the relevance of retrieval demonstrations.

3 The Proposed Approach

3.1 Problem Definition

Zero-Shot NER. We assume an unlabeled corpus $\mathcal{U} = \{x_i\}_{i=1}^{N}$ with the entity label set $\mathcal{T} = \{t_j\}_{j=1}^{M}$. Our goal is to perform zero-shot NER using LLMs in low-resource scenarios without any annotated data. Given a input test sentence x^q, we prompt the LLM $\mathcal{M}$ to generate corresponding entity predictions for x^q:

$$y_i = \arg\max_{y} P_{\mathcal{M}}(y \mid T, S, x^q), \tag{1}$$

where $P_{\mathcal{M}}(\cdot)$ represents the output probability of LLMs, T is the task description prompt, and S denotes pseudo ICL demonstrations retrieved from self-annotated data for x^q. We obtain the predicted results of named entities $y_i = \{(e_i^j, t_i^j)\}_{j=1}^l$: e is the entity span, which is a sequence of tokens from x; t is the corresponding entity type, which belongs to the predefined entity label set; l is the number of the predicted entity span.

Type-Related Feature Extraction. Type-related features are tokens that are strongly associated with entity types and relevant to the input sentence. Given an input sentence $x \in \mathcal{U}$ with entity types $\mathcal{T}_x$, the goal of TRF extraction is to identify all TRFs $\mathcal{R}$ that are related to the input sentence x for all entity types in $\mathcal{T}$. We calculate the mutual information between word $\mathbf{w}$ and entity class c to describe this relation:

$$I(\mathbf{w}; c) = \sum_{\mathbf{w} \in \mathcal{V}} \sum_{c \in \mathcal{C}} p(\mathbf{w}, c) \log \frac{p(\mathbf{w}, c)}{p(\mathbf{w}) \cdot p(c)} \tag{2}$$

where $\mathcal{V}$ is the vocabulary, $\mathcal{C}$ is the entity class set, and $p(\cdot)$ is the distribution probability. We select the top k tokens with the highest mutual information scores as the candidate set of TRFs.

3.2 Self-enhancement Framework

Two-Stage Self-annotation for Unlabeled Data. For each unlabeled example $x_i \in \mathcal{U}$, we generate predicted annotations using LLMs with zero-shot prompting. This process is formulated as follows:

$$y_i = \arg\max_y P_{\mathcal{M}}(y \mid T_{\mathbf{e}}, x_i), \tag{3}$$

In the second stage, we generate point entities from the self-annotated data. Formally, let $\mathcal{E}_t = \{e_{t,1}, e_{t,2}, \cdots, e_{t,|\mathcal{E}_t|}\}$ be the entity set for corresponding entity type $t \in \mathcal{T}$ from self-annotated data. For each entity e_t of entity type t, we can use the text-embedding model such as `bge-m3` to obtain its embedding representation:

$$\mathbf{e}_t = \text{Encoder}(e_t) \in \mathbb{R}^d, \tag{4}$$

where d is the embedding size, which generally depends on the encoder used. For `bge-m3`, the size is 1024. Let $\mathbf{E}_t = \{\mathbf{e}_{t,1}, \mathbf{e}_{t,2}, \cdots, \mathbf{e}_{t,|\mathcal{E}_t|}\}$ be the corresponding entity representation set. For entity type t, we use the following procedure to obtain a_t entity centers with classic K-Means clustering algorithm:

$$(\mathbf{c}_{t,1}, \mathbf{c}_{t,2}, \cdots, \mathbf{c}_{t,a_t}) = \text{K-Means}(\mathbf{E}_t, a_t), \tag{5}$$

Then, we take $\mathbf{c}_{a_t}$, which is called a point entity as typical examples of entity spans to make LLMs generate predictions again:

$$y_i = \arg\max_y P_{\mathcal{M}}(y \mid T_{\mathbf{e}}, x_i, \mathbf{c}_{a_t}). \tag{6}$$

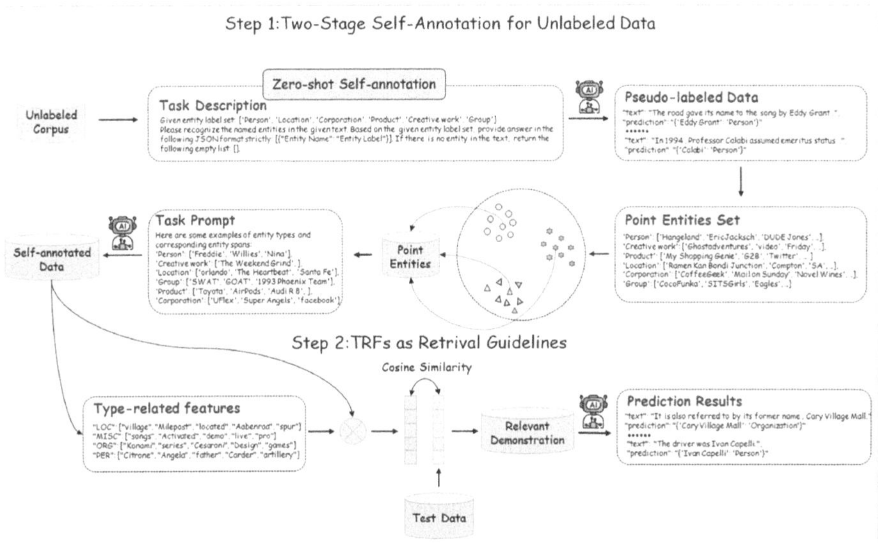

Fig. 2. The overview of the proposed self-enhancement framework for zero-shot NER with LLM.

To better utilize point entities as auxiliary information, we investigate the three following select strategies. (1)*Random selection from clusters.* For each cluster, we select a single entity to serve as its representative across the entire cluster. (2)*Similarity-based selection.* We choose the corresponding point entity c_t with the highest similarity to the cluster center.(3)*Density-based selection.* We identify the point entity with the highest local density within the cluster as its representative (Fig. 2).

TRFs as Retrieval Guidelines. We generate TRFs from the self-annotated data. We define $\mathcal{U}_i$ as a set that contains all sentences where entities of the i-th type appear. However, the mutual information criteria may favor tokens that are highly associated with $\mathcal{U} \setminus \mathcal{U}_i$ rather than with $\mathcal{U}_i$. Thus, we introduce a filtering condition as follows:

$$\frac{C_{\mathcal{U}}(\mathbf{w}_m)}{C_{\mathcal{U}_i}(\mathbf{w}_m)} < \tau \tag{7}$$

where $C_{\mathcal{U}_i}(\mathbf{w}_m)$ represents the count of the m-gram $\mathbf{w}_m$ (for simplicity, only 1-gram texts are considered here) in $\mathcal{U}_i$, and τ is an m-gram threshold hyperparameter. By applying this filtering criterion, we ensure that the frequency of the token w in $\mathcal{U}_i$ is significantly higher than its frequency in other sets ($\mathcal{U} \setminus \mathcal{U}_i$).

Traditional methods typically employ the K-Nearest Neighbors (KNN) algorithm to retrieve the K sentences with the highest similarity scores. However, the similarity between sentences is often noisy, and the retrieved examples may not be relevant to the test sentence and have the same entity types, which affects

LLMs' performance in NER tasks. Sentences with similar TRFs often have very similar context structures, which means they exhibit local relevance and involve entities of the same type.

Specifically, We take attention mechanism to integrate the representation of TRFs into sentence embedding. We take these novel embeddings to employ the KNN algorithm to retrieve the demonstrations:

$$\alpha_i = \frac{\exp(\mathbf{q}_i^\top \mathbf{e}_x/\tau)}{\sum_{i=1}^{M} \exp(\mathbf{q}_i^\top \mathbf{e}_x/\tau)}, \qquad \mathbf{e}_x^{\text{new}} = \sum_{i=1}^{M} \alpha_i \cdot \mathbf{q}_i \tag{8}$$

$$N_K(S_{\text{demo}}) = \underset{S_{\text{demo}}}{\arg\max} \cos(\mathbf{e}_{\text{demo}}^{\text{new}}, \mathbf{e}_{\text{test}}) \tag{9}$$

where $\mathbf{q}$ and $\mathbf{e}$ are the embeddings of TRFs and input sentences, τ is the temperature. Because we focus on the strict zero-shot scenarios, we do not perform any operations on the test sentences. In short, we compute the similarity between the original representations of test sentences and the novel representations of demonstration sentences.

4 Experiments

In this section, we evaluate our proposed self-enhancement framework for the zero-shot NER task. We consider a prediction correct only when both the boundary and the type of the predicted entity exactly match those of the ground truth, specifically in terms of precision, recall and the micro-F1 score.

4.1 Experiment Settings

Datasets. The data, such as aviation equipment failure maintenance manuals, are relatively sensitive, as they involve confidentiality. For our general-domain and domain-specific experiments, we consider four commonly used datasets, namely the CoNLL03 [20], WikiGold [21], WNUT17 [22] and GENIA [23], where GENIA is a professional dataset in the biological field (Table 1).

Table 1. Statistics of datasets used in the experiments.

Dataset	CoNLL03	WikiGold	WNUT-17	GENIA
Train	14,382	1,422	4,403	16,692
Test	3,453	274	1,287	1,854

Similar to related methods [15,17], we randomly sample 500 examples from the training set as the unlabeled corpus for two-stage self-annotation, and repeatedly sample 300 examples three times from the test set to calculate the average performance. But WikiGold has a test set of only 247 examples.

Baselines. To effectively evaluate our method, we compare our framework with the following baselines:(i)**Vanilla** employs a straightforward and common prompting strategy that directly asks LLMs to extract entity labels from input texts.(ii)**TSMV** applies the self-consistency strategy to obtain a SC score for reliable annotation selection.(iii)**ICL with gold labeled demonstrations** indicates the upper bound of performance.

4.2 Experimental Results

To explore the performance of the proposed self-enhancement framework on weaker LLMs, we conduct experiments on the Qwen2.5-Instruct-7B model (Table 2 and Fig. 3).

Table 2. Experimental results micro-F1(%) on the above datasets.

Method	CoNLL03	WikiGold	WNUT-17	GENIA
Vanilla	61.86	58.74	35.55	19.19
TSMV	63.91	62.05	37.48	17.18
Gold(3-shot)	78.16	65.51	42.83	54.06
Gold(5-shot)	78.64	66.78	42.92	54.94
Ours	66.87	66.61	44.08	42.79

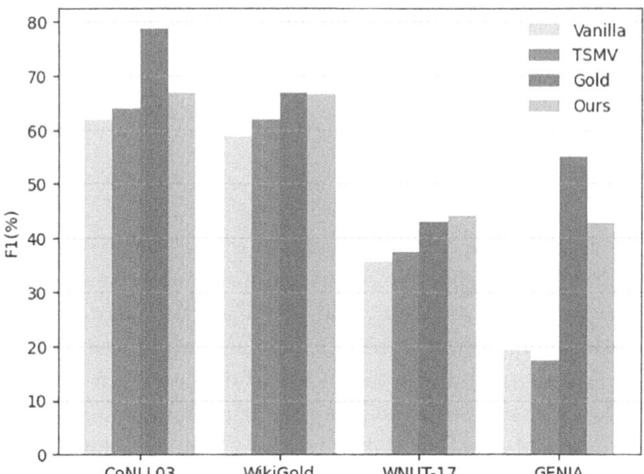

Fig. 3. Experimental results (F1) with LLMs on CoNLL03, WikiGold, WNUT-17, and GENIA.

4.3 Ablation Study

To analysis the individual contributions of each component to our framework, this study conducted ablation experiments on each module separately (Table 3).

Table 3. The results of ablation study.

Method	CoNLL03	WikiGold	WNUT-17	GENIA
Vanilla	61.86	58.74	35.55	19.19
Analysis of Point Entities Selection				
Random	58.9	59.02	**43.33**	25.7
Similarity-based	**64.18**	**63.4**	40.66	24.99
Density-based	58.61	63.26	41.81	**26.72**
Analysis of Pseudo-sample Retrieval				
Random	65.6	64.64	42.11	42.31
KATE	65.2	65.19	42.32	42.60
Ours	**66.87**	**66.61**	**44.08**	**42.79**

Point entities provide more detailed and clear entity definitions for LLMs, which improve the performance of zero-shot NER with LLMs. For these datasets where the micro-F1 value of zero-shot NER with LLMs is relatively high, the extracted entities have higher similarity. Therefore, similarity-based selection methods can more easily obtain representative point entities to improve the performance of self-annotation.

The experimental results validate the effectiveness of TRFs, as incorporating TRFs into the representation of sentences significantly improves the relevance of retrieved examples.

5 Conclusion

In this paper, we introduce a self-enhancement framework to improve the performance of LLMs in the zero-shot NER task on unlabeled data. Our approach can be deployed directly in the aviation field to efficiently extract information such as key fault entities, component numbers, and abnormal states, providing structured inputs for subsequent fault diagnosis and prediction models to improve the intelligence level of the entire aviation health management system.

Based on experiments on four datasets, representative entity points obtained by the cluster algorithm can enhance the LLMs' understanding of entity spans and entity types. Type-related features help retrieve relevant ICL demonstrations. In future work, we will improve the zero-shot information extraction ability of LLMs via cross-domain transfer learning in low-resource scenarios such as the aviation equipment failure maintenance manuals.

Acknowledgement. This work was supported by the Fund of Shanghai Engineering Research Center of Civil Aircraft Health Monitoring.

References

1. Wei, J., et al.: Chain-of-thought prompting elicits reasoning in large language models. In: Advances in Neural Information Processing Systems, vol. 35, pp. 24824–24837 (2022)
2. Brown, T., et al.: Language models are few-shot learners. In: Advances in Neural Information Processing Systems, vol. 33, pp. 1877–1901 (2020)
3. Yin, H., et al.: An generative entity relation extraction model based on UIE for legal text. In: International Conference on Web Information Systems and Applications, pp. 91–99. Springer (2024)
4. Liu, J., et al.: What makes good in-context examples for GPT-3? arXiv preprint: arXiv:2101.06804 (2021)
5. Li, J., Wang, J., Zhang, Z., Zhao, H.: Self-prompting large language models for zero-shot open-domain QA. arXiv preprint: arXiv:2212.08635 (2022)
6. Scarlatos, A., Lan, A.: RetiCL: sequential retrieval of in-context examples with reinforcement learning. arXiv preprint: arXiv:2305.14502 (2023)
7. Kossen, J., Gal, Y., Rainforth, T.: In-context learning learns label relationships but is not conventional learning. arXiv preprint: arXiv:2307.12375 (2023)
8. Wang, S., et al.: GPT-NER: named entity recognition via large language models. arXiv preprint: arXiv:2304.10428 (2023)
9. Yan, F., et al.: LTNER: large language model tagging for named entity recognition with contextualized entity marking. In: International Conference on Pattern Recognition, pp. 399–411. Springer (2025)
10. Ashok, D., et al.: PromptNER: prompting for named entity recognition. arXiv preprint: arXiv:2305.15444 (2023)
11. Wang, Z., et al.: Generalizing few-shot named entity recognizers to unseen domains with type-related features. arXiv preprint: arXiv:2310.09846 (2023)
12. Jiang, G., et al.: P-ICL: point in-context learning for named entity recognition with large language models. arXiv preprint: arXiv:2405.04960 (2024)
13. Xie, T., Li, Q., Zhang, J., Zhang, Y., Liu, Z., Wang, H.: Empirical study of zero-shot NER with ChatGPT. arXiv preprint: arXiv:2310.10035 (2023)
14. Wang, X., et al.: Self-consistency improves chain of thought reasoning in language models. arXiv preprint: arXiv:2203.11171 (2022)
15. Xie, T., et al.: Self-improving for zero-shot named entity recognition with large language models. arXiv preprint: arXiv:2311.08921 (2023)
16. Wang, A., et al.: ReverseNER: a self-generated example-driven framework for zero-shot named entity recognition with large language models. arXiv preprint: arXiv:2411.00533 (2024)
17. Tong, Z., et al.: EvoPrompt: evolving prompts for enhanced zero-shot named entity recognition with large language models. In: Proceedings of ACL, pp. 5136–5153 (2025)
18. Xiang, W., et al.: ChatIE: zero-shot information extraction via chatting with ChatGPT (2024). https://arxiv.org/abs/2302.10205
19. Wang, Z., et al.: A cooperative multi-agent framework for zero-shot named entity recognition. arXiv preprint: arXiv:2502.18702 (2025)

20. Sang, E., et al.: Introduction to the CoNLL-2003 shared task: Language-independent named entity recognition. arXiv preprint: cs/0306050 (2003)
21. Balasuriya, D., et al.: Named entity recognition in Wikipedia. In: Proceedings of the 2009 Workshop on the People's web Meets NLP: Collaboratively Constructed Semantic Resources (People's Web), pp. 10–18 (2009)
22. Derczynski, L., et al.: Results of the WNUT2017 shared task on novel and emerging entity recognition. In: Proceedings of the 3rd Workshop on Noisy User-generated Text, pp. 140–147 (2017)
23. Ohta, T., Tateisi, Y., Kim, J.D., Mima, H., Tsujii, J.: The GENIA corpus: an annotated research abstract corpus in molecular biology domain. In: Proceedings of the Human Language Technology Conference, pp. 73–77. Morgan Kaufmann Publishers Inc. San Francisco (2002)

DM-ERC: Emotion Recognition in Conversations Based on Large Language Models for Hidden Information Deep Mining

Zhinan Gou(✉), Yuxin Chen, Mengyao Jia, and Siyu Liu

School of Management Science and Information Engineering, Hebei University of Economics and Business, Shijiazhuang 050061, China
gouzhinan@hueb.edu.cn

Abstract. Emotion Recognition for Conversations (ERC), a key research area in Natural Language Processing (NLP), has been increasingly implemented in human-computer interaction (HCI) systems. Existing approaches employ graph-based or sequence models to extract features from utterances, yet these features are often insufficient for emotional reasoning tasks. With the growing adoption of Large Language Models (LLMs), researchers have integrated speaker identification techniques to boost model performance. However, the hidden information about the current dialogue is not fully exploited. In this work, we introduce a novel framework enabling large language models to deeply mine context and speaker hidden information (DM-ERC). First, speaker latent commonsense is generated by Llama2-Chat from historical utterances and injected into the large language model to improve its emotional dynamics understanding. Next, emotion prediction experiments are conducted on the pre-trained LLM to anticipate future emotional trends. Furthermore, a similar utterance retrieval module is integrated to assist reasoning during emotion recognition. Extensive experiments validate that the framework outperforms current baselines on three publicly available datasets.

Keywords: Emotion recognition for conversations · Large language models · Hidden information

1 Introduction

With ongoing advancements in Artificial Intelligence (AI) [1], ERC, a key research area in NLP, has intersected with multiple interdisciplinary fields. For example, within social media ecosystems [2], ERC supports public sentiment monitoring, user profiling, and optimization of intelligent customer service systems by analyzing emotions in user-generated content. Researchers have traditionally modeled utterances in a sequential manner, incorporating global and

S. Yang et al. (Eds.): WISA 2025, LNCS 16299, pp. 567–578, 2026.
https://doi.org/10.1007/978-981-95-6426-2_50

local information to aid emotion recognition [3]. However, these approaches place excessive reliance on local utterances and fail to capture long-range contextual information. Graph-based methods address this limitation by constructing graph-specific edges and nodes to effectively link historical information with speaker features [4]. Furthermore, several models utilize supervised contrastive learning to enhance discrimination between similar emotional categories [5]. Since the advent of large language models, their robust generative and inference capabilities have become a focal point of AI research. Some researchers employ them in multiple-assisted tasks to improve the accuracy of emotion recognition [6]. However, all of these methods focus on the surface characteristics of the speakers and do not dig deeper to exploit the hidden information.

In this paper, we propose a new framework for deep mining of hidden information by applying prompt engineering techniques for large language models (DM-ERC). First, we guide Llama2-Chat to reason and generate "the reaction of listener" based on historical information, incorporating this common sense into the LLM for pre-training. Second, the pre-trained model is jointly trained on emotion prediction and recognition tasks to derive the final LLM. In addition, we add similarity retrieval of the target discourse to emotion recognition.

The main contributions of this work are set out below:

- DM-ERC achieves deep mining of contextual information from the listener's point of view, effectively capturing the dynamic emotional cues of the dialogue context.
- We predict speaker emotions from historical dialogues, thereby providing effective support for the emotion recognition task.
- We conduct extensive experiments on three publicly available datasets, MELD, IEMOCAP, and EmoryNLP, which demonstrate the effectiveness of the proposed model.

2 Related Work

As a result of the increased demand from society [7], more and more researchers are committed to the study of ERC tasks. This field aims to infer emotional states by comprehensively analyzing the textual content [8], vocal intonation, and facial expressions of a speaker [9]. In this paper, we only study textual modals. DialogRNN [4] utilizes multiple recurrent neural networks to dynamically model discourse. To enhance the perception of conversational context, DialogCRN [10] adeptly captures and synthesizes emotional cues within dialogues by employing multiple sophisticated inference modules. After the graph neural network-based approach is proposed, the information of remote utterances is effectively utilized. RGAT [11] pays attention to the location information of each discourse, encodes it and integrates it into the graph. DAG-REC [12] utilizes a directed acyclic graph to model global and local statements. Models like TODKAT [13] and COSMIC [14] also incorporate commonsense knowledge to better capture the mood of the target discourse. In addition, InstructERC [6] utilizes LLM for the first time to unify ERC tasks in the Seq2Seq paradigm.

3 Methodology

DM-ERC consists of two critical stages: "Commonsense Generation and Injection" and "Emotion Prediction and Recognition". Subsequent to the aforementioned two stages of training, the final LLM is utilized for the emotion recognition task. Additionally, we propose a similar statement retrieval method to assist inference. The general framework of the DM-ERC is shown in Fig. 1 .

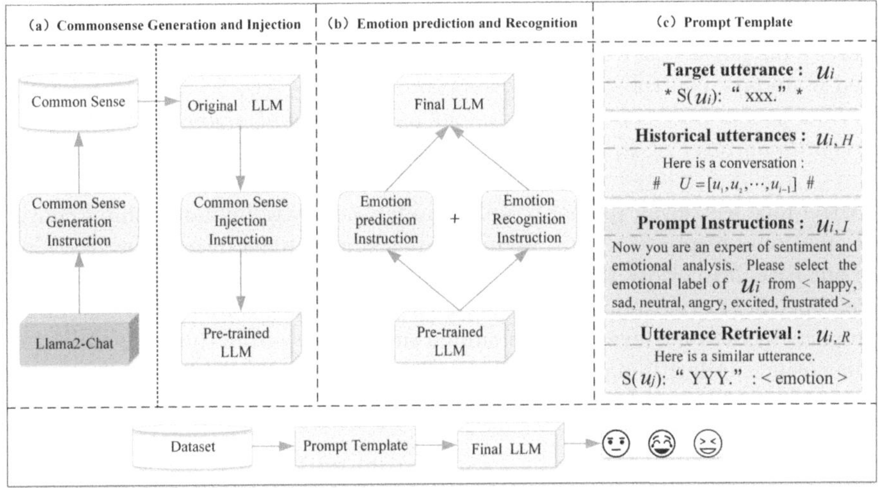

Fig. 1. The overall architecture of DM-ERC.

In ERC, each conversation consists of several utterances $U = [u_1, u_2, \ldots, u_n]$, where n represents the number of utterances. Each utterance is associated with an emotion label $Y = \{y_1, y_2, \ldots, y_m\} \in O$, where O denotes the predefined set of emotion categories. Additionally, define $S(\cdot)$ as a function mapping utterance u_i to its corresponding speaker identity information, where $S(u_i)$ denotes the speaker identity of utterance u_i.

3.1 Prompt Template and Retrieval

With the proliferation of LLMs, prompt engineering is now the dominant interaction paradigm. Building effective instructions can bridge the gap when applying LLMs to specific subtasks. Therefore, we enable LLMs to efficiently complete assigned tasks by designing accurate prompt templates for the ERC task. The detailed prompt template contains four parts: target utterance, history utterances, prompt instructions, and utterance retrieval, as shown in Fig. 1.

Target Utterance. To better utilize the features of the speaker, we synchronously embed the identity information of the speaker when inputting the target utterance. Moreover, we adopt the special marker "*" to emphasize the target utterance.

History Utterances. Deep mining of context-hidden information enables effective capture of dynamic emotion cues. To model contextual dependencies, we introduce hyperparameter w to quantify historical window size. Specifically, during input encoding, we prefix and suffix historical utterances with special marker "#", explicitly bounding contextual boundaries for LLM reasoning.

Prompt Instructions. We initiate by engineering role-specific identities for LLMs, activating domain knowledge, tonal patterns, and reasoning logics of assigned personas to generate human-aligned precise responses. To enhance task accuracy, output is constrained within a pre-defined label scope through prompt-based boundary control.

"Now you are an expert of sentiment and emotional analysis. Please select the emotional label of u_i *from* $\langle$ *happy, sad, neutral, angry, excited, frustrated* $\rangle$*."*

Utterance Retrieval. In contextually similar scenarios, humans exhibit consistent emotional responses. Thus, we integrate a similar utterance retrieval module into our ERC framework to ground LLM reasoning with real-world response patterns. Specifically, we employ Sentence-BERT to semantically encode the target utterance u_i, then retrieve the most similar utterance r_d from the dialogue corpus D via cosine similarity matching. The retrieved r_d is fed into LLM as contextual augmentation for emotion reasoning. The process of this abstraction is shown below:

$$r_d = \underset{r_j \in \mathcal{D}}{\operatorname{argmax}} \, \mathrm{SBERT}(u_i, r_j) \tag{1}$$

In summary, when performing emotion recognition, the prompt template entry can be abbreviated to x_i:

$$x_i = \left[u_i;\ u_{i,H};\ u_{i,I};\ u_{i,R} \right] \tag{2}$$

where $[;]$ denotes text linking, u_i, $u_{i,H}$, $u_{i,I}$, $u_{i,R}$ denotes target utterance, historical utterances, prompt instructions, utterance retrieval.

3.2 Commonsense Generation and Injection

Modeling Speaker Emotional Dynamics constitutes a core challenge in human-computer interaction systems, where accurate emotion tracking enables adaptive and empathetic responses. In the Emotion Recognition in Conversation (ERC) task, most existing models focus on surface-level features of contextual information while neglecting deep contextual mining and utilization. Conversely, a growing body of research leverages external knowledge graphs to enrich utterance representations, thereby enhancing model performance through semantic augmentation. The most widely adopted resource is the ATOMIC knowledge graph, which not only captures objective facts but also specializes in causal reasoning patterns inherent in human commonsense. The ATOMIC knowledge graph encompasses nine relation types. For ERC tasks, we focus specifically on mental states, including oReaction (emotional response of the listener), xReaction (emotional reaction of the speaker), and xIntent (intention of the speaker). However, relying solely on ATOMIC knowledge is insufficient to capture deeply hidden contextual information embedded in historical dialogues, thereby impeding the model's ability to track emotional dynamics clues. In recent years, the generative and inference capabilities of LLMs bring about a significant shift in the field of NLP. Therefore, we stimulate the LLM to generate hidden common sense by designing a prompt template and inject the generated common sense into the LLM to help it perform the emotion recognition task.

Commonsense Generation. Llama2-Chat exhibits exceptional performance in dialogue generation, contextual understanding, and multi-turn interaction, making it widely applicable to intelligent customer service, virtual assistants, and educational tutoring scenarios. Inspired by the ATOMIC relationship type, we utilize Llama2-Chat to deeply mine hidden information based on historical discourse and generate listener response common sense. To make the generated common sense more accurate, we limit the output to no more than 10 words. The prompt instruction for generating common sense is as follows:

"Now You are an expert who is good at using common sense for reasoning. Based on the above historical utterances, please use common sense to infer the reaction of potential listeners in $*u_i*$ *in no more than ten words of output:"*

Commonsense Injection. Although LLM has good reasoning capabilities, there are some limitations when targeting specific ERC tasks. To enable LLM to learn dynamic cues in dialogue scenarios, we inject the aforementioned generated commonsense knowledge to activate their perceptual capabilities for emotion-aware reasoning. To fit in with common sense, we use similar prompt instructions for injection:

"Now You are an expert who is good at using common sense for reasoning. Based on the above historical utterances, please use common sense to infer the reaction of potential listeners in $*u_i*$ *in no more than ten words of output:,"target": "*t_i*"*

Where t_i denotes the common sense of the potential listener response corresponding to u_i. Meanwhile, the input x_i^c to the prompt template and the loss function for this task are as follows:

$$x_i^c = \left[u_{i,H};\, u_{i,I}^c \right] \tag{3}$$

$$L_c = \sum_i^N - \log P\left(t_i \mid x_i^c, \theta_c\right) \tag{4}$$

3.3 Emotion Prediction and Recognition

In a dialogue, the complex relationships between speakers can have a significant impact on the words and emotions of subsequent speakers. Previous studies solely rely on speaker perception for modeling while neglecting dialogue structural information. To address this problem, DualGATs [15] utilize nodes and edges specific to graph neural networks to focus on both contextual structure and speaker perception. To further leverage speaker features, MPLP [16] integrates contextual information and speaker experience knowledge, embedding them as prompts into the decoder for enhanced dialogue modeling. However, all of the above approaches focus only on the representational information of the existing text. In the ERC task, most of the models focus on utilizing the target discourse for inference and recognition, while ignoring the powerful predictive capabilities in generative models. The essence of prediction lies in inferring future events based on historical information. In the commonsense generation task, we leverage this principle to elicit latent responses of the listener from LLM. Through commonsense injection, LLM develop preliminary contextual understanding of dialogue scenarios, enhancing their readiness for emotion reasoning. Therefore, we design an emotion prediction task to further mine the historical information to support the emotion recognition task. It implicitly captures the interplay between dialogue and emotional affect. In this work, we simultaneously utilize emotion recognition and emotion prediction tasks to fine-tune LLM for enhanced performance in dialogue emotion modeling. The prompt instructions and overall input for the emotion prediction task are as follows:

"Now you are expert of sentiment and emotional analysis. Based on the above historical utterances, the next utterance is spoken by <S(ui)>, please predict the emotion states of $S(u_i)$ *from* $\langle\, y_1, y_2, \ldots, y_m \,\rangle$*:"*

The loss function for the emotional prediction task is as follows:

$$L_e = \sum_{i}^{N} -\log P\left(\mu_i \mid x_i^e, \theta_e\right) \tag{5}$$

where μ_i denotes the emotional label token of the text label y_i corresponding to the formatted input discourse x_i. The loss function of the emotion recognition task is as follows:

$$L_{main} = \sum_{i}^{N} -\log P\left(\mu_i \mid x_i, \theta\right) \tag{6}$$

4 Experiments

4.1 Datasets and Evaluation Indicators

Extensive experiments are conducted on three publicly available datasets using DM-ERC. Detailed statistics of the datasets are presented in Table 1.

MELD [17], a multimodal dataset derived from dialogues in the TV series Friends, contains 1,400 dialogues and 13,000 utterances for emotion recognition research. Emotional categories include joy, anger, fear, disgust, sadness, surprise, and neutral.

Table 1. Statistics for the three datasets.

	MELD	IEMOCAP	EmoryNLP
C:Train	1038	108	713
C:Val	114	12	99
C:Test	280	31	85
U:Train	9989	5163	9934
U:Val	1109	647	1344
U:Test	2610	1623	1328
classes	7	6	7

IEMOCAP [18], a dyadic dialogue dataset, involves 8 speakers in the training set and 2 speakers in the test set for emotion recognition tasks. Emotional tags in IEMOCAP are happy, sad, frustrated, angry, excited, and neutral.

EmoryNLP [19], another dataset derived from Friends, primarily consists of single-turn dialogues and focuses on speaker identity and contextual dependency for emotion analysis. This dataset includes verbal expressions classified into seven emotional types: neutral, joy, peaceful, powerful, fear, anger, sad.

To ensure consistency with previous methods, accuracy (Acc) and weighted-F1 (W-F1) [20] are employed as the evaluation metrics for emotion recognition tasks. Due to the severe class imbalance problem in the EmoryNLP and MELD datasets, weighted-F1 is more responsive to model performance. In addition, we utilize LLaMA2 as the model underlying this paper.

4.2 Baseline Model

- **DialogueGCN** [21] is a graph convolutional network model that constructs dialogue graph structures to model multi-turn interactions and contextual dependencies in conversational data, enabling effective representation learning for dialogue understanding tasks.
- **COSMIC** [14] pioneers a framework incorporating ATOMIC and COMET to leverage commonsense knowledge for refining conversational states, explicitly modeling contextual relationships across intrinsic, extrinsic, and intentional dimensions.
- **TODKAT** [13] integrates a dedicated module for topic identification, which pairs with domain-specific commonsense knowledge from a contextual repository to boost conversational proficiency through profound contextual insights.
- **CoG-BART** [20] utilizes the BART pretrained encoder-decoder framework as its core structure, with a supplementary response generation task integrated to enhance contextual comprehension capabilities.
- **InstructERC** [6] transforms emotion recognition from traditional frameworks to a LLM-based generative framework. It integrates dialogue information via retrieval templates, models role relationships through speaker recognition and emotion prediction tasks, and unifies emotion labels for real-world application adaptation.

4.3 Experimental Results and Analysis

A comparison of the performance of DM-ERC with other state-of-the-art 5 baseline models on the three datasets is shown in Table 2.

Table 2. The main results on three datasets.

Dataset	MELD	IEMOCAP		EmoryNLP
Models	W-F1	Acc	W-F1	W-F1
DiaglogueGCN	63.02	64.49	64.18	36.43
COSMIC	65.03	65.42	65.16	38.49
TODKAT	65.24	61.11	61.33	38.69
CoG-BART	65.02	66.02	65.96	37.33
InstrutERC	69.15	71.68	71.39	41.37
DM-ERC	**69.47**	**72.91**	**72.75**	**42.48**

Experimental results demonstrate that LLM-based models significantly outperform other paradigms in emotion recognition challenges, validating the effectiveness of generative frameworks. Pre-trained large language models (LLMs) understand contextual information and speaker features from multi-faceted perspectives, enabling sophisticated reasoning in emotion recognition tasks. MELD, featuring a complex contextual structure, exposes limitations in GNN-based DialogueGCN and RNN-based COSMIC, as their simple architectures fail to fully exploit contextual information for emotion recognition. However, DM-ERC can deeply mine contextual information in a reasonable generative manner to capture dynamic cues, and W-F1 improves by 0.32 relative to InstructERC. On the IEMOCAP dataset, Transformer-based CoG-BART achieves promising results by incorporating supervised contrastive learning and auxiliary tasks for emotion reasoning. DM-ERC improves the model by 1.23 and 1.36 in Acc and W-F1, respectively, by designing two different auxiliary tasks. Additionally, in the extremely class-imbalanced EmoryNLP dataset, DM-ERC improves W-F1 by 1.11 relative to state-of-the-art models. This demonstrates the effectiveness of the similar utterance retrieval module. In conclusion, DM-ERC deeply mines historical utterance information through two auxiliary tasks, enabling effective emotion recognition in dialogue contexts.

4.4 Ablation Study

In this section, we perform ablation experiments to analyze the effectiveness of the auxiliary tasks, as shown in Table 3.

Table 3. The W-F1 results of ablation study on three datasets.

	MELD	IEMOCAP	EmoryNLP
DM-ERC	**69.47**	**72.75**	**42.48**
w/o C	68.02	70.96	40.92
w/o E	68.84	71.06	41.15
w/o R	69.19	72.43	42.01

Based on the experimental results, we find that the performance of the model degrades most significantly across the three datasets when the common sense injection task is removed (denoted as w/o C). This demonstrates the importance of grasping dynamic emotion cues in dialogue context for emotion reasoning. When the emotional prediction task is removed (denoted as w/o E), a more significant impact is observed on the IEMOCAP and EmoryNLP datasets. This validates the hypothesis of dialogue context-emotion influence correlations in emotion-aware dialogue systems. When the similar statement retrieval module is removed (denoted as w/o R), a slight decrease in the performance of the model is observed across the three datasets. This finding suggests that individuals in analogous situational contexts tend to demonstrate consistent affective patterns. In summary, each task and module in DM-ERC plays a crucial role in ERC.

5 Conclusion

In this paper, we propose a novel framework leveraging multi-auxiliary tasks to enhance emotion recognition capabilities in large language models. Latent listener responses serve as prompts to elicit commonsense generation from LLMs, deeply mining contextual information and speaker features to enable LLMs to grasp contextual emotion dynamics cues. By implicitly capturing dialogue-emotion influence relationships, the emotion prediction task facilitates subsequent reasoning in LLMs for emotion-aware dialogue processing. Retrieval of similar utterance provides substantial support for emotion recognition tasks in dialogue scenarios. Future work will further optimize DM-ERC by exploring multimodal approaches for emotion recognition, integrating audio-visual information to enhance accuracy in contextual emotion analysis.

Acknowledgments. This work is partly supported by Hebei Natural Science Foundation (grant number No. F2023207003), Higher Education Teaching Reform Research and Practice Project of Hebei Province (grant number No. 2023GJJG187), Hebei University of Economics and Business Educational Research Project (grant number No. 2024JYQ09), and Hebei University of Economics and Business Scientific Research and Development Program (grant number No. 2024YB23).

Disclosure of Interests. The authors declare that they have no known competing financial interests or personal relationships that could have appeared to influence the work reported in this paper.

References

1. Schuff, H.: Human-centered explainable artificial intelligence for natural language processing (2024). https://nbn-resolving.org/urn:nbn:de:bsz:93-opus-ds-141966
2. Cai, M., Wang, D., Feng, S., Zhang, Y.: PECER: empathetic response generation via dynamic personality extraction and contextual emotional reasoning. In: IEEE International Conference on Acoustics, Speech and Signal Processing, ICASSP 2024, Seoul, Republic of Korea, 14–19 April 2024, pp. 10631–10635. IEEE (2024). https://doi.org/10.1109/ICASSP48485.2024.10446914
3. Hazarika, D., Poria, S., Mihalcea, R., Cambria, E., Zimmermann, R.: ICON: interactive conversational memory network for multimodal emotion detection. In: Riloff, E., Chiang, D., Hockenmaier, J., Tsujii, J. (eds.) Proceedings of the 2018 Conference on Empirical Methods in Natural Language Processing, Brussels, Belgium, October 31–November 4, 2018, pp. 2594–2604. Association for Computational Linguistics (2018). https://doi.org/10.18653/V1/D18-1280
4. Majumder, N., Poria, S., Hazarika, D., Mihalcea, R., Gelbukh, A.F., Cambria, E.: DialogueRNN: an attentive RNN for emotion detection in conversations. In: The Thirty-Third AAAI Conference on Artificial Intelligence, AAAI 2019, The Thirty-First Innovative Applications of Artificial Intelligence Conference, IAAI 2019, The Ninth AAAI Symposium on Educational Advances in Artificial Intelligence, EAAI 2019, Honolulu, Hawaii, USA, January 27–February 1, 2019, pp. 6818–6825. AAAI Press (2019). https://doi.org/10.1609/AAAI.V33I01.33016818
5. Hu, D., Bao, Y., Wei, L., Zhou, W., Hu, S.: Supervised adversarial contrastive learning for emotion recognition in conversations. CoRR **abs/2306.01505** (2023). https://doi.org/10.48550/ARXIV.2306.01505
6. Lei, S., Dong, G., Wang, X., Wang, K., Wang, S.: InstructERC: reforming emotion recognition in conversation with a retrieval multi-task LLMs framework. CoRR **abs/2309.11911** (2023). https://doi.org/10.48550/ARXIV.2309.11911
7. Chatterjee, A., Gupta, U., Chinnakotla, M.K., Srikanth, R., Galley, M., Agrawal, P.: Understanding emotions in text using deep learning and big data. Comput. Hum. Behav. **93**, 309–317 (2019). https://doi.org/10.1016/J.CHB.2018.12.029
8. Wang, C., Zhang, Y., Yao, S.: Text-independent speaker verification based on mutual information disentanglement **14094**, 309–318 (2023). https://doi.org/10.1007/978-981-99-6222-8_26
9. Chen, F., et al.: VLP: a survey on vision-language pre-training. Int. J. Autom. Comput. **20**(1), 38–56 (2023). https://doi.org/10.1007/S11633-022-1369-5
10. Hu, D., Wei, L., Huai, X.: DialogueCRN: contextual reasoning networks for emotion recognition in conversations. In: Zong, C., Xia, F., Li, W., Navigli, R. (eds.) Proceedings of the 59th Annual Meeting of the Association for Computational Linguistics and the 11th International Joint Conference on Natural Language Processing, ACL/IJCNLP 2021, (Volume 1: Long Papers), Virtual Event, 1–6 August 2021. pp. 7042–7052. Association for Computational Linguistics (2021). https://doi.org/10.18653/V1/2021.ACL-LONG.547

11. Ishiwatari, T., Yasuda, Y., Miyazaki, T., Goto, J.: Relation-aware graph attention networks with relational position encodings for emotion recognition in conversations. In: Webber, B., Cohn, T., He, Y., Liu, Y. (eds.) Proceedings of the 2020 Conference on Empirical Methods in Natural Language Processing, EMNLP 2020, Online, 16–20 November 2020, pp. 7360–7370. Association for Computational Linguistics (2020). https://doi.org/10.18653/V1/2020.EMNLP-MAIN.597
12. Shen, W., Wu, S., Yang, Y., Quan, X.: Directed acyclic graph network for conversational emotion recognition. In: Zong, C., Xia, F., Li, W., Navigli, R. (eds.) Proceedings of the 59th Annual Meeting of the Association for Computational Linguistics and the 11th International Joint Conference on Natural Language Processing, ACL/IJCNLP 2021, (Volume 1: Long Papers), Virtual Event, 1–6 August 2021. pp. 1551–1560. Association for Computational Linguistics (2021). https://doi.org/10.18653/V1/2021.ACL-LONG.123
13. Zhu, L., Pergola, G., Gui, L., Zhou, D., He, Y.: Topic-driven and knowledge-aware transformer for dialogue emotion detection. In: Zong, C., Xia, F., Li, W., Navigli, R. (eds.) Proceedings of the 59th Annual Meeting of the Association for Computational Linguistics and the 11th International Joint Conference on Natural Language Processing, ACL/IJCNLP 2021, (Volume 1: Long Papers), Virtual Event, 1–6 August 2021. pp. 1571–1582. Association for Computational Linguistics (2021). https://doi.org/10.18653/V1/2021.ACL-LONG.125
14. Ghosal, D., Majumder, N., Gelbukh, A.F., Mihalcea, R., Poria, S.: COSMIC: commonsense knowledge for emotion identification in conversations. In: Cohn, T., He, Y., Liu, Y. (eds.) Findings of the Association for Computational Linguistics: EMNLP 2020, Online Event, 16–20 November 2020. Findings of ACL, vol. EMNLP 2020, pp. 2470–2481. Association for Computational Linguistics (2020). https://doi.org/10.18653/V1/2020.FINDINGS-EMNLP.224
15. Zhang, D., Chen, F., Chen, X.: DualGATs: dual graph attention networks for emotion recognition in conversations. In: Rogers, A., Boyd-Graber, J.L., Okazaki, N. (eds.) Proceedings of the 61st Annual Meeting of the Association for Computational Linguistics (Volume 1: Long Papers), ACL 2023, Toronto, Canada, 9–14 July 2023, pp. 7395–7408. Association for Computational Linguistics (2023). https://doi.org/10.18653/V1/2023.ACL-LONG.408
16. Zhang, T., Chen, Z., Zhong, M., Qian, T.: Mimicking the thinking process for emotion recognition in conversation with prompts and paraphrasing. In: Proceedings of the Thirty-Second International Joint Conference on Artificial Intelligence, IJCAI 2023, Macao, SAR, China, 19–25 August 2023. ijcai.org. (2023). https://doi.org/10.24963/ijcai.2023/699
17. Poria, S., Hazarika, D., Majumder, N., Naik, G., Cambria, E., Mihalcea, R.: MELD: a multimodal multi-party dataset for emotion recognition in conversations. In: Korhonen, A., Traum, D.R., Màrquez, L. (eds.) Proceedings of the 57th Conference of the Association for Computational Linguistics, ACL 2019, Florence, Italy, July 28–August 2, 2019, Volume 1: Long Papers, pp. 527–536. Association for Computational Linguistics (2019). https://doi.org/10.18653/V1/P19-1050
18. Busso, C., et al.: IEMOCAP: interactive emotional dyadic motion capture database. Lang. Resour. Evaluation **42**(4), 335–359 (2008). https://doi.org/10.1007/S10579-008-9076-6
19. Zahiri, S.M., Choi, J.D.: Emotion detection on TV show transcripts with sequence-based convolutional neural networks. vol. abs/1708.04299 (2017). http://arxiv.org/abs/1708.04299

20. Li, S., Yan, H., Qiu, X.: Contrast and generation make BART a good dialogue emotion recognizer. In: Thirty-Sixth AAAI Conference on Artificial Intelligence, AAAI 2022, Thirty-Fourth Conference on Innovative Applications of Artificial Intelligence, IAAI 2022, The Twelveth Symposium on Educational Advances in Artificial Intelligence, EAAI 2022 Virtual Event, February 22–March 1, 2022, pp. 11002–11010. AAAI Press (2022). https://doi.org/10.1609/AAAI.V36I10.21348
21. Ghosal, D., Majumder, N., Poria, S., Chhaya, N., Gelbukh, A.F.: DialogueGCN: a graph convolutional neural network for emotion recognition in conversation. In: Inui, K., Jiang, J., Ng, V., Wan, X. (eds.) Proceedings of the 2019 Conference on Empirical Methods in Natural Language Processing and the 9th International Joint Conference on Natural Language Processing, EMNLP-IJCNLP 2019, Hong Kong, China, 3–7 November 2019, pp. 154–164. Association for Computational Linguistics (2019). https://doi.org/10.18653/V1/D19-1015

LFT4POI: Multimodal POI Recommendation via Large Language Model Fine-Tuning

Chuanchang Zhang[1], Yuchen Zheng[1], Xuan Pan[2], Sihan Xu[3](✉), and Xiangrui Cai[1]

[1] TMM, DISSec, NDST, College of Computer Science, Nankai University, Tianjin, China
2120250706@mail.nankai.edu.cn, zhengyuchen@dbis.nankai.edu.cn, caixr@nankai.edu.cn

[2] School of Artificial Intelligence, Tiangong University, Tianjin, China
panxuan@tiangong.edu.cn

[3] DISSec, NDST, College of Cryptology and Cyber Science, Nankai University, Tianjin, China
xusihan@nankai.edu.cn

Abstract. Multimodal point-of-interest (POI) recommendation seeks to enhance recommendation performance by leveraging diverse types of POI information, such as textual descriptions, user reviews, and images. Recently, Large Language Models (LLMs) have demonstrated remarkable capabilities in various recommendation tasks. However, to the best of our knowledge, there has been no prior work exploring the application of LLMs to multimodal POI recommendation. In this paper, we propose a novel method, i.e., LFT4POI, that incorporates LLM fine-tuning to effectively model and integrate multimodal POI data for improved recommendation performance. Specifically, LFT4POI introduces a multitask LLM fine-tuning framework for large language models, combining contrastive learning and supervision LLM fine-tuning to align the model with POI-specific contexts. After fine-tuning, the large language model generates enriched POI embeddings, which are fused with sequential patterns from a conventional POI recommendation model (e.g., MMPOI) for final recommendations. Experiments on NYC and TKY datasets demonstrate superior performance: our method outperforms seven baselines including MMPOI, STGCN, and STGN with an average 4% improvement across all the metrics such as Acc@20 and MRR. Ablation studies further confirm the contributions of each component, specifically multimodal fusion and LLM fine-tuning.

Keywords: Multimodal POI Recommendation · LLM Fine-tuning · Contrastive Learning

S. Yang et al. (Eds.): WISA 2025, LNCS 16299, pp. 579–590, 2026.
https://doi.org/10.1007/978-981-95-6426-2_51

1 Introduction

With the rapid development of mobile Internet, various intelligent devices and real-time positioning technology, location-based Social Networks (LBSNs) emerged as The Times require [1]. Unlike ordinary social networks, LBSNs breaks through the time and space limitations of traditional social networks and builds a new type of social graph with geographical locations as the bond. Point of Interest (POI) recommendation, as the core task of LBSNs, has become the key point for improving the user experience. Many recent papers on POI recommendation [9–11,14], aim to use historical behavior data of previous users to help users find POIs that may be of interest.POI recommendations not only facilitate the exploration of new things, but also provide a new basis for the business growth of enterprises.

Most of the current studies only use the historical check-in data of users for POI recommendation. Although these studies have made considerable progress, but lack of associated with POI using modal data. By introducing multimodal data including basic POI information, user reviews, and images into POI recommendation, as shown in Fig. 1, the system can obtain additional information, thereby improving recommendation effectiveness. At the same time, it alleviates the problem of data sparsity [17].

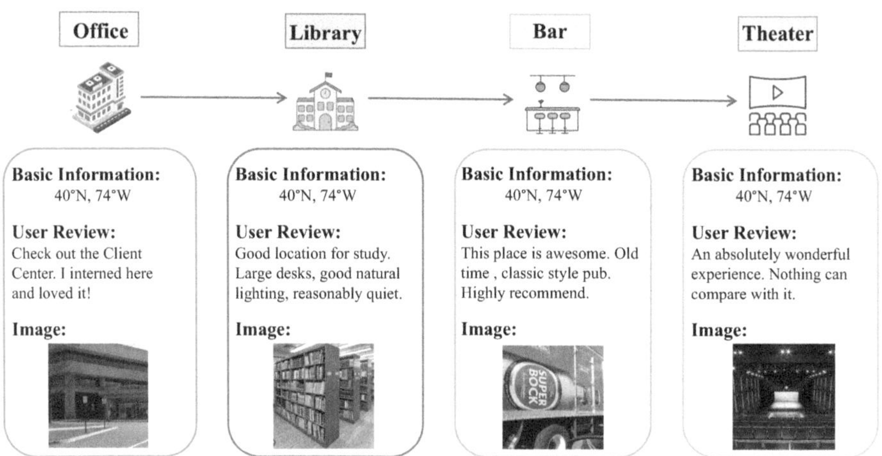

Fig. 1. User trajectory with multimodal data.

In recent years, large language models have shown powerful capabilities in semantic understanding and knowledge reasoning, which provides new opportunities for the optimization of POI recommendation systems. Large language models can directly give recommendation results by constructing prompt, or generate high-quality embeddings by using the check-in data and POI data of users. The current large language model is basically trained on the general data set. By fine-tuning large language models, it can be made closer to the scene of POI recommendation and improve the recommendation effect.

To this end, we propose a multi-task framework for fine-tuning large language models by using multi-modal POI data and user check-in records. The framework consists of two parts: a contrastive learning module and a supervised LLM fine-tuning module. The contrastive learning (CL) module improves the recommendation effect by learning the collaborative relationship between related pairs, and the supervision LLM fine-tuning (SFT) module prevents the model from relying too much on specific data and resulting in poor generalization ability after only fine-tuning for the main task scene. The contributions of this paper are as follows:

(1) For the first time, we applied large language models to multimodal POI recommendations and generated high-quality embeddings through large language models.
(2) We propose a multi-task LLM fine-tuning framework to fine-tune the large language model, and make the large language model closer to the scenarios recommended by POI through the CL module and the SFT module.
(3) We conduct experiments on Foursquare-NYC and Foursquare-TKY datasets. The recommendation accuracy is improved by 4% on average compared with the baselines. Ablation studies further confirm the contributions of each component, specifically multimodal fusion and LLM fine-tuning.

2 Related Work

This section introduces the work related to this paper, including traditional POI recommendation methods and POI recommendation methods based on large language model.

2.1 Traditional POI Recommendation

The Next POI Recommendation refers to predicting the POI of a user's next check-in by using their historical check-in data. These works [2,12,23] mainly focus on the personalized modeling of users using Markov chains for the recommendation of the next point of interest.

In recent studies, recurrent neural network and self-attention mechanism are mostly used for Next POI recommendation. Several researchers [24,26] used recurrent neural networks to model the historical behavior sequence of users and capture the timing relationship of users' access to POI for POI recommendation. However, none of these methods can solve the problem of long-term dependencies in time series. Therefore, methods such as ST-GCN, LSTPM, and PLSPL [13, 16,25] emerged. These methods all attach importance to the hiding of users' long-term and short-term preferences on the recommendation effect and learn users' long-term and short-term preferences in the model.

The latest research on Next POI recommendation combines graph representation learning technology to capture group preference patterns and complex spatio-temporal interaction patterns for Next POI recommendation by constructing trajectory flow graph [19], introducing hypergraph structure [18] or adaptively generating geographic dependence graph [15].

Traditional POI recommendation methods predominantly rely on check-in sequences of users to model their preference patterns, often neglecting other potentially valuable multi-source information, e.g., POI descriptions, reviews, and images. This limitation hinders the comprehensive modeling of user preferences and affects the overall recommendation performance.

2.2 LLM-Based POI Recommendation

With the rise of large language models, many papers [8,20–22] apply large language models to recommendation systems. At present, there are also some papers that apply large models to the field of POI recommendation. These studies can mainly be divided into two categories. One type of research mainly utilizes the powerful semantic understanding ability of large models to extract the hidden information in the POI access sequence and directly recommend POI. Feng et al. [5] proposed the LLMmove method, which uses large models to analyze users' long-term preferences, short-term preferences, geographical locations, and sequence patterns for recommendation. Li et al. [7] utilized a pre-trained large language model to extract rich contextual information from the raw data and integrate common sense knowledge for POI prediction. On the other hand, the research utilizes large models to extract the features of multimodal information, construct the embedding of POI, and indirectly recommend POI. Xu et al. [17] proposed the MMPOI model, which uses the pre-trained large model to extract the features of each mode of multi-mode data, and recommends the next POI by combining the proposed method of multi-mode trajectory flow graph.

Unfortunately, existing POI recommendation approaches based on large language models still fall short in capturing the deep semantic alignment between users and POI contexts.

3 LFT4POI

To solve these problems, we propose a multimodal POI recommendation method based on large language models. While introducing multimodal data, the large language model is fine-tuned to make it suitable for the POI recommendation scenario.

Figure 2 shows the overall architecture of LFT4POI. Specifically, it can be divided into three stages. First, the multimodal POI dataset is processed. Specifically, a multimodal large model is employed to convert non-textual data into text format, extracting key information to construct prompts. These prompts are then input into a large language model for encoding. Next, the pre-trained large model is fine-tuned. This stage incorporates contrastive learning alongside supervised LLM fine-tuning to better align the model with POI recommendation scenarios. Subsequently, the fine-tuned large model is used to generate feature

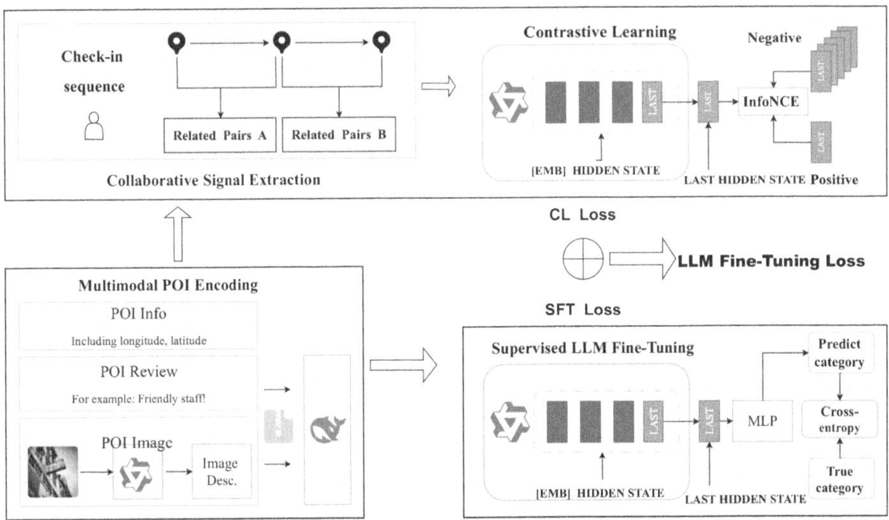

Fig. 2. Overview of LFT4POI.

embeddings corresponding to the multimodal POI data within the training set. Finally, the POI recommendation is completed by integrating these embeddings into the POI recommendation model.

3.1 Multimodal POI Encoding

The multimodal POI data utilized in this study encompasses basic POI information, user reviews, and images. For image data, we employ Qwen-VL-Max to generate corresponding textual descriptions, thereby aligning the semantic representations across modalities. Subsequently, to address the inherent redundancy in both user reviews and the generated image descriptions, we leverage DeepSeek-V3 to extract their key informational content. Finally, we apply a POI information compression template to obtain the POI embedding.

Table 1. POI Information Compression Template

Components	Contents
Instruction	Compress the poi information into one word for recommendation and generate its category
Input POI	{Name: poi_id, Latitude: poi_latitude, Longitude: poi_longitude, Review: poi_review, Image: poi_image}
Output Guidance	The compression word is: [EMB]. The category is:

We adopt a unified POI information compression template to derive the POI embedding. To leverage the generative capabilities of autoregressive large language models, we designed this template (as shown in Table 1), which capitalizes on the LLM's powerful semantic understanding ability. The template comprises three components: Instructions, Input POI data, and Output Guidance.

In this compression template, [EMB] functions as a specially designated token. The multimodal embedding of the POI is derived by extracting the final-layer hidden state corresponding to [EMB] from the LLM's output. Given that the selected large language model operates autoregressively, each token generated is conditioned on preceding context. Consequently, when the [EMB] token is produced, its final-layer hidden state comprehensively encapsulates all POI information, thus serving as its representation.

3.2 Collaborative Signal Extraction

Generally speaking, pre-trained large models are fine-tuned through instructions or optimized from user feedback through reinforcement learning, but these methods cannot obtain the collaborative signals between data. These collaborative signals are crucial for the recommendation of POI, containing information such as the common transfer patterns of the group. To extract the collaborative signals among the data, we proposed the concept of related pairs and used contrastive learning to extract the collaborative limit numbers among the data.

Ralated Pair Construct. In POI recommendations, related pairs are defined as adjacent combinations of POIs in the same user access sequence. After extracting related pairs from all historical records, high-score related pairs need to be calculated and screened. To prevent bias caused by active users dominating the scores, the contribution of each user is normalized during the calculation. The specific related pair fraction calculation formula is:

$$s_{p_i \to p_j} = \sum_{u=1}^{U} \frac{count_u(p_i \to p_j)}{N_u} \tag{1}$$

where N_u denotes the length of user $u's$ POI sequence.

Contrastive Learning. After identifying all the realted POI pairs, they are grouped into batches of size B. In this paper, we denote the hidden representation of the i-th POI as p_i and its related POI's representation as p_i^+. Using the standard InfoNCE loss function for contrastive learning, the fine-tuning loss for this stage is defined as follows:

$$L_{cl} = -\frac{1}{2B} \sum_{i=1}^{2B} \log \frac{e^{sim(p_i, p_i^+)/\tau}}{\sum_{j \in [2B] \setminus \{i\}} e^{sim(p_i, p_j)/\tau}} \tag{2}$$

3.3 Supervised Fine-Tuning

In addition to contrastive learning, a supervision module needs to be designed during the fine-tuning process to prevent the model from being overly dependent on specific data and having poor generalization ability after being fine-tuned only for the main task scenario. This paper designs the category generation task of POI as the supervision task for the fine-tuning of large language models, and uses multi-objective learning to improve the quality of POI embedding.

The loss function of this part adopts the cross-entropy loss function. In this paper, the true category of POI is defined as y_i, and the prediction result given by the LLM is $\hat{y}_i$. Thus, the loss of the supervision LLM fine-tuning obtained is:

$$L_{sft} = -\sum_{i=1}^{C} y_i \log(\hat{y}_i), \tag{3}$$

Finally, define a weighted hyperparameter α, combine the losses of the two parts, and obtain the final fine-tuning loss as follows:

$$L = \frac{L_{cl} + \alpha L_{sft}}{1 + \alpha}, \tag{4}$$

3.4 POI Recommendation

After fine-tuning the large language model, we obtained multimodal embeddings for each POI. These multimodal embeddings were fused with sequential patterns from a conventional POI recommendation model (e.g., MMPOI). Within this enhanced MMPOI framework, the multimodal representations replaced the original Sentence-BERT embeddings. Components like the POI embedding model were consequently restructured into fully connected layers to effectively learn from these fused representations, and the resulting model was deployed for POI recommendation.

4 Experiments

4.1 Experimental Settings

Datasets. We conducted experiments on two public datasets: Foursquare-NYC and FourSquare-TKY, and on this basis, added the multimodal data of POI collected in the MMPOI paper, including the basic information, comments and pictures of POI. Finally, we divided the data set into the training set and the test set in an 8:2 ratio. Table 2 shows the statistical information of the dataset.

Table 2. Metrics of Foursquare-NYC and Foursquare-TKY Datasets

Metric	Foursquare-NYC	Foursquare-TKY
Number of Users	1,080	2,275
Number of Locations	4,637	4,943
Total Check-ins	115,134	326,932

Our Models. In our experiment, we used Qwen2.5-7B-Instruct as the base large language model. During the construction process of the relevant note pairs, we set the upper limit of the co-occurrence score u to 1 and the lower limit l to 0.005. Meanwhile, we set the temperature parameter t to 3, set the dimension d embedded in POI to 128, and set α to 0.01 in Eq. (4).

Evaluation Metrics. To comprehensively evaluate the performance of next POI prediction, we employ two widely adopted ranking metrics: Hit Ratio (HR@K) and Mean Reciprocal Rank (MRR), each providing distinct perspectives on recommendation quality. HR@K, a standard metric in POI recommender systems, measures whether the user's next check-in location appears within the top-K recommendations. MRR, commonly used in ranking tasks, evaluates the prioritization of correct answers by averaging the reciprocal rank of the first relevant item across all test samples.

Baselines. To evaluate the effect of the model, we select the following model as the baseline model:

- **FPMC** [12]**:** A model integrating Markov chains and matrix factorization, which can model the long-term preferences of users and capture the dependencies of adjacent POIs by using the transition matrix.
- **LSTM** [6]**:** A variant improved upon RNN networks, which solves the vanishing gradient problem of traditional RNNs through a gating mechanism and excels at processing long sequence data.
- **PRME** [4]**:** A model integrating metric space embedding and Euclidean distance modeling, which utilizes the distance between vectors to capture the transition dependency between POIs.
- **ST-RNN** [3]**:** A model integrating the spatio-temporal transition matrix and the standard RNN, which can model the spatio-temporal constraints of sign-in records and capture the dependencies therein.
- **STGN** [25]**:** A variant based on LSTM, adding temporal gates and spatial gates to LSTM to model short-term preferences and long-term preferences.
- **STGCN** [25]**:** STGCN is a variant of STGN. Building upon STGN, it innovatively couples the input gate with the forget gate.
- **MMPOI** [17]**:** The state-of-the-art multimodal POI recommendation method, which constructs a multimodal trajectory flow graph and a geographic trajectory flow graph.

The above-mentioned baseline model covers multi-level technical routes ranging from traditional matrix factorization to deep spatio-temporal networks, and from individual behavior modeling to group pattern mining. Among them, FPMC and PRME represent the traditional methods based on statistical learning, LSTM and its variants (ST-RNN, STGN, STGCN) belong to the time series modeling schemes under the deep learning framework, while MMPOI reflect the cutting-edge combination of graph neural networks and trajectory mining.

4.2 Performance Comparison

In this subsection, experiments were conducted on the public datasets NYC and TKY, comparing the model proposed in this paper with seven other models. The results of the comparison are shown in Table 3, demonstrating the performance of the model proposed in this paper and the baseline model on four metrics: Acc@5, Acc@10, Acc@20, and MRR. In the table, the performance of LFT4POI is highlighted in bold, while the baseline with the best performance is highlighted with an underline. From the table, we have the following conclusions:

The LFT4POI model we proposed significantly outperforms the current state-of-the-art POI recommendation method MMPOI on both the two public datasets, Foursquare NYC and Foursquare TKY. Specifically, on the New York (NYC) and Tokyo (TKY) datasets, LFT4POI improved by an average of 3.74% and 4.22% respectively in all evaluation metrics. We believe that the significant improvement in performance mainly stems from two aspects: First, large language models can extract richer feature representations from the multi-modal information of POI, thereby enhancing the effect of POI recommendation; Second, the pre-trained large language model fine-tuned by the multi-task framework is better adapted to specific POI recommendation scenarios. Furthermore, the experimental results fully verify the superiority of the POI recommendation algorithm based on large models.

Table 3. Performance Comparison on NYC and TKY Datasets

Model	NYC Dataset				TKY Dataset			
	Acc@5	Acc@10	Acc@20	MRR	Acc@5	Acc@10	Acc@20	MRR
FPMC	0.2126	0.2970	0.3323	0.1701	0.2045	0.2746	0.3450	0.1344
LSTM	0.2719	0.3283	0.3568	0.1857	0.2728	0.3277	0.3598	0.1834
PRME	0.2236	0.3105	0.3643	0.1712	0.2278	0.2944	0.3560	0.1786
ST-RNN	0.2923	0.3622	0.4502	0.2198	0.3022	0.3577	0.4753	0.2212
STGN	0.3381	0.4122	0.5017	0.2598	0.3391	0.3848	0.4514	0.2422
STGCN	0.3425	0.4279	0.5214	0.2788	0.3453	0.3927	0.4763	0.2504
MMPOI	<u>0.4154</u>	<u>0.5581</u>	<u>0.6605</u>	<u>0.2559</u>	<u>0.4555</u>	<u>0.5525</u>	<u>0.6331</u>	<u>0.3095</u>
LFT4POI	**0.4671**	**0.5942**	**0.6832**	**0.2995**	**0.4981**	**0.5926**	**0.6703**	**0.3583**
Improve	**+5.17%**	**+3.61%**	**+2.27%**	**+4.36%**	**+4.26%**	**+4.01%**	**+3.72%**	**+4.88%**

4.3 Ablation Study

In this subsection, we conduct ablation studies to verify the effectiveness of the key modules within the proposed fine-tuning framework for large language models. Specifically, we aim to isolate and evaluate the contribution of the Contrastive Learning (CL) module and the Supervision Fine-Tuning(SFT) module by comparing the performance of different framework variants. To achieve this, we designed experiments utilizing several configurations of the LLM fine-tuning framework, thereby validating the impact of specific module combinations on the model's fine-tuning performance.

The performance of the full proposed framework is compared against the following variants:

(1) w/o CL: A variant of LFT4POI utilizing only the SFT module.
(2) w/o CL & SFT: A fine-tuning-free variant of LFT4POI.

From Fig. 3, we draw the following conclusions. First, fine-tuning the large language model through our supervision LLM fine-tuning module (SFT) yields noticeable performance gains over the fine-tuning-free variant, demonstrating SFT's effectiveness in enhancing multimodal POI embeddings. Second, the complete framework—integrating both the CL module and SFT module—significantly outperforms the variant using SFT alone. This indicates that both proposed components synergistically improve multimodal embedding quality, consequently boosting the model's predictive performance.

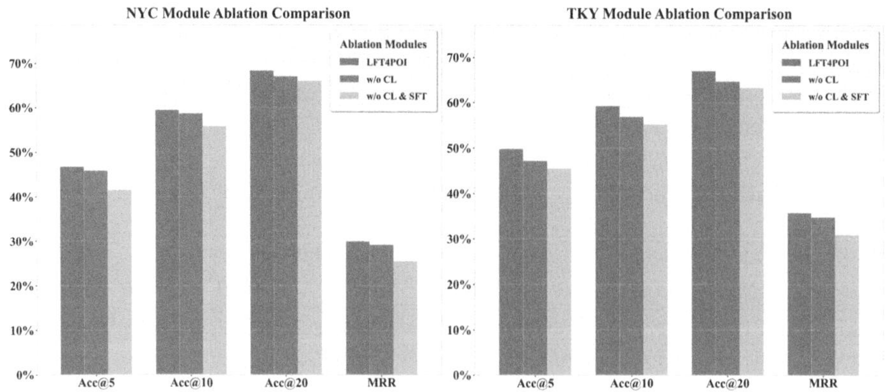

Fig. 3. Ablation Studies.

4.4 Generalization Study

The proposed LFT4POI can be also integrated with other conventional sequence recommendation model. To investigate the generalization of LFT4POI, we conduct further experiments by combine LFT4POI with GetNext [19]. We can observe from Table 4 that our method improves the recommendation performance as well. The results demonstrate that LFT4POI, through LLM fine-tuning, effectively learns high-quality POI representations, which enhance the performance of downstream recommendation tasks. Note that we present the results of GetNext separately from Table 3, as the dataset used in this GetNext differs from that used in MMPOI.

Table 4. Comparison between GetNext and LFT4POI on NYC and TKY Datasets

Model	NYC Dataset				TKY Dataset			
	Acc@5	Acc@10	Acc@20	MRR	Acc@5	Acc@10	Acc@20	MRR
GetNext	0.4579	0.5384	0.6021	0.3001	0.5391	0.6381	0.7097	0.3964
LFT4POI	**0.4648**	**0.5696**	**0.6398**	**0.3146**	**0.5530**	**0.6514**	**0.7226**	**0.4108**
Improve	**+0.69%**	**+3.12%**	**+3.77%**	**+1.45%**	**+1.39%**	**+1.33%**	**+1.29%**	**+1.44%**

5 Conclusion

This paper introduces a novel multimodal POI recommendation framework leveraging large language models. We propose a multi-task LLM fine-tuning approach using contrastive learning and supervised LLM fine-tuning modules to adapt pre-trained LLMs for POI recommendation scenarios. The fine-tuned LLM is then integrated into the MMPOI architecture to enhance recommendation performance. Experimental validation on Foursquare's NYC and TKY datasets demonstrates the efficacy of our approach.

Acknowledgments. This work was supported in part by the National Key Research and Development Program of China under Grant 2022YFB3103202, and in part by the Natural Science Foundation of Tianjin, China, under Grant 23JCQNJC01960.

References

1. Bao, J., Zheng, Yu., Wilkie, D., Mokbel, M.: Recommendations in location-based social networks: a survey. GeoInformatica **19**(3), 525–565 (2015). https://doi.org/10.1007/s10707-014-0220-8
2. Cheng, C., Yang, H., Lyu, M.R., King, I.: Where you like to go next: successive point-of-interest recommendation. In: IJCAI, pp. 2605–2611 (2013)
3. Deng, B., Qu, B., Wang, P., Yang, D., Fankhauser, B., Cudre-Mauroux, P.: Replay: modeling time-varying temporal regularities of human mobility for location prediction over sparse trajectories. IEEE Trans. Mob. Comput. 1–14 (2025)
4. Feng, S., Li, X., Zeng, Y., Cong, G., Chee, Y.M., Yuan, Q.: Personalized ranking metric embedding for next new poi recommendation. In: IJCAI, pp. 2069–2075 (2015)

5. Feng, S., Lyu, H., Li, F., Sun, Z., Chen, C.: Where to move next: zero-shot generalization of llms for next poi recommendation. In: IEEE CAI, pp. 1530–1535 (2024)
6. Graves, A., Graves, A.: Long short-term memory. Neural Comput. 37–45 (2012)
7. Li, P., de Rijke, M., Xue, H., Ao, S., Song, Y., Salim, F.D.: Large language models for next point-of-interest recommendation. In: SIGIR, pp. 1463–1472 (2024)
8. Li, X., Chen, B., Hou, L., Tang, R.: Ctrl: connect collaborative and language model for ctr prediction. ACM Trans. Recommender Syst. (2023)
9. Lim, N., Hooi, B., Ng, S.K., Goh, Y.L., Weng, R., Tan, R.: Hierarchical multi-task graph recurrent network for next poi recommendation. In: SIGIR, pp. 1133–1143 (2022)
10. Pan, X., Cai, X., Song, K., Baker, T., Gadekallu, T.R., Yuan, X.: Location recommendation based on mobility graph with individual and group influences. IEEE Trans. Intell. Transp. Syst. **24**(8), 8409–8420 (2022)
11. Pan, X., Cai, X., Zhang, J., Wen, Y., Zhang, Y., Yuan, X.: STMG: spatial-temporal mobility graph for location prediction. In: Jensen, C.S., et al. (eds.) Database Systems for Advanced Applications. DASFAA 2021. LNCS, vol 12681, pp. 667–675. Springer, Cham (2021). https://doi.org/10.1007/978-3-030-73194-6_45
12. Rendle, S., Freudenthaler, C., Schmidt-Thieme, L.: Where to move next: zero-shot generalization of llms for next poi recommendation. WWW, pp. 811–820 (2010)
13. Sun, K., Qian, T., Chen, T., Liang, Y., Nguyen, Q.V.H., Yin, H.: Where to go next: modeling long-and short-term userpreferences for point-of-interest recommendatio. AAAI **34**(1), 214–221 (2020)
14. Sun, S., Cai, T., Yan, F., Ju, S.: Two-stage enhancement for recommendation systems based on contrastive learning. In: WISA, vol. 14883, pp. 150–162 (2024)
15. Wang, Z., Zhu, Y., Wang, C., Ma, W., Li, B., Yu, J.: Adaptive graph representation learning for next poi recommendation. In: SIGIR, pp. 393–402 (2023)
16. Wu, Y., Li, K., Zhao, G., Qian, X.: Personalized long-and short-term preference learning for next poi recommendation. IEEE Trans. Knowl. Data Eng. **34**(4), 1944–1957 (2020)
17. Xu, Y., Cong, G., Zhu, L., Cui, L.: Mmpoi: a multi-modal content-aware framework for poi recommendations. WWW, pp. 3454–3463 (2024)
18. Yan, X., et al.: Spatio-temporal hypergraph learning for next poi recommendation. In: SIGIR, pp. 403–412 (2023)
19. Yang, S., Liu, J., Zhao, K.: Getnext: trajectory flow map enhanced transformer for next poi recommendation. In: SIGIR, pp. 1144–1153 (2022)
20. Zhai, J., et al.: Actions speak louder than words: trillion-parameter sequential transducers for generative recommendations. In: ICML, pp. 58484–58509 (2024)
21. Zhang, C., et al.: Notellm: a retrievable large language model for note recommendation. WWW, pp. 170–179 (2024)
22. Zhang, C., et al.: Notellm-2: multimodal large representation models for recommendation. In: KDD, pp. 2815–2826 (2025)
23. Zhang, J.D., Chow, C.Y., Li, Y.: Lore: exploiting sequential influence for location recommendations. In: SIGSPATIAL, pp. 103–112 (2014)
24. Zhang, Y., et al.: Sequential click prediction for sponsored search with recurrent neural networks. AAAI **28**(1) (2014)
25. Zhao, P., et al.: Where to go next: a spatio-temporal gated network fornext poi recommendation. IEEE Trans. Knowl. Data Eng. **34**(5), 2512–2524 (2020)
26. Zhu, Y., et al.: What to do next: modeling user behaviors by time-lstm. IJCAI **17**, 3602–3608 (2017)

Adapting Language Models to Text Matching Based Recommendation Systems

Haidong Xin, Sen Mei, Zhenghao Liu(✉), Xiaohua Li, Minghe Yu, Yu Gu, and Ge Yu

School of Computer Science and Engineering, Northeastern University, Shenyang, China
liuzhenghao@mail.neu.edu.cn

Abstract. The **T**ext m**A**tching based **S**equen**T**ial r**E**commendation model (TASTE) offers a self-contained recommendation system by mapping items and users into an embedding space and recommending items through text matching. To enhance its performance, this paper proposes TASTE$^+$, which adapts language models to sequential recommendation via pretraining. TASTE$^+$ follows the instruction tuning method by designing two pretraining tasks, Masked Item Prediction (MIP) and Next Item Prediction (NIP), which enable language models pretrained on recommendation datasets to better capture text matching signals from user-item interaction sequences. Experimental results on the Yelp and Amazon Product datasets demonstrate that TASTE$^+$ enhances user-item text matching by assigning more balanced attention to both prompt tokens and item IDs. This enables the model to construct more accurate user representations by attending more evenly to recent interaction history. This work demonstrates the effectiveness of language model pretraining in boosting text matching-based recommendation systems.

Keywords: Sequential Recommendation · Text Matching · Language Model Pretraining

1 Introduction

Sequential recommendation systems play a vital role in web applications by dynamically suggesting the next item to meet users' needs and mitigate information overload [5,34,51]. Platforms like Amazon and Yelp utilize user browsing and purchase histories to enhance user experience through personalized recommendations. Traditional approaches typically encode items as randomly initialized IDs and model user behavior via neural architectures to capture item dependencies [4,14,18,36]. However, such ID-based systems often suffer from issues like popularity bias and cold-start problems [1,2,26,52].

Recently, some work [26] introduces the **T**ext m**A**tching based **S**equen**T**ial r**E**commendation (TASTE) model, which represents users and items as text and models their relevance through text matching [21,26]. TASTE constructs textual

S. Yang et al. (Eds.): WISA 2025, LNCS 16299, pp. 591–605, 2026.
https://doi.org/10.1007/978-981-95-6426-2_52

prompts using item IDs and attributes, providing semantic clues to capture user-item dependencies [24]. To address the sequence length limitation of language models, it adopts an attention sparsity mechanism [3] that segments interaction histories into sessions and encodes them separately, reducing computational cost. While pretraining has shown promise in adapting language models to sequential recommendation [11,34], fully exploiting its benefits in text-matching-based settings remains an open challenge.

In this paper, we propose the TASTE$^+$ model, which employs language model pretraining to improve the effectiveness of text matching based sequential recommendation. Specifically, TASTE$^+$ uses the Yelp and Amazon Product datasets for pretraining language models. It follows the instruction-tuning method [6,39] and introduces two pretraining tasks to adapt language models to the sequential recommendation task: Masked Item Prediction (MIP) and Next Item Prediction (NIP). The MIP task trains the model to better capture item dependencies within user-item interactions by randomly masking an item in the interaction history and asking the model to predict the masked item. The NIP task, on the other hand, trains the model to predict the next item in the sequence, which directly aligns with the objective of sequential recommendation. This task helps to better align the representations of user-item interaction history and items in the embedding space.

Our experiments show that TASTE$^+$ outperforms previous sequential recommendation baselines [34,42,51] on Yelp and Amazon Product datasets. Our analyses demonstrates the effectiveness of our pretraining strategy, which brings about 3% improvements to the text matching based recommendation system. Compared to item ID-based recommendation systems, TASTE$^+$ enhances performance by addressing the cold-start problem and reducing popularity bias. Moreover, TASTE$^+$ improves text matching between users and items by assigning more balanced attention weights to both prompt tokens and item IDs. This enables the model to learn more effective user representations that better reflect the user's most recent interactions.

2 Related Work

Sequential recommendation systems aim to predict the next item based on user-item interaction histories. Traditional methods represent items using randomly initialized IDs and learn sequential patterns through matrix factorization [12,31], CNNs [36,45], RNNs [14,25], GNNs [4,47], and self-attention architectures [18,34]. To enhance representation learning, recent works incorporate item attributes and side information [48,51], using attention mechanisms to reduce noise and highlight relevant features [23,42]. However, both ID-based and attribute-enhanced models typically rely on randomly initialized embeddings, making it difficult to generalize to unseen or long-tail items. As a result, the cold-start problem remains a significant challenge.

A growing body of research has turned to fully text-based item modeling in recommendation systems as an alternative to ID-based approaches [10,15,21,43,

46]. These methods leverage pretrained language models to encode item content directly, bypassing randomly initialized embeddings that heavily depend on rich interaction data [20,29,33,49]. For instance, some work [21] proposes an ID-free recommendation framework based entirely on text matching, demonstrating its advantage in mitigating cold-start and enabling cross-domain generalization. CoWPiRec [43] enhances item representations by incorporating collaborative filtering signals into pretrained models, improving transferability and robustness in sparse settings. While MoRec [46] contrasts ID-based and modality-based representations, showing each excels under different data conditions, they treat the two modalities separately. In contrast, TASTE [26] unifies item IDs and content in a text embedding space, treating IDs as prompt tokens to better capture user-item relevance within language models.

To further enhance the performance of text-matching-based sequential recommendation systems, recent studies have explored adapting pretrained language models [9,30,35] to recommendation scenarios. While these models are originally trained with general language modeling objectives, prior research has identified a mismatch between such objectives and the requirements of recommendation-specific text matching tasks [16]. To bridge this gap, several works have proposed tailored pretraining strategies for more effective item ID prediction [11,34]. Bert4Rec [34] represents items using randomly initialized embeddings and adopts masked language modeling [9] to pretrain both self-attention layers [37] and item embeddings. Zhou et al. [51] further introduce an auxiliary task to jointly learn item identifiers and attribute embeddings, aiming to better capture the interactions between users and item features. P5 [11] takes a different approach by representing items as natural language descriptions and framing multiple recommendation tasks as sequence-to-sequence problems. It continually pretrains the T5 model [30] to generate item descriptions that match users' future preferences. However, the challenge of fully leveraging language model pretraining within text-matching-based sequential recommendation systems remains underexplored, limiting their potential to capture fine-grained user-item alignment.

3 Methodology

This section introduces our TASTE$^+$ model, which is illustrated in Fig. 1. We first introduce the text-matching-based recommendation framework, which verbalizes users and items and encodes them into a unified textual representation space for relevance modeling (Sect. 3.1). Then, we describe two pretraining objectives that enhance the language model's adaptation to recommendation tasks (Sect. 3.2).

3.1 Text Matching Based Sequential Recommendation Modeling

Given a user-item interaction history $\mathcal{H} = \{v_1, v_2, \ldots, v_{t-1}\}$, the sequential recommendation task aims to recommend the item v_t to satisfy the user's needs at t-th time step. TASTE$^+$ utilizes the pretrained language model T5 [30] to encode

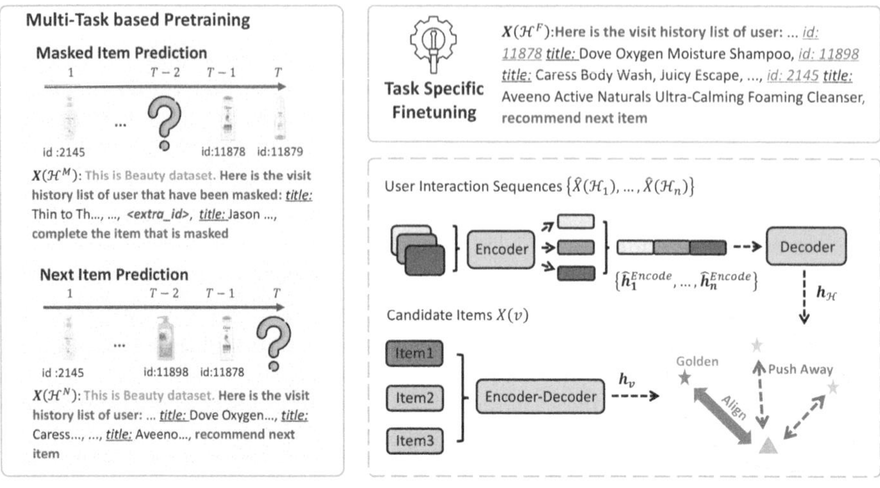

Fig. 1. The Framework of Our TASTE$^+$ Model.

the user-item interaction history $\mathcal{H}$ and candidate item v, and establishes user-item relevance through text matching of their representations.

Text Representation. Each item v can be verbalized using its item ID and k attributes <Attr> with the following template:

$$X(v) = \text{id}: v(\text{ID}) \ldots < \text{Attr} >_k : v(< \text{Attr} >_k) \tag{1}$$

where $v(\text{ID})$ and $v(<\text{Attr}>_i)$ denote the textual description of the item ID and its i-th attribute. The inclusion of the ID serves as a prompt, helping the model learn matching signals beyond attribute semantics.

The user-item interaction history $\mathcal{H}$ is verbalized as: *Here is the visit history list of user:* $X(\mathcal{H})$ *recommend next item*, where

$$X(\mathcal{H}) = X(v_{t-1}); \ldots ; X(v_1) \tag{2}$$

is the reversed concatenation of previously interacted items. This format helps pretrained language models understand sequential recommendation behaviors [6, 11] and prioritizes recent interactions during truncation.

Long Text Encoding with Attention Sparsity. In real-world scenarios, user histories are often long, which introduces challenges for pretrained language models due to sequence length limits [9,30]. To encode longer user-item interaction texts $X(\mathcal{H})$ effectively, we introduce an attention sparsity mechanism inspired by the Fusion-in-Decoder (FiD) architecture [17].

Specifically, $X(\mathcal{H})$ is split into n sub-sequences $\hat{X}(\mathcal{H}) = \{\hat{X}(\mathcal{H})_1, ..., \hat{X}(\mathcal{H})_n\}$, each containing m tokens and representing session-based interactions [22,41]. Each sub-sequence is independently encoded using the T5-Encoder:

$$\hat{h}_i^{\text{Encode}} = \text{T5-Encoder}(\hat{X}(\mathcal{H})_i). \tag{3}$$

The resulting encoded sub-sequences are concatenated to form the history representation:

$$\hat{h}^{\text{Encode}} = \{\hat{h}_1^{\text{Encode}}; \hat{h}_2^{\text{Encode}}; ...; \hat{h}_n^{\text{Encode}}\}. \tag{4}$$

This approach avoids attention computation across session boundaries, reducing complexity from $O(n^2)$ to $O(n)$ and enabling efficient long text encoding.

The T5-Decoder then takes the fused representation and the first token embedding h_0^{Decoder} to generate the final user embedding:

$$h_{\mathcal{H}} = \text{T5-Decoder}(\hat{h}^{\text{Encode}}, h_0^{\text{Decode}}), \tag{5}$$

where cross-attention helps the model attend to relevant user-item sessions and extract matching signals across the entire interaction history.

Relevance Estimation. The item v is encoded using the same model:

$$h_v = \text{T5}(X(v)). \tag{6}$$

The relevance between $h_{\mathcal{H}}$ and h_v is measured using the dot product followed by a softmax over candidate items:

$$P(v|\mathcal{H}) = \text{Softmax}_v(h_{\mathcal{H}} \cdot h_v), \tag{7}$$

where $\cdot$ denotes the dot product.

Training. The model is trained by minimizing the following loss:

$$\mathcal{L}_{\text{Finetune}} = \text{CrossEntropy}(P(v|\mathcal{H}), v^*), \tag{8}$$

where v^* denotes the ground truth item that the user interacts with at the t-th time. We apply in-batch negatives and randomly sampled negatives [13,32] for contrastive learning to enhance training.

3.2 Adapting Language Models to Sequential Recommendation via Pretraining

Existing pretrained language models [9,30] typically rely on language modeling objectives for pretraining. However, prior research [16] has highlighted a significant gap between language model pretraining and the requirements of text matching tasks. To bridge this gap, we pretrain T5 model [30] on both Yelp and Amazon Product datasets using the instruct tuning method [6,39]. The pretraining method $\mathcal{L}_{\text{Pretrain}}$ helps language models to capture the relevance signals between the representations of items and users using two tasks, Masked Item Prediction (MIP) and Next Item Prediction (NIP):

$$\mathcal{L}_{\text{Pretrain}} = L_{\text{MIP}} + L_{\text{NIP}}. \tag{9}$$

Masked Item Prediction. Given the user-item interaction sequence $\mathcal{H}$, we corrupt its text representation $X(\mathcal{H})$ by randomly masking and then ask language models to map the user representations to the masked representation. Our masked item prediction mechanism guides the language model to better understand the context of user-item interactions, learn the semantics of masked items and better model the dependency among user-interacted items.

Specifically, we randomly mask an item in the user-item interaction $\mathcal{H}$ and then verbalized user interaction sequence, which is similar to Eq. 2:

$$X(\mathcal{H}_{mask}) = X(v_{t-1}); ...; < \text{extra\_id} >; ...; X(v_1), \tag{10}$$

where <extra_id> is a special token of T5 that denotes the masked item. Then we use the following instruction to guide language models to predict the masked items: *This is* $\mathcal{D}_{name}$ *dataset. Here is the visit history list of user that have been masked:* $X(\mathcal{H}_{mask})$*, complete the item that is masked.* $\mathcal{D}_{name}$ indicates the name of the dataset that the instance comes from. During pretraining, we optimize the language model to predict the masked item using the loss function L_{MIP}:

$$L_{\text{MIP}} = \text{CrossEntropy}(P(v|\mathcal{H}_{mask}), v^*_{\text{mask}}), \tag{11}$$

where v^*_{mask} denotes the ground truth item that is masked in the user-item interaction sequence.

Next Item Prediction. The next item prediction task teaches the language model to establish the relevance between users and items by aligning the representation of the next interacted item with the representation of user-item interaction.

We use the same method to verbalize user-item interaction history $\mathcal{H}$ (Eq. 2) and get its representation $X(\mathcal{H})$. Then we use the instruction to demonstrate the next item prediction task: *This is* $\mathcal{D}_{name}$ *dataset. Here is the visit history list of user:* $X(\mathcal{H})$*, recommend next item.* $\mathcal{D}_{name}$ indicates the name of the dataset that the instance comes from. We use the loss function L_{NIP} to optimize the language model to predict the next item during pretraining:

$$L_{\text{NIP}} = \text{CrossEntropy}(P(v|\mathcal{H}), v^*_{\text{next}}), \tag{12}$$

where v^*_{next} denotes the ground truth item that is interacted with users in the next step.

Table 1. Statistics of Datasets.

Dataset	Finetune			Pretrain (Train)		Pretrain (Dev)	
	Train	Dev	Test	NIP	MIP	NIP	MIP
Beauty	131,413	22,363	22,363	225,685	148,704	19,412	19,412
Sports	189,543	35,598	35,598	225,685	140,889	19,412	19,412
Toys	109,361	19,412	19,412	225,685	145,516	19,412	19,412
Yelp	225,685	30,499	30,499	225,685	164,687	19,412	19,412
Total	–	–	–	902,740	599,796	77,648	77,648

4 Experimental Setup

This section provides details on the datasets, evaluation metrics, baseline models, and implementation details used in our experiments.

Dataset. We evaluate TASTE$^+$ on four sequential recommendation datasets: Yelp[1], and Amazon Beauty, Sports, and Toys [28][2]. These datasets cover location and product recommendations. Data statistics are shown in Table 1.

We follow Recbole [50] for data preprocessing and maintain consistent settings with prior work [26,42,51] to ensure fair comparisons. Users and items with fewer than five interactions are removed, and all interactions are treated as implicit feedback. After filtering, each sample is a user interaction sequence $\mathcal{H} = \{v_1, v_2, \ldots, v_T\}$. We adopt the leave-one-out evaluation [34,42], where the test set predicts v_T from $v_{1:T-1}$, the dev set predicts v_{T-1} from $v_{1:T-2}$, and the training set uses $v_{1:i-1}$ to predict v_i for $1 < i < T - 1$.

For pretraining, we construct NIP and MIP datasets using the same four domains. We set the NIP training/dev sizes to the largest/smallest among the four datasets, resampling where needed to balance representation. For MIP, we filter out sequences with fewer than three items and randomly mask one item in the remaining sequences as the prediction target.

Evaluation Metrics. Following DIF-SR [42], we use Recall@10/20 and NDCG@10/20 to evaluate recommendation performance. All results are reported under the full ranking setting [8,19], and statistical significance is assessed using a permutation test with $P < 0.05$.

Baselines. We compare TASTE$^+$ with several representative sequential recommendation models. GRU4Rec [14] uses RNNs to model interaction sequences, while SASRec [18] and Bert4Rec [34] adopt self-attention for user preference modeling. To evaluate the use of side information, we include S^{3}Rec [51], NOVA [23], and DIF-SR [42], which incorporate item attributes via self-supervised learning, graph modeling, or attention mechanisms. We further compare with language-model-based approaches. We further compare TASTE$^+$ with language-model-based approaches. T5-ID [11] generates the next item ID from text-encoded user histories, while T5-DPR employs a dense retrieval framework by encoding both user and item texts using a T5 encoder. CoWPiRec [43] enhances pretrained language models with collaborative filtering signals to improve item representation learning. TASTE [26] integrates item IDs and content into a unified text embedding space by treating IDs as prompt tokens, enabling more effective modeling of user-item relevance within the language model.

Implementation Details. We implement our models using OpenMatch [27,44], initializing TASTE$^+$ with the T5-base checkpoint from Huggingface Transformers [40]. Item texts are truncated to 32 tokens, and user-item interaction

[1] https://www.yelp.com/dataset.
[2] http://jmcauley.ucsd.edu/data/amazon/.

Table 2. Overall Performance of Sequential Recommendation Models. Underlined scores indicate the best-performing results among baselines. †, §, ‡ indicate statistically significant improvements over DIF-SR†, T5-DPR§ and T5-ID‡.

Dataset	Metrics	GRU4Rec	Bert4Rec	SASRec	S^{3}Rec	NOVA	DIF-SR	T5-DPR	T5-ID	CoWPiRec	TASTE	TASTE$^+$
Beauty	Recall@10	0.0530	0.0529	0.0828	0.0868	0.0887	0.0908	0.0716	0.0785§	0.0837	0.1030	**0.1071**†‡§
	Recall@20	0.0839	0.0815	0.1197	0.1236	0.1237	0.1284	0.1082	0.1138§	0.1280	0.1550	**0.1586**†‡§
	NDCG@10	0.0266	0.0237	0.0371	0.0439	0.0439	0.0446	0.0345	0.0440§	0.0405	0.0517	**0.0528**†‡§
	NDCG@20	0.0344	0.0309	0.0464	0.0531	0.0527	0.0541	0.0437	0.0529§	0.0517	0.0649	**0.0657**†‡§
Sports	Recall@10	0.0312	0.0295	0.0526	0.0517	0.0534	0.0556	0.0329	0.0464§	0.0588	0.0633	**0.0694**†‡§
	Recall@20	0.0482	0.0465	0.0773	0.0758	0.0759	0.0800	0.0554	0.0689§	0.0892	0.0964	**0.1056**†‡§
	NDCG@10	0.0157	0.0130	0.0233	0.0249	0.0250	0.0264	0.0159	0.0252§	0.0300	0.0338	**0.0356**†‡§
	NDCG@20	0.0200	0.0173	0.0295	0.0310	0.0307	0.0325	0.0216	0.0308§	0.0376	0.0421	**0.0447**†‡§
Toys	Recall@10	0.0370	0.0533	0.0831	0.0967	0.0978	0.1013	0.0805‡	0.0754	0.0502	0.1232	**0.1249**†‡§
	Recall@20	0.0588	0.0787	0.1168	0.1349	0.1322	0.1382	0.1243‡	0.1065	0.0739	0.1789	**0.1825**†‡§
	NDCG@10	0.0184	0.0234	0.0375	0.0475	0.0480	0.0504	0.0375	0.0421§	0.0272	**0.0640**	0.0610†‡§
	NDCG@20	0.0239	0.0297	0.0460	0.0571	0.0567	0.0597	0.0485	0.0499	0.0331	**0.0780**	0.0756†‡§
Yelp	Recall@10	0.0361	0.0524	0.0650	0.0589	0.0681	0.0698	0.0460	0.0450	0.0657	0.0738	**0.0768**†‡§
	Recall@20	0.0592	0.0756	0.0928	0.0902	0.0964	0.1003	0.0740	0.0745	0.0959	0.1156	**0.1192**†‡§
	NDCG@10	0.0184	0.0327	0.0401	0.0338	0.0412	0.0419	0.0250‡	0.0235	0.0416	0.0397	**0.0419**‡§
	NDCG@20	0.0243	0.0385	0.0471	0.0416	0.0483	0.0496	0.0320	0.0309	0.0497	0.0502	**0.0526**†‡§

sequences are limited to 512 tokens, split into two parts to apply sparse attention (Eq. 3). For pretraining, we use a batch size of 32, learning rate of 1e−4, and in-batch negatives; the best checkpoint is selected based on validation loss. Fine-tuning uses the Adam optimizer with a 1e−4 learning rate, 0.1 warm-up proportion, batch size 8, and 9 randomly sampled negatives per example.

5 Experimental Results

In this section, we first evaluate the recommendation performance of TASTE$^+$ and conduct ablation studies. Then we analyze the attention mechanism and the recommendation behaviors of TASTE$^+$.

5.1 Overall Performance

The recommendation performance of TASTE$^+$ across all datasets is detailed in Table 2. In comparison to item ID embedding based recommendation models, such as Bert4Rec, TASTE$^+$ nearly doubles recommendation accuracy by leveraging text-matching signals to better capture the relevance between users and items. Additionally, TASTE$^+$ outperforms DIF-SR, which incorporates item attributes as side information to enhance user behavior modeling. DIF-SR relies on embedding item attributes and employs complex mechanisms to decouple and integrate side information into item representations. In contrast, TASTE$^+$ simplifies the process by verbalizing user and item data using a general template and leveraging the pretrained attention mechanisms of T5 to align textual representations. This highlights the inherent advantages of employing pretrained language models in recommendation tasks. Furthermore, TASTE$^+$ outperforms

Table 3. Ablation Study.

Dataset	Metrics	Pretraining					Fine-Tuning	
		Zero-Shot	w/ MIP	w/ NIP	Plus w/o Instruct	Plus	TASTE	TASTE$^+$
Beauty	Recall@10	0.0009	0.0341	0.0546	0.0450	0.0582	0.1030	**0.1071**
	Recall@20	0.0016	0.0536	0.0907	0.0741	0.0927	0.1550	**0.1586**
	NDCG@10	0.0005	0.0157	0.0246	0.0217	0.0271	0.0517	**0.0528**
	NDCG@20	0.0007	0.0206	0.0337	0.0290	0.0358	0.0649	**0.0657**
Sports	Recall@10	0.0004	0.0182	0.0234	0.0263	0.0288	0.0633	**0.0694**
	Recall@20	0.0008	0.0313	0.0415	0.0454	0.0467	0.0964	**0.1056**
	NDCG@10	0.0001	0.0081	0.0111	0.0122	0.0134	0.0338	**0.0356**
	NDCG@20	0.0002	0.0114	0.0156	0.0170	0.0179	0.0421	**0.0447**
Toys	Recall@10	0.0011	0.0345	0.0762	0.0490	0.0789	0.1232	**0.1249**
	Recall@20	0.0015	0.0570	0.1160	0.0792	0.1209	0.1789	**0.1825**
	NDCG@10	0.0004	0.0154	0.0370	0.0237	0.0376	**0.0640**	0.0610
	NDCG@20	0.0005	0.0211	0.0470	0.0313	0.0482	**0.0780**	0.0756
Yelp	Recall@10	0.0003	0.0395	0.0425	0.0461	0.0478	0.0738	**0.0768**
	Recall@20	0.0007	0.0622	0.0672	0.0698	0.0702	0.1156	**0.1192**
	NDCG@10	0.0001	0.0215	0.0243	0.0258	0.0275	0.0397	**0.0419**
	NDCG@20	0.0002	0.0271	0.0305	0.0317	0.0331	0.0502	**0.0526**

the prior state-of-the-art model, TASTE, illustrating that the proposed pretraining method effectively adapts language models to the text based sequential recommendation task.

5.2 Ablation Study

As shown in Table 3, we explore the recommendation effectiveness of language models using different pretraining methods. Compared with the Zero-Shot model, both pretraining tasks, Masked Item Prediction (MIP) and Next Item Prediction (NIP), achieve significant improvements, especially the NIP task. It shows the important role of language model pretraining, which learns more effective representations of users and items for the recommendation task. The recommendation performance is further improved when we use both tasks to pretrain language models. During multi-task pretraining, the task instruction shows its effectiveness by indicating both the task objective and the corresponding dataset. After finetuning with the specific dataset, TASTE$^+$ achieves much better performance, confirming the effectiveness of our pretraining method in adapting language models to different recommendation tasks.

5.3 Attention Mechanism of Text Matching Based Recommendation Systems

In this experiment, we explore the attention mechanism of TASTE and TASTE$^+$. The cross-attention score distributions of T5 [30] are shown to further under-

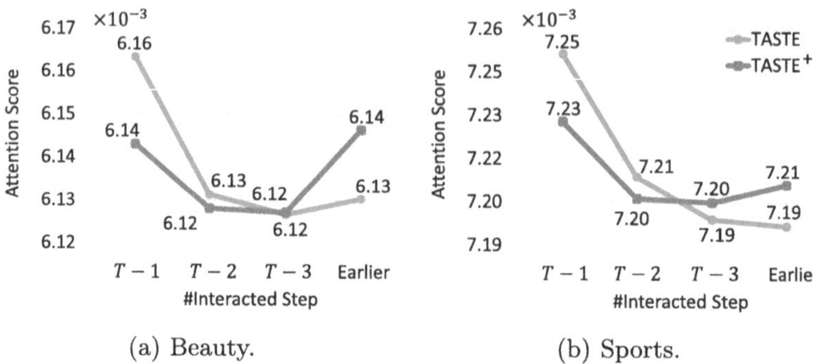

(a) Beauty. (b) Sports.

Fig. 2. Attention Distributions of User-Item Interactions.

stand the semantic learning mechanism of these text based recommendation systems.

Attention Distribution of Items. As shown in Fig. 2, we first show the attention distribution of the items at different user-interacted steps.

Overall, on Amazon Product datasets (Beauty), the last user-interacted items always win the most attention weights from the TASTE model, showing that the last user-interacted items are crucial to recommending the next items. Compared to the TASTE model, TASTE$^+$ achieves a more balanced attention distribution across all user interactions. While it still emphasizes the last user-interacted items for capturing short-term interests, it also gives more attention weight to earlier user-interacted items. This balance helps in reflecting the user's long-term preferences more effectively. Moreover, on the Yelp dataset, the gap in attention weights among different user-item interaction steps remains lower than the gap in other datasets. It indicates that user's previously visited restaurants almost play the same important role in characteristic the user preference. Compared to the TASTE model, the TASTE$^+$ model, while maintaining a balance between long-term and short-term preferences, increases dependence on the last user-interacted items, which can better reflect the user's interests and preferences in the recommended session.

Attention Weights on Different Tokens. We further explore the attention weights on different types of tokens assigned by TASTE$^+$, and compare them with those assigned by the TASTE model. We divide the token types into three types according to the textual representation of users (Eq. 2), which include prompt tokens (id, v(ID) and $<\text{Attr}>_i$), ID tokens (id and v(ID)), and text tokens ($v(<\text{Attr}>_i)$).

Figure 3 illustrates a comparison of the average cross-attention scores on different types of tokens assigned by the TASTE$^+$ model. The average cross-attention scores for the tokens of different types are computed. For both models, the experimental results show that the prompt tokens consistently receive almost the same attention scores as texts, showing that these prompt tokens are informative to learn the user representations. These prompt tokens gather infor-

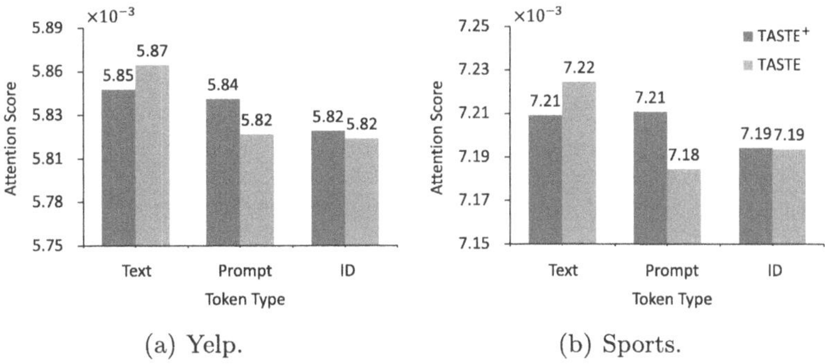

(a) Yelp. (b) Sports.

Fig. 3. Attention Distribution of Different Types of Tokens.

mation in shallow layers and are extracted in the deeper layer, which is also observed in previous work [38]. Moreover, both models also assign more attention to ID tokens. It demonstrates that these item ID tokens also learn valuable information to better represent users. Compared to the TASTE model, TASTE$^+$ assigns more attention weights to prompt tokens and IDs. This illustrates that our pretraining mechanism helps language models to better encode semantics using these prompt and ID tokens [7], to build a more effective recommendation system.

5.4 Recommendation Behavior Analyses on TASTE$^+$

This experiment evaluates the effectiveness of TASTE and TASTE$^+$ by analyzing their recommendation behaviors in different testing scenarios.

Long-Tailed Items. We evaluate the recommendation performance of TASTE$^+$ on these items with different user-interacted frequencies, as shown in Fig. 4. The items are categorized into two groups: long-tail items, which have lower interaction frequencies, and others that are more frequently interacted with by users. TASTE$^+$ demonstrates notable improvements over T5-ID on long-tail items, highlighting its few-shot capability in learning representations of items using their identifiers and attributes. This effectively alleviates the "Cold Start" problem [20,33] often encountered in recommendation systems. TASTE$^+$ extends pretrained language models to long-tail items, enabling self-contained recommendation systems that model user-item relevance through text matching.

Popularity of Recommended Items. We investigate the recommendation behaviors of TASTE under different item modeling strategies, comparing four models: T5-ID, TASTE w/o ID, TASTE, and TASTE$^+$. T5-ID uses randomly initialized item embeddings to directly predict item IDs. TASTE w/o ID and TASTE adopt a two-tower structure to encode item attributes, with TASTE additionally including item IDs. TASTE$^+$ further introduces pretraining to adapt language models for recommendation.

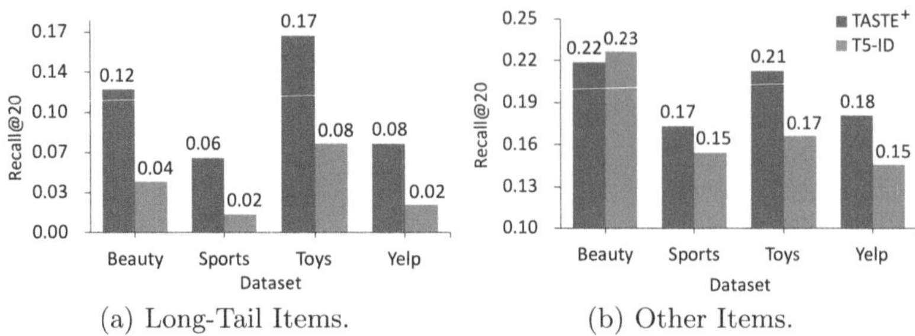

(a) Long-Tail Items. (b) Other Items.

Fig. 4. Recommendation Effectiveness on Items with Different User-Interacted Frequencies.

Table 4. Recommendation Behaviors with Different Item Verbalization Methods.

Dataset	Metrics	T5-ID	TASTE w/o ID	TASTE	TASTE$^+$
Beauty	Dist-1	**0.1456**	0.1385	0.1398	0.1333
	Dist-2	**0.5003**	0.4452	0.4641	0.4524
	Bleu-4	0.0267	0.0321	0.0322	**0.0332**
	Recall-5	0.0509	0.0540	**0.0639**	0.0635
	Popular	52%	29%	31%	**26%**
Sports	Dist-1	**0.1740**	0.1571	0.1623	0.1532
	Dist-2	**0.5668**	0.4917	0.5193	0.4986
	Bleu-4	0.0130	0.0153	0.0163	**0.0181**
	Recall-5	0.0304	0.0326	0.0403	**0.0428**
	Popular	62%	36%	33%	**31%**
Toys	Dist-1	**0.1728**	0.1564	0.1601	0.1568
	Dist-2	**0.5268**	0.4493	0.4697	0.4677
	Bleu-4	0.0467	0.0566	**0.0588**	0.0541
	Recall-5	0.0507	0.0636	**0.0793**	0.0779
	Popular	38%	29%	29%	**27%**
Yelp	Dist-1	**0.1402**	0.1363	0.1351	0.1239
	Dist-2	**0.4367**	0.4184	0.4177	0.4187
	Bleu-4	0.0479	**0.0542**	0.0540	0.0513
	Recall-5	0.0276	0.0465	0.0469	**0.0502**
	Popular	46%	**30%**	**30%**	34%

As shown in Table 4, T5-ID exhibits strong popularity bias, with nearly 49.5% of its recommendations being popular items. TASTE significantly mitigates this issue by leveraging text-based representations, reducing the popular item ratio by 20% on average. It also improves diversity compared to TASTE w/o ID, indicating that integrating item IDs enhances recommendation variety. TASTE$^+$ decreases bias on Amazon datasets while slightly increasing it on Yelp. This phenomenon shows that popular restaurants attract attention, while prevalent products among most people may not be tailored to individual user preferences.

6 Conclusion

In this paper, we propose the TASTE$^+$ model, which utilizes two pretraining tasks, Masked Item Prediction and Next Item Prediction to adapt language models to the sequential recommendation tasks. TASTE$^+$ achieves further improvements over TASTE on both Yelp and Amazon Product datasets, showing its effectiveness in learning more tailored representations of users and items for recommendation. The experimental results show that TASTE$^+$ has the ability to alleviate the cold start problem and popularity bias of existing recommendation systems.

Acknowledgment. This work is partly supported by the National Natural Science Foundation of China (No. 62206042), and the Fundamental Research Funds for the Central Universities (No. N25ZLL045).

References

1. Abdollahpouri, H., Mansoury, M.: Multi-sided exposure bias in recommendation. ArXiv preprint (2020)
2. Abdollahpouri, H., Mansoury, M., Burke, R., Mobasher, B.: The unfairness of popularity bias in recommendation. ArXiv preprint (2019)
3. Beltagy, I., Peters, M.E., Cohan, A.: LongFormer: the long-document transformer. CoRR (2020)
4. Chang, J., et al.: Sequential recommendation with graph neural networks. In: Proceedings of SIGIR, pp. 378–387 (2021)
5. Chen, X., et al.: Sequential recommendation with user memory networks. In: Proceedings of the Eleventh ACM International Conference on Web Search and Data Mining, pp. 108–116 (2018)
6. Chung, H.W., et al.: Scaling instruction-finetuned language models. CoRR (2022)
7. Clark, K., Khandelwal, U., Levy, O., Manning, C.D.: What does BERT look at? An analysis of BERT's attention. In: Proceedings of the 2019 ACL Workshop BlackboxNLP: Analyzing and Interpreting Neural Networks for NLP, pp. 276–286 (2019)
8. Dallmann, A., Zoller, D., Hotho, A.: A case study on sampling strategies for evaluating neural sequential item recommendation models. In: Proceedings of Fifteenth ACM Conference on Recommender Systems, pp. 505–514 (2021)
9. Devlin, J., Chang, M.W., Lee, K., Toutanova, K.: BERT: pre-training of deep bidirectional transformers for language understanding. In: Proceedings of NAACL-HLT, pp. 4171–4186 (2019)
10. Ding, H., Ma, Y., Deoras, A., Wang, Y., Wang, H.: Zero-shot recommender systems. CoRR (2021)
11. Geng, S., Liu, S., Fu, Z., Ge, Y., Zhang, Y.: Recommendation as language processing (RLP): a unified pretrain, personalized prompt & predict paradigm (P5). In: Proceedings of Sixteenth ACM Conference on Recommender Systems, pp. 299–315 (2022)

12. He, R., McAuley, J.J.: Fusing similarity models with Markov chains for sparse sequential recommendation. In: Proceedings of IEEE 16th International Conference on Data Mining, pp. 191–200 (2016)
13. He, X., Liao, L., Zhang, H., Nie, L., Hu, X., Chua, T.: Neural collaborative filtering. In: Proceedings of WWW, pp. 173–182 (2017)
14. Hidasi, B., Karatzoglou, A., Baltrunas, L., Tikk, D.: Session-based recommendations with recurrent neural networks. In: Proceedings of ICLR (2016)
15. Hou, Y., Mu, S., Zhao, W.X., Li, Y., Ding, B., Wen, J.: Towards universal sequence representation learning for recommender systems. In: Proceedings of KDD, pp. 585–593 (2022)
16. Hu, X., Yu, S., Xiong, C., Liu, Z., Liu, Z., Yu, G.: P3 ranker: mitigating the gaps between pre-training and ranking fine-tuning with prompt-based learning and pre-finetuning. In: Proceedings of the 45th International ACM SIGIR Conference on Research and Development in Information Retrieval, pp. 1956–1962 (2022)
17. Izacard, G., Grave, E.: Leveraging passage retrieval with generative models for open domain question answering. In: Proceedings of EACL, pp. 874–880 (2021)
18. Kang, W., McAuley, J.J.: Self-attentive sequential recommendation. In: Proceedings of IEEE International Conference on Data Mining, pp. 197–206 (2018)
19. Krichene, W., Rendle, S.: On sampled metrics for item recommendation. In: Proceedings of KDD, pp. 1748–1757 (2020)
20. Lam, X.N., Vu, T., Le, T.D., Duong, A.D.: Addressing cold-start problem in recommendation systems. In: Proceedings of the 2nd International Conference on Ubiquitous Information Management and Communication, pp. 208–211 (2008)
21. Li, J., et al.: Text is all you need: learning language representations for sequential recommendation. In: Proceedings of KDD, pp. 1258–1267 (2023)
22. Li, J., Ren, P., Chen, Z., Ren, Z., Lian, T., Ma, J.: Neural attentive session-based recommendation. In: Proceedings of CIKM, pp. 1419–1428 (2017)
23. Liu, C., Li, X., Cai, G., Dong, Z., Zhu, H., Shang, L.: Non-invasive self-attention for side information fusion in sequential recommendation. CoRR (2021)
24. Liu, P., Yuan, W., Fu, J., Jiang, Z., Hayashi, H., Neubig, G.: Pre-train, prompt, and predict: a systematic survey of prompting methods in natural language processing. ACM Comput. Surv. **9**, 1–35 (2023)
25. Liu, Q., Wu, S., Wang, D., Li, Z., Wang, L.: Context-aware sequential recommendation. In: Proceedings of IEEE 16th International Conference on Data Mining, pp. 1053–1058. IEEE (2016)
26. Liu, Z., et al.: Text matching improves sequential recommendation by reducing popularity biases. In: Proceedings of the 32nd ACM International Conference on Information and Knowledge Management, pp. 1534–1544 (2023)
27. Liu, Z., Zhang, K., Xiong, C., Liu, Z., Sun, M.: OpenMatch: an open source library for neu-IR research. In: Proceedings of SIGIR, pp. 2531–2535 (2021)
28. McAuley, J.J., Targett, C., Shi, Q., van den Hengel, A.: Image-based recommendations on styles and substitutes. In: Proceedings of SIGIR, pp. 43–52 (2015)
29. Pan, F., Li, S., Ao, X., Tang, P., He, Q.: Warm up cold-start advertisements: improving CTR predictions via learning to learn ID embeddings. In: Proceedings of SIGIR, pp. 695–704 (2019)
30. Raffel, C., et al.: Exploring the limits of transfer learning with a unified text-to-text transformer. J. Mach. Learn. Res. 140:1–140:67 (2020)
31. Rendle, S.: Factorization machines. In: Proceedings of The 10th IEEE International Conference on Data Mining, pp. 995–1000 (2010)
32. Rendle, S., Freudenthaler, C., Gantner, Z., Schmidt-Thieme, L.: BPR: Bayesian personalized ranking from implicit feedback. CoRR (2012)

33. Schein, A.I., Popescul, A., Ungar, L.H., Pennock, D.M.: Methods and metrics for cold-start recommendations. In: Proceedings of SIGIR, pp. 253–260 (2002)
34. Sun, F., et al.: BERT4Rec: sequential recommendation with bidirectional encoder representations from transformer. In: Proceedings of CIKM, pp. 1441–1450 (2019)
35. Sun, S., Cai, T., Yan, F., Ju, S.: Two-stage enhancement for recommendation systems based on contrastive learning. In: International Conference on Web Information Systems and Applications, pp. 150–162. Springer (2024)
36. Tang, J., Wang, K.: Personalized top-n sequential recommendation via convolutional sequence embedding. In: Proceedings of the Eleventh ACM International Conference on Web Search and Data Mining, pp. 565–573 (2018)
37. Vaswani, A., et al.: Attention is all you need. In: Proceedings of NeurIPS, pp. 5998–6008 (2017)
38. Wang, L., et al.: Label words are anchors: An information flow perspective for understanding in-context learning. arXiv preprint arXiv:2305.14160 (2023)
39. Wang, Y., et al.: Self-instruct: aligning language model with self generated instructions. arXiv preprint (2022)
40. Wolf, T., et al.: HuggingFace's transformers: state-of-the-art natural language processing. CoRR (2019)
41. Wu, S., Tang, Y., Zhu, Y., Wang, L., Xie, X., Tan, T.: Session-based recommendation with graph neural networks. In: Proceedings of AAAI, pp. 346–353 (2019)
42. Xie, Y., Zhou, P., Kim, S.: Decoupled side information fusion for sequential recommendation. In: Proceedings of SIGIR, pp. 1611–1621 (2022)
43. Yang, S., et al.: Collaborative word-based pre-trained item representation for transferable recommendation. In: 2023 IEEE International Conference on Data Mining (ICDM), pp. 728–737. IEEE (2023)
44. Yu, S., Liu, Z., Xiong, C., Liu, Z.: OpenMatch-v2: an all-in-one multi-modality PLM-based information retrieval toolkit. In: Proceedings of SIGIR, pp. 3160–3164 (2023)
45. Yuan, F., Karatzoglou, A., Arapakis, I., Jose, J.M., He, X.: A simple convolutional generative network for next item recommendation. In: Proceedings of the Twelfth ACM International Conference on Web Search and Data Mining, pp. 582–590 (2019)
46. Yuan, Z., et al.: Where to go next for recommender systems? ID- vs. modality-based recommender models revisited. In: Proceedings of SIGIR, pp. 2639–2649 (2023)
47. Zhang, M., Wu, S., Yu, X., Liu, Q., Wang, L.: Dynamic graph neural networks for sequential recommendation. TKDE (2022)
48. Zhang, T., et al.: Feature-level deeper self-attention network for sequential recommendation. In: Proceedings of IJCAI, pp. 4320–4326 (2019)
49. Zhang, Y., et al.: Language models as recommender systems: evaluations and limitations. In: Proceedings of I (Still) Can't Believe It's Not Better! NeurIPS 2021 Workshop (2021)
50. Zhao, W.X., et al.: RecBole: towards a unified, comprehensive and efficient framework for recommendation algorithms. In: Proceedings of CIKM, pp. 4653–4664 (2021)
51. Zhou, K., et al.: S3-rec: self-supervised learning for sequential recommendation with mutual information maximization. In: Proceedings of CIKM, pp. 1893–1902 (2020)
52. Zhu, Z., He, Y., Zhao, X., Caverlee, J.: Popularity bias in dynamic recommendation. In: Proceedings of KDD, pp. 2439–2449 (2021)

Enabling Intelligent Support for SunDB via Knowledge Graphs and LLMs

Peng Ren(✉), Yong Zhang, Chunxiao Xing, and Yuxuan Li

BNRist, DCST, RIIT, Tsinghua University, Beijing 100084, China
{renpeng,zhangyong05,xingcx,liyuxuan0120}@tsinghua.edu.cn

Abstract. As data volumes surge and digital transformation accelerates, enterprises increasingly expect database support to be fast, precise, and intelligent. We present SunDB.ai, an intelligent database support platform that tackles core challenges in this domain by combining a high-precision knowledge graph with an LLM-driven natural-language interface [1]. It features an intuitive, user-friendly question–answer interface with strong interpretability and traceability, and integrates intelligent data processing tools with the ticketing system for seamless coordination. By enabling rapid and precise resolution of user issues, the platform significantly enhances service quality. A prototype demonstration highlights its strong performance, ease of use, and tangible benefits in real-world applications.

Keywords: Knowledge Graph · Retrieval-Augmented Generation (RAG) · Natural Language Processing (NLP) · SunDB · Large Language Models (LLMs) · Ticketing System

1 Introduction

As enterprise digital infrastructure continues to evolve, databases have become the backbone of data storage and management. With this growing centrality comes increasing complexity in maintaining and supporting database systems. Conventional technical support approaches—typically based on static documentation and manual troubleshooting—are often inadequate in addressing user issues in a timely and context-sensitive manner. These limitations make it difficult for organizations to keep pace with rapidly changing business requirements. Consequently, there is a rising demand for intelligent, adaptive, and explainable support systems that can deliver accurate assistance at scale [2]. Developing such systems has become a key strategy for enhancing operational efficiency and sustaining competitive advantage.

2 System Architecture

The SunDB Intelligent Technical Support System adopts a modular microservice architecture comprising the following core components (Fig. 1):

S. Yang et al. (Eds.): WISA 2025, LNCS 16299, pp. 606–609, 2026.
https://doi.org/10.1007/978-981-95-6426-2_53

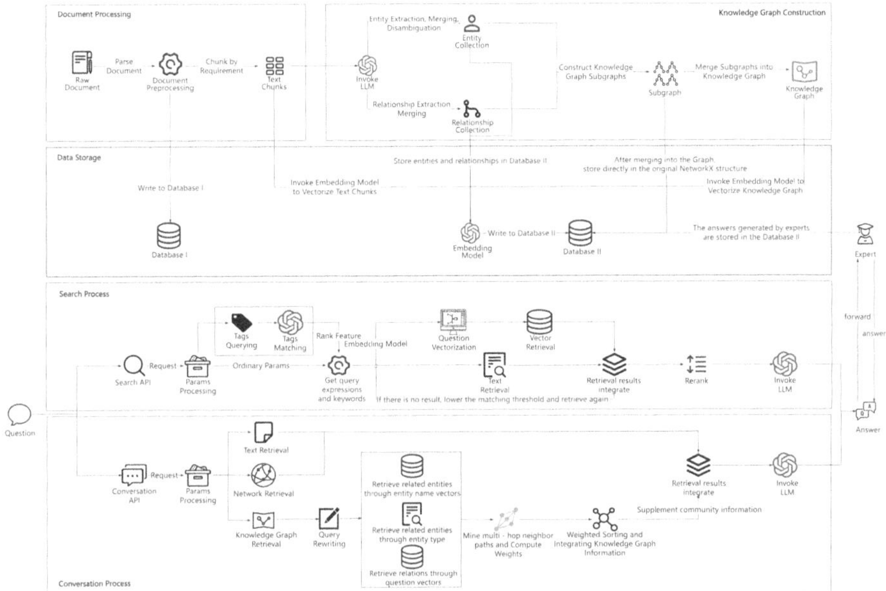

Fig. 1. System architecture consisting of a knowledge base and a QA module.

User Interface Module: Provides a user-friendly and intuitive natural language interface that enables users to submit technical queries and receive feedback seamlessly.

Knowledge Base Module: This module supports the ingestion and processing of diverse data types and document formats. It integrates advanced document parsing techniques to extract structured information from unstructured sources. Leveraging large language models (LLMs), the system further transforms document content into structured knowledge graphs. Embedding models are then used to convert these graphs into vector representations for efficient retrieval. The module also includes a suite of tools for managing multiple knowledge bases.

Knowledge Retrieval Module: Combines a range of retrieval strategies, including vector search algorithms, reranking techniques, graph-based reasoning, and prompt templates. This module ensures high-efficiency, high-precision retrieval while filtering out irrelevant information, laying the foundation for accurate and responsive answer generation.

Model Management Module: Integrates various models required by the system, including large language models, embedding models, reranking models, and their corresponding APIs. This design enables flexible model switching and modular integration based on specific task needs.

Ticket System Integration Module: For complex or unresolved issues, the system automatically generates and routes support tickets to domain experts. Once an expert responds, the answer is automatically processed and incorporated into the knowledge base, enabling continuous enrichment of system knowledge and closing the loop between human expertise and automated support.

3 Core Techniques

3.1 RAG-Based Intelligent Interaction Mechanism

Traditional QA systems often rely on fixed rules or basic natural language processing techniques to interpret user input, which limits their ability to handle complex queries or ambiguous semantics. To address these shortcomings, our system integrates large language models, a vector database, and a knowledge graph to construct a Retrieval-Augmented Generation (RAG)–based interaction framework. This design significantly enhances the system's natural language understanding capabilities and enables accurate intent recognition even in challenging scenarios.

3.2 Intelligent Ticket Interface with Expert-In-The-Loop

Conventional ticket systems typically rely on manual triage or rule-based routing, which often proves inefficient when handling complex or time-sensitive issues. To address these limitations, we design an intelligent ticket interface tightly coupled with an expert-in-the-loop knowledge platform. The system adopts a hybrid workflow: routine queries are answered directly via the knowledge base, while complex or novel issues are escalated to human experts. Expert responses are then automatically distilled and incorporated back into the knowledge base, enabling continuous knowledge refinement and significantly reducing response latency for recurring issues.

3.3 Systematic Extraction Method for Multi-Source and Heterogeneous Data

Traditional data extraction and integration approaches often rely on unimodal processing pipelines, which are inadequate for handling unstructured content originating from diverse sources. These limitations frequently result in information loss and incomplete data fusion. To overcome these challenges, we develop a multimodal information parsing engine that systematically incorporates advanced techniques across modalities processing techniques across modalities. The engine enables unified extraction and integration of heterogeneous data formats, significantly improving processing efficiency and enriching the underlying knowledge base. This, in turn, enhances the coverage and quality of responses generated by the QA system.

3.4 Precision RAG Enhanced by Knowledge Graphs

Traditional retrieval systems often rely on keyword matching, which fails to capture complex semantic relationships. To address these challenges, we propose a high-precision Retrieval-Augmented Generation (RAG) approach enhanced by knowledge graphs. Unlike prior methods, our system constructs and leverages structured knowledge graphs to identify fine-grained semantic associations and contextual cues embedded in user queries. This significantly improves both retrieval relevance and response generation accuracy. Moreover, the system incorporates a dynamic learning mechanism that continuously refines the structure and content of the knowledge graph based on user feedback and emerging domain knowledge.

4 System Demonstration Workflow

During the system demonstration, users can intuitively experience the platform's interaction flow and response behavior. The process consists of the following steps:

1. **Document Ingestion**: The user uploads a document containing various forms of complex information. The system extracts relevant data and stores it in the knowledge base.
2. **User Query Input**: The user enters a specific technical question through the interactive interface.
3. **Knowledge Retrieval**: The system retrieves relevant solutions by precisely locating matching entries from the knowledge base and knowledge graph.
4. **Answer Generation**: The retrieved content is composed into a natural language response with explicit source attribution.

5 Demonstration Environment

Due to data privacy considerations, the system is deployed within the internal infrastructure of the SunDB host organization. The demonstration runs on a single NVIDIA A100 GPU.

6 Application

The SunDB Intelligent Technical Support Platform has been officially deployed within the pre-sales and post-sales support teams of Beijing Kolan Software Systems Co., Ltd. The system assists in efficiently resolving both customer-reported issues during database usage and technical problems encountered by engineers during development. In real-world operation, the platform has demonstrated strong performance by significantly reducing the workload of support teams, accelerating issue resolution, and markedly improving customer satisfaction and the overall quality of technical support services.

References

1. Khan et al.: Unifying Large Language Models and Knowledge Graphs for Question Answering: Emerging Paradigms, VLDB (2024)
2. Rowe, F., et al.: Explainable artificial intelligence in information systems: a review of the status quo and future research directions. Electron. Markets (2023). https://doi.org/10.1007/s12525-023-00644-5

Security

TrustSched: A Blockchain-Enhanced Distributed Scheduling Framework for Trusted Synthetic Data Generation

Ding Sheng, Zheming Ye, Qi Xu, Yanqin Yang(✉), Zhao Zhang, and Cheqing Jin

East China Normal University, Shanghai, China
{dsheng,zhmye,qxu}@stu.ecnu.edu.cn,
{yqyang,zhzhang,cqjin}@dase.ecnu.edu.cn

Abstract. The growing demand for synthetic data in privacy-sensitive domains necessitates scheduling frameworks that balance generation efficiency with verifiable traceability. Existing approaches often struggle to maintain secure auditability while scaling across distributed environments. This paper presents TrustSched, a distributed scheduling framework integrating blockchain-backed integrity proofs for trusted synthetic data generation. By employing a Master-Executor architecture with slot-epoch scheduling, the system achieves parallelized data synthesis while enforcing hierarchical Merkle tree commitments for tamper-evident traceability. A batched blockchain submission mechanism decouples evidence anchoring from task execution, optimizing both computational throughput and on-chain efficiency. Experimental validation confirms the framework's ability to maintain robust security guarantees without compromising synthesis performance, demonstrating its suitability for enterprise applications requiring auditable data generation.

Keywords: Data Synthesis · Distributed Task Scheduling · Blockchain

1 Introduction

In recent years, the rapid advancement of artificial intelligence (AI) and big data technologies has significantly accelerated their adoption across various domains, including healthcare [1], finance [2] and education [3]. However, this surge in AI development has also brought to light two increasingly pressing challenges: data scarcity and privacy risks. On the one hand, high-quality labeled data is often difficult and expensive to obtain, particularly in specialized or emerging fields, where expert annotation is required or data samples are inherently limited. On the other hand, many real-world datasets contain sensitive personal or proprietary information, making it difficult or even legally restricted to share or utilize them directly for model training.

To address these challenges, the field of synthetic data generation has rapidly gained momentum. In particular, the rise of powerful generative models—such

S. Yang et al. (Eds.): WISA 2025, LNCS 16299, pp. 613–624, 2026.
https://doi.org/10.1007/978-981-95-6426-2_54

as generative adversarial networks has enabled the creation of realistic and statistically representative data that mimics the distribution of real-world samples. These synthetic datasets can be used to augment limited real data, replace sensitive datasets to protect privacy, and improve model robustness by mitigating issues like class imbalance or distributional bias. As such, synthetic data has emerged as a promising solution for enabling data-driven AI development in domains where traditional data acquisition is constrained.

While synthetic data addresses challenges such as data scarcity, privacy protection, and annotation costs, large-scale deployment introduces new difficulties. Generative models like diffusion models and GANs are computationally intensive and memory-heavy, requiring substantial GPU resources and generating significant storage overheads. Consequently, data owners often lack the capacity to meet large-scale synthetic data demands independently. A common solution is to offload generation tasks to distributed third-party platforms, which offer greater scalability, concurrent execution, and reduced local infrastructure burden. However, outsourcing synthetic data generation also introduces fundamental challenges that must be addressed.

Consistency Overkill for Synthetic Workloads. Existing distributed computing frameworks like Spark [4] are fundamentally mismatched with synthetic data workloads. Their strong consistency and frequent synchronization mechanisms introduce excessive overhead, while synthetic tasks are primarily compute-bound with minimal data exchange requirements. This misalignment calls for lighter-weight, asynchronous communication models that prioritize computational throughput over strict consistency.

Lack of Traceability. Current synthetic data pipelines offer little visibility into the internal workflow of third-party platforms. As generation algorithms are often treated as proprietary black boxes, data owners lack the ability to monitor execution, verify intermediate outputs, or audit the generation process. This absence of procedural oversight becomes particularly problematic in the event of anomalies or quality concerns, where tracing back to the source of errors is essential. Furthermore, for privacy-sensitive and regulated domains, the inability to reconstruct or verify the synthesis path hinders accountability, validation, and compliance auditing.

To address these challenges, we propose TrustSched, a distributed scheduling framework designed for trusted synthetic data generation. Our solution combines lightweight task orchestration with blockchain-based verification to achieve efficient and auditable execution. Specifically, our contirbution of this paper includes:

1. A multi-layer scheduling model that enables lightweight task delegation and dynamic load balancing while eliminating strong consistency requirements to optimize computational efficiency for synthetic data workloads
2. A blockchain-based verification mechanism combining Merkle Tree commitments with batched on-chain evidence archival, providing immutable audit trails while maintaining low overhead through asynchronous batch processing

3. We propose TrustSched framework, a Master-Executor architecture that integrates lightweight task orchestration with blockchain verification. It ensures trusted execution through hierarchical Merkle proof validation and dynamic slot-epoch scheduling.

2 Related Work

2.1 Data Synthesis

Data synthesis technologies have made significant strides, with methods like Generative Adversarial Networks (GANs) and diffusion models showing considerable promise in addressing data scarcity and enhancing privacy protection. CTGAN [5], through its conditional generation mechanisms and mode-specific normalization techniques, effectively models mixed-type features and resolves class imbalance issues in tabular data, making it a critical tool for privacy-sensitive data generation in healthcare and finance. Diffusion models, based on iterative denoising processes, excel in synthesizing high-fidelity data, particularly in image generation, and have been extended to tabular data (e.g., Tab-DDPM) [6]. However, challenges remain in computational cost and generation efficiency. While these methods exhibit distinct advantages in generation quality, further research is needed to address key challenges such as privacy-utility trade-offs, thereby advancing broader applications of data synthesis in academia and industry. In addition to algorithmic advancements, distributed platforms for large-scale data synthesis have emerged. For example, SiloFuse [7] introduces a distributed architecture that enables high-quality data synthesis across multiple nodes while preserving privacy. This approach enhances communication efficiency and reduces training rounds, making it suitable for decentralized data synthesis tasks.

2.2 Blockchain

Blockchain technology has emerged as a foundational approach to establish trust in decentralized traceability systems by ensuring data integrity and transaction transparency. While public blockchains provide censorship-resistant frameworks, their proof-of-work mechanisms suffer from severe scalability limitations, making them impractical for enterprise applications requiring high throughput. This performance bottleneck has driven the adoption of permissioned blockchains, which employ controlled participation and optimized consensus protocols. Byzantine Fault Tolerant (BFT) algorithms such as Practical Byzantine Fault Tolerance (PBFT) [8] exemplify this shift, replacing energy-intensive mining with deterministic voting to achieve sub-second finality and improved transaction throughput. A notable implementation is FISCO-BCOS [9], a leading permissioned blockchain platform that integrates modular multi-chain expansion, privacy-preserving transactions, and configurable smart contracts. At the data layer,

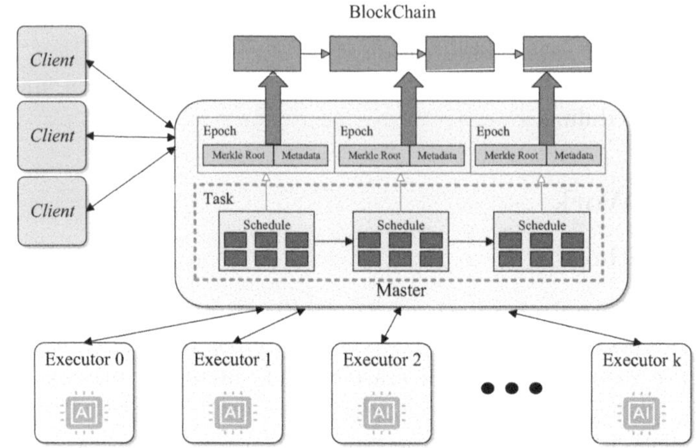

Fig. 1. System architecture diagram.

Merkle Trees [10] play a critical role by cryptographically aggregating transactions into a hierarchical structure, enabling efficient integrity verification without requiring full dataset storage. This lightweight auditing mechanism, combined with enterprise-grade consensus, forms a dual-layer architecture that meets the core requirements of supply chain traceability: immutable ledgers for non-repudiation, parallel processing for operational efficiency, and hierarchical hashing for scalable verification.

3 System Overview

The architecture of TrustSched is designed as illustrated in Fig. 1. TrustSched employs a Master-Executor architecture to enable scalable and verifiable distributed computing.

Master. The Master orchestrates the entire lifecycle of task execution, including task scheduling, collecting and evidence archiving. Upon receiving user tasks, it decomposes them into manageable slots according to real-time node capacity, assigns and monitors their execution, and enforces timeout thresholds through prioritized tracking. At regular intervals, it redistributes expired slots to maintain system throughput and updates metadata for completed tasks. During task execution, the Master continuously aggregates evidence data, batches them efficiently, and submits them onto the blockchain through lightweight smart contract interactions, ensuring an immutable audit trail while minimizing on-chain overhead. When users request retrieval, the Master verifies the health of candidate nodes, retrieves validated data slots, merges redundant chunks based on size optimization strategies, cross-verifies data integrity against blockchain-anchored records, and delivers the final datasets to users. Throughout the process, it

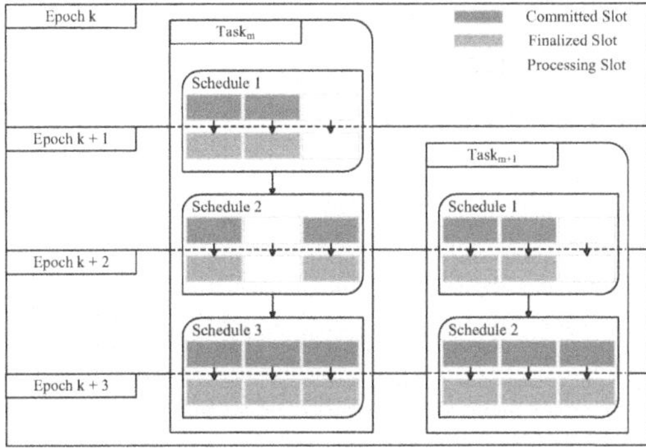

Fig. 2. The scheduling model.

logs essential metadata to support fault recovery and maintain service resilience against node failures or network disruptions.

Executor. Each Executor node operates as an independent computing unit, performing three core functions. Upon receiving task assignments from the Master, the node first verifies whether it has sufficient computational resources to handle the workload. During execution, while generating synthetic data, the node constructs a Merkle tree structure to aggregate the hash values of data blocks, ultimately producing an integrity proof (represented as a Merkle root). Upon task completion, the node submits these compact Merkle root proofs to the Master while retaining local copies for redundancy. This design ensures both the autonomous operation of the node and its effective participation in the verifiable computation framework through the Merkle root's hierarchical authentication mechanism.

4 Task Scheduling

4.1 Slot-Epoch Model

Before introducing the scheduling algorithm of TrustSched, we first define the core abstractions underlying the system design. In TrustSched, any user-initiated synthetic data generation request is defined as a *Task*, which specifies the generative model to use, the target data volume, and other quality parameters. In practical deployments, multiple Tasks from different users may arrive concurrently.

Each Task is conceptually decomposed into a collection of *Slots* to enable fine-grained parallelism. Let T denote the target result set of a Task, and let n represent the requested size specified by the user. The system ensures that $|T| \geq n$. A Slot is defined as a set of generated samples, formally denoted as

$$S_i = \{s_{i_1}, s_{i_2}, \ldots, s_{i_m}\}.$$

If the union of a set of Slots $S_1, \ldots, S_p$ satisfies the requirement

$$\left| \bigcup_{j=1}^{p} S_j \right| \geq n,$$

then these Slots together constitute a valid completion of the Task.

Within TrustSched, each Slot transitions through a sequence of states during execution:

- **Committed:** The Slot has been generated by a node and the preliminary result has been submitted to the system.
- **Finalized:** The Slot has undergone blockchain-based attestation, establishing an immutable record of the result.

This Slot-based abstraction allows each Task to be represented as the union of multiple Slots generated over different scheduling rounds. Consequently, it becomes possible to reconstruct the *execution trajectory* of a Task, providing a fine-grained audit trail for synthetic data generation. Importantly, this trajectory is inherently *time-dependent*: a Task may complete certain Slots during an initial time window t_1 and additional Slots during a subsequent window t_2. Only by aggregating the contributions across all relevant time windows can the complete Task result be obtained. In realistic settings, the system typically handles multiple concurrent Tasks. That is, within a given time window, Slots corresponding to different Tasks (e.g., T_1 and T_2) may simultaneously reach completion stages. To systematically capture the temporal structure of slot generation and scheduling across Tasks, TrustSched introduces the notion of an *Epoch*. Following the Epoch model, progress is synchronized at fixed intervals, during which committed Slots are batch transitioned to finalized status, and unfinished work is rescheduled accordingly. This design ensures consistent state progression while facilitating scalable and traceable task execution across distributed nodes.

Figure 2 illustrates the scheduling process: During Epoch k, a user submits Task m. The task is initially divided into three slots allocated to three nodes. By the end of this epoch, two slots complete synthesis and report to the master node, while one remains in progress. In Epoch $k+1$, the committed slots transition to the finalized state. Persistent uncommitted slots trigger a second scheduling round, where the remaining workload is re-partitioned into three new slots assigned to available nodes. When partial progress persists into Epoch $k+2$, a third scheduling iteration begins. Concurrently, the framework supports multi-task coordination, as demonstrated by the introduction of a new Task $m+1$ during Epoch $k+2$.

4.2 Scheduling Algorithm

The workflow of Algorithm 1 operates through phased epoch transitions anchored by blockchain verification. Initialization (line 1) establishes the remaining data volume $R = R_0$, initial epoch counter, and empty pending slot set $\mathcal{P}$.

Algorithm 1. Task Dynamic Scheduling

Require: R_0: Initial data volume, $L_{\max}$: Max load per node
1: $R \leftarrow R_0$, $epoch \leftarrow 1$, $\mathcal{P} \leftarrow \emptyset$
2: **while** $R > 0$ **do**
3:     **if** $\mathcal{P} \neq \emptyset$ **then**
4:         $\mathcal{F} \leftarrow \text{FinalizeSlots}(\mathcal{P})$
5:         $\text{AppendToChain}(\mathcal{F}, epoch - 1)$
6:         $R \leftarrow R - |\mathcal{F}|$
7:         $\mathcal{P} \leftarrow \mathcal{P} \setminus \mathcal{F}$
8:     **end if**
9:     $N \leftarrow \text{GetAvailableNodes}()$
10:     **if** $N = \emptyset$ **then**
11:         $\mathcal{P} \leftarrow \emptyset$
12:         **wait** next epoch
13:         $epoch \leftarrow epoch + 1$
14:         continue
15:     **end if**
16:     Allocate A using $\min(\lfloor R/|N| \rfloor, L_{\max})$ strategy
17:     $\mathcal{C} \leftarrow \text{ParallelExecute}(A)$
18:     $\mathcal{P} \leftarrow \mathcal{C}$
19:     $epoch \leftarrow epoch + 1$
20: **end while**

The main loop (line 2) iterates until complete task fulfillment ($R \leq 0$), implementing a two-phase progression per epoch. At each epoch boundary, the system first processes *previously committed slots* through blockchain finalization (lines 3-7). The FinalizeSlots function verifies the validity of pending slots in $\mathcal{P}$ and transitions them to the finalized state. Then the finalized slot set $\mathcal{F}$ is then recorded on the blockchain (line 5) along with the integrity proofs. This enforces a mandatory one-epoch delay between slot commitment and blockchain finalization, where slots generated in epoch t attain immutable status only during epoch $t+1$. The remaining workload R updates strictly based on blockchain-confirmed results (line 6), ensuring consensus-aligned progress tracking.

Node availability checks (lines 10-14) provide fault tolerance by resetting $\mathcal{P}$ and pausing scheduling when no nodes are available. Upon node availability, the allocation strategy (line 16) distributes workload using:

$$A[i] = \begin{cases} \min\left(\left\lfloor \frac{R}{|N|} \right\rfloor + (R \mod |N|),\ L_{\max}\right) & \text{for } i = 1 \\ \min\left(\left\lfloor \frac{R}{|N|} \right\rfloor,\ L_{\max}\right) & \text{for } i > 1 \end{cases}$$

This formulation guarantees $\sum A \leq \min(R, |N| \times L_{\max})$ while maintaining load imbalance ≤ 1 unit between nodes. Newly generated slots from ParallelExecute (line 17) enter the *Committed* state, buffered in $\mathcal{P}$ (line 18) for next-epoch finalization. The epoch counter increments (line 19) to complete the cycle, establishing temporal isolation between slot generation and finalization phases.

5 On-Chain Evidence Archiving

5.1 MHT-Based Data Model

Based on the definitions of Task and Slot introduced in Sect. 4.1, we further formalize the notion of data integrity within TrustSched.

Each *Slot* S_i consists of a set of generated data elements: $S_i = \{s_{i_1}, s_{i_2}, \ldots, s_{i_m}\}$. To verify the completeness and integrity of a Slot, these elements are organized into a Merkle tree. Let $\mathsf{Merkle}(S_i)$ denote the Merkle root computed over all elements in S_i:

$$r_i = \mathsf{Merkle}(S_i),$$

where r_i is referred to as the *Slot root.*

Building upon this structure, a *Task* T is composed of a collection of Slots $\{S_1, S_2, \ldots, S_p\}$. The integrity of the entire Task is ensured by constructing a second-level Merkle tree over the Slot roots. Let

$$\mathsf{Roots}(T) = \{r_1, r_2, \ldots, r_p\}$$

denote the set of Slot roots, then the *Task root* r_T is defined as:

$$r_T = \mathsf{Merkle}(\mathsf{Roots}(T)).$$

Additionally, given a Task T, composed of Slot roots $\{r_1, r_2, \ldots, r_p\}$, and a corresponding Task root r_T, the integrity of T is verified by the Merkle Proof over the Slot roots. The verification function is defined as:

$$\mathsf{VerifyTask}(\{r_1, r_2, \ldots, r_p\}, r_T) = \begin{cases} 1, & \text{if } r_T = \mathsf{Merkle}(\{r_1, r_2, \ldots, r_p\}); \\ 0, & \text{otherwise.} \end{cases}$$

Through this hierarchical Merkle tree construction, TrustSched ensures that data integrity can be efficiently and securely verified. Moreover, by leveraging Merkle proofs, TrustSched enables fine-grained verification: the integrity of individual data elements within a Slot or specific Slots within a Task can be independently validated without reconstructing the entire dataset. In TrustSched, the collection of roots—including both the Slot roots and the Task root—is regarded as the *Evidence* of the Task, serving as a compact and verifiable commitment to the integrity of its associated data. This design allows partial data corruption or loss, either within Slots or across Slots, to be precisely detected and localized through corresponding Merkle proof verification.

5.2 On-Chain Evidence Submission

To ensure traceability while minimizing on-chain overhead, TrustSched adopts an asynchronous evidence submission mechanism. Rather than uploading each piece of evidence immediately after its generation, evidences—such as Slot roots or Task roots—are temporarily accumulated in a local evidence pool. Submission

Algorithm 2. Asynchronous On-chain Evidence Submission

1: Initialize evidence buffer $\mathcal{E} \leftarrow \emptyset$
2: **upon** receiving a new evidence e
3: Append e to $\mathcal{E}$
4: **while** true **do**
5: **if** $|\mathcal{E}| \geq \mathsf{BATCH\_SIZE}$ **or** $t_{elapsed} \geq \mathsf{TIME\_LIMIT}$ **then**
6: Package $\mathcal{E}$ into a batch transaction
7: Submit the batch to the blockchain
8: Reset $\mathcal{E} \leftarrow \emptyset$
9: Update the timestamp of last submission
10: **end if**
11: **end while**

to the blockchain is then performed in a batched manner, triggered either by reaching a threshold in batch size or by the expiration of a timeout.

Upon receiving a new evidence item, the system appends it to the evidence buffer $\mathcal{E}$. Meanwhile, an asynchronous submission process runs in the background, continuously monitoring the buffer status. Whenever either the number of buffered evidences reaches a predefined batch size threshold $\mathsf{BATCH\_SIZE}$ or the elapsed time $t_{elapsed}$ since the last submission exceeds a timeout threshold $\mathsf{TIME\_LIMIT}$, the buffered evidences are packaged and submitted to the blockchain. The detailed procedure is summarized in Algorithm 2.

This design effectively decouples evidence collection from on-chain submission, significantly reducing per-evidence transaction overhead and improving overall blockchain throughput by amortizing submission costs across multiple evidences. Furthermore, by bounding the maximum delay before evidence submission, TrustSched maintains a timely commitment to the blockchain, achieving a balance between auditability, system responsiveness, and resource efficiency.

5.3 Integrity Verification of Synthetic Data

Upon receiving the synthetic data associated with a task T, the user verifies its integrity through a two-stage Merkle proof validation process:

- **Slot-level verification:** For each received slot S_i, the user performs the following steps:
  1. Compute the local Merkle root:
  $$r'_i \leftarrow \mathsf{Merkle}(S_i)$$
  2. Retrieve the corresponding on-chain Merkle root:
  $$r_i^{\text{chain}} \leftarrow \text{BlockchainLookup}(S_i)$$
  3. Compare the local root with the on-chain root:
  $$r'_i \stackrel{?}{=} r_i^{\text{chain}}$$

If the equality check fails, the slot S_i is flagged as corrupted. This enables fine-grained detection and isolation of faulty data segments without affecting the validation of other slots.

- **Task-level verification:** After individually verifying all slots, the user verifies the overall task structure as follows:
  1. Aggregate the locally computed slot roots to reconstruct the task-level Merkle root:
  $$r'_T \leftarrow \mathsf{Merkle}(\{r'_1, r'_2, \ldots, r'_p\})$$
  2. Retrieve the blockchain-anchored task root:
  $$r_T^{\text{chain}} \leftarrow \text{BlockchainLookup}(T)$$
  3. Compare the reconstructed task root with the on-chain task root:
  $$r'_T \stackrel{?}{=} r_T^{\text{chain}}$$
  Successful verification confirms that all received slots are correctly assembled and that the global task integrity is maintained.

This hierarchical verification ensures both localized error detection at the slot level and comprehensive consistency verification at the task level, providing strong guarantees of data authenticity with efficient fault recovery capabilities.

6 Evaluation

The experiments were conducted on a server equipped with an Intel Xeon Gold-6330 2.00 GHz CPU (56 cores, 112 threads), 500 GB RAM, and an NVIDIA RTX A6000 GPU (48 GB memory), running Ubuntu 20.04.2 LTS. A 4-node blockchain network was deployed using FISCO-BCOS [9] for performance evaluation.

We conducted a series of experiments to evaluate the performance of our phased scheduling system for tabular data synthesis, using CTGAN [5] as the baseline model. In the experimental setup, we configured 5 synthetic virtual nodes and set the maximum slot capacity to 1,000 rows per scheduling cycle. Figure 3a demonstrates the relationship between synthetic task data size and execution time, revealing that execution times remain stable within the same scheduling phase despite varying data volumes. This validates the effectiveness of our static resource allocation strategy in maintaining load balancing across nodes for synthetic tasks within moderate data ranges. Furthermore, Fig. 3b illustrates a linear correlation between data volume and the slot collection-aggregation time, confirming the system's stability in data processing and metadata management. To investigate scalability, Fig. 3c shows the impact of parallel node count on system efficiency with a fixed task size of 5,000 rows. The results indicate optimal efficiency when using 5–10 parallel nodes, as this range balances synthesis acceleration through reduced scheduling iterations against the overhead from increased blockchain metadata recording. Notably, excessive nodes amplify slot quantity and associated metadata, partially offsetting performance gains. These findings highlight the importance of carefully configuring both the maximum slot capacity per scheduling cycle and the number of parallel nodes to optimize the trade-off between computational efficiency and blockchain overhead in practical deployments.

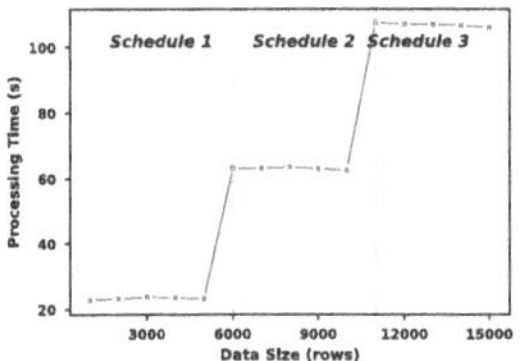

(a) Relationship between data size and processing time.

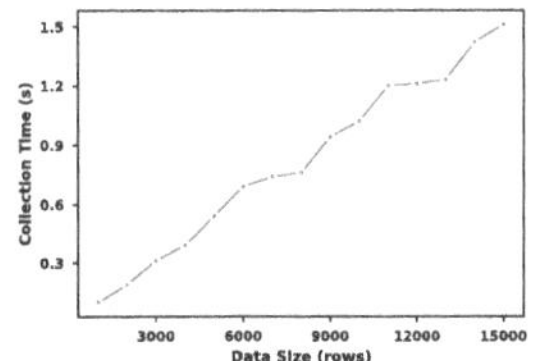

(b) Relationship between data size and collection time.

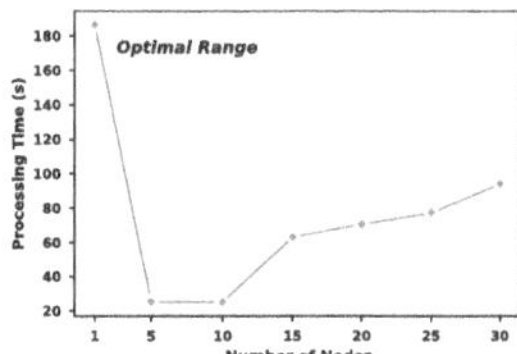

(c) Relationship between node number and processing time.

Fig. 3. Performance evaluation of scheduling.

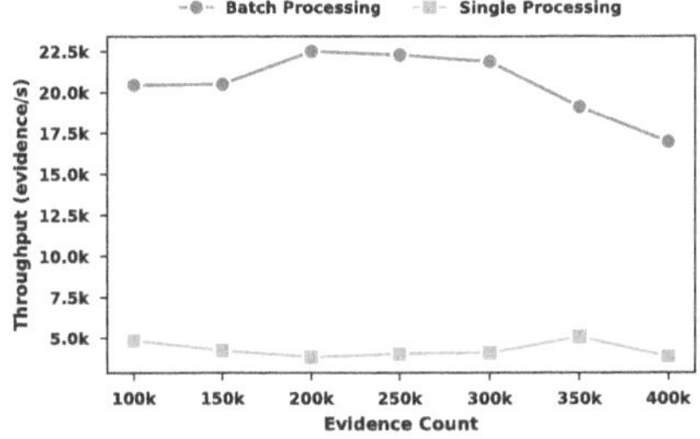

(a) Throughput comparison between batch and single processing.

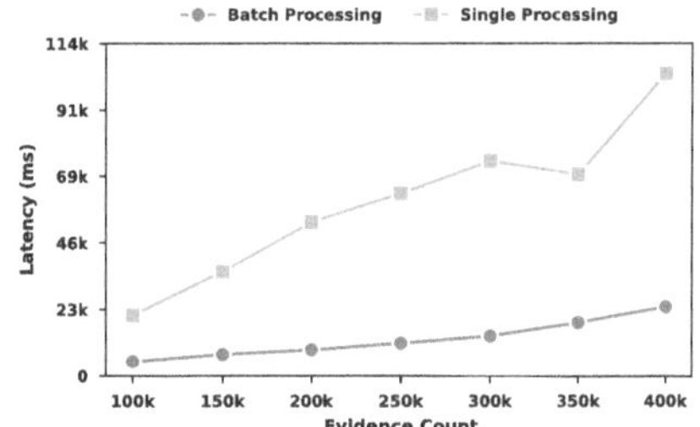

(b) Latency comparison between batch and single processing.

Fig. 4. Performance evaluation of batch processing vs single processing.

The experimental results in Fig. 4 demonstrate that our asynchronous batch processing method significantly outperforms synchronous single processing in both throughput and latency. The batch approach achieves up to 4.4× higher throughput and 4.2× lower latency, with the performance advantage becoming more pronounced as evidence volume increases. These results validate the effectiveness of our design for large-scale blockchain evidence storage.

7 Conclusion

This paper presents TrustSched, a distributed scheduling framework for trusted synthetic data generation, which harmonizes computational efficiency, verifiable integrity, and scalable task execution. By integrating hierarchical Merkle proof validation with dynamic slot-epoch scheduling, the framework achieves robust privacy-utility trade-offs while ensuring tamper-evident traceability through blockchain-anchored evidence. The Master-Executor architecture, coupled with asynchronous batch processing, minimizes on-chain overhead and maintains load balancing across distributed nodes. Experimental results demonstrate the system's ability to adaptively manage concurrent tasks, mitigate node failures, and optimize resource utilization. The hierarchical verification mechanism enables granular integrity checks at both slot and task levels, fostering trust in decentralized synthetic data workflows. Future work will explore adaptive scheduling policies for dynamic load variations and further enhancements to support cross-institutional collaboration at scale.

Acknowledgements. This work was supported by National Key Research and Development Program of China (No.2023YFC3304700), Shanghai "Science and Technology Innovation Action Plan" Project (No.24BC3201100) and Program of Shanghai Academic/Technology Research Leader (No.23XD1401100).

References

1. Giuffrè, M., Shung, D.L.: Harnessing the power of synthetic data in healthcare: innovation, application, and privacy. NPJ Digit. Med. **6**, 186 (2023)
2. Assefa, S.A., Dervovic, D., Mahfouz, M., Tillman, R.E., Reddy, P., Veloso, M.: Generating synthetic data in finance: opportunities, challenges and pitfalls. In: Proceedings of the First ACM International Conference on AI in Finance, pp. 1-8. (2020)
3. Chen, J., Meng, D., Gao, X., Zhang, L., Kong, C.: AGNE: Attentional Graph Convolutional Network Embedding for Knowledge Concept Recommendation in MOOCs. In: WISA 2024, pp. 463-475. Springer, (2024)
4. Zaharia, M., et al.: Resilient distributed datasets: a fault-tolerant abstraction for in-memory cluster computing. In: 9th USENIX symposium on networked systems de-sign and implementation (NSDI 12), pp. 15-28 (2012)
5. Xu, L., Skoularidou, M., Cuesta-Infante, A., Veeramachaneni, K.: Modeling tabular data using conditional GAN. Adv. Neural Inf. Process. Syst. **32** (2019)
6. Kotelnikov, A., Baranchuk, D., Rubachev, I., Babenko, A.: TabDDPM: Modelling tabular data with diffusion models. In: International Conference on Machine Learning, pp. 17564–17579. PMLR, (2023)
7. Shankar, A., Brouwer, H., Hai, R., Chen, L.: SiloFuse: Cross-silo synthetic data generation with latent tabular diffusion models. In: 2024 IEEE 40th International Conference on Data Engineering (ICDE), pp. 110–123. IEEE (2024)
8. Castro, M., Liskov, B.: Practical byzantine fault tolerance. In: OsDI, pp. 173–186 (1999)
9. Li, H., et al.: Fisco-bcos: An enterprise-grade permissioned blockchain system with high-performance. In: Proceedings of the International Conference for High Performance Computing, Net-working, Storage and Analysis, pp. 1–17 (2023)
10. Becker, G.: Merkle signature schemes, Merkle trees and their cryptanalysis. Ruhr-University Bochum, Technical report 12, 19 (2008)

MP-PBFT: PBFT Consensus Algorithm Based on Multidimensional Reputation Score and Probabilistic Election

Jiahao Wang[1], Sanyuan Wang[1], Xin Shi[1], Xueqing Zhao[1](✉), Yun Wang[2], and Guigang Zhang[2]

[1] School of Computer Science, Xi'an Polytechnic University, Shaanxi Key Laboratory of Clothing Intelligence, Xi'an, China
{240721126,230721113}@stu.xpu.edu.cn, {shixin,zhaoxueqing}@xpu.edu.cn
[2] Institute of Automation, Chinese Academy of Sciences, Beijing, China
{y.wang,guigang.zhang}@ia.ac.cn

Abstract. To address the issues of high communication overhead, unfair node election, and insufficient security in Byzantine Fault Tolerance (BFT)-based consensus algorithms in blockchain systems, this paper proposes a multidimensional reputation score and probabilistic election-based PBFT algorithm, termed MP-PBFT. A node reputation scoring mechanism is designed, evaluating historical behavior, participation level, and honesty, and employing a progressive reward-and-punishment strategy for dynamic scoring. This mechanism allows high-reputation nodes to be adaptively and dynamically selected for key roles, improving consensus efficiency while maintaining security. Furthermore, to overcome centralization tendencies of highest-score selection, a probabilistic election mechanism is introduced for critical nodes. By randomly selecting the advantages of high-reputation nodes and the participation opportunities of low-reputation nodes, honest behavior is encouraged and fairness is improved. Simulation experiments validate MP-PBFT's performance in communication efficiency, election fairness, and security. Compared to traditional PBFT, MP-PBFT reduces communication overhead by 65% while ensuring fairness and robustness. The proposed approach demonstrates significant advantages in resisting collusive attacks and improving the network's fault tolerance through the integration of multidimensional reputation scoring and probabilistic elections.

Keywords: Blockchain · PBFT · Consensus Mechanism · Credit Score · Probabilistic Elections

1 Introduction

Blockchain is a decentralized distributed ledger [1], initially applied to digital currencies such as Bitcoin. In a blockchain, each block contains a set of transaction records, which are processed using cryptographic hash algorithms [2] and

S. Yang et al. (Eds.): WISA 2025, LNCS 16299, pp. 625–637, 2026.
https://doi.org/10.1007/978-981-95-6426-2_55

linked to the hash value of the previous block, forming a chain-like structure. Each node maintains a complete copy of the ledger, and consensus mechanisms are used to ensure agreement on valid transactions and complete data. The application areas of blockchain continue to expand, from digital currencies [3] to smart contracts [4] , with its technology now permeating numerous industries.

Currently, three mainstream architectures have emerged, public blockchains [5], consortium blockchains [6], and private blockchains [7]. These three types of blockchains each have unique characteristics in terms of application scenarios, forming diverse landscape of blockchain technology. Various consensus algorithms have emerged, with common ones including Proof of Work (PoW) [8], Proof of Stake (PoS), and Practical Byzantine Fault Tolerance (PBFT) [9]. PoW is the algorithm commonly used in public blockchains like Bitcoin, where nodes compete for the right to record transactions by solving complex mathematical problems. While PoW offers high security, it is energy-intensive and inefficient. PoS determines which node packages the block based on the number of tokens held by the node. Compared to PoW, PoS is more energy-efficient. However, PoS is prone to power centralization, potentially leading to the "rich get richer" issue. The PBFT consensus algorithm can achieve consensus even if up to one-third of nodes are malicious, demonstrating strong fault tolerance. However, its static node evaluation mechanism lacks incentives for honest behavior, leading to resource waste and efficiency limitations, master nodes pose single-point-of-failure risks, and view switching causes high latency that impacts system stability, O(n) communication overhead limits its scalability in large-scale networks.

To address this, this paper proposes a PBFT consensus algorithm based on multi-dimensional reputation scoring and probabilistic election, referred to as MP-PBFT (Multi-Dimensional Reputation and Probabilistic Election-Based Practical Byzantine Fault Tolerance Algorithm). This algorithm aims to resolve inherent issues in PBFT algorithms, such as insufficient node dynamism, inefficient view switching, and excessive communication overhead. The innovative features of the proposed MP-PBFT include,

1) This paper proposes a multi-dimensional node reputation scoring mechanism that comprehensively considers the historical behavior, participation, and honesty of nodes. Through a gradual reward and punishment strategy, the reputation value of nodes is dynamically adjusted to avoid the problems of node homogenization and staticization in the PBFT algorithm.

2) This paper designs a probability-based node election mechanism. This mechanism balances the advantages of high-reputation nodes with the participation opportunities of low-reputation nodes through probability-based elections, thereby enhancing network fairness and decentralization.

3) Introduced a group threshold mechanism, dividing nodes into multiple groups and setting independent thresholds for each group. Within each group, consensus is reached through threshold signatures, reducing the numbers of communication within the group and thereby improving consensus efficiency.

2 Problem Description

The PBFT algorithm is a consensus mechanism that ensures consistency in distributed systems even in the presence of Byzantine nodes. It achieves consensus through three phases of message exchange, the preparation phase, the readiness phase, and the commitment phase. An operation must be approved by more than two-thirds of the nodes to be executed, ensuring that the system remains consistent even if one-third of the nodes fail or act maliciously. Although PBFT has strong fault tolerance capabilities, its communication complexity is O(nÂš), and when the number of nodes is large, communication and computational overhead increase significantly. Therefore, PBFT is more suitable for scenarios with a small number of nodes and high security requirements, such as consortium blockchains and private blockchains.

To address issues in the PBFT consensus algorithm, such as communication overhead, node activity, and reward and punishment mechanisms, existing optimization methods primarily focus on the following areas. Firstly, Sheng Gao et al. [10] proposed a reputation assessment mechanism based on the EigenTrust model. By calculating reputation scores based on a node's historical transactions, high-reputation nodes are selected to form a consensus group, optimizing a single master node into a master node group. Group signatures and mutual supervision are utilized to reduce the probability of view changes, thereby enhancing fault tolerance while effectively reducing communication overhead. Secondly, Yunhua He et al. [11] designed a blockchain truthful incentive mechanism for distributed P2P applications. By rewarding nodes that actively participate in consensus and penalizing nodes that passively participate or attempt to act maliciously, addressing the pain point of "low node activity" in PBFT. Furthermore, Guangxia Xu et al. [12]proposed combining PBFT with an optimized DPoS mechanism, using a share voting mechanism to elect consensus nodes, significantly reducing communication complexity and improving consensus efficiency. In addition, Houshyar Honar Pajooh et al. [13] proposed a network self-clustering technique suitable for IoT scenarios. High-reliability seed nodes are first selected as clustering centers to divide large-scale IoT [14] node networks into multiple independent subsets. Agent nodes are then screened from each subset, with only agent nodes participating in the consensus process. This significantly reduces cross-node communication interaction volume and effectively addresses the scalability challenges of PBFT in multi-node scenarios.

Improvements to the PBFT consensus algorithm primarily focus on reducing communication overhead, enhancing efficiency, and strengthening system security. By implicitly reducing the number of consensus nodes through a grouping mechanism, communication complexity can be significantly reduced. In the selection of critical nodes, a probability-based election mechanism weighted by reputation is introduced to replace the fixed master node selection method, enabling the system to avoid the deterministic allocation of critical nodes, thereby reducing the likelihood of prediction and attack, and effectively mitigating single-point failure risks. By introducing a reputation scoring mechanism that dynamically adjusts a node's reputation value based on its behavioral performance, the system incentivizes honest participation and limits the influence of malicious nodes, thereby improving consensus efficiency and system security.

3 MP-PBFT Algorithm Design

3.1 Summary

MP-PBFT is an improved Byzantine fault-tolerant consensus algorithm that optimizes the PBFT algorithm to address issues such as high communication overhead, insufficient node dynamism, and low view switching efficiency in applications. As shown in Fig. 1, the algorithm establishes a multi-dimensional node reputation scoring mechanism that comprehensively considers key indicators such as historical consensus performance, node participation, and honesty to achieve a comprehensive assessment of node behavior. The algorithm designs a dynamic adjustment strategy based on node behavior, dynamically adjusting node reputation values through a progressive reward and punishment mechanism. Good behavior is rewarded exponentially, while malicious behavior faces progressively increasing penalties, thereby achieving rapid removal of malicious nodes. Additionally, the MP-PBFT algorithm introduces a probability-based node election mechanism, ensuring that high-reputation nodes have a higher election probability while providing low-reputation nodes with moderate opportunities for consensus participation. To optimize network communication efficiency, the algorithm employs a grouping strategy, dividing the network into multiple consensus groups and processing transaction messages through a combination of intra-group and inter-group consensus.

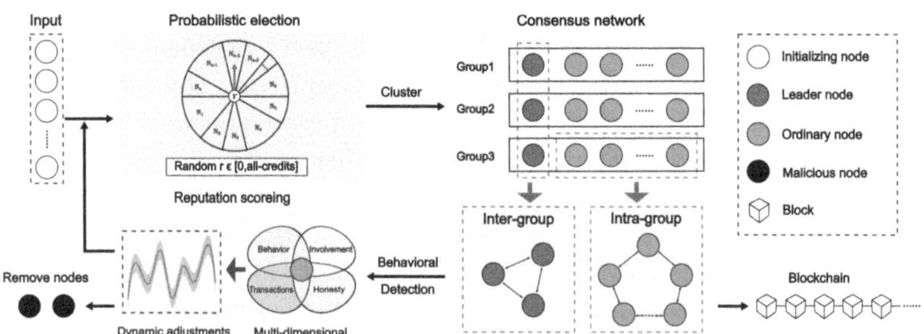

Fig. 1. MP-PBFT Architecture Diagram.

3.2 Multi Dimensional Node Reputation Scoring Mechanism

Although the PBFT algorithm has certain advantages in terms of fault tolerance, the lack of a node reward and punishment mechanism makes it difficult to fully stimulate node participation, thereby affecting consensus efficiency. To address this issue, the MP-PBFT algorithm has designed a multi-dimensional node reputation scoring mechanism. This mechanism consists of a multi-dimensional scoring system and a dynamic adjustment strategy. The former comprehensively evaluates node status by integrating four dimensional indicators (historical behavior, participation, honesty, and transaction-related indicators). The latter implements rewards and penalties based on nodes' consensus performance, thereby fostering a healthy network environment that rewards good behavior and penalizes bad behavior.

Multi-dimensional Scoring System. The PBFT algorithm struggles to comprehensively assess node reliability. To address this, the MP-PBFT algorithm introduces a multi-dimensional reputation scoring mechanism that combines four core dimensions, historical behavior, participation, honesty, and transaction-related factors. Specifically, this includes, 1) Historical behavior, which evaluates a node's behavior record during historical consensus cycles, such as voting completion rate, signature validity, and abnormal behavior. Exponential penalties will be imposed for malicious behaviors such as forged messages or double signatures. 2) Participation, quantifying a node's activity level in the consensus process, including participation frequency, response time, and online duration. Nodes with high participation typically respond more quickly, thereby accelerating reputation growth. 3) Integrity, measuring whether a node adheres to consensus rules and whether it engages in malicious behaviors such as message tampering or collusion attacks. Reputation points will be dynamically deducted based on the severity of violations, and nodes with no violation records will receive progressive rewards. 4) Transaction-related, Reflects a node's performance in transaction processing, including transaction volume, verification accuracy, and response to abnormal transactions. Nodes that successfully process multiple transactions will receive reputation boosts, and nodes that can quickly detect and report abnormal behavior will also receive additional rewards.

Dynamic Adjustment Strategies. MP-PBFT employs a progressive reward and punishment strategy. If a node consistently achieves consensus over multiple rounds, its reputation value will grow exponentially. Conversely, if a node exhibits malicious behavior continuously, it will be penalized, and its reputation value will decrease arithmetically. Simultaneously, the reputation values of all nodes are dynamically adjusted. The cumulative reputation value of a node is given by formula 1:

$$R_t^i = R_{t-1}^i + (a_t^i \times c_t^i) \tag{1}$$

In this context, i denotes a node, and t represents the consensus round. R_t^i indicates the reputation value of node i at the t-th round, while R_{t-1}^i denotes the reputation value of node i at the t-1-th round, c_t^i represents the number of consecutive executions of the same type of behavior by node i up to the t-th round, starting with an initial value of 0. The update rule for the number of consecutive behaviors is shown in formula 2:

$$c_t^i = \begin{cases} c_{t-1}^i + 1 & \text{if } t \geq 1 \text{ and } a_t^i - a_{t-1}^i \\ 1 & \text{else} \end{cases} \tag{2}$$

Among them, $a_t^i \in \{+1, -1\}$ indicates the type of behavior exhibited by node i in round t (+1 for honest, −1 for malicious), while c_{t-1}^i denotes the number of consecutive rounds prior to t-1 during which node i engaged in the same type of behavior. The behavior of nodes is regularly evaluated, and their reputation scores are updated to ensure long-term participation and honesty. This approach prevents nodes from receiving excessively high scores based on short-term good behavior, thereby stopping malicious nodes from gaining unfairly high reputation values through temporary performance.

3.3 Probability-Based Node Election Mechanism

The PBFT algorithm uses a polling method to replace the primary node, lacking dynamic adaptability and ignoring node performance and trustworthiness. This may result in inefficient or potentially malicious nodes periodically assuming the role of leader, thereby impacting consensus efficiency. To address this, MP-PBFT designs a probability-based node election mechanism, where all nodes initially have a credibility value of 100, which continuously changes during the consensus process. Each group elects a leader through a probability-based election mechanism, where the election probability is proportional to the node's credibility value. A random factor r is introduced to enhance election fairness and prevent nodes from occupying critical roles for extended periods. The system re-elects the group leader every 5 rounds, and the master node is randomly selected from the group leaders every 10 rounds. The higher a node's reputation value, the greater its probability of being selected as a critical node. Assuming there are n nodes in the system, the node set is$\{n_1, n_2, \ldots, n_n\}$. Each node n_i is associated with a reputation value$w_i \geq 0$. The total sum of all nodes' reputation values is calculated as $\sum_{i=1}^{n} w_i$. Under this mechanism, the probability P_i of node i being selected is given by formula 3:

$$P_i = \frac{w_i}{\sum_{i=1}^{n} w_i} \tag{3}$$

As shown in Table 1, the algorithm utilizes a reputation value-weighted probabilistic random selection mechanism to balance the fairness and efficiency of node election. First, the total sum of all nodes' reputation values, denoted as W, is calculated. Then, a uniformly random number r is generated within the interval $[0, W]$. By linearly traversing the node set and accumulating each node's reputation value, the node corresponding to the first cumulative value that exceeds r is selected as the key node. If no node is selected due to floating-point errors, the last node is chosen as a fault tolerance measure. The algorithm ensures that the probability of a node being elected is strictly proportional to its reputation value, $P_i = \frac{w_i}{W}$.

3.4 Node Grouping Mechanism

To improve the consensus efficiency of the MP-PBFT algorithm in distributed systems, grouping mechanism has been introduced. All nodes are divided into

Table 1. Key node election algorithm.

Algorithm 1: Key node election algorithm

Input: node set nodes = $\{n_1, n_2, \ldots, n_n\}$, where each node n_i has a reputation value w_i stored in array`[i][j]` // Node set and their reputation values

Output: selected key node n_i // The final selected key node

1: initialize total reputation value $W \leftarrow 0$
2: **for each** node reputation value w_i in array`[i][j]` **do**
3: $W \leftarrow W + w_i$ // Calculate the total reputation of all nodes
4: **end for**
5: $r \leftarrow$ random number in the range $[0, W]$ // Generate a random number r
6: initialize cumulative weight $C \leftarrow 0$
7: **for each** node n_i and reputation w_i in array`[i][j]` **do**
8: $C \leftarrow C + w_i$ // Gradually accumulate reputation values
9: **if** $r \leq C$ **then** // Select node if number falls in its reputation interval
10: **return** node n_i as the key node
11: **end if**
12: **end for**
13: **return** nodes$[n]$

k groups based on their reputation values, with different nodes in each group assuming different roles. The allocation method uses integer division and modulo operations to ensure fairness in grouping. Nodes within a group conduct independent message consensus within the group to ensure intra-group consensus, while the group leader node is responsible for collecting intra-group consensus results and coordinating with other group leader nodes.

In the grouping mechanism, MP-PBFT uses threshold signatures to achieve consensus within the group, with the threshold set to a majority of nodes agreeing. The group leader is responsible for collecting the consensus results within the group. If consensus cannot be reached within a group, it is invalid, ensuring that consensus can be reached quickly within the group with fewer participating nodes. Only when a majority of groups agree is the consensus considered to have been reached. This grouping mechanism eliminates the need for communication between all nodes, instead relying on intra-group communication and leader coordination, significantly reducing communication overhead.

3.5 MP-PBFT Algorithm Consensus Process

The consensus process of the MP-PBFT algorithm is carried out through 9 stages. Each stage is designed based on message transmission and consensus protocols among nodes. Figure 2 illustrates the message transmission during the consensus process, detailing each stage's message flow. The consensus process is as follows, 1) Request phase, The client sends transaction information to the master node. After verifying its validity, the master node adds it to the transaction

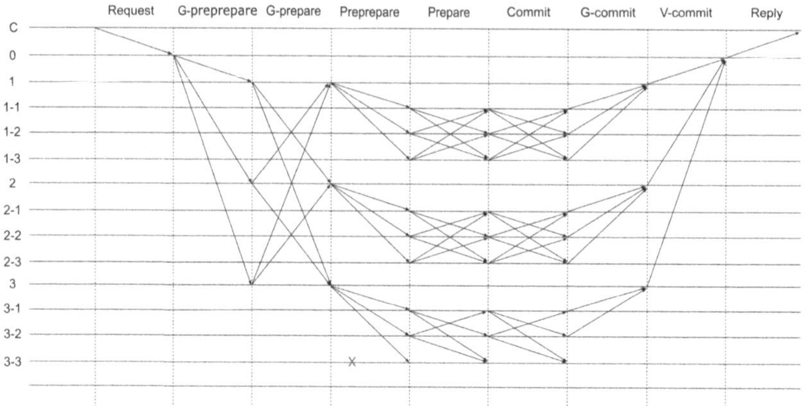

Fig. 2. MP-PBFT consensus process.

queue. 2) G-preprepare phase, The master node broadcasts the <G-preprepare, m, v, t_1, S_i > message to all group leader nodes. Upon receiving it, the group leaders transition to the pre-prepare state, where m represents transaction information, v is the master node, t_1 is the timestamp, and S_i is the signature. 3) G-prepare phase, Each leader node broadcasts the <G-prepare, m, v,t_2, S_i> message to other leader nodes. Upon receiving it, they verify consistency. If successful, they transition to the prepare state. 4) Preprepare phase, Leader nodes broadcast the <Preprepare, m, v,t_3, S_i > message to group members. After verifying validity, members transition to the preprepare state. 5) Prepare phase, Members within the group broadcast the <Prepare, m, v, t_4, S_i> message to each other. After verification, they transition to the prepare state until sufficient signatures are received. 6) Commit phase, Members continue to broadcast the <Commit, m, v, t_5, S_i > message. After receiving sufficient commit messages, they transition to the confirm state, indicating that consensus has been reached within the group. 7) G-commit phase, Members broadcast <G-commit, m, v, t_6, S_i> message to the group leader. After receiving sufficient commit messages from the group, the group leader transitions to the confirmed state. 8) V-commit phase, The group leader broadcasts the <V-commit, m, v, t_7, S_i > message to the main node. After receiving sufficient confirmation messages from the group leader, the main node transitions to the final confirmed state. 9) Reply phase, The master node sends the <V-commit, m, v, t_8, S_i> message to the client, returning the consensus result.

4 Experiments and Analysis

4.1 Node Reputation Values and Consensus Rounds

In the multi-dimensional node reputation scoring mechanism, the system adopts a progressive reward and punishment strategy. Honest nodes earn points each

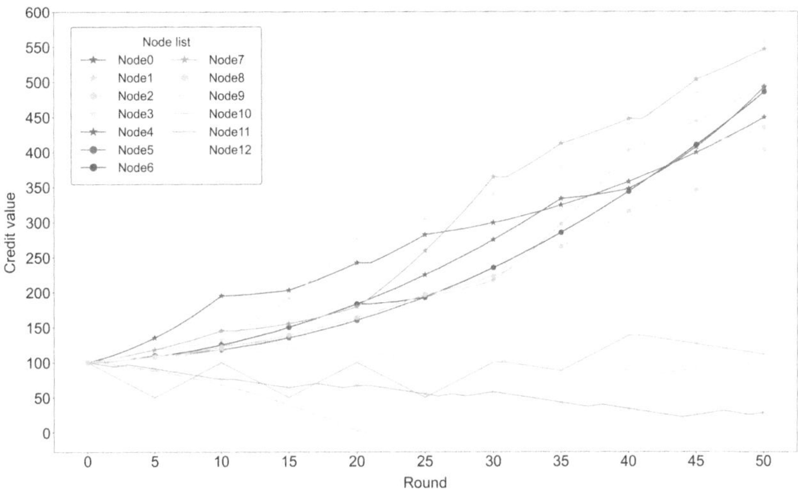

Fig. 3. Node Reputation Values and Consensus Rounds.

time they successfully reach consensus, and enjoy additional weighted rewards when serving as master nodes. Malicious nodes are penalized each time they violate the rules, and repeated violations trigger exponential point deductions.

Figure 3 shows the changes in reputation scores of 13 nodes after 50 rounds of consensus. As the number of consensus rounds increases, honest nodes see their reputation scores rise due to adhering to consensus rules and responding promptly. Among these, honest nodes that have served as master nodes have higher reputation scores than other nodes. Conversely, malicious nodes experience a continuous decline in their reputation scores over time. Node 9 in the figure exhibits persistent malicious behavior, with its reputation score declining at the fastest rate, dropping to 0 by the 21st round and being removed from the consensus network. This dynamic reputation scoring mechanism effectively punishes malicious behavior, ensuring the reliability of the network.

4.2 Communication Overhead Analysis

Communication overhead refers to the communication volume generated between nodes during the network consensus process multiplied by the size of the information. Here, the size of the information is set to a fixed value for ease of calculation and comparative analysis. The MP-PBFT consensus algorithm achieves staged consensus by reducing the number of consensus nodes and optimizes the consistency protocol of the PBFT algorithm. Since the communication process during the request phase is identical between the MP-PBFT optimized algorithm and the PBFT consensus algorithm, to simplify parameter settings and highlight core differences, the comparison will exclude communication overhead during the request phase. Assuming that the total number of nodes is n and the number of nodes participating in the consensus group is k, the total communication overhead of the MP-PBFT algorithm is shown in formula 4:

$$C_{MP-PBFT} = 2k + k \times (k-1) + 2k \times \left(\frac{n}{k} - 1\right) + 2k \times \frac{n}{k} \times \left(\frac{n}{k} - 1\right)$$
$$= k^2 + 2\frac{n^2}{k} - k \approx O(n^{\frac{4}{3}}) \tag{4}$$

Figure 4 compares the communication overhead of PBFT, RPBFT and the MP-PBFT algorithm proposed in this paper under different network scales. The results show that when the network size is small, the communication overhead of the three algorithms is similar. However, as the network size increases, the communication overhead of MP-PBFT is significantly lower than that of PBFT and RPBFT. This advantage stems from the grouping mechanism and threshold signature technology adopted by MP-PBFT, which significantly reduce the total message volume per consensus round. In a network with 60 nodes, MP-PBFT achieves a substantial reduction in communication overhead compared to PBFT. Experiments demonstrate that in large networks, MP-PBFT exhibits significantly superior communication efficiency compared to other algorithms.

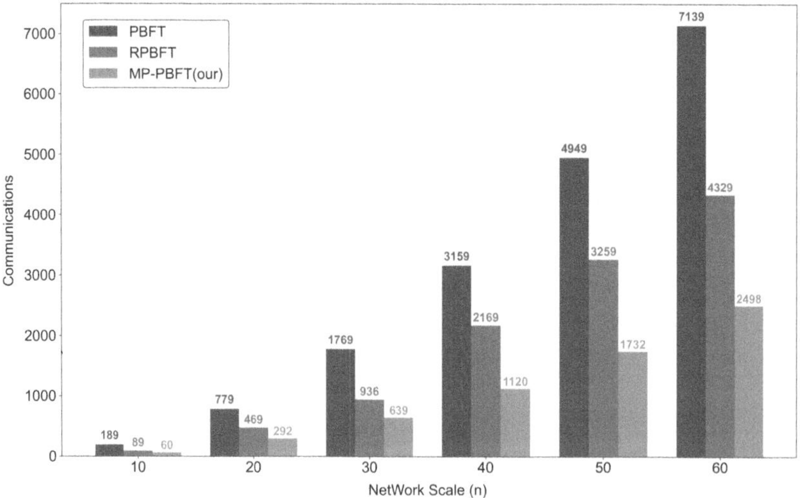

Fig. 4. Communication Overhead Comparison.

4.3 Analysis of Node Election Fairness

In distributed consensus algorithms, the fairness of the election mechanism directly impacts the system's long-term stability. MP-PBFT combines a multi-dimensional node reputation scoring mechanism with a probability-based node election mechanism to enhance efficiency while ensuring all nodes have equal opportunities to participate. Figure 5 illustrates the probability distribution of nodes being selected as key nodes during a 50-round consensus process.

As shown in Fig. 5(a), nodes 3 and 7 are selected as group leaders more frequently, primarily due to their significantly higher reputation values compared to other nodes. Figure 3 shows that node 3's reputation value grows rapidly after the 10th round of consensus, while node 7's reputation value grows even faster after the 20th round, leading to their frequent selection as group leaders, thereby validating the characteristic that nodes with higher reputation values are more likely to be selected. Figure 5(b) shows the probability distribution of each node being selected as a leader node. Node 3 has a higher probability of being selected as a leader node, indicating that its leadership frequency and high probability advantage contribute to the election of leader nodes. However, despite having the lowest probability of being selected as a leader, node 4 was still selected as a leader node. This outcome is attributed to the probabilistic election mechanism of MP-PBFT, which allows nodes with varying reputation values to participate in the election of key nodes with a reasonable probability, thereby enhancing the system's fairness and level of decentralization.

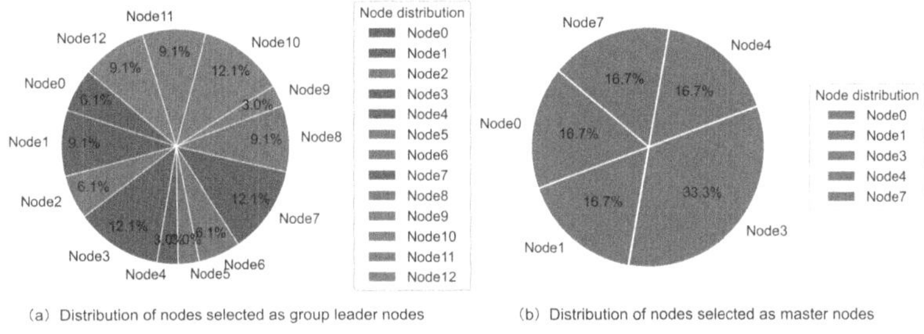

(a) Distribution of nodes selected as group leader nodes (b) Distribution of nodes selected as master nodes

Fig. 5. Fairness of node election.

5 Conclusion

This paper proposes an MP-PBFT algorithm that combines a multi-dimensional node reputation scoring mechanism with a probabilistic node election mechanism, aiming to address the shortcomings of the PBFT algorithm in terms of communication overhead and election fairness. Simulation results show that MP-PBFT outperforms PBFT in terms of communication efficiency, reducing communication overhead by approximately 65% in large-scale networks. The probabilistic election mechanism effectively prevents centralization issues, thereby enhancing the system's fairness. Future research will focus on optimizing consensus efficiency from three aspects. Firstly, adjusting the grouping mechanism to reduce redundant communication and improve consensus speed. Secondly, introducing a "group-internal redundancy and cross-group backup" strategy based on the existing fault-tolerance mechanism to enhance the system's resistance to attacks. Finally, further refining the reputation scoring mechanism to enhance the network's overall fault tolerance capability and reliability.

Acknowledgment. This work was supported by the National Social Science Foundation of China Art Project (No.23EH232), the Key Research and Development Program of Shaanxi Province in 2023 (No. 2023-ZDLGY-48), the 2025 Public Digital Cultural Service Project (GGSZWHFW2025-004), Research Center for Culture & Sci-Tech Integration Innovation, Key Research Base of Humanities and Social Sciences of Hubei Province, the Project of Public Digital Cultural Services (No.WK2023001), Xi'an Polytechnic University's 2024 Undergraduate Education and Teaching Reform Research Key Project (No. 24JGZD13) and Shaanxi Province University Young Outstanding Talents Support Program.

References

1. Rauchs, M., et al.: Distributed ledger technology systems: a conceptual framework. Available at SSRN 3230013 (2018)
2. Velmurugadass, P., Dhanasekaran, S., Anand, S. S., Vasudevan, V.: Enhancing blockchain security in cloud computing with IOT environment using ECIES and cryptography hash algorithm. Mater. Today Proc. 37,2653–2659 (2021)
3. Walter Engert and Ben Siu-Cheong Fung: Central bank digital currency: motivations and implications. Technical report, Bank of Canada Staff Discussion Paper (2017)
4. Schär, F.: Decentralized finance: on blockchain-and smart contract-based financial markets. FRB of St, Louis Review (2021)
5. Zhao, X., et al.: An ethereum-based image copyright authentication scheme. In: International Conference on Web Information Systems and Applications, pp. 724–731. Springer (2022)
6. Dib, O., Brousmiche, K., Durand, A., Thea, E., Hamida, E. B.: Consortium blockchains: overview, applications and challenges. Int. J. Adv. Telecommun, **11**(1), 51–64 (2018)
7. Pahlajani, S., Kshirsagar, A., Pachghare, V.: Survey on private blockchain consensus algorithms. In: 2019 1st International Conference on Innovations in Information and Communication Technology (ICIICT) (2019)
8. Pongnumkul, S., Siripanpornchana, C., Thajchayapong, S.: Performance analysis of private blockchain platforms in varying workloads. In: 2017 26th international conference on computer communication and networks (ICCCN), pp. 1–6. IEEE (2017)
9. Castro, M., et al.: Practical byzantine fault tolerance. In OsDI **99**, 173–186 (1999)
10. Gao, S., Tianyu, Yu., Zhu, J., Cai, W.: T-pbft: an eigentrust-based practical byzantine fault tolerance consensus algorithm. China Commun. **16**(12), 111–123 (2019)
11. He, Y., et al.: A blockchain based truthful incentive mechanism for distributed p2p applications. IEEE access **6**, 27324–27335 (2018)
12. Xu, G., Liu, Y., Khan, P.W.: Improvement of the DPOS consensus mechanism in blockchain based on vague sets. IEEE Trans. Industrial Inf., **16**(6), 4252–4259 (2019)

13. Pajooh, H.H., Rashid, M., Alam, F., Demidenko, S.: Multi-layer blockchain-based security architecture for internet of things. Sensors, **21**(3), 772 (2021)
14. Oktian, Y.E., et al.: Patchman: firmware update delivery service over the blockchain for IOT environment. In: International Conference on Information Security Applications, pp. 131–142. Springer (2023)

An Incentive Mechanism Defending Against Blockchain Selfish Denial-of-Service Attack

Shujie Zhang, Qinglin Yang, Yuan Zhou, Yaoyao Zhang, Pengdeng Li, Chenlu Zhuansun, Yuan Liu(✉), and Zhihong Tian

Guangdong Key Laboratory of Industrial Control System Security, Cyberspace Institute of Advanced Technology/Huangpu Research School of Guangzhou University, Guangzhou University (Huangpu), Guangzhou 510006, China
{zhangshujie,wendribs,yaoaoz}@e.gzhu.edu.cn,
{yangqinglin,pdli,chenluzhuansun,yuanliu,tianzhihong}@gzhu.edu.cn

Abstract. This paper focuses on defending against a novel attack targeting PoW-based blockchains—namely, the Selfish Denial-of-Service (SDoS) attack. SDoS requires as little as 19.6% of the total computational power to earn revenue exceeding its proportional share. Alarmingly, as the revenue-to-cost ratio of Bitcoin mining has recently approached 1, SDoS attacks can easily induce all rational honest miners to voluntarily cease mining altogether. To defend SDoS, this paper proposes a game-based strategy to defend against the SDoS attack. This method restricts the strategy whereby a node publishes only the block header after finding a new block and later reveals the full private chain immediately after a competing block is observed. Experimental results show that the proposed method not only reduces the attacker's revenue, thereby encouraging voluntary termination of the attack, but also mitigates the impact of the SDoS attack and increases the revenue of honest miners.

Keywords: Selfish Mining · PoW Blockchain · Incentive Mechanisms

1 Introduction

In recent years, most studies [1], have focused on blockchain applications, while relatively little attention to its fundamental security issues.

Bitcoin, based on the Proof-of-Work (PoW) consensus, remains the cryptocurrency with the largest market capitalization. As of March 29, 2025, its price reached USD 112,788[1], making it an increasingly attractive target for financially motivated blockchain attacks.

Several attacks allow adversaries to earn rewards disproportionate to their computational power. In selfish mining [2], a miner withholds and strategically releases blocks to gain revenue exceeding its fair share, becoming profitable with only 25% of the network's hash power. The Selfish Mining-based

S. Zhang and Q. Yang—Equal contribution.

[1] https://www.binance.com/en/price/bitcoin.

S. Yang et al. (Eds.): WISA 2025, LNCS 16299, pp. 638–649, 2026.
https://doi.org/10.1007/978-981-95-6426-2_56

Denial-of-Service (SDoS) attack [3] targets PoW blockchains by withholding and strategically releasing blocks while broadcasting block headers to signal hidden ones. Upon receiving these headers, rational miners assess the attack and decide whether to continue or suspend mining based on expected revenue. Notably, an SDoS attacker can gain profits with only 19.6% of the network's hash power [3].

The principle of SDoS is illustrated in Fig. 1. Initially, both the attacker and honest miners mine after block $N-1$. The arrow above $N-1$ indicates the honest miner's position, while the one below shows the attacker's position (*state* 0). When the attacker discovers a block N' after $N-1$, they publish only its header H and continue mining on N', whereas honest miners keep mining on $N-1$ (*state* 1). If the attacker then finds another block $N+1'$ and honest miners find none, the attacker keeps it hidden (*state* 2). However, if honest miners publish a new block N, the attacker—seeing a one-block lead—immediately releases N' and $N+1'$, causing honest miners to switch to $N+1'$ while N becomes orphaned (event $2-M$).

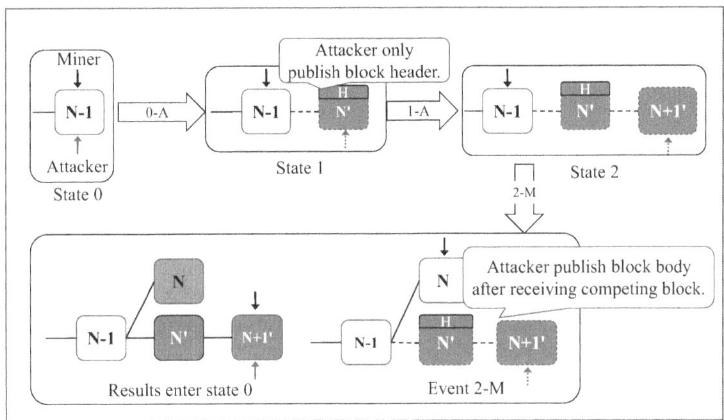

Fig. 1. The demonstration of SDoS attack principle.

In extreme cases, when Bitcoin's cost-to-price ratio reaches 1, rational miners will suspend operations to avoid losses, leaving only the attacker mining on the blockchain. Alarmingly, the current ratio is already near 1[2], suggesting that under current conditions, an SDoS attack could easily drive honest miners to stop mining.

Few studies have examined SDoS attacks in depth. Most only mention them in introductions or related work. Bai *et al.* [4] use SDoS merely for system modeling, while several others [5–8] propose new attack strategies without discussing mitigation. Wang *et al.* [9] detect selfish mining using features such as fork height, mining rewards, and block production rate. Wang *et al.* argue [10] that high-reputation nodes should gain greater profits to deter inactivity, though this

[2] en.macromicro.me/charts/29435/bitcoin-production-total-cost.

only mitigates the attack without addressing honest nodes' profitability. Kadu et al. [11] apply Q-learning to enable optimal miners to reduce latency and enhance throughput, but their method lacks defenses against SDoS attacks.

In short, no defense method has been reported in the literature to counter this threat. Defending against SDoS attacks in blockchain systems presents several significant challenges: i) The attack behavior closely resembles rational or delayed mining activity, making it difficult to distinguish from legitimate actions. ii) Rooted in game-theoretic strategies, SDoS exploits the incentive structure of blockchain systems, requiring defense mechanisms to go beyond simple detection and instead reshape economic incentives. iii) Additionally, ensuring that defenses are incentive-compatible is crucial to avoid deterring honest miners.

In response to this gap, we propose a novel defense mechanism against SDoS attacks and rigorously evaluate its effectiveness through simulation experiments. The main contributions of this work are summarized as follows.

- We propose an SDoS defense strategy tailored for honest miners and provide a preliminary analysis of its effectiveness.
- We prove that the proposed strategy satisfies the incentive compatibility (IC) constraints and demonstrate that its utility exceeds that of a system without defense.
- The simulation results demonstrate that the defense strategy not only increases the utility of honest miners under SDoS attacks but also encourages the widespread adoption of the strategy among honest miners.

The structure of the paper is as follows. Section 2 introduces the model of mining and game-theoretic. Section 3 elaborates on the defense strategy and analyzes the revenue events under defense. Section 4 provides a theoretical analysis of the defended system. Section 5 reports the simulation results. Finally, Sect. 6 concludes the study.

2 Mining Model and Game-Theoretic Model

To facilitate the analysis , this paper models the system in a similar way to [3]. Different from it, we take partial delay into account. Secondly, we integrate the impact of block reception time into the choice of branch. Finally, we leverage the number of blocks that the attacker's private chain leads the defense strategy chain to distinguish the system states.

1) Participants: Assuming that the system consists of one attacker, A, and several honest miners $P_1, P_2, \ldots, P_n$. Let the hash power ratios of the attacker A and each honest miner P_i be denoted as α_A and α_i, respectively, such that $\alpha_A + \sum_{i=1}^{n} \alpha_i = 1$.

2) Delay: To simplify the analysis, we consider transmission delay, but ignore the delays caused by the network structure and different network locations, meaning that when a block is published, all participants receive the block at the same time after a certain delay.

3) Utility: This analysis considers only the block rewards, which account for the predominant portion of miners' revenue. Let K denote the reward of a block, let λ denote the initial block generation rate of the system, and let c_i represent the mining cost per unit of mining power and time for a miner P_i. Then, under normal conditions, the expected revenue per unit time for miner P_i is $R_i = \alpha_i \lambda K$, and the corresponding unit time mining cost is $C_i = \alpha_i c_i$. The unit time utility U_i of an honest miner P_i is expressed by $U_i = \lambda K - c_i$.

4) Profitability Factor w_i: Let w_i present the profitability factor of P_i, and $w_i = R_i/C_i = \lambda K/c_i$. If there is no SDoS attack, then for any honest miner, P_i will have $w_i > 1$. Otherwise, P_i would have no incentive to mine because of no utility.

5) Action of Honest Miners: Assuming that all honest miners are rational and adopt the strategy that maximizes their expected revenue. The strategy set for honest miners is $v = \{D, E\}$, where D represents the defense strategy and E represents the evasive strategy.

Miners adopting the defense strategy are referred to as D miners, while those adopting the evasive strategy are referred to as E miners. D miners will blacklist blocks exhibiting characteristics of SDoS attack, along with their descendants. E miners will mine only when the attacker A does not lead.

6) Choice of Branch: Upon receiving a block, a miner cannot anticipate whether competing blocks at the same height will subsequently appear. Moreover, by the time such competing blocks are received, the miner has already invested a certain amount of computational effort. Therefore, when multiple blocks of the same height are received, honest miners tend to mine after block they received first.

7) System States: The chain supported by the D miners is called defense chain, and the private chain maintained by the attacker is called attacker chain. The system state is defined by the number of blocks that attacker chain leads defense chain, resulting in a set of system states: $\{0, 0', 1, 2, \ldots, n\}$. State 0 represents the normal condition where all participants are mining on the same block, while state $0'$ represents a situation where defense chain and attacker chain are both the longest chains.

8) Events: To distinguish miners employing different strategies, the notations $0-A$, $0-D$, and $0-E$ are used to represent the events where the attacker A, D miners, and E miners find a block in state 0, respectively. Similar notations are used to represent events occurring in other states. The event set is $\{0-A, 0-D, 0-E, 0'-A, 0'-D, 0'-E, 1-A, 1-D, 2-A, 2-D, n-A, n-D\}$.

3 Defense Strategy

The proposed defense strategy blacklists blocks, along with their descendant blocks, that are initially published with only their headers and subsequently fully published only after the appearance of competing blocks, as illustrated in

Fig. 2. Under the SDoS protocol, an attacker first releases only the header of a discovered block, withholding the full block content until a rival block has been observed. When the D miners receive the header of block N', they begin monitoring the timing of its corresponding body, as illustrated in State 1. If they subsequently receive the full body of N' and its child block $N' + 1$ after having already received the competing block N, then both N' and $N' + 1$ are blacklisted, as demonstrated in event $2 - D$.

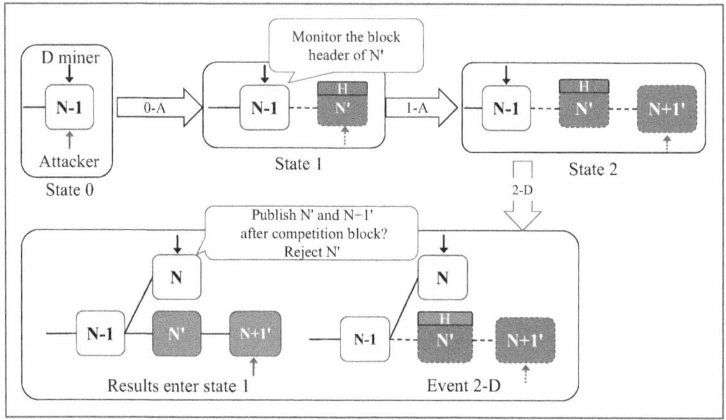

Fig. 2. Core of the defense strategy.

3.1 Defense Strategy Design

By analyzing the attack strategy of the SDoS attacker described in [3], it is observed that the attacker, upon initiating the attack, first publishes the block header and delays the release of the body until a competing block appears. Based on this behavioral characteristic, a defense strategy is developed for honest miners, which adds to the blacklist any blocks that exhibit this behavior. The specific algorithm for the defense strategy is provided in Algorithm 1.

Initially, D miners initialize t_1, t_2, t_3, a flag variable $delayedReleaseFlag$, and the blacklist, respectively, as shown in lines 1–2 of Algorithm 1. When receiving a block header, D miners mark the corresponding block as N' and record the time t_1 of receipt, as shown in lines 3–4. Once receiving a full block, they mark the block as N, as shown in lines 5–6. If $t_1 > 0$ and N is a competing block of N' (a different block at the same height), D miners record the time of receiving N as t_2, like shown in lines 7–10. If N and N' are the same block, they first records the time t_3 of receiving N. If $t_1 < t_2 < t_3$, indicating that N' was fully released after the competing block appeared, the miner sets $delayedReleaseFlag$ to 1, as shown in lines 11–16. If N is the child of N' and the value of $delayedReleaseFlag$ is 1, both N' and N are added to the blacklist, as they were published after the competing block, and then t_1, t_2, t_3, and $delayedReleaseFlag$ are reset, as shown in lines 17–20. In the case where no block header has been received, if the parent block of N is in the blacklist, N will also be added to the blacklist, as shown in lines 21–23. Finally, the honest miner shall execute mine behavior following the longest earliest chain that is not in the blacklist, as shown in line 24.

Algorithm 1. Defense Strategy.

1: Initialization:
2: $t_1, t_2, t_3 \leftarrow 0$,$delayedReleaseFlag \leftarrow 0$,$blacklist \leftarrow null$
3: When receive a header of a block:
4: Mark the block as N' and record the receive time as t_1
5: When receiving a whole block:
6: Mark the block as N
7: **if** $t_1 > 0$ **then**
8: **if** N, N' are Different blocks of equal height **then**
9: Record the receive time of N as t_2
10: **end if**
11: **if** N, N' are the same blocks **then**
12: Record the receive time of N as t_3
13: **if** $t_1 < t_2$ & $t_2 < t_3$ **then**
14: $delayedReleaseFlag \leftarrow 1$
15: **end if**
16: **end if**
17: **if** N is the son of N' & $delayedReleaseFlag == 1$ **then**
18: Add N', N to blacklist
19: $t_1, t_2, t_3 \leftarrow 0$,$delayedReleaseFlag \leftarrow 0$
20: **end if**
21: **else if** N's parent is in blacklist **then**
22: Add N to blacklist
23: **end if**
24: Mining on the longest and earliest non-blacklist chain

3.2 Revenue Events of Attacker

Assume that the hash rate of the defense strategy, α_m, is greater than that of the attacker A, α_A; otherwise, the attacker would be capable of launching a 51% attack.

If attacker A does not modify the SDoS strategy, then whenever the attacker leads by two or more blocks, all of these blocks will be added to the blacklist. Since the hash rate of the defense strategy is greater than that of attacker A, the attacker's local private chain will eventually be overtaken, and the blocks will become orphaned.

When the attacker leads by one block, a block mined by an honest miner will result in a fork. If the attacker subsequently finds another block after his own (denoted as N'), both blocks mined by the attacker will be added to the blacklist. If the attacker fails to find a new block after N', then, because no honest miner mines after N', the chain containing N' will be surpassed, and

N' will eventually become an orphan block. In summary, if the attacker does not modify the SDoS strategy, all attack blocks will become orphan blocks, and attacker A will earn zero revenue.

Suppose the attacker modifies the SDoS strategy and publishes a block earlier than the anticipated release of a competing block. In that case, they must publish the block body immediately after releasing the header, as the average block time of the defense strategy is shorter than that of the attacker. Consequently, any miner who finds a block will publish it immediately, resulting in the system operating as it would under normal conditions without an SDoS attack.

In short, if attacker does not modify the SDoS strategy, the revenue will be zero. If attacker modify the SDoS strategy, the system operates as if it were in a normal, attack-free state.

3.3 Revenue Events of Honest Miners

After defense, honest miners who find blocks at any time have a high probability of receiving block rewards and a small probability of losing rewards because of blacklisted block, so we quantify this probability as following.

1) Situation of Blacklisted Honest Blocks: An honest miner P_i finds block N' and publish the block header first, and during transmission of the block body, if any other honest miner publish a competing block N of the same height, then block N' will be added to the blacklist.

2) Block Header Average Existence Time θ: BIP 152 allows nodes publish block header information firstly, and publish lack information latter. Values of block header existence time, θ, range from 0 to $\frac{1}{\lambda}$, and the worst case is $\frac{1}{\lambda}$, the time a block found in system.

3) Probability of Publishing Block Headers p_h: The probability of publishing a block header, p_h, ranges from 0 to 100% , and the worst case is is 100%.

4) False Negative Rate $q_D^f, q_{\bar{A}}^f$: This paper models the case where the block mined by honest miner P_i is incorrectly added to the blacklist as follows: P_i mines the block at most θ seconds earlier than other honest miners and publishes the block header first. When the mining difficulty after defense stabilizes, we denote the system's block generation rate, including orphan blocks, as λ'. Recall that $\alpha_{D(v)}$ represents the total hash power of D miners when honest miner P_i selects strategy v. For the block mined by P_i, when the honest mining hash power is $\alpha_{D(v)}$, we denote the probability of the honest block being blacklisted as q_D^f. When the honest mining hash power is $1 - \alpha_A$, we denote the probability of the honest block being blacklisted as $q_{\bar{A}}^f$. Under normal conditions, where the block mining time for a single miner follows an exponential distribution, we use the formulas Eq.(1) to estimate the probability of the honest block being blacklisted.

$$\begin{cases} q_D^f = \dfrac{\alpha_i}{\alpha_{D(v)}}(1 - e^{-(\alpha_{D(v)} - \alpha_i)\lambda'\theta})p_h, \\ q_{\bar{A}}^f = \dfrac{\alpha_i}{1 - \alpha_A}(1 - e^{-(1 - \alpha_A - \alpha_i)\lambda'\theta})p_h. \end{cases} \quad (1)$$

5) Revenue Event of Honest Miners:For honest miners, after defending, if they mine a block in states 1, 2, or n (where $n > 2$), they have a probability of q_D^t of receiving a block reward. When they mine a block in states 0 or $0'$, they have a probability of $q_{\bar{A}}^t$ (i.e., the probability of not being added to the blacklist) of receiving a block reward. The values of q_D^t and $q_{\bar{A}}^t$ are calculated as $q_D^t = 1 - q_D^f$ and $q_{\bar{A}}^t = 1 - q_{\bar{A}}^f$, respectively.

4 Theoretical Analysis

This section focuses on the expected utility of honest miners choosing different strategies after defense. It calculates the expected utilities U_i^D and U_i^E for honest miner P_i when choosing the defense and evasive strategies, respectively, and proceeds as follows: First, define the Markov chains after the defense is implemented. Secondly, solve for the state probabilities using the Markov chains. Third, calculate the utility under different strategies using the state probabilities and revenue events. Finally, construct and calculate the critical profitability factor w_i^* (the profitability factor when $U_i^D = U_i^E$) to determine the strategy with the higher utility for P_i.

1) Markov chain after defense. Figure 3 depicts the Markov chain after defending against the SDoS attack. $\alpha_D(v)$ represents the hash power ratio of miners who choose strategy v, including the honest miner P_i, with the calculation formula provided in Eq. (2), and α_D represents the hash power ratio of miners who choose the defense strategy, excluding the honest miner P_i. Furthermore, it is required that $\alpha_D > \alpha_A$ to prevent attacker A from launching a 51% attack.

$$\alpha_D(v) = \begin{cases} \alpha_D + \alpha_i & , v = D, \\ \alpha_D & , v = E. \end{cases} \tag{2}$$

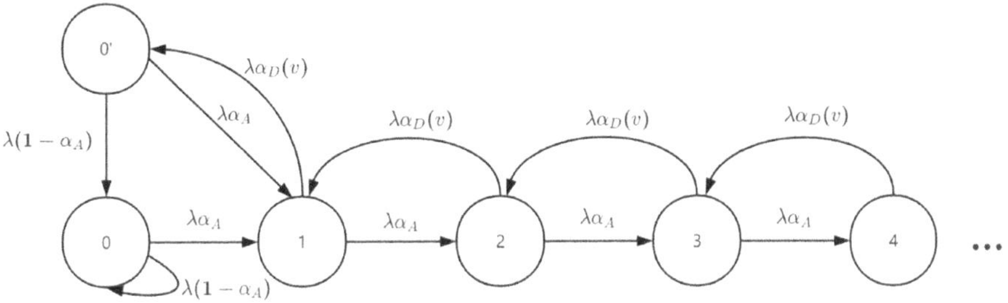

Fig. 3. The evolution of Markov chain after defense.

2) State probability after defense. $p_0^v, p_{0'}^v, p_1^v, p_2^v, p_n^v$ represent the probabilities of the system being in states $\{0, 0', 1, 2, n\}$ (where $n > 2$) under different strategies chosen by honest miner P_i. Assume that the block generation rate (including orphan blocks) of the system after stabilization through defense is denoted as λ'. According to Eq. (3), the state probabilities of the Markov chain in Fig. 3 are given by Eq. (4).

$$\begin{cases} p_0^v = (1-\lambda'\alpha_A)p_0^v + \lambda'(1-\alpha_A)p_{0'}^v, \\ p_{0'}^v = (1-\lambda')p_{0'}^v + \lambda'\alpha_D(v)p_1^v, \\ p_1^v = (1-\lambda'\alpha_A - \lambda'\alpha_D(v))p_1^v + \lambda'\alpha_A p_0^v + \lambda'\alpha_A p_{0'}^v + \lambda'\alpha_D(v)p_2^v, \\ p_2^v = (1-\lambda'\alpha_A - \lambda'\alpha_D(v))p_2^v + \lambda'\alpha_A p_1^v + \lambda'\alpha_D(v)p_3^v, \\ p_n^v = (1-\lambda'\alpha_A - \lambda'\alpha_D(v))p_n^v + \lambda'\alpha_A p_{n-1}^v + \lambda'\alpha_D(v)p_{n+1}^v,\ n > 2, \\ p_0^v + p_{0'}^v + p_1^v + \sum_{i=2}^{\infty} p_i^v = 1. \end{cases} \tag{3}$$

$$\begin{cases} p_1^v = \frac{1}{\frac{\alpha_D(v)}{\alpha_A}+1+\frac{\alpha_A}{\alpha_D(v)-\alpha_A}}, \\ p_0^v = \frac{\alpha_D(v)(1-\alpha_A)}{\alpha_A}p_1^v, \\ p_{0'}^v = \alpha_D(v)p_1^v, \\ p_2^v = \frac{\alpha_A}{\alpha_D(v)}p_1^v, \\ p_n^v = (\frac{\alpha_A}{\alpha_D(v)})^{n-1}p_1^v, n > 2. \end{cases} \tag{4}$$

3) Mining difficulty after defending When P_i chooses different strategies, in states 0 and $0'$, the total active mining power in the system is 1, and in other states, it is $\alpha_A + \alpha_D(v)$. Therefore, the expected active mining power α_* can be calculated using Eq. (5).

$$\alpha_* = p_0^v + p_{0'}^v + (1 - p_0^v - p_{0'}^v)(\alpha_A + \alpha_D(v)). \tag{5}$$

The valid block generation rate of honest miners is denoted as r_H, and according to the revenue events (see Sect. 3.3) and the state probabilities, r_H can be calculated by $r_H^v = ((p_0^v + p_{0'}^v)(1-\alpha_A)q_{\overline{A}}^t + (1 - p_0^v - p_{0'}^v)\alpha_D(v)q_D^t)/\alpha_*$.

The initial mining difficulty is $\frac{1}{\lambda}$, and the mining difficulty after defending is $\frac{1}{\lambda'}$. Then, λ' is calculated by $\lambda' = \lambda / r_H^v$.

4) Utility after defense. After defending, the revenue of an honest miner P_i choosing different strategies is given by Eq. (6), the cost by Eq. (7), and the utility by Eq. (8).

$$\begin{cases} R_i^D = (p_0^D q_{\overline{A}}^t + p_{0'}^D q_{\overline{A}}^t + \sum_{i=1}^{\infty} p_i^D q_D^t)\alpha_i \lambda' K, \\ R_i^E = (p_0^E q_{\overline{A}}^t + p_{0'}^E q_{\overline{A}}^t)\alpha_i \lambda' K. \end{cases} \tag{6}$$

$$\begin{cases} C_i^D = \alpha_i c_i, \\ C_i^E = (p_0^E + p_{0'}^E)\alpha_i c_i. \end{cases} \tag{7}$$

$$\begin{cases} U_i^D = \dfrac{R_i^D - C_i^D}{\alpha_i} = (p_0^D q_{\overline{A}}^t + p_{0'}^D q_{\overline{A}}^t + \sum_{i=1}^{\infty} p_i^D q_D^t)\lambda' K - c_i, \\ U_i^E = \dfrac{R_i^E - C_i^E}{\alpha_i} = (p_0^E q_{\overline{A}}^t + p_{0'}^E q_{\overline{A}}^t)\lambda' K - (p_0^E + p_{0'}^E)c_i. \end{cases} \tag{8}$$

5) Critical profitability factor w_i^*. To find the more profitable strategy for P_i, which may have different c_i, this section uses the critical profitability factor w_i, the value of the profitability factor w_i when $U_i^D = U_i^E$, as defined in Eq. (9). If $w_i > w_i^*$, it means that P_i choosing the defense strategy will achieve a higher utility than the evasive strategy.

$$w_i^* = \frac{1 - p_0^E - p_{0'}^E}{p_0^D q_{\overline{A}}^t + p_{0'}^D q_{\overline{A}}^t + \sum_{i=1}^{\infty} p_i^D \overline{q}_D^f - p_0^E q_{\overline{A}}^t - p_{0'}^E q_{\overline{A}}^t}. \tag{9}$$

5 Simulated Experiments

Our experiments simulate two stages: i) start defending, and ii) all honest miners defend. The simulated experiment was run on a Windows 11 system with an Intel(R) Core(TM) i5-13500H CPU, using MATLAB version 9.14.0. The experiment makes the worst-case assumption, simulating with $\theta = \frac{1}{\lambda'}$ and $p_h = 100\%$, and facilitates the calculation of q_D^t and $q_{\overline{A}}^t$. Note that at any time, the defense strategy's mining power, $\alpha_D(v)$, must exceed that of attacker A to prevent a 51% attack.

5.1 Start Defending

When attacker launches SDoS attack, honest miners have two choices without defending: constantly mining or mining only when attacker does not lead, which is called evasive strategy. If miners who always mine find that the defense strategy is more profitable than always mining, they will switch to the defense strategy, and the system will begin defending. The result shown in Fig. 4 confirms this.

Figure 4 shows the ratio of the revenue from P_i's defense strategy to the revenue from P_i's always-mining strategy in the SDoS attack. The ratio increases monotonically with the mining power of the attacker, α_A, and decreases monotonically with the mining power of other miners using the defense strategy, α_D, excluding P_i. Specifically, the ratio is always no less than 1, otherwise there will be gray areas. For any P_i who chooses always mining, since the mining cost of both strategies is equal, they will switch to defense strategy because of higher utility.

5.2 All Honest Miners Defend

After the system begins defending, there are two choices for honest miners: defense strategy or evasive strategy. If an evasive strategy honest miner P_i finds that the defense strategy is more profitable, P_i will switch to the defense strategy. The result shown in Fig. 5 confirms this.

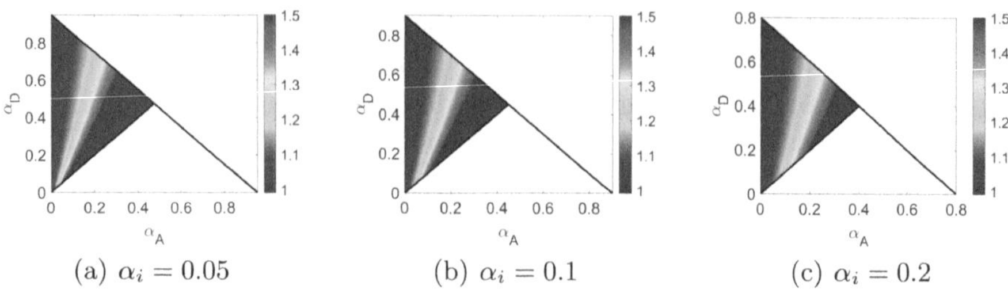

(a) $\alpha_i = 0.05$ (b) $\alpha_i = 0.1$ (c) $\alpha_i = 0.2$

Fig. 4. Ratio of P_i's defense strategy revenue to always mining revenue under SDoS.

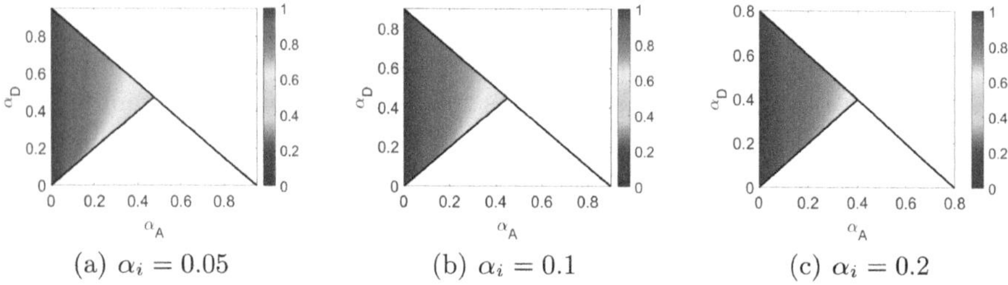

(a) $\alpha_i = 0.05$ (b) $\alpha_i = 0.1$ (c) $\alpha_i = 0.2$

Fig. 5. Critical profitability factor w_i^* after defense strategy implementation.

The critical profitability factor w_i^* is used to find the more profitable strategy between defense strategy and evasive strategy for each P_i. If $w_i > w_i^*$, it proof that defensive strategy pays better. The result is shown in Fig. 5. w_i^* decreases monotonically with α_A and increases monotonically with α_D. Specifically, w_i^* is never greater than 1 in any case; otherwise, there will be gray areas, meaning that the defense strategy is more profitable than the evasive strategy for any P_i, since $w_i > 1 > w_i^*$. The defense strategy satisfies the Incentive Compatibility (IC) constraint for any honest miners P_i.

6 Conclusion

This paper proposes a defense strategy against the SDoS attack and evaluates its effectiveness. The strategy not only compels the attacker to terminate the attack but also increases the mining utility of honest miners, thereby incentivizing their active participation in defense. Simulation results show that the strategy enhances the utility of miners who continue mining, which in turn encourages other honest miners to adopt the defense. Ultimately, all honest miners remain engaged in mining, while the attacker ceases the SDoS attack due to zero revenue.

Acknowledgments. This work was supported in part by the GuangDong Basic and Applied Basic Research Foundation under Grant 2025B1515020022; in part by the National Natural Science Foundation of China under Grant T2522011.

References

1. Yu, Z., Dong, X., Wen, Z., Nie, T.: A blockchain-based dynamic symmetric searchable encryption scheme for sharing elderly health data. In: International Conference on Web Information Systems and Applications, pp. 389–402. Springer (2024)
2. Kwon, Y., Kim, D., Son, Y., Vasserman, E., Kim, Y.: Be selfish and avoid dilemmas: Fork after withholding (faw) attacks on bitcoin. In: Proceedings of the 2017 ACM SIGSAC conference on computer and communications security, pp. 195–209 (2017)
3. Wang, Q., et al.: Selfish mining-based denial-of-service attack: Sdos. IEEE Trans. Inf. Forensics Secur. **17**, 3335–3349 (2022)
4. Bai, Q., Yuedong, X., Liu, N., Wang, X.: Blockchain mining with multiple selfish miners. IEEE Trans. Inf. Forensics Secur. **18**, 3116–3131 (2023)
5. Wang, Q., et al.: Optimal selfish mining-based denial-of-service attack. IEEE Trans. Inf. Forensics Secur., **19**, 835–850 (2023)
6. Zhang, J., Zha, C., Zhang, Q., Ma, S.: A denial-of-service attack based on selfish mining and sybil attack in blockchain systems. *IEEE Access* (2024)
7. Shang, J., Lu, T., Zhao, P.: Sim: achieving high profit through integration of selfish strategy into innocent mining. IEEE Trans. Netw. Serv. Manag. (2024)
8. He, Z., et al.: Nurgle: exacerbating resource consumption in blockchain state storage via mpt manipulation. In: 2024 IEEE Symposium on Security and Privacy (SP), pp. 2180–2197. IEEE (2024)
9. Wang, Y., et al.: A detection method against selfish mining-like attacks based on ensemble deep learning in IOT. IEEE Internet of Things Journal (2024)
10. Wang, Z.: A decentralized oracle network constructed from weighted schnorr multisignature. IEEE Trans. Netw. Serv. Manag. (2025)
11. Kadu, R., Kadu, A.: An efficient incremental learning bioinspired model for improving QOS of blockchain-based large-scale wireless networks. In: 2024 3rd International Conference for Innovation in Technology (INOCON), pp. 1–9. IEEE (2024)

High-Performance GPU Implementation of Key Generation in Post-quantum SNOVA Digital Signature Algorithm

QingZheng Wang[1], ZhanPeng Shi[1], QingQuan Tan[2], and Dong Wang[1](✉)

[1] Henan University, Henan 475000, China
juliawdd@henu.edu.cn
[2] Wuhan University, Wuhan 430040, China

Abstract. SNOVA is a post-quantum digital signature scheme based on multivariate quadratic equations. It can effectively resist quantum attacks. However, SNOVA faces computational efficiency bottlenecks in practical deployment due to the extensive matrix operations required during key generation. To address this issue, this paper proposes a GPU-accelerated scheme for SNOVA key generation. First, we adopt a multi-dimensional grid mapping strategy to optimize thread allocation. We design specialized 2D and 3D grid configurations to achieve efficient parallelization of matrix operations. Second, we design a three-tier memory optimization system that effectively reduces memory access latency. Finally, we construct a batch processing framework for key generation. Through pipeline and reuse mechanisms, we implement batch key generation for SNOVA. Experimental results demonstrate that compared to the basic CPU implementation, the key generation speed achieves up to 12.30× improvement. In batch processing mode, GPU-SNOVA achieves up to 3.0× improvement in key generation time compared to the AVX2 version, and up to 58.44× improvement compared to the CPU version, meeting the requirements of high-throughput applications.

Keywords: Post-quantum digital signature · Multivariate quadratic equations · SNOVA · GPU

1 Introduction

Digital signatures serve as a core component of Public Key Infrastructure (PKI). They play a crucial role in addressing security issues such as data integrity, identity authentication, and non-repudiation. Digital signatures are widely applied in domains including e-commerce, e-government, and blockchain systems. However, the rapid development of quantum computing poses fundamental threats to traditional digital signature systems. Existing digital signature schemes are primarily based on mathematical hard problems such as integer factorization and elliptic curve discrete logarithm [1]. Quantum computers can efficiently solve these problems through quantum algorithms like Shor's algorithm [2], creating

S. Yang et al. (Eds.): WISA 2025, LNCS 16299, pp. 650–662, 2026.
https://doi.org/10.1007/978-981-95-6426-2_57

systematic risks to the security of current cryptographic and digital signature systems [3]. To address quantum threats, the National Institute of Standards and Technology (NIST) initiated the post-quantum cryptography standardization process in 2016 to evaluate next-generation post-quantum cryptographic algorithms [4]. Among numerous candidate schemes, signature schemes based on multivariate quadratic equations have received significant attention. Most of these schemes are derived from the Oil and Vinegar scheme proposed by Patarin in 1997 [5]. SNOVA is an improved Unbalanced Oil and Vinegar (UOV) signature scheme [7] proposed by Wang et al. [6] in 2022. By significantly reducing public key size, SNOVA effectively alleviates the key storage problem of multivariate cryptographic systems. It has become the latest candidate scheme for NIST post-quantum cryptography standardization. However, SNOVA involves extensive intensive matrix operations during the key generation phase. This affects its practical application in large-scale computational scenarios requiring high concurrent processing. GPUs serve as important platforms for massive parallel computing. They can distribute these intensive matrix operations across numerous threads for parallel execution. However, there is currently no GPU-accelerated SNOVA key generation scheme available. In this work, we design specialized CUDA kernel functions for SNOVA key generation. We adopt multi-dimensional grid configurations to achieve GPU parallelization of the key generation process. Through memory optimization techniques, we reduce memory access latency and implement a batch processing architecture. Our main contributions can be summarized as follows:

- We adopt a dynamic multi-dimensional grid configuration strategy. We implement efficient parallel execution of matrix operations through 2D and 3D grid configurations. We design a batch processing framework for key generation. Through pipeline processing and resource reuse mechanisms, we achieve batch key generation. This significantly improves GPU resource utilization and overall throughput.
- We construct a three-tier memory optimization system. This system includes shared memory caching, constant memory storage, and coalesced global memory access. By preloading frequently accessed GF(16) operation tables and matrix data, we substantially reduce memory access latency.
- We propose the first GPU-accelerated SNOVA key generation scheme. Through rigorous testing, our GPU-based SNOVA key generation achieves 1.93× to 12.30× speedup compared to CPU implementations. In batch processing mode, our GPU-based key generation achieves up to 3.0× speedup compared to the state-of-the-art AVX2 version [16]. Compared to the CPU version, it achieves up to 58.44× speedup.

2 Related Work

The SNOVA algorithm was proposed by Wang et al. [6] in 2022. It is a variant of the UOV scheme based on non-commutative rings [7]. Subsequently, Ikematsu et al. [8] and Li et al. [9] conducted security analyses of SNOVA in 2024. However, there is currently no research on GPU implementation of SNOVA key generation. Current academic research on GPU acceleration of post-quantum signature algorithms mainly focuses on several mainstream algorithms standardized by NIST. Zhao et al. [10] proposed a multi-level parallel framework for CRYSTALS-Dilithium. This framework includes three parallel levels: inter-block, intra-thread, and inter-thread parallelism. The research focuses on optimizing computation-intensive Number Theoretic Transform (NTT) operations. It adopts shared memory and loop unrolling techniques to significantly improve algorithm performance. Shen et al. [11] further proposed a high-throughput GPU implementation of Dilithium. They optimized the sequential operations of rejection sampling, improved memory access patterns of NTT, and minimized resource consumption of hash functions and inner product computations. This achieved effective performance improvement of the algorithm. Wang et al. [12] proposed the CUSPX implementation for SPHINCS+. They constructed a three-level parallel framework: cross-tree parallelism, intra-node parallelism, and inter-node parallelism. The research designed Merkle tree construction algorithms supporting arbitrary-scale parallel computation. They proposed multiple load balancing strategies to improve resource utilization, achieving significant performance improvements. Hao et al. [17] proposed a blockchain-based and attribute-based dynamic verifiable and searchable encryption framework to address privacy issues in medical data sharing. Their work offers a valuable perspective for potential future applications of our SNOVA acceleration scheme in distributed healthcare environments. All of the above provide beneficial references for GPU implementation of SNOVA.

3 Preliminaries

3.1 Notation

$\mathbb{F}_q$ is defined as the finite field of order q. The parameters l and rank represent the dimensions of the matrix ring. R represents an $l \times l$ matrix ring composed over the finite field. The variables v, o, and S represent the number of vinegar variables, oil variables, and symmetric matrices respectively. The parameter n equals $v + o$. The parameter m represents the number of equations. X is a vector and it represents vinegar variables. T represents the invertible linear mapping in SNOVA. F represents the central mapping. P represents the public key mapping. Matrix addition over the finite field is represented by $\oplus$. Matrix multiplication over the finite field is represented by $\otimes$.

3.2 Basic Concepts

MQ Problem. The Multivariate Quadratic (MQ) problem can be defined as follows. Given a system of m multivariate quadratic polynomials with n variables over a finite field GF(q) of order q, denoted as $S = \{Q_1(\boldsymbol{x}), ..., Q_m(\boldsymbol{x})\}$, the goal is to find a vector $\boldsymbol{x} = (x_1, \ldots, x_n)$ such that $Q_1(\boldsymbol{x}) = \cdots = Q_m(\boldsymbol{x}) = 0$. This

Algorithm 1. SNOVA GenKey

Input: public and private seeds of SNOVA $(S_{public}, S_{private})$
Output: $pk = (S_{public}, P_i^{22}), sk = (T^{12}, F_i^{11}, F_i^{12}, F_i^{21})(for 0 \leq i \leq m)$
1: $S_{out} \leftarrow Hash_{Shake256}(S_{private})$
2: $T^{12} \leftarrow GenF_q(S_{out})$; m $\leftarrow$ 0
3: $GenerateEntries\left(P_i^{11}, P_i^{12}, P_i^{21} for 0 \leq i \leq m\right) \leftarrow GenF_q(Hash_{AES128}(S_{public}))$
4: Compute P_i^{22} and $F_i^{11}, F_i^{12}, F_i^{21}$
5: $pk \leftarrow (S_{public}, P_i^{22})$ (for $0 \leq$ i $\leq$ m)
6: $sk \leftarrow (T^{12}, F_i^{11}, F_i^{12}, F_i^{21})$(for $0 \leq$ i $\leq$ m)
7: **return** (pk, sk)

problem is NP-hard [13]. The security of the SNOVA algorithm is based on the MQ problem.

3.3 SNOVA

SNOVA is a digital signature scheme based on multivariate quadratic equations. It is a variant of the unbalanced oil and vinegar signature scheme [7]. SNOVA differs from the UOV scheme in several key aspects.

First, SNOVA is based on non-commutative rings, specifically matrix rings. Second, it adopts the "random alignment" technique proposed by Petzoldt et al. [14,15]. This technique further compresses the size of the public key.

The algorithm maintains the classical oil-vinegar variable structure [5]. Additionally, it introduces random matrices A_α, B_α, $Q_{\alpha,1}$, and $Q_{\alpha,2}$. These matrices perform mixing transformations on the central mapping to eliminate sparsity. The resulting central mapping takes the following form:

$$F_i = \sum_{\alpha=0}^{l^2+l-1} A_{i,\alpha} \cdot \left(\sum X^T Q_{i,\alpha,1} F_{i'} Q_{i,\alpha,2} X\right) \cdot B_{i,\alpha} \tag{1}$$

As shown in Algorithm 1, the generation of the F matrix and P matrix requires multiple nested loop iterations. These loops traverse through each element position to systematically construct both matrices. The nested loop structure ensures that all matrix elements are properly initialized according to the SNOVA specification requirements. Algorithm 2 implements the signature generation process for SNOVA. It uses a 16-byte salt generated by AES256-CTR to increase signature complexity. The process has two steps: construct a Gaussian matrix, then perform matrix elimination.Algorithm 3 describes the signature verification process.

4 Method

This chapter presents a detailed optimization scheme for porting the SNOVA signature algorithm from CPU implementation to GPU. The work focuses on comprehensive parallelization and optimization of SNOVA's key generation process.

Algorithm 2. SNOVA Sign

Input: pk,Message digest, $S_{public}, S_{private}, Salt$

Output: Signature sig

1: $signed_{hash} \leftarrow SHAKE256(S_{public} \parallel digest \parallel salt)$
2: $hash_{gf16} \leftarrow ConvertToGF16(signed_{hash}), retry_{count} \leftarrow 0$
3: **repeat**
4: $retry_{count} \leftarrow retry_{count} + 1$
5: $vinegar \leftarrow SHAKE256(S_{private} \parallel digest \parallel salt \parallel retry_{count})$
6: Compute Left,Right Matrix and Fvv
7: $Gauss \leftarrow BuildLinearSystem(Left, Right, F^{12}, F^{21}, hash_{gf16}, Fvv)$
8: $solution \leftarrow GaussianElimination(Gauss)$
9: **until** solution exists
10: $signature \leftarrow ApplyAffineTransform(X_{vinegar}, solution, T^{12})$
11: **return** $sig \leftarrow ConvertToBytes(signature)||salt$

Algorithm 3. SNOVA Verify

Input: pk,Signature sig,Message digest

Output: Accept or Reject

1: Compute $signature_{bytes}$, $salt$ and $signed_{hash}$
2: $X \leftarrow ConvertToGF16Matrix(signature_{bytes})$
3: $Left, Right \leftarrow ComputeTransforms(Aalpha, Balpha, Qalpha1, Qalpha2, X)$
4: $P \leftarrow CombineMatrices(P^{11}, P^{12}, P^{21}, P^{22})$
5: $computed_{hash} \leftarrow EvaluatePublicMap(Left, Right, P)$
6: $hash_{bytes} \leftarrow ConvertToBytes(computed_{hash})$
7: **if** $hash\_bytes = signed\_hash$ **then**
8: **return** Accept
9: **else**
10: **return** Reject
11: **end if**

We design a complete CUDA acceleration framework to enhance algorithm performance. The framework employs algorithm parallelization and memory optimization techniques. These techniques significantly improve the overall performance of the algorithm.

4.1 Overall Architecture

The architecture is a hierarchical optimization and heterogeneous collaborative design architecture. The architecture consists of three layers: CPU layer, transmission layer, and GPU layer. The CPU layer is responsible for algorithm flow control and data management. It includes CUDA device constant initialization, memory management, and concurrent coordination of multiple CUDA streams. The CPU layer executes tasks such as seed expansion and random number generation. The transmission layer serves as a data bridge between CPU and GPU.

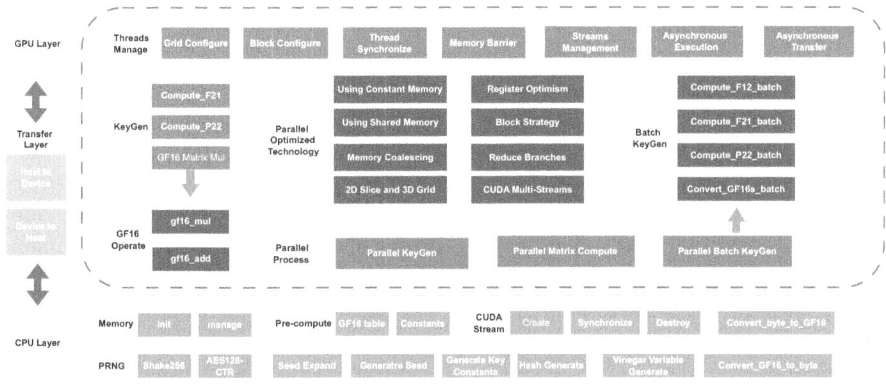

Fig. 1. Overall Architecture Diagram.

It implements efficient bidirectional data transfer mechanisms. The GPU layer focuses on intensive matrix operations for key generation. It processes tasks such as matrix computations over finite fields, polynomial evaluations, and hash function computations in parallel. In terms of memory optimization, we design a three-tier memory optimization system. This system includes shared memory caching, constant memory storage, and coalesced global memory access. We adopt parallel optimization techniques such as block computation and asynchronous data transfer. These techniques achieve GPU acceleration of SNOVA algorithm key generation. Figure 1 illustrates the specific architecture design.

4.2 GPU-Accelerated Single Key Generation Implementation

The SNOVA key generation process represents one of the most computationally intensive components in the entire algorithm. We migrate the computation-heavy F matrix and associated matrix calculations to GPU execution for enhanced performance. According to Algorithm 1, the F matrix consists of three submatrices with F^{11},F^{12},F^{21}. The calculation is as follows:

$$F_i^{12}[j][k] += \sum_{j=0}^{v-1} \sum_{k=0}^{o-1} \sum_{index=0}^{v-1} (P_i^{11}[j][index] \times T^{12}[index][k]) \tag{2}$$

$$F_i^{21}[j][k] += \sum_{j=0}^{o-1} \sum_{k=0}^{v-1} \sum_{index=0}^{v-1} (T^{12}[index][j] \times P_i^{11}[index][k]) \tag{3}$$

We establish three dedicated CUDA streams for F matrix computation. These streams handle data transfer and summation calculations respectively. This design achieves task-level parallelism across the computation pipeline.

Matrix multiplication operations over finite fields utilize lookup queries through the mt4b multiplication table and inv4b inverse element table. We store both lookup tables in the GPU's constant memory space. This placement significantly enhances lookup efficiency during computation.

Algorithm 4 demonstrates our multi-level parallel strategy for optimizing F matrix operations. The grid layer employs a (m, v) two-dimensional layout structure. Each thread block manages computation for specific (i, j) positions in the output matrix. This arrangement ensures balanced load distribution across processing units. Each block contains 128 threads configured for collaborative computation. Individual threads process multiple elements along the third dimension. We preload frequently accessed P^{11} and T^{12} data into shared memory. This approach substantially reduces global memory access requirements. Our computational optimizations incorporate several key strategies. Loop unrolling reduces operational overhead. Register-based local variables accumulate intermediate computation results. Data layout optimization enables coalesced memory access patterns.

The computation of P^{22} follows a similar approach to F matrix operations. The calculation proceeds as follows:

$$P_i^{22}[j][k] += \sum_{j=0}^{v-1} \sum_{k=0}^{o-1} \sum_{index=0}^{v-1} (T^{12}[index][j] \times (P_i^{11}[j][index] \times T^{12}[index][k]) + P_i^{21}[j][index] \times T^{12}[index][k]) \tag{4}$$

We implement a three-dimensional CUDA grid structure for parallel computation of the P^{22} matrix. This implementation reuses optimization strategies developed for F matrix computation. Upon completion of calculations, we employ a specialized kernel function `ConvertToBytes`. This function efficiently packs pairs of 4-bit $\mathbb{F}_{16}$ elements into byte format for storage and transmission.

4.3 GPU Implementation for Batch Key Generation

To fully utilize the parallel computing capabilities of GPUs, this research designs a batch processing scheme for key generation. This batch processing scheme can generate up to 2500 key pairs in parallel within a single GPU call. This approach significantly improves the overall throughput of key generation.

The F matrix and matrix generation processes consume the most time during key generation. We design dedicated CUDA kernels to address these computationally intensive operations. These kernels employ a three-dimensional grid structure. Each thread block is responsible for processing one matrix block within a specific batch. We achieve acceleration by parallelizing the innermost loop calculations. We adopt thread coarsening techniques in our implementation. Each thread handles multiple consecutive computational tasks. The number of elements processed by each thread is calculated using the following formula, blockDim.x denotes the size of a thread block in the x-dimension:

$$elements_{per_{thread}} = \frac{(total_{elements} + blockDim.x - 1)}{blockDim.x} \tag{5}$$

Algorithm 4. GPU-Accelerated SNOVA KeyGen

Input: public and private seeds of SNOVA $(S_{public}, S_{private})$

Output: $pk = (S_{public}, P_i^{22}), sk = (T^{12}, F_i^{11}, F_i^{12}, F_i^{21})(for 0 \leq i \leq m)$

1: $T^{12} \leftarrow GenF_q(HashShake256(S_{private}))$

2: $GenerateEntries(P_i^{11}, P_i^{12}, P_i^{21} for 0 \leq i \leq m) \leftarrow GenF_q(Hash_{AES128}(S_{public}))$

3: $sharedT^{12}[v \times o \times rank^2] \leftarrow$ Load T^{12} to SharedMemory

4: $sharedP^{11}[v \times rank^2] \leftarrow$ Load $P^{11}[i][j]$ to SharedMemory

5: **Launch Kernel Compute F**$12 <<\leq (m, v), o, stream1 >>>$

6: $i \leftarrow blockIdx.x, j \leftarrow blockIdx.y, k \leftarrow threadIdx.x$

7: $F^{12}[i][j][k] \leftarrow \sum_{index=0}^{v-1} P^{11}[i][j][index] \otimes T^{12}[index][k]$

8: **Launch Kernel Compute F**$21 <<\leq (m, v), o, stream2 >>>$

9: $i \leftarrow blockIdx.x, k \leftarrow blockIdx.y, j \leftarrow threadIdx.x$

10: $F^{21}[i][j][k] \leftarrow \sum_{index=0}^{v-1} T^{12}[index][k] \otimes P^{11}[i][index][k]$

11: **Launch Kernel Compute P**$22 <<\leq (m, o), o, stream >>>$

12: $i \leftarrow blockIdx.x, j \leftarrow blockIdx.y, k \leftarrow threadIdx.x$

13: $P^{22}[i][j][k] \leftarrow \sum_{index=0}^{v-1}(T^{12}[index][j] \otimes F^{12}[i][index][k]) \oplus (P^{21}[i][j][index] \otimes T^{12}[index][k]))$

14: **Launch Kernel Convert To Bytes** $<<< CTgrid_size, CTblock_size, stream >>>$

15: $pk \leftarrow (S_{public}, P_i^{22}), sk \leftarrow (T^{12}, F_i^{11}, F_i^{12}, F_i^{21})$

We use `cudaMemcpyAsync` for asynchronous data transfer to hide data transmission latency. While the current batch undergoes matrix computation, data for the next batch can be transferred to the GPU in parallel. This approach achieves overlap between computation and communication operations.

The detailed implementation of batch key generation is presented in Algorithm 5

5 Experimental Results

5.1 Experimental Setup

We conducted performance evaluations on two distinct hardware setups. The first configuration used an NVIDIA RTX 4090 GPU with 24 GB GDDR6X memory. The second configuration employed an NVIDIA RTX 3060 GPU with 12 GB GDDR6 memory. Both systems used the same CPU configuration for consistent comparison. The CPU was an Intel Xeon Gold 5418Y processor with a base frequency of 2.0 GHz and 32 cores. This provided a reliable baseline for performance comparison. Table 1 shows the detailed hardware specifications.

5.2 Testing Methodology

Our experiments cover all standard parameter sets proposed by the authors of the SNOVA algorithm [16]. The values in each parameter set sequentially

Algorithm 5. GPU-Accelerated Batch Key Generation

Input: Message digest M, $S_{public}/S_{private}$key seeds, batch size B, salt
Output: batch key pairs
1: $T^{12} \leftarrow GenF_q(HashShake256(S_{private}))$
2: $GenerateEntries(P_i^{11}, P_i^{12}, P_i^{21} for 0 \leq i \leq m) \leftarrow GenF_q(Hash_{AES128}(S_{public}))$
3: $sharedSize \leftarrow (v \times o + v) \times l^2$
4: $blocksBatch \leftarrow 200, totalBlocks \leftarrow blocksBatch \times B$
5: **Launch compute F12 batch kernel** $<<< totalBlocks, 64, sharedSsize >>>$
6: $batchId \leftarrow blockIdx.x/blocksBatch$
7: **Load** $T^{12}[batchId]$ **and** $P^{11}[batchId]$ **into** SharedMemory
8: **for** each(i , j) to **this block do**
9:     **for** each(k , index) to **(o - 1 , v - 1) do**
10:         $result \leftarrow result \oplus P^{11}[batchId][i][j][index] \otimes T^{12}[batchId][index][k]$
11:         $F^{12}[batchId][i][j][k] \leftarrow P^{12}[batchId][i][j][k] \oplus result$
12:     **end for**
13: **end for**
14: **Launch compute F21 batch kernel** $<<< totalBlocks, 64, sharedSsize >>>$
15: Similar computation with transposed indexing for F^{21} batch matrices
16: **Launch compute P22 batch kernel** $<<< (m, o, B), o >>>$
17: $batchId \leftarrow blockIdx.z, i \leftarrow blockIdx.x, j \leftarrow blockIdx.y, k \leftarrow threadIdx.x$
18: **for** $i \leftarrow 0$ to $v - 1$ **do**
19:     $temp1 \leftarrow T^{12}[batchId][index][j] \otimes F^{12}[batchId][i][index][k]$
20:     $temp2 \leftarrow P^{21}[batchId][i][j][index] \otimes T^{12}[batchId][index][k]$
21:     $P^{22}[batchId][i][j][k] \leftarrow P^{22}[batchId][i][j][k] \oplus temp1 \oplus temp2$
22: **end for**
23: **Launch convert GF16s to bytes batch kernel** $<<< (gridX, B), 256 >>>$
24: Convert batch P^{22} matrices form GF16 format to byte arrays
25: **Transfer results from device to host and construct public keys**
26: **Retrun batch key pairs**

represent the sizes of v, o, q, and l. We implemented comprehensive testing to evaluate GPU parallelization effectiveness. The evaluation consisted of two main components. First, we conducted single-operation performance comparison tests. Second, we performed batch processing performance evaluations. The batch processing tests included comparative analysis with AVX2-optimized versions [16].

5.3 Single Key Generation Performance

Table 2 presents the speedup results for single key generation operations. On the RTX 3060 platform, different parameter sets achieved speedup ratios rang-

Table 1. Hardware Configuration of Test Environment

Component	RTX 3060	RTX 4090
Device	GeForce RTX 3060	GeForce RTX 4090
SM Count	28	128
Core Count	3,585	16,384
Global Memory	12 GB	24 GB
Host CPU	Xeon Gold 5418Y	Xeon Gold 5418Y
CPU Cores	24	24
CPU Frequency	2.00 GHz	2.00 GHz
Operating System	Ubuntu 20.04	Ubuntu 20.04
CUDA Version	12.4	12.4

Table 2. Latency (ms) and Speedup Ratios for Single Key Generation on Different GPUs Compared to CPU Reference Implementation

Level	Parameter Set	CPU (Ref)	RTX 3060	RTX 4090
I	SNOVA_37_17_16_2	5.15	1.12 (4.6×)	1.10 (4.68×)
	SNOVA_25_8_16_3	1.70	0.71 (2.39×)	0.70 (2.43×)
	SNOVA_24_5_16_4	1.44	0.74 (1.95×)	0.69 (2.09×)
III	SNOVA_56_25_16_2	23.83	2.92 (8.16×)	2.89 (8.25×)
	SNOVA_49_11_16_3	10.66	2.08 (5.13×)	1.78 (5.99×)
	SNOVA_37_8_16_4	7.85	1.73 (4.54×)	1.47 (5.34×)
	SNOVA_24_5_16_5	2.34	1.21 (1.93×)	0.9 (2.6×)
V	SNOVA_75_33_16_2	73.17	6.75 (10.84×)	5.95 (12.30×)
	SNOVA_66_15_16_3	35.12	4.03 (8.71×)	3.93 (8.94×)
	SNOVA_60_10_16_4	29.69	3.83 (7.75×)	3.68 (8.07×)
	SNOVA_29_6_16_5	4.71	1.60 (2.94×)	1.26 (3.74×)

ing from 1.93× to 10.84×. The largest parameter set, SNOVA_75_33_16_2, achieved a 10.84× performance improvement. Performance further improved on the RTX 4090 platform. The same parameter set reached a speedup ratio of 12.30× on this higher-end GPU.

5.4 Batch Processing Performance

We implemented batch processing mode for key generation to better utilize GPU parallel computing advantages. We compared our implementation with highly optimized AVX2 versions [16]. For experimental comparison, we used the largest parameter set SNOVA_75_33_16_2. Key generation's most time-consuming components are F matrix and matrix generation processes. These two matrices

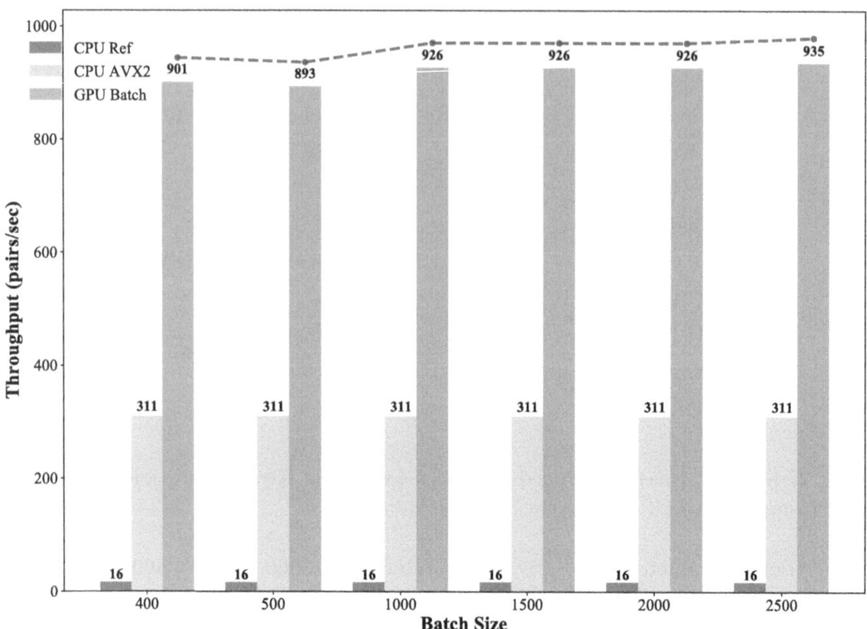

Fig. 2. Performance Comparison of SNOVA Batch Processing Implementations.

are also key components of the key. Therefore, we measured execution time by calculating the total duration of these two processes only.

The batch processing test generated 2500 key pairs simultaneously. The RTX 4090 platform demonstrated exceptional performance in this scenario. The RTX 4090 completed the entire batch within 2675 ms. This resulted in an average of 1.07 ms per key pair. This represents a 58.44× improvement compared to sequential CPU generation. As shown in Fig. 2, our GPU batch processing implementation achieved approximately 3.0× throughput improvement. This comparison was made against the current state-of-the-art AVX2-optimized version [16].

6 Conclusion

In this work, we present the first GPU-accelerated key generation scheme for SNOVA. We develop specialized optimization mechanisms for matrix operations in the finite-field GF(16). Our implementation incorporates a three-tier memory optimization framework. In addition, we establish an efficient batch parallel processing architecture. Our approach successfully improves the performance of single core operations in key generation by a factor ranging from 1.93× to 12.30×. In batch processing mode, the throughput achieves remarkable acceleration of up to 58.44× compared to sequential CPU execution. Furthermore, our implementation demonstrates a 3.0× performance advantage over the current state-of-the-art AVX2-optimized implementation. This scheme effectively addresses the computational efficiency bottleneck in SNOVA key generation. Our work facilitates the practical deployment of post-quantum signature algorithms in large-scale real-world scenarios. It also advances the engineering applications of post-quantum cryptographic solutions. In future work, we plan to further refine the GPU acceleration scheme for SNOVA.

References

1. Menezes, A.: The discrete logarithm problem. In: Elliptic Curve Public Key Cryptosystems. The Springer International Series in Engineering and Computer Science, LNCS, vol. 234, pp. 49–59. Springer, Boston, MA (1993). https://doi.org/10.1007/978-1-4615-3198-2_4
2. Shor, P.W.: Algorithms for quantum computation: discrete logarithms and factoring. In: Proceedings 35th Annual Symposium on Foundations of Computer Science, pp. 124–134. IEEE (1994)
3. Shor, P.W.: Polynomial-time algorithms for prime factorization and discrete logarithms on a quantum computer. SIAM Rev. **41**(2), 303–332. SIAM (1999)
4. Alagic, G., et al.: Status report on the third round of the NIST post-quantum cryptography standardization process. US Department of Commerce, National Institute of Standards and Technology (2022)
5. Patarin, J.: The oil and vinegar signature scheme. Presented at the Dagstuhl Workshop on Cryptography (1997)
6. Wang, L.-C., Tseng, P.-E., Kuan, Y.-L., Chou, C.-Y.: A simple noncommutative UOV scheme. Cryptology ePrint Archive (2022)
7. Kipnis, A., Patarin, J., Goubin, L.: Unbalanced oil and vinegar signature schemes. In: Stern, J. (eds) Advances in Cryptology – EUROCRYPT '99. EUROCRYPT 1999. LNCS, vol. 1592, pp. 206–222. Springer, Berlin, Heidelberg (1999). https://doi.org/10.1007/3-540-48910-X_15
8. Ikematsu, Y., Akiyama, R.: Revisiting the security analysis of SNOVA. In: Proceedings of the 11th ACM Asia Public-Key Cryptography Workshop, pp. 54–61 (2024)
9. Li, P., Ding, J.: Cryptanalysis of the SNOVA signature scheme. In: Saarinen, M.J., Smith-Tone, D. (eds.) Post-Quantum Cryptography. PQCrypto 2024. LNCS, vol. 14772, pp. 79–91. Springer, Cham (2024). https://doi.org/10.1007/978-3-031-62746-0_4
10. Chen, Z., et al.: An optimized GPU-based acceleration of CRYSTALS-Dilithium. In: ICASSP 2025-2025 IEEE International Conference on Acoustics, Speech and Signal Processing (ICASSP), pp. 1–5. IEEE (2025)
11. Shen, S., Yang, H., Dai, W., Zhang, H., Liu, Z., Zhao, Y.: High-throughput GPU implementation of Dilithium post-quantum digital signature. IEEE Trans. Parallel Distrib. Syst. IEEE (2024)
12. Wang, Z., Dong, X., Chen, H., Kang, Y., Wang, Q.: CUSPX: efficient GPU implementations of post-quantum signature SPHINCS+. IEEE Trans. Comput. IEEE (2024)
13. Garey, M.R., Johnson, D.S., et al.: A guide to the theory of NP-completeness. In: Computers and Intractability, pp. 37–79. Freeman (1990)
14. Petzoldt, A.: Selecting and reducing key sizes for multivariate cryptography. tuprints (2013)

15. Petzoldt, A., Thomae, E., Bulygin, S., Wolf, C.: Small public keys and fast verification for multivariate quadratic public key systems. In: Preneel, B., Takagi, T. (eds.) Cryptographic Hardware and Embedded Systems – CHES 2011. CHES 2011. LNCS, vol. 6917, pp. 475–490. Springer, Berlin, Heidelberg (2011). https://doi.org/10.1007/978-3-642-23951-9_31
16. Wang, L.-C., Chou, C.-Y., et al.: SNOVA. NIST Post-Quantum Cryptography Project (2024). https://csrc.nist.gov/csrc/media/Projects/pqc-dig-sig/documents/round-2/spec-files/snova-spec-round2-web.pdf
17. Hao, S., Dong, X., Wen, Z., Nie, T.: Enhancing medical data sharing with an attribute-based dynamic verifiable searchable encryption scheme using blockchain. In: Jin, C., Yang, S., Shang, X., Wang, H., Zhang, Y. (eds.) Web Information Systems and Applications. WISA 2024. LNCS, vol 14883. Springer, Singapore (2024). https://doi.org/10.1007/978-981-97-7707-5_33

Author Index

S. Yang et al. (Eds.): WISA 2025, LNCS 16299, pp. 663–666, 2026.
https://doi.org/10.1007/978-981-95-6426-2

The manufacturer's authorised representative in the EU is Springer Nature Customer Service Centre GmbH, Europaplatz 3, 69115 Heidelberg, Germany. If you have any concerns regarding our products, please contact ProductSafety@springernature.com

Printed and bound by CPI Group (UK) Ltd, Croydon, CR0 4YY
07/07/2026
02160917-0019